International Conference on Superconductivity and Magnetism 2008

AA008829

(ICSM 2008)

Journal of Physics: Conference Series Volume 153

Side-Antalya, Turkey
25-29 August 2008

Editors:

A. Gencer
G. Grasso
B. Aktas
I. Belenli
M. A. Aksan
M. I. J. Probert
M. E. Yakinci
A. Bozbey

ISBN: 978-1-61738-380-9

Printed from e-media with permission by:

Curran Associates, Inc.
57 Morehouse Lane
Red Hook, NY 12571

Some format issues inherent in the e-media version may also appear in this print version.

Printed by Curran Associates, Inc. (2010)

Institute of Physics
Dirac House, Temple Back
Bristol BS1 6BE UK

Phone: 44 1 17 929 7481
Fax: 44 1 17 920 0979

techtracking@iop.org

Additional copies of this publication are available from:

Curran Associates, Inc.
57 Morehouse Lane
Red Hook, NY 12571 USA
Phone: 845-758-0400
Fax: 845-758-2634
Email: curran@proceedings.com
Web: www.proceedings.com

TABLE OF CONTENTS

SUPERCONDUCTIVITY

Magnetoresistance Properties of MgB_2 Thick Films on YSZ Substrate Using Spray Pyrolysis Method 1
S Altin, M A Aksan, Z D Yakinci, M Özabaci, Y Balci, M E Yakinci

The Chaotic Points and XRD Analysis of Hg-based Superconductors 6
Ö Aslan, Z Güven Özdemir, S S Keskin, Ü Onbaşli

Loss Analysis of a 1 MW Class HTS Synchronous Motor 15
S K Baik, Y K Kwon, H M Kim, J D Lee, Y C Kim, H J Park, W S Kwon, G S Park

Pinning Properties of Bi-2212 Single Crystal Whiskers 24
S Altin, M A Aksan, Y Balci, M E Yakinci

Anomalous Ferromagnetic Behaviour of Y_2O_3 and CuO Nanoparticles in $YBa_2Cu_3O_y$ Superconductor 31
M Basoglu, K Öztürk, Ş Çelik, C R M Grovenor, E Yanmaz

Parametric Dynamic Analysis of a Superconducting Bearing System 39
A Cansiz, U C Hasar, Ö Gundogdu, B Ates Çam

Investigation of Effect of x in $Yb_{1-x}Gd_xBa_2Cu_3O_{7-y}$ Superconducting Structures on Upper Critical Magnetic Field and Coherence Length 47
Ş Çelik, K Öztürk, E Yanmaz

Electrodynamic Behaviour of a Two-junction Quantum Interferometer Enclosing a Superconducting Nanocylinder 55
R De Luca

Pinning Behavior of Al, Mn and Ag Substitutions in YBCO TTSMG Bulk Superconductors 62
P Diko, V Antal, M Kaňuchová, M Šefčiková, J Kováč

Enhancing the Transport Properties of Polycrystalline MgB_2 Superconductors 68
M Gharaibeh, I M Obaidat, B A Albiss, M K Hasan (Qaseer)

Studies on Structural and Magnetic Properties of Melt Textured Growth $YBa_2Cu_3O_{7-x}$ Superconducting Bulk Materials 73
U S Gokay, M Tepe

Influence of Annealing Temperature on the Microstructure and Mechanical Properties of MgB_2 82
O Gorur, M Nursoy, C Terzioglu, A Varilci, I Belenli

Artificial Pinning Centers Using the Barrier Layer of Ordered Nanoporous Alumina Templates 95
X Hallet, S Mátéfi-Tempfli, M Mátéfi-Tempfli, S Michotte, L Piraux, J Vanacken, V V Moshchalkov

Optimization of the Pr Concentration in $Y_{1-x}Pr_xBCO$ Films Prepared by Pulsed Laser Deposition 102
M Irjala, H Huhtinen, P Paturi, Anuj Kumar, V P S Awana, A V Narlikar, R Laiho

Effect of Oxygen Annealing on Magnetic Performance of $(Nd,Eu,Gd)Ba_2Cu_3O_y$ with a Lamelar Pinning Nanostructure 106
M Jirsa, M Rames, M Muralidhar, P Diko

Dependence of the Structural, Electrical and Magnetic Properties of the Superconductive YBCO Thin Films on the Deposition Rate 113
A B Karci, M Tepe, H Sozeri

Response Analysis of Free-membrane Transition-edge Detectors with Thin Substrates 121
Alireza Kokabi, Amirhosein Khoshaman, Sasan Razmkhah, Seyed Mehdi Hoseini, Ali Moftakharzadeh, Ali Bozbey, Mehdi Fardmanesh

DC V-I Characteristics of a HTS Tape for a 600kJ SMES in an Oblique External Magnetic Field 129
H J Kim, S K Baik, K C Seong, Zhuyong Li, YongHu Ma, K Ryu

Critical Current Anisotropy of MgB_2 Tapes 136
P Kováč

Revisiting the $Y_{1-x}Pr_xBa_2Cu_3O_{7-\delta}$ System- CuO_2 Plane Disorder Induced Flux Pinning 142
Anuj Kumar, Arpita Vajpayee, Shiva Kumar, V P S Awana, H Kishan

Dynamic Fatigue Behaviour of Ag-doped Bi-2212 Textured Thin Rods 149
M A Madre, Sh Rasekh, J C Diez, A Sotelo

Effect of Partial Substitution of Ca by 4f Elements on Dissipative Processes in Bi:2223 Superconductors 155
D Marconi, G Stiufiuc, A V Pop

Improvement of Superconducting Properties of MgB_2 by Changing the Argon Ambient Pressure and Sintering Conditions 160
Burcu Savaskan, K Öztürk, Ş Çelik, E Yanmaz

Influence of Gold Diffusion-doped on Phase Formation, Superconducting and Microstructure Properties of $Bi_{1.8}Pb_{0.35}Sr_{1.9}Ca_{2.1}Cu_3O_y$ Superconductors ... 169
O Ozturk, M Akdogan, C Terzioglu, A Gencer

Low-field Ac Susceptibility Study of the Flux Motion in $Bi_2Sr_2CaCu_2O_y$ Thin Films ... 186
A V Pop, S Manolache, I Matei, V Pop

Electrothermal Performance and Growth of Single Phase in Barium Doped Bi-2223 High-T_c Superconductors ... 192
M Anis-ur-Rehman

Field- and Current Controlled Switching Between Vortex States in a Mesoscopic Superconductor ... 204
A Rydh, T Golod, V M Krasnov

Superconducting Properties of Noncentrosymmetric $Li_2(Pt_{1-x}Pd_x)_3B$ Superconductors ... 208
H Takeya, S Kasahara, M ElMassalami, K Hirata, K Togano

Transport and Ac Susceptibility Measurements on Silver Wrapped Pellets of $Bi_{1.7}Pb_{0.35}Sr_{1.9}Ca_{2.1}Cu_3O_y$ Superconductors ... 212
C Terzioglu, D Yegen, A Varilci, I Belenli

Theory of Nernst Effect in Layered Superconductors ... 222
B D Tinh, B Rosenstein

Ammonium Nitrate Melt Technique for the Synthesis of Oxide Compounds ... 229
Ugur Topal

Experimental and Numerical Study of Influence of Ferromagnetic Cover on Critical Current of BiSCCO-2223/Ag Tape Superconductor ... 234
M Vojenčiak, J Šouc, F Gömöry

Superconductor Ferromagnet Bilayers; Experimental Investigation of Spin Polarized Transport Across the Interface ... 240
S K Wanchoo, John J, V C Bagwe, S P Pai, A M Narsale, R Pinto

The Metal-insulator Transition in High-T_c Cuprates–An Unusual Quantum Transition ... 247
W Winkler

Magnetic Torque Transferring Study for Bulk High-T_c Superconductors and Permanent Magnets ... 283
A Wongsatanawarid, A Suzuki, H Seki, M Murakami

Reel-to-reel Continuous Deposition of $Ce_xZr_{1-x}O_2$ Single Buffer Layer for YBCO Coated Conductors ... 291
Jie Xiong, Bowan Tao, Wenfeng Qin, Xiao Feng, Xiaoke Song, Fei Zhang, Yanrong Li

Influence of Pr Substitution on Acoustic Gruneisen Parameter and Lattice Vibrational Anharmonicity Contribution in Ceramic Superconductors ... 300
A K Yahya

The Influence of Cooling Rates on Microstructure and Mechanical Properties of $Bi_{1,6}Pb_{0,4}Sr_2Ca_2Cu_3O_y$ Superconductors ... 306
H A Cetinkara, M Yilmazlar, O Ozturk, M Nursoy, C Terzioglu

Effect of Pr-doping on Magnetic and Transports in $YBa_2Cu_3O_{7-\delta}$ Single Crystal Superconductor ... 318
J C Zhang, Z P Qin, G Jin, M Q Chen, X Yao, C B Cai, S X Cao

MAGNETISM

The Structure and Magnetic Properties of $Fe_{1.2}In_{1.87}Se_4$ Single Crystals ... 323
J Hasani Barbaran, G G Guseinov, Y I Aliyev, G M Agamirzoeva

Kinetics of First Stage of Formation Magnetite Nanoparticles Prepared by Chemical Precipitation Method ... 329
R A Ali-zade

Localized Magnetic Polaritons in Antiferromagnetic Superlattice with Impurity ... 336
R T Askerbeyli (Tagiyeva), B Tanatar

Rashba Spin-orbital Splitting in Kane Type Quantum Anti-wires ... 343
A M Babayev, S Cakmaktepe, D Turkoz Altug

Magnetism in Wide Band Gap Semiconductors Implanted with Non-magnetic Ions ... 354
R P Borges, M M Cruz, R C da Silva, F Bern, P Venezuela, M Dionízio Moreira, A T Costa, M Godinho

Strain and Interface Effects on the Magnetic and Transport Properties of $La_{0.7}Ca_{0.3}MnO_3$/CaO Multilayers ... 360
A Casaca, R P Borges, P Ferreira, A Saraiva, M A Rosa, R C da Silva, W C Nunes, S Magalhães, M Godinho

Magnetic and Magnetocaloric Properties of Monovalent Substituted $La_{0.65}Ca_{0.3}M_{0.05}MnO_3$ (M=Na, Ag, K) Perovskite Manganites ... 370
W Cheikh-Rouhou Koubaa, M Koubaa, A Cheikhrouhou

Development of Halbach Magnet for Portable NMR Device ... 375
N Doğan, R Topkaya, H Subaşi, Y Yerli, B Rameev

Observation of Room Temperature Ferromagnetism in ZnTe:Cr Films Grown Onto Glass Substrate by Thermal Evaporation Method 383
D Soundararajan, D Mangalaraj, D Nataraj, L Dorosinskii, J Santoyo-Salazar

Production, Characterization and Magnetic Properties of Bismuth-substituted Yttrium Iron Garnet Layers Prepared by Sol-gel Process 395
M Erol, Y Ozturk, I Avgin, E Celik

The Effect of NaCl Prepared by Ultra-sonic Vibration on Sintering of Annealed FePt Nanoparticles 402
M Farahmandjou, S A Sebt, S S Parhizgar, P Aberomand, M Akhavan

The Effect of Milling Speed on the Magnetic Properties of Mechanically Alloyed Fe-45%Ni Powders 408
K H Gheisari, S Javadpour, J T Oh

EPR Study of Mn-implanted Single Crystal TiO_2 414
S Güler, B Rameev, R I Khaibullin, O N Lopatin, B Aktaş

Effect of Annealing Temperature and Substitution of Zr-Cu on Magnetic Properties of Strontium Hexaferrite Nanoparticles 420
Muhammad Javed Iqbal, Muhammad Naeem Ashiq, Pablo Hernandez-Gomez

Analysis of the Series Resistance and Interface State Densities in Metal Semiconductor Structures 427
G Güler, Ö Güllü, Ş Karataş, Ö F Bakkaloğlu

Galvanomagnetic Effects in $A_2^V B_3^{VI}$ 434
T G Kerimova, N A Abdullayev, K M Mustafayeva

Structural, Static and Dynamic Magnetic Studies of Evaporated CO_xCr_{1-x}/Si (100) and Co_xCr_{1-x}/Glass Thin Films 439
A Kharmouche

Superconductivity, Antiferromagnetism and Ferromagnetism in Periodic Table of D.I. Mendeleev 450
L Koroleva, T M Khapaeva, M V Lomonosov

Room-temperature Ferromagnetism in Novel Mn-doped $ZnSiAs_2$ Chalcopyrite 463
S F Marenkin, V M Novotortsev, I V Fedorchenko, S A Varnavskiy, L Koroleva, D M Zashchirinskii, T M Khapaeva, R Szymczak, B Krzymanska, V Dobrowolski, L Kilanski

First-principle Calculation of GaAs Electronic Structure by Doping with Mn and P 468
Z B Li, X Wang, Y Zhu, M W Ran, K L Yao

Correlation Between Crystallographic Texture and the Degree of $L1_0$-ordering in Post-annealed Ag/CoPtBilayers and Comparison with Ag/CoPt Nanocomposites 473
E Manios, D Stamopoulos, I Panagiotopoulos, D Niarchos

Dielectric and Magnetic Properties of YIG/PMMA Nanocomposites 485
Anjum Qureshi, Ayhan Mergen, B Aktaş

Optically Induced Hall Effect in Semiconductors 493
M Idrish Miah, E Mac A Gray

Observation of Magnetocaloric Effect in the $LaMn_{1.9}Fe_{0.1}Si_2$ Compound at Low Fields in the Vicinity of Phase Transition Around Room Temperature 501
Onur Nane, Hasan Ağil, Berdan Özkurt, Bekir Özçelik, Ali Gencer

Magnetostriction in Orthorhombic Manganites 508
Iliya Radulov, Vassil Lovchinov, Dimitar Dimitrov, Andrey Apostolov

Study of Ferromagnetism in Mn Doped ZnO Dilute Semiconductor System 515
R K Singhal, M S Dhawan, S K Gaur, S N Dolia, Sudhish Kumar, T Shripathi, U P Deshpande, Y T Xing, Elisa Saitovitch, K B Garg

Another Alternative to the Solid State Reaction Method to Synthesize Nanocrystalline YIG: Ammonium Nitrate Melt Technique 527
H Sozeri, N Ghazanfari

Magnetic and Fluorescence Properties of Cobalt Implanted Hydrogels 533
H Sozeri, A Gelir, E Alveroğlu, M Tulun, Y Yilmaz

FMR Studies of Half Metallic Ferromagnetic Thin Films Co_2MnSn and Co_2MnGe 539
R Yilgin, S Kazan, B Rameev, B Aktaş, K Westerholt

Ferromagnetism in Mn-implanted Ge/Si Nanostructure Material 543
I T Yoon, C J Park, T W Kang

SUPERCONDUCTIVITY – LATE SUBMISSION

Meissner Effect in Films of Ropes of Boron-doped Single-walled Carbon Nanotubes; Correlation with Applied Pressure and Boron-doped Multi-walled Nanotubes 550
M Matsudaira, J Nakamura, T Shimizu, M Özabaci, J Haruyama, J Reppert, A M Rao, T Eguchi, T Nishio, Y Hasegawa, H Sano, Y Iye

Author Index

PREFACE

The International Conference on Superconductivity and Magnetism (ICSM2008) was held at the congress centre of Ankara University in Side, Antalya, between 25–29 August 2008. The conference was the first conference on the combined fields of superconductivity and magnetism organized in Turkey at international level, and it had broad international participation from 42 countries, with registered delegates numbering over 400. A quarter of the attendees were research students. The conference attracted many of the best known leading scientists and experts in the field of superconductivity and magnetism from all over the world. The scientific program involved the presentation and discussion of 336 papers, classified as 65 invited, 81 oral and 190 posters. Submission of papers for the proceedings was on a volunteer basis and we therefore had nearly half of the presented papers, i.e. 30 submitted invited papers, peer-reviewed by *Superconductor Science and Technology*, and 85 submitted contributing papers, peer-reviewed by the organizers through processes administered by the Editorial Board and Scientific Committee. Reviews were conducted by expert referees at professional level and with the scientific standards expected of a proceedings journal issue published by IOP Publishing. The invited papers on superconductivity and magnetism with superconductivity were considered and processed for *Superconductor Science and Technology* by IOP itself. Although there are missing papers from some of the plenary speakers, we believe that this special issue of *Superconductor Science and Technology* (SUST) and the corresponding issue of *Journal of Physics: Conference Series* (JPCS) reflect most of the booming research in the fields of superconductivity and magnetism. We are very pleased to have worked with IOP on the conference proceedings, with special thanks to Dr Tom Miller and Dr Graham Douglas. Based on a refereed evaluation of all the papers and posters submitted, about 93 papers were selected for publication in the *Journal of Physics: Conference Series* and *Superconductor Science and Technology*.

The immensely rich and diverse scientific program started with Professor D Hampshire's opening plenary on 26 August 2008 and sessions were mostly chaired by the invited speakers. The topics included:

- New superconductors
- Theory and applications of MgB_2 and boride superconductors
- Hybrid magnetic-superconducting systems
- High temperature superconductors: theory and applications
- Superconducting devices and applications
- Oxypnictide superconductors
- Superconducting sources of THz-radiation

- Vortex dynamics
- Spintronics
- Superconductivity and magnetism at nanoscale
- Multiferroic materials
- Manganities
- Materials and fabrication techniques
- Conventional superconductors
- Microscopic theories of high temperature superconductors
- Device physics and thin films.

We believe that this conference, with the above topics, has provided a forum where the many matters of interest to the superconductivity and magnetism community have been debated. We are of the opinion that the conference has been a very successful one and we hope that the subsequent conferences in this field will constitute a series of fruitful meetings, by growing to become larger global events with greater success in bridging the gap between the best scientists, engineers, exhibitors and participants from countries underrepresented in science and technology.

The delegates and companions also enjoyed the social program in the touristic and historical places of the surrounding area, during the excursion time after 4.30pm on conference days. The welcome reception was held at the historical site of the Apollo Temple in Side and sponsored by the Municipality of Side; a second welcome party was also organized at the congress centre in the evening of 25 August 2008, for late-arriving participants. The social program also continued in the evenings with a variety of different entertainment.

The meeting was financially supported by generous contributions from Turkish research institutions, and an Italian firm, Columbus Superconductors SpA. The list of contributors is as follows:

- TÜBİTAK
- Boren
- Columbus Superconductors
- Ankara University Research Fund.

The following small research companies also contributed:

- Boronsan
- IMTEK
- Zeiss
- Nanomagnetics
- Optomek
- Side Municipality
- Belek Municipality.

It is a pleasure to thank them all for their pivotal role in the organization of the conference. We would also like to thank the speakers, the participants, the exhibitors and the Ankara University staff and students for their efforts which made ICSM2008 a very successful meeting. The list of people involved in the organization is given below. More information about the conference can be found at http://www.icsm2008side.org. It is due to the immense demand from most of the attendees that we are planning to organize the second International Conference on Superconductivity and Magnetism (ICSM2010) for April 2010, in Antalya. Further details will be made clear and circulated soon. We will be very happy to host the next conference and look forward to welcoming you again.

Conference Director
A Gencer *Ankara University*

Conference Co-directors
B Aktas *GYTE, Gebze*
M E Yakinci *Inonu University*

Scientific Committee
D Abukay *IYTE, İzmir*
M Altunbas *KTU, Trabzon*
I Askerzade *Ankara University*
I Belenli *AIBU, Bolu*
G Grasso *Columbus Superconductors SpA, Genova*

N Guclu *G O Pasa University*
K Kiymac *Çukurova University*
S Nezir *Kirikkale University*
A Oral *Sabanci University*
B Ozcelik *Çukurova University*
Y Oner *IYTE, İzmir*
L Ozyuzer *IYTE, İzmir*
O Uzun *G O Pasa University*
E Yanmaz *KTU, Trabzon*—Coordinator
C Terzioglu *AIBU, Bolu*

International Advisory Board
M Tachiki *Tokyo University*
E Aksu *TAEK*
H Akalin *Renko*
O Atiker *Boronsan*
R Cubitt *Institut Laue-Langevin, Grenoble*
G Crabtree *Argonne National Laboratory*
A Elmali *Ankara University*
C Ferdeghini *CNR INFM, Genova*
R Flukiger *Genova University*
W Goldacker *FZ Karlsruhe*
C E Gough *Birmingham University*
R Huebener *Tubingen University*
K Kadowaki *NIMS, Tsukuba University*
M A R LaBlanch *UO, CA*
E Inger *Boren*
E Ozdas *Hacettepe University*
H Ozkan *METU, Ankara*
R Penko *ASG*
V Vlasov *Argonne National Laboratory*

Scientific Programme Committee
E H Brandt *Max-Planck-Institute, Stuttgart*
W Goldacker *FZ Karlsruhe*
C Ferdeghini *CNR INFM, Genova*
D M Hampshire *Durham University*
Y Oner *ITU, Instanbul*
B Aktas *GYTE, Gebze*
B Ramaev *GYTE, Gebze*
E Ozdas *Hacettepe University*—Coordinator
M A Aksan *Inonu University*—Abstract Handling

Local Organizing Committee
S Safran *Ankara University*—General Secretary
E Ertekin *Ankara University*—Registrar

This is a co-publication with *Superconductor Science and Technology*. The International Conference On Superconductivity and Magnetism (ICSM2008) took place at The Congress Centre of Ankara University in Side-Antalya, Turkey, between 25–29 August 2008. The bulk of the papers, after peer review, are published in *Journal of Physics: Conference Series*. However a selection of papers, are published separately in a special issue of *Superconductor Science and Technology*.

All papers published in this volume of *Journal of Physics: Conference Series* have been peer reviewed through processes administered by the proceedings Editors. Reviews were conducted by expert referees to the professional and scientific standards expected of a proceedings journal published by IOP Publishing.

Magnetoresistance properties of MgB_2 thick films on YSZ substrate using spray pyrolysis method

S. Altin[1], M. A. Aksan[1], Z. D. Yakıncı[2], M. Özabacı[1], Y. Balcı[1] and M. E. Yakıncı[1]

[1]İnönü Üniversitesi, Fen Edebiyat Fakültesi, Fizik Bölümü, Süperiletkenlik Araştırma Grubu, 44280-Malatya
[2]İnönü üniversitesi,Sağlık Hizmetleri M.Y.O, 44280-Malatya

E-mail: saltin@inonu.edu.tr

Abstract. The MgB_2 thick films have been fabricated on the polycrystalline YSZ substrates using spray pyrlosis technique. Structural/microstructural, electrical and magnetic properties were investigated by XRD, SEM-EDX and R-T measurements. The films fabricated showed dense and homogenous structural formation with strong grain connections. The electrical properties of MgB_2 thick films were investigated under different magnetic fields. It was seen that resistive transition broadened with increasing the applied magnetic fields. It was believed that the broadening is due to the thermally activated flux flow (TAFF). The activation energy, $U(H)$, of the MgB_2 films was calculated using Arhenius law.

Introduction

The superconductivity at 39 K discovered in MgB_2 among binary alloys attracted much interest in its fabrication techniques and practical applications [1]. MgB_2 does not exhibit weak-link at grain boundaries. Low anisotropy, high critical current density, J_c, due to strongly linked grains, long coherence length and high critical fields ensure rapid processing in making thin/thick films, wires or nanowires and single crystal. Especially, the low processing temperature in thin/thick film formation offers important electronic applications such as junction. However, MgB_2 films processed in the low processing temperature show T_c values between 34 and 37 K, which are relatively low values compared to the T_c value of pure MgB_2 material. The incorporation of MgO could be the possible reason for this observed low T_c.

H.J.Kim et al reported that the vortex phase diagram of MgB_2 films is similar to High-T_c systems and similarity of phase diagrams of MgB_2 and HT_c materials leads us to explain the pinning properties in MgB_2 easily. MgB_2 films show 3D vortex glass state of flux vortices, which is possible origin of high current density, J_c [2,3].

In this paper, we have fabricated MgB_2 thick films on the YSZ substrate using spray prylosis technique. The effects of high and low heat processing temperature on the structural/microstructural and electrical properties were investigated by XRD, SEM and R-T. Flux dynamics of MgB_2 thick films on YSZ substrate are studied by temperature dependence of resistance under magnetic fields up to 4 T. The activation energy, $U(H)$ was calculated using Arrhenius law.

Experimental

High purity MgB_2 powder was mixed in absolute ethanol with concentration 0.01mol/liter. The solution with MgB_2 were mixed for 0.2 h and poured into nozzle container. The tip of the nozzle is 0.1 mm diameter. YSZ substrates were put in a glass container and sprayed with the solution. The spraying process was repeated four-five times. Fabricated films were heated at 600^0C and 900^0Cfor 24 h under Ar atmosphere in the molybdenum tube.

The films' microstructure and phase analysis were characterized by XRD (Rigaku RadB) and SEM (Leo Evo 40 VPX). Electrical properties of the films fabricated were investigated by temperature dependence of resistance under under magnetic fields between 0-4 T using VSM (Crygonic) magnetometer.

Results and Discussions

XRD patterns of the MgB_2 thick films fabricated are presented in Figure 1. Films heated treated at 600^0C and 900^0C for 24 h showed the single phase MgB_2. The crystal symmetry of both films was found to be hexagonal and the unit cell parameters of the films heat treated at 600^0C and 900^0C for 24 h were calculated to be a=3.0881 and 3.0898 Å, and c=3.5341 and 3.5359 Å, respectively, which are close to previously obtained results [4,5]. However, it was found that the peak intensities of the film heat treated at 900^0C increased. B Although the main phase was obtained to be MgB_2, some MgO impurity phases were detected due to oxidation of films which are believed to be due to small amounts of oxygen in the non-spectroscopic Ar gas used during the heat treatment cycles.

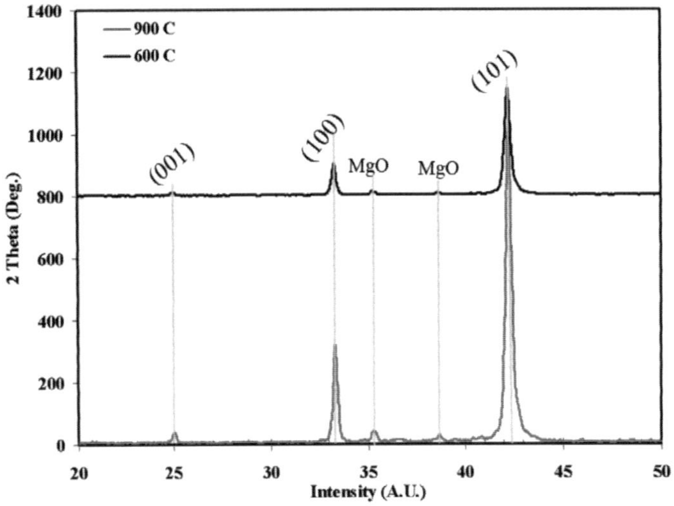

Figure 1. XRD pattern of films fabricated at 900^0 C and 600^0C.

Surface morphology of the films and microanalysis of the phases have been investigated by SEM and EDX analysis. The microstructure of the films fabricated at 600^0C and 900^0C for 24 h is shown in Figure 2.a and b, respectively. It was found that the film fabricated at 600^0C consisted mainly of the MgB_2 phase, Figure 2.a, which is consistent with XRD pattern. The white colored grains on the surface of the film were identified to be the MgO phase. The surface morphology changed with increasing heat treatment temperature from 600^0C to 900^0C. More compact grain formation without porosity was observed, Figure 2.b. This is typical structure of MgB_2 material which was previously obtained by many research groups [6].

Figure 2. SEM image of MgB2 films fabricated at a) 600^0C and b) 900^0C.

Temperature dependence of resistance in MgB_2 films heat treated at 600^0C and 900^0C was given in Figure 3. The films fabricated under low temperature heat treatment condition showed superconducting transition at 37 K, T_c, and the zero resistance temperature, T_0, was reached at 20 K, Figure 3.a. The sample shows a metallic behavior in the normal state. However, the magnetic field influenced strongly T_c and T_0. Transition broadened with increasing the applied magnetic field and T_c and T_0 decreased with the applied magnetic field as given Table 1. Transition broadening ($\Delta T = T_c - T_0$) was found to be 17 K at the magnetic field of 0 T and increased with increasing the applied magnetic field.

Resistance versus temperature curves of the MgB_2 films heat treated at 900^0C for 24 h were shown in Figure 3.b. It was observed that the T_c and T_0 values at 0 T was 38 K and 33 K, respectively. Although the superconducting transition at zero magnetic field is sharp (ΔT=5.6 K), the resistive broadening in R-T with increasing the magnetic field curves was obtained, Table 1. The broadening effect is similar to that in HT_c systems. Thermally activated flux flow (TAFF) model is generally used to explain such a broadening behavior in the HT_c materials [7]. According to the TAFF model, flux lines are depinned with the help of thermal activation while $T_{irr} < T < T(H_{c2})$. During the experiments, the applied current ensured that only small driving forces exist in the vortex system of MgB_2 and therefore we believed that the TAFF theory is suitable to interpret the magnetoresistance data.

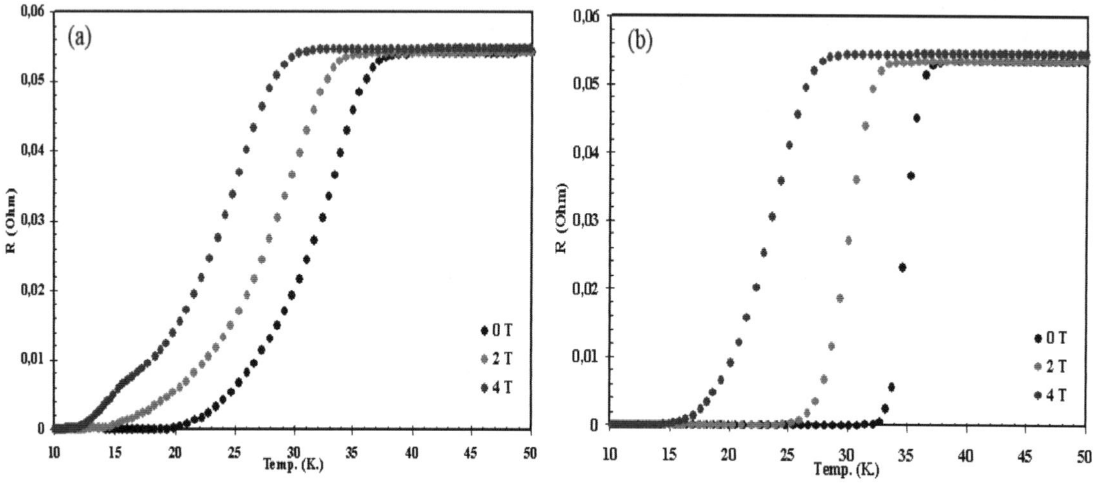

Figure 3. Temperature dependence of resistance of MgB_2 films at under magnetic field of 0, 2 and 4 T. MgB_2 films heat treated at a) 600^0C for 24 h and b) 900^0C for 24 h.

Table 1. Superconducting Properties of films fabricated with different heat treatment cycles

Heating Temp.(^{0}C)	**T_c(K)**			**T_0(K)**			**ΔT(K)**			**$U(H,T)/k_B$**		
	Applied Magnetic Field											
	0 T	**2 T**	**4 T**	**0 T**	**2 T**	**4 T**	**0 T**	**2 T**	**4 T**	**0 T**	**2 T**	**4 T**
600	37.7	33.8	29.1	20.8	12.3	12.8	16.9	17.5	11.3	220	111	79
900	38.3	34.8	29.7	32.7	25.3	15.3	5.6	9.5	14.4	2230	760	185

The activation energy *U (H,T)* plays an important role for the determination of the flux dynamics in the superconducting materials. *U (H,T)* also reveals the height of the energy barriers of superconducting materials which control the vortex motion in the vortex liquid [8]. In the Thermally Activated Flux Flow (TAFF) region, the thermally activated resistivity is defined as [9],

$$R = R_{\exp}\left[-\frac{U(J,H,T)}{k_B T}\right] \quad (1)$$

where ρ_0 is the pre-exponent factor, k_B the Boltzmann constant and *U* the activation energy for the flux motion which depends on the temperature, *T*, the current density, *J*, and the applied magnetic field, *H*. At the constant current, the dependence of *U (H,T)* is believed to give information about the dissipation mechanism. *U (H,T)* can be given as temperature independent in the narrow temperature region of the low resistivity part. The activation energy *U(H)*, can be calculated from the slope of the plot of ln(*R*) versus *1/T* as shown in figure 4. The calculated activation energy values of the films fabricated under high and low temperature heat treatment conditions are given in Table 1.

The activation energy, *U(H)*, decreased rapidly with increasing magnetic field. The variation of *U(H)* with the applied magnetic field is explained as an increase of the magnetic field and also flux trapping. When the magnetic field was increased, the fluxes from the externally applied magnetic field penetrate into the samples and are trapped there leaving the internal field almost unchanged.

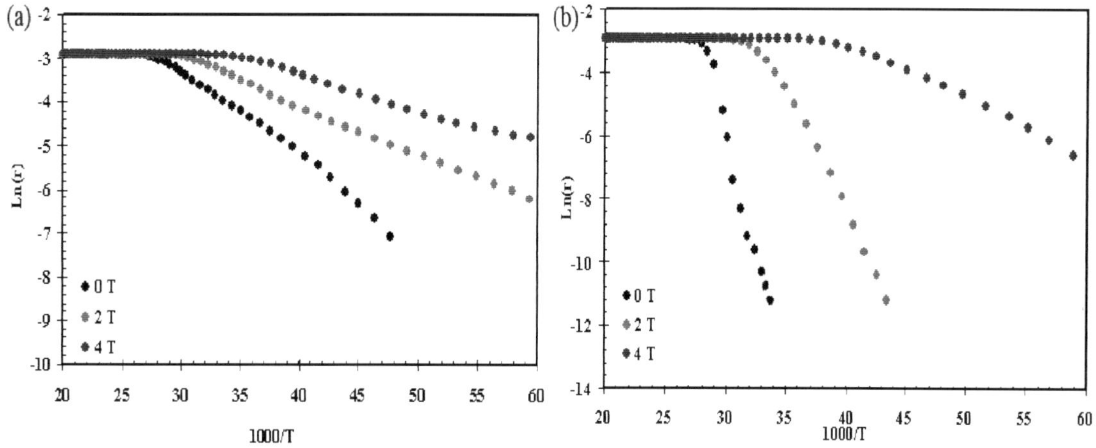

Figure 4. ln(R) versus 1000/T plots of MgB_2 films fabricated at a) 600^0C 24 h and b) 900^0C 24 h.

Conclusion

In this study, we have fabricated MgB_2 thick film on the YSZ substrate by using Spray Pryloysis technique. The films fabricated were heat treated under low (600^0C) and high temperature (900^0C) conditions. The XRD and SEM analysis showed that heat treatment temperature played an important role on the crystallization and grain connections in the films. R-T measurement pointed out that the film processed at low temperature showed the broadened superconducting transition. This problem was solved with high temperature heat cycle. Activation energy, $U(H)$ was calculated using Arrhenius law. U(H) decreased strongly with increasing the magnetic fields which is attributed to the flux motion.

Acknowledgement

This research was supported by Bor Araştırma Enstitüsü (BOREN) under contract no 2008.Ç 0171.

References.

[1] Nagamatsu, Nakagawa N, Muranaka T, Zenitan Y and Akimitsu J 2001 *Nature* **410** 63
[2] Senatore C, Polichetti M, Zola D, Di Matteo T, Giunchi G and Pace S 2003 *Supercond. Sci. Technol.* **16** 183
[3] Kusevic I, Babic E, Husnjak O, Soltanian S. Wang X L and Dou S X 2004 *Solid state Commun.* **132** 761
[4] Rogacki K, Battlog B, Karpinski J, Zhigadlo N D, Schuck G, Kazakov S M, Wagli P, Puzniak R, Wisniwseki A, Carbone F, Brinkman A and Van Der Marel D 2006 *Phys. Rev. B* **73** 174
[5] Bernardini F and Massidda S 2006 *Phys. Rev. B* **74** 014513
[6] Naito M and Ueda K 2004 *Supercond. Sci. Technol.* **17** R1
[7] Kes P H, Aarts J, Vanden Berg J, Van der Beek C J and Mydosh J A 1989 *Supercond. Sci. Technol.* **1** 242
[8] Masui T, Lee S and Tajima S 2003 *Physica C* **383** 299
[9] Li H L, Ruan K Q, Li S Y, Yu Y, Wang C Y and Cao L Z 2003 *Physica C* **386** 560

The Chaotic Points and XRD Analysis of Hg-Based Superconductors

Ö Aslan[1], Z Güven Özdemir[2], S S Keskin[3] and Ü Onbaşlı[4]

[1] Anatürkler Educational Consultancy and Trading Company, Orhan Veli Kanık Cad., 6/1, Kavacık 34810 Beykoz, İstanbul, Turkey.
[2] Physics Department, Yıldız Technical University, Davutpaşa Campus, Esenler 34210, İstanbul, Turkey.
[3] Dept. of Environmental Eng., University of Marmara, Ziverbey, 34722, İstanbul, Turkey.
[4] Physics Department, University of Marmara, Rıdvan Paşa Cad. 3. Sok. 85/12 Goztepe, Istanbul, Turkey.

E-mail: ozdenaslan@yahoo.com

Abstract. In this article, high T_c mercury based cuprate superconductors with different oxygen doping rates have been examined by means of magnetic susceptibility (magnetization) versus temperature data and X-ray diffraction pattern analysis. The under, optimally and over oxygen doping procedures have been defined from the magnetic susceptibility versus temperature data of the superconducting sample by extracting the Meissner critical transition temperature, T_c and the paramagnetic Meissner temperature, T_{PME}, so called as the critical quantum chaos points. Moreover, the optimally oxygen doped samples have been investigated under both a.c. and d.c. magnetic fields. The related a.c. data for virgin(uncut) and cut samples with optimal doping have been obtained under a.c. magnetic field of 1 Gauss. For the cut sample with the rectangular shape, the chaotic points have been found to occur at 122 and 140 K, respectively. The Meissner critical temperature of 140 K is the new world record for the high temperature oxide superconductors under normal atmospheric pressure. Moreover, the crystallographic lattice parameters of superconducting samples have a crucial importance in calculating Josephson penetration depth determined by the XRD patterns. From the XRD data obtained for under and optimally doped samples, the crystal symmetries have been found in tetragonal structure.

1. Introduction

It is well known that, the superconducting materials have a phase transition from normal state to superconducting state at the Meissner transition temperature, T_c. The most common property of the superconductivity is the diamagnetic response to the applied magnetic field. In addition to diamagnetic response, some superconductors exhibit a simultaneous paramagnetic response under a weak applied magnetic field [1-3]. This paramagnetic behaviour is called as the Paramagnetic

[1] To whom any correspondence should be addressed.

Meissner Effect (PME) and it can be observed within a specific temperature interval with the maximum value at the paramagnetic Meissner temperature, T_{PME}. At this temperature, the direction of the orbital current changes in the momentum space. Since both temperatures represent the transition from one to another state of being, T_c and T_{PME} are considered as the critical quantum chaos points of the superconducting sample [4]. The superconducting system, which is a system with two interacting electrons called as Cooper pairs, is considered to be the best material media displaying the chaotic behaviour [5-7]. In this study, the concept of quantum chaos has been discussed in the context of some critical temperatures defined at the superconducting state.

The investigated mercury based copper oxide layered superconductor, which is one of the high temperature superconductors, has the highest critical parameters such as Meissner transition temperature, T_c, critical current density, J_c, etc. The primitive cell of the mercury cuprate superconductors given in figure 1 has three superconducting copper oxide planes separated by insulating layers and this structure is considered as an intrinsic Josephson junction array. As is known, the superconductivity occurs in the copper oxide planes which form intrinsic structural layers. The origin of the superconductivity is based on the harmony which is extended to all copper oxide layers along the c-axis via coupling at the Josephson plasma frequency, ω_p [8].

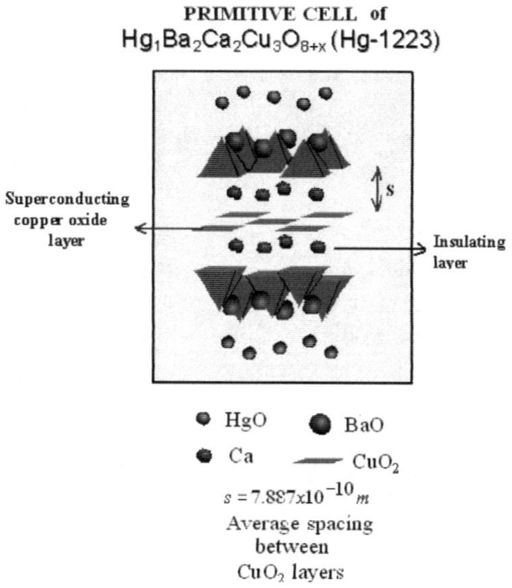

Figure 1. The primitive cell of Hg-1223 superconductor.

Due to the highest critical parameters of the bulk superconducting $HgBa_2Ca_2Cu_3O_{8+x}$ (Hg-1223) samples, the determination of some electrodynamics parameters such as Josephson penetration depth, plasma frequency and the anisotropy factor has a great importance for both theoretical and various advanced technological applications. To calculate these electrodynamics parameters, the average spacing of copper oxide bilayers, s, is required to be measured. In this work, the bulk superconducting Hg-1223 samples have been examined by means of critical quantum chaotic points and X-ray diffraction (XRD) patterns. The critical chaotic points T_c and T_{PME} have been observed on both a.c. and d.c. magnetic susceptibility versus temperature data of mercury based superconductors [9].

2.Theoretical

The concept of the chaos can be defined as the transition from one state to another state where the probability density of the system, which is sensitive to the initial conditions [10], changes via temperature. In this point of view, the superconducting system is one of the best samples to understand the unexpected chaotic transitions. Superconducting systems, which exhibit the second order phase transition, possess some critical chaos points as defined above. According to Evangalou, the phenomenon of the critical quantum chaos is observed in the quasi periodic systems, the systems with two interacting electrons and the fractal matrices [5]. Furthermore, the superconductors investigated, in which phonon mediated attractive electron-electron interaction leads to form quasi-particles, namely Cooper pairs[11-14], constitute a natural laboratory for searching and observing quantum critical chaos points [4].

The crystal structure of the mercury based copper oxide layered high temperature superconductor is used to determine the electrodynamics parameters and can be obtained from the XRD data. The average spacing of the adjacent copper oxide layers, *s*, which is obtained from the c-axis value of the primitive cell given by the XRD data, is a crucial parameter to calculate Josephson penetration depth, λ_j. Josephson penetration depth is inversely proportional to the square root of *s* value of the sample investigated [15-18].

From the view of the point group symmetry (crystal symmetry), the crystal structure of cuprates can be divided into two categories; tetragonal and orthorhombic lattices. Some cuprates such as $La_{2-x}Sr_xCuO_4$, $Tl_2Ba_2CaCu_2O_8$, $HgBa_2CaCu_2O_6$ (Hg-1212), and $HgBa_2Ca_2Cu_3O_{8+x}$ (Hg-1223) have the tetragonal crystal structure [19-21]. Since the crystal structure of the superconducting sample affects the critical quantum chaos points of the superconducting samples, the determination of lattice parameters is a necessary step to make an analogy between crystal structure and quantum chaotic points.

3. Experimental

In this work, the high T_c mercury based cuprates with different oxygen doping rates have been investigated by means of magnetic susceptibility versus temperature data and X-ray diffraction pattern analysis.

3.1. The Analysis of Temperature Dependence of Magnetization

In this study, the under, optimally and over oxygen doping procedures have been investigated by referring to T_c and T_{PME} temperatures extracted from the magnetic susceptibility versus temperature data taken by Quantum Design SQUID susceptometer model MPMS-5S. In all of the magnetization measurements, the magnetic field has been applied to the superconducting specimen along the *c*-axis.

The under [22], optimally [9] and over-oxygen doped [23] samples, which are very durable superconductors obtained by pressing under 1 ton weight [24]. The virgin samples have been cut by diamond saw in the rectangular shape of 4*x*2*x*1 mm. Over-doped samples have been obtained by the oxygen annealing procedure that has been performed in a furnace kept at 250° C for five hours and repeated several times [9,23].

In this article, Hg based superconducting samples have been investigated under both a.c. and d.c. magnetic fields [9,23]. The magnetic moment versus temperature data of the non-oxygen doped (as prepared) superconducting sample and the optimally doped sample obtained from the same batch have been given in figure 2 [25].

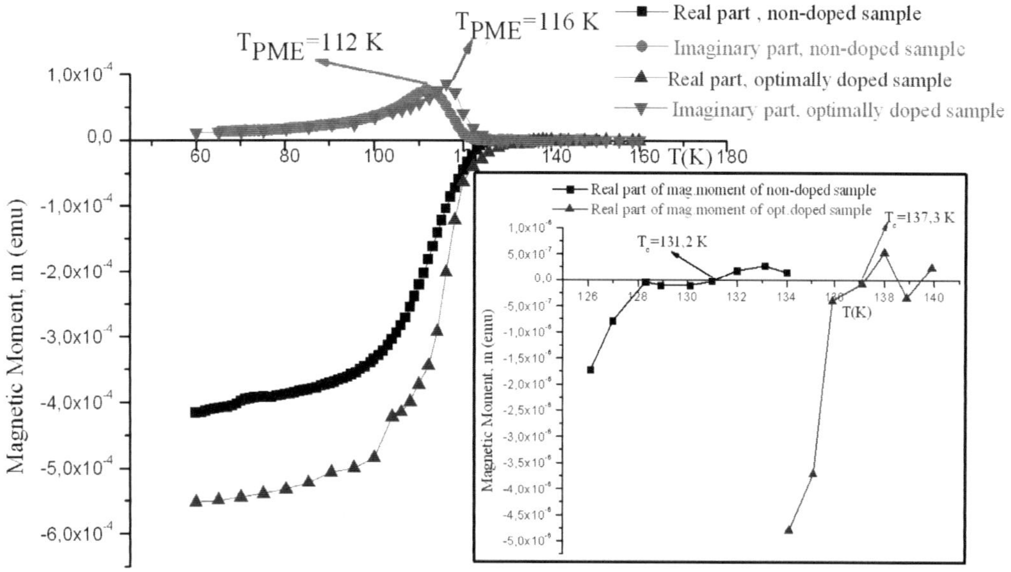

Figure 2. Magnetic moment versus temperature curves of the non-doped and optimally oxygen doped samples under a.c. magnetic field of 1 Gauss.

The related a.c. data for optimally doped virgin (uncut) and cut samples have been given in figure 3. Both data have been taken under a.c. magnetic field of 1Gauss with 1 kHz frequency.

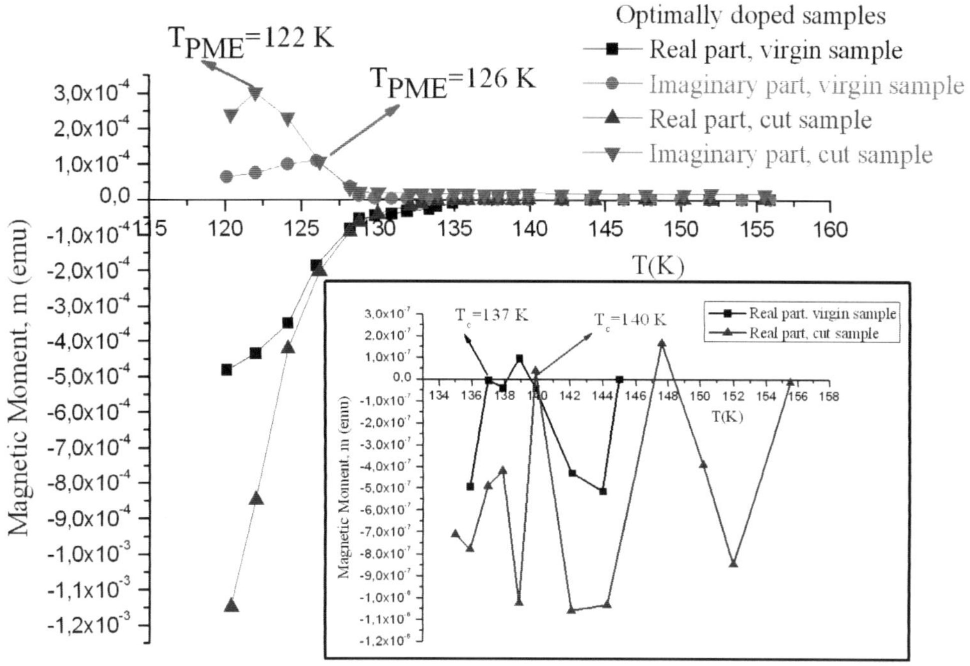

Figure 3. Magnetic moment versus temperature curves of the virgin (uncut) and cut samples under a.c. magnetic field of 1 Gauss.

As seen in figure 3, the paramagnetic Meissner and the critical Meissner chaotic temperatures of uncut samples have been determined as 126 and 137 K, respectively. For cut sample with rectangular shape, T_{PME} and T_c have been found as 122 and 140 K, respectively. To our knowledge, the Meissner critical temperature of 140 K is the new world record of the high temperature superconductors under normal atmospheric pressure.

In a previous study, both T_c and T_{PME} of the optimally doped cut sample have been determined as 137.7 K under zero field and 1 Gauss of d.c. magnetic field [9]. The chaotic points of the under and optimally doped samples have been observed from magnetic moment versus temperature graphic given in figure 4.

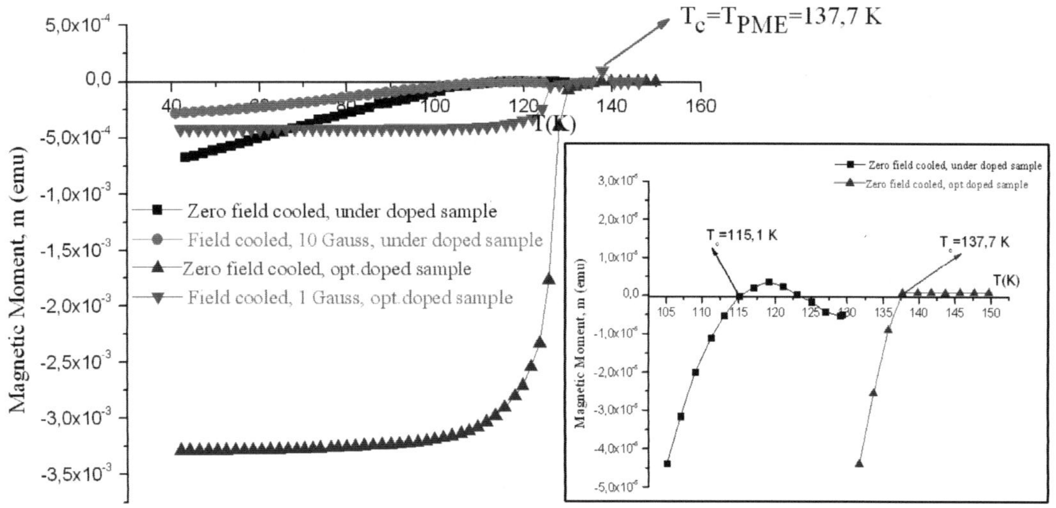

Figure 4. Magnetic moment versus temperature curves of under and optimally doped samples under d.c. magnetic field between zero and 10 Gauss.

In magnetic moment versus temperature curve of cut sample taken under zero and 1 Gauss of d.c. magnetic field, the paramagnetic Meissner effect has been observed under field cooled.

3.2. X-Ray Diffraction (XRD) Pattern Analysis and the Lattice Parameters

The XRD pattern, from which the lattice parameters of the sample have been calculated, has been obtained from a Rigaku Model XRD device. The average spacing of copper oxide bilayers, "s", by which the Josephson penetration depth values has been calculated, has been determined by means of the translation vectors of the mercury cuprate superconductor samples. The XRD patterns of the optimally doped and under-doped samples have been shown in figure 5 and figure 6, respectively. The lattice parameters for them have been given in table 1 and table 2, respectively [22].

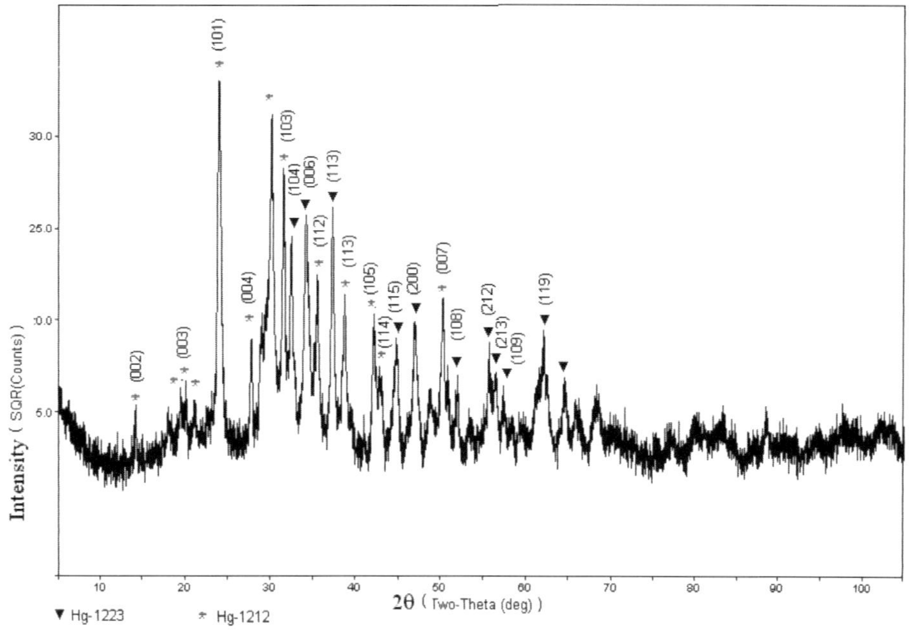

Figure 5. XRD Pattern and (*hkl*) planes (Miller indices) of optimally doped sample in 50 minutes counting. * indicates Hg-1212 and ▼ indicates Hg-1223, respectively.

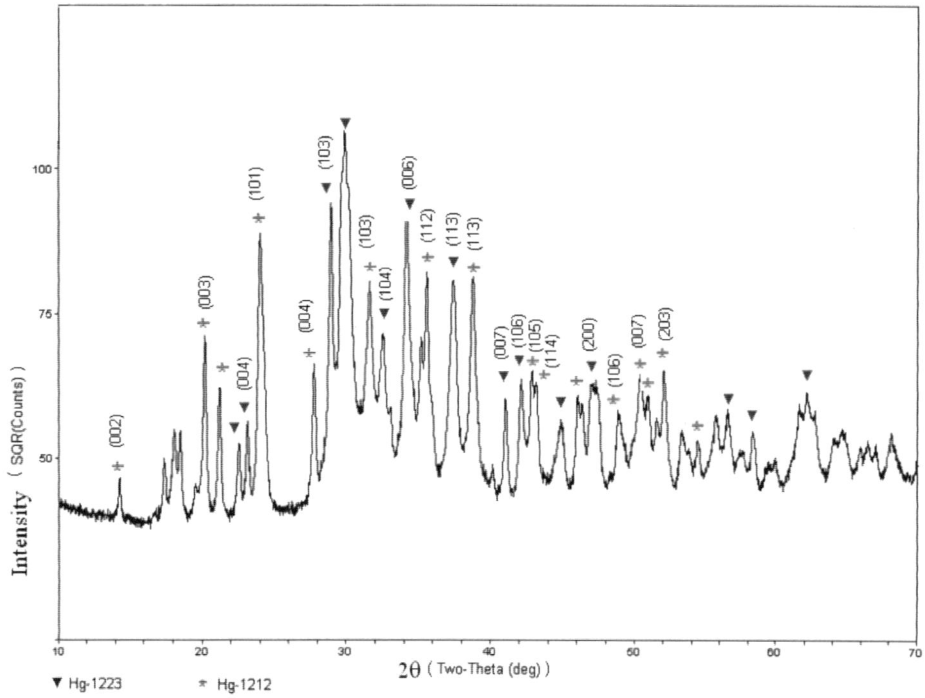

Figure 6. XRD Pattern and (*hkl*) planes (Miller indices) of under-doped sample in 7 hours counting. * indicates Hg-1212 and ▼ indicates Hg-1223, respectively.

Table 1. The lattice parameters of optimally doped sample calculated from the XRD data given in figure 5.

a-axis	*b*-axis	*c*-axis
3.8684 Å	3.8684 Å	15.7182 Å

Table 2. The lattice parameters of under-doped sample calculated from the XRD data given in figure 6.

a-axis	*b*-axis	*c*-axis
3.8328 Å	3.8328 Å	15.7452 Å

The average spacing of the mercury based sample, "s", has been derived from the *c* parameter of the sample. The average spacing values of the optimally and the under-doped samples have been calculated as 7.8591 Å and 7.8726 Å, respectively. Both of the average spacing values have been used to determine the Josephson penetration depth of the superconductors.

The existence of the intrinsic Josephson Effect in unconventional superconductors has indicated that these materials are natural superlattices of Josephson junctions formed on the atomic structure. The spatial period of the superlattices is only 15 Å, so Josephson junctions are densely packed in the intrinsic structure [26]. In this point of view, the mercury based copper oxide layered superconductors include a natural superlattice due to the magnitude of the translation vector which is about 15 Å.

4.Results and Discussion

Since the charge carriers of the most high temperature copper oxide superconductors are the electron holes (hole), there is a strong correlation between the hole concentration and the Meissner critical temperature. The density of the charge carriers in the superconducting sample can be increased by the procedure of oxygen annealing. In other words, optimally oxygen doping procedure results in the coupling of the uncoupled electrons in the hole type of conductivity. Therefore, over oxygen process to the superconducting sample causes extra electron doping to the system which results in some anti-symmetric contamination of the symmetric state [27].

In this study, X-ray data taken on optimally and under-doped samples have been clearly shown a mixed phase of $HgBa_2CaCu_2O_{6+x}$ (Hg-1212) and Hg-1223.According to the data taken on both samples, the crystal symmetries have been found to be tetragonal structure with a space group of P4/mmm. It has been determined that the superconducting plane (*ab*-plane) of the optimally doped sample is larger than that of the under-doped sample. However, the lattice parameter along the *c*-axis of the optimally doped sample is 0.027 Å shorter than the other one. Recalling the fact that the reduction in c-axis parameter increases the quantum tunnelling probability, so that the T_c of the optimally doped sample is higher than that of the under-doped sample. It has been first reported in this article that the optimally oxygen doped cut sample has the highest T_c of 140 K ever obtained among the superconducting samples prepared under normal atmospheric pressure. Furthermore, an appropriate oxygen annealing has been found to be very important in stabilizing the Hg-cuprate samples prepared at atmospheric pressure. This result has been confirmed by dc-SQUID measurements performed on mercury based samples which have been kept in air for several months after being synthesized [28].

According to our experimental studies, it has been determined that deficiency of oxygen doping reduces the T_c by about 20 K for bulk mercury based sample. Also, it has been observed that the critical quantum chaos points of the optimally doped sample are higher than that of non-doped samples. As is known, the geometric shape of the superconducting sample affects T_c and T_{PME}. While T_{PME} of virgin sample is higher than that of cut sample, T_c of virgin sample is lower than that of cut sample.

5. Conclusions

In this study, the investigation of the variation of the tunnelling probability in high temperature superconductors depending on oxygen content and geometry of the sample have been realized. It is also shown that this work displays a correlation between XRD data and quantum chaos points of the superconducting sample as an indication of a bridge between the momentum and Cartesian spaces.

The determination of chaotic points has a crucial importance for technological applications of the superconductors. Hence, the prediction of chaotic points of a superconducting system enables the technologists to figure out the reliable working temperature interval for construction of superconducting devices.

6. References

[1] Nielsen A P, Cawthorne A B, Barbara P, Wellstood F C, Lobb C J, Newrock R S and Forrester M G 2000 *Physical Review B* **62** 14380.

[2]Braunish W, Knauf N, Kataev V, Neuhausen S, Grutz A, Kock A, Roden B, Khomskii D and Wohlleben D 1992 *Phys.Rev.Lett.* **68** 1908.

[3]Braunish W, Knauf N, Bauer G, Kock A, Becker A, Freitag B, Grutz A, Kataev V, Neuhausen S, Roden B, Khomskii D and Wohlleben D 1994 *Phys.Rev.B* **48** 4030.

[4] Onbasli Ü, Ozdemir Z G and Aslan O 2008 *Chaos, Solitons and Fractals* submitted.

[5] Evangelou S N 2001 *Physica B* **296** 62.

[6]Waintal X, Weinmann D and Pichard J L 1999 *The European Physical Journal B* **7** 451-456.

[7] Bogomolny E, Gerland U and Schmit C 1999 *Physical Review E* **59** R1315-R1318.

[8] Lawrence W E, Doniach S 1971 *Proceedings of the 12th International Conference on Low Temperature Physics*, ed E. Kanda (Kyoto/Academic Press of Japan) p 361.

[9] Onbasli Ü, Wang Y T, Naziripour A, Tello R, Kiehl W and Hermann A M 1996 *Phys. Status Sol. (b)* **194** 371.

[10] James Gleick, Chaos: Making a New Science, Penguin Books, USA, 1987

[11] Aoki R, Murakami H, Kita T, Shirai M, Nishio Y, Svistunov V M, Dyachenko A I and Afanassyev D N 1996 *Physica B* **210-1120** 172.

[12] Egami T, Chung J H, McQueeney R J, Yethiraj M, Mook H A, Frost C, Petrov Y, Dogan F, Inamura Y, Arai M, Tajima S and Endoh Y 2002 *Physica B* **316-317** 62.

[13] Tsudo N and Shimada D 2003 *Physica C* **390** 95.

[14] Egami T 2006 *Journal of Phys. and Chem. of Sol.* **67** 50.

[15]Gough C E 1998 *The Gap Symmetry and Fluctuations in High-Tc Superconductors* Nato ASI Series, Series B: Physics 371 ed J. Bok, G. Deutscher, D. Pavuna and S.A. Wolf (New York: Plenum Press,) p465.

[16] Ketterson J B and Song S N 1999 Superconductivity (United Kingdom: Cambridge University Press) 86-87

[17] Özdemir Z G, Aslan Ö and Onbaşlı Ü 2006 *Journal of Phys. and Chem.Sol.* **67** 453.

[18] Özdemir Z G, Aslan Ö and Onbaşlı Ü 2007 *The Seventh International Conference on Vibration Problems ICOVP 2005 Springer Proceedings in Physics* vol 111, ed E İnan and A Kırış (Berlin/Heidelberg /Springer Verlag) p 377

[19] Li M S 2003 *Physics Report* **376** 133.

[20] Tsuei C C and Kirtley J R 2000 *Reviews of Modern Phys.* **72** 969.

[21] Wagner J L, Hunter B A, Hinks D G and Jorgensen J D 1995 *Physical Review B* **51** 407.

[22]Aslan Ö 2007 *Investigation of the Symmetries and the Breakages in Relativistic and Non-Relativistic Regions in High Temperature Superconductors* PhD Thesis Marmara University Institute of Pure and Applied Sciences.

[23] Onbaşlı Ü 2000 *Physica C* **332** 333.

[24]Özdemir Z G, Aslan Ö and Onbaşlı Ü 2008 ICSM submitted.

[25]Özdemir Z G 2007 *Determination of Electrical and Magnetic Properties of Mercury Based CuO_2 Layered Superconductors* PhD Thesis Marmara University Institute of Pure and Applied Sciences.

[26] Ustinov A V 1998 *Physica D* **123** 315.

[27] Ketterle W 2007 Quantum Mechanics At The Crossroads, ed J Evans and A S Thorndike (Berlin/Heidelberg: Spinger Verlag) p 159-182

[28] Onbaşli Ü, Öztürk S and Hasçiçek Y 1998 *Applied Superconductivity Conference, (ASC, 98)*, California, USA.

7. Acknowledgements

We would like to thank to our colleagues at the XRD Laboratory of Faculty of Engineering at Marmara University. The work has been supported by the Scientific Research Projects Commission of Marmara University with the project number of KPS-FEN-080805-0175.

Loss analysis of a 1 MW class HTS synchronous motor

S. K. Baik[1,4], Y. K. Kwon[1], H. M. Kim[1], J. D. Lee[1], Y. C. Kim[2], H. J. Park[2], W. S. Kwon[2] and G. S. Park[3]

[1]Superconducting Devices & Cryogenics Group,
Korea Electrotechnology Research Institute, P.O. BOX 20, Chang Won, Korea
[2]System Tech. Research Team, Doosan Heavy Industries & Construction Co., Ltd., Chang Won, 641-792, Korea
[3]School of Electrical and Computer Engineering, Pusan National University, Busan, 609-735, Korea

E-mail: skbaik@keri.re.kr

Abstract. The HTS (High-Temperature Superconducting) synchronous motor has advantages over the conventional synchronous motor such as smaller size and higher efficiency. Higher efficiency is due to smaller loss than the conventional motor, so it is important to do loss analysis in order to develop a machine with higher efficiency. This paper deals with machine losses those are dissipated in each part of a HTS synchronous motor. These losses are analyzed theoretically and compared with loss data obtained from experimental results of a 1 MW class HTS synchronous motor. Each machine loss is measured based on IEEE 115 standard and the results are analyzed and considered based on the manufacturing of the test machine.

1. Introduction

A 1 MW class superconducting synchronous rotating machine with HTS field coil is considered for test and analysis. This machine was designed to have the rotating speed with 3600 rpm that is the ordinary speed of turbine generator and industrial motors applied to blowers, pumps and compressors installed in large plants. The developed HTS synchronous machine has been tested while connected to a driving and load induction machine with 1.1 MW rating. As the first step to a synchronous rotating machine test, the object 1 MW HTS motor has been rotated by the driving induction motor while the three phase armature (stator) coils are open-circuited and short-circuited at the terminal. Through armature open circuit test, excitation voltage was measured at different field current and rotor speed. This excitation voltage is induced at the armature coils of a synchronous rotating machine by rotating magnetic field of the field coil and does an important role in output characteristics of a synchronous rotating machine. During this test open-circuit losses had been measured at each test condition and open-circuit core losses were derived from them. Through armature short circuit test, short-circuit losses had been measured at each test condition and stray load loss was derived from them. These two tests are basic tests as generator operation of a synchronous machine without any connected electrical load. After these two tests, the object HTS machine was driven by an inverter and the efficiency as a motor was measured under several load conditions applied by the 1.1 MW induction machine operated as a generator.

2. The object 1 MW class HTS synchronous motor

The developed HTS synchronous machine is composed of Bi-2223 HTS field coils cooled by conduction heat transfer of liquid neon and copper armature coils cooled by water passing through them. The Figure 1 shows the construction of the developed 1 MW HTS synchronous machine and the Table 1 shows the major specifications.

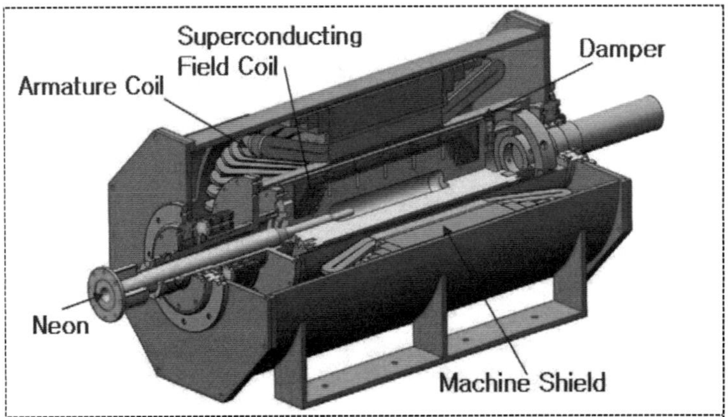

Figure 1. Construction of the developed 1 MW HTS synchronous machine

Liquid neon is supplied from outside of the rotor through a long stainless steel pipe connected to a cold head of a GM cryocooler which liquefies gas neon evaporated from inside of the rotor. This 2 phase closed loop cooling mechanism is called thermosiphon that has been applied to several HTS motors developed worldwide [1] [2].

Table 1. Major specifications of the object 1 MW HTS synchronous motor

Rating Capacity	1 MW
Rotating Speed	3600 rpm
Frequency	60 Hz
Poles	2
Armature Terminal Voltage	3300 V
Power Factor	1.0
Synchronous Reactance	0.13 p.u
Field Coil Current	150 A
Field Coil Operating Temperature	30 ~ 35 K
Field Coil Turns	3348
Armature Slots	36
Armature Coil Turns	48 turns/phase
Axial Length × Height	2.4 × 1.2 m

Journal of Physics: Conference Series **153** (2009) 012003 doi:10.1088/1742-6596/153/1/012003

3. Armature open and short circuit test

As a basic test of a synchronous machine, the object 1 MW HTS machine has been rotated by a 1.1 MW induction machine shown in Figure 2 while the three phase armature coils open- and short-circuited as shown in Figure 3 and Figure 4. During these two tests called "Armature Open Circuit Test" and "Armature Short Circuit Test", the open circuit voltage, E_{af}, and the short circuit current, I_a, were measured while the 1 MW HTS machine was operating as a generator driven by the dynamo 1.1 MW induction machine [3].

Figure 2. Armature open circuit (OCC) and short circuit (SCC) test result

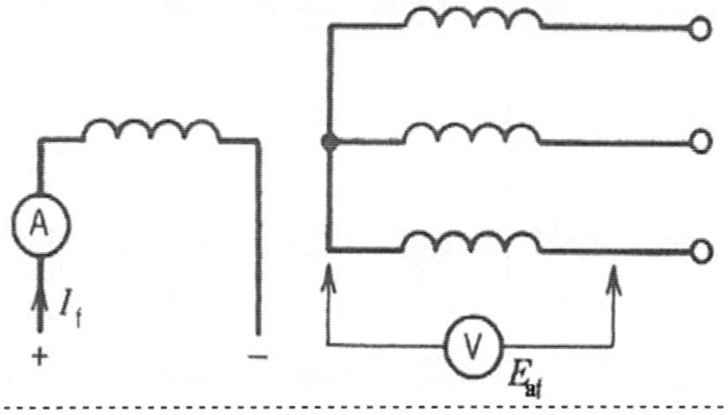

Figure 3. Circuit for armature open circuit test

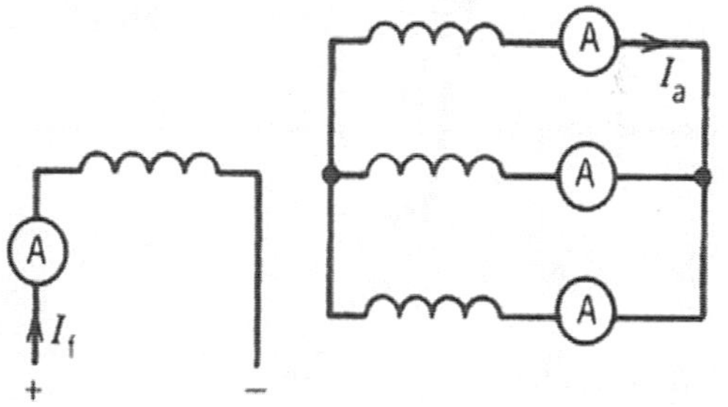

Figure 4. Circuit for armature short circuit test

The armature open circuit voltage was measured at various rotor speeds up to 1800 rpm, the half of 3600 rpm rating speed and field currents up to the rating current of 150 A. Due to vibration problem of the dynamo system coupled to the HTS machine, it was impossible to do tests at 3600 rpm rating speed. The results are shown as OCC (Open Circuit Characteristics) curves in Figure 5 at each different field current and rotor speed. The open circuit voltage in Figure 5 is three phase line-to-line terminal voltage (Root Mean Square(RMS) value) that is $\sqrt{3}$ times of the armature phase voltage (RMS value), E_{af}, shown in Figure 3. This voltage is also called "excitation voltage" and directly proportional to the rotor speed (angular frequency, ω) because the time variation of the armature coil flux linkage by the field coil excitation is directly proportional to the rotor speed. Other parameters to decide the excitation voltage are mutual inductance between the field and the armature coils and the field current, I_f, by the relation of the equation (1). In case of the conventional iron-cored synchronous machine, the open circuit voltage is saturated due to magnetic saturation of iron-core as the field current increases. However, the developed 1 MW machine shows non-saturated (straight) OCC curves as shown in Figure 5 because no ferro-magnetic material (iron) is used for the rotor and the armature teeth [4]. Moreover, the measured waveforms of open circuit voltages are sinusoidal as shown in Figure 6 because there are non-magnetic FRP (Fiber Reinforced Plastic) armature teeth instead of iron teeth to support the armature coils.

From the OCC test result, the maximum mutual inductance between the armature and the field coils can be calculated from the equation (1), which is 41.74 mH and almost same with a calculated value, 43.24 mH, from 3 dimensional Finite Element Analysis [5].

$$E_{af} = \frac{\omega L_{af} I_f}{\sqrt{2}} \qquad (1)$$

Where, ω is the angular frequency in rad/sec

I_f is field coil current

L_{af} is maximum mutual inductance between field and armature coils

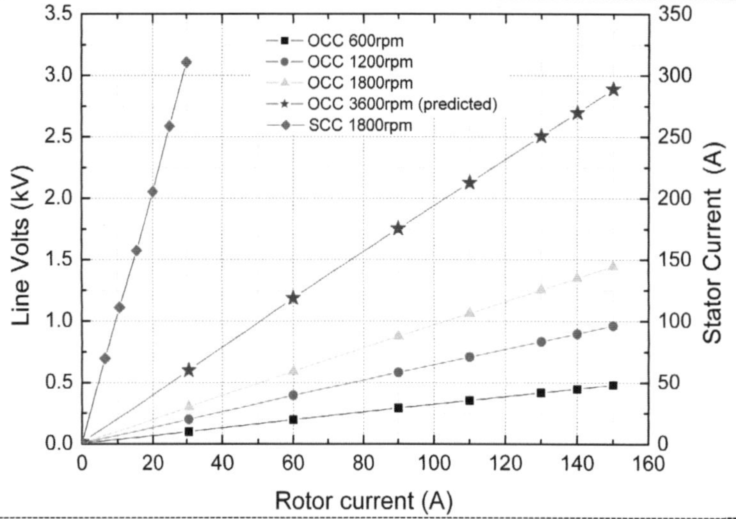

Figure 5. Armature open circuit (OCC) and short circuit (SCC) test result

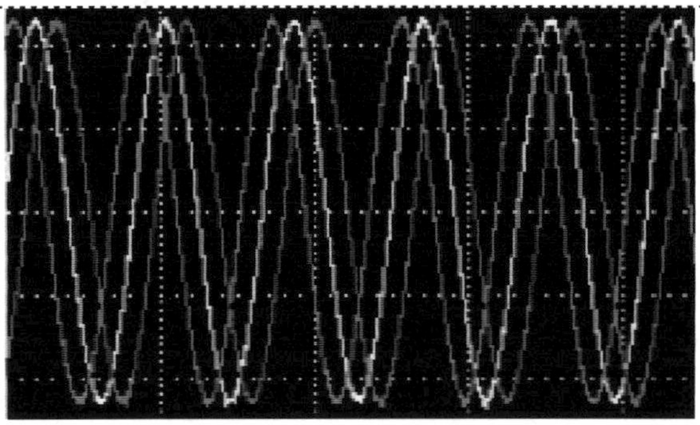

Figure 6. Measured armature open circuit voltage waveforms

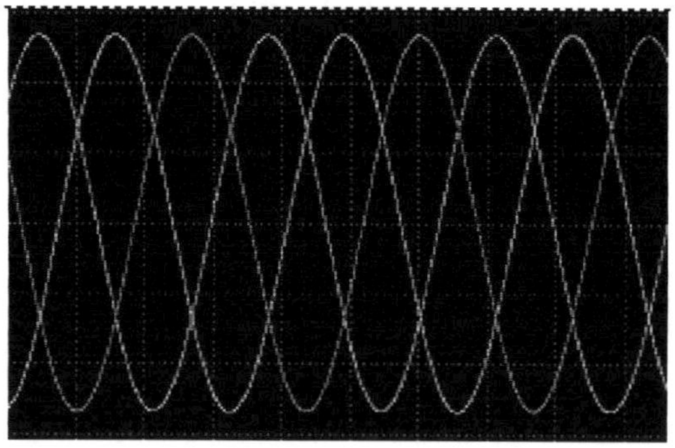

Figure 7. Measured armature short circuit current waveforms

The armature short circuit current was measured at 1800 rpm, the half of 3600 rpm rating speed and several field currents up to 30 A. The result is also shown as SCC (Short Circuit Characteristics) curve in Figure 5 at each different field current and very sinusoidal short circuit current waveforms are also measured by an oscilloscope as shown in Figure 7. The armature short circuit current is related with other parameters as the equation (2). The armature open circuit voltage, E_{af}, is 173.6 V per phase measured from OCC test and the short circuit current, I_a, is 310.8 A measured from the SCC test with the same field current of 30 A and rotor speed of 1800 rpm and the armature coil resistance per phase, R_{ph}, is 0.0965 Ω measured at 20°C. By using these parameters the synchronous reactance, X_s, is 0.5502 Ω at 1800 rpm, so the synchronous inductance, L_s, can be calculated from the equation (3), which is 2.92 mH and the synchronous reactance at the rating speed of 3600 rpm is 1.1008 Ω calculated from the equation (3) by substitution of 2 times of the angular frequency, ω, at 1800 rpm [4].

$$\mathrm{I_a} = \frac{E_{af}}{\sqrt{R_{ph}^2 + X_s^2}} \tag{2}$$

Where, I_a is the armature coil short circuit current during SCC test
E_{af} is the excitation voltage per phase during SCC test
R_{ph} is armature coil resistance per phase
X_s is synchronous reactance

$$\mathrm{L_s} = \frac{X_s}{\omega} \tag{3}$$

4. Open circuit core loss

Table 2. Core loss calculation from open circuit characteristics test (Generator mode)

Field Current [A]	Open Circuit Loss (HTS Machine Input Power) [kW]			Core Loss [kW]		
	600 rpm	1200 rpm	1800 rpm	600 rpm	1200 rpm	1800 rpm
0	0.346	0.702	1.027	0	0	0
30	0.348	0.856	1.491	0.002	0.154	0.464
60	0.614	1.656	3.114	0.268	0.954	2.087
90	0.908	2.708	5.414	0.562	2.006	4.387
110	1.12	3.638	7.344	0.774	2.936	6.317
130	1.419	4.714	9.6	1.073	4.012	8.573
140	1.597	5.329	10.916	1.251	4.627	9.889
150	1.817	6.107	12.492	1.471	5.405	11.465

During the open circuit test, input power to the HTS machine was measured at each field current and rotating speed as shown in Table 2. Core loss of a synchronous machine is generated at armature iron core by alternating magnetic field due to rotating magnetic field of rotor, which consists of hysteresis and eddy-current losses. In case of superconducting synchronous machine, core loss is generated at the machine shield stacked with silicon steel sheet. The open circuit loss at 0 A field current (with zero excitation) is generated by windage and friction loss due to rotation, so the core loss shown in Table 2 is calculated by subtracting this loss at 0 A field current from the open circuit loss. Figure 8 shows the core loss according to line-to-line armature terminal voltage [3].

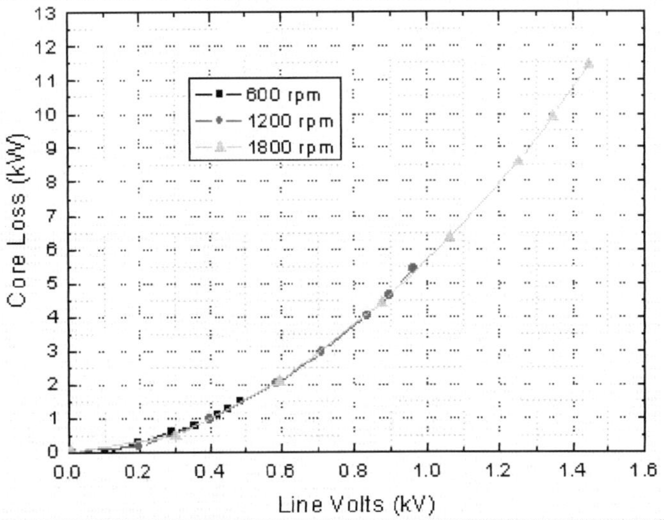

Figure 8. Core loss curves from open circuit test

5. Short circuit loss and stray load loss

Table 3. Short circuit characteristics test data (Generator mode, 1800 rpm)

Field Current [A]	Armature Current [A]	Armature Coil Temperature (Coil Center) [□] R	S	T	HTS Machine Input Power [kW]	Short Circuit Loss [kW]
0	0.432	23.8	23.4	23.9	1.163	0.136
6.5	69.76	25.5	25.2	25.7	2.569	1.542
10.5	111.19	26.6	26.0	26.5	4.807	3.78
15.3	157.52	28.1	27.8	28.2	8.569	7.542
20	205.41	31.4	31.1	31.5	14.058	13.031
25	258.62	37.4	37.2	37.6	22.131	21.104
30	310.76	45.3	45.6	45.9	32.402	31.375

During the short circuit test, input power to the HTS machine was also measured at each field current and 1800 rpm rotating speed as shown in Table 3. The short circuit loss is calculated by subtraction of windage and friction loss, which is 1.027 kW at 1800 rpm and zero excitation from Table 2, from the

HTS machine input in Table 3. Stray load loss consists of the losses arising from non-uniform current distribution in copper and the additional core losses produced in the iron by distortion of the magnetic flux by the load current. Stray load loss during the short circuit test can be calculated from the equation (4) with the data in Table 2 and Table 3. The core loss is 0.464 kW from Table 2 of the open circuit test result and the armature I^2R loss is calculated to 27.958 kW with the armature coil resistance per phase of 0.0965 Ω measured at 20°C and the armature short circuit current of 310.76 A at 1800 rpm and 30 A field current. So the stray load loss, 2.935 kW, is calculated by subtracting the core loss and the armature I^2R loss from the short circuit loss, 31.375 kW, at this condition. Figure 9 shows the short circuit loss and the stray load loss curves obtained from the test [3].

$$\text{Stray load Loss} = \text{Short circuit Loss} - \text{Core Loss} - \text{Armature } I^2R \text{ Loss} \tag{4}$$

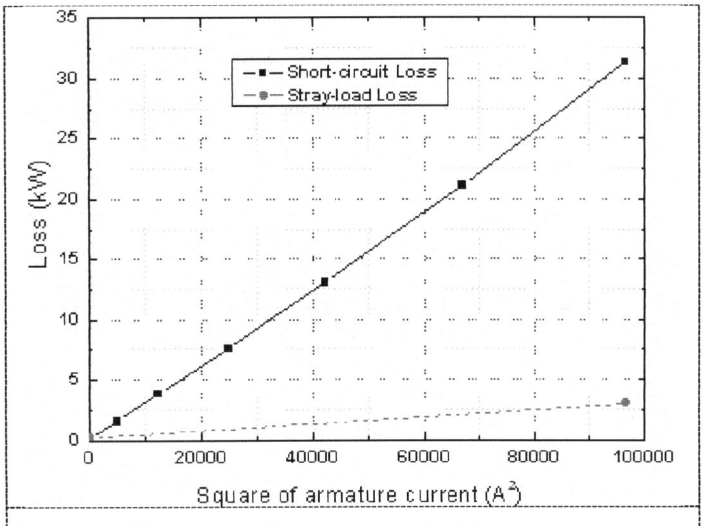

Figure 9. Short circuit loss and stray load loss curves

6. Efficiency measurement at load test as a motor

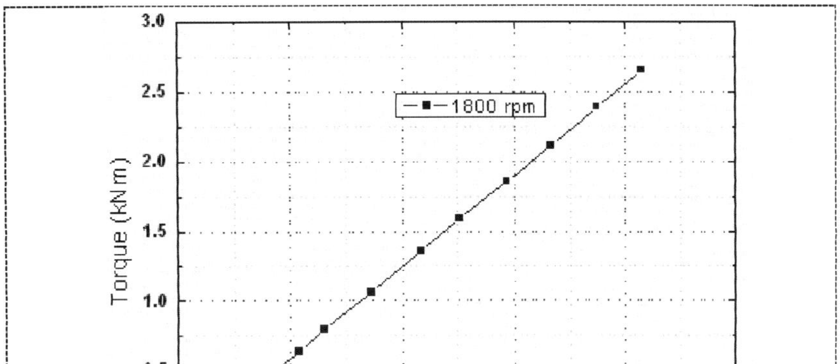

Figure 10. Generated torque according to armature current (1800 rpm)

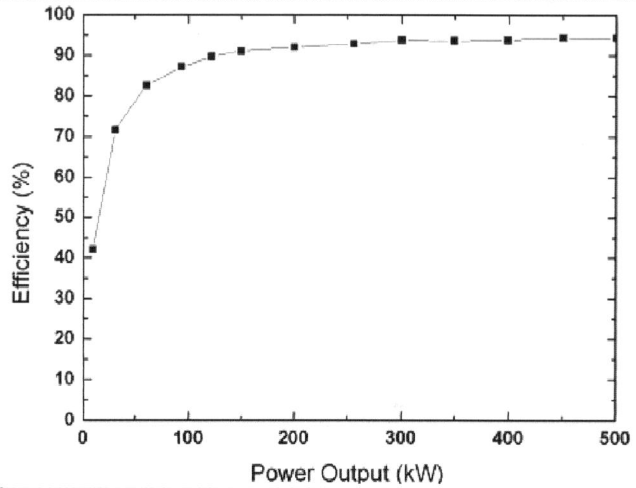

Figure 11. Efficiency variation according to output power at 1800 rpm motor operation

Figure 10 shows test result of generated torque according to armature current at 1800 rpm as a motor operation. Due to vibration problem this test could not be done at the rating speed of 3600 rpm, but the test machine generated rating torque, 2650 Nm, with 207.2 A armature current and 1475 V armature line-to-line voltage at 1800 rpm. Based on these values, per unit synchronous reactance is 0.13 p.u, which is the same with the design value. At this load torque condition, losses can be summarized as Table 4. During this test HTS motor was rotating at 1800 rpm by an adjustable speed drive. Figure 11 shows efficiency variation at partial loads during load test at 1800 rpm. The output powers are calculated from the measured torques multiplied by the angular frequency, 188.5 rad/sec, at 1800 rpm rotating speed. It is confirmed that the HTS motor shows high efficiency even at small loads on the contrary of conventional motor having low efficiency at small load. This result comes from the fact that the conventional synchronous motor has the same field coil excitation I^2R loss regardless of power output, so this loss becomes relatively larger and reduces largely the efficiency of the conventional motor at small load situation. The overall efficiency including a cryocooler power consumption is 93.17 % at 1800 rpm and 2650 Nm during motor operation. We could not test at 3600 rpm rating speed due to vibration problem, but it is expected that the developed HTS motor will show higher efficiency at 3600 rpm rating speed.

Table 4. Loss analysis at 1800 rpm and 2650 Nm during motor operation

Losses	[kW]	Percentage
Core loss	11.465	31.26 %
Armature I^2R loss	12.429	33.89 %
Stray load loss	1.305	3.56 %
Windage and friction loss	1.027	2.80 %
Field coil cooling loss	7.5	20.45%
Etc	2.953	8.04 %
total	36.679	100 %

7. Conclusion

The generated losses of the developed 1 MW class HTS machine were calculated from the basic tests of a synchronous machine based on the IEEE standard 115. Although the HTS machine could not be tested at 3600 rpm rating speed, efficiency was measured at each applied load condition up to the full load torque during synchronous motor operation. In this full torque condition, the generated losses were analyzed based on the results obtained from the basic tests such as OCC and SCC tests. The core loss among the generated losses is too high to bring a test result showing distinguished high efficiency compared with a conventional synchronous machine with the same specification. It is estimated that this high core loss even in the armature open circuit condition is due to single-stranded armature coil made of hollow conductor applied for a purpose of water-cooling. Too large eddy current might be induced in this single-stranded copper conductor. Another reason to decrease efficiency is et cetera (Etc) loss shown in Table 4, which might come from large armature current harmonics generated from the driving inverter.

Acknowledgments
This research was supported by a grant from Center for Applied Superconductivity Technology of the 21st Century Frontier R&D Program funded by the Ministry of Education, Science and Technology, Republic of Korea.

8. References

[1] Kalsi S S, Weeber K, Takesue H, Lewis C, Neumueller H –W and Blaugher R D 2004 *Development Status of Rotating Machines Employing Superconducting Field Windings Proc. IEEE* vol **92**, NO 10 p 1688–1704

[2] Frank M, Frauenhofer J, Hasselt P van, Nick W, Neumueller H –W and Nerowski G 2003 *Long-Term Operational Experience with First Siemens 400 kW HTS Machine in Diverse Configurations IEEE Trans. Applied Superconductivity* vol **13** p 2120–2123

[3] The Institute of Electrical and Electronics Engineers Inc. 1995 *IEEE Guide: Test Procedures for Synchronous Machines (IEEE Std 115-1995)* p 27–38

[4] Fitzgerald A E, Kingsley Charles Jr. and Umans Stephen D *Electric Machinery McGRAW-HILL International Editions 5th Edition* p 114–229

[5] Baik S K, Kwon Y K, Kim H M, Lee E Y, Lee J D, Kim Y C, Moon T S, Park H J, Lee C H and Kwon W S 2008 *Excitation Voltage Estimation of HTS Motor via Magnetic Energy Calculation Proceedings of the 39th Korean Institute of Electrical Engineers Summer Conference* p 870–871

International Conference On Superconductivity and Magnetism (ICSM2008) IOP Publishing
Journal of Physics: Conference Series **153** (2009) 012004 doi:10.1088/1742-6596/153/1/012004

Pinning Properties of Bi-2212 Single Crystal Whiskers

S Altin[1], **M A Aksan, Y Balci, M E Yakinci**
Inonu Universitesi, Fen Edebiyat Fakultesi, Fizik Bolumu, *Süperiletkenlik Arastirma Grubu,* 44280, Malatya, Turkey

E-mail: saltin@inonu.edu.tr

Abstract. Bi-2212 single crystal whiskers have been fabricated on the surface of the bulk glass BSCCO pieces prepared by the melt-quenching process. The characterizations of the whiskers were made using XRD, SEM-EDX, M-T, M-H analysis. Pinning force calculations were performed using magnetic hysteretic loops taken at different temperatures and results obtained were analyzed using scaling law equation. The whiskers fabricated showed higher pinning at low temperature. The pinning force at 10 K was calculated to be $3{,}83\text{x}10^{11}$ N/m^3 under 2 T applied magnetic field. The results obtained showed that a decrease on the flux pinning force decreases the critical current density of whiskers.

1. Introduction

Technological developments lead scientists to the fabrication of HT_c materials in single crystal form with micro-nano dimensions which is essential to use in technological devices. Since discovery of high temperature (HT_c) superconducting cuprates, many studies have been done on fabrication of the HT_c materials in different dimensions by using different preparation techniques. The superconducting properties and technological applications of the fabricated materials have been widely investigated.

The $Bi_2Sr_2CaCu_2O_y$ (Bi-2212) whiskers have been discovered by Matsubara [1]. The single crystalline nature of the Bi-2212 whiskers has drawn great attention among the researchers which tried to produce the materials in different dimensions by using different preparation methods [2-4]. The fabrication of flexible whiskers without grain boundary and crystal defect is necessary for scientific researches and then integration to the devices [5, 6]. In future, it is believed that superconducting single-crystal whiskers will be most used materials in the technological applications such as micro/nano electronic switches, high current applications and Josephson junction based devices [7,8].

Flux pinning is one of the crucial problems in the development of HT_c superconductors of high current density. In 1973, Krammer first suggested a mechanism of flux motion characterized by synchronous shear of the vortex line around flux lines which are too strongly pinned to be broken. The model was further developed owing to new insight into the physics of flux pinning, especially for the HT_c superconductor cuprates [9, 10].

Grain boundaries, stacking fault, crystal defects and impurities behave as the primary flux-pinning centers in HT_c superconductors. The applied temperature is very important for high current and high pinning force. At high temperature, thermal energy causes melting of vortex lattice while in

[1] Inonu Universitesi, Fen Edebiyat Fakultesi fizik Bolumu, 44280, Malatya

the region of low temperature, HT_c materials generally show strong pinning. It is well known that HT_c materials in small dimensions also exhibit strong pinning. The behavior of the vortex state in Bi-2212 remains still controversial againist many theories being purposed to explain experimentally observed results, such as collective pinning, vortex glass, flux lattice melting, flux creep and scaling law. Studies on pinning properties of HT_c materials showed that the scalling law can be easily applied to explain the pinning force F_p (H). Flux pinning, $F_p(H)=\mu_0 HJ_c(H)$, shows a characteristic field dependence proportional to scaling law $h^m(1\text{-}h)^n$ where h is the reduced field H/H^* and H^* is the normalization field [11-13].

In this study, we fabricated the Bi-2212 whiskers using conventional glass-ceramic technique. Structural and microstructural properties of the whiskers fabricated have been characterized with XRD and SEM-EDX. The pinning properties were investigated by using the scalling law and experimental results obtained were compared with theoretical results.

2. Experimental Procedure

$Bi_2Sr_2CaCu_2O_y$ (Bi-2212) whiskers were grown on BSCCO glass plates using the glass-ceramic technique. In this study, the starting composition was selected to be $Bi_3Sr_2Ca_2Cu_3O_y$ (Bi-3223). An appropriate amount of high purity (at least 99.9%) Bi_2O_3, $SrCO_3$ and CaCO3 CuO oxide powders was mixed in an agate mortar for 3–5 h to attain the starting composition. The mixture was melted in an α-Al_2O_3 crucible at 1200^0C for 2-3 h in a high temperature furnace and then molten material were quenched between two cold coppers plates. Thus, rapidly quenched, dark, shiny and approximately 1–3mm thick amorphous materials were obtained. The whiskers were grown on the glass plates in oxygen atmosphere at 850^0C for 120 h

Structural analysis of both main matrix and the whiskers fabricated were characterized by X-ray difractometer (XRD). The Rigaku RadB system having CuK_α radiation (λ=1.5405 Å) and scan rate of 3^0/min were used during XRD analysis. Unit cell parameters were calculated by using XRD data obtained. Surface morphology of the whiskers and the main matrix were carried out by using a Leo Evo40VP scanning electron microscope (SEM).

Magnetization measurements were investigated by using Cryogenic Q-3398 vibrating sample magnetometer (VSM). Superconducting critical trainsition temperature, T_c, of the whiskers fabricated was determined with M-T experiments. Pinning force calculation was performed by magnetic hysteretic loops taken at different temperatures and directions and the data obtained was fitted to scaling law equation. The results were compared with experimentally and theoretically obtained results.

3. Results and Discussion

X-ray diffraction (XRD) pattern of the whisker and the main matrix is showed in Figure 1. The whisker consisted of pure Bi-2212 phase without impurites. It was found that the whisker was oriented along (00*l*) direction. The crystal symmetry was obtained to be tetragonal and unit cell parameters to be a(=b)=5.4049Å and c= 30.6017Å which corresponds to the previously obtained values for Bi-2212 single crystal materials [14,15]. For the main matrix, it was seen that intensity of diffraction peaks decreased significantly, compared to the XRD peaks for the whiskers. Some Bi-2212 peaks were detected, however, large number of impurity phases such as $Cu_3Sr_2O_5$, $Bi_4Sr_8Cu_5O_x$ was found in the main matrix. We believe that ionic diffusion from main matrix to whiskers takes place during the whisker formation and this ionic diffusion causes the multiphased main matrix.

Figure 2 shows the surface morphology of whiskers and the main matrix. It was observed that whiskers have grown on the local nucleation centers on the surface of the main matrix and almost perpendicular to surface, Figure 2.a. This suggests that whiskers grow by the time dependent nucleation-growth process. Figure 2.b shows the cross-section of a whisker. The SEM observations revealed very smooth surface formation with no grain boundaries on the direction of the growth. Any impurity or substructures were not observed, indicating a clear and perfect microstructural formation EDX analysis showed that composition of the whisker perfectly corresponded to the Bi-2212 phase.

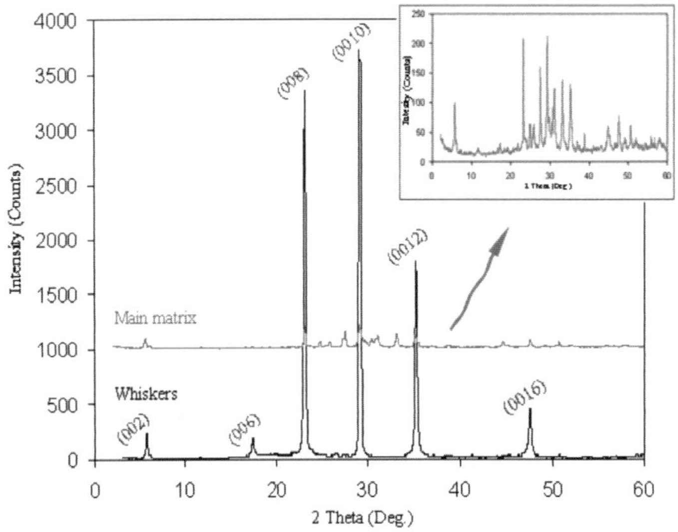

Figure 1. X-ray diffraction of whiskers and main matrixes.

Figure 2. a) Main matrix with whiskers and **b)** whisker surface

morphology.

Temperature dependence of magnetization of both the whisker and the main matrix is shown in Figure 3. A sharp drop in the magnetization curve was obtained, suggesting a good diamagnetic property and single crystal nature of the whisker fabricated. The T_c value for the whisker was found to be 89.4 K. However, the main matrix has a transition indicating intergranular connectivity. The diamagnetic signal increased and the T_c value decreased. It was seen that the impurity concentration influenced the onset T_c of the main matrix. The T_c value for the main matrix was obtained to be 76.8 K, which is attributed to the existence of weakly coupled grains.

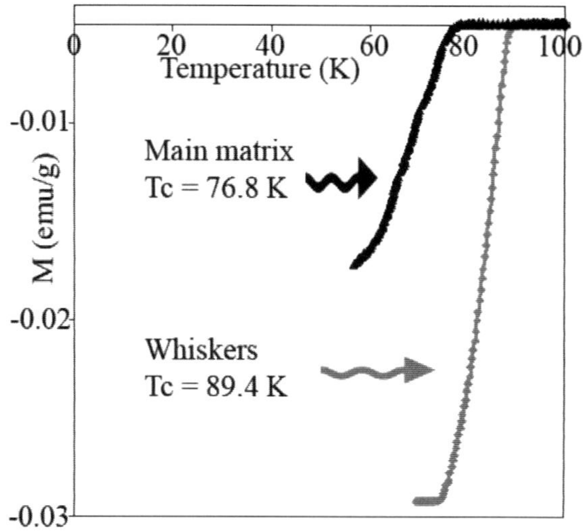

Figure 3. M-T results of patterns produced

Under a fixed magnetic field, when transport current density, J, is passed through a HTc superconductor in the mixed state, there is an electromagnetic interaction between the electrons carrying the current and fluxes themselves.A Lorentz force acts on each flux at right angles to both the direction of the flux penetration and the transport current. Two regimes can take place depending on the competition between the Lorentz and pinning forces: flux flow when the Lorentz force is larger than the pinning and/or formation of pinning when the pinning force is stronger than the Lorentz force. The fluxes remain stationary and the sample continues to have zero resistance, as long as there is a pinning force at least equal to the Lorentz force acting to move the fluxes across the sample.However, when the Lorentz force acting on the flux exceeds the pinning force, the fluxes begin to move across the sample. The energy dissipated by the drag or viscous force opposing this motion is supplied by the transport current. Thus, when fluxes are in motion, the material becomes resistive [16]. In HTc superconductors containing various defects sucah as grain boundaries and impurities, vortices are pinned by the defects.

Change of the pinning force of whiskers and the main matrix with applied field was calculated using $F_p(H)=\mu_0 H J_c(H)$, where H magnetic field and $J_c(H)$ the critical current density calculated by using Bean equation. The obtained maximum $F_p(H)$ values and the corresponding $J_c(H)$ and H values were presented in Table 1. The maximum $F_p(H)$ value was found at 10 K for all the materials. It is obvious that $F_p(H)$ in the ab-axis of the whisker was higher than that of the c-axis. For the main matrix, it was determined that the $F_p(H)$ value was approximately 100 times lower than that of both c-axis and ab-axis of the whisker.

We also fitted to pinning force obtain from M-H curves by using scaling law. According to scaling law, the pinning force was given below equation [17];

$$f = F \cdot h^m (1-h)^n \quad (1)$$

where $h = H / H^*$ and H* is normalization field determined by using equation $J_c(H=H^*)=0$, *F, m ,n* are constants determined pinning force. The parameters obtained fitting of experimentally calculated pinning forces are shown in table 2.

Table 1. Materials produced maximum pinning force, critical current density at constant temperature.

Materials	Temperature (K)	Max. Pinning Force(N/m^3)	Magnetic Field (Tesla)	Critical Current Density (A/cm^2)
Bi-2212 Whisker (perp. to c-axis)	10	$3.13x10^{11}$	2.0	$0.2x10^6$
Bi-2212 Whisker (perp. to ab-axis)	10	$3.83x10^{11}$	5.5	$1.3x\ 10^6$
Main Matrix	10	$1.17x10^9$	1.5	$2.7x10^3$
Bi-2212 Whisker (perp. to c-axis)	20	$2.39x10^9$	0.3	$8x10^4$
Bi-2212 Whisker (perp. to ab-axis)	20	$2.73x10^{10}$	1.0	$5x10^5$
Main Matrix	20	$1.40x10^8$	0.3	$1.1x10^3$

Variation of the pinning force calculated by using equation (1) with the magnetic field applied along the c-axis of the whisker is shown in Figure 4.a. The pinning force rapidly increased at low magnetic field values. Maximum pinning force was obtained to be $3.13x10^{11}$ N/m^3 under 2.0 T at 10 K. The pinning force decreased exponantially with increasing the magnetic field and also temperature. The magnetic field (applied along a-b axis) dependence of the pinning force is shown in Figure 2.b. Maximum pining force value was found to be $3.83x10^{11}$ N/m^3 under 5.5 T at 10 K. The F_p^{max} in the a-b axis was higher than that of the c-axis, indicating that vortices on the ab-plane were pinned stronger than the c-axis. We believed that this can be due to low dimensionality of a-b axis. Figure 4.c shows the pinning force variation with the applied magnetic field for the main matrix. Maximum pining force value was found to be $1.17x10^9$ N/m^3 under 1.5 T at 10 K. It is clear that the variation of the pinning force with the applied magnetic field for the c-axis of the whisker displayed similar behavior as the main matrix. The pinning force in the main matrix reduces due to high density of impurity and so low critical current density is attained [18]. It was concluded from the results obtained that Bi-2212 single crystal whiskers showed strong pinning force and the impurites in the main matrix decreased the critical current density, J_c, and pinning properites.

The experimental results were fitted to Scaling law and the results are given in Table 2. As a result of the fitting process, the normalization field for the c-axis and the a-b axis of the whisker and main matrix was determined and compared with *H** values. It was found that the a-b axis shows strongest pinning and highest *H** value.

Table 2. The parameters obtained scaling law

Materials	Temperature (K)	m	n	F	$h(f_{pmax})$	H* (T)
Bi-2212 Whisker c-axis	10	0,850	2,93	280.1	0,25	11
Bi-2212	10	0,700	2,39	980,1	0,79	7,1

Whisker ab-axis						
Main Matrix	10	0,235	5,09	0,067	0,18	5

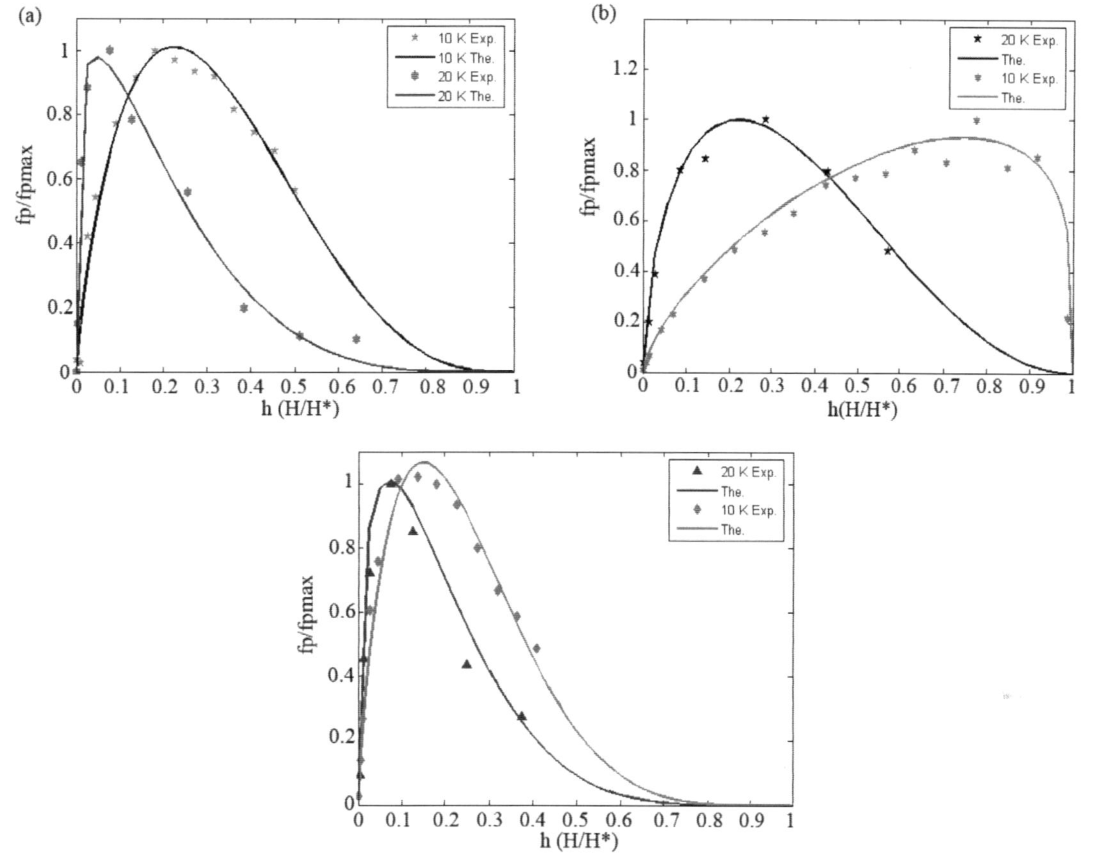

Figure 4. Pinning force vs. applied field perpendicular l to a) c-axis, ab-axis of whiskers and c) main matrix

4. Conclusions

In this study, we have fabicated the single-crystal Bi-2212 whiskers and investigated pinning properties of the whisker fabricated. XRD and SEM investigations showed that the whisker fabricated was oriented along the c-axis. The whiskers have grown on the local centers by the nucleation-growth process. Pinning force was calculated using Scaling law. Maximum pinning force value was found to be $3.83x10^{11}$ N/m^3 under 5.5 T at 10 K for the ab-axis of the whisker. However, it was seen that the pinning force values decreased to $3.13x10^{11}$ N/m^3 under 2.0 T at 10 K and $1.17x10^9$ N/m^3 under 1.5 T at 10 K for the c-axis and the main matrix, respectively. High pinning force is believed to be due to low anisotropy of the ab-axis. This clearly indicates that vortices were pinned on the ab-plane.

5. Acknowledgement

This work was supported by the Inonu universitesi Bilimsel Araştırma fonu with the project no 2008-07.

References

[1] Matsubara I, Kageyama H, Tanigawa H, Ogura T, Yamashita H and Kawai T 1989 *Jpn. J. Appl. Phys.* **28** L1121
[2] Tange M , Amano T , Yokoshima M , Nishizaki S And Yoshizaki R 2004, *Physica C* **412-414** 610
[3] Takano Y, Hatano T, Fukuyo A , Ishii A , Arisawa S, Tachiki M and Togana M 2001 *Supercond. Sci. Technol.* **14** 765
[4] Matsubara I , Funahashi R , Ueno K and Ishikawa H , 2000 *Meter. Res. Bull.* **35** 441
[5] Zhou Y Q, Chen Z J, Jin H, Zheng P and Wang W H 1999 *J. Cryst. Growth* **204** 289
[6] Jayevel R, Mochiku T, Ooi S and Hirata K 2001 *J. Cryst. Growth* **229** 339
[7] Takano Y, Hatano T, Fukuyo A, Ishii A, Ohmori M, Arisawa S, Togano K and Tachiki M 2002 *Pys. Rev. B* **65** 140513
[8] http://www nims go jp/eng/
[9] Cheggour N and Hampshire D P , 1999 *J. Appl . Phys.* **86** 552
[10] Okamoto H, Irie F, Kiss T, Inoue M and Kanazawa M 1999 *Supercond. Sci. Technol.* **12** 1102
[11] Wördenweber R 1999 *Rep. Prog. Phys.* **62** 187
[12] Goffman M F , Herbsommer J A and De la Cruz F 1998 *Phys. Rev. B* **57** 3663
[13] Chorin A J 1988 *Commun. Math. Phys.* **114** 167
[14] Bean C P 1962 *Phys. Rev. Lett.* **8** 250
[15] Okamura K, Kiuchi M, Otabe E S, Yasuda T, Matsushita T and Okayasu S 2004 *Supercond. Sci. Technol.* **17** S20
[16] Okamoto H, Irie F, Kiss T, Inoue M and Anazawa K 1999 *Supercond. Sci. Technol.* **12** 1102
[17] Albrecht J, Djupmyr M and Br¨uck S 2007 *J. Phys.: Condens. Matter* **19** 216211
[18] Babu N H, Iida K and Cardwell D A 2007 *Supercond. Sci. Technol.* **20** S141

Anomalous ferromagnetic behaviour of Y_2O_3 and CuO nanoparticles in $YBa_2Cu_3O_y$ superconductor

M. Basoglu[1], K. Öztürk[1], Ş. Çelik[2], C.R.M. Grovenor[3] and E. Yanmaz[1]

[1]Department of Physics, Faculty of Arts & Sciences, Karadeniz Technical University, 61080-Trabzon, Turkey
[2]Department of Physics, Faculty of Arts & Sciences, Rize University, Rize, Turkey
[3]Department of Materials, University of Oxford, Parks Road, OX1 3PH, Oxford, UK.

E-mail: mehmet_basoglu@ktu.edu.tr

Abstract. Anomalous ferromagnetic behaviour of the Y_2O_3 and CuO components in $YBa_2Cu_3O_y$ (Y123) was observed after severe reduction of particle size. The particle size reduction was performed in following way: Y_2O_3, $BaCO_3$ and CuO were ground for different times to reduce the particle size to the nanoscale. It was found that the superconducting phase could not be obtained without heat treatment, but a weak ferromagnetic behaviour of the mixture was observed at liquid and room temperatures. . In order to understand which compound is responsible for this weak ferromagnetism, the Y_2O_3, $BaCO_3$ and CuO powders were separately ground and M-H loops taken at 290K. The results indicated that both Y_2O_3 and CuO powders showed weak ferromagnetism, but $BaCO_3$ showed anti ferromagnetism. It is thought that energy transfer to the powder particles by grinding results in severe plastic deformation of the particles to produce dislocations, vacancies and atomic disorder. We discuss then the observations in the context of recent studies of the magnetic properties of nanoscale oxide materials.

1. Introduction

Most physical and chemical properties of solid matter change when the particle size is decreased to the nanoscale [1]. There are several methods for the production of nano powders; conventional methods for particle size reduction like milling, grinding, jet milling, crushing, and air micronization [2-4]. Advantages of reducing of the particle size in ceramic materials include increased reaction rates during sintering and improvements in both microstructure and properties even of reduced processing temperatures [5].

The aim of this work was to reduce the particle size to the nanoscale of material with a nominal composition of $YBa_2Cu_3O_y$ using mechanical alloying. We were then interested to characterize the differences in structural and magnetic properties between mixtures of Y_2O_3, $BaCO_3$ and CuO in stoichiometric ratios subjected directly grinding without any heat treatment and powders of pre-calcined $YBa_2Cu_3O_y$. The primary aim was to explore the feasibility of preparing nanoscale YBCO powder directly by mechanical alloying of oxide precursors, but we have found it necessary to study the magnetic properties of both kinds of samples in some detail.

2. Experimental details

100 grams batches of Y_2O_3 (Aldrich 99.99%), $BaCO_3$ (Strem Chemicals 99.999%) and CuO (Alfa Aesar 99.0%) powder were mixed together for 1h in a molar ratio 1:2:3 using a mortar machine. The mixture was transferred back to the mortar machine and ground for 10, 20, 40, and 60h in air. These samples will be referred to below as A10, A20, A40 and A60.

Solid state phase transformations in the powder samples were studied using a Differential Thermal Analysis (DTA) model NETZSCHl. Structural analyses were carried out in an X-ray diffractometer (XRD), Rikago Dmax/III, which emits X-rays from a CuKα source at a wavelength of 0.154nm. Analysis of the particle size of the powders was monitored by a high resolution JEOL 840F Scanning Electron Microscopy (SEM) working at 5kV, and Energy Dispersive X-ray analyses (EDX) were performed using a JEOL 6300 SEM working at 10kV. The magnetic properties (M-H loops) of samples were determined using a Vibrating Sample Magnetometer (VSM) on the same system.

3. Results and Discussion

In order to see any solid state phase transformations in the powders during heating to determine appropriate processing temperatures, DTA measurements were performed for samples series A (Fig. 1). It is clearly seen that an exothermic peak at around 330°C and two endothermic peaks at around 800°C and 970°C were observed for all samples. We believe that the peak at around 330°C corresponds to the strain release temperature for the lattice damage induced during grinding. It can also be seen that for the 10h sample, A10, the peak intensity at 330°C is very low and the peak is very broad, but that this peak becomes sharper and higher gradually for series A samples with increasing grinding time, consistent with the residual strain in the particles increasing with additional grinding. The higher endothermic peaks correspond to the formation of liquid phase and the peritectic temperature of Y123 respectively.

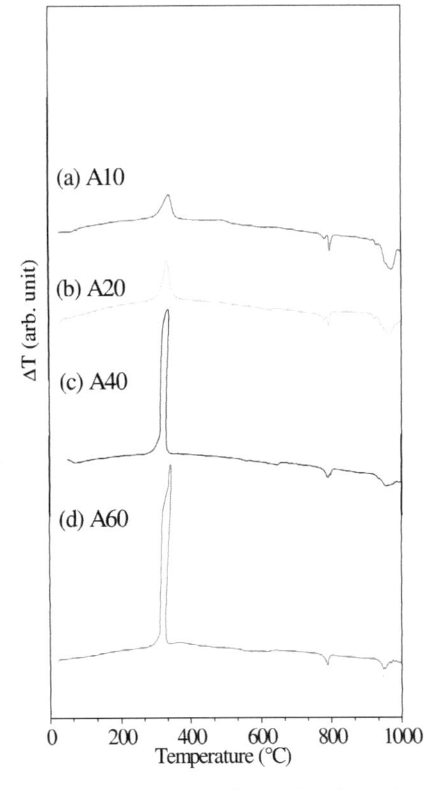

Figure 1. DTA curves of samples in series A

It is well known that a substantial energy transfer to powder particles takes place by grinding action. This process results in plastic deformation of the particles which introduces strain and lattice defects. With the continued grinding action the joining of particles develops lamellar structures. Initially, grinding process flattens ductile powders into flaky/ plate-like particles. As the particles are repeatedly ground, the lamellar structure is continually getting thinner until a steady state is established and the layers cannot become any thinner [6]. An optical photograph of powder A60 is given in Fig.2 which shows a flaky morphology as discussed.

Figure 2. Shows an optical photograph of sample A60.

The XRD patterns of samples from series A, which are as-ground without any heat treatment, are given in Fig.3. These profiles indicate no formation of the superconducting high-T_c phase by grinding a mixture of Y_2O_3, $BaCO_3$ and CuO. The peaks shown in Fig.3 were indexed to CuO, Ba_2CuO_3, $Ba_2Cu_2O_5$, $Ba_3YCu_2O_{6.5}$ and a few unknown phases. As a result, we can conclude that the mortar grinder used in this work reduced the particle size, but an appropriate condition could not be found to form the high-T_c phase by direct mechanical alloying.

An SEM/EDX analysis was performed to obtain information on the particle size and determination of the second phases present in the samples. Fig.4 shows SEM images of powder samples of A60. The image of A60 indicates an average particle size much less than 1μm, but some thin fibres between the particles are shown in same image. In addition, EDX analysis indicated four different phases are formed in sample A60 during the grinding process. These phases are Y-rich, Cu-rich and two different compositions of Y, Ba and Cu oxides. No evidence of Y123 was found.

At first, we attempted to check the superconducting properties of all these samples by a simple method using the repulsive Meissner effect which allows the suspension of a superconductor above a permanent magnet. We found that samples from series A which were not subjected to any heat treatment, except as a result of grinding, showed no Meissner effect even after 60h grinding. In contrast, all samples, especially samples after long grinding times, showed a weak attractive behaviour to the magnet both after cooling liquid nitrogen temperature and also, more surprisingly, at room temperature there is still a strong attraction between the samples and the magnet. These preliminary results suggested that all the A samples showed ferromagnetic rather than diamagnetic properties.

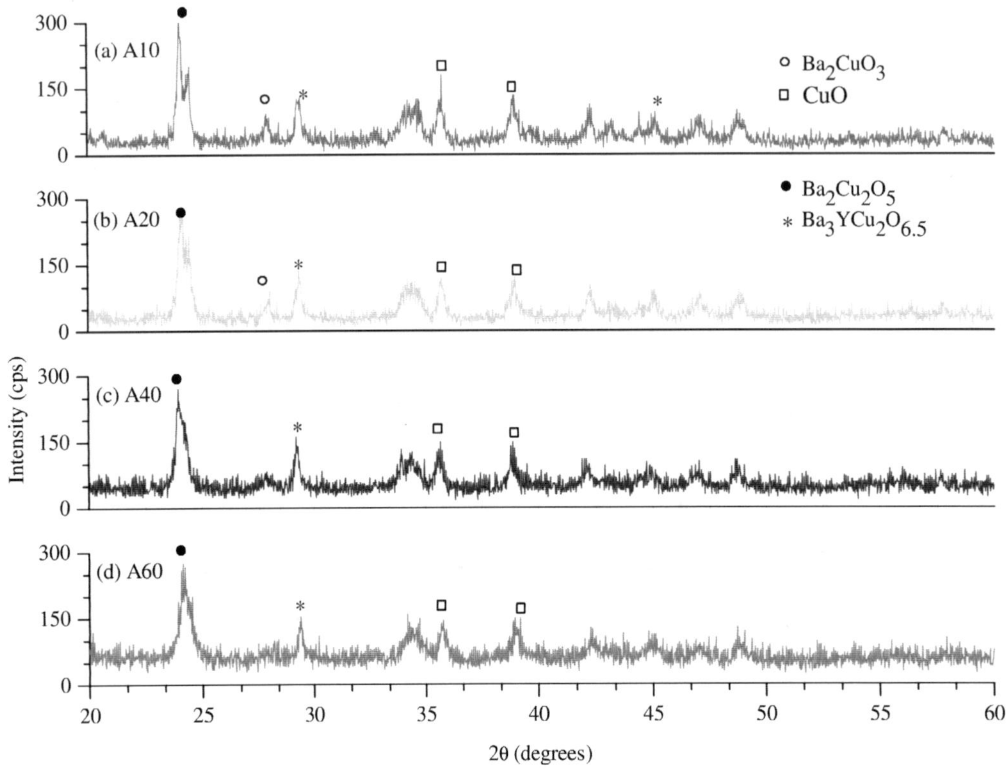

Figure 3. Shows XRD spectrums of samples series A, which were ground for 10, 20, 40 and 60h in air without any heat treatment.

Figure 4. Shows SEM images of powder samples of A60.

In order to determine the magnetic behaviour of our samples in more detail, the M-H loops of sample A10, A40 and A60 at 77K were taken, and are shown in Fig. 5. It can be clearly seen that the M-H loops indicate the hysteresis typical of ferromagnetic behaviour. The remenance magnetisation values of hysteretic loop increased when the milling time increased. The coercive force (H_c) and remenant magnetization (M_r) values were determined at 77K for A60 sample to be respectively -87.013 Oe and 0.156 emu/cm^3 and values of other samples are as indicated in Fig. 5.

At this point, we decided to find out which compound is responsible for the ferromagnetic character after long grinding times. 10g of the starting Y_2O_3, $BaCO_3$ and CuO powders were separately subjected to grinding for 20 hours under same condition as all the samples described above. XRD and M-H measurements were then performed on each compound. Firstly, the peak intensities in the X-ray diffraction patterns decreased (not included here). Secondly, Y_2O_3 and CuO powders after 20 h grinding were attracted to the magnet at 290 K as we seen before, but $BaCO_3$ powders did not show any unusual behaviour. Then we measured the M-H loops of the three compounds before (insets of Fig. 6) and after 20h grinding, as shown in Fig. 6. As can be seen in the insets of Fig. 6, none of starting compounds show ferromagnetism. After 20h grinding, however, both the Y_2O_3 and CuO powders showed weak ferromagnetic hysteresis at 290K but the $BaCO_3$ powders showed a typical antiferromagnetic behaviour with the superimposed to a diamagnetic signal. The remenant magnetization (M_r) and coercive force (H_c) values for ground Y_2O_3, CuO are indicated in Fig. 6.

In summary, samples prepared by grinding a mixture of Y_2O_3, $BaCO_3$ and CuO (series A) showed no sign of the superconducting Y-123 phase in the XRD spectra and no sign of a typical Meissner effect when cooled by liquid nitrogen. From the DTA analysis, a strong peak at around 330°C corresponds to a possible strain release temperature since the peak intensity is increasing with the grinding time. This clearly indicates that strain in the particles is induced during grinding. We conclude that mechanical alloying of starting oxides is not an effective process to produce nanoscale Y123 powder, partly because of the excess strain introduced after long grinding times. In addition long grinding times also cause decomposition of prereacted Y123 powder, and restoring of the Y123 phase by re-sintering was not possible. After 20h grinding oxide mixtures, prereacted Y123 or CuO or Y_2O_3 separately all showed weak ferromagnetic hysteresis at 290K.

We believe that the XRD, SEM and magnetic measurements all provide evidence that extensive grinding of all these samples results in the formation of nanoscale CuO and/or Y_2O_3. Previous work has shown that weak ferromagnetism has frequently been discussed in nano scale CuO [7-8], but not yet in Y_2O_3.

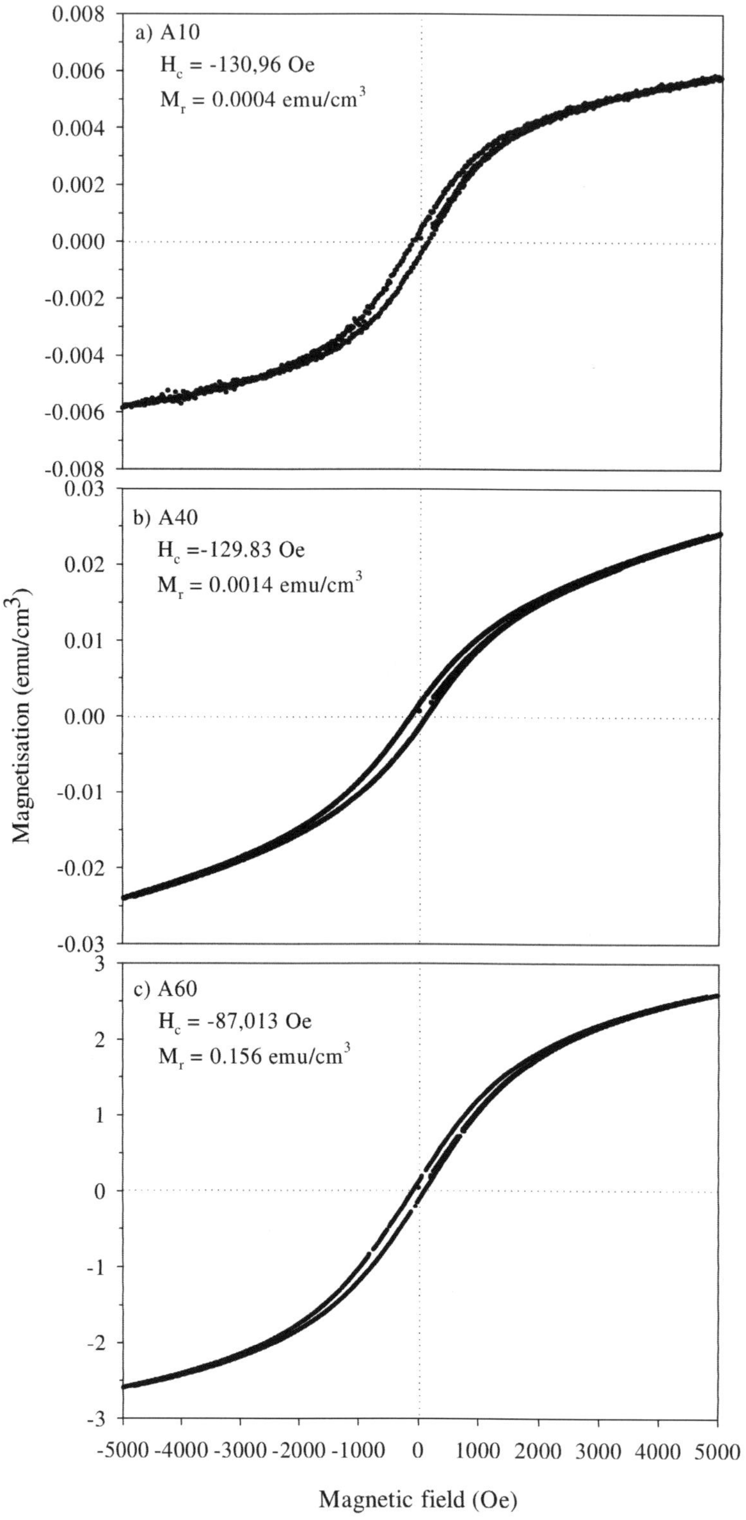

Figure 5. The M-H loops measured at 77 K for samples (a) A10 (b) A40 and (c) A60.

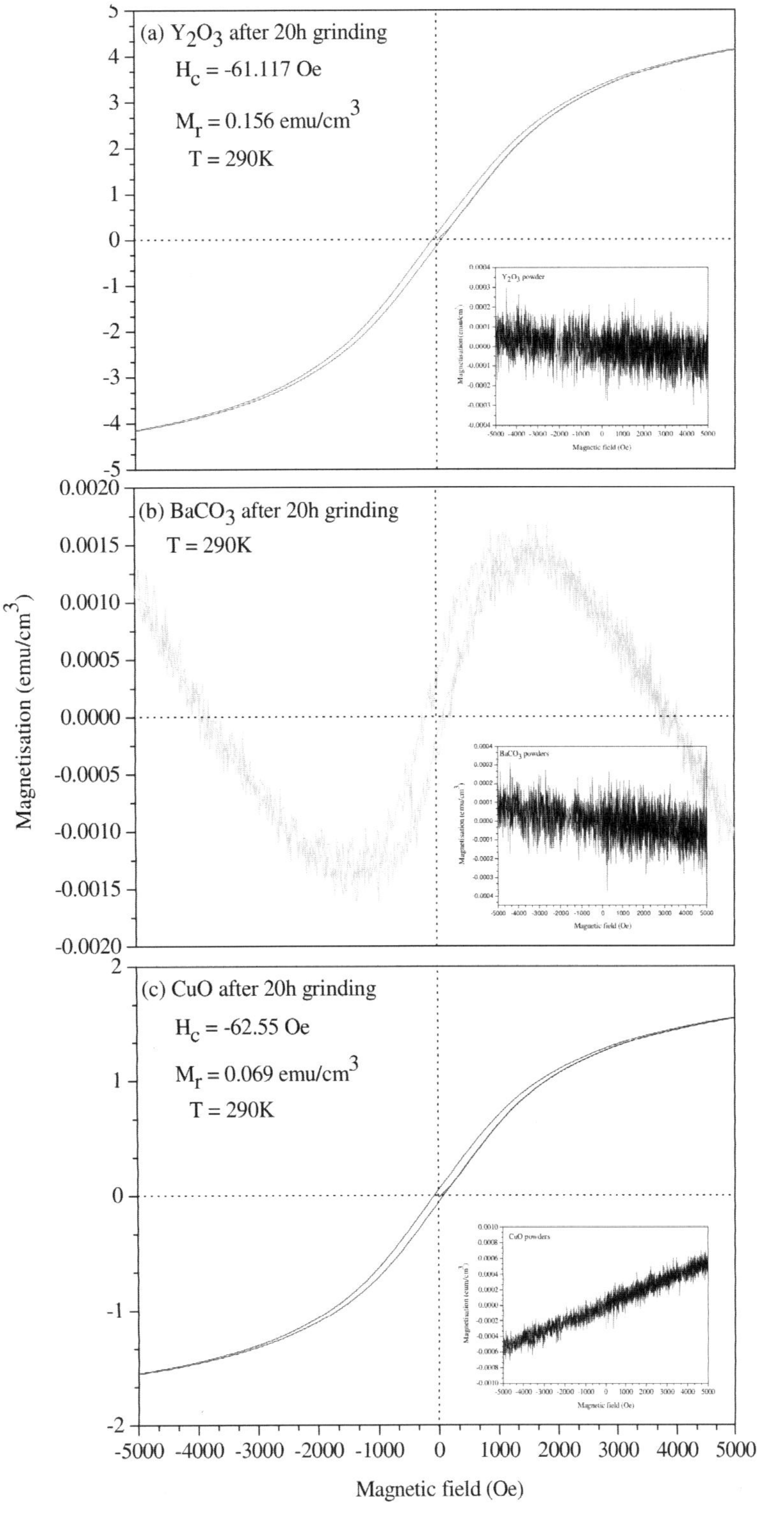

Figure 6. The M-H loops of individual Y_2O_3, $BaCO_3$ and CuO after 20h grinding measured at temperature 290 K. The insets of each curve indicate the M-H loops of powders before grinding.

References

[1] Ayyup P, Palkar V R, Chattopadhyay S and Multani M 1995 *Phys. Rev. B* **51** 6135
[2] Murty B S and Ranganathan S 1998 *Int. Mat. Rev* **43** 3
[3] Diko P 2006 *Physica C* **445** 323
[4] Perner O, Eckert J, Häßler W, Fischer C, Müller K H, Fuchs G, Holzapfel B and Schultz L 2004 *Supercond. Sci. Technol.* **17** 1148
[5] Delogu F, Cocco G 2006 *J. Alloys and Compd.* **420** 246
[6] Koch C C 1998 *Mater. Sci. Engin. A* **244** 39
[7] Seehra M S and Punnoose A 2003 *Solid State Commun.* **128** 299
[8] Xiao H M, Zhu L P, Liu X M and Fu S Y 2007 *Solid State Commun.* **141** 431

Parametric dynamic analysis of a superconducting bearing system

A Cansiz[1,3], U C Hasar[1], Ö Gundogdu[2] and B Ates Çam[1]
[1]Electrical and Electronics Engineering Department, Ataturk University, Erzurum, Turkey
[2]Mechanical Engineering Department, Ataturk University, Erzurum, Turkey

Abstract. The dynamics of a disk-shaped permanent-magnet rotor levitated over a high-temperature superconductor is studied. The interaction between the rotor magnet and the superconductor is modelled by assuming the magnet to be a magnetic dipole and the superconductor as a diamagnetic material. In the magneto-mechanical analysis of the superconductor part, the frozen image concept is combined with the diamagnetic image and the damping in the system was neglected. The interaction potential of the system is the combination of magnetic and gravitational potential. From the dynamical analysis, the equations of motion of the permanent magnet are stated as a function of lateral, vertical and tilt directions. The vibration behaviour of the permanent magnet is analyzed with a numerical calculation obtained by the non-dimensionalized differential equations for small initial impulses.

Introduction

Since the discovery of high temperature superconductors (HTS), many engineering applications have emerged. These applications are generally based on the interactions between the permanent magnets (PM) and superconductors [1-3]. The possibilities of engineering applications were improved by the theoretical and experimental studies [3]. One of the most important issues in the levitation applications is the stability of the levitated PM over the HTS. The response of the superconductor to the magnetic field of the PM was widely studied from an experimental point of view. However, modelling the motion of the PM/HTS systems is not easy due to the complicated hysteresis in the superconductor [4-7]. Some deficiencies also occur such as the tilt on the levitated magnet (the angle between the rotational symmetry of the PM with its vertical direction). These deficiencies cause non-linear drag on the levitated PM which in turn results in instabilities.

Studying the levitation for real geometries generally requires the combination of electromagnetic and mechanical analysis. The aim of this paper is to study the dynamics of a tilted PM levitated over a superconductor. The study performed by Sugiura et al. [8] used the frozen image concept to define the interaction potential for the PM/HTS system and also discussed the lateral, vertical and tilt effect. In order to provide a more complete dynamical analysis, vertical, radial and tilt characteristics were studied.

[3] Tel: 90 442 2314877, Fax: 90 442 2360957,
E-mail: acansiz@atauni.edu.tr

Magneto-mechanical analysis

The configuration basically consists of vibration permanent magnet levitated over the high temperature superconductor. A disk shaped NdFeB PM levitated over a melt-textured YBCO. The PM disk has a diameter of 25.4 mm, height of 6.35 mm, mass of 35.6 gram and axially polarized with magnetization of 1.1 Tesla.

The governing equation of the PM levitated over the HTS is not simple due to the complicated magnetic interaction potential. This is because in actual geometries the total potential depends on both gravitational and magnetic forces. The magnetic part of the interaction for this system can be approximated in terms of the frozen-image model [4]. In this approach the PM is assumed as a magnetic dipole and HTS as the material that consists of the combination of flux pinning and diamagnetic property. In this study, the frozen-image model does not account for hysteresis in the HTS, which would result in a transfer of some part of the moment of the diamagnetic-image into a small frozen-flux moment as the PM moves closer to the HTS surface. For quasi-static motions, where the hysteresis is not a major effect, it was proven that the frozen-image concept successfully predicts the interaction force between the PM and HTS. In addition, prediction of some of the characteristics of the system by the frozen image model in the dynamical case is expected.

The interaction of the PM and superconductor system can be represented for a first order approximation by the forces between the dipoles and their images. In this system, the first dipole is represented by the PM and second dipole by its image. The total potential represents the system of the dipole and its diamagnetic mirror image. When the HTS is field cooled with the PM at a particular field cooling height (h), two images appear: one is a frozen-image and the other one is a diamagnetic-image. In this configuration, while the frozen image is fixed in h, the diamagnetic image is able to move with the PM. Thus, the PM has the distance of $(h+z)$ from the surface of the HTS, which is called as levitation height. The frozen and diamagnetic images are equal in magnitude and are opposite in direction to each other. Thus, the frozen image model states that the net interaction forces between the PM and its frozen and diamagnetic images are zero.

When the dipole from the PM and its frozen image is considered, $\mathbf{m}_1$ is the dipole from the PM and this time $\mathbf{m}_2$ is the dipole from the frozen image of the PM [3, 4]. By taking HTS surface at zero potential, with M and H as the mass and the half height of the PM, respectively. Defining $A_1=[r^2+(2h+H+z)^2]^{1/2}$ and $A_2=r\sin\theta+(2h+H+z)\cos\theta$, the total interaction potential of the system can be written as

$$U_{\text{tot}}=\beta\left[\frac{1+\cos^2\theta}{2(2h+H+z)^3}+\frac{\cos\theta}{A_1^3}-\frac{3A_2(2h+H+z)}{A_1^5}\right]+Mg(h+H/2+z) \tag{1}$$

where θ is the angle between rotational magnetization symmetry axis of PM with the vertical direction and $\beta=\mu_0 m_1 m_2/4\pi$. In order to simplify the geometry of the system, the origin was chosen to be the centre of mass, which is the sum of field cooling height h (distance from the surface of the superconductor to the bottom of the PM) and half of the PM height. The kinetic energy of the levitated PM above the superconductor is also given as

$$T=\frac{1}{2}M(\dot{r}^2+\dot{z}^2)+\frac{1}{2}I\dot{\theta}^2 \tag{2}$$

where I is the radial moment of inertia and the dot over each variable, from here on, presents the time derivative of that variable. The kinetic energy also consists of the spatial components since the PM is not fixed in one point. From the Lagrangian function of the PM/HTS system the equations of motion of the system given as

$$\ddot{r}+\frac{3\beta}{M}\left[\frac{5rA_2(2h+H+z)}{A_1^7}-\frac{A_3}{A_1^5}\right]=0 \tag{3}$$

$$\ddot{z}+g+\frac{3\beta}{M}(B_2+B_3)=0 \tag{4}$$

$$\ddot{\theta}+\frac{\beta}{I}\left[\frac{-\sin\theta\cos\theta}{(2h+H+2z)^3}-\frac{\sin\theta}{A_1^3}-\frac{3A_4(2h+H+z)}{A_1^5}\right]=0 \tag{5}$$

where

$$B_1=\frac{5rA_2(2h+H+z)}{A_1^7}-\frac{A_3}{A_1^5} \tag{6}$$

$$B_2=-\frac{(1+\cos^2\theta)}{(2h+H+2z)^4} \tag{7}$$

$$B_3=-\frac{2(2h+H+z)\cos\theta+A_2}{A_1^5}+\frac{5A_2(2h+H+z)^2}{A_1^7} \tag{8}$$

and

$$A_3=r\cos\theta+(2h+H+z)\sin\theta \tag{9}$$

$$A_4=r\cos\theta-(2h+H+z)\sin\theta. \tag{10}$$

Here, the double dot over each variable, from here on, presents the second time derivative of that variable.

The equations given in (3), (4) and (5) represent the centre of mass motion of the PM. The PM is assumed to be cooled in the coordinates of $(r,\theta,z)=(0,0,h)$. This assumption provides the magnetization of the PM being aligned purely in the vertical direction so that the magnetization of the frozen image can only occur in the vertical direction.

The equations from (3) to (5) are coupled non-linear differential equations and an analytical solution is not possible. It is possible to characterize the PM motion from the numerical calculation obtained by using the Runge-Kutta method. The theoretical analysis of the various vibrational frequencies such as vertical, radial and tilt will be next in discussion. In this analysis first we will directly solve the equation of the system with the Runge-Kutta method and compare the results with non-dimensionalized equations solved again with Runge-Kutta method.

Non-dimensionalization of the equation set

Non-dimensionalization can be explained as replacing units with suitable variables and for these units to be removed partially or completely from mathematical equality. For example, if a system owns a resonance frequency length or a time consistency then non-dimensionalization can rearrange values owned by this system. Non-linear equation applications can be given as an example for this situation. Non-dimensionalization ensures the determining of the characteristic of the system used by the natural units of the system. In other words, non-dimensionalization is used to determine parameters which are actually necessary for the analysis of the system.

The non-dimensionalized equations set are derived with the proper variable changes. Bearing in the mind that $H_0 = h + H/2$ and by converting $r^* = r/H_0$, $z^* = z/H_0$ and $t^* = t\omega_0 = t\sqrt{K/M}$ the parametric terms are given as

$$K = \frac{3\mu_0 m_1 m_2}{16\pi H_0^5} \qquad g_A = \frac{M}{H_0 K} g. \tag{11}$$

At first we considered starred equations, and then we omitted star superscripts in the equations. By some arrangements, below equations are derived:

$$\ddot{r} + 4\left[\frac{5C_2(2+z)}{C_1^{7/2}} - \frac{C_3}{C_1^{5/2}}\right] = 0 \tag{12}$$

$$\ddot{z} + 4\left\{-\frac{(1+\cos^2\theta)}{16(1+z)^4} + \left[\frac{5C_2(2+z)^2}{C_1^{7/2}} - \frac{C_4}{C_1^{5/2}}\right] + g_A\right\} = 0 \tag{13}$$

$$\ddot{\theta} + \frac{M}{K}\frac{\beta}{I}\frac{1}{H_0^3}\left[\frac{\sin\theta\cos\theta}{8(1+z)^3} + \frac{\sin\theta}{C_1^{7/2}} + \frac{3C_5(2+z)}{C_1^{5/2}}\right] = 0 \tag{14}$$

where

$$C_1 = r^2 + (2+z)^2 \tag{15}$$

$$C_2 = r[r\sin\theta + (2+z)\cos\theta] \tag{16}$$

$$C_3 = r\cos\theta + (2+z)\sin\theta \tag{17}$$

$$C_4 = r\sin\theta + 3(2+z)\cos\theta \tag{18}$$

$$C_5 = r[r\cos\theta - (2+z)\sin\theta]. \tag{19}$$

Discussion

Obtaining the numerical solutions of the equation of motion for the PM levitated above the superconductor provides an insight for the stability analysis. Because of the non-linear nature of the effective force components of the magnetic systems, the magnetic bearings act like non-linear systems. In order to find physical solutions of such systems, the linearization procedure is generally used. One of the common techniques for the stability analysis is to use the non-dimensionalization under the initial constant impulse applied to the system. From the numerical calculations, the model predicted the resonance frequency with a reasonable accuracy for a typical vertical vibration amplitude of the PM as a function of time.

Although the impulse, which is applied vertically, does not affect the other directions, when a small displacement is applied to the PM either from radial or tilt, there exists vibrational modes in the directions of radial and tilt and also negligibly small amount in the vertical. Figure 1 shows the theoretical vibration spectrum, where the displacement was applied in the radial direction. The appearance of both radial and tilt vibration in the same frequency spectrum indicates the strong cross-couplings among these directions. This indicates that all the vibration frequency modes are actually the combination of pure vibration frequency modes such as vertical, lateral and even the angular directions. However, in the vertical direction this agreement is not too strong, and the agreement gets worse especially in the tilt direction. This difference in vertical direction is also due to the hysteresis effect, which acts as damper in the system. This is because the hysteresis occurs more in the vertical direction than that of the radial, so the difference in between the theory and the experiments is high for the vertical direction compared to the radial. In addition, as shown in figure 2, a non sinusoidal shape of the radial vibration amplitude as a function of time indicates that there is actually radial and tilt displacement of the permanent magnet.

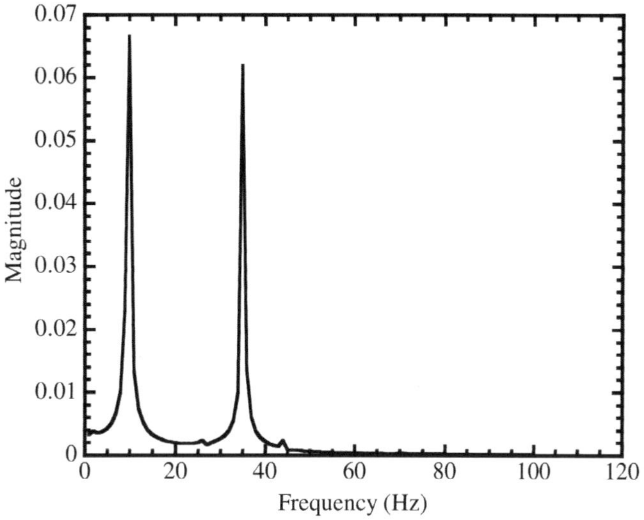

Figure 1. Resonance frequencies in the radial and tilt directions from the FFT analysis.

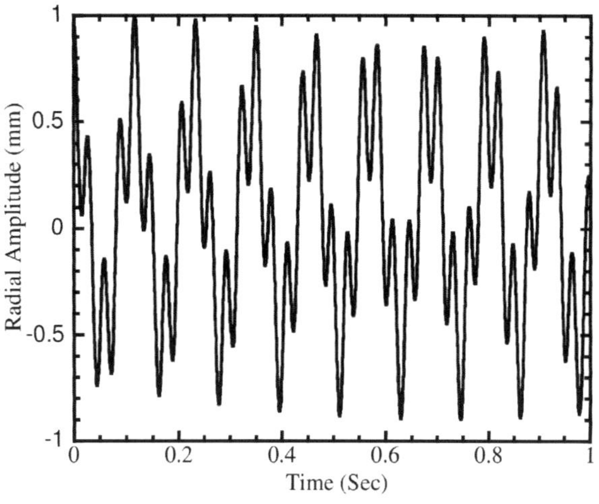

Figure 2. The amplitude of vibration in radial and tilt directions.

From the dynamical point of view, the response of the damping forces is not dominant compared to the forces due to effective potential of the system. For this reason, the analysis presented in the model is assumed to be reliable for studying the motion along the vertical, radial and tilt. The eccentricity between the magnetic centre and the mass centre gets larger if the inhomogeneities from the magnet and superconductor are too much. This behaviour can be observed in figure 3 that the resonance frequency as a function of cooling height gets unrealistic for the case of very low and very high cooling heights. This can be explained in terms of the frozen image model assumption that the permanent magnet was assumed as a dipole which creates unrealistic results if the dipole is not far away from its image.

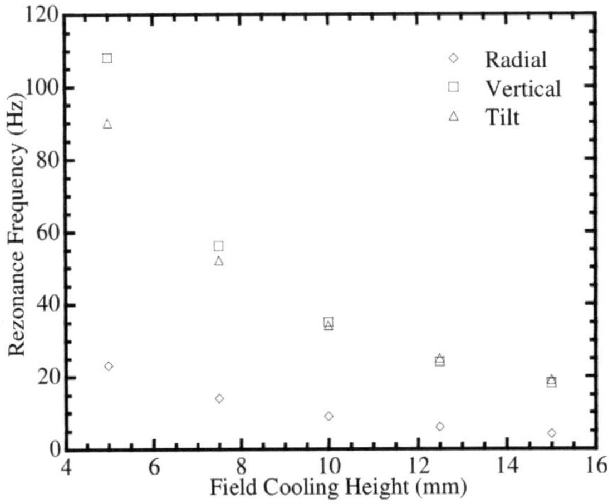

Figure 3. Resonance frequency as a function of field cooling height.

The amplitude of the radial and tilt vibration modes as a function of cooling height shows non linear behaviour. As shown in figure 4, the amplitude is changing with the cooling height in a parabolic function. This may be attributed to frozen image model that the dipole acts in appropriate cooling height. However a real explanation of this behaviour still needs further theoretical analysis.

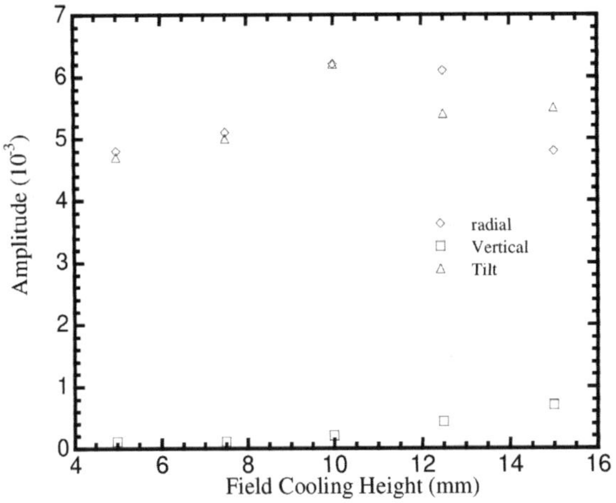

Figure 4. Amplitude of vibration as a function of field cooling height.

Preliminary results obtained for non-dimensionalized equations show that vertical vibration amplitude as a function of time is scaled. As shown in figure 5, the vertical resonance frequency predicted for 10 mm cooling height is 5 Hz, which is far different than the one obtained in direct solution shown in figure 3. In order to explain these unexpected results, the set of equations needs to be analyzed by Taylor series of expansion. At this stage, these preliminary results also show that the non-dimensionalization must be carried out with the Taylor expansion of the equations of the motion to investigate initial condition dependence on the parametric equations.

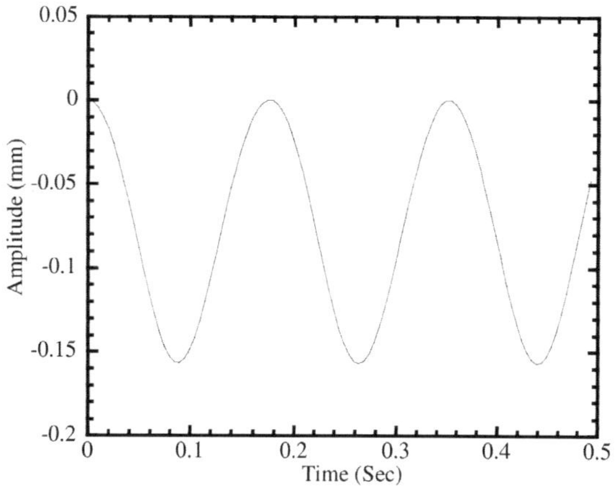

Figure 5. Resonance frequency in the vertical direction for non-dimensional equation set.

Conclusion

In this paper, the dynamics of a levitated PM over a superconductor was studied. The dynamic behaviour of the PM was investigated with the combination of mechanical and electromagnetic considerations. The governing equations of the motion of the PM levitated over the HTS were simplified by approximating the magnetic part of the interaction for this system in terms of the advance frozen-image model approach. In this approach, the PM was assumed to be a dipole and the HTS as a diamagnet, in which the image dipole resides.

In conclusion, for designing a superconducting bearing for applications, the geometric and magnetic considerations must be taken into account. The theoretical results show non-sinusoidal shape of the radial vibration amplitude as a function of time. This indicates that there are actually radial and tilt displacements of the permanent magnet. Also, the non-linear behaviour of the amplitude of radial resonance frequency still needs further theoretical analysis, either considering frozen image model or other existing models. In order to explain these unexpected results, the set of equations needs to be analyzed by Taylor series of expansion. At this stage, these preliminary results also show that the non-dimensionalization must be carried out with the Taylor expansion of the equations of the motion to investigate initial condition dependence on the parametric equations.

Acknowledgments

This work was partially supported by Ataturk University Research Grant 167 and U S Department of Energy.

References

[1] Hull J R 2000 *Supercond. Sci. Technol.* **13** R1
[2] Day A C, Strasik M, McCrary K E, Johnson P E and Gabrys J W 2002 *Supercond. Sci. Technol.* **15** 838
[3] Cansiz A, Campbell A M and Coombs T A 2003 *Physica* C **390** 305
[4] Hull J R and Cansiz A 1999 *J. Appl. Phys.* **86** 6396
[5] Navau C and Sanchez A 2002 *Physica* C **372** 1474
[6] Qin M J, Li G, Liu H K, Dou S X and Brandt E H 2002 *Phys. Rev. B* **66** 24516
[7] Campbell A M 2002 *Supercond. Sci. Technol.* **15** 759
[8] Sugiura T, Aoyagi T, Ura H and Yoshizawa M 2003 *IEEE Trans. Appl. Supercond.* **13** 2247

Journal of Physics: Conference Series **153** (2009) 012007 doi:10.1088/1742-6596/153/1/012007

Investigation of effect of x in $Yb_{1-x}Gd_xBa_2Cu_3O_{7-y}$ superconducting structures on upper critical magnetic field and coherence length

Ş. Çelik[1], K. Öztürk[2] and E. Yanmaz[2]
[1] Rize University, Faculty of Arts and Sciences, Department of Physics, Rize, Turkey.
[2] Karadeniz Technical University, Faculty of Arts and Sciences, Department of Physics, Trabzon, Turkey.

E-mail: sukrucelik@yahoo.com

Abstract. The resistivity measurements of the $Yb_{1-x}Gd_xBa_2Cu_3O_{7-y}$ superconducting samples prepared by using the solid-state reaction technique for x=0.0, 0.2, 0.4, 0.6, 0.8 and 1.0 were performed in QD-PPMS system under different magnetic fields up to 5T in zero field cooling regime. The upper magnetic field $H_{c2}(T=0)$ for each sample was calculated from 50% of the normalized resistivity (R_n) by the extrapolation $H_{c2}(T)$ to T=0K and the coherence lengths $\xi(T=0)$ of the samples were calculated from the $H_{c2}(0)$ values. The examination of the effects of x in the composition on $H_{c2}(0)$ showed us that the upper critical magnetic field decreased from 186.31T for x=0.0 to 37.99T for x=1.0 with the increasing of content x from x=0.0 to 1.0. Using the content x in the composition, the upper critical magnetic field can be controlled and this can be used in superconducting application especially as a superconducting relays.

1. Introduction

Superconducting materials, such as YBCO and BSCCO, must operate at high currents and magnetic fields for many practical applications. Typical large scale applications of them are magnets, power transmission lines, transformers and generators because superconductors have smaller power losses compared to the metals. For these applications, one of the important properties of the superconducting materials is the upper critical magnetic field. The upper critical magnetic field $H_{c2}(0)$ that the material keep its superconductivity is the huge magnetic field at zero temperature for that material. In the higher magnetic field than $H_{c2}(0)$, the material loses its superconductive properties at zero temperature. Many of the studies up to now have been on the development of the transition temperature T_c, $H_{c2}(0)$ and the critical current density J_c of the superconducting structures.

Some researchers studied the effects of the substitution of the rare earth atom with Y in the polycrystalline or crystal superconducting structure of $YBa_2Cu_3O_7$ (Y123). In the studies, the substitution of RE atoms (RE: Nd, Sm, Eu, Gd) in RE123 structure showed that they are effective in achieving high value of critical current density (J_c) in a high magnetic field region accompanied by the secondary peak effect [1]. While the substitution of rare earth atoms (except Ce, Pr, Tb, Pm) for Y in $YBa_2Cu_3O_{7-y}$ (Y-123) preserves the 90K superconductivity [2], the same substitution changes the critical current density J_c and the upper critical magnetic field H_{c2}. One of the factors affecting J_c and H_{c2} is the fabrication processes such as calcinating and sintering processes [3]. When the sintering temperature of the superconducting structure is taken very lower than the peritectic decomposition temperature, the structure has the weak-link interaction or its superconductivity disappear. While the strong and weak interactions in the same superconducting sample are under control, the upper critical magnetic field of the sample can be assigned at the desired value in the specific range of magnetic field which is possible.

Our previous study was about the effects of content x in $Y_{1-x}Yb_{x/2}Gd_{x/2}Ba_2Cu_3O_7$ (Y, Yb, Gd)123 structure on the upper critical magnetic field and coherence length [10]. It was found in our previous study that Yb123 phase in this composition has stronger interactions and Gd123 has weaker interactions compared to Y123. Owing to three interactions in the structure, the upper critical magnetic field was changed from 84T to 122T with the content x.

1 Corresponding author: e-mail: sukrucelik@yahoo.com, Phone: +90 4642236126, Fax: +90 4642235376.

In this study, we investigated the effect of Gd123 and Yb123 structures without Y123 phase in the superconducting material of $Yb_{1-x}Gd_x\ Ba_2Cu_3O_7$ (Yb, Gd)123 for x=0.0, 0.2, 0.4, 0.6, 0.8 and 1.0. Because the peritectic decomposition temperature of Gd123 structure is 975°C [4] and Yb123 is 910°C [5], the sintering temperature of the sample was selected as 890°C to fabricate Yb123 with strong interaction and Gd123 with weak interaction in the same composition. Using the content x, the amounts of both interactions were thought to change and to control upper critical magnetic field of the sample.

Although numerous studies have been undertaken on the pinning properties of melt processed Yb123 [6] with finely dispersed Y211, the effect of x in $Yb_{1-x}Gd_xBa_2Cu_3O_7$ polycrystalline structure on the activation energy and the *c*-lattice parameter [7, 8], the effect of x of $Y_{1-x}Gd_{x/2}Yb_{x/2}Ba_2Cu_3O_7$ superconducting structures on the activation energy [9] and the upper critical magnetic field [10], the effect of Gd diffusion into Y123 [11, 12], and Yb diffusion into Y123 [13], no studies on the effect of x in such $Yb_{1-x}Gd_xBa_2Cu_3O_{7-y}$ polycrystalline superconducting structure on the upper critical magnetic field and the coherence length have been reported.

2. Experimental

The polycrystalline $Yb_{1-x}Gd_xBa_2Cu_3O_{7-y}$ ((Yb, Gd)123) superconducting samples were produced by solid state reaction method for the six different x values. After stoichiometric mixtures of Yb_2O_3, Gd_2O_3 $BaCO_3$ and CuO for each composition of x were thoroughly ground and calcined at 900ºC in air for 10 h, each mixture was ground again, and pressed into a pellet form with 13mm diameter under 250 MPa pressure, and then each of the samples for x=0, 0.2, 0.4, 0.6, 0.8 and 1 was sintered at 890ºC for 24h in air. Finally, all the samples together were subjected to oxygenation process at 400ºC for 2h, followed to 300 ºC at the rate of -1 ºC /min under the oxygen flowing and then they were cooled to the room temperatures in air. Samples were cut into bar-shape with identical dimensions with the size of 1.5x2.0x6.0 mm^3. The sintering and oxygenation temperatures of the samples were obtained from the Differential Thermal Analysis (DTA) model NETZSCH1.

The standard four-probe resistivity of samples were measured by using a physical properties measurement system (Quantum Design PPMS system) under various magnetic fields such as 0, 1, 2, 3, 4 and 5T in the zero field cooling regime (ZFC). In order to remove the trapped magnetic field inside the samples, 0.1T magnetic field was applied for a short time at 100K and then decreased to zero field before ZFC regime.

3 Results and Discussions

The measurements of $H_{c2}(T)$ for the HTSCs have been limited to temperatures near T_c because H_{c2} in these materials rapidly exceeds accessible laboratory magnetic fields when the temperature is reduced to only nine-tenths of T_c [14]. Thus the values of $H_{c2}(0)$ and $\xi(0)$ can be calculated from such data require either a large extrapolation to T=0K using a presumed functional form for H_{c2} or calculation of H_{c2} from the slope $(dH_{c2}/dT)_{Tc}$ [15] using the Werthamer-Helfand-Hohenberg (WHH) expression $H_{c2}(0)=0.7Tc(dH_{c2}/dT)_{Tc}$ [14]. In this paper, the extrapolation method was used to calculate $H_{c2}(0)$.

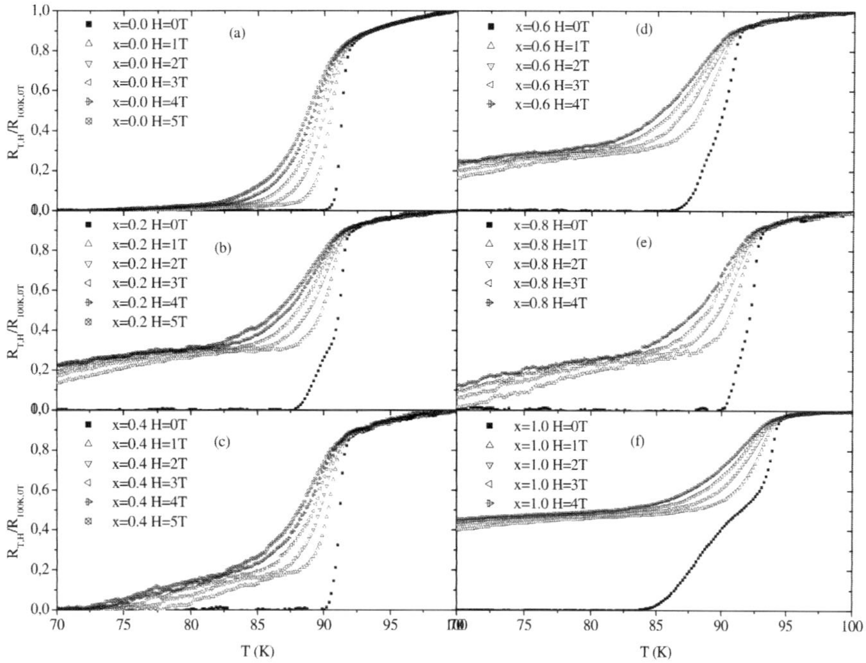

Figure 1. Shows the normalized resistivity versus temperature plots under different magnetic fields up to 5T for the sample a) x=0.0, b) x=0.2, c) x=0.4, d) x=0.6, e) x=0.8 and f) x=1.0

Using the R-T plots in Figure 1, of which the detailed explanations were given in our previous paper in ref [7], the upper critical magnetic field can be calculated from the expression below [16, 17]:

$$H_{c2}(T) = \frac{\phi_0}{2\pi\xi^2(T)} \tag{1}$$

Where ϕ_0 is the flux quantum (ϕ_0=2,067833636x10^{-15}Tm^2), and $\xi(T)$ is the coherence length at the temperature T [16, 17]. The dependency of ξ on temperature is given by:

$$\xi(T) = \xi_0\left(1-\frac{T_c}{T}\right)^{-1/2} \tag{2}$$

Where ξ_0 is the coherence length ($\xi(0)$) at the temperature of T=0K. When the Equation (2) is replaced with ξ in the Equation (1), the upper critical magnetic field at temperature close to T_c can be rewritten as below:

$$H_{c2}(T) = H_{c2}(0)\left(1-\frac{T}{T_c}\right) \tag{3}$$

In the expressions, $H_{c2}(T)$ and $H_{c2}(0)$ is the upper critical magnetic field at T and T=0K, respectively. The Equation (3) can be rewritten as follow:

$$H_{c2}(T) = H_{c2}(0) - H_{c2}(0)\frac{T}{T_c} \tag{4}$$

There are some ideas in previous researchers to define the transition temperature T in the Equation (4). In the study of Müller and his friends [19], they used the magnetic field and the temperature values (as H_{10}, H_{50} and H_{90}) were defined when the resistance of the sample is 10, 50 and 90% of the normal-state resistance R_n, respectively. After the comparison of the upper critical magnetic fields of 10, 50 and 90% of R_n with the fields obtained from dc magnetization and ac susceptibility measurements, they showed us that H_{90} agrees with the result obtained from magnetization, whereas H_{c2} which determined from susceptibility coincides with the H_{10} for MgB_2. On the other hand, in the study of Ando and his friends [18], they associated H_{50} with the mean-field upper critical magnetic field H_{c2}.

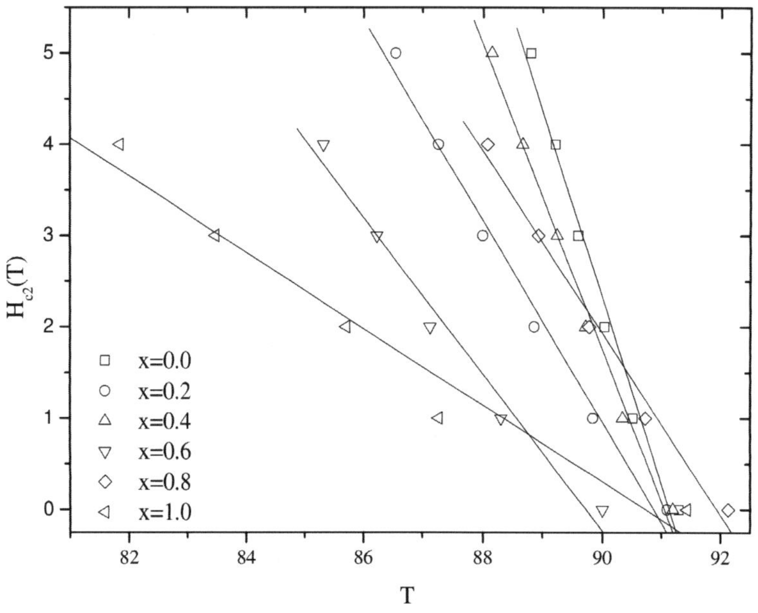

Figure 2. Shows the H_{c2}(T) versus T_c(H) calculated from R-T plots for the samples under different magnetic fields.

In this paper, we calculated the upper critical magnetic field such as H_{c2}(T) and T values at which the normalized resistivity equal to 50% of R_n using the R-T plots in Figure 1, and H_{c2}(T) versus T plots were shown in Figure 2. In this figure, each dot corresponds to the upper critical magnetic field for the samples at the different critical transition temperature T_c under different applied magnetic fields. After the linear fittings of the plots in Figure 2, $H_{c2}(0)$ values for the samples were calculated using the extrapolation of H_{c2}(T) to T=0K. $H_{c2}(0)$ versus content x was shown in Figure 3.

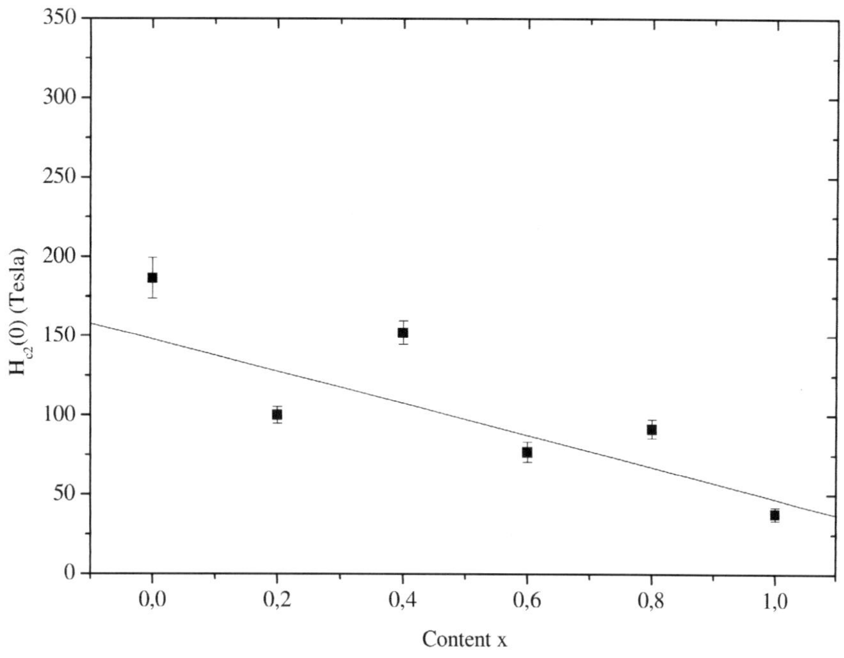

Figure 3. Shows $H_{c2}(0)$ values versus content x. Solid line in the plot is the linear fitting.

When the Figure 3 was generally examined, it was resulted that the value of the $H_{c2}(0)$ is roughly decreasing with the increasing of the content x. While content x is increasing, not only is the amount of content Yb decreasing, but also that of content Gd is increasing. So, it is resulted in decreasing of $H_{c2}(0)$ value with increasing of content x. This says us that Yb123 structure is more resistive than Gd123 structure against to the higher magnetic fields in this study. In fact, this is not unusual result of Gd123 structure due to the sintering temperature. Because Yb123 liquidation temperature which is 910°C [5], the sintering temperature of the composition was estimated as 890°C and also this is very low sintering temperature for Gd123 structure. While this temperature is very suitable for Yb123, it is also near enough below sintering temperature of Gd123 structure. The decreasing of $H_{c2}(0)$ with the increasing of the content x shows us that this sintering temperature resulted the weak interaction between grains in Gd123 and also strong interaction in Yb123 structures.

The separation from the linearity in the plot in Figure 3 is thought to originate from the extrapolations which are used to calculate $H_{c2}(0)$. In the extrapolation of plots in Figure 3 to calculate $H_{c2}(0)$, only five different magnetic fields up to 5T were used. Higher magnetic fields than 5T is not possible in our laboratory. On the other hand, when the Figure 3 is examined, it is clear to be seen that the strongest structure against to the magnetic field is x=0.0, corresponding to Yb123, and the weakest one is x=1.0, corresponding to Gd123 structure. While the $H_{c2}(0)$ value of Y123 thin single superconducting crystal sample with B||c and B⊥c are 250 T and 120 T respectively [20], that value for our polycrystalline superconducting samples in this study is decreasing from 186 T to 38 T with the increasing of content x. Indeed, the upper critical magnetic field of Gd123 is not around 38 T and more much than 38 T. But, we had to estimate the sintering temperature as 890 °C to apply the same sintering temperature for all the samples in this study. We aimed to produce the sample which has Yb123 superconducting structure with the strong interaction and Gd123 with the weak interaction between grains. This temperature produced Gd123 with very weak-link interaction in the composition as low as it can keep its superconductivity, on the other side, produced Yb123 with very strong interaction. The effects of both structures in the same composition in this study resulted in the variation of the upper critical magnetic field from 186T to 38 T. While the strong interaction tries to make H_{c2} of the sample to be higher, the weak interaction does it to be lower. The greatnesses of two interactions are originated from the quantities of Yb123 and Gd123 structures in the composition. Because these quantities are related to x, the content x can be used to control the upper critical magnetic field. This idea can be understood from the Figure 3. In the figure, there is nearly linear decreasing with the increasing of the content x.

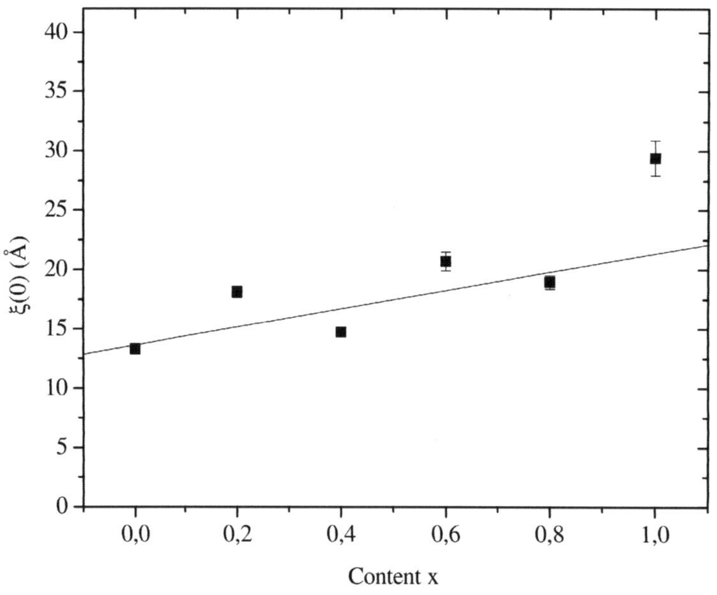

Figure 4. Shows the coherence length ξ(0) versus content x. Solid line in the plot is the linear fitting.

As for the coherence length ξ, it is one of the characteristic parameters of the superconductors. It is a difficult quantity to measure directly, however, it was thus calculated from the expression $H_{c2}(0)=\Phi_0/2\pi\xi^2$ and ξ - content x plot is shown in Figure 4. The coherence length is about 16 Å for $YBa_2Cu_3O_{7-y}$ [21,22], 13.6 Å for Tl-2223 [23,24], 9.7 Å for Bi-2223 and 9±1 Å for Bi-2212 [25], and this value in this study is varying roughly linear from ~13 Å to ~29 Å with increasing of the content x from x=0.0 to x=1.0. The coherence length can be controlled by the content x. The separation from the linearity is originated from the separation in $H_{c2}(0)$ versus content x plot in Figure 3.

From the linear fittings of the plots in Figure 3 and Figure 4, the relationships between $H_{c2}(0)$ and ξ(0) with content x and the inverse relations are defined below:

$H_{c2}(0)=A+B.x$ (in T) x is defined in the range of 0-1.0 (5a)

$\xi(0)=A'+B'.x$ (in Å) x is defined in the range of 0-1.0 (5b)

$x=A''+B''H_{c2}(0)$ $H_{c2}(0)$ is defined in the range of 38T-186T (6a)

$x=A'''+B'''.\xi(0)$ $\xi(0)$ is defined in the range of 13Å-29Å (6b)

Table 1. The constants used in the Eq(5a, 5b, 6a and 6b) with their errors.

Constant	Value	Error	Constant	Value	Error
A	147.8606	±5.3237	B	-10.0602	±0.7329
A′	13.6241	±0.3387	B′	7.7089	±0.7729
A′′	11.3526	±2.2226	B′′	-0.0591	±0.0189
A′′′	-0.5496	±0.3591	B′′′	0.0547	±0.0181

The constants in Equation (5) and Equation (6) are listed in Table 1. Using the content x in $Yb_{1-x}Gd_xBa_2Cu_3O_{7-y}$ structure, and the constants in Table 1, $H_{c2}(0)$ or $\xi(0)$ of the sample with any content x in the range of x=0.0 – 1.0 can be calculated from Equation (5a) and Equation (5b) respectively. On the other hand, it can be calculated that which sample with the desired $H_{c2}(0)$ or $\xi(0)$ values in the specific range should have the content x using Equation (6a) and Equation (6b) respectively.

4. Conclusion

Using the standard solid-state-reaction technique, the nominal composition of $Yb_{1-x}Gd_xBa_2Cu_3O_{7-y}$ superconductor samples for x=0.0, 0.2, 0.4, 0.6, 0.8 and 1.0 were prepared. The upper critical magnetic field $H_{c2}(T)$ at the temperature T=T for the samples were calculated by the midpoint of the normalized resistivity values obtained from the resistivity measurements in the different magnetic fields up to 5T under zero field cooling regime and the upper critical magnetic field $H_{c2}(0)$ at the temperature T=0K was found by the extrapolation $H_{c2}(T)$ to T=0K. The coherence length values were calculated by using the upper critical magnetic field values at the temperature T=0K.

The calculations presented us that while the coherence length is 16 Å for Y123, 13.6 Å for Tl2223, 9.7 Å for Bi2223and ~9 Å for Bi2212, this value in this study varies from ~13 Å to ~29 Å and $H_{c2}(0)$ is varying from 186 T to 38 T with increasing content x from x=0.0 to 1.0.

In our study it was found that there are two interactions in the composition and they are the strong interaction in Yb123 and the weak interaction in Gd123 structure. The quantities of both interactions were changed with the content x because the amount of Yb123 is decreased and that of Gd123 is increased with the increasing of the content x. The quantities of Yb123 and Gd123 structures in the composition supplied us to control the upper critical magnetic field in the identical range of 37T-186T. Consequently, using the content x in $Yb_{1-x}Gd_xBa_2Cu_3O_{7-y}$ structure, $H_{c2}(0)$ or $\xi(0)$ of the sample with any content x between 0.0 – 1.0 can be calculated from the Equation (5a) and Equation (5b) respectively. On the other hand, it can be calculated that which sample with the desired $H_{c2}(0)$ or $\xi(0)$ values in the specific range should have the content x using Equation (6a) and Equation (6b) respectively. By this way, we don't need to produce and measure the sample with any other content x in the range of 0.0-1.0 to find $H_{c2}(0)$ and $\xi(0)$. In addition, this study showed us that the upper critical magnetic field can be controlled using the content x in our compositions and this controlling of the upper critical magnetic field can be used in the application of superconductivity as an electronic superconducting switches.

References

[1] Murakami M 1992 *Supercond. Sci. Technol.* **5** 185

[2] Tarascon J M, McKinnon W R, Greene L H, Hull G W and Vogel E M 1987 *Phys. Rev. B* **36** 26

[3] Yamani Z and Akhavan M 1997 *Supercond. Sci. Technol.* **10** 427
[4] Nariki S, Sakai N, Inoue K and Murakami M, 2003 *Physica C* **392** 468
[5] Won C, Kinder L R, Gim Y, Fan Y, Coulter J Y, Maley M P, Foltyn S R, Peterson D E and Jia Q X, 1999 *Appl. Supercond.* **9** (2) 1575
[6] Mochida T, Sakai N, Yoo S I and Murakami M, 2002 *Physica C* **366** 229
[7] Çelik Ş, Öztürk K and Yanmaz E, 2008 *J. Alloys Compd.* **456** 1
[8] Çelik Ş, Thesis Ph D, Karadeniz Technical University, June 2006, Trabzon, Turkey.
[9] Çelik Ş, Öztürk K, Çevik U and Yanmaz E, 2008 *J. Alloys Compd.* **460** 79
[10] Çelik Ş, Öztürk K and Yanmaz E, 2008 *J. Alloys Compd.* **458** 30
[11] Öztürk K, Çelik Ş, Çevik U and Yanmaz E 2007 *J. Alloys Compd.* **433** 46
[12] Öztürk K, Çelik Ş and Yanmaz E, 2008 *J. Alloys Compd.* **462** 19
[13] Öztürk K, Çelik Ş, Çevik U and Yanmaz E, 2008 *J. Alloys Compd.* **456** 34
[14] Osofsky M S, Soulen R J, Wolf S A, Broto J M, Rakoto J M, Ousset J C, Coffe G, Askenazy S, Pari P, Bozovic I, Eckstein J N and Virshup G F, 1993 *Phys. Rev. Lett.* **71** 14
[15] Helfand E, Werthamer N R and Hohenberg P C, 1966 *Phys. Rev.* **147** 295
[16] Thinkam M, 1996 *Introduction to Superconductivity*, Second Ed. McGraw-Hill New York 118
[17] Shigeta I, Abiru T, Abe K, Nishida A and Matsumoto Y, 2003 *Physica C* **392** 359
[18] Ando Y, Boebinger G S, Passner A, Schn L F, Kimura T, Okuya M, Watauchi S, Shimoyama J, Kishio K, Tamasaku K, Ichikawa N and Uchida S, 1999 *Phys. Rev. B* **60** 17.
[19] Müller K H, Fuchs G, Handstein A, Nenkov K, Narozhnyi V N and Eckert D, 2001 *J. Alloys Compd.* **322** L10
[20] Sekitani T, Miura N, Ikeda S, Matsuda Y H and Shiohara Y, 2004 *Physica B* **346** 319
[21] Welp U, Kwok W K, Crabtree G W, Vandervoort K G and Liu J Z, 1989 *Phys. Rev. Lett.* **62** 1908
[22] Hao Z, Clem J R, McElfresh M W, Civale L, Malozemoff A P and Holtzberg F, 1991 *Phys. Rev.* B **43** 2844
[23] Brandstatter G, Sauerzopf F M, Weber H W, Ladenberger F and Schwarzmann E, 1994 *Physica C* **235** 1845
[24] Thompson J R, Ossandon J G, Christen D K, Chakoumakos B C, Sun Y R, Paranthaman M and Brynestad J, 1993 *Phys. Rev. B* **48** 14031
[25] Li Q, Suenaga M, Hikata T and Sato K, 1992 *Phys. Rev. B* **46** 5857

Electrodynamic behaviour of a two-junction quantum interferometer enclosing a superconducting nanocylinder

R De Luca[1]
Dipartimento di Matematica e Informatica, *University of Salerno*,
Via Ponte don Melillo, 84084 Fisciano (SA), ITALY

E-mail: rdeluca@unisa.it

Abstract. The electrodynamic response of a two-junction symmetric quantum interferometer, containing two equal *0*-junctions and enclosing a superconducting nanocylinder, is studied by means of a first order perturbation analysis of the dynamical equations obtained for the composite system in the absence of noise. It is shown that, when an appropriate number of fluxons is trapped in the inner superconducting cylinder, the system behaves like a π-SQUID.

1. Introduction

Two-junctions quantum interferometer models have been adopted as model systems of Superconducting QUantum Interference Devices (SQUIDs) [1-2] for a long time and are now being considered, for their peculiar properties, as possible candidates in realizing elementary memory cells in quantum computing [3-4]. The interest in these systems, however, rests also upon their intrinsic analytical properties. In particular, it has been shown [5-6] that a perturbation analysis, aimed to solving the coupled nonlinear ordinary differential equations governing the dynamics of the two superconducting phase differences in the device, ϕ_1 and ϕ_2, gives rise to a *single-junction equivalent model*, for small values of the parameter $\beta = \dfrac{LI_{J0}}{\Phi_0}$, where L is the inductance of a single branch of the device, I_{J0} is the maximum Josephson current of the two equal junctions, and Φ_0 is the elementary flux quantum. This model was already known in the literature in the rather simple case of $\beta = 0$. In the latter case, indeed, a symmetric quantum interferometer containing identical junctions can be described by a differential equation analogous to the one governing the dynamics of a single junction. In this equivalence, however, we need to specify that the superconducting phase difference ϕ across the equivalent junction is given by the average of the two superconducting phases and that the effective current-phase relation is of the conventional type $I = I_J \sin\phi$, with a maximum Josephson junction I_J modulated by the externally applied normalized magnetic flux ψ_{ex} as follows: $I_J = I_{J0}\cos(\pi\psi_{ex})$.

In case the parameter β is finite a different analysis of the dynamical system is required. For small values of the parameter β, indeed, one can first realize that, by a mere change in variables, the

[1] To whom any correspondence should be addressed.

system can be analyzed in terms of the average superconducting phase ϕ and of the fluxon number ψ, which represents the total magnetic flux linked to the superconducting loop normalized to the elementary flux quantum Φ_0. This change of variables is necessary also to discriminate between *slow* and *fast* dynamics in the system. Indeed, for small values of the parameter β, the flux number ψ possesses a much more rapid dynamical behaviour than the average superconducting phase ϕ. In this way, the dynamical equation for ψ, in which a coupling term in $\cos\phi$ appears, can be solved as if the coupling term would be a fixed parameter. When the solution of this equation is found by a perturbation analysis to first order in β and the result is substituted in the dynamical equation for the slow variable ϕ, one obtains [6, 7] the following nonlinear ordinary differential equation

$$\frac{\mathrm{d}\phi}{\mathrm{d}\tau} + \cos(\pi\psi_{ex})\sin\phi + \pi\beta\sin^2(\pi\psi_{ex})\sin 2\phi = \frac{i_B}{2}, \qquad (1)$$

where $\tau = \dfrac{2\pi R I_{J0}}{\Phi_0}$, R being the resistive parameter of the junctions, and i_B is the bias current injected in the device, normalized to I_{J0}. As one can see from equation (1), the single junction equivalent model described above is recovered by setting $\beta = 0$. Carrying out the perturbation analysis to first order in β, however, we notice the appearance of second harmonic term in the effective current-phase relation for the equivalent junction. This term is not due to intrinsic junction properties, since the current-phase relation (CPR) for each junction is of conventional type. On the contrary, the second harmonic sine term appears because of the inductively mediated electrodynamic coupling between the junctions.

In what follows we shall carry out a similar perturbation analysis for a two junction quantum interferometer containing two identical conventional *0*-junctions and a superconducting nanocylinder entrapping magnetic flux as shown in figure 1. We shall notice that, already at a stage where the dynamical equations is written for the superconducting phases ϕ_1 and ϕ_2, one can argue that the composite system could behave like a π-SQUID, which is realized by substituting one of the two *0*-junctions with a π-junction in a conventional superconducting quantum interference device. This behaviour is obtained by virtue of flux quantization into the superconducting nanocylinder. We finally exhibit the reduced model for this system and explore the possibility of realizing an experiment which could confirm the prediction made in the present work.

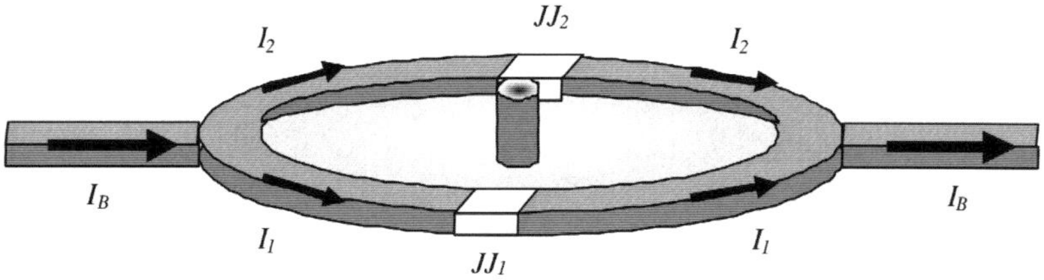

Figure 1. Schematic representation of the composite system.

2. The dynamical model

In the present section we first find the dynamical equations for the superconducting phase differences ϕ_1 and ϕ_2 across the two junctions in the interferometer enclosing a superconducting nanocylinder with entrapped magnetic flux. Starting from these equations and following the analysis described in references [5] and [6], the reduced model for the composite system is derived.

The Resistively Shunted Junction (RSJ) model [1-2] is used to describe the dynamics of the two identical overdamped junctions, which symmetrically interrupt the interferometer superconducting loop. In this way, the dynamical equations in the absence of noise are written as follows:

$$\frac{\mathrm{d}\phi_1}{\mathrm{d}\tau}+\sin\phi_1=i_1=\frac{i_B+i_S}{2}, \tag{2a}$$

$$\frac{\mathrm{d}\phi_2}{\mathrm{d}\tau}+\sin\phi_2=i_2=\frac{i_B-i_S}{2}, \tag{2b}$$

where $i_S=\dfrac{i_1-i_2}{2}$ is the circulating current. In order to express the above equations in terms of the variables ϕ and ψ, we first proceed to the following transformation of variables: $\phi=\dfrac{\phi_1+\phi_2}{2}$, $\xi=\dfrac{\phi_1-\phi_2}{2}$. Furthermore, by fluxoid quantization [1], we may express ξ in terms of ψ as follows: $\xi=\pi(n-\psi)$, where n is an integer. Let us now express the currents in terms of all fluxes in the system. For this purpose, let us define the geometric fluxes $\psi_{ex}=\dfrac{\mu_0 HS}{\Phi_0}$ and $\psi'_{ex}=\dfrac{\mu_0 HS_0}{\Phi_0}=\dfrac{r^2}{R^2}\psi_{ex}$, where H is the applied field and S and S_0 are, respectively, the total circular area of radius R of the region enclosed by the interferometer loop and the superconducting cylinder section of radius r. On the other hand, the flux numbers ψ and $\psi'=k$, k being an integer owing to flux quantization in the superconducting cylinder, represent the actual flux linked to S and S_0, respectively. By elementary electrodynamics and simple algebraic manipulations, we finally write the dynamical equations for the composite system in terms of ϕ and ψ in the following way [8]:

$$\frac{\mathrm{d}\phi}{\mathrm{d}\tau}+(-1)^n\cos(\pi\psi)\sin\phi=\frac{i_B}{2}, \tag{3a}$$

$$\pi\frac{\mathrm{d}\psi}{\mathrm{d}\tau}+(-1)^n\cos\phi\sin(\pi\psi)+\frac{\psi}{2\tilde{\beta}}=\frac{1}{2\tilde{\beta}}\left[\left(1-\frac{r^2}{R^2}\frac{\tilde{\beta}}{\gamma}\right)\psi_{ex}+k\frac{\tilde{\beta}}{\gamma}\right]. \tag{3b}$$

where $\tilde{\beta}$ and γ are the characteristic parameters of the system. Indeed, denoting by $2L$ and L' the self inductance of the interferometer loop and of the inner superconducting cylinder, respectively, and by $2M$ the value of the mutual inductance between these two elements of the composite system, we may define: $\tilde{\beta}=\dfrac{lI_{J0}}{\Phi_0}$, with $l=\dfrac{LL'-M^2}{L'}$, and $\gamma=\dfrac{mI_{J0}}{\Phi_0}$, with $m=\dfrac{LL'-M^2}{M}$. In this way, $\tilde{\beta}$ is analogous to the usual interferometer parameter β, while the ratio $\dfrac{\tilde{\beta}}{\gamma}=\dfrac{M}{L}$ takes account of the electrodynamic interaction between the inner superconducting cylinder and the interferometer loop.

Notice also that equation (3a) is formally identical to the corresponding equations for a superconducting quantum interferometer, while only equation (3b) differs, when $M \neq 0$, from the corresponding dynamical equation for the flux number of an ordinary superconducting quantum interferometer.

3. Some properties of the composite system

In the present section we investigate the electrodynamic properties of the composite system. We start by analyzing some characteristic features coming from equation (3b). We first notice that this equation clearly reduces to the usual expression for a superconducting quantum interferometer when we set $M = 0$. In this case, indeed, the parameter $\tilde{\beta}$ goes to β and the ratio $\frac{\tilde{\beta}}{\gamma}$ vanishes. Interesting new behaviour of the composite system is thus expected only for $M \neq 0$. In order to simplify the analysis, let us assume $r << R$ in such a way that we can make use of the approximated mutual inductance expression

$$M \approx \frac{\mu_0 \pi r^2}{4R}. \tag{4}$$

By taking $L' \approx \frac{\mu_0 \pi r^2}{d}$, where *d* is the cylinder height, we have that $\frac{\tilde{\beta}}{\gamma} = \frac{M}{L'} \approx \frac{d}{4R}$. By choosing *d* of the same order of magnitude of *r*, we notice that $\frac{r^2}{R^2}\frac{\tilde{\beta}}{\gamma} \approx O\left(\frac{r^3}{R^3}\right)$ and equation (3b) can be written as follows

$$\pi \frac{\mathrm{d}\psi}{\mathrm{d}\tau} + (-1)^n \cos\phi \sin(\pi\psi) + \frac{\psi}{2\tilde{\beta}} \approx \frac{1}{2\tilde{\beta}}\left(\psi_{ex} + \frac{d}{4R}k\right). \tag{5}$$

The above equation, apart from a redefinition of the characteristic parameter β ($\beta \to \tilde{\beta}$), presents a shift in the normalized external magnetic flux ψ_{ex}, which, as specified in reference [8], can give rise to π-SQUID-like behaviour of the composite system if this shift is equal to a half integer number.

First of all, let us notice that, in order to obtain an integer number *k* of fluxons inside the inner superconducting cylinder, we need to apply a magnetic field H_N before cooling the cylinder itself below its critical temperature. Therefore, we have:

$$k = \Omega\left[\frac{\mu_0 H_N S_0}{\Phi_0}\right], \tag{6}$$

where Ω is a nonlinear operator, which gives the closest integer value of the quantity inside the square brackets. Therefore, if we chose to trap a given number *k* of fluxons inside the superconducting cylinder, we need to have:

$$\left(k - \frac{1}{2}\right)\frac{\Phi_0}{S_0} < \mu_0 H_N < \left(k + \frac{1}{2}\right)\frac{\Phi_0}{S_0}. \tag{7}$$

In this way, in order to get π-SQUID-like behaviour of the composite system, we may set $\frac{d}{4R}k = \frac{2s+1}{2}$, with *s* integer, so that equation (7) can be recast in the following form:

$$\left[\frac{2R}{d}(2s+1)-\frac{1}{2}\right]\frac{\Phi_0}{S_0} < \mu_0 H_N < \left[\frac{2R}{d}(2s+1)+\frac{1}{2}\right]\frac{\Phi_0}{S_0}. \quad (8)$$

We should notice, however, that the range of applied magnetic field values, obtained for various integers s, need to be compatible with the trapping capabilities of the superconducting cylinder. Therefore, one needs to compute the saturation number n_S corresponding to a given hole size, giving the maximum number of trapped fluxons in a cylindrical cavity. In type-II superconductors. For example, in the limit of $r >> \lambda$, with λ the penetration depth of the superconducting material, it can be shown that [9]

$$n_S \approx \frac{r^2}{3\sqrt{3}\xi\lambda}, \quad (9)$$

where ξ is the coherence length of the superconducting cylinder. In this way, since we can only take values of k less than n_S, we might introduce the following constraint to equation (8):

$$k = (2s+1)\frac{2R}{d} \le \frac{r^2}{3\sqrt{3}\xi\lambda}. \quad (10)$$

For a practical realization of the system, let us choose $s = 0$, $R = 1.0\,\mathrm{mm}$, $d = 10.0\,\mu\mathrm{m}$, giving $k = 200$. We need to find a hollow cylindrical type-II superconductor whose geometry and whose properties are such to satisfy equation (10). By choosing, for instance, Niobium at temperatures close to $4.2\,\mathrm{K}$, with $\lambda \approx 40\,\mathrm{nm}$ and $\xi \approx 9.0\,nm$, and by taking $r = 800\,\mathrm{nm}$, we get $n_S \approx 342$, which makes the constraint in (10) compatible with the quantities chosen above. We can thus safely evaluate the field range in which trapping of 200 fluxons in the superconducting cylinder is possible:

$$0.411\,\mathrm{T} < \mu_0 H_N < 0.412\,\mathrm{T}. \quad (11)$$

4. The reduced model for the composite system

We now consider again equations (3a) and (3b), in order to get a single junction equivalent model of the composite system. Let us first set $f = \frac{d}{4R}k$, so that the dynamical equations for the variables ϕ and ψ can be written as follows:

$$\frac{d\phi}{d\tau} + (-1)^n \cos(\pi\psi)\sin\phi = \frac{i_B}{2}, \quad (12a)$$

$$\pi\frac{d\psi}{d\tau} + (-1)^n \cos\phi\sin(\pi\psi) + \frac{\psi}{2\tilde{\beta}} = \frac{1}{2\tilde{\beta}}[\psi_{ex} + f]. \quad (12b)$$

Let us now assume that the parameter $\tilde{\beta}$ is small enough to allow a perturbation solution of equation (12b) for the variable ψ, so that [6,7]:

$$\psi \approx \psi_0 + \tilde{\beta}\psi_1 = \psi_{ex} + f - 2\tilde{\beta}(-1)^n \sin[\pi(\psi_{ex} + f)]\cos\phi. \quad (13)$$

By substituting (13) in (12), we get the following single-junction equivalent dynamical equation

$$\frac{d\phi}{d\tau}+(-1)^n\cos[\pi(\psi_{ex}+f)]\sin\phi+\pi\tilde{\beta}\sin^2[\pi(\psi_{ex}+f)]\sin 2\phi=\frac{i_B}{2}, \qquad (14)$$

the only difference with respect to the usual model is the *f*-shift of the applied magnetic flux.

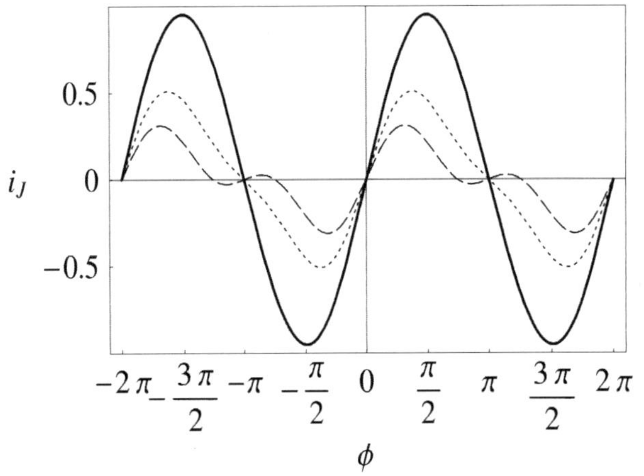

Figure 2. Current-phase relation of the equivalent single-junction model of the composite system with n even, $\tilde{\beta}=0.05$, $\psi_{ex}=0.1$ and with $f=0$ (solid line), $f=0.25$ (dotted line), $f=0.333$ (dashed line).

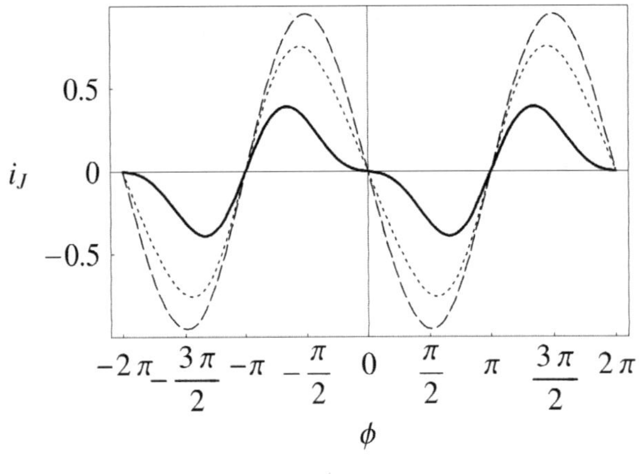

Figure 3. Current-phase relation of the equivalent single-junction model of the composite system with n even, $\tilde{\beta}=0.05$, $\psi_{ex}=0.1$ and with $f=0.5$ (solid line), $f=0.666$ (dotted line), $f=1.0$ (dashed line).

We might then exhibit the effective CPR of this device, for various values of *f*, as in figures 2 and 3. In particular, in figure 2 we show the effective superconducting current $i_J=(-1)^n\cos[\pi(\psi_{ex}+f)]\sin\phi+\pi\tilde{\beta}\sin^2[\pi(\psi_{ex}+f)]\sin 2\phi$ versus the average phase difference

ϕ for n even, $\tilde{\beta} = 0.05$, $\psi_{ex} = 0.1$ and for various values of the parameter *f*. The same is done in figure 3, for the same values of the parameters n, $\tilde{\beta}$ and ψ_{ex}, and for still different values of f.

5. Conclusions

We have studied the electrodynamic behaviour of a composite system consisting of a two-junction symmetric quantum interferometer enclosing a concentric superconducting cylinder of nanoscopic dimensions. It has been shown that the physical properties of this system can be described by a single-junction equivalent model with two characteristic parameters: $\tilde{\beta}$, which is the analogous of the parameter β for d. c. SQUIDs and *f*, proportional to the number of trapped fluxons in the inner cylinder. It has been noted that, for specific numbers of trapped fluxons inside the inner superconducting cylinder, the composite system behaves like a π-SQUID. Furthermore, deviation of the CPR from the usual sinusoidal behaviour is detected mainly for two reasons. First of all, when the characteristic parameter $\tilde{\beta}$ is different from zero, a second harmonic term appears in the effective CPR of the single-junction equivalent model. This feature is also present in two-junction quantum interferometers. It has also been shown that, for $\tilde{\beta} \neq 0$, by keeping the externally applied field fixed, the CPR may be different from its sinusoidal behaviour by making the parameter *f* vary.

The composite system has been studied in the absence of noise. Further work will be devoted to the analysis of the physical properties of the system in the presence of thermal fluctuations.

6. References

[1] Barone A, Paternò G 1982 *Physics and Applications of the Josephson Effect* (New York: Wiley)
[2] Likharev K K 1986 *Dynamics of Josephson Junctions and Circuits* (Amsterdam: Gordon and Breach)
[3] Bocko M F, Herr A M and Feldmann M J 1997 *IEEE Trans Appl. Supecond.* **7** 3638
[4] Majer J B, Paauw F G, ter Haar A C J, Harmans C J P M and Mooij J E 2005 *Phys. Rev. Lett.* **94** 090501
[5] Grønberg-Jensen N, Thomson D B, Cirillo M and Cosmelli C 2003 *Phys. Rev.* **B 67** 224505
[6] Romeo F and De Luca R 2004 *Phys. Lett.* **A 238** 330
[7] De Luca R, Fedullo A and Gasanenko V A 2007 *Eur. Phys. J.* **B 58** 461
[8] De Luca R 2008 *Physica* **C 468** 932
[9] Denisov D V, Shantsev D V, Galperin J M and Johansen T H 2007 *arXiv:1086v3[cond-mat.supr-con]*

Pinning behavior of Al, Mn and Ag substitutions in YBCO TSMG bulk superconductors

P. Diko, V. Antal, M. Kaňuchová, M. Šefčiková, J. Kováč

Institute of Experimental physics SAS, Watsonova 47, 04001 Košice, Slovak Republic

Corresponding author: dikos@saske.sk

Abstract. Peak effect in dependences of critical current density on applied magnetic field caused by nanosize chemical pinning centers was studied in $Y_{1.5}Ba_2(Cu,M)_3O_y$, for M = Ag, Mn, and Al with CeO_2 technological addition. The behavior of Mn, Ag and Al dopants was analyzed in relation to sample microstructure and oxygenation heat treatments respectively. It is shown that formation of $Cu_2Mn_3Ba_6O_y$ compound and low negative partition coefficient of Ag between the solid and melt are responsible for declination from expected pinning behavior for Mn and Ag substituents.

1. Introduction

As a consequence of low coherent length (≈ 3 nm at 77 K), the most effective pinning centers in RE123superconductors should have nanosize dimensions. For example optimally irradiated Y123 bulk superconductors can reach 3 times higher trapped fields due to effective nanosize pinning centers induced by irradiation damage [1]. There are successful attempts to introduce nanosize pinning centers in the form of ex-situ [2] or in-situ [3] created nanoparticles or through introduction of nanosize pinning centers by chemical substitutions in the RE123 lattice [4]. In the case of RE123 bulk superconductors the most problematic step in the incorporation of nanoparticles is the pushing of the particles with the size smaller than critical by the growing crystal. It seems therefore sensible to create nanosize pinning centers in these bulks through substitution of atoms in the RE123 lattice. It is characteristic for TSMG RE123, especially for Y123, that small technological addition of Pt or Ce is necessary for their successful fabrication. These elements can react with added substitutions and consequently misrepresent their influence on the pinning behavior. For example in the TSMG Y123 bulks with Pt or Ru addition the phase $Ba_4Cu_2PtO_y$ type is formed and this phase can incorporate Cu substituents [5, 6]. All reported studies on TSMG YBCO bulks wit Cu substitutions were done on the systems with Pt addition [4, 7, 8]. Therefore in our program we used Ce addition where no reaction with Cu substitutions has been reported.

2. Experimental details

YBCO bulk single-grain samples with a nominal composition $Y_{1.5}Ba_2(Cu,M)_3O_y$, for M = Ag, Mn and Al were fabricated by top-seeded melt-growth process (TSMG). The stoichiometry of the M elements in $Y_1Ba_2(Cu_{1-x}M_x)_3O_y$ was chosen in the range x = 0.0025 to 0.05. $YBa_2Cu_3O_7$, Y_2O_3, CeO_2 and M-oxide powders were mixed in appropriate amounts, milled in a friction mill and pressed in to cylindrical pellets 20 mm in diameter. The as grown samples were cut in two halves along the {100} plane. The microstructure of the sample was analyzed by an optical microscope and by scanning electron microscope (SEM). For magnetization measurements, the plate-like samples (2 x 2 x 0.5) mm^3 in size were cut from the a-growth sectors of the single-grain bulks in a distance 1 mm from the seed. The samples were oxygenated in flowing oxygen at 400 °C for 150 hours. Magnetization properties of the samples were measured at 77 K by a vibrating sample magnetometer under fields applied parallel to the *c*-axis. The critical current density, J_c, was calculated from the width of magnetization hysteresis loops based on the extended Bean model. The critical temperatures, T_c, were determined from the zero field and field cooling curves with an applied external magnetic field 2 mT. The transition temperatures were obtained from the mid-point of the magnetization transition curves (50% magnetization) from the superconducting state to the normal state.

3. Results and discussion

In the case of Al dopant the peak effect appeared from the lowest used dopant concentration x = 0.0025 while for Mn the strongest peak effect was observed for x = 0.02 and for Ag the peak effect was observed only at the highest dopant concentration x = 0.05 (Figure 1). According to Ishii [8] in the

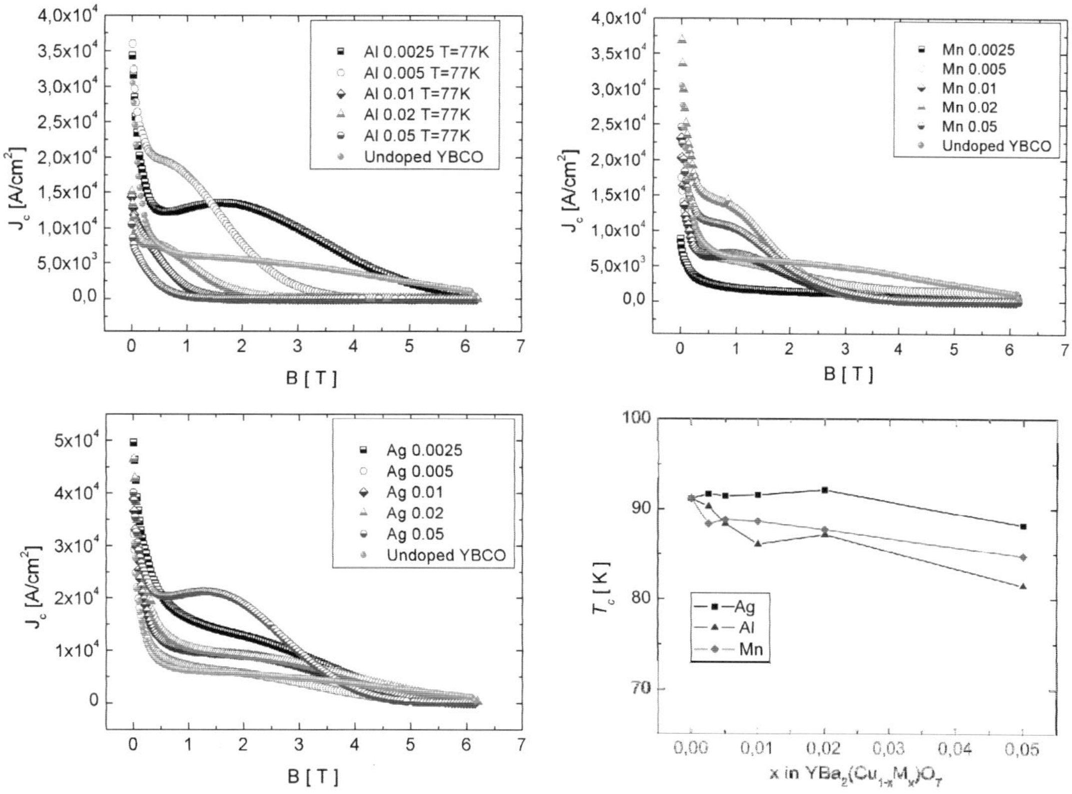

Fig. 1. Dependences of critical current densities, J_c, on applied magnetic field, *B*, and transition temperature, T_c, on dopant concentration. Peak effect appears at the lowest Al concentration, starts at higher Mn concentration and is observed only at the highest Ag concentration. Decreasing of T_c with dopant concentration evidences substitutions in Y123 lattice.

RE123 melt processed samples with RE element, which do not substitute Ba (for example Y and Dy) the peak effect should be observed at concentration of dopant leading to the distance between randomly distributed dopant atoms in the CuO_2 plane equal to 2 coherent lengths what is about 6 nm at 77 K. In our case this distance between substituents is reached for x = 0.0013 when they substitute Cu

in CuO chains. This level of doping can correspondent to our lowest used concentration as some dopant atoms can substitute also in Y211 phase what will decrease their concentration in Y123 phase. The decreasing of T_c with dopant nominal concentration confirms that Cu atoms in Y123 crystal lattice were partially substituted. Some differences in T_c decreasing in comparison with polycrystalline ceramics may be caused by redistribution of dopants between Y123 and Y211 phases, inhomogeneous distribution of dopant caused by solidification process as well as due to formation of secondary phases rich in dopant.
The peak effect caused by Al and Mn was not observed so far in the TSMG bulk YBCO superconductors and Ag and Mn behavior is different because both these elements exhibit peak effect at concentrations higher than expected. Further we will analyze more in detail Ag, Mn and Al doping separately.

Ag substitution. The peak effect caused by Ag addition in TSMG YBCO was reported by Nakashima [9] but the concentration of Ag in the sample with peak effect was not given. Important information, which can explain why the peak effect in our case was observed at much higher nominal dopant concentration, was obtained from conducted microstructural observation. We found that in the middle of the samples, at the beginning of solidification, the microstructure does not contain any Ag phase at any Ag nominal concentration and consists only from Y123 crystal and Y211 particles and well developed gray barium cerate phase (Fig. 2 (a)).

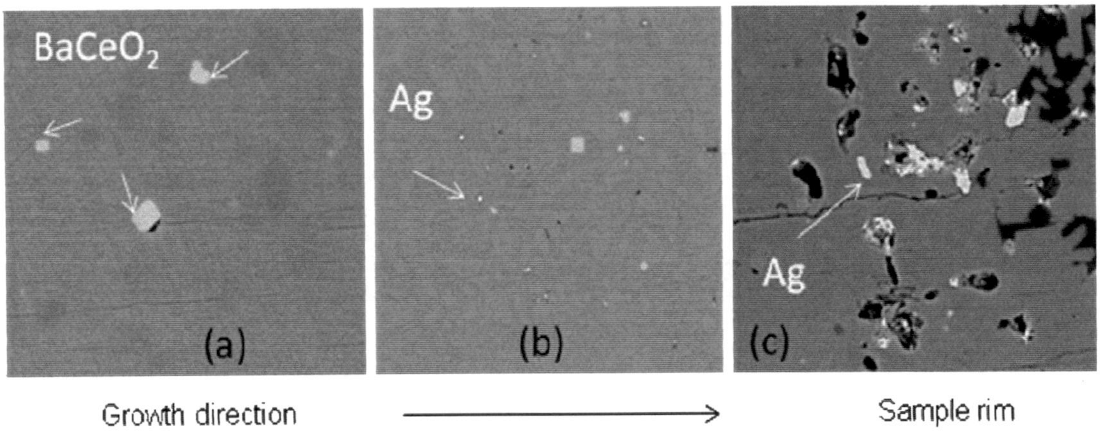

Fig. 2. SEM micrographs of a/c-cross section of Ag doped samples. Besides Y123 and Y211 phases only grey barium cerate, $BaCeO_2$, rectangular particles were present at the beginning of solidification (a). Later, white globular Ag particles are trapped (b) but the most Ag is concentrated as separated phase at the sample rim (c).

After some distance from the beginning of solidification the Ag phase start to appear in the structure in the form of globular particles (Fig. 2 (b)) as it was confirmed by EDAX analysis. The distance from the seed, when the first Ag particles appeared, was longer for the samples with lower nominal Ag concentration. At the lowest nominal Ag concentration the Ag particles were observed only at the end of solidification, at the rim of the sample where they solidified together with the rest of melt (Fig. 2 (c)). This experimental result confirms that there is some partition coefficient controlling ratio of Ag concentration between the solid Y123 crystal, C_{AgS}, and the nominal Ag concentration in melt, C_{Ag0}.

$$k_{Ag} = C_{AgS} / C_{Ag0} \qquad (1)$$

From the ratio between nominal Ag concentration at which the peak effect was observed (x = 0.05) and the Ag concentration in the Y123 crystal at which the peak should appear (x = 0.0013) we can conclude than the Ag distribution coefficient, k_{Ag}, is somewhere around 0.026 (0.0013/0.05). During solidification of the system with negative partition coefficient the dopant element is distributed inhomogenously with the lowest concentration in the solid, C_{AgS} , at the beginning of solidification and increasing concentration with the relative distance from the beginning of solidification, λ, according to equation [10]:

$$C_{AgS} = k_{Ag}\, C_{Ag0}\, (1-\lambda)^{kAg\ -1} \qquad (2)$$

Dependence of Ag concentration on the relative distance from the seed calculated from equation (2) for used starting nominal Ag concentrations is presented in Fig. 3.

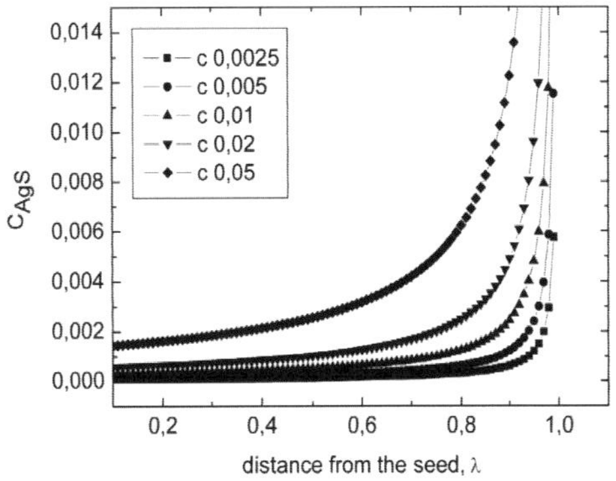

Fig. 3. Calculated Ag concentration in the Y123 crystal, C_{AgS} versus relative distance from the seed, λ.

We measured magnetization and T_c of the samples separated at the beginning of solidification where much lower concentration of Ag was substituting Cu in the Y123 lattice than the nominal Ag concentration. Only at the nominal concentration x = 0.05 the concentration of Ag solved in the Y123 crystal at the beginning of solidification reached value which generated the effective distance of Ag atoms for pinning. As the solubility of Ag in the Y123 phase is limited [11] after some solidification length the Ag concentration in the Y123 crystal exceeds this limit and all the rest of Ag diluted in the melt is pushed by growing crystal. When the Ag concentration in the melt overcomes the solubility limit, Ag rich phase star to precipitate in the form of droplets in the melt. These droplets are trapped by growing Y123 crystal and later, during cooling from the crystal growth temperature, they solidify [12].

Mn substitution. Microstructure analysis revealed that in the studied system with Mn addition the compound with composition close to $Cu_2Mn_3Ba_6O_y$, measured by EDAX, was formed (Fig. 4). Formation of this compound consumes a part of added Mn, but some Mn also substitute Cu in the Y123 lattice what is evidenced by the influence of Mn addition on transition temperature as well as appearance of peak effect at higher Mn concentrations (Fig. 1).

Al substitution. T_c decreasing with Al concentration which was measured (Fig. 1) is comparable with that measured on polycrystalline ceramics [13]. Microstructure analysis has shown that Al addition does not form any compound with Y, Ba, Cu or Ce in the system, therefore we can suppose that the Al partition coefficient is close to one and Al is mostly diluted in the Y123 phase. The peak effect, which we observed at the lowest Al concentrations (Fig. 1), also confirms that Al substitution in the Y123 lattice is close to nominal concentration. Al^{3+} substitutes CuO chains in the Y123 lattice and should behave similarly to Fe and Co substitutions. Both these elements prefer higher coordination number than two and therefore at annealing in Ar at higher temperatures they shear deficit oxygen atoms and form clusters [14]. Additional heat treatment of the samples with Al doping in argon at 800 °C for 1 hour caused significant changes in both T_c and pinning behavior. In the concentration range for x from 0.0025 to 0.02 we observed recovery of transition temperature to the values found in the sample without Al substitution (Fig. 5) as well as extinction of peak effect and

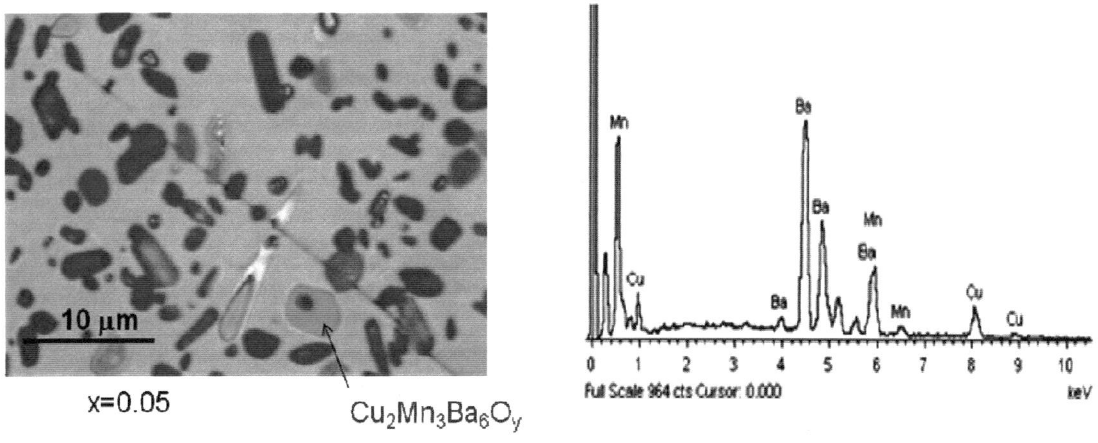

Fig. 4. Y211 particles and particles of $Cu_2Mn_3Ba_6O_y$ phase in the Mn doped sample with x = 0.05 and EDAX spectrum of Mn rich phase.

its appearance at the highest Al concentrations (Fig. 5). If we suppose clustering of Al atoms and effective mean distance between Al clusters 6 nm we can expect about 20 Al atoms in the cluster.

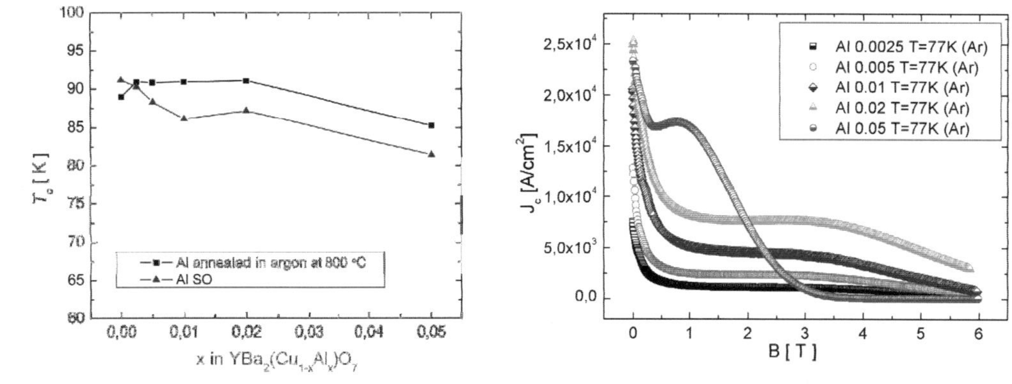

Fig. 5. Critical temperature, T_c, recovered and peak effect shifted to the highest Al concentration after preannealing in Ar at 800 °C.

4. Conclusions

The study of Cu substitutions in TSMG YBCO bulk superconductors with CeO_2 technological addition showed that the peak effect can be induced by Al, Mn or Ag doping. In the case of Al, the peak effect appeared from the lowest used dopant concentration x = 0.0025. We suggest that Al behaves similarly as Fe and Co e.g. under annealing in low oxygen atmosphere it has a tendency to form clusters. Mn addition is partially bound by $Cu_2Mn_3Ba_6O_y$ phase and only partially substitutes Cu in the Y123 lattice. Ag is distributed highly inhomogenously in the sample due to low Ag partition coefficient between solid and melt, which was estimated to be about 0.026. Very low Ag partition coefficient causes that peak effect appeared only at the highest nominal Ag concentration x = 0.05.

Acknowledgements
This work has been supported by VEGA project No.- 2/7052/27, APVV projects No. 51-061505 and LPP-0334-06, EU network NESPA, and by the Centre of Excellence of Slovak Academy of Sciences NANOSMART.

References

[1] Weinstein R, Sawh R, Ren Y, Parks D 1998 *Materials Science and Engineering* **B53** 38
[2] Hari Babu N, Kambara M, Shi Y, Cardwell DA, Tarrant CD and Schneider KR 2003*Physica C* **392–396** 110
[3] Puig T, Gutiérrez J, Pomar A, Llordés A, Gázquez J, Ricart S, Sandiumenge F and Obradors X 2008 *Supercond. Sci. Technol.* **21** 034008
[4] Shlyk L, Krabbes G, Diko P, Nenkov K, Fuchs G 2004 *Inst. Phys. Conf. Ser.* **181** 1401
[5] Diko P, Shlych L, Krabbes G 2003 *Physica C* **390** 143
[6] Diko P, Stöver G, Krabbes G 2002 *Supercond. Sci. Technology* **15** 217
[7] Zhou YX, Scruggs S and Salama K 2006 *Supercond. Sci. Technol.* **19** S556
[8] Ishii Y, Yamazaki Y, Nakashima T, Ogino H, Shimoyama J, Horii S, Kishio K 2008 *Mat. Sci. Eng. B* **151** 69
[9] Nakashima T, Ishii Y, Katayama T, Ogino H, Horii S, Shimoyama J and Kishio K 2008 *J. Phys.: Conf. Ser.* **97** 012007.
[10] Tiller WA 1970 Chapter 9 Solidification in Physica Metallurgy edited by R.W. Cahn, Nord-Holland Publishing Company – Amsterdam-London, p. 416.
[11] Wiesner U, Krabbes G, Uelzen M, Magerkurth C, Plewa J and Altenburg H 1998 *Physica* C **294** 17
[12] Diko P, Krabbes G, Wende C 2001 *Supercond. Sci. Technol.* **14** 486
[13] Tarascon JM et al. 1988 *Phys. Rev. B* **37** 7458
[14] Renevier HP et al. 1989 *Physica C* **162-164** 51

Enhancing the Transport Properties of Polycrystalline MgB_2 Superconductors

M Gharaibeh[1], I M Obaidat[2], B A Albiss[1], and M K Hasan (Qaseer)[1]

[1] *Superconductivity & Magnetic Measurements Laboratory, Physics Department, Jordan University of Science and Technology, 22110 Irbid, Jordan*

[2] *Department of Physics, United Arab Emirates University, Al-Ain 17551, United Arab Emirates*

E-mail: magh@just.edu.jo

Abstract: MgB_2 polycrystalline superconducting specimens were irradiated with 10 MR and 20 MR doses of γ-rays. An increase in the normal state resistance and a broadening of the resistive transition to the superconducting state were observed with increasing γ-irradiation dose within the temperature range of 28 – 31 K. The change in the critical temperature, T_c was found to be insignificant after irradiation. The critical current density, J_c was found to be enhanced almost at all temperatures after irradiation. The enhancement of J_c was found to be linear with the irradiation dose at all temperatures. These results are explained by the effects of γ-rays and temperature on the migration of MgO impurities into the grains and the oxygen segregation in the grain boundaries.

1. Introduction

Magnesium diboride MgB_2, superconductors have become the subject of an enormous interest since their first discovery in 2001 [1]. The mechanism of superconductivity in MgB_2 was found to be electron phonon mediated [2, 3]. MgB_2 superconductors were also found to have two superconducting gaps [4, 5]. Compared to high temperature superconductors (HTSCs), MgB_2 has a lower critical temperature $T_c \sim 39$ K, a simple hexagonal structure (probable absence of problems associated with weak links in HTSCs, a lower anisotropy, a relatively low cost of production and a high critical current density, J_c. All these properties make MgB_2 promising material for applications in the temperature range 20 - 30 K.

Irradiation of MgB_2 superconductors alerts the lattice structure and induces defects (point, columnar, or cluster defects) which greatly affect the superconducting properties [6-10]. Shinde *et a.l* have found only marginal enhancement in J_c after irradiating MgB_2 films by 200 MeV Ag ions [6]. Okayasu *et al.* have found no enhancement in J_c in 2.5 MeV electron irradiated MgB2 [7]. Recently, Okayasu *et al* have conducted a comparison between the effect of boron-ion and neutron irradiations on bulk MgB_2 superconductors [8]. They have found that J_c decreased as a result of irradiations in both cases. They have attributed the decrease in J_c to the decrease in pinning in the case of B-irradiation and the degradation of superconductivity in the case of n-irradiation. Sözeri *et al* have studied the effect of γ-irradiation on the magnetic properties of MgB_2 samples of different geometry at low magnetic fields, namely the paramagnetic Meissner effect (PME) [9]. They have found that PME was lowered with increasing γ-irradiation dose and disappeared completely with a dose of 20 MR. They have concluded that PME is not geometrical in origin but it has to do with the intrinsic properties of the material. Sekkina *et el.* have studied the DC resistivity over the temperature range 36-43 K under γ-irradiation doses up to 40 MR [10]. They have found that T_c was not affected by γ-irradiation for doses up to 20 MR, but there was a slight change at a dose of 40 MR. They have also found that doses of 10 and 20 MR enhanced the structure quality of MgB_2 via the depression of the un-reacted Boron and MgO minor secondary phases, furthermore the minor phases were not noticed at the inter-grain layers with increasing γ-irradiation dose. In the present study we report on the effects of γ-irradiation on the normal state resistance and on the behavior of J_c in polycrystalline MgB_2 superconductors.

2. Experimental procedures

Stoichiometric mixture of Mg and B was made from Mg turnings and finely powdered boron. The mixture was grinded and compacted under a pressure of 100 MPa into a pellet with 10 mm in diameter and 1 mm in thickness. The pellet was then wrapped in a tantalum foil and heat treated in a programmable three-zone tubular furnace in flow of a commercially available gas mixture of 92% Ar and 8% H_2. The pellet was held at 600 °C for 2 hours, followed by 800 °C for 2 hours and finally at 950 °C for one hour, then it is cooled down to room temperature. The produced pellet was grinded for 1 hour, pressed again into a pellet under a pressure of 200 MPa. This pellet is heated at 950 °C for 2 hours followed by a slow cooling to room temperature at the rate of 1 °C/min. The resulting pellets were cut into rectangular specimens approximately 1mm thick with a diamond saw. Based on X-ray diffraction studies, the samples used in the present study were found to be nearly single hexagonal MgB_2 crystalline phase with insignificant traces of secondary phases and small amounts of MgO impurities possibly due to the use of commercial gas mixtures. SEM studies revealed a granular microstructure of the bulk polycrystalline material with the grain sizes ranging from 10 to 30 μm.

The Current–Voltage *(I–V)* characteristics and resistance versus temperature, *R(T)* curves were measured using a standard four point dc-method. The sample was cooled in a closed cycle helium cryostat (Leybold 320) under a vacuum of 10^{-3} mbar. The voltage was measured by a sensitive nano-voltmeter (Keithley 2182) with accuracy better than 10 nV. The current was applied by a programmable constant current source (Keithley 2245) with an accuracy of 1 μA. The *I–V* curves were measured over the temperature range 28-40 K in zero applied magnetic fields. The critical current densities were determined using an electric field criterion of 1 μV/cm. The temperature was measured with a calibrated platinum resistance sensor with an accuracy of ± 0.1 K. The electrical probes were attached to a rectangular-shaped samples (~ 8×4×1 mm^3) using silver paste. Gamma irradiation of the samples was carried out at room temperature by a Co^{60} gamma ray source at a dose rate of 0.5 MR/h. The irradiation of the samples was conducted in a gamma research irradiator instrument (Gamma-facility PX-γ-30, Issledovatelj, Techsnabexport, Moskivia) in Jordan Atomic Energy Commission, Amman, Jordan. The instrument is spherically shaped and has a sample capacity of 4400 cm^3 and uses ^{60}Co as an irradiation source.

3. Results and Discussion

Here we present experimental results on the effect of the γ-irradiation on the superconducting properties of the MgB_2 polycrystalline samples. Figure 1 shows the temperature dependence of resistance for all samples, before and after irradiation with γ-rays with doses of 10 and 20 MR. We can see that the non-irradiated sample displays a typical metallic behavior in the normal state followed by a sharp transition to the superconducting state. We also see that γ-irradiation has a slight effect on the superconducting transition temperature, T_c where it is found to range nearly between 38 and 39 K. The figure also shows a small broadening of the transition region due to γ-irradiation. The broadening region becomes larger as the irradiation dose is increased. This behavior of resistance might be an indication of thermal fluctuation in the order parameter [11]. The critical current densities were determined from the *I-V* characteristic curves using an electric field criterion of 1 μV/cm. Figure 2 is a plot of the critical current density, J_c versus temperature, measured in zero applied magnetic fields before and after γ-irradiation. For the non-irradiated sample, we can see that the decay in J_c is faster at lower temperatures but at higher temperatures, the decay becomes slower and J_c becomes nearly zero above 31 K. After the sample was irradiated with γ-rays, the behavior of J_c as function of temperature is some what similar to the non-irradiated case; however, there is an enhancement in the value of J_c with γ-irradiation in the whole temperature range of 28 – 31 K. Above 31 K no significant effect of γ-irradiation on J_c was observed at both irradiation doses. Figure 3 is a plot of J_c as a function of γ-irradiation dose at different temperatures. As can be seen in the figure, there is a linear increase in the value of J_c with irradiation dose at all temperatures. Figure 4 shows the change in the critical current density, ΔJ_c which is calculated as the difference between the values of J_c after irradiation and before irradiation. We can see that ΔJ_c becomes smaller as the irradiation dose increases. This indicates that as the temperature increases, the effect of γ-irradiation on J_c becomes less significant. This limiting effect of γ-irradiation dose on the critical current density has been reported also for several HTSCs irradiated by heavy ions and by γ-rays [12-15].

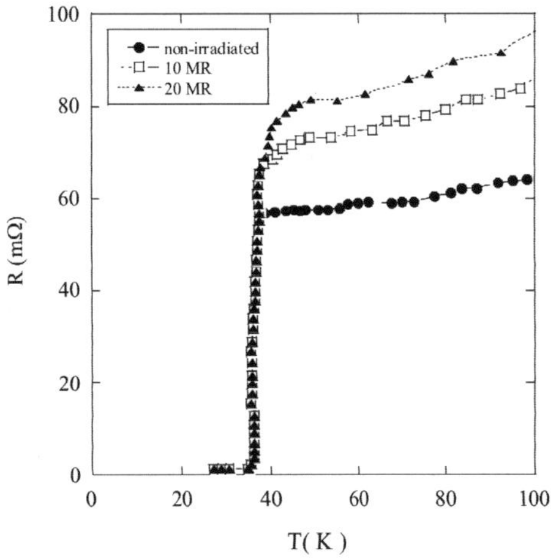

Figure 1. The temperature dependence of resistance, $R(T)$ at zero applied field for all samples, before and after γ-irradiation with two different doses.

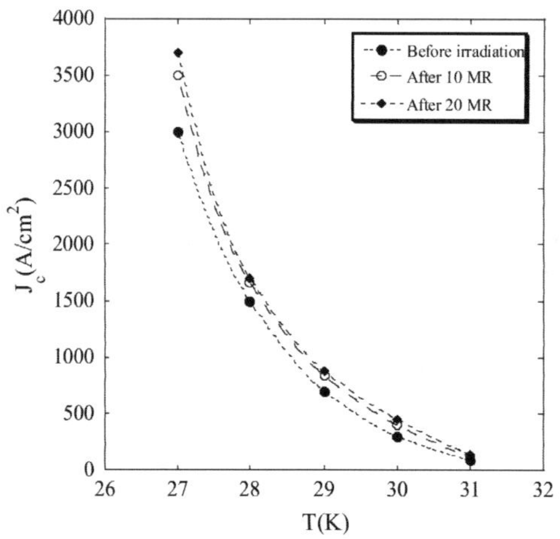

Figure 2. The temperature dependence of the critical current density, J_c (T), for the non-irradiated and γ-irradiated MgB_2 polycrystalline superconductors measured in zero applied magnetic fields.

The results presented here in this study demonstrated a clear effect of γ-irradiation on the electrical transport properties in MgB_2 polycrystalline superconductors. There was a slight effect on the superconducting transition temperature which is quite different from the recent findings in Ref. [10]. The lattice alterations expected from the γ-irradiation might be the cause in the observed increase in the normal state resistance and the broadening of the resistive transition to the superconducting state. The critical current density was clearly enhanced by γ-irradiation. The present MgB_2 polycrystalline sample has small amounts of MgO impurities and possibly small traces of secondary Mg-B phases. Sekkina *et el.* have reported the disappearance of the MgO and B minority phases between the MgB_2 inter-grain layers in MgB_2 polycrystalline samples after γ-irradiation dose of 20 *MR* [10]. Zhu *et al.* have shown that MgO can exist inside MgB_2 grains as stacking faults and created micro strains [16].

As the MgO penetrate inside the MgB_2 grains with increasing γ-irradiation dose, the lattice structure of MgB_2 is altered. This might explain the observed increase in the normal state resistance and broadening of the resistive transition after irradiation. Xue *et al.* have reported that the transport properties of MgB_2 are dominated by oxygen related defects, these defects arise from the existence of MgO inside the MgB_2 grains [17]. Oxygen, unlike some other elements, is present in MgB_2 as an unintentional impurity due to its high reactivity with MgB_2. Typically, oxygen rich precipitates form in the bulk of MgB_2 are expected to enhance the vortex flux pinning and thus to enhance the critical current density [18]. Moreover, it was found that oxygen can segregate at grain boundaries of MgB_2 where it contributes significantly to the vortex flux pinning [19]. Thus we suggest that there are two main effects caused by γ-rays that have a role on enhancing J_c. The first effect is the migration of MgO inside the MgB_2 grains which improves the grains connectivity and also causes oxygen related defects in the grains hence the enhancement in J_c. The other effect is oxygen segregation in the grain boundaries where it increases the vortex flux pinning which leads to an enhancement of J_c. It is also clear from the results presented in figure 4 that the temperature plays a role in controlling these two mechanisms. The level of enhancement of J_c is believed to depend on the specific details of the properties of the granular material.

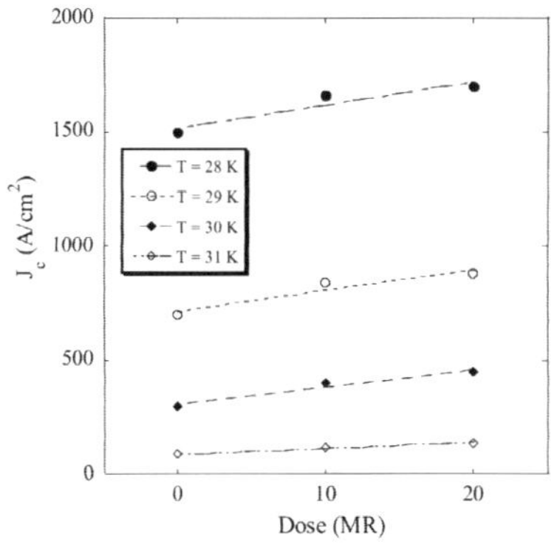

Figure 3. The critical current density, J_c as a function of γ-irradiation dose at different temperatures.

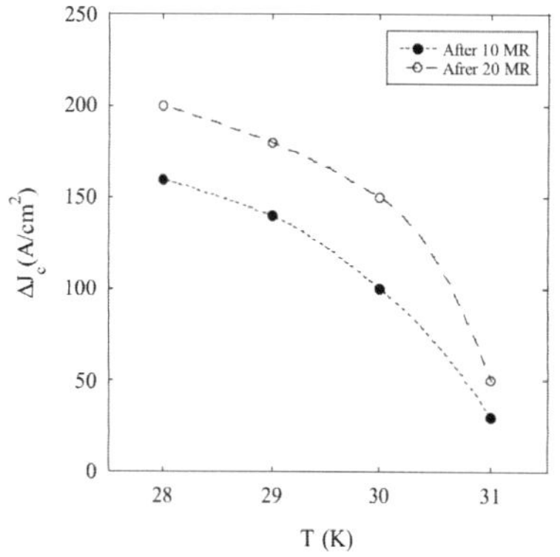

Figure 4. The change of the critical current density, ΔJ_c as a function of temperature.

4. Conclusions

MgB_2 polycrystalline superconducting specimens were irradiated with γ-rays. We have found that γ-rays have nearly insignificant effect on the critical temperature but a broadening of the resistive transition to the superconducting state were observed with increasing γ-irradiation dose. A noticeable increase in normal resistance was also observed after irradiation. The critical current density, J_c was found to increase almost linearly with the dose of γ-rays. The enhancement of J_c becomes less significant as the temperature increases. The behavior of J_c in this material was attributed to two different mechanisms caused by γ-rays as the dose or temperature is increased.

Acknowledgment

This work was financially supported by Jordan University of Science and Technology and by the Research Affairs at the UAE University under a contract no. 03-02-2-11/07.

References:

[1] Nagamatsu J, Nakagawa N, Muranaka T, Zenitani Y, and Akimitsu J 2001 *Nature* **410** 63
[2] Bugoslavsky Y, Perkins G K, Qi X, Cohen L F and Caplin A D 2001 *Nature* **410** 563
[3] Finnemore D K, Ostenson J E, Bud'ko S L, Lapertot G and Canfield P C 2001 *Phys. Rev. Lett.* **86** 2420
[4] Choi H J, Roundy D, Sun H, M. Cohen M L, Louie S G 2002 *Nature* **418** 758
[5] Liu A Y, Mazin I I, Kortus J, 2002 *Phys. Lett.* **87** 87005
[6] Shinde S R, Ogale S B, Higgins J, Choudhary R J, Kulkarni V N, Venkatesan T, Zheng H, Ramesh R, Pogrebnyakov A V, Xu S Y, Li Q, Xi X X, Redwing J M, and Kanjilal D 2004 *App. Phys. Lett.* **84** 2352
[7] Okayasu S, Sasase M, Hojou K, Chimi Y, Iwase A, Ikeda H, Yoshizaki R, Kambara T, Sato H, Hamatani Y, and Maeda A 2002 *Physica C* **382** 104
[8] Okayasu S, Sataka M, Ikeda H, Yoshizaki R 2005 *Physica C* **426-431** 360
[9] Sözeri H, Dorosinskii L, Topal U, Ercan I 2004 *Physica C* **408-410** 109
[10] Sekkina M M A, Elsabawy K M 2002 *Physica C* **377** 411
[11] Kim H, Kim K, Kang W N, Lee H S, Kim K H P, Choi E, Kim H J, Kim H, Lee S, Mun M 2004 *Physica C* **408-410** 72
[12] Obaidat I M, Goeckner H P, and Kouvel J S Submitted to *Physica C.*
[13] Gandini A, Weinstein R, Parks D, Sawh R P, and Dou S X 2001 *IEEE Transactions on Applied Superconductivity*, Vol. **13** 2
[14] Samoilov A V, Feigel_man M V, Konczykow ski M, Holtzberg F, 1996 *Phys. Rev. Lett.* **76** 2798
[15] Kito H, Iyo A, Hirai M, Crisan A, Tokumoto M, Okayasu S, Sasase M, Sataka M, Ihara H, Tanaka Y 2003 *Physica C* **392-396** 181.
[16] Zhu Y, Wu L, Volkov V, Li O, Gu G, Moodenbaugh A R, Malac M, Suenaga M, and Tranquada J 2001 *Physica C* **356** 239
[17] Xue Y Y, Meng R L, Lorenz B, Meen J K, Sun Y Y, Chu C W 2002 *Physica C* **377** 7
[18] Liao X X, Serquis A, Zhu Y T, Huang J Y, Civale L, Peterson D E, Mueller M, and H. F. Xu 2003 *J. Appl. Phys.* **93** 6208
[19] Larbalestier D, Gurevich A, Feldmann D M, and Polyanskii A 2001 *Nature* **414** 368

Studies on structural and magnetic properties of melt textured growth $YBa_2Cu_3O_{7-x}$ superconducting bulk materials

U S Gokay[1], M Tepe[1]
1 Physics Department, Ege University, Bornova, Izmir, Turkey

E-mail: ulasgokay@gmail.com, mustafa.tepe@ege.edu.tr

Abstract. In this study, production methods and the structural and magnetic properties of melt-textured growth of $YBa_2Cu_3O_{7-x}$ bulk superconductors are discussed. During the research; various temperatures for melt texturing are tried for the optimization of the production of bulk YBCO. With the help of AC magnetic susceptibility measurements, some insights for the produced samples are handed out including critical phase transition temperatures and the effects of crystal growth inside the YBCO samples. XRD measurements showed bigger and greater number of crystal growth in a correlation with the increasing temperature and the microphotographs of the samples verified successive growth of YBCO crystals inside the samples. Sharper phase transition is reached with the increasing texturing temperature. Top seeding for the selected crystal orientation has also been tried at the 1070°C for the texturing temperature, but the resulting melt textured sample was not different with the only-textured sample in the means of magnetic properties. Optimization over the texturing temperature found out to be between 1070°C and 1080°C while this result only depends on magnetic and structural properties.

Introduction

Developments on the high temperature superconductivity have taken a great advance over the last 20 years. The discovery of the high-Tc oxide superconductor, named as Y-Ba-Cu-O system [1] brought a great deal on the area of superconductivity. These new type of superconductors have many new possibilities which are useful to design them in desired shapes and contain useful mechanical and electromagnetic properties. Producing this type of superconductors and optimization of the fabrication conditions have been a great concern and been overcome by several approaches. Flywheel energy storage systems [2-3], benefiting magnetic levitation properties [4] and a good value of critical current density are all among the required properties of the bulk samples. However, the quality of the bulks depend on many factors; including grain size, crystal orientation and the other properties such as secondary phases, homogeneity of the sample and microcracks have critical effects. In this case; obtaining good quality YBCO bulk superconductors is at most importance. One of the important and useful approaches to fabricate bulk YBCO is the melt-textured growth [5-6].

In this study production method of melt-textured growth has been experimented for obtaining bulk YBCO superconductors. Magnetic properties and structural properties, together with x-ray diffraction and microphotography have been applied to get characterization and optimization of the samples. Peritectic decomposition temperature has been changed to get the difference of crystal growth among the samples while other parameters kept fixed.

Experimental details

Throughout this study the method of Melt-Textured-Growth [7] of YBCO bulks has been experimented. Experimentation consisted of some predefined temperature steps which are crucial to crystallization in the means of peritectic decomposition temperature. These temperatures are selected with the help of previous studies [8-9], so that resulting YBCO bulks expected to differ in the means of crystal sizes, crystal orientation and samples' magnetic properties. These samples differed among their decomposition temperature followed as in this order: 1030 °C, 1050 °C, 1070 °C, 1080 °C, 1090 °C. Other variables; such as size, mass and the homogeneity of the samples, thermal cycle with further oxygenation at which melt-textured growth has taken place are tried to keep fixed while the decomposition temperature is changed.

All samples are prepared according to the method of solid state reaction by required chemicals which are weighted by a 4-digit weightier in stoichiometric proportions and mixed for a suitable time for getting homogenous mix. This mixture then subjected to calcination process which is required to vaporize CO_2 from the mixture. After the calcination, mixture is grounded, mixed homogenously and pelleted to get disks of 8mm in diameter and 1mm in thickness which will be used in sintering process. Sintering process is performed at 920°C under air atmosphere for all samples for a suitable time period. Sintering process is required for getting a good melted and homogenous mixture. These disks then called as 'sintered samples', are then ready for Melt-Textured-Growth. For all the thermal cycles, a Euroterm horizontal tubular furnace is used. Sintered samples are subjected to a thermal gradient via entering them into the hot zone of the furnace with a speed of 3cm/min at the peritectic decomposition temperature. At this temperature two phases of Y211 and Y123 are melted and mixed so that a well mixed liquid phase is reached. After this temperature plateau, samples are subjected to a slow cooling rate at which the first Y123 crystal seeding starts. This cooling ends at the temperature of 980°C and samples are subjected to this temperature for a long period so that the stabilization of the growth crystals and further crystal formation are enhanced. After this dwell, samples are cooled to the room temperature. Further oxygenation at 600 °C is needed in case of getting orthorhombic structure which supplies property of superconductivity.

Results and Discussion

Numerous microphotography and XRD measurements are performed to engage into the inner structure of the samples. Due to the superconductive properties of the YBCO crystal form, a predefined and good growth of layers should be provided during the crystal growth. This layer formation consists of a growth along the *c*-axis and also a periodicity along the *a-b* layer. Also the texturing of the samples which is the main goal of this study should be significant with the peaks along the *c*-axis.

XRD measurements are performed with an X'pert Pro Cu-K_α X-ray diffractometer and results are tabulated as a graph of intensity versus 2θ. Intensity peaks are expected to show well developed Y123 crystal and Y211 phase signals along the *a-b* plane with *c*-axis between the angles of 20° to 60°. The first sample is shown at Fig. 1 which has crystal growth temperature of 1030 °C shows numerous intensities from all degrees which are close to each other and many peaks are mixed in the means of signal competence. There are some peaks for the growth along the (00*l*) but none of them reflect strong peaks for the *c*-axis growth. Also the other peaks of *a-b* plane are also mixed and no strong signal is recorded. There are also $(040)^*$ and $(151)^*$ peaks which may result from Y211 phase. This situation shows a bad crystal growth or no crystal growth at all. On the other hand, the sample having crystal growth temperature of 1050 °C shows many strong and clear peaks for the *c*-axis growth as seen in Fig. 2. Characteristic orthorhombic structure peaks [10] of (103), (013), (213), and (116) are also seen in the diffraction pattern. Growth along the *a-b* plane can also be seen providing the periodic crystal structure. There is also a low melting point phase signal as $(002)^*$ peak of CuO. The sample with crystal growth temperature of 1070 °C as seen in Fig. 3 states some (00*l*) peaks along the *c*-axis. Intensity data is similar to the previous one in the means of peak positions and coherence of characteristic orthorhombic structure peaks. Intensities are clean and strong so that there has been a good crystal growth in *c*-axis.

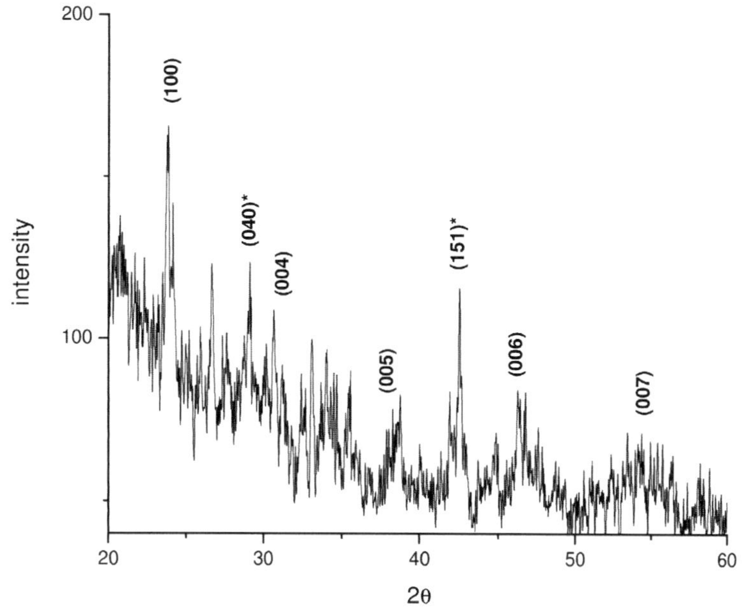

Figure 1. XRD pattern of the sample with crystal growth temperature of 1030 °C

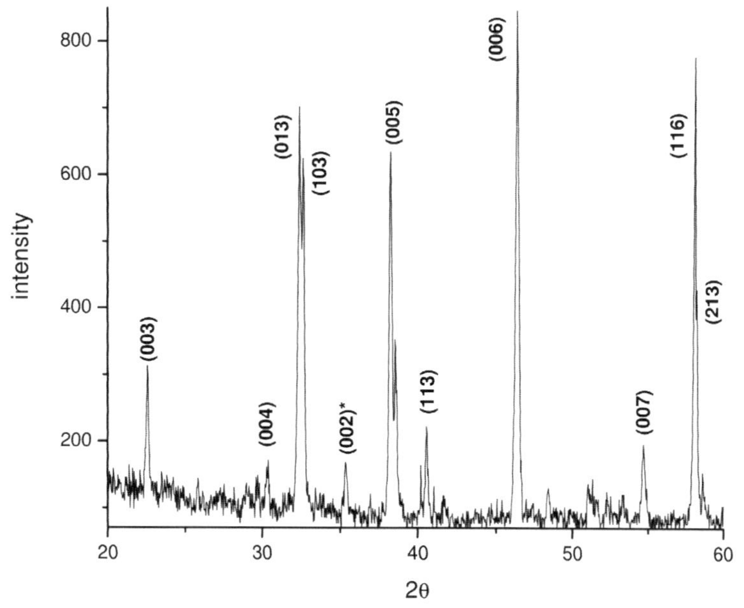

Figure 2. XRD pattern of the sample with crystal growth temperature of 1050 °C

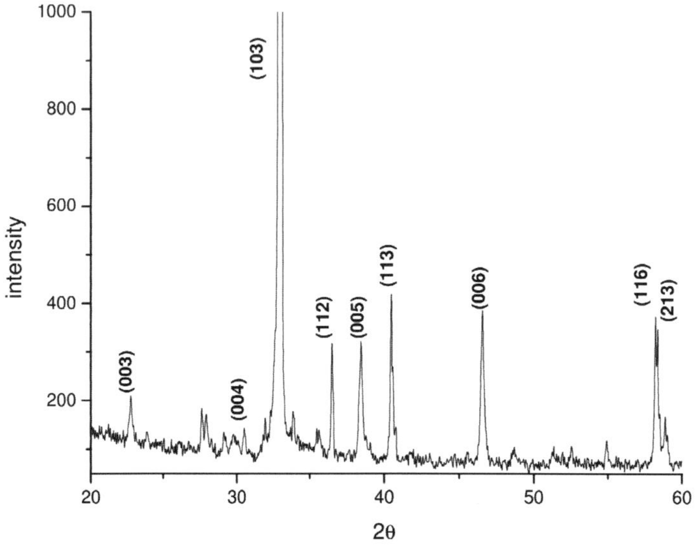

Figure 3. XRD pattern of the sample with crystal growth temperature of 1070 °C

Samples are polished by using diamond paste and microphotographs are taken in a few different magnifications with a polarized light microscope. As stated by previous studies [11], melt-textured growth induces non-uniform heating. Sample's surfaces mostly reach maximum temperature during the thermal cycle and also the interior may not reach the surface temperature above the peritectic decomposition. The microstructure and phase distribution may be effected with this non-homogenous heating and resulting structure may contain many cracks. Corollary, all samples in this study regardless of their crystal growth temperature show highly porous structure with cracks and many miss aligned plate-like structures. In figure 4, the sample with crystal growth temperature of 1070 °C can be seen in high magnification. There seems to be 3 different planar regions; the region at the right bottom corner of the image and the others are the upper top corner of the image and the regions of white color. These color difference may be resulted by the reflection of the light with different angles from the *a-b* plane periodicity of Y123 phase but also there seems some low melting phase regions which have reflections in bluish color. Regions which are between 50 μm to 100 μm show many cracks in various directions. They may be resulted with the further oxygen annealing and with the temperature differences [11]. Figure 5 show the image of the sample with crystal growth temperature of 1030 °C. Porous structure is visible and also no sign periodicity in *a-b* plane. Figure 6 shows the image of the sample with crystal growth temperature of 1050 °C. There seems to be different planes again with pores and cracks. Sample may have *a-b* plane growth for the periodicity in different regions and these regions seem to be at around 500 μm in size.

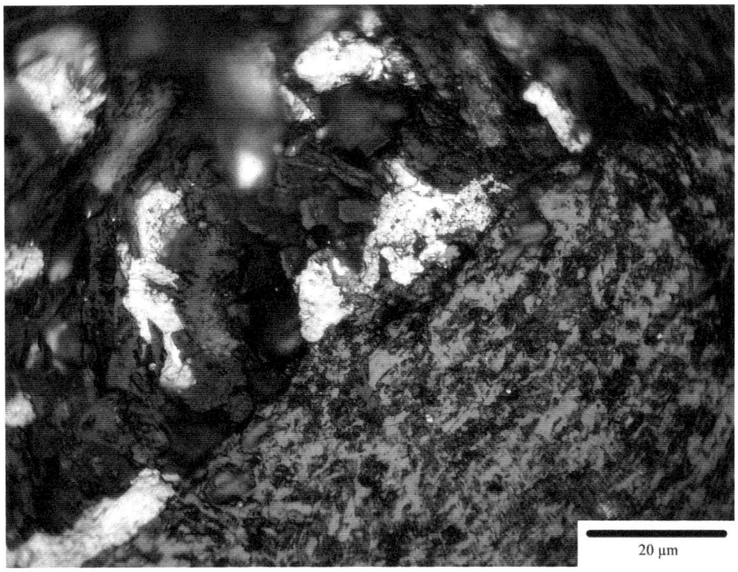

Figure 4. Microphotograph of the sample with crystal growth temperature of 1070 °C

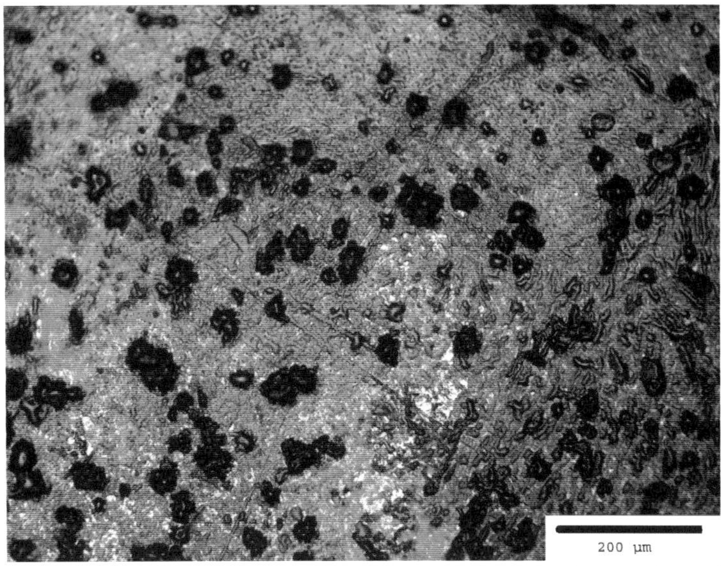

Figure 5. Microphotograph of the sample with crystal growth temperature of 1030 °C

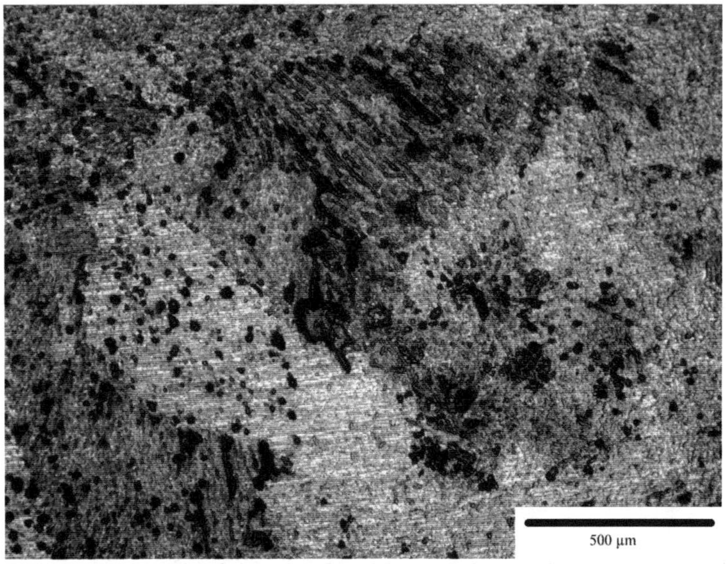

Figure 6. Microphotograph of the sample with crystal growth temperature of 1070 °C

AC magnetic susceptibility measurements are mostly used for qualifying samples for the superconductivity. Since the material expresses perfect diamagnetism property of shielding the entire applied magnetic field and exhibiting the susceptibility of exact -1 in SI units, this property can be used to determine the onset temperature of the critical transition and its temperature [12]. AC magnetic susceptibility consists of two parts of real and imaginary. Real part of AC magnetic susceptibility is in phase with the applied field and represents superconductivity transition from normal conductivity. This part also contains the information of inter and intra granular interactions of the crystalline structure of the sample [13-14]. The first peak like bending part occurs at a temperature slightly below the critical transition temperature indicating the first response of the sample to the applied field as vortices [15]. This response is called as inter-granular interaction at which the superconductive grains of the crystalline structure behave as small cylinders and start to trap the applied field. The second response of the sample occurs at a temperature lower than the first interaction, and called as intra-granular interaction. At this stage, the motion of trapped magnetic flux creates a Josephson coupling [16] and shows itself as a bend-like formation in the real part of the AC susceptibility. On the other hand, the imaginary part of the AC susceptibility reflects the magnetic loss of from the applied field to establish the ordered state of the superconductivity.

AC magnetic susceptibility measurements of the samples are performed by a lock-in magnetic field supplier of SR530 Lock-in Amplifier under vacuum atmosphere with Keithley 224 Current source and Keithley 196 System DMM nanovoltmeter. Results can bee seen at Fig. 3. These measurements are performed by using liquid nitrogen as a cooler and samples are subjected to AC magnetic field of three different magnetic field values.

In figure 7, onset phase transition temperatures range from 89.8 K for 1070 °C to 89.2 K for 1030 °C. This difference can be neglected and it can be said that samples have nearly same critical transition temperature. Whereas, ΔT transition range differs among samples. The sample of 1030 °C has a respectively wider range of 3.3 K for phase transition; on the other hand the sample of 1090 °C has a narrow transition range of 1.1 K. Figure 8 shows the middle amplitude applied AC magnetic field results for the samples. Phase transition temperatures range from 90.2 K for 1080 °C to 90.0 K for 1030 °C. ΔT transition range differs as 8.5 K for 1030 °C to 2.5 K for 1090 °C. Figure 9 shows the high amplitude applied AC magnetic field responses for the samples. Phase transition temperatures range from 90.2 K for 1050 °C to 90.7 for 1030 °C. ΔT transition ranges from 9.2 K for 1030 °C to 3.9 K for 1090 °C. Increasing field amplitude resulted with broadened and smoother phase transition in the means of temperature as expected. Also at the corresponding imaginary parts, a decrease in the temperature of the peak and a general widening of the peaks are indicated which are predicted by the bean critical state model [17]. In figure 8, it's clearly seen that intra-inter granular interactions took place for the sample of 1030 °C and this interaction can also be seen at the figure 9 for the sample with crystal growth temperature of 1090 °C. But these measurements are not sufficient to correctly interpret the magnetic response of the samples in the means of flux dynamics. On the other hand, measurements show that at all the applied field values, samples with crystal growth temperature of 1070 °C and 1080 °C have the higher onset transition temperature and also narrow transition range, which may state that the optimization may took place between 1070 °C and 1080 °C. Figure 9, which is the measurement with the highest applied field value, this optimization is clearly indicated. Real part of the magnetic response of all the samples except 1030 °C took place between the response of the 1070 °C and 1080°C indicating that optimization may be in that range.

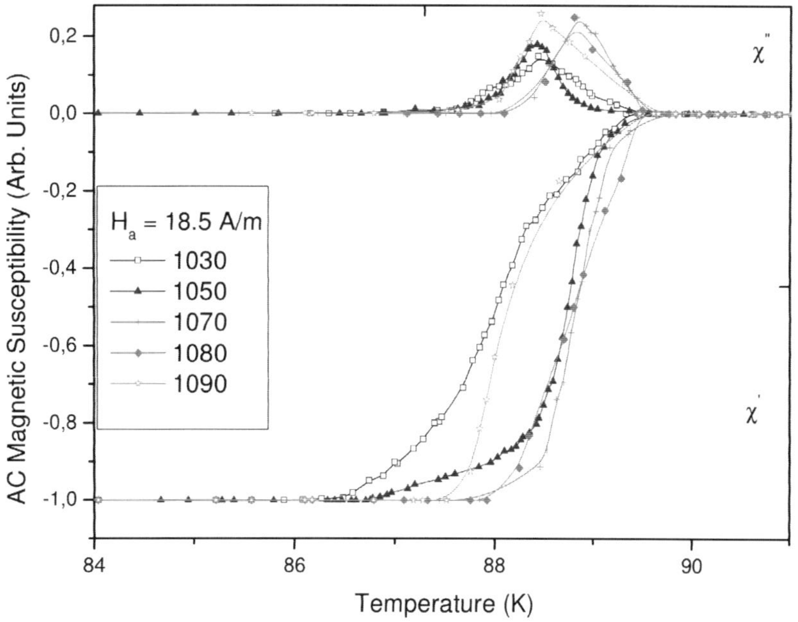

Figure 7. AC Magnetic Susceptibility Measurement of all the samples at applied magnetic field of H_a = 18.5 A/m.

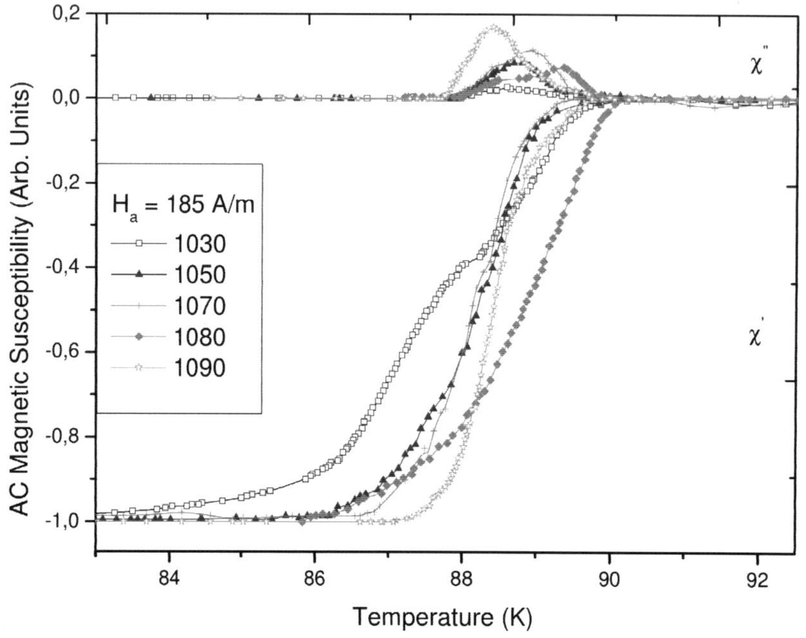

Figure 8. AC Magnetic Susceptibility Measurement of all the samples at applied magnetic field of H_a = 185 A/m.

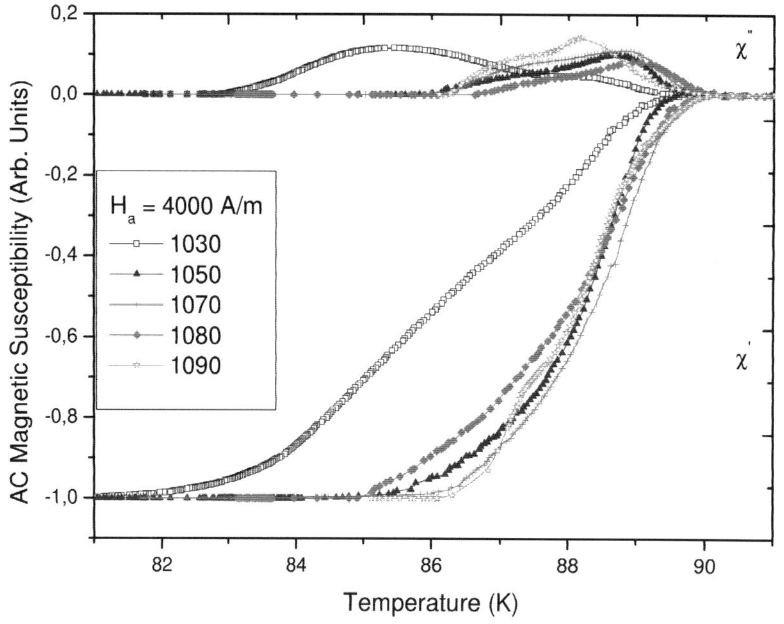

Figure 9. AC Magnetic Susceptibility Measurement of all the samples at applied magnetic field of H_a = 4000 a/m.

4. Conclusion

Through this study, an optimisation over the peritectic decomposition temperature of the bulk YBCO has been experimented by using melt-textured growth method. XRD results and microphotography of the samples show a linear correlation with higher peritectic temperature as increasing *a-b* plane periodicity and *c*-axis alignment. AC magnetic susceptibility measurements reflected a narrow superconductive transition temperature window for the samples of higher peritectic temperature. These results show the importance of the slow cooling rate part of the MTG cycle. Higher the peritectic temperature resulted with more time for the crystal formation with fixed cooling rate. A more homogenic mixture of the YBCO may supply the possibility of less porous structure and also less temperature differences during the cycles may result with fewer cracks in the samples. However, MTG cycle resulted with crystal formation in various samples respectively.

5. Acknowledgements

This study was supported by the Ege University science foundation under the project number of 2008/FEN/060.

6. References

[1] Wu M K, Ashburn J R, Torng C J, Hor P H, Meng Rl, Gao L, Huang ZJ, Wang YQ and Chu C W 1987 *Phys. Rev. Lett.* **58** 908
[2] Bornemann H J 1994 *J. Appl. Supercond.* **2** 439
[3] Murakami M 1992 *Supercond. Sci. Technol.* **185** 5
[4] Wang J, et al. 2002 *Physica C* **809** 378
[5] Jin S, Tiefel T H, Sherwood R C and Davis M E 1988 *Appl. Phys. Lett.* **52** 2074
[6] Yang W M, Zhou L, Feng Y, Zhang P X, Chen S K, Wu M Z, Zhang C P, Wang J R, Du Z H, Wang F Y, Yu Z M, Wu X Z, Gawalek W and Gornert P 1998 *Physica C* **307** 217
[7] Jın S, Tıefel T H, Sherwood R C, Vandover R B, Davis M E, Kammlott G W and Fastnacht RA 1988 *Phys. Rev. B* **37** 7850
[8] Leblond C, Monot I, Provost J and Desgardin G 1999 *Physica C* **311** 211
[9] Reddy E S, Babu N H, Shi Y and Cardwell D A 2005 *J. Europ. Ceramic Soc.* **25** 2935
[10] R. Loh 1998 *Ceramics International* **24** 559-563
[11] Giovannelli F and Monot-Laffez I, 2002 *Supercond. Sci. Technol.* **15** 1193
[12] Opagiste C, Couach M, Khoder A, Triscone G, Jorda J L and Cohenadad M T 1994 *Physica B* **194** 1809
[13] Ilonca G 2001 *Int. J. Inorg. Matter.* **3** 763
[14] Clem J R 1988 *Physica C* **50** 153
[15] Huth M, Schmitt M and Adrian H 1991 *Physica C* **178** 203
[16] Muller K H, MacFarlane J C and Driver R 1989 *Physica C* **158** 69
[17] Bean C P 1962 *Phys. Rev. Lett.* **8** 250

Influence of annealing temperature on the microstructure and mechanical properties of MgB_2

O Gorur[1], M Nursoy[2], C Terzioglu[2], A Varilci[2], I Belenli[2]

[1] Department of Physics, Faculty of Arts and Sciences, Rize University, 53100 Rize,Turkey.
[2] Department of Physics, Faculty of Arts and Sciences, Abant Izzet Baysal University, 14280 Bolu,Turkey.

E-mail: gorur@ktu.edu.tr

Abstract: We have investigated the effect of annealing temperature on the microstructure, magnetic and mechanical properties of MgB_2 superconducting samples employing X-ray diffraction (XRD), scanning electron microscopy (SEM), ac susceptibility and Vickers microhardness measurements. XRD patterns and SEM micrographs are used to obtain information about lattice parameters and grain size, respectively. These measurements indicate that MgB_2 grain size, lattice parameters, and critical temperature are increased, and grain connectivity is improved, with increasing the sintering temperature up to 850°C. It is also observed that the Vickers microhardness of the samples is dependent of the sintering temperature and applied load. In addition, we calculate the load dependent mechanical properties of MgB_2 samples such as the Young's modulus, yield strength, and fracture toughness. The possible reasons for the observed improvements in microstructure, superconducting and mechanical properties due to annealing temperature are discussed.

Keywords: Annealing temperature; Microstructure; Lattice parameters; Ac susceptibility; Microhardness; Fracture toughness; Young's modulus; Yield strength.

1. Introduction

The improvement of the superconducting properties of MgB_2 is a decisive goal to enable its potential applications [1-6]. MgB_2 has been an attractive candidate for various applications due to its relatively high critical current density (J_c), critical temperature (T_c), and coherence length. It is generally known that chemical doping or substitution, and preparation conditions, play very important roles in the properties of high-T_c and conventional superconductors. A number of substitutions into the MgB_2 system have been tried [7-12]. In these investigations, variations in the lattice parameters, microstructure, critical temperature and critical current density with the chemical composition were observed. Variations in these properties were also obtained by changing conditions of sample preparation [13-21]. In a previous study of the present samples, Varilci [22] investigated the influence of magnetic field on hysteretic ac losses in bulk MgB_2 by using Hall probe ac susceptibility. MgB_2 powder is commercially available. Usually its quality is not sufficiently good. The critical temperature of MgB_2 powder prepared in laboratory is slightly higher. On the other hand, even with the same

1 Osman Gorur, Department of Physics, Faculty of Arts and Sciences, Rize University, 53100 Rize-Turkey.

preparation technique, different research groups observed different values of the critical temperature and the critical current density [15,23].
It is very important to investigate the microstructure and mechanical properties as a function of annealing temperature. To our knowledge, no detailed study of the effect of annealing temperature on mechanical properties of MgB_2 has been published in the literature. In this paper, we report the effects of annealing temperature on the microstructure, superconducting and mechanical properties of MgB_2 by performing XRD, SEM, ac susceptibility and microhardness measurements.

2. Experimental details

A commercially available MgB_2 powder (Alfa Aesar) was used. The powder was pressed into pellets under a pressure of 750 MPa at room temperature and the pellets were wrapped with a Ta foil then placed in a quartz tube for annealing. They were annealed at temperatures of 650 °C, 750 °C, 850 °C and 950 °C for 1 hour in flowing Ar gas [22]. These bulk samples were rectangular bars with approximate dimensions of 10x3x1 mm^3. They were analyzed by XRD, SEM, ac susceptibility and microhardness measurements. The MgB_2 pellets will be hereafter denoted as M650, M750 , M850 , and M950, to indicate their annealing temperature (650 °C, 750 °C, 850 °C, and 950 °C, respectively).
The phase composition of the samples were characterized by XRD method using a Jeol Multiflex XRD with Cu K_α radiation (λ = 1.5418 Å) in the range 2θ = 30°-60° with a scan speed of 3°/min and a step increment of 0.02° at room temperature. Phase purity and lattice parameters were determined from these XRD patterns. The accuracy in determining the lattice parameters (a and c) was ± 0.0001 Å. The surface morphology and microstructure of the samples were investigated by SEM (JEOL 6390-LV).
Measurements of ac susceptibility were made in the closed cycle refrigerator, using our homemade susceptometer equipped with a lock-in amplifier. A Si diode, attached to the cold head, was used to read the temperature of the sample. The temperature was stabilized with an accuracy of 1 mK. Measurements were taken in the temperature range of 30-45 K with a heating rate of 0.5 K/min using Lake Shore 332S-T1 temperature controller. The real and imaginary components of ac susceptibility were collected with a driving field having amplitude of $H_{app.}$ = 960 A/m and a frequency of 10 Hz. Since the signal generated from the Hall probe was very weak, Tegam model 73 precision ratio transformations was used for voltage amplifier. Details of experimental system were described elsewhere [22,25].
To characterize the mechanical properties of the annealed samples, microhardness measurements were performed with a digital microhardness tester (Instron Series 2100) at room temperature. A Vickers pyramidal indenter with different loads (0.245, 0.490, 0.980, 1.960, and 2.940 N) and a single loading time of 10 s was applied and the diagonals of indentation were measured with an accuracy of ± 0.1 μm. Indentations were made at different parts of the samples' surface such that the distance between any two indentations was more than two times the diagonal of the indentation mark to avoid surface effects and work hardening. An average of 5 readings at different locations of specimen surfaces was taken to obtain reasonable mean values for each load.
Conventional Vickers microhardness measurements consist of applying a load F on the test material via a geometrically defined indenter and after the indenter is removed, measuring the characteristic dimension, d, of the resultant impression. The Vickers microhardness, H_v, was estimated using the following equation

$$H_v = 1.8544\left(F/d^2\right) \tag{1}$$

where F is the load applied (in N) and d the mean diagonal length of the indentation impression (in μm). The scatter in the values of d did not exceed 3%.

In most materials, the elastic modulus (Young's modulus) E is related to the Vickers microhardness (apparent) H_v by the relation [26]

$$E = 81.9635 H_v, \tag{2}$$

and yield strength Y is related to the hardness by the relation [27,28]

$$Y \approx H_v/3. \tag{3}$$

In this work, fracture toughness of the annealed MgB_2 samples was measured by Vickers indentation method. Several semi-empirical formulas have been used on the basis of crack types in certain materials [29-33]. Fracture toughness was calculated from the following equation [29]

$$K_{IC} = 0.0016\sqrt{E/H_v}\,(F/a^{3/2}) \tag{4}$$

where K_{IC} is the fracture toughness in MPa.m$^{1/2}$, F is the applied load in N, E is the Young's modulus in GPa, H_v is the hardness in GPa and a is the crack length in mm.

3. Results and discussion

3.a Lattice parameters and microstructure

The x-ray diffraction patterns taken from the surface of MgB_2 samples (M650, M750, M850 and M950) are shown in Fig. 1. Some of the Miller indices are indicated in the figure. The XRD pattern of sample M950 showed, besides MgB_2, the presence of a small amount of elemental B (Boron) peaks [17,19]. No traces of MgO or degradation compounds such as MgB_6 or MgB_{12} were observed. The presence of B peaks in the XRD pattern of sample M950 (while being nonexistent in XRD patterns of the other samples) indicates that 950°C is too high for heat treating MgB_2, and may have caused segregation of elemental boron through sublimation of Mg. Since we did not observe any MgO peaks in the XRD patterns, annealing MgB_2 pellets under flowing Ar gas was sufficient enough to avoid partial oxidation of Mg. For samples M650, M750 and M850, only MgB_2 phase was detected [17,19]. The intensity of all peaks increased while the position of the peaks shifted towards smaller angles with increasing the annealing temperatures up to 850 °C then decreased at 950 °C. The increase of the intensity of the peaks and the decrease of the FWHM with increasing the annealing temperature from 650 to 850 °C indicate that crystallite size becomes larger as confirmed by SEM investigations. On the other hand, the XRD patterns revealed that all samples show polycrystalline structure with (100), (101), (002.) and (110) peaks.

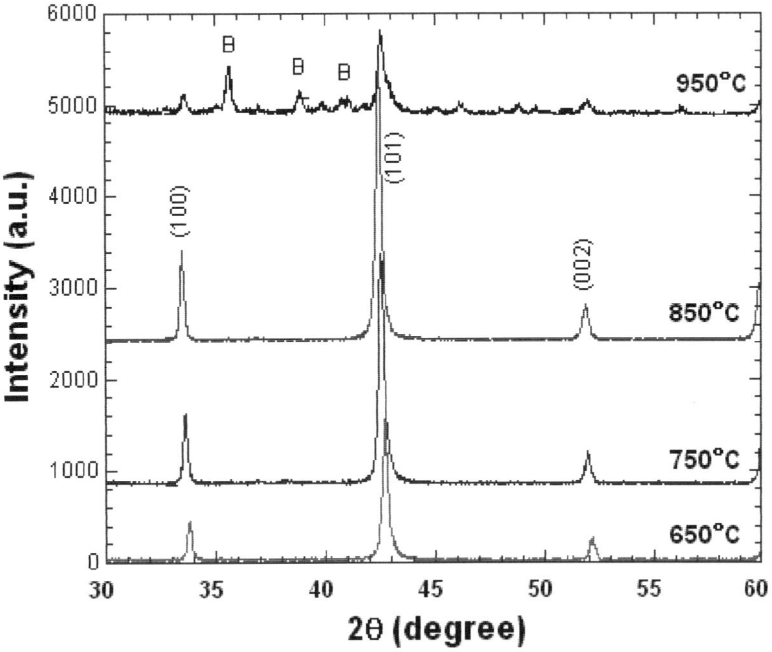

Figure 1. XRD patterns for samples M650, M750, M850 and M950.

The dependence of the lattice parameters on annealing temperature was determined by least-square fits to the position of high angle Bragg peaks ($30° < 2\theta < 60°$) and the results are provided in Table 1. An appreciable change in the lattice parameters of the samples was found within the experimental limit of 0.0001 Å. The in-plane B-B distance, proportional to the lattice parameter *a*, increased slightly from 3.0701 to 3.0844 Å with increasing the annealing temperature up to 850 °C. Moreover, the inter-plane B-B distance, proportional to the lattice parameter *c*, increased significantly from 3.2522 to 3.5285 Å with increasing the annealing temperature up to 850 °C. The lattice parameter *c* of sample M850 was found to be 3.5285 Å. On the other hand, it decreased to 3.5090 Å for sample M950. These lattice parameters are reasonably consistent with other work [34].

Table 1. The lattice parameters *a* and *c* the M950, M850, M750 and M650 samples.

Samples	*a* (Å)	*c* (Å)
M950	3.0752	3.5090
M850	3.0844	3.5285
M750	3.0723	3.3522
M650	3.0701	3. 2522

The surface morphology of the four MgB_2 samples was studied by SEM (Fig. 2). The microstructure of the sample annealed at 850 °C is remarkably different from that of samples annealed at 650 and 750 °C.

(a)

(b)

(c)

(d)

Figure 2. SEM micrographs of samples (a) M650, (b) M750, (c) M850 and (d) M950.

A broad grain size distribution can be seen for the sample annealed at 850 °C. The grain size is about 100-120 nm for sample M850. It is observed that the grain connectivity is enhanced greatly with increasing the annealing temperature from 650 to 850 °C. The surface of sample M850 is also denser. M650 and M750 have a non-uniform surface appearance with smaller grains. From Fig. 2, one can say that the grains in samples M650, M750 are randomly oriented and poorly connected. Grain size of sample M950 is similar to that of M850 but M950 shows more signs of partial melting due to higher annealing temperature. We propose that sublimation of Mg occurred at the grain boundaries where the internal energy of the crystal is higher. As a result, accumulation of segregated boron also occurred at the grain boundaries. This finding is supported by ac susceptibility measurements as explained below.

Journal of Physics: Conference Series **153** (2009) 012012 doi:10.1088/1742-6596/153/1/012012

3.b Critical temperature by ac susceptibility measurements

The magnetic ac susceptibility measurements as a function of temperature were made in the temperature range between 35 and 45 K with a driving field having amplitude of $H_{app.}$ = 960 A/m and a frequency of 10 Hz. Fig. 3 shows ac susceptibility data as a function of temperature for the four MgB_2 samples annealed at different temperatures.

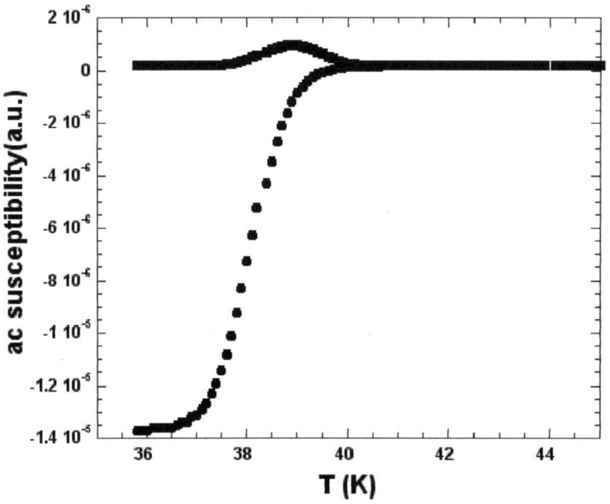

(a)

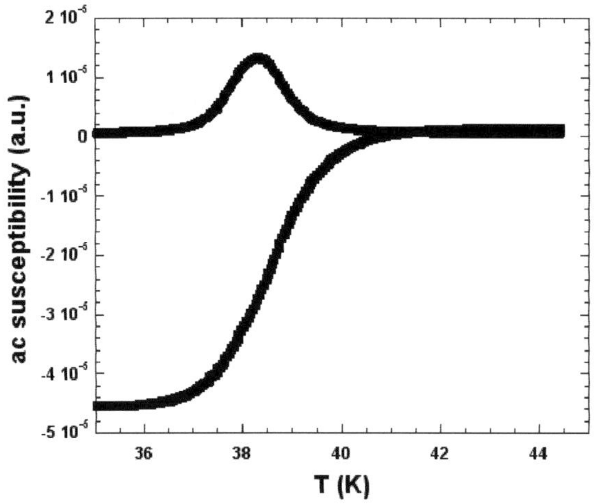

(b)

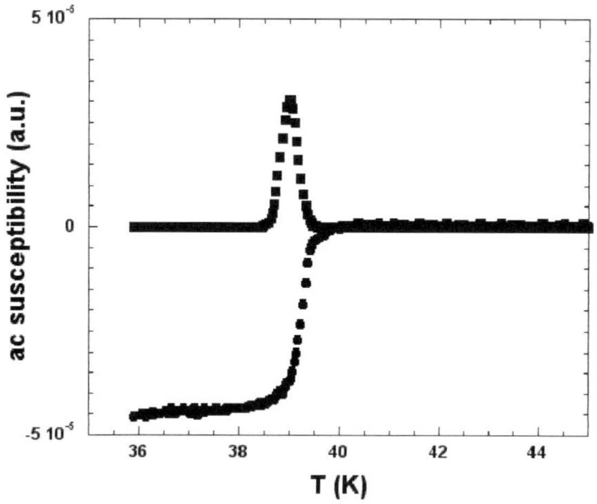

(c)

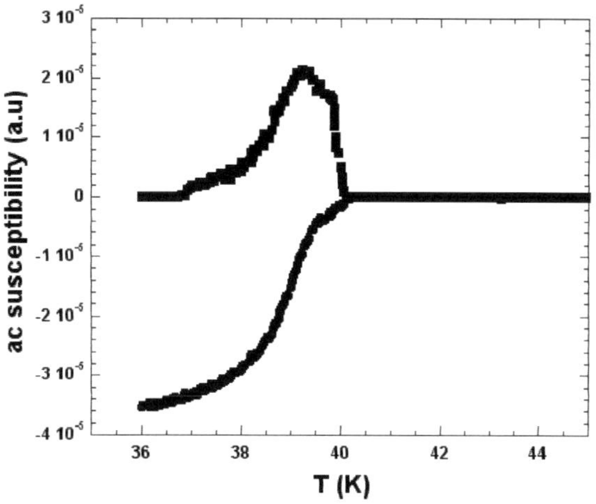

(d)

Figure 3. Temperature dependence of ac susceptibility for samples (a) M650, (b) M750, (c) M850 and (d) M950.

As seen from Fig. 3, sample M850 has a very sharp transition $(\Delta T_c \approx 1 \quad K)$. All the other samples exhibit wider transitions $(\Delta T_c \approx 3K)$. M650 shows a lower T_c due to low annealing temperature.

As the annealing temperature is increased to 750 °C, T_c increased but the transition was still broad. When the annealing temperature increased to 950 °C, the transition widened again. The onsets of T_c of all samples are found to be about 40 K. The broadening could be related to non stockiometric phases and/or poor sintering between the grains. The shape of the ac susceptibility curves of sample M950 is different from that of the other samples, showing a double step-like behaviour. Double step transitions may indicate the presence of other phases at the grain boundaries indicating that Mg deficient MgB_2 has lower transition temperature. This is consistent with the appearance of boron peaks in the XRD pattern of sample M950.

3.c Mechanical properties

In order to investigate the effect of the annealing temperature on the mechanical properties of MgB_2 samples, we conducted the Vickers microhardness tests. We measured the diagonal length as a function of the test load. The Vickers microhardness (apparent) values, calculated for different applied loads by using Eq. (1), are listed in Table 2.

Table 2. Load dependent micro hardness, elastic modulus, yield strength and fracture toughness of MgB_2 samples annealed at 650, 750, 850 and 950 °C.

Sample	*Load (N)*	$H_v(GPa)$	$E(GPa)$	$Y(GPa)$	$K_{IC}(MPa\sqrt{m})$
M650	0.245	1.011	82.86	0.337	4.005
	0.490	0.682	55.90	0.227	3.806
	0.980	0.537	43.96	0.179	3.787
	1.960	0.502	41.14	0.167	3.539
	2.940	0.494	40.45	0.165	3.874
M750	0.245	2.568	210.5	0.856	8.321
	0.490	1.228	100.7	0.409	6.301
	0.980	0.863	70.70	0.288	5.883
	1.960	0.791	64.81	0.264	5.142
	2.940	0.780	63.94	0.260	4.762
M850	0.245	3.824	313.4	1.275	12.55
	0.490	1.674	137.2	0.558	8.488
	0.980	1.065	87.33	0.355	7.348
	1.960	0.952	78.00	0.317	5.711
	2.940	0.944	77.37	0.315	5.352
M950	0.245	1.450	118.9	0.483	6.830
	0.490	0.917	75.06	0.305	6.349
	0.980	0.721	59.11	0.240	5.728
	1.960	0.658	53.96	0.219	5.420
	2.940	0.628	51.80	0.208	5.142

Fig. 4 displays the variation of microhardness as a function of applied load for samples M650, M750, M850 and M950. These MgB_2 samples have higher mechanical strength than the other high-T_c ceramic superconductors (YBaCuO and Bi(Pb)SrCaCuO) [5,6,11,35].

The variation of microhardness with load has similar shape irrespective of the annealing temperature although its value for a given load is different depending on the annealing conditions. We have observed that the microhardness values increased with increasing annealing temperature from 650 to 850 °C, and decreased at 950 °C. At 950 °C, the strength of bonding between the grains decreased and consequently the microhardness degraded. The rapid variation of microhardness was observed with increasing applied load from 0.245 to 1.000 N. The reason for this behaviour is due to mechanically weak grain boundaries [36]. It is also obvious from Fig. 4 that the Vickers microhardness is load dependent for all samples; the calculated microhardness decreased non-linearly as the applied load increased until 1 N, and then attained saturation. The reason of this behaviour is explained elsewhere [37]. This non-linearity has also been observed for *Bi-Pb-Sr-Ca-Cu-O* [38-40] and for MgB_2 [5,6], and is known as indentation size effect (ISE). To account for this effect, several relationships between the applied load and the resulting indentation size have been suggested [41-44]. However, we are not going to analyze the ISE in this paper, but will focus on the load dependent mechanical properties. Quin et al. [45] investigated the variation of the Vickers microhardness as a function of the indentation load for a variety of ceramic materials. They observed that such hardness-load curve shows distinct transition to a plateau and concluded that this plateau corresponds to the intrinsic hardness value of the material. In this study, this plateau is reached at an applied load of about 1 N. It is evident that the value of microhardness of sample M850 is larger than that of the other samples. This result is consistent with XRD, SEM and ac susceptibility measurements.

The value of load dependent *E*, *Y*, and K_{IC} were calculated for each load by using Eqs. (2) and (4), and summarized in Table 2. As seen in this table, the load dependent *E*, *Y* and K_{IC} increase significantly with increasing annealing temperature and decreasing loads.

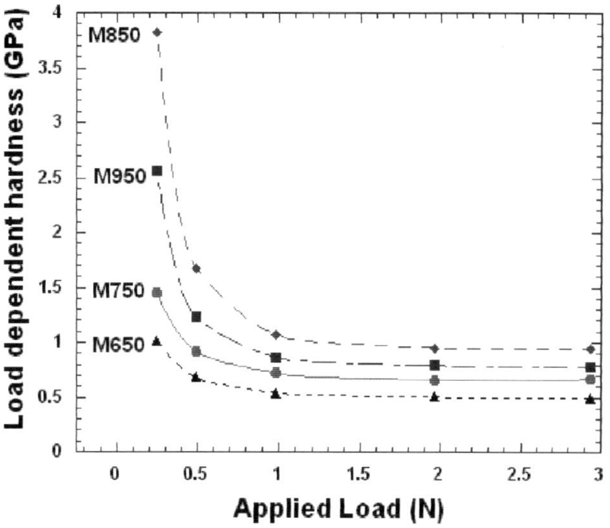

Figure 4. Load dependent microhadness versus applied load for the samples.

Similar changes in the yield strength, elastic modulus and fracture toughness were reported in the literature [46-48]. An increase in K_{IC} corresponds to an increase in the average surface energy as proposed from the hardness calculations. One should point out that the apparent microhardness, Young's modulus, yield strength, and fracture toughness of the samples in the present work indicate strong dependency on applied load and weak dependency on annealing temperature. It was observed that the mechanical properties of the samples are improved with increasing the annealing temperature from 650 to 850 °C (Fig. 5). At 950 °C, however, the mechanical properties showed a significant decrease. This indicates that the optimum annealing temperature to harden MgB_2 material is around 850 °C.

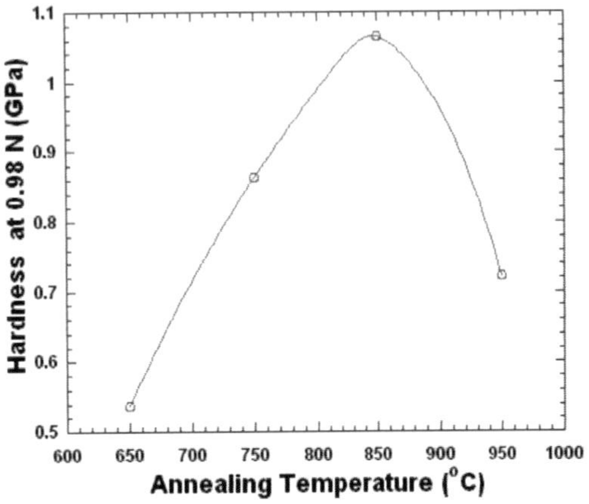

(a)

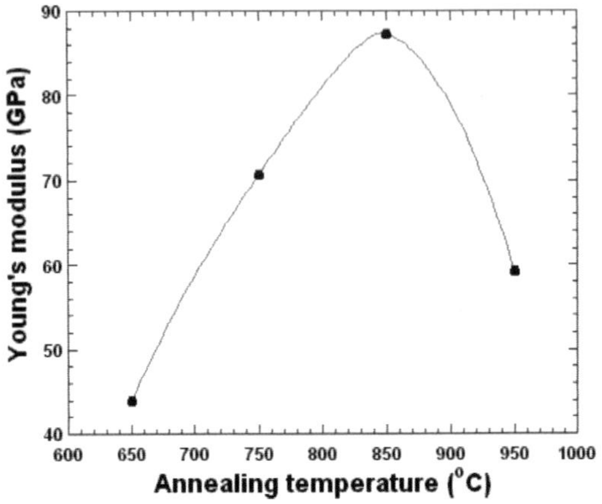

(b)

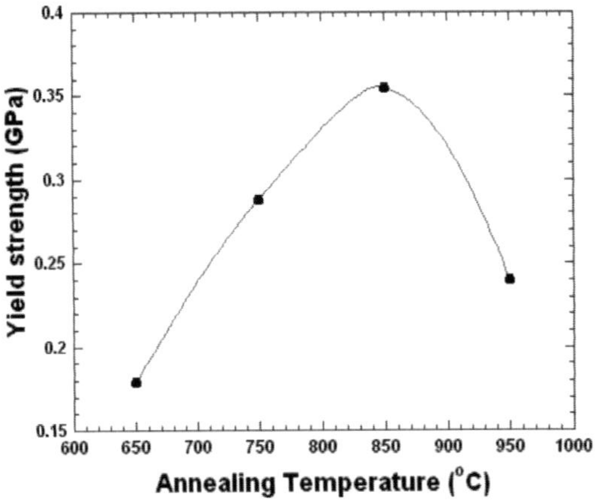

(c)

Figure 5. (a) Hardness, (b) Elastic modulus, and (c) Yield strength versus annealing temperature at 0.98 N.

4. Conclusions

Our measurements showed that an increase in the annealing temperature up to 850^{0}C improved the microtructure, mechanical and superconducting properties of MgB_2 bulk samples. XRD and SEM examinations show that the lattice parameters and grain connectivity increase, and the number and size of voids at the sample surface decrease with increasing the annealing temperature from 650 to 850 °C. As a result, load dependent microhardness, elastic modulus, yield strength and fracture toughness values increase with increasing the annealing temperature. It was also found that the hardness of MgB_2 samples is greater than that of high-T_c ceramic superconductors (YBaCuO and Bi(Pb)SrCaCuO). The improvement of the superconducting properties (T_c and transition width) with increasing the annealing temperature is due to the modification of grain boundaries together with a better crystallinity and larger grains. The improvement of the mechanical properties can be observed by the increase of microhardness with increasing annealing temperature. The mechanical properties enhanced with increasing the annealing temperature, up to 850 °C and degraded at 950 °C. This behaviour is ascribed to the strength of the bonds between grains. The higher annealing temperature raised the reactivity for the formation of MgB_2 phase. We conclude that the best annealing temperature of these MgB_2 samples is 850 °C.

Acknowledgements
One of the authors, Ahmet Varilci, would like to thank to the Turkish State Planning Organization (DPT) for its financial support (project No: 2004K120200).

References

[1] Sou H L Benedence C Dhall M Musolino N Genound J Y Flükiger R 2001 *Appl. Phys. Lett.* **79** 3116
[2] Anderson N E Straszheim W E Bud'ko S L Canfield P C Finnemore D K Suplinskas R J 2003 *Physica C* **390** 11
[3] Zhao Y Cheng C H Feng Y Machi T Huang D X Zhou L Koshizuka N Mukrakami M 2003 *Physica C* **386** 581
[4] Vilke R H T Bud'ko S L Canfield P C Kramer M J Wu YQ Finnemore D K Suplinskas R J Marzik J V Hannahs S T 2005 *Physica C* **418** 160
[5] Kolemen U 2006 *J. Alloys Comp.* **425** 429
[6] Kolemen U Uzun O Aksan M A Güçlü N Yakıncı E 2006 *J. Alloys Comp.* **415** 294
[7] Xu X L Guo J D Wang Y Z Wang X 2003 *Materials Letters* **58** 142
[8] Tampieri A Celotti G Sprio S Rinaldi D Baruccai G Caciuffo R 2002 *Solid State Comm.* **121** 497
[9] Gencer A Kılıç A Okur S Güçlü N Özyüzer L and Belenli I 2005 *IEEE Transactions on Applied Superconductivity* **15(2)** 3352
[10] Kazakov S M Angst M Karpinski J Fita I M Puzniak R 2001 *Solid State Comm.* **119** 1
[11] Hong-xia L Hong-wei S Gou-xing L Chang-ping C 2005 *Ceramic International* **31** 105
[12] Güçlü N 2007 *Materials Chemistry and Physics* **101** 470
[13] Kolesnikov N N Kulakov M P 2001 *Physica C* **363** 166
[14] Takano H Kokubo H Kinami T Amakai Y Murayama S 2007 *Journal of Magnetism and Magnetic materials* **310** e134
[15] Wu T Yau J K F Cai Y M Cui Y G Gu D W Yuan P F Yuan G Q Shen L J Jin X 2003 *Physica C* **386** 638
[16] Yamamoto A Shimoyama J Ueda S Katsura Y Iwayama I Horii S Kishio K 2005 *Physica C* **426-431** 1220
[17] Qing-rong F Chinping C Jun X Ling-wen K Xin C Yong-zhong W Yan Z Zheng-xiang G 2004 *Physica C* **411** 41
[18] Matsmoto A Kumakara H Kitaguchi H Fuji H Tugano K 2002 *Physica C* **382** 207
[19] Aswal D K Sen S Singh A Chandrasekhar Rao T V Vyas J C Gupta L C Sahni VC 2001 *Physica C* **363** 149
[20] Xu G J Pinholt R Bilde-Sorensen J Grivel J C Abrahamsen A B Andersen N H 2006 *Physica C* **434** 67
[21] Abe H Naito M Nogi K Matsuda M Miyake M Ohara S Kondo A Fukui T 2003 *Physica C* **391** 211
[22] Varilci A 2007 *Supercond. Sci. Technol.* **20** 397
[23] Buzea C Yamashita T 2001 *Supercond, Sci, Technol.* **14** R115
[24] Cullity B D 1978 Element of X-ray diffraction, Addition – Wesley, Reading, MA

[25] Yegen D Varilci A Yilmazlar M Terzioglu C Belenli I 2007 *Physica C*
[26] Veerender C Dumke V R Nagabhooshanam M 1994 *Phys. Status Solid A* **144** 199
[27] McClintock F A Argon A S 1996 *Mechanical Behaviour of Materials*, Addison-Wesley, Reading, MA, (p. 455)
[28] Tabor D 1951 *The Hardness of Metals*, Clarendon, Oxford
[29] Anstis GR Chantikul P Lawn B R and Marshall D B 1981 *J. Am. Ceram. Soc.,* **64** 533-538
[30] Evans A F Wilshaw T R 1976 *Acta Met.* **24** 939
[31] Liang K M 1990 *J. Mater. Sci.* **25** 207
[32] Austis G R 1994 *J. Am. Ceram.* **64** 533
[33] Habib K A Saura J J Ferrer C Damra M S Gimenez E Cabedo L 2006 *Surface& Coatings Technology* **201** 1436.
[34] Drozd V A Gabovich A M Gierłowski P Pezkała M Szymczak H 2004 *Physica C* **402** 325
[35] Beilin V Dul'kin E Yashchin E Galstyan E Lapides Y Tsindlekht M Felner I Roth M 2004 *Physica C* **405** 70
[36] Ling H C and Yan M F 1988 *J. Appl. Phys.* **64** 1307
[37] Rabier J Denanot M F 1990 *J. Less Common Met.* **223** 164
[38] Yilmazlar M Cetinkara H A Nursoy M Ozturk O and Terzioglu C 2006 *Physica C* **442** 101
[39] Khalil S M 2001 *J. Phys Chem Solids* **62** 457
[40] Murakami A Katagiri K Noto K Kasaba K Sohoji Y Muralidhar M Sakai N and Murakami M 2002 *Physica C* **378-381** 794
[41] Fröhlinch F Grau P Grellmann W 1997 *Phys. Status Solidi* **42** 79
[42] Li H Bradt R C 1993 *J. Mater. Sci.* **22** 917
[43] Hays C Kendall E G 1973 *Metallography* **6 (4)** 275
[44] Gong J Wu J Guan Z 1999 *J. Eur. Ceram. Soc.* **19** 2625
[45] Quinn J B Quinn G D 1997 *J. Mater. Sci.* **32** 4331
[46] Khalil SM 2005 *Smart Mater. Struct.* **14** 804
[47] Veerender C Dumke V R Nagabhooshanam M 1994 *Phys. Status Solid A* **144** 199
[48] Goyal A Funkenbusch P D Kroeger D M Burns S J 1992 *J. Appl. Phys.* **71** 1363

Artificial pinning centers using the barrier layer of ordered nanoporous alumina templates

X Hallet[1], S Mátéfi-Tempfli[1], M Mátéfi-Tempfli[1], S Michotte[1], L Piraux[1], J Vanacken[2], V V Moshchalkov[2]

[1] Unité de Physico-Chimie et Physique des matériaux, Université Catholique de Louvain, Place Croix du Sud 1, 1348 Louvain-la-Neuve, Belgium

[2] Institute for Nanoscale Physics and Chemistry, Katholieke Universiteit Leuven, Celestijnenlaan 200D, 3001 Leuven, Belgium

E-mail: xavier.hallet@uclouvain.be

Abstract. The barrier layer of self-ordered anodized aluminium oxide, which is grown from an aluminium foil, has been revealed by a selective chemical etching of the remaining aluminium. The surface obtained in this way consists of a triangular lattice of bumps with 100nm spacing, and heights of approximately 50nm. Using this surface as a template for controlling the pinning in thin superconducting films, superconducting Nb was deposited with different thicknesses and under different deposition angles. The evaporation under a 30° angle shows an asymmetric pinning potential composed of two triangular lattices having different pinning strengths. Matching effects are observed up to 1T. Matching effects are also maintained at relatively low temperature.

1. Introduction

When a current is applied to a superconductor in the mixed state, vortices move perpendicularly to this current and generate energy loss. This dissipation weakens the superconducting state and the critical current is generally reached much below the theoretical pair-breaking current. Vortex pinning has therefore been extensively studied in the last decades because of the technological and scientific interest.

Abrikosov has shown [1] that in defect free superconductors, the vortices tend to form a triangular lattice with a lattice constant depending on the applied magnetic field (equation 1). Therefore, random pinning sites do not provide an optimized vortex lattice pinning. Since the development of lithographic techniques, several artificially created lattices of pinning sites have been investigated [2-7]. The commensurability of the vortex lattice with the periodic pinning array (PPA) generates a strong stabilization of the vortex lattice. This results in an increase of the critical current, a decrease of the magnetoresistance and a reduction of the flux creep. More recently, magnetic defects attracted much attention. The interaction between the superconductor and the magnetic field of the dots produces asymmetric matching effects [6,8] and field-induced superconductivity [9].

Lithography enables to create a tremendous variety of PPA's. However, this technique is very expensive and time-consuming. Nowadays, matching effect from lithographically introduced PPA's was demonstrated only to about hundreds of gauss. However, practical applications usually require much higher fields.

A breakthrough was made by Welp et al. [10] with the use of anodized aluminium oxide (AAO) templates to imprint a triangular lattice of holes in a thin Nb film increasing the first matching field by about more than one order of magnitude compared to previous lithographic techniques results. Another self-assembled approach was proposed by Vinckx et al. [11]. In that case, a triangular lattice of polystyrene spheres was used to modulate the thickness of a Nb thin film inducing a hexagonal lattice of pinning sites with matching effects up to 90 mT.

We present in this paper, a novel method to make hexagonal lattices of pinning centres with a very small lattice parameter. Highly ordered AAO templates are created by a two step anodization of a high purity Al foil. The oxide barrier layer is then revealed providing a triangular lattice of bumps (almost half-spheres) on which the evaporation of a thin Nb layer is made under different angles. This induces a periodic thickness modulation of the Nb film and thus creates a periodic lattice of pinning sites.

Pinning centres form a hexagonal arrangement since they are supposed to be localised on the interstitial sites of the bump lattice. This configuration of pinning centres allows us to increase by two the number of pinned vortices compared to the configuration obtained when the opposite face (with the holes) is chosen such as in [10,12]. In our case, the evaporation under an angle provides an interesting asymmetric pinning potential, which can be seen as two interpenetrating triangular pinning lattices of different strength. They can thus pin twice as many vortices before the first matching field is reached. The formation of undesirable interstitial vortices is thus delayed to higher fields. Moreover, the highly ordered structure obtained in this way remarkably shows matching effects up to 1T.

2. Experimental

The non-supported AAO templates used in this work were fabricated by a two step anodization process (Figure 1). An ultra-pure (99.999 %) aluminium foil (500μm) is first partially anodized in 0.3M oxalic acid at 2°C under a constant voltage of 40V (fig. 1a) leading to a mean interpore distance of 100nm (fig.1b). The first anodization step was performed to a depth of 50μm to promote ordering. Then this first anodized layer is selectively removed with a mixture of H_3PO_4 6wt% and chromic acid 1.8wt% at 60°C (fig. 1c). A second anodization step is then applied (fig. 1d) in order to obtain a highly ordered template.

The pores form a triangular lattice of long channels closed at one end by round-shaped Al_2O_3 barrier layers. The selective removal of the unanodized aluminium with $CuCl_2$ reveals a nanoporous template with contrasting surfaces (fig. 1e). Indeed, the top surface presents a triangular lattice of holes, while the bottom surface shows a triangular lattice of bumps. These two lattices have the same periodicity because the bumps represent the barrier layer at the bottom of the pores.

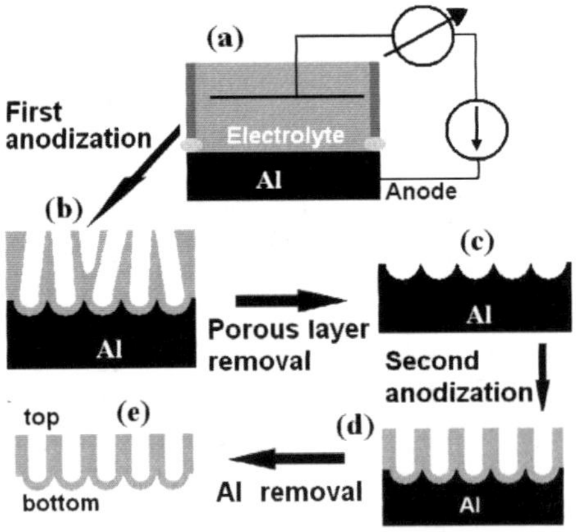

Figure 1. (a) Anodization system (b) Aluminium + alumina layer after the first anodization (c) Aluminium layer after the Al_2O_3 removal (d) After the second anodization (e) Aluminium oxide template after the aluminium removal

The use of triangular lattices of holes to create antidots in thin superconducting films has been reported recently [10,12]. In this present paper, we present results obtained using the bottom face of the template (conjured barrier layer) to modulate laterally thin superconducting films of Nb. These films were deposited using molecular beam epitaxy (MBE) at room temperature at a working pressure of 10^{-10} torr and an evaporation rate of 1.5 $\text{Å}s^{-1}$. The SEM image in figure 2 shows the bottom surface of a highly ordered sample but yet not a perfectly one. Vacancies and deformed bumps are observed throughout the sample at the boundaries of the ordered domains having different orientations. Average domain size [13] is about 1 μm², which is enough for the reported behaviour.

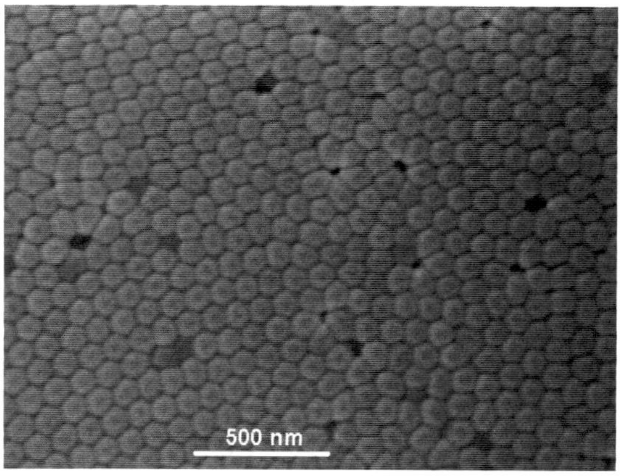

Figure 2. SEM image of the bottom surface (bumps) of a highly-ordered alumina template

Six different Nb films were deposited on the same bottom surface. Two different nominal thicknesses (30 and 50nm) and three different angles (0°, 30° and 60° from normal) were tried. Real thicknesses are obtained by multiplying the evaporated thicknesses by the cosine of the corresponding evaporation angle.

3. Results and discussion

We first discuss the thickness modulation under the different evaporation conditions in order to underline the importance of angular evaporation on our template. Then, we present and discuss the experimental results for the best samples that are those evaporated under 30° (30 and 50 nm nominal thickness).

The bottom surface looks like a triangular lattice of half spheres of radius of ~50nm. The hemispheres are "slightly" deformed into a hexagonal shape. The deepest points are the interstitial sites between 3 adjacent bumps. These sites form a hexagonal lattice. The number of these sites is twice the number of the bumps. Between two interstitial sites, there is a small maximum which is in fact a saddle point.

In the case of a perpendicular evaporation, the thickness of the deposited layer is theoretically the same everywhere on the surface. Figure 3a (upper panel) is a SEM image of the AAO surface. A transversal cut of the 50 nm thick layer along the dotted line is represented on the down scheme. There is no thickness modulation meaning that the volume of the superconductor occupied by a vortex is the same wherever the vortex is placed. However, the surface induces a shape modulation of the superconducting film. At the interstitial sites, the Nb layer has a sharp angle, which can be penetrated more easily by an applied magnetic field. Vortices will tend to go on these sites where the vortex energy is lowered. Therefore, the layer forms a hexagonal lattice of pinning centres. The matching field expected for this hexagonal lattice can be easily calculated. Indeed, the number of interstitial sites is twice the number of bumps. But the first matching field for a triangular lattice (bumps) is equal to:

$$H_1^{\Delta} = \frac{2\Phi_0}{\sqrt{3} \cdot d^2} \cong 237mT \tag{1}$$

where d ≈ 100nm is the bumps spacing and Φ_0=h/2e is the magnetic flux quantum. Therefore, the first matching field in the case of interstitial pinning is simply twice this value:

$$H_1 = 2 \cdot H_1^{\Delta} \cong 474mT \tag{2}$$

Figure 3a (top) represents the location of the vortices on the AAO surface in the case of perpendicular evaporation (0°). Vortices most favourable location is indicated. The pinning strength in this case is weak (sharp angle part is small compared to the total thickness).

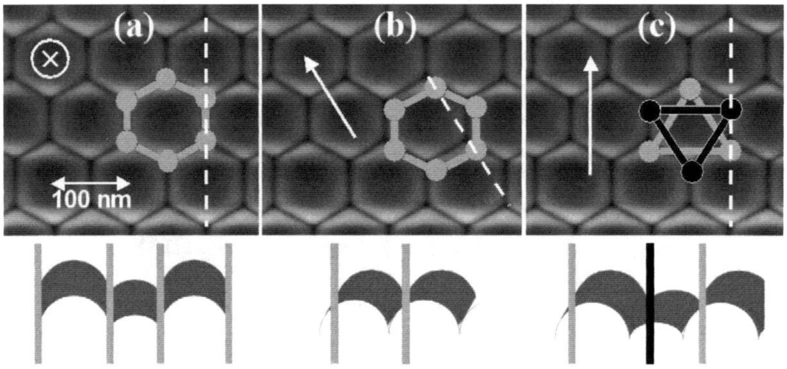

Figure 3. SEM micrograph with a schematic representation of the vortex configuration (upper frame) and simulation of the layer thickness along the dotted line (lower panel) for (a) perpendicular evaporation (0°) (b) 30° evaporation in the direction of a row of bumps (c) 30° evaporation in the direction of the side of the hexagonal unit cell. Most favourable sites for vortices are shown in grey and the other pinning sites in black.

The evaporation under an angle produces a thickness modulation, which depends on the direction of evaporation. As seen in figure 2, the bump lattice does not present a unique direction. In fact, the template is split in domains of tens of bumps, which have their own orientations. Therefore, it is not possible to match the evaporation direction with the same lattice direction for all the domains.

The thickness modulation will depend on the orientation of the evaporation axis compared to the hexagonal lattice. Because of the symmetry of the system we only have to consider directions spread over 30°. The figures 3b and c represent two extremes of the considered angular distribution. The first direction is located on a row of bumps (indicated by the arrow in Fig. 3b – upper frame). The other direction is along one side of a hexagonal unit cell (Fig. 3c – upper frame). It is enough to consider only the set of directions in the 30° angle formed by the two extreme ones. As a first approximation, we guess that the thickness modulation in an intermediate direction will be close to the average of the two principal directions. Figure 3b (lower frame) represents the superconducting film thickness (along the dotted line) in the case of an evaporation in the direction of the bumps row. One can see that the angular evaporation induces a pronounced thickness modulation. Indeed, the thickness is proportional to the cosine of the angle between the normal to the surface and the direction of evaporation (30°). Moreover shadow effects also play an important role. Vortex pinning sites will be situated close to every interstitial site. In this case, all pinning sites are equivalent and therefore the pinning lattice has a hexagonal symmetry.

If the evaporation is performed along the direction of a hexagon side, the six interstitial sites are not equivalent anymore. Indeed, half of the sites are situated behind a bump and the other half are behind a small bump in the middle of a hexagon side (saddle point). The simulated layer deposited along this direction is given in figure 3c (lower panel). It shows that the layer thickness is now strongly varying, which induces an asymmetric vortex pinning potential. In this case, the hexagonal lattice of pinning sites consists of a double triangular lattice of different pinning strength. In figure 3c (upper panel), the energy of the vortices is indicated with their colour. Gray means a low energy (strong pinning) and black a higher energy (weaker pinning).

The behaviour of the whole sample should be somewhat intermediate between these two limiting cases. Therefore, the pinning lattice should consist of two interpenetrating triangular lattices with different pinning strength. If the pinning strengths of the two lattices are similar, the pinning pattern will behave like a hexagonal lattice. On the other hand, if for one of the lattices, pinning strength is negligible compared to the other, it will behave more like a single triangular lattice.

From this moment, we will only discuss the results obtained with 30° evaporation. Critical current measurements have been performed on the 30 nm (nominal) sample (Figure 4a). In this case, the first matching field would correspond to the filling of both triangular pinning lattices forming together a hexagonal lattice. Well-pronounced matching effects are observed every 0.2T. This is clearly visible on the differentiated curve (t=0.83 of figure 4a) up to $H_{5/2}$ = 1T (figure 4b). This result is consistent with theoretical [14] as well as experimental [7,11] results obtained on hexagonal lattice of pinning centres. Indeed, these works showed that periodic anomalies (of different amplitude) appear every half-integer matching field, which corresponds here to 0.2T. In particular, $H_{1/2}$ = 0.2T corresponds to the filling of the strongest triangular lattice. As the triangular arrangement of vortices is known to be the most stable [1], $H_{1/2}$ is well-pronounced and even more than H_1 as predicted theoretically [14]. One can also observe $H_{1/4}$ = 0.1T, which is only predicted for hexagonal pinning lattice. The matching effect is more pronounced at high temperature which is due to a decrease of intrinsic pinning but an important matching effect is maintained at low temperatures (down to t=0.75) contrary to lithographically made samples.

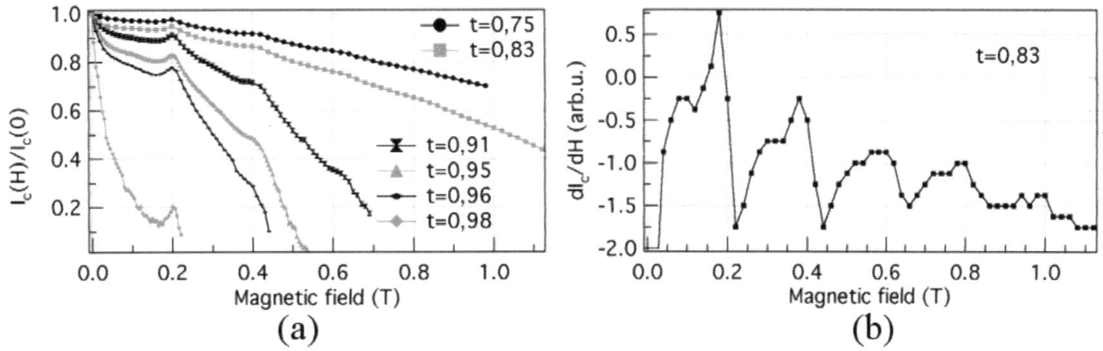

Fig. 4: 30nm (nominal) Nb film deposited under 30° on bumps of AAO: (a) Normalized critical current versus magnetic field curves for several reduced temperatures (t= T/T_c). The critical current is determined from the onset of a measurable resistance. At high temperature only $H_{1/2}$ is visible. At low temperature, matching effects are observed every 0.2T. (b) Differentiated curve of figure 4a at t=0.83. One can see apparent periodic behaviour up to 1T. $H_{1/4}$ = 0.1T is also clearly seen.

The figure 5a shows the magnetoresistance curves of our sample (50nm nominal deposited under 30°) for reduced temperatures from 0.89 to 0.99 by step of 0.01. Several matching effects are clearly seen in the entire range of reduced temperature. As observed previously on the 30nm thick sample, periodic anomalies are present every 0.2T. This periodicity corresponds to a triangular lattice of bumps spaced by 108nm, which is consistent with our anodization conditions.

As seen on figure 4b, the 50nm layer also shows a weak fractional matching effects at 0.1T corresponding to $H_{1/4}$. The matching effects are more pronounced for a 50nm nominal thickness than for 30nm.

At lower temperature (figure 5b) matching effects are preserved (zoom of the dashed frame of figure 5a). The matching effects are still seen every 0.2T at very low resistance. The commensurability effect at the first and second matching field results in a drop in the magnetoresistance curves. At the third matching field the commensurability effect appears as a change in the slope of the magnetoresistance curves. The sharp increase of the resistance from the first matching field is consistent with the appearance of interstitial vortices situated at the centre of the hexagon. The matching effect observed at $H_{3/2}$ = 0.6T then corresponds to the filling of each hexagon centre producing a triangular lattice. The configurations at the different matching field can be found in [14].

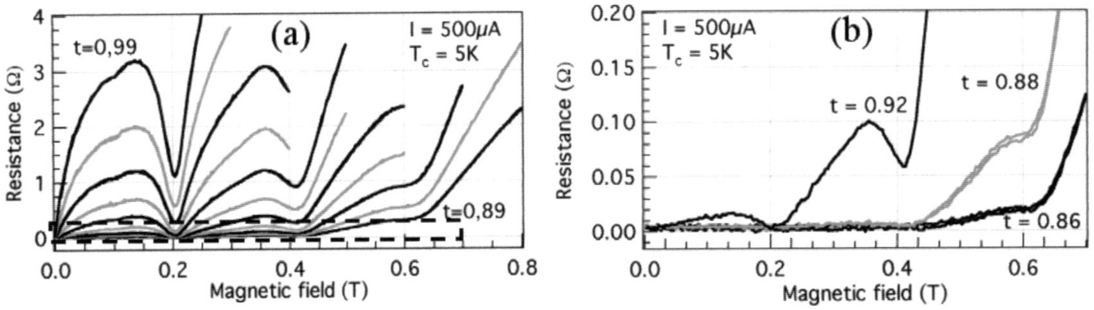

Fig. 5: Magnetoresistance curves at different reduced temperatures for the 50nm (nominal) Nb film evaporated under 30° on bumps of AAO. (a) A strong decrease of the resistance is observed at $H_{1/2}$ and H_1. (b) Zoom on the dashed frame of figure 5a. Matching effects are present at relatively low temperature.

4. Conclusion

An original new technique leading to a laterally modulated thin film has been presented. We used self-ordered alumina templates prepared by a two step anodization of an Al foil. Two modulated surfaces on each side of the template are obtained and both can be used for superconducting vortices pinning. The top surface of the template contains a triangular array of holes, while the bottom surface shows a triangular array of bumps. The latter (bottom surface) is revealed after the removal of the remaining Al substrate and was used here as a substrate for the evaporation of a thin Nb layer. Two thicknesses and three different angles were compared. The depositions under 30° from normal show matching effects up to ~1T, which is a big improvement of the magnetic field range for matching effects compared to previous works [11,12]. The periodic matching effects observed are consistent with a quasi-hexagonal lattice of pinning centres that is composed of two interpenetrating triangular sublattices of different pinning strength. This results from the particular thickness modulation obtained using angle deposition over such a template.

This work was supported by the Interuniversity Attraction Pole Program (P6/42) - Belgian State - Belgian Science Policy. X. H. acknowledges financial support of the Fund for Training in Research in Industry and Agriculture. S.M. is a post-doctoral researcher of the National Fund for Scientific Research. The authors wish to thank Bas Opperdoes for the MBE growth of the Nb thin films.

References

[1] Abrikosov A A 1957 *Sov. Phys. JETP* **5** 1174
[2] Daldini O, Martinoli P, Olsen J L and Berner G 1974 *Phys. Rev. Lett.* **32** 218
[3] Fiory A T, Hebard A F and Somekh S 1978 *Appl. Phys. Lett.* **32** 73
[4] Baert M, Metlushko V V, Jonckheere R, Moshchalkov V V and Bruinseraede Y 1995 *Phys. Rev. Lett.* **74** 3269 ()
[5] Bezryadin A and Pannetier B 1995 *J. Low Temp. Phys.* **98** 251
[6] Morgan D J and Ketterson J B 1998 *Phys. Rev. Lett.* **80** 3614

[7] Wu T C, Wang J C, Horng L, Wu J C and Yang T J 2005 *J. Appl. Phys.* **97** 10B102
[8] Lange M, Van Bael M J, Bruinseraede Y and Moshchalkov V V 2003 *Phys. Rev. Lett.* **90** 197006
[9] Lange M, Van Bael M J, Moshchalkov V V and Bruinseraede Y 2002 *J. Magn. Magn. Mat.* **240** 595
[10] Welp U, Xiao Z L, Jiang J S, Vlasko-Vlasov V K, Bader S D, Crabtree G W, Liang J, Chik H and Xu J M 2002 *Phys. Rev. B* **66** 212507
[11] Vinckx W, Vanacken J and Moshchalkov V V 2006 *J. Appl. Phys.* **100** 044307
[12] Vinckx W, Vanacken J, Moshchalkov V V, Mátéfi-Tempfli S, Mátéfi-Tempfli M, Michotte S and Piraux L 2006 *Eur. Phys. J. B* **53** 199
[13] Mátéfi-Tempfli S, Mátéfi-Tempfli M and Piraux L 2008 *Thin Solid Films* **516** 3735
[14] Reichhardt C and Olson Reichhardt C J 2007 *Phys. Rev. B* **76** 064523

Optimization of the Pr concentration in $Y_{1-x}Pr_xBCO$ films prepared by pulsed laser deposition

M. Irjala[1,2], H. Huhtinen[1], P. Paturi[1], Anuj Kumar[3], V. P. S. Awana[3], A. V. Narlikar[1,4] and R. Laiho[1]

[1] Wihuri Physical Laboratory, Department of Physics, FIN-20014 University of Turku, Finland
[2] Graduate School of Materials Research, Turku, Finland
[3] Superconductivity Division, National Physical Laboratory, New Delhi-110012, India
[4] UGC-DAE Consortium for Scientific Research, University Campus, Khandwa Road, Indore-452017, MP, India

E-mail: matti.irjala@utu.fi

Abstract. A series of studies has been carried out on thin films which were prepared by pulsed laser deposition from micro-grained targets of $Y_{1-x}Pr_xBa_2Cu_3O_{7-\delta}$ ($Y_{1-x}Pr_xBCO$) to determine the doping effect of Pr (at Y site) in concentrations between x = 0 - 0.20. Earlier results on bulk materials show that the Pr substitution can significantly enhance the value of $J_c(B)$ at cost of T_c. Our X-ray diffraction results confirmed the good quality of the films without any impurity phases. Magnetometric studies show the systematic decrease of $J_c(B)$ when the Pr concentration is increased.

1. Introduction

Since the discovery of $YBa_2Cu_3O_{7-\delta}$ (YBCO) and other high temperature superconductors (HTS), cationic substitutions (such as Pr, Ca, Ni, etc.) have been widely studied, because the response of the normal and superconducting state to the substitutional disorder has given new insights and information about these materials (see e.g. [1]).

The interest in such studies has recently risen, because despite the decrease of the superconducting critical temperature, T_c, other superconducting properties such as critical current density, $J_c(B)$, can be improved [2, 3, 4, 5, 6]. The doping concentrations for the improving superconducting effects are, however, relatively small. For example, Zn, Ni, Co, Ga, Er and Ho, when doped in small concentrations of 0.1 - 5% depending on the sites they occupy, can enhance $J_c(B)$, which is important for high magnetic field and critical current applications.

Pr presents an intriguing case, because it isostructurally replaces Y in YBCO over the entire range from x = 0 to 1, giving rise to a dramatic depression in T_c [7, 8, 9], and at $x \geq 0.55$, no superconductivity is observed. Moreover, the magnetic insulator $PrBa_2Cu_3O_{7-\delta}$ is a compound which has a single ortorhombic 123 structure, but it does not show superconductivity [10, 11, 12, 13]. Nevertheless, Pr has been found to increase the value of $J_c(B)$, e.g. in bulk [14, 15, 16] and melt-textured samples [17]. In the bulk samples [16], $J_c(B)$ increases with increasing Pr concentration up to x = 0.08 and decreases beyond this level. Because of the similarity of the ionic radii between Y and Pr, there is no remarkable change in the lattice parameters between pure and doped samples. It is proposed that the reason for the enhanced

$J_c(B)$ for Pr-doped bulk samples is due to formation of rows of oxygen vacancies in CuO_2 planes, which provide strong pinning interaction with pancake vortices formed in the same planes [16].

In this paper, we report the effect of Pr^{3+} substitution at the Y^{3+} site in $Y_{1-x}Pr_xBCO$ (x = 0 - 0.20) thin films, prepared by pulsed laser deposition (PLD), whereas earlier studies on films have been made on samples with small concentrations ($x \leq 1\%$) [18].

2. Experimental details

The $Y_{1-x}Pr_xBCO$ (x = 0 - 0.20) films were deposited on $SrTiO_3$ (001) substrates with PLD. The deposition targets were prepared as described previously [19, 20, 21, 22, 23]. Detailed description of the deposition process can be found in [24].

The crystallinity of the film samples was characterized by x-ray diffraction measurements (XRD), using CuK_α radiation for 2θ scans and texture measurements with Shultz geometry. The grain size, surface roughness and thickness of the films were measured by atomic force microscopy (AFM).

Magnetic measurements were performed with a Physical Properties Measurement System (PPMS), at temperatures between 10 - 300 K in fields up to 8 T. The critical current densities $J_c = 3\Delta M/2R$ of the films were calculated from the opening of the hysteresis loop by using the Bean's critical state model for disc-shaped samples [25, 26]

3. Results and discussion

The first optimization of the PLD parameters was done with films deposited from pure YBCO target. The substrate temperature was varied from 620°C to 730°C and the best film quality was obtained at 670°C. Other deposition parameters, such as laser fluence, atmosphere, cooling process of films etc. were unchanged due to earlier optimization [24].

All the films were highly c-axis-oriented, and they had the amount of the a-axis-oriented material less than 2% without any clear dependence on the Pr content. The $Y_{1-x}Pr_xBCO$ (212)/(122) XRD peaks show no effect of Pr on twinning and straining of the films. Also 2θ scans show only typical YBCO (001) peaks without any signs of impurities. Therefore, it is clear that Pr replaces Y in the lattice over the whole range of $x \leq 0.20$ and the structural quality of the samples is good. The AFM data showed that on the contrary to the unaligned polycrystalline pellets prepared using the solid state method [15], grain size in the films and the surface roughness did not have correlation with the Pr concentration.

As shown in figure 1, the values of $J_c(B)$ decreased at all the measured temperatures (a) and in external magnetic field densities (b) when the Pr concentration was increased. The insert in figure 1(a) shows the decrease of T_c with the increasing Pr concentration. The values of T_c are roughly the same as in the sintered bulk samples [16]. However, the $J_c(B)$ does not behave as in the earlier studies on different type of samples [14, 17, 15, 16], where $J_c(B)$ was found to increase at some specific Pr concentrations. The same tendency is also seen in figure 2, where the $J_c(B)$ values decrease as the Pr concentration increases in all fields.

The accomodation field, B^*, defined with the criterion $J_c(B^*)/J_c(B(0) = 0.9$ [27, 28] was found to decrease as the Pr concentration was increased, giving the highest value of 35 mT in a pure sample at 10 K. This value agrees with those obtained from similar targets [29]. The B^* values decrease with increasing doping, reaching 14 mT for $x = 0.20$ sample at 10 K. This behaviour is similar at all measured temperatures. The field dependence of J_c above B^* is often described by a power law expression $J_c(B) \sim B^\alpha$ with α varying from -0.50 to -0.60 [30, 29, 31]. The values of α are in the typical range for all the samples up to 70 K, varying from -0.40 to -0.60 without any dependence on the Pr concentration, excluding the case of $x = 0.20$ where α decreases rapidly from -0.45, observed at 10 K to -0.72 at 60 K, indicating a rapid decrease of $J_c(B)$ with temperature.

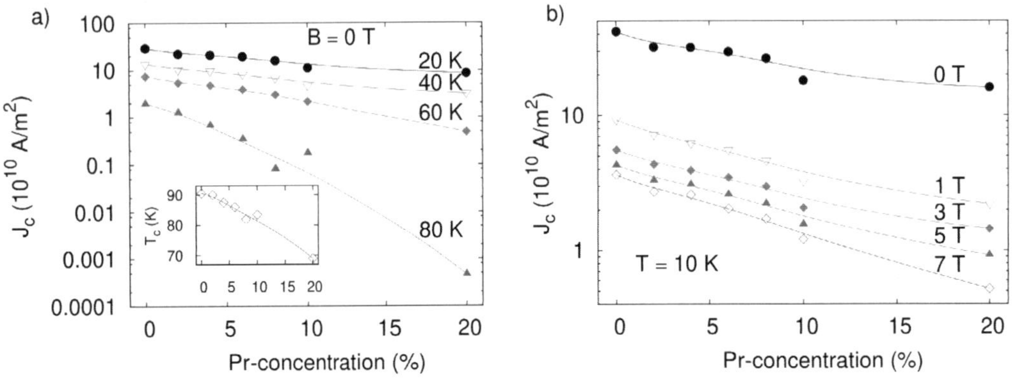

Figure 1. a) Dependence of $J_c(B)$ for all the film samples on the Pr concentration for $B = 0$ T at different temperatures and b) at 10 K in different external fields. Inset of (a) shows the dependence of T_c on Pr concentration. Lines are for guiding the eye.

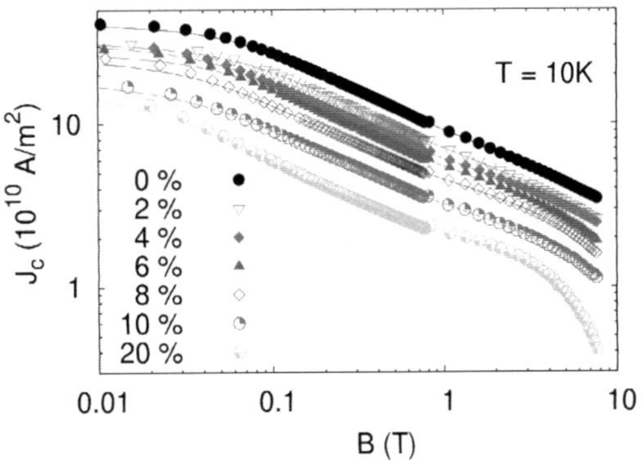

Figure 2. Dependence of $J_c(B)$ in samples with $x = 0$ - 0.20 at 10 K. The step seen in all the signals below 1 T is due to the measurement sequence. Lines are for guiding the eye.

Because doping with Pr did not improve the superconducting quality of the $Y_{1-x}Pr_xBCO$ films, we performed a second optimization, using the $Y_{1-x}Pr_xBCO$, $x = 0.08$ PLD target. Though this dopant level gave the highest $J_c(B)$ for bulks [16], we did not come up with a 8% film sample holding greater $J_c(B)$ than the pure YBCO film. E.g., at 10 K and 0 T the pure film sample has $J_c(B) = 4.2 \cdot 10^{11}\mathrm{A/m^2}$ and the best $x = 0.08$ film sample (deposited at the substrate temperature 710°C) has $J_c(B) = 2.8 \cdot 10^{11}\mathrm{A/m^2}$, whereas in the $x = 0.08$ bulk samples $J_c(B)$ was three times higher than in the pure sample [16].

It is possible that the clear difference in the SC properties on the dopant level, between the polycrystalline bulk samples and the epitaxial thin films, can be explained by the effect of surface pinning [32]. Due to the formation of defects at the interior of thin films, they become rough on the atomic scale, and since the coherence length is small, the barriers for flux entry and flux expulsion are broken or weakened. This gives rise to decrease in $J_c(B)$ of thin films when doped with Pr. Anyhow, our structural measurements did not indicate any dependence between surface roughness and Pr concentration. Thus, we propose that this difference is at least partly due to the grain boundary critical current density $J_c^{gb}(B)$, which is due to increased misorientation angle between the grains. This effect has been reported many times [33, 34] and Ca^{2+} doping is well known to increase $J_c^{gb}(B)$. Furthermore, the ionic radii of Pr and Ca are roughly the same, and it has been suggested lately that the ionic radius is more important than the valence of the dopant in creation of high $J_c^{gb}(B)$ [35]. This is in agreement of the enhancement of $J_c(B)$ only in the polycrystalline samples, but lacks from our epitaxial thin film samples.

4. Conclusions

In this work, we optimized Pr-doped YBCO films. Even though the structural quality of the samples was good and the dependence of T_c on Pr concentration was as expected, neither $J_c(B)$ nor B^* increased in the doped films. We suggest that this is at least partly due to the grain boundary critical current effect, which increases J_c in the polycrystalline samples and is absent in thin film samples.

References

[1] Narlikar A 1989-2006 *Studies of High Temperature Superconductors vol 1 to 50* (New York: Nova Science)
[2] Feng Y, Pradhan A K, Zhao Y, Wu Y, Koshizuka N and Zhou L 2001 *Physica C* **357-360** 799
[3] Wang Z H, Zou X W, Fang J, Yang T, Tang Y L, Huang Z, Qiu L, Chen J L, Zhang H and Ding S Y 2002 *Supercond. Sci. Technol.* **15** 183
[4] Gonzalez M T, Hari-Babu N and Cardwell D A 2002 *Supercond. Sci. Technol.* **15** 1372
[5] Shlyk L, Krabbes G, Fuchs G, Nenkov K and Schüpp B 2005 *Supercond. Sci. Technol.* **18** S10
[6] Zhou Y X, Scruggs S and Salama K 2006 *Supercond. Sci. Technol.* **19** S556
[7] Xu Y and Guan W 1992 *Phys. Rev. B* **45** 3176
[8] Neumeier J J and Maple M B 1992 *Physica C* **191** 158
[9] Tomkowicz Z 1999 *Physica C* **320** 173
[10] Soderholm L, Zhang K, Hinks D G, Beno M A, Jorgensen J D, Segre C U and Schuller I K 1987 *Nature* **328** 604
[11] Liang J K, Xu X T, Xie S S, Ran G H, Shao X Y and Duan Z G 1987 *J. Phys. B* **69** 137
[12] Dalichaouch Y, Torikachvili M S, Early E A, Lee B W, Seaman C L, Yang K N, Zhou H and Maple M B 1988 *Solid State Commun.* **65** 1001
[13] Malik S K, Jhans H, Dhar S K, Awana V P S, Yelon W B, Rhyne J J, Prasad R and Soni N C 1995 *Phys. Rev. B* **52** 13006
[14] Harada T and Yoshida K 2002 *Physica C* **383** 48
[15] Paulius L M, Almasan C C and Maple M B 1993 *Phys. Rev. B* **47** 11627
[16] Huhtinen H, Awana V P S, Gupta A, Kishan H, Laiho R and Narlikar A V 2007 *Supercond. Sci. Technol.* **20** S159
[17] Ren H T, Taylor K N R, Chen Y J, Xia J A and Qing H 1993 *Physica C* **216** 447
[18] Barnes P N, Kell J W, Harrison B C, Haugan T J, Varanasi C V, Rane M and Ramos F 2006 *Appl. Phys. Lett.* **89** 012503
[19] Narlikar A V, Rao C V N and Agarwal S K 1989 *Studies of High Temperature Superconductors vol 1, ed A Narlikar* (New York: Nova Science)
[20] Narlikar A V, Gupta A, Samanta S B, Chen C, Hu Y, Wondre F, Wanklyn B M and Hodby J W 1999 *Phil. Mag. Part B* **79** 717
[21] Awana V P S and Narlikar A V 1994 *Phys. Rev. B* **49** 6353
[22] Awana V P S, Tulapurkar A, Malik S K and Narlikar A V 1994 *Phys. Rev. B* **50** 594
[23] Sedky A, Gupta A, Awana V P S and Narlikar A V 1998 *Phys. Rev. B* **58** 12495
[24] Huhtinen H, Järvinen J, Laiho R, Paturi P and Raittila J 2001 *J. Appl. Phys.* **90** 1521
[25] Bean C P 1962 *Phys. Rev. Lett.* **8** 250
[26] Conner L W and Malozemoff A P 1991 *Phys. Rev. B* **43** 402
[27] Peurla M, Paturi P, Stepanov Y P, Huhtinen H, Tse Y Y, Bódi A C, Raittila J and Laiho R 2006 *Supercond. Sci. Technol.* **19** 767
[28] Cai C, Holzapfel B, Hänisch J, Fernandez L and Schultz L 2004 *Phys. Rev. B* **69** 104531
[29] Peurla M, Huhtinen H and Paturi P 2005 *Supercond. Sci. Technol.* **18** 628
[30] Klaassen F C, Doornbos G, Huijbregtse J M, van der Geest R C F, Dam B and Griessen R 2001 *Phys. Rev. B* **64** 184523
[31] van der Beek C J, Konczykowski M, Abal'oshev A, Abal'osheva I, Gierlowski P, Lewandowski S J, Indenbom M V and Barbanera S 2002 *Phys. Rev. B* **66** 24523
[32] Bean C P and Livingston J D 1964 *Phys. Rev. Lett.* **12** 14
[33] Ivanov Z G, Winkler P A N D, Alarco J A, Claeson T, Stepantsov E and Tzalenchuk A Y 1991 *Appl. Phys. Lett.* **59** 3030
[34] Heinig N F, Redwing R D, Nordman J E and Larbalestier D C 1999 *Phys. Rev. B* **60** 1409
[35] Klie R F, Buban J P, Varela M, Franceschetti A, Jooss C, Zhu Y, Browning N D, Pantelides S T and Pennycook S J 2005 *Nature* **435** 475

Effect of Oxygen Annealing on Magnetic Performance of (Nd,Eu,Gd)$Ba_2Cu_3O_y$ with a Lamelar Pinning Nanostructure

M. Jirsa[1], M. Rames[1], M. Muralidhar[2], P. Diko[3]

[1] Institute of Physics ASCR, Na Slovance 2, CZ-182 21 Prague 8, Czech Republic
[2] SRL/ISTEC, 10-13, Shinonome 1-chome, Koto-ku, Tokyo 135-0062, Japan
[3] Institute of Experimental Physics SAS, Watsonova 47, SK- 04353 Kosice, Slovakia

Corresponding author: jirsa@fzu.cz

Abstract. The nanometer-scale correlated pinning disorder filling channels between regular twin boundaries in the melt-textured (Nd,Eu,Gd)$Ba_2Cu_3O_y$ composite is studied. Aligned with regular twin boundaries but tightly packed on nm-scale, this pinning medium is capable to significantly enhance pinning performance at high magnetic fields. We show that oxygen annealing plays a principal role in the formation mechanism of this defect type. It seems that the nanoscale lamelar structure is extremely sensitive to the oxygenation procedure and to the terminal temperature.

1. Introduction

In the LRE-$Ba_2Cu_3O_y$ compounds (LRE-123, LRE=light rare earth, Nd, Sm, Eu, Gd) much better vortex pinning environment can be created than in $YBa_2Cu_3O_y$ (Y-123) and thus they can carry significantly higher engineering currents [1-3]. Therefore, the LRE-123 bulks can trap the same magnetic field as Y-123 [4] but at higher temperatures or a higher magnetic field at the same temperature. It is commonly due to the additional source of point-like pins consisting of LRE/Ba solid solution nanometer scale clusters. Amount of this type of substitution is usually kept in an optimal range by growing the material in a reduced oxygen atmosphere [1-3] or by a slight excess of Cu over Ba in the starting powder [5]. This kind of pinning enhances mainly the secondary peak, the intermediate field range of J_c-B dependence. The low-field part of it is usually increased by "large" particles of secondary phases (in the micron and sub-micron scales). The smaller the secondary phase particles size, the stronger is the effect [6,7]. In the case of nanometric nonsuperconducting particles the J_c enhancement is so strong that extends up to intermediate fields and at high temperatures (where the secondary peak becomes weak) it passes over the whole field range up to the irreversibility field.

In binary and ternary LRE-123 compounds variation in the mutual LRE ratio offers a natural tool how to manipulate with the local environment (say tensions) in the superconducting matrix. Due to different sizes of LRE ions their positions and concentrations are not exactly equivalent. It has been found that in a narrow range of the Nd:Eu:Gd ratio (around 33:38:28) the superconducting matrix is frustrated (in the positive sense from our point of view) so that a periodical correlated nanostructure appears [1]. It has either a zig-zag shape or, in the better case, it is planar, aligned with the regular twin boundaries and filling the channels between them. As the period of this new structure is typically a few nm, just in the range of the vortex core size, no wonder that a very strong pinning effect appears. This effect enhances critical currents at high magnetic fields and presents itself as a hump or shoulder on the $J_c(B)$ curve The currents enhancement persists up to substantially higher magnetic fields so that the irreversibility field can be increased by factor 2-3 [1]. While the typical irreversibility field value at 77 K, both in Y-123 and LRE-123 samples, commonly ranges from 5 to 7 Tesla, the value obtained in the above-described material reached about 15 Tesla [1]. Such kind of defects is in any case very attractive for all applications operating at high magnetic fields. The recent extensive study of the material composition and electromagnetic performance homogeneity [8] confirmed a close correlation between LRE ratio and the type of pinning in the melt-textured $(Nd_{0.33}Eu_{0.38}Gd_{0.28})Ba_2Cu_3O_y$ samples.

In the present work we refer on the effect of oxygenation on the pinning structure of these samples.

2. Experimental details

The $(Nd_{0.33}Eu_{0.38}Gd_{0.28})Ba_2Cu_3O_y$ "NEG-123", doped by 0.035 mol% ZnO, 0.5 mol% Pt, 5 mol% Gd-211, and 10 wt% Ag, was produced from high-purity commercial powders of Nd_2O_3, Eu_2O_3, Gd_2O_3, $BaCO_3$ and CuO mixed in nominal composition. The starting powders were three times thoroughly ground and calcined at 880°C for 24 h with intermediate grinding. Sintering was carried out at 900 °C for 15 h at Ar/0.1% O_2. ZnO was added to improve growth control of the compound, and silver to enhance mechanical strength of the material. The well-mixed powders were pressed into pellets 20 mm in diameter and 10 mm thick, and consolidated by cold isostatic pressing to 200 MPa. The sample was grown from MgO seed in Ar/0.1% O_2 atmosphere, at the gas flow of 300 ml/min [1]. After that the pellet was oxygenated in the following way: the pellet was during 2 hours heated up to 600 °C, there kept for 2 hours. Then, it was cooled in 12 h to 500 °C, further cooled in the half rate to 400 °C, and cooled further with the rate 2 °C/h to 300 °C. There the sample was kept for 150 h and finally left cooling down to room temperature. The final dimensions were 16 mm diameter and 8 mm height. The RINT2200 (Rigaku) high-resolution automated x-ray powder difractometer identified the sample to be single-phase, with less than 3 mol% Gd-211. The chemical profile analysis of the final product, based on the x-ray emission spectrometry and the x-ray mapping image analysis was made on the cut interface running across the pellet diameter, just next to the positions of the samples for magnetic measurements. The analysis showed a negligible fluctuation of Cu and Ba, irrespective of the growth sector. For copper it was 24.38 ± 0.16 at%; for barium 15.21 ± 0.14 at%. The variation in the LRE

Fig.1. (left) Top and bottom sides of the investigated NEG-123 pellet. The image manifests the perfect growth (indicated by the perfectly developed growth sectors boundaries, especially on the pellet bottom), the growth being facilitated by a tiny addition of ZnO; (right) Optical microscope image of the material showing a perfect structure of regular twin boundaries. The bright spot is a small silver drop embedded in the superconducting matrix.

ions content was much higher, reaching 2.69 ± 0.13 at% for Nd, 3.55 ± 0.29 at% for Eu, and 2.11 ±0.16 at% for Gd. A notable spatial dependence was found for Nd and Eu. Nd content was significantly higher in the pellet center and continuously decreased towards the pellet circumference, while Eu content varied in opposite way, though much less [8]. The average final composition was $(Nd_{0.32}Eu_{0.43}Gd_{0.25})Ba_{1.8}Cu_{2.9}O_y$, showing a higher excess of Eu over Gd than the initial one. Just an opposite tendency should be expected, due to the small amount of Gd-211 added, that can on the other hand, be the reason for the slight depletion of Ba and Cu with respect to the stoichiometric concentration. The average content of zinc was only 0.014 at%, less than half of the input value. Major part of Zn evidently evaporated during the pellet processing, after having made its work in the perfect crystal growth initiation. The average amount of silver incorporated into the superconducting matrix (without the silver droplets) was about 0.044 at%. No degradation of magnetic properties due to zinc or silver was observed [8]. The critical temperature, identified from ac-susceptibility, was 92 K.

The pellet form and the material morphology are presented in Figs. 1 and 2. Figure 1 left shows top and bottom of the pellet, while the figure on the right presents the optical microscopy image of the a-b plane of the pellet, characterized by nice equidistant regular twins. The bright spots, nearly regularly distributed over the sample volume, are silver micro-droplets embedded in the superconducting matrix. They are very important for mechanical stiffness of the material, as without silver or another type of impregnation the compound is very crispy. A more complex insight into the material morphology is in Fig. 2. One can see typical horizontal cracks along a-b planes. Twin boundaries, running along the vertical *c*-axis, possess the same character on both sides from the growth sector (GS) boundary running along the image diagonal from the left upper corner to the right bottom one. The boundary is "decorated" by Gd-211 particles agglomerated along this boundary. The Gd-211 particles are much smaller than Ag droplets. Magnetic measurements on the samples cut off the growth sector boundaries exhibit an equally high central peak as the samples cut just at the boundary. As the central peak is partly due to "large" secondary phase particles, the magnetic data indicate that a reasonable amount of Gd-211 is distributed over the whole pellet volume. As the optical microscopy was not able to detect them, we conclude that these particles were only a few tens of nm small inside the growth sectors and thus invisible by optical microscopy. Dark and bright large spots

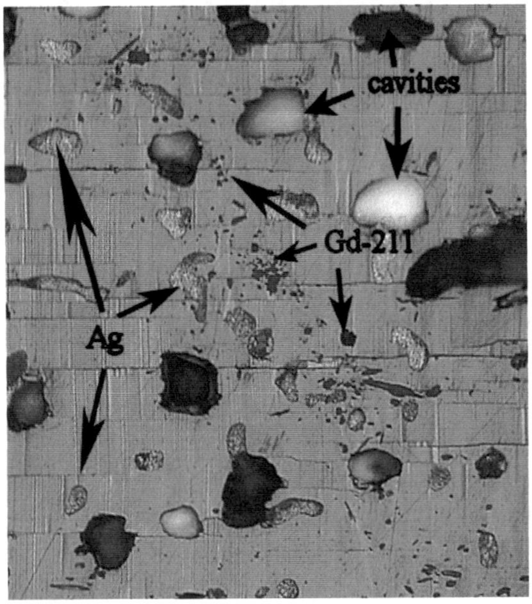

Fig. 2. An optical microscope image showing structure of the pellet at the position of growth sector boundary in the a-c plane. The a-b planes are horizontal, c-axis goes upwards.

are cavities in the material. No special growth difference can be seen by this technique in the two growth sectors (c-axis GS at the lower left corner, a-axis GS at the upper right corner). The material quality was very good.

The nanoscale lamellas filling channels between adjacent regular twin boundaries were extensively characterized in Refs. [1,9]. In the same work a correlation between the nanoscale lamellar pinning structure and the special shoulder on the magnetic hysteresis loop (MHL) and the associated $J_c(B)$ dependence was established. The pinning enhancement at high magnetic fields was manifested not only by the shoulder but also by a significant increase of irreversibility field. When speaking about shoulder, we mean that the high-field inflexion point is shifted significantly up-wards from its regular position close behind the secondary peak, so as the part of the curve above the secondary peak is convex instead of the common concave shape.

The samples for magnetic measurements were of the typical dimensions a x b x c = 1.5 x 1.5 x 0.5 mm^3. Critical current was estimated from magnetic measurements using the extended Bean critical state model for a rectangular sample [10], $J_c = 2\Delta m/[Va(1-a/3b)]$, where V is the sample volume, a and b are the transverse dimensions ($b \geq a$), and Δm is the difference between the upper and lower branch of the MHL. Note that this formula is strictly valid only for a thin plate magnetized in the plane and does not take into account demagnetizing effects. Irrespective of that, it is commonly used also for the configuration with magnetic field normal to the plane, valid only at high magnetic fields. For sake of compatibility and simplicity we keep this practice, too. The magnetization loops were measured at 77 K by means of the PPMS (Quantum Design) vibrating sample magnetometer, with magnetic field sweeping at the rate of 12 mT/s, the field being directed along c-axis of the sample.

Altogether, 45 samples cut from the pellet at different positions were investigated. Here we report on the behaviour of only a few of them that were subjected to a simultaneous thermal treatment. The numbers 1 to 15 in the sample denomination indicate the sample position along the pellet height starting from the top surface (each column was cut to 15 samples) and by characters "C" for the pellet centre and "P" for its periphery.

3. Experimental results

Most of the samples exhibited $J_c(B)$ curve similar to that of sample 15P in Fig. 3 (a). This $J_c(B)$ curve characterizes a corporate effect of nanoscale lamellas and twin planes. Remarkable is the shoulder at about 7-8 Tesla shifting the irreversibility field up-wards, out of our magnetic field range (limited at 9 Tesla). Above the low-field peak the curve is rather flat, what we attributed to the effect of regular twins [11]. This type of curve could not be fitted by any of the commonly used theoretical schemes [12,13].

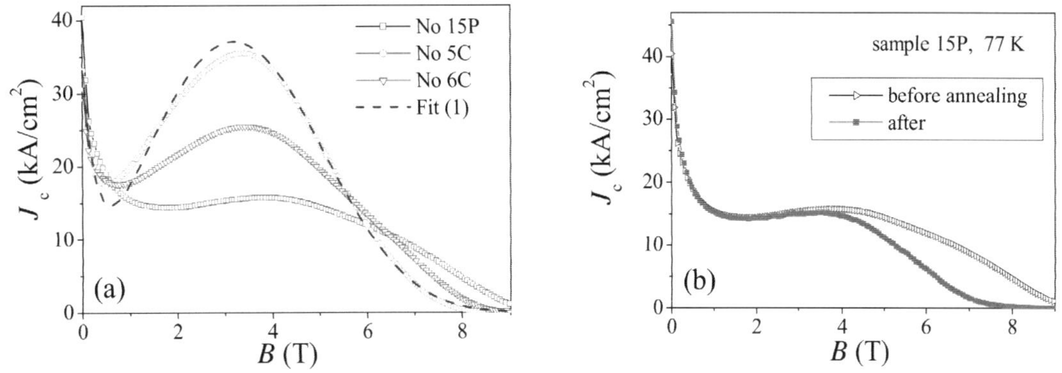

Fig. 3. (a) $J_c(B)$ dependences of three samples cut from the same $(Nd_{0.33}Eu_{0.38}Gd_{0.28})Ba_2Cu_3O_y$ pellet. The dashed curve in shows the best fit by the phenomenological model function (1) for twin-free RE-123 samples. (b) Change of the $J_c(B)$ dependence of sample 15P after annealing at 320 °C for 7 days. The high-field pinning effect disappeared.

From the whole set of samples only two behaved in a different manner. The exceptional ones were two adjacent samples 5C and 6C cut from the pellet center. On the $J_c(B)$ curve of sample 5C the high-field shoulder is missing (the curve shape above the secondary peak is mostly concave, resembling thus the common response of bulk RE-123 samples). Although not evident on the first glance, the secondary peak is somewhat flattened due to the channeling effect of regular twins. The flattening is evident from comparison with the model function (dash curve) for twin-free samples [13,14],

$$J_c(B) = J_{c1}\exp(-B/B_L) + J_{c2}\, b \exp[(1-b^n)/n], \quad (1)$$

where $b=B/B_{max}$, B_{max} and J_{c2} are coordinates of the second peak, J_{c1} is magnitude of the first peak, B_L is the field scale of the central peak decay, and n is the scaling factor of the second peak. Sample 5C was exceptional in the context of all others by exhibiting a well-resolved secondary peak and no high-field shoulder. We found that this magnetic "anomaly" correlated with a notable departure of the Eu:Gd elemental ratio value from the pellet average [8]. Sample 6C showed also a pronounced peak at intermediate fields and the curve above the secondary peak was mostly linear, neither convex, nor concave, representing thus a compromise between the two typical courses.

One of the natural explanations of the anomalous behavior of these two samples was an insufficient oxygenation. Therefore we oxygenated these two samples simultaneously, together with reference ones from both the pellet center and periphery, at 320 °C for 7 days. We expected a general change of J_c of all samples in intermediate fields (the secondary peak area) due to reduction of oxygen deficiency. For samples 5C and 6C an increase of pinning at high magnetic fields was expected. Fig. 3 (b) shows what happened: $J_c(B)$ of the representative sample 15P at low and intermediate fields practically did not change, while at high fields it transformed from convex to concave with a significant reduction of irreversibility field. Such a change was really unexpected. $J_c(B)$ characteristics

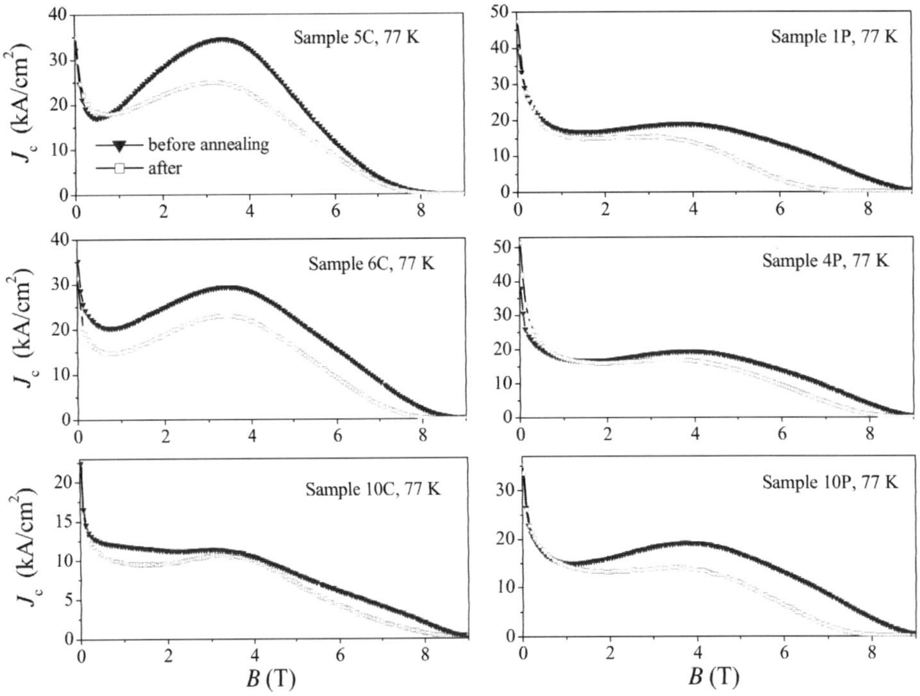

Fig. 4. $J_c(B)$ dependences of six samples cut from the same $(Nd_{0.33}Eu_{0.38}Gd_{0.28})Ba_2Cu_3O_y$ pellet, measured at 77 K by a vibrating sample magnetometer before (full symbols and after (open symbols) annealing in flowing oxygen at 320 °C for 7 days. Samples in the left column are from the pellet center; those in the right column are from the pellet circumference.

of practically all other samples changed in a similar manner, as documented in Fig. 4. In general, the changes were: (i) a systematic decrease of the secondary peak height and (ii) disappearance of the high-field shoulder accompanied by a decrease of irreversibility field, B_{irr}. J_c at low fields did not practically change. This indicates that the annealing did not practically touch the "large" particles distribution but in most cases reduced concentration of point-like oxygen vacancies. On the other hand, the decrease of J_c and B_{irr} indicates that the lamellar nanostructure was destroyed. This, however, implies that oxygenation plays a basic role in creation (and destruction) of the lamellas. This conclusion is in accord with the fact that the lamellas are aligned with regular twin boundaries, which appear first when the material transforms from tetragonal to orthorhombic structure, i.e. during the first material oxygenation. To what extent this process is reversible, is a subject of just running experiments.

Lets now turn back to the samples from the pellet center. Although J_c of these samples also decreased, the reduction was mainly in the intermediate field range and the change of B_{irr} was not much remarkable. Each sample experienced slightly different change of J_c. Worth mentioning is the practically unchanged irreversibility field of sample 5C, the sample that already before annealing exhibited no high-field shoulder and the lowest B_{irr} of all the investigated specimens. This fact again supports our conclusion that the irreversibility field enhancement in the investigated samples is due to the special pinning nanostructure created thanks to, or at least with help of, a special oxygenation treatment.

4. Conclusions

A melt-textured pellet of $(Nd_{0.33}Eu_{0.38}Gd_{0.28})Ba_2Cu_3O_y$ with a small amount of secondary phase particles and a tiny amount of ZnO was investigated. The material exhibited a pronounced enhancement of J_c at high magnetic fields and a substantial increase of irreversibility field. This was attributed to a correlated nanoscopic pinning disorder composed of nanoscale lamellas filling channels between regular twin planes. In the attempt to explain behavior of two samples from the pellet center, deviating from the average of the 45-sample batch magnetically studied, these two samples together with four reference ones were annealed in flowing oxygen at 320 °C for 7 days. In general, this simple oxygenation process resulted in suppression of J_c at intermediate and high magnetic fields. This indicates that oxygenation process is a basic agent in formation of the nanoscale lamellas. Further experiments are under way to find to what extent the creation and suppression of the lamellas is reversible and how the process can be optimized.

Acknowledgements

This work was accomplished under the support of Grant Agency of the Czech Republic (GACR 202/08/0722), one of the authors (MM) acknowledges the grant No. 19510125 (C) of JSPS (Japan).

References

[1] M. Muralidhar, N. Sakai, N. Chikumoto, M. Jirsa, T. Machi, M. Nishiyama, Y. Hu, M. Murakami, *Phys. Rev. Letters* **89** (2002) 237001
[2] M. Muralidhar, N. Sakai, M. Jirsa, N. Koshizuka, M. Murakami, *Appl. Phys. Lett.* **85** (2004) 3504-3506
[3] M. Muralidhar, N. Sakai, M. Jirsa, M. Murakami, N. Koshizuka, *Supercond. Sci. Technol.* **18** (2005) L9-L12
[4] M. Tomita and M. Murakami, *Nature* **421** (2003) 517-520
[5] Y. Q. Cai, X. Yao, A. Hu, X. J. Wu, M. Jirsa, S. Xu, G. Jin, *Supercond. Sci. Technol.* **19** (2006) S506-S509
[6] M. Muralidhar, N. Sakai, M. Jirsa, M. Murakami, *Appl. Phys. Lett.* **83** (2003) 5005-5007
[7] M. Muralidhar, N. Sakai, M. Jirsa, M. Murakami, N. Koshizuka, *Supercond. Sci. Technol.* **16** (2003) L46-L48

[8] M. Jirsa, M. Rameš, K. Jurek, M. Muralidhar, P. Das, M. R. Koblischka, T. Wolf, *Mater. Sci. Engin.* B **151** (2008) 25-30
[9] M. Muralidhar, N. Sakai, A. Wu, M. Murakami, M. Jirsa, T. Nishizaki, N. Kobayashi, *Supercond. Sci. Technol.* **15** (2002) 1357-1363
[10] D. X. Chen and R. B. Goldfarb, *J. Appl. Phys.* **66** (1989) 2489
[11] M. Jirsa, M. R. Koblischka, T. Higuchi, M. Murakami, *Phys. Rev. B* **58** (1998) R14771-R14774
[12] E. J. Kramer, *J. Appl. Phys.* **44** (1973) 1360; D. Dew Hughes, *Philos. Mag.* **30** (1974) 293
[13] M. Jirsa, L. Půst, D. Dlouhý, M.R. Koblischka, *Phys. Rev.* **B 55** (1997) 3276-3284
[14] M. Jirsa, T. Nishizaki, N. Kobayashi, M. Muralidhar, M. Murakami, *Phys. Rev.* B **70** (2004) 024525

Dependence of the Structural, Electrical and Magnetic Properties of the Superconductive YBCO Thin Films on the Deposition Rate

A.B. Karci[1], M.Tepe[1], H. Sozeri[2]

1 Department of Physics, Ege University, Bornova 35100, Izmir, Turkey.

2 TÜBİTAK, Institute of National Meteorology, Gebze 41470, Kocaeli, Turkey.

belkis.karci@ege.edu.tr

Abstract. In this study, YBCO thin films on single crystal $LaAlO_3$ (100) substrates have been grown using DC inverted cylindrical magnetron sputtering technique and the effect of the deposition rate on these films is investigated. Three different deposition rates are used to produce superconducting YBCO thin films with 150 nm of thickness on (100) $LaAlO_3$ single crystal substrate at 780 ^{0}C. The samples are analyzed in detail by means of XRD, R-T, χ-T, M-H and AFM characterizations and also the critical current densities (J_c) are derived from the magnetic hysteresis curves using the modified Bean formula [1]. The critical current density at 50 K was found to be in the range of 3.10^7 A/m^2 to $8.\ 10^7$ A/m^2 with a deposition rate between 2nm/min and 1.2nm/min. A correlation has been obtained so that as the film deposition rate increases, the surface smoothness and crystalline quality of the films significantly deteriorate, resulting in a significant decrease in J_c.

1. Introduction

After the discovery of high temperature superconductivity [2], enormous effort has been put forward to produce thin film forms using superconductive materials with epitaxial structures. A great percentage of these studies have been focused on determining the required physical growth conditions of thin films for electronic device applications and optimization of these conditions [3].

Among the high temperature superconductors, $YBa_2Cu_3O_{7-\delta}$ (YBCO) is a widely used material which has a transition temperature over the boiling point of liquid nitrogen. Further advantages of YBCO over others are non-poisonous components, high phase stability, high quality of crystallization, high flux pinning, low surface resistance and single step deposition oxygenation in-situ process [3]. The main disadvantage of YBCO is the strong dependence of its crystal structure and superconducting properties on the level of oxygenation during the growth process. This critical dependence on the oxygen content determines a transition from tetragonal phase which is non-superconductive to orthorhombic phase which is superconductive. There are various techniques for epitaxial growth of YBCO thin films, such as chemical vapor deposition (CVD) [4], pulsed laser deposition (PLD) [5], molecular beam epitaxy (MBE) [6], laser molecular beam epitaxy (L-MBE) [7], DC- and RF–sputtering [8,9]. The widely used sputtering technique of producing large area films in industry should be viable for the industrial production of YBCO thin films. As of now, this technique is a popular method in laboratories for small-scale fabrication of YBCO films [10]. Li *et al.* [11] showed that films produced by inverted cylindrical magnetron sputter system (ICMS) are better than those produced by non-magnetron inverted cylindrical sputtering at the same conditions. Non-magnetron system bombards the film surface with energetic ions so that there occurs re-sputtering. Geerk *et al.* [12] also expressed that during the planar magnetron sputtering; film surface is bombarded by negative ions having the same amount of energy of argon ions which bombards the target surface. As a result, ICMS system is very suitable for obtaining high quality and homogenous thin films and this system is widely used.

There are many parameters which affect the superconducting properties of YBCO thin films. These are substrate preparation, target condition, gas mixture, growth temperature and deposition rate. Degardin *et al.* [13] have reported that the substrate preparation is a major step for the elaboration of good quality superconducting films. The choice of the substrate material is of primary importance because the crystal formation of the coated thin films parallels the crystal structure of the substrate. Therefore, there should be accordance between the substrate and the crystal lattice of high temperature superconductor together with a similar thermal expansion and also no chemical interaction on the boundary surface [3]. Taniwaki *et al.* have reported that the change of substrate temperature affects the quality of crystallization of thin films [14]. Also, Selinder *et al.* reported that sputtering pressure effects the film crystallization [15], and Liu *et al.* reported that the increase of deposition rate will negatively affect the superconducting transition properties, and also will increase the lattice disorder concentration and make a rough surface in the resulting YBCO thin films [16]. And also, Avci *et al* reported the variations in YBCO thin film properties for different substrate coatings [17].

In this study, we systematically investigated the YBCO thin film preparation on substrates of single crystal (100) oriented $LaAlO_3$ by ICMS technique with different deposition rates. Other parameters were kept fixed at optimized values derived in previous studies [17, 18]. The YBCO samples were studied by structural (XRD, AFM), electrical (R-T) and magnetic (χ-T and M-H) measurements.

2. Experimental

$YBa_2Cu_3O_{7-\delta}$ thin films of 150 nm thickness were deposited onto 10x10x0.5 mm^3 (100) $LaAlO_3$ substrates by using DC Inverted Cylindrical Magnetron Sputtering (ICMS) system. The sputtering device, as well as the YBCO target, was commercially purchased from Hitec-Materials [19]. The substrate surface cleaning is a very important process for the epitaxial thin film growth. In this study, the substrates were cleaned with ultrasonic cleaner in acetone for 1 hour and subsequently in isopropyl alcohol for 1 hour. After cleaning, the substrate was glued to the substrate holder of sputter system with silver paste. The thin film deposition was carried out at a fixed temperature of 780 ^{0}C. Argon and oxygen gas pressures were maintained at 0.15 mbar during the deposition. In order to obtain different deposition rates 1.2, 1.6 and 2.0 nm/min. for different samples, we varied the plasma power from 55 to 75 watts, simultaneously adjusting the deposition times for obtaining a fixed film thickness of 150 nm.

After the deposition, the crystalline structure and surface morphologies of the YBCO thin films were analyzed by x-ray diffraction (XRD) and atomic force microscopy (AFM). The superconducting transition temperatures of the YBCO thin films were measured by dc four-probe point method. Current reversal technique was used to eliminate the effects of voltage offsets. The ac magnetic susceptibility (χ-T) and magnetization (M-H) measurements have been carried out with commercial SQUID magnetometer (Quantum Design Inc., MPMS-5). The samples were attached to the sample holder so that their plane was perpendicular to the applied magnetic field during the measurements. M-H loops were measured at 50 K and critical current densities (J_c) were calculated based on the extended Bean critical state formula [20], $J_c = 20\Delta M/[a(1-a/3b)]$ where ΔM is the width of magnetization loop in emu/cm^3, a and b $(a<b)$ are dimensions of the YBCO samples in cm.

3. Results and Discussion

3.1. R-T Measurement

Temperature versus resistance change of the YBCO thin films which were deposited on the $LaAlO_3$ substrates with the rates of 2.0, 1.6 and 1.2 nm/min are shown at the Fig. 1. The transition temperature of all the samples (T_{c0}) is at around 90 K but the value of the ΔT changes. Transition to superconductive phase occurs at a narrower ΔT with the decreasing deposition rates. Very small positive value of R(0) resistance or the exact zero value of R(0) states that the normal phase of the samples show better metallic properties [17]. The different resistance values of the different deposition rates of YBCO superconductive thin films at normal phases can be seen at Figure 1 clearly. The low value of room temperature resistance and bigger ratio of R(300)/R(100) show that there will be a greater critical transition temperature [21]. Samples with the deposition rates of 2.0, 1.6 and 1.2 nm/min in that order have ratios of R(300)/R(100) as 2.83, 3.09, 3.13. It is clearly seen that the deposition rate of 1.2 nm/min has the highest ratio with the smallest normal state resistance, so that it supplies the highest zero resistance temperature T_c (R=0).

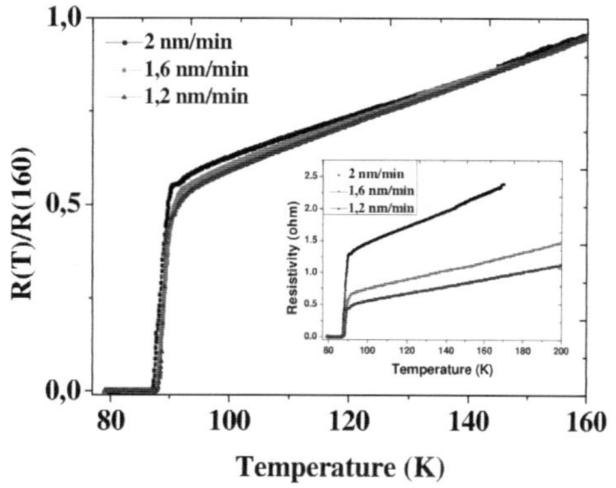

Fig. 1. Resistance vs. temperature curves of the $YBa_2Cu_3O_{7-\delta}$ YBCO thin films by different deposition rates

3.2. Susceptibility Measurement

Susceptibility measurement is widely used for the determination and characterization of the polycrystalline HTSCs. Usually two peaks is observed [22]. In Fig. 2 and 3, responses of reel (χ') and imaginary (χ'') AC magnetic susceptibilities can be seen depending on the different magnetic field (1 Oe and 4 Oe) with frequency values of 80 Hz, 586 Hz and 986 Hz. It is clearly seen from Fig. 2 that as the deposition rate and AC magnetic field amplitudes increase, χ''(T) shifts to lower temperature values and broadens over temperature. It is well known that the amount of the shift as a function of the field amplitude is proportional to the strength of the pinning force. The sharpness of these curves proves the homogeneity of the films whereas the increasing deposition rates show a loss at the susceptibility curves. So, it can be safely said that slower deposition rates result in the best homogeneous film. And also it can be seen at Fig. 2 that increasing magnetic field causes a wider region for the transition and increasing deposition rate results in a more sensitive behavior for the magnetic field. On the other hand, in figure 3, it is clearly seen that samples are not affected from changing frequency that much of magnetic field so that there has been no significant change at the sharpness of the curves. Many researchers have observed that the AC magnetic susceptibility of high T_c superconductors is weakly dependent on the frequency [23-25] and depends strongly on the AC magnetic field amplitude H_{ac} [26-28]. And also, by looking at the figure 3, it's clearly seen that as frequency increases, AC magnetic susceptibility curves slightly shift to the higher temperature.

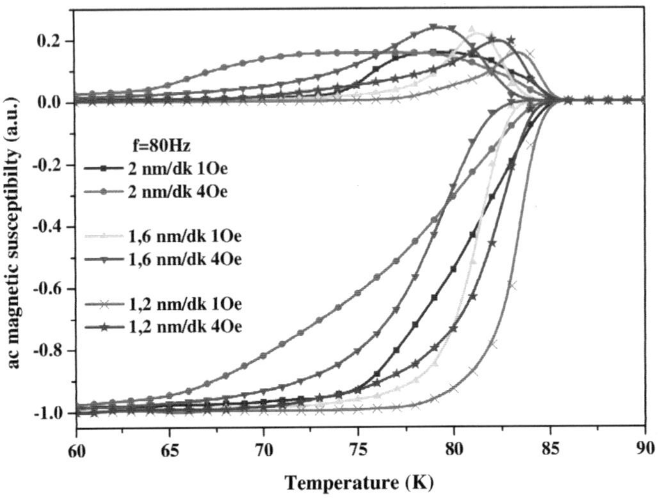

Figure 2. Magnetic susceptibility vs. temperature curves of YBCO thin films on different deposition rate for 1 and 4 Oe magnetic field

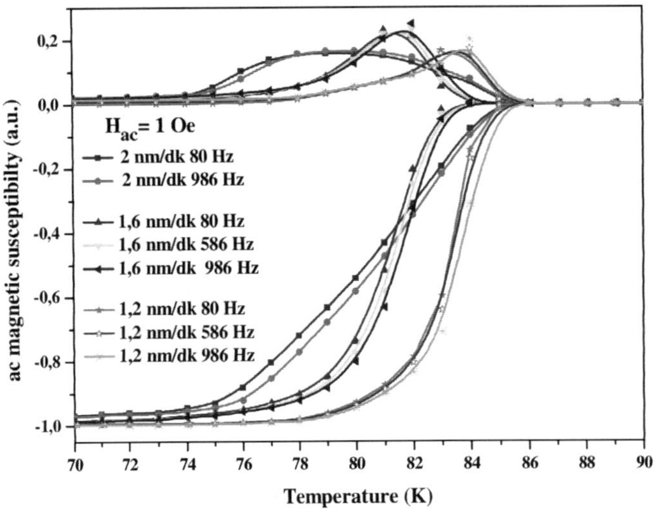

Figure 3. Magnetic susceptibility vs. temperature curves of YBCO thin films on different deposition rate for different frequencies

3.3. Magnetization Measurement

At the Fig. 4, the magnetization curves of the samples at the 50K can be seen. It is clearly seen that the sample with the slowest deposition rate traps more magnetic field. In a relation with these results, calculated critical current density values depending on the Bean formula can be seen at the Fig 5. At the brief sight of the critical current densities at 50K, samples with the deposition rate of 1.2nm/min are significantly different with that of the samples with the deposition rate of 2nm/min.

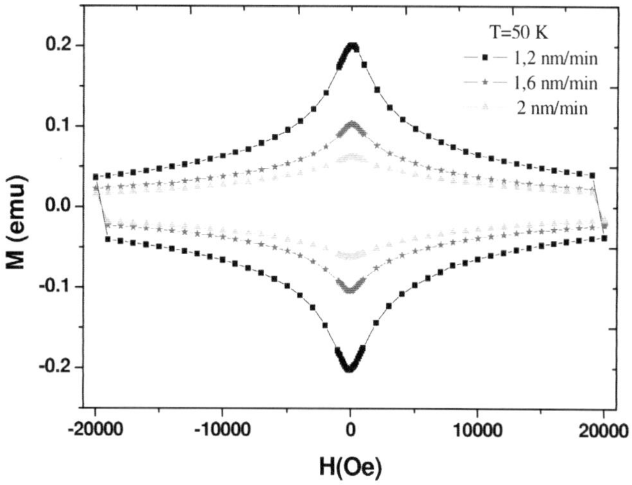

Figure 4. M(H) curves of YBCO thin films on different deposition rates

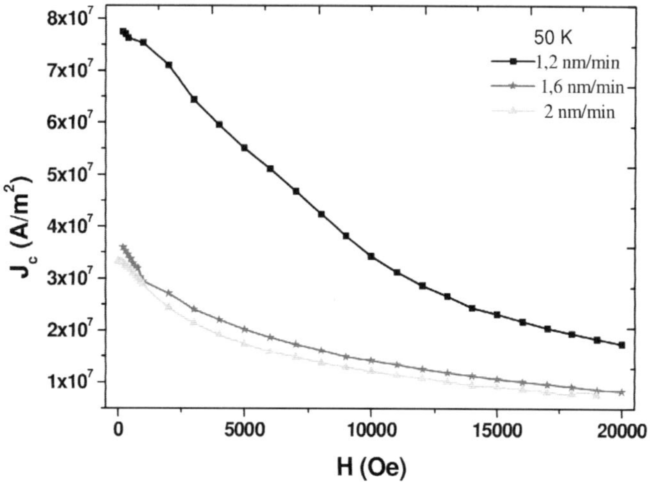

Figure 5. J_c curves of YBCO thin films on different deposition rates

3.4. XRD Measurement

In figure 6, XRD results of the thin films in the range of $2\theta=10^0$-60^0 can be seen. YBCO structure deposited as a thin film can have two different alignments due to its anisotropic formation. In this case, a right angle alignment with the substrate surface is called as c-axis alignment. There are many causes which affect the alignment of the YBCO thin films and the most important is the structure of the substrate used and its alignment. $LaAlO_3$, $SrTiO_3$ and MgO substrates with (100) alignment are mostly used substrates [21]. Also, the quality of the crystallization of thin films strongly depends on the consistency of the substrate lattice. The accordance between the substrate and films in the means of lattice, improve the crystallization of the films [21]. As seen at the figure. 6, produced thin films have c-axis alignment. The sample grown on (100) single crystal $LaAlO_3$ substrate with a relatively low deposition rates of 1.2 nm/min shows a highly aligned *c*-axis orientation. The peak intensities and the amount of *c*-axis orientation are decreasing as the deposition rate increases to 2nm/min Full width at half maximum of the (005) peak show the growth along the c-axis. With smaller FWHM of the films there will be better c-axis alignment according to substrate [16] and this value has a most important role on the superconductive properties of the films [21]. In our samples, the YBCO (005) peaks had FWHM values equal to 0.35°.

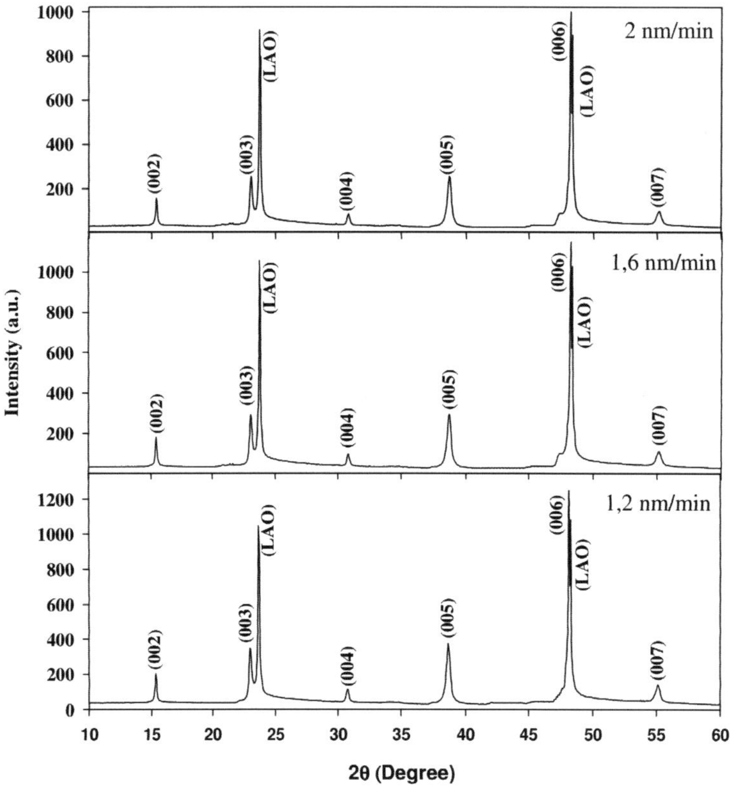

Figure 6. XRD measurement of samples

3.5. AFM Measurement

Our investigation has yielded a correlation between the deposition rates the surface morphology parameters, such as the surface roughness parameter, by employing AFM studies. Higher deposition rate creates more energetic ions so that more massive parts can be plucked from the target. Higher deposition rate and shorter time for the deposition causes non homogeneous distribution of these parts on the substrate. This situation creates rougher surface. Three dimensional images and analysis of the roughness can be seen at Fig. 7. Samples deposited with the rates of 2nm/min, 1.6 nm/min. and 1.2 nm/min. have surface roughness values of 6.071 nm, 5.459 nm and 4.944 nm. In this analyze, in accordance with the previous ones [29], 1.2 nm/min deposition rate yields with the decreased surface roughness.

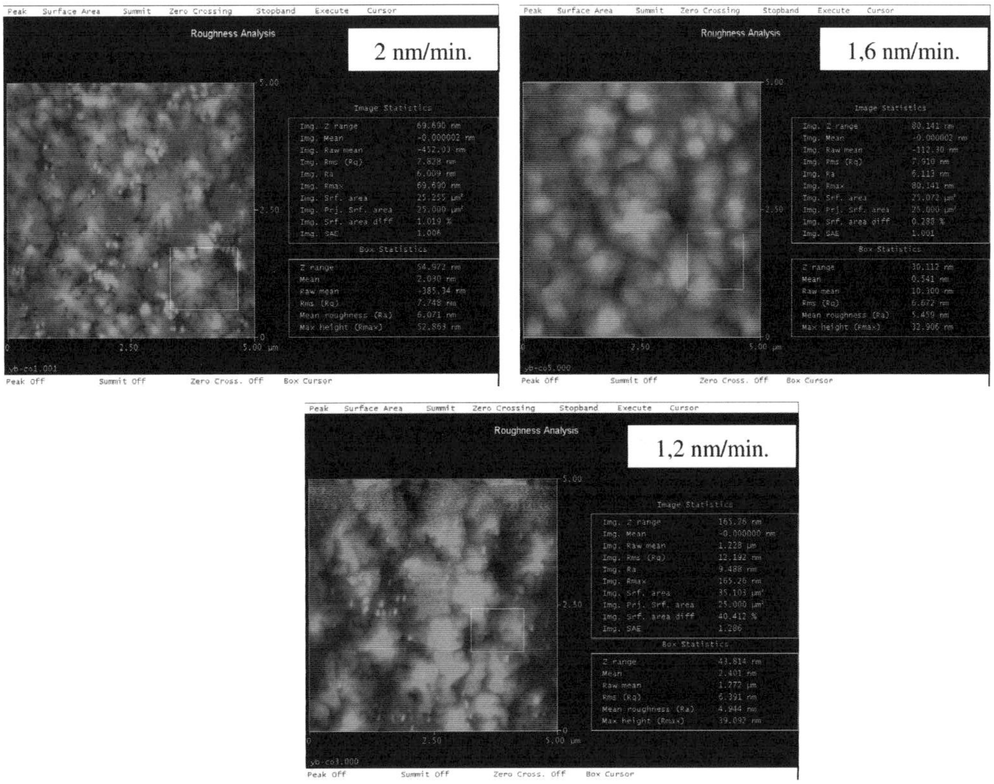

Figure 7. Surface roughness images of YBCO thin films with different deposition rates.

4. Conclusion

In this study, we grew thin films using ICMS technique on $LaAlO_3$ substrates. We have selected the deposition rate parameter, which is very important among the other parameters such as the choice of substrate and its temperature, total oxygen pressure during the deposition and partial pressure of the oxygen. We obtained three different samples. Upon experimenting with the structural and magnetic properties of the samples, we have found that although all the samples showed respectively useful good properties, the sample with low deposition rate showed a smoother surface and higher *c*-axis orientation and also contained more capacity for flux trapping which is accompanied by more resistance to the applied magnetic field.

5. Acknowledgements

This study was supported by the Ege University science foundation under the project number of 2007/FEN/38.

References

[1] Gyorgy E. M., Van Dover R. B., Jackson K. A., Schneemeyer L. F., and Waszczak J. V. 1989 *Appl. Phys. Lett.* **55** 283

[2] Bednorz J.G. and Müller K. A. 1986 *Z. Phys.* B **64** 189

[3] Wördenweber R. 1999 *Supercond. Sci. Technol.* **12** 86

[4] Kashima N., Niwa T., Nagaya S., Onabe K., Saito T., Muroga T., Miyata S., Watanabe T., Yamada Y. 2004 *Physica C* **412-414** 944-947

[5] Develos-Bagarinao K., Yamosaki H., Nakagawa Y., Endo K. 2004 *Physica C* **412-414** 1286-1290

[6] Nakamura T., Inada H., Iiyama M. 1998 *Applied Surface Science* **130-132** 576-581
[7] Koinuma H., Konda N., Nishino J., Ohtomo A., Kubota H., Kawasaki M., Yoshimoto M. 1997 *Applied Surface Science* **109-110** 514-519

[8] De Santis A., Grimaldi G., Gambardella U., Pace S., Cucolo A.M., Cucolo M.C., Boffa V., Celentano G., Fabbri F. 2000 *Physica C* **340** 225-229
[9] Jian Yang J., Shi D., Wang X, Liu A., Yuan G 2000 *Physica C* **341-348** 2499-2500
[10] Kim H. S., Chung K. C., Lee B. S., Lim S. M. and Youm D. 2003 *Supercond. Sci. Technol.* **16** 760-767
[11] Li Y. R., Liu X. Z. and Tao B. W. 2001 *Journal of Superconductivity: Incorporating Novel Magnetism* **14** 3
[12] Geerk J., Linker G. and Meyer O. 1992 *Journal of Superconductivity* **5** 4
[13] Degardin A. F., Castel X., Achani M., Guilloux M., Cariston E., Roelens Y., Carru J-C., Perrin A., Kreisler A. J. 2000 *Physica C* **341-348** 1993-1994
[14] Taniwaki M. and Hirose T. 2002 *Journal of Crystal Growth* **235** 384-388
[15] Selinder T. I., Larsson G., Helmersson U. and Rudner S. 1991 *Supercond. Sci. Technol.* 379-381
[16] Liu X. Z., Li Y. R., Tao B. W., Luo A., He S. M. 2002 *Physica C* **371** 133-138
[17] Avci I., Tepe M., Abukay D. 2004 *Solid State Communications* **130** 357-361
[18] Avci I., Akram R., Bozbey A., Tepe M. and Abukay D 2007 *IEEE Transactions On Applied Superconductivity* **17** 2
[19] Dr. Ing. Keschtkar Gmbh&Co. KG. G. Braun Str. 2-4, 7500 Karlsruhe, Germany
[20] Masoliver A. P. *Critical Currents and Dissipation of Grain Baundary Networks in Coated Conductors* (Barcelona: Department de Materials Magnetics i Superconducts ICMAB-CSIC)
[21] Singh R. K. and Kumar D. 1998 *Mat. Sci. Eng.* **R22** 113
[22] Bahgat A. A., Shaisha E. E., Saber M. M. 2007 Physica B 399 70-76
[23] Nikolo M., Goldfarb R. B. 1989 *Phys. Rev. B* **39** 6615.
[24] Prozorov R., Shaulov A., Wolfus Y., Yeshurun Y. 1995 *Phys. Rev. B* **52** 12541.
[25] Müller K. H. 1990 *Physica C* **168** 585
[26] Chenagues D.X., Rao K.V. 1989 *Cryogenics* **29** 800
[27] Stoppard O., Gugan D. 1995 *Physica C* **241** 375.
[28] Wurlitzer M., Lorenz M., Zimmer K., Esquinazi P. 1992 *Phys.Rev. B* **45** 11816
[29] Braˇnescu M., Vailionis A., Huh J., Moldovan A., Socol G. 2007 *Applied Surface Science* **253** 8179–8183

Response Analysis of Free-Membrane Transition-Edge Detectors with Thin Substrates

Alireza Kokabi[1], Amirhosein Khoshaman[1], Sasan Razmkhah[1], Seyed Mehdi Hoseini[2], Ali Moftakharzadeh[1], Ali Bozbey[3], Mehdi Fardmanesh[1]
[1]Electrical Engineering Department, Sharif University of Technology, Tehran, Iran;
[2]Department of Physics, Sharif University of Technology, Tehran, Iran;
[3]Department of Quantum Engineering, Nagoya University, Furo-cho, Chikusaku, Nagoya 464-8603, Japan

Corresponding Author: Mehdi Fardmanesh: Fardmanesh@sharif.edu

Abstract. In this work, we have investigated and presented the bolometric response of free-standing membrane type of superconductive transition-edge detectors at micro-scale and submicron-scale substrate thicknesses. The technique applied here is based on the finite-size scaling method which is now extensively used for the calculation of thermal parameters of materials at low dimensions when the significance of surface energy contribution is not negligible compared to that of the total bulk energy. This method is used to evaluate the variation of specific heat per unit volume and heat conductivity and hence its effect on the total heat capacitance and conductance of the substrate versus its thickness. By using the thermal parameters obtained from the applied method we calculated the IR response in these types of the devices. This response when inspected at low thicknesses of substrates showed anomalistic behavior with respect to previous response analyses. Discussed in this paper, we also obtained the optimized substrate thickness for maximum responsivity of the device. The obtained results are compared to previous response calculations and experimental measurements.

1. Introduction

Superconductive bolometers have attracted much attention because of their high sensitivity at long wavelengths. Three basic types of these detectors have been introduced: the bolometer on a solid substrate, the bolometer on a thermal filament and the bolometer on a membrane. Measurements indicated that the free-membrane type exhibits more sensitivity than the other two basic types of bolometric detectors [1]. The empirical result compares very well with previous theoretical analyses, which have been derived in [2], where the IR response of bolometric detectors is presented as follow,

$$r_v = \frac{\eta I}{(G + j\omega C)} \tag{1}$$

where G is the total thermal conductance (steady-state and oscillating conductivity), I is DC bias current, η is the fraction of the incident power absorbed by the bolometer and C is the total heat capacitance observed by the film. For the free-membrane type, these two parameters are considered to be less than that of the other types. This leads to a higher value of response for the free-membrane type. Therefore, the theoretically predicted sensitivity for the free-membrane type is more than the thermal filament and solid substrate types as confirmed by experiments.

It can be shown that the application of a thinner substrate would result in a lowering of C and G and therefore more sensitivity. It seems that we can increase the responsivity to any desired value by thinning the substrate. This might be true if the specific heat per volume and thermal conductivity of the substrate remains constant, while its thickness is decreased. But empirical work shows that this assumption is not right and indicates the variation of thermal parameters of substrates with respect to their thickness especially for very thin substrates. It has been shown that by decreasing the substrate thickness the heat conductivity increases in the range of micrometer [3] and decreases in the range of hundreds of nanometers [4] through two different mechanisms. Theoretically this dependency of specific heat and thermal conductivity on the dimensions of a material is described by finite size effects [5]. In the adopted method, we consider the total energy of a substance as a sum of its surface and bulk energies. When the dimensions are large, the surface energy is usually much less than the bulk one and so its effect on the total energy is negligible. However, when the substance is miniaturized, this effect becomes more pronounced, and the term of surface energy becomes dominant in comparison to the bulk energy. Thus, for small size bodies we cannot neglect the surface energy in the calculation of the total energy. It is known that the specific heat is the rate of variation of total energy with respect to temperature. Also, the thermal conductivity per volume in electrically isolated solids is linearly related to specific heat through the sound velocity and phonon mean free path in the substrate [6]. Knowing these relations of thermal parameters to total energy, we conclude that for small dimensions, surface effects are required in the calculation of specific heat per unit volume and thermal conductivity as it is done for total energy. As we will illustrate in the next section, the inclusion of surface effects in the calculation of thermal parameters for a thin substrate caused it to become dependent on substrate thickness which is confirmed by the measurements as we mentioned before. In this work, we obtained the dependencies of specific heat per unit volume and thermal conductivity on substrate thickness and used them to calculate the total heat capacitance and thermal conductance versus substrate thickness. Then these parameters are substituted in equation (1) to obtain the IR response for membrane type of superconductive bolometers. The results are also applied for the optimization of the detectivity of this type of detectors.

2. Theoretical Model

The calculations were performed on a thin layer crystal substrate with a square surface. Thus, the surface energy was as important as the volume energy when the thickness of substrate was brought down. The total energy of the substrate could be considered as the addition of the energies of the volume and the surface,

$$E_{\text{tot}} = E_{\text{V}} + E_{\text{S}} \tag{2}$$

Here, E_{V} and E_{s} are the total volume and surface energies. E_{V} is a function of the substrate volume and E_{S} is a function of its surface area. Dividing both sides of above equation by the volume of the substrate leads to the finite size scaling method where the energies per volume are described as below [5],

$$f_{\text{tot}} = f_{\text{V}} + \frac{2}{d} f_{\text{S}} + \ldots \tag{3}$$

where d is the substrate thickness and f_{V}, f_{S}, f_{tot} are the energies per unit volume as discussed above and the factor of two comes from taking into account the two possible polarizabilities for the two-dimensional method discussed in the following to calculate the $U_{2\text{D}}$.

These energies are calculated by using phonon heat transfer equations and Debye model at low temperatures [6, 7]. For calculating the two dimensional density of states, periodic boundary conditions are assumed over N^2 primitive samples within a square plane of side L, which calls for the quantization of the momentum vector elements on the accompanying plane coordinates, x and y, i.e. they can only take the following values,

$$0, \pm(\frac{2\pi}{L}), .., (\frac{N \times 2\pi}{L}) \tag{4}$$

Consequently, there is only one value of **K** per unit area $(2\pi/L)^2$ on k-plane and $(2\pi/L)^2$ allowed values of **K** per unit area of k-plane for each polarization and each branch. The area of the specimen, S is L^2. Hence the total number of modes with a wavevector with a magnitude is less than K for each polarization in a circular area of radius K on this plane is $N=SK^2/4\pi$.
By direct differentiation with respect to frequency, $D(\omega)$ the density of states for each polarization is:

$$D(\omega) = \frac{S}{2\pi} K \frac{dK}{d\omega} \tag{5}$$

which, by using Debye approximation, is reduced to be $S\omega/2\pi v$, where v is the speed of sound in the material and is assumed constant.

By neglecting the contributions of phonon modes with frequencies higher than the Debye frequency, ω_D which is the maximum frequency that makes up the crystal, the two-dimensional internal energy, U_{2D} is found by integration of $D(\omega)$ over ω-space to the boundary value ω_D by considering the fact that a set of identical harmonic oscillators in thermal equilibrium take a Plank distribution. The result divided by the area of the substrate turns out to be,

$$U_{2D} = \frac{\hbar}{d\pi v^2} \int_0^{\omega_D} \frac{\omega^2 d\omega}{\exp(\frac{\hbar\omega}{k_B T})} \tag{6}$$

where the two-dimensional specific heat per unit area is found by differentiation of the U_{2D} with respect to temperature, at constant surface area, S.
To evaluate U_{3D}, a three-dimensional cubic space with edges of length L is assumed, the same strategy approached above leads to the quantization of K_x, K_y and K_z to only the following values,

$$0, \pm(\frac{2\pi}{L}), ..., (\frac{N \times 2\pi}{L}) \tag{7}$$

Hence there is only one value of **K** per unit volume $(2\pi/L)^3$ in k-space and $(2\pi/L)^3$ allowed values of **K** per unit volume of k-space for each polarization and branch. The volume of the specimen, V is L^3. Hence the total number of modes corresponding to a wavevector with magnitude less than K for each polarization in a spherical volume of radius K is,

$$N = \left(\frac{L}{2\pi}\right)^3 \times (\frac{4\pi K^3}{3}) \tag{8}$$

Therefore, the three-dimensional density of states after utilizing Debye's approximation turns out to be,

$$D(\omega) = \frac{V\omega^3}{2\pi^2 v^3} \tag{9}$$

Neglecting the contributions of the modes with frequencies higher than Debye frequency ω_D and using the Plank distribution function, and for the sake of brevity assuming the phonon velocity independent of the polarization, the result is multiplied by a factor of three to yield the three-dimensional internal energy as,

$$U_{3D} = \frac{3\hbar}{4\pi^2 v^3} \int_0^{\omega_D} \frac{\omega^3 d\omega}{\exp(\frac{\hbar\omega}{k_B T}) - 1} \tag{10}$$

Now the specific heat per unit volume and area can be found in two and three dimensions as follow,

$$c_{V2D} = \left(\frac{\partial U_{2D}}{\partial T}\right)_S \tag{11}$$

$$c_{V3\mathrm{D}} = \left(\frac{\partial U_{3\mathrm{D}}}{\partial T} \right)_V \tag{12}$$

The total heat capacitance could be assumed as the sum of the two latter heat capacitances,

$$c_V = c_{V3D} + \frac{2}{d} c_{V2\mathrm{D}} \tag{13}$$

By using 1D model, diffusion length is calculated as below,

$$L_f = \left(\frac{D}{\pi f} \right) \tag{14}$$

where f is the modulation frequency and $D=k/cV$ is the thermal diffusivity of this substrate and k is thermal conductivity of material. The thermal conductance, G and heat capacitance, C of the substrate are:

$$G_\mathrm{s} = \frac{k_\mathrm{s} S}{L_f} \tag{15}$$

$$C_\mathrm{s} = c_{V\mathrm{s}} S L_f \tag{16}$$

where S is the substrate area. Substrate thickness in membrane type bolometers is typically less than the thermal diffusion length and therefore for calculation of G and C, thermal diffusion length is replaced by substrate thickness. We also know from the 1D model that $G=G_t+G(0)$ where G_t is the thermal conductance at modulation frequency and $G(0)$ is the steady-state thermal conductance due to lateral thermal diffusion. G_t is the series of two thermal conductances, first the substrate conductance and second the boundary conductance between the back of the active area of the substrate and the cold-head. So the total thermal conductance at modulation frequency is zero because in the membrane type the substrate is floating behind it there is no thermal contact. The only thermal conductance in this type of bolometer is the steady-state conductance, $G(0)$ which corresponds to lateral heat flow through the substrate and it is also taken to be constant because it is mainly determined by the thermal boundary resistance of substrate to cold-head (we consider that the lateral length of the substrate is more than thermal diffusion length in a way that lateral heat flow at the modulation frequency can be neglected). By these assumptions responsivity is calculated using equation (1)

3. Results and Analyses

Here we present the results of our calculation for substrate with cubic active area. In figure 1 the specific heat per unit volume versus the thickness of the substrate is plotted. This figure shows that the specific heat per unit volume of the substrate at large thicknesses is approximately constant. This is in line with our theoretical analysis because as mentioned previously, when the thickness is large we expect that the surface has little effect and so the dependency on thickness is thought to be negligible. Below 100μm, specific heat varies with respect to the substrate thickness. This behaviour of specific heat is also expected because at this range of thickness, the surface energy is comparable to the bulk energy. Therefore, decreasing one dimension and so the total volume of the substrate makes the surface energy more significant with respect to bulk energy. In [8] similar results were obtained for thin substrates.

The heat capacitance versus substrate thickness is depicted in figure 3. As illustrated, heat capacitance decays until a specific thickness, called "new thickness" in this work. Below the new thickness, heat capacitance is moderately constant. The new thickness is strongly related to substrate material.

Variation of time constant with respect to the substrate thickness is presented in figure 4. Also in this figure, we observed two different behaviours below and above the new thickness. This behaviour is consistent with a previously reported measurement on the time constant of NTC thermal sensors, which has analogous principle of operation and similar thermal model to that of transition-edge sensors [11].

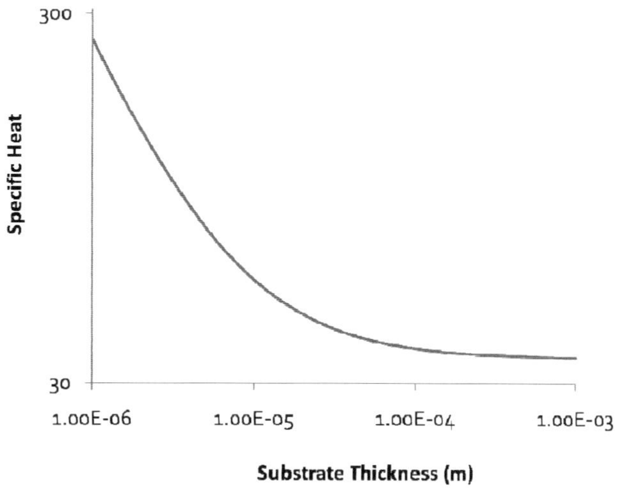

Figure 1. Heat capacity versus thickness.

In figure 2 the frequency-dependent response of the bolometer versus the substrate thickness is drawn at an excitation frequency of 4Hz. For this calculation, thermal boundary resistance at the substrate-cold-head of the samples, RSC is taken to be $6.82KW^{-1}cm^{2}$ which is found experimentally for MgO crystal lattice connected to gold cold-head [2]. Using this value for RSC and neglecting thermal resistance of substrate, which is usually one order of magnitude below RSC, total thermal conductance, G calculated to be approximately $14.6Wmm^{2}K^{-1}$.

It is noteworthy to see that as the thickness of the substrate diminishes the responsivity is enhanced until it reaches a fairly constant value at thicknesses of about 10μm. Thus thinning the substrate is desirable until an particular thickness of substrate and after that, the responsivity would not improve.

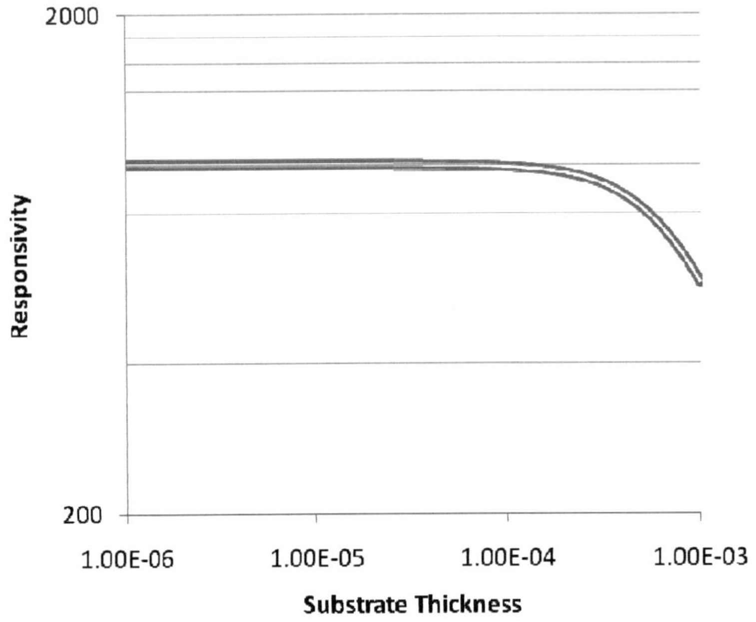

Figure 2. Frequency response of transition edge superconducting edge versus substrate thickness.

The three-dimensional plot of the frequency dependent response versus substrate thickness and excitation frequency of the bolometer is drawn in figure 3. Similar to figure 2, in this figure it is observed that responsivity saturates below a particular substrate thickness. As it is illustrated in this figure, this impact of the substrate thickness which was also predicted by the finite size scaling method is more tangible as the frequency is decreased. It should also be noted that in addition to saturating responsivity at small value of substrate thickness, it also become constant for thicker substrates at near-to-DC modulation frequencies. This is attributed to slight effect of the heat capacitance at low frequencies which is the only thickness dependent parameter in the calculation. Saturation of response at low frequencies for different substrate thicknesses is previously reported [8].

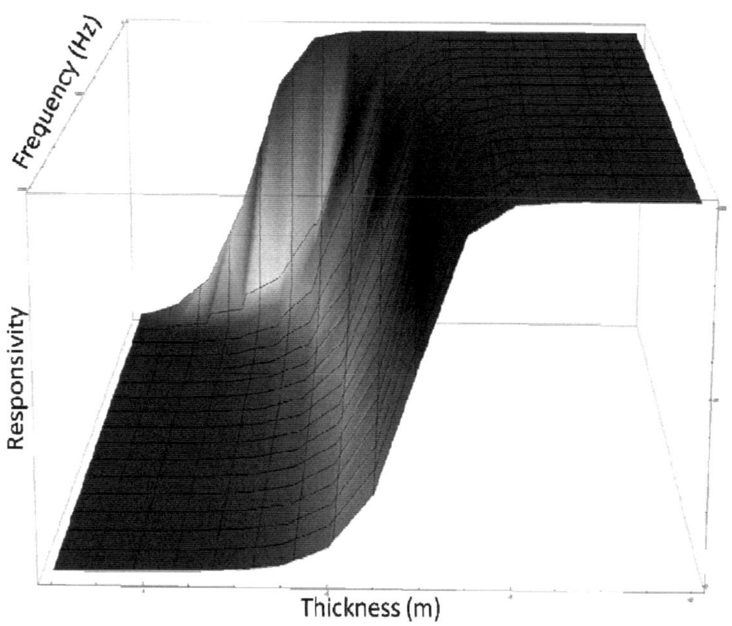

Figure 3. The three-dimensional plot of response versus frequency and substrate thickness.

4. Conclusions

The magnitude of response in superconductive edge transition bolometers is contingent on thickness of the substrate material. As thickness is lowered the response shows an unexpected behaviour which relates to its surface energy. By using finite size scaling method and calculating surface energy the response behaviour can be explained as a function of substrate thickness.

Acknowledgments

This work was supported in part by the National Elite Foundation, Tehran, Iran.

References

[1] Khrebtov I A and Tkachenko A D 1999 *Opticheskii Zhurnal* **66** 79
[2] Fardmanesh M 2001 *Appl. Opt.* 40 1080
[3] Cahill D G, Katiyar M and Abelson J R 1994 *Phys. Rev. B* **50** 6077
[4] Lee S M and Cahill D G 1997 Heat transport in thin dielectric films *J. Appl. Phys.* **81** 2590
[5] Binder K 1976 *in Phase Transitions and Critical Phenomena*, edited by Domb C and Green M S vol 5B (Academic Press)
[6] Kittel C 1976 Introduction to Solid State Physics (Wiley, New York)
[7] Huang K 1987 Statistical Mechanics (John Wiley & Sons) 2nd Edition
[8] Palasantzas G 2002 *Solid State Communications* **122** 523

[9] Lakew B, Brasunas J C, Pique´ A, Fettig R, Mott B, Babu S and Cushman G M 2000 *Physica C* **329** 69
[10] Lakewa B, Brasunas J C, Aslam S and Pugel D E 2004 *Sensors and Actuators A* **114** 36
[11] Lee M and Yoo M 2002 *Sensors and Actuators A* **96** 97

DC V-I Characteristics of a HTS tape for a 600kJ SMES in an Oblique External Magnetic Field

H. J. Kim[1], S. K. Baik[1], K. C. Seong[1], Zhuyong Li[2], YongHu Ma[2] and K. Ryu[2]

[1]Korea Electrotechnology Research Institute, Changwon, 641-120 Korea
[2]Chonnam National University, Gwangju, 500-757 Korea

E-mail: hjkim@keri.re.kr

Abstract. The critical current and n-value of the HTS tape, which are important for the characteristic analysis of the 600kJ HTS SMES system, were studied, and based on the results of the study, the DC V-I characteristics of a model coil fabricated by the same HTS tape of a 500 m length conductor were analyzed. The result show that the measured critical current and n-value of the conductor for parallel field are constant in external magnetic fields less than about 0.2, whereas the characteristics decreased for the whole range of the oblique magnetic field. The critical current of the coil, which was calculated based on the results of the study on a short sample conductor, corresponded to the obtained value, which means that the critical current and n-value of a conductor for an oblique magnetic field in an arbitrary direction in the HTS SMES system can be estimated only by using the DC V-I characteristics of a short sample conductor for parallel and perpendicular magnetic fields.

1. Introduction

Since the HTS (high temperature superconducting) conductor was discovered, which has high stability for system applications with an about 100 K critical temperature; its commercialization has not been realized because it is very expensive in comparison to the NbTi wire. The development of the HTS SMES system with more than hundreds of kJ, however, is under way technologically, satisfying the level of about 100 A/mm^2 [1], [2] which is the conventional operational current density of the SMES system fabricated with an NbTi wire, thanks to the development of the HTS tape technology. Therefore, the 600kJ HTS SMES system was fabricated for characteristic evaluation in this study, and the system is still being developed.

The critical current and n-value of the HTS tape for an oblique magnetic field in an arbitrary direction are very important for the analysis of the characteristics of a 600 kJ HTS SMES system, such as the critical current, flux flow resistance, and AC loss from the charge and discharge, but they have not been studied sufficiently for various reasons, including the high anisotropy of the HTS tape [3]. Therefore, a short sample was fabricated using a high current conductor that was selected for the 600kJ HTS SMES system, and the DC V-I characteristics of the HTS tape, which are important for the analysis of the SMES system characteristics, were examined. Based on the measurement results, the DC V-I characteristics of a model coil fabrication of the same HTS tape of a 500m long were analyzed and examined.

2. Experimental

Fig. 1 shows a diagram of the HTS tape and the magnetic field around it in a short sample conductor. The external magnetic field applied to the conductor surface at an angle of θ consists of the components perpendicular to the conductor ($B_{\perp}$) and parallel to the conductor ($B_{//}$).

Table 1 shows the main specifications of the conductor that was chosen for the HTS SMES system, and the parameters that were measured for the coil that was fabricated in this study.

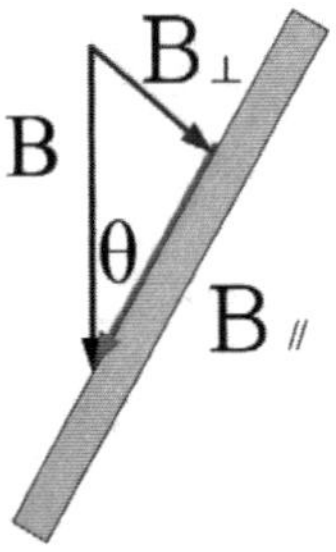

Figure 1. Diagram of a magnetic field applied to the conductor

Liquid neon is supplied from outside of the rotor through a long stainless steel pipe connected to a cold head of a GM cryocooler which liquefies gas neon evaporated from inside of the rotor. This 2 phase closed loop cooling mechanism is called thermosiphon that has been applied to several HTS motors developed worldwide [1] [2].

Table 1. Main Specifications of the Conductor and Model Coil

Conductor	Width	4.57 mm
	Thickness	0.77 mm
	Bi-2223/Ag/Solder/Brass	28/42/5/25
	Critical current @ 77 K	280 A
Model Coil	Inner diameter	500 mm
	Outer diameter	698 mm
	Height	15.4 mm
	Total number of turns	262 turns
	Inductance	69 mH
	Critical current @ 77 K	123 A
	Stored energy	522 J

A short sample conductor with voltage taps with an 80 mm distance between them was fixed on an FRP(Fiberglass Reinforced Plastics) plate to examine the sample's DC V-I characteristics, which are important for the HTS tape applications, under an oblique magnetic field, as in Fig. 1. The FRP plate was securely fixed to avoid mechanical deformation due to the magnetic force generated between the high critical currents (280A) on the selected HTS tape, as in Table 1, and the external magnetic field to minimize the angle error of the external magnetic field during the experiment. To prove the results of the experiment that was conducted on the sample conductor, a model coil was fabricated using the same material and an about 500 m HTS tape to examine the sample's DC V-I characteristics experimentally.

The critical current and n-value are the main parameters among the DC V-I characteristics of the HTS tape. n-value is expressed as Equation [4].

$$\frac{E}{E_c} = \left(\frac{J}{J_c}\right)^n \tag{1}$$

where E and E_c are the measured electric field for the conductor and 1 μV/cm, respectively; J and J_c are the measured current density and critical current density, respectively; and N is the index of the DC V-I curve, which is called n-value.

To measure the DC V-I curve for the sample conductor under an external magnetic field, a core-type background magnet with a magnet constant of 25 mT/A and an opening of 100 mm (length) × 45 mm (width) × 8 mm (height), and 30A/50V power, was used for magnetization. 600A/8V power was used for the DC current on the sample conductor, and the voltage of voltage taps attached to the sample conductor was measured with a nano-voltmeter.

In addition, 1000A/50V power supply was used for the DC V-I characteristic evaluation of the model coil in Table 1, and the evaluation of the sample conductor and model coil was performed at the liquid nitrogen (77 K).

3. Experiment Results and Discussions

Fig. 2 shows the typical DC V-I characteristics from the measurement where oblique magnetic fields of various magnitudes were applied to the HTS tape at θ = 60, as in Fig. 1. The voltage in Fig. 2 is that between the voltage taps attached to the HTS tape. The distance between them is 80 mm.

As shown in Fig. 2, the voltage signal measured at around 8 μV (1 μV/cm), which is conventionally used for the definition of the HTS critical current, was relatively good, but the noise increased as it approached 0.8 μV, or 10% of such voltage.

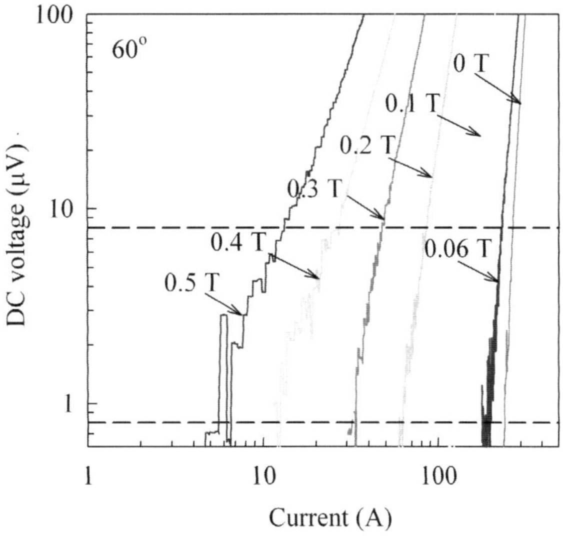

Figure 2. DC V-I characteristics of the HTS tape under a 60 degree oblique magnetic field

Fig. 3 shows the critical current of the HTS tape and the dependence of the n-value on the external magnetic field in Eq. 1, determined by the DC V-I data measured from the application of the external magnetic field at θ = 0 (parallel magnetic field) and θ = 90 (perpendicular magnetic field) to the HTS tape. In Fig. 3, the symbols and full lines represent the experiment results and the results calculated by the interpolated functions, respectively. The interpolated function (f) for the critical current of the conductor and the n-value is expressed as an exponential function, as in Equation (2).

$$f(B) = K + \alpha \exp^{-\beta B} + \gamma \exp^{-\delta B} + \varepsilon \exp^{-\eta B} \quad (2)$$

Where f is the interpolated function for the critical current or n-value; B is the parallel or perpendicular magnetic field; and κ, α, β, γ, **δ**, ε, and η are the coefficients, which are presented in Table 2.

Table 2. COEFFICIENTS OF THE INTERPOLATED FUNCTION

		Coefficients						
f		κ	α	β	γ	δ	ε	η
θ =0°	Ic	0	107	1.01	113	1.01	113	1.01
	n	7.40	6.30	5.02	4.55	5.01	3.83	5.02
θ =90°	Ic	-15.7	106	5.54	104	5.55	102	5.55
	n	1.93	4.35	6.69	4.27	6.69	4.17	6.66

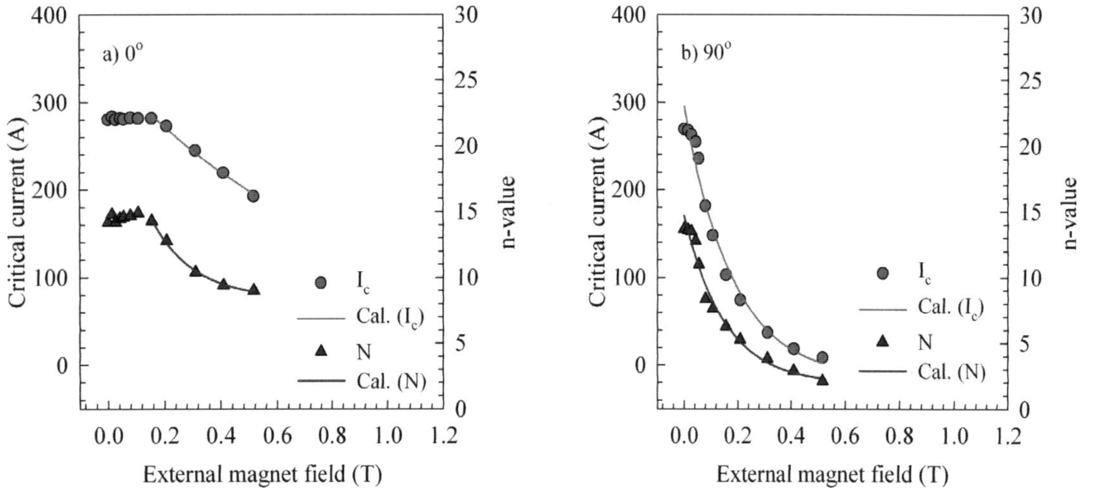

(a) 0 degree external magnetic field (b) 90 degree external magnetic field

Figure 3. Critical current of the HTS tape and dependence of the n-value on the magnetic field under a 0/90 degree external magnetic field

The critical current and n-value that were measured for the magnetic field parallel to the surface of the HTS tape at θ = 0 showed stable characteristics regardless of the up to about 0.2T external magnetic fields, and then decreased as the magnetic field increased, as shown in Fig. 3. Meanwhile, in the case of the magnetic field perpendicular to the HTS tape surface at θ = 90, the critical current and n-value decreased rapidly for the whole external magnetic field as the magnetic field increased. In the region where the critical current and n-value decreased according to the magnetic field, the experiment results corresponded well to the results obtained using Eq. 2 and the coefficients in Table 2.

Fig. 4 shows the critical current and dependence of the n-value on the external magnetic field measured from the application of the oblique magnetic field at θ = 10, 30, and 60 to the HTS tape.

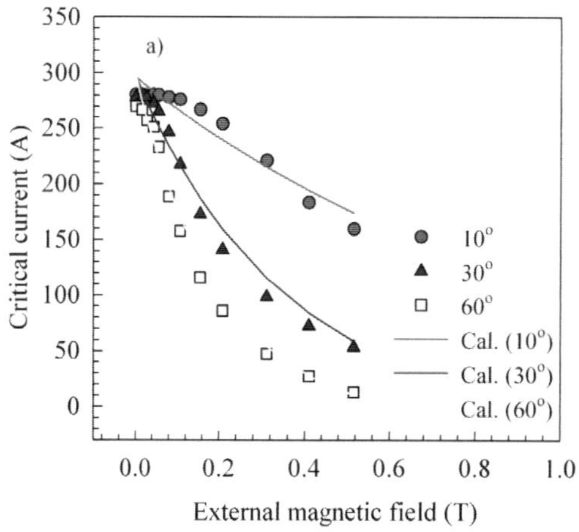

(a) Critical current

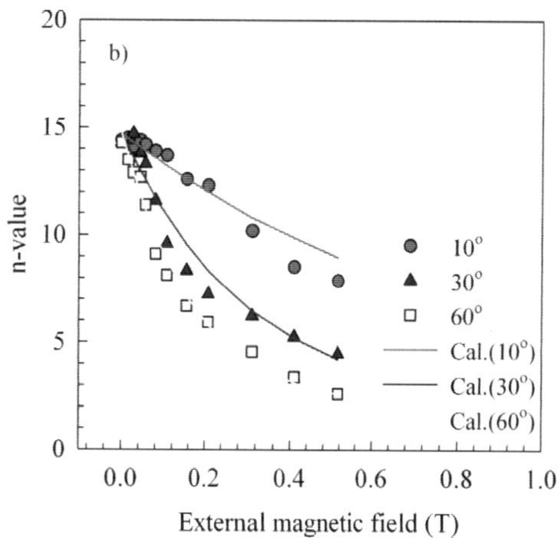

(b) n-value

Figure 4. Critical current of the HTS tape and dependence of the n-value on the magnetic field under a 10/30/60 degree oblique magnetic field

In Fig. 4, the symbols represent the experiment results, and the full lines represent the results calculated by the substitution of the perpendicular component into Equation (2) after analyzing the oblique magnetic field that was applied to the conductor, as in Fig. 1, into the components either parallel or perpendicular to the conductor's surface. The coefficients for Equation (2) were taken from the case of the perpendicular magnetic field at $\theta = 90$ in Table 2.

The critical current and n-value measured in the oblique magnetic field in Fig. 4 tend to decrease as the magnetic field become larger throughout the external magnetic field area as in the case of Fig. 3 (b). Those values decreased more rapidly when the angle between the HTS tape's surface and the external magnetic field increased. The critical current and n-value, which were calculated by considering only the perpendicular magnetic field for each oblique magnetic field (10°, 30°, and 60°), corresponded well to the experiment results. This indicates that the critical current and n-value for the oblique magnetic field in an arbitrary direction can be estimated using the experiment results from the parallel and perpendicular magnetic fields and Equation (2), if the angle between the HTS tape's surface and the external magnetic surface, and the magnitude of the external magnetic field, are known.

To prove the results of the experiment that was conducted on the 80 mm sample conductor, a pancake model coil was fabricated using the same material and an about-500-m-long HTS tape to examine its DC V-I characteristics, and the results are shown in Fig. 5. The voltage taps attached to the HTS tape were used to measure the voltage. The critical current for the fabricated coil is 123 A, which can be found in Fig. 5.

The magnetic field in the coil must be calculated before the calculation of the critical current in Fig. 5. The magnetic-field application program COMSOL was used for such calculation. Fig. 6 shows the axially symmetrical cross-section of the coil that was used in the numerical analysis and the calculation results, respectively. In the critical-current-density calculation in Fig. 6(b), the critical current density of the HTS tape in the coil for θ of 10 or more was calculated by substituting the perpendicular component of the magnetic field calculated at each point and the coefficients for $\theta = 90$ in Table 2 into Eq. 2, based on the results from the short-sample conductor in Fig. 4. Meanwhile, when θ was less than 10, the parallel component of the magnetic field calculated at each point, and the coefficients for $\theta = 0$, were substituted into Fig. 2 to calculate the critical current density of the HTS tape in the coil.

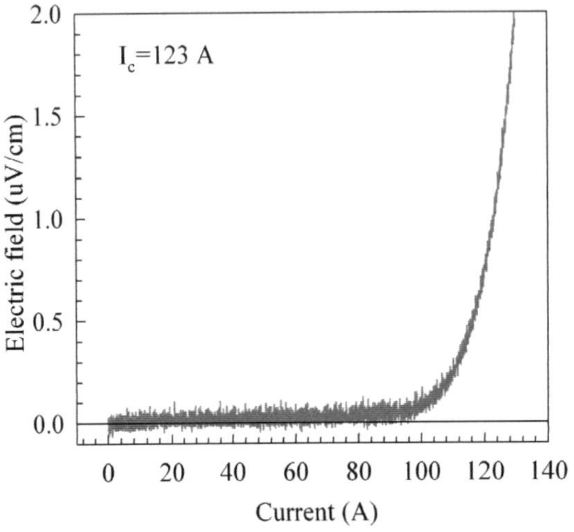

Figure 5. DC V-I characteristics of the model coil

From the critical current density distribution in the model coil winding in Fig. 6(b), the critical current density that was calculated at around the mid-plane at half the height of the winding was much higher than the coil current density (35.0 A/mm^2), but the farther it was from the plane, the lower the value was. The minimum critical current density at around the mid-plane of the upper and lower surface was 30.5 A/mm^2, which was about 90% of the coil current density of 35.0 A/mm^2, as shown in Fig. 5. Therefore, the two values corresponded to each other relatively well.

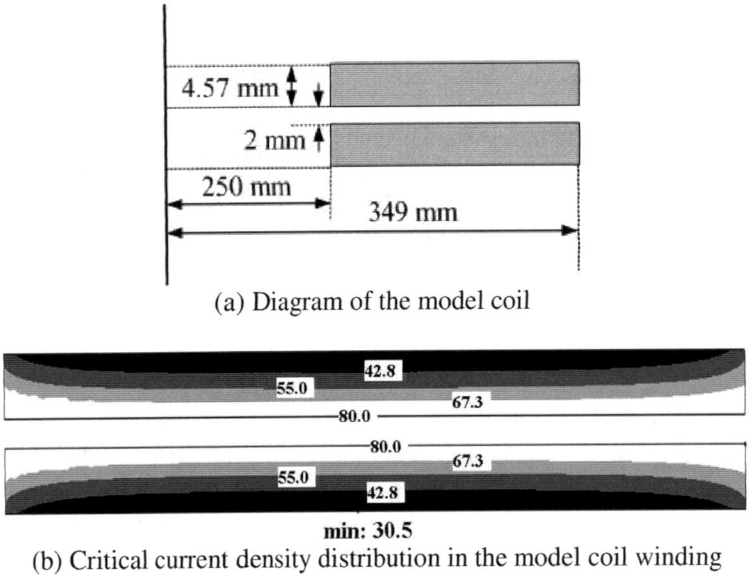

Figure 6. Diagram of the coil and the numerically analyzed result

4. Conclusion

A high current sample conductor selected for the fabricate of a 600kJ HTS SME system and a model coil was fabricated to examine the sample's DC V-I characteristics, and the experiment results were applied to the model coil.

The result show that the measured critical current and n-value of the conductor for parallel field are constant in external magnetic fields less than about 0.2, whereas the characteristics decreased for the whole range of the oblique magnetic field. The critical current of the coil, which was calculated based on the results of the study on the short conductor, corresponded to the measured value, which means that the critical current and n-value of a conductor for an oblique magnetic field in an arbitrary direction in the HTS SMES system can be estimated only by using the DC V-I characteristics of a short-sample conductor for parallel and perpendicular magnetic fields.

5. References

[1] Ryu K, Kim H J, Seong K C, Cho J W, Kim S W, Kwon Y K, and Ryu K S 2002 *IEEE Trans. Appl. Supercond.* **12** 796

[2] Kim H J, K. C. Seong, J. W. Cho, S. W. Kim, Y. K. Kwon, and K. Ryu 2003 *IEEE Trans. Appl. Supercond.* **13** 1863

[3] M. A. Young, J. A. Demko, M. J. Gouge, M. O. Pace, J. W. Lue, and R. Grabovickic 2003 *IEEE Trans. Appl. Supercond.* **13** 2964

[4] Iwasa Y, *Case Studies in Superconducting Magnets*. New York: Plenum Press, pp. 306-308, 1994.

Journal of Physics: Conference Series **153** (2009) 012019 doi:10.1088/1742-6596/153/1/012019

Critical current anisotropy of MgB_2 tapes

P Kováč[1]
Institute of Electrical Engineering of Slovak Academy of Sciences, 84104 Bratislava, Slovakia

E-mail: Pavol.Kovac@savba.sk

Abstract. Flattened MgB_2 superconductors have an advantage of increased filament density and consequently high transport current density. On the other side, flat rolling is introducing the texture into MgB_2, which leads to critical current anisotropy. Therefore, MgB_2 tapes prepared from different precursor powders (ex-situ, in-situ and mechanical alloying) deformed in several metallic sheaths were measured in the external magnetic field of variable orientations. Critical current anisotropy of these composite tapes are compared and discussed.

1. Introduction

It was shown that flattening of low temperature superconductors (e.g. NbTi and Nb_3Sn) introduces the pinning anisotropy and consequently different I_c values are measured for parallel and perpendicular external field with respect to the flat tape surface [1-2]. In the case of MgB_2 manufactured by powder-in-tube (PIT) technique, the filament (powder) density is strongly influenced by the applied deformation. Could rolling produces higher core density in comparison to drawing, which allows reaching considerably higher transport current densities. Therefore, flat rolling is preferred deformation for high current MgB_2 [3-4]. On the other side, flat rolling leads to critical current anisotropy: higher I_c's are measured in the parallel orientation of external field with the tape surface. I_c anisotropy of MgB_2 tapes made by *ex-situ* process was presented by several authors [5-7]. Kumakura et al. have published critical current anisotropy for Cu-Ni and stainless steel sheathed tapes and explained it by some c-axis grain alignment of the MgB_2 layer [5]. Lezza et al. have shown a strong I_c anisotropy dependence on the grain size [6]. X-ray analysis of the degree of texturing indicated that samples are weakly textured, which correlates with the relative hardness of MgB_2 and the used metallic sheath. Flattened four-core *ex-situ* MgB_2-W/Fe wire with small aspect ratio b/a = 2.19 has shown also anisotropy above $\mu_0 H = 3$ T [7]. I_c anisotropy of *in-situ* MgB_2 tapes [8-9] and the tapes made of mechanically alloyed powders were also measured [10-11]. Recently, the anisotropic property of highly textured thin MgB_2 layers formed by molecular beam epitaxy was examined by Yamamoto et al. [12]. Surprisingly, the measured I_c anisotropy of well-textured MgB_2 thin layers is smaller than those observed for MgB_2 tapes made by PIT. The reason of high I_c anisotropy in PIT MgB_2 tapes is not yet fully understood. The aim of the work reported here is to show the scale of I_c anisotropy for differently made MgB_2 tapes (ex-situ, in-situ, mechanical alloying) using variable metallic sheaths, additions and also at increased conductor temperatures (≈ 20K) interesting for future applications.

[1] Pavol Kováč, Pavol.Kovac@savba.sk

2. Experimental

Single- and multi-core flat composite superconductors containing MgB_2 filaments of variable aspect ratio *b/a* (width/thickness) and size have been made by powder-in-tube technique (PIT) [7-11]. Different precursor powders (in-situ, mechanical alloying) and metallic sheath materials (Fe, Nb, Ta, Ti) were used [8,10]. Mechanical alloying (MA) is a special variant of the in-situ technique [3]. During the high energy milling a partial reaction to MgB_2 ($\approx$ 30%) takes place. Particle size of MA powder is much smaller (5 – 20 nm) than for typical ex-situ (0.1-100 μm) and/or in-situ (1-20 μm) rotes. The aspect ratio *b/a* of rolled tapes is ranging from 3.5 to 11, which influences the degree of grain alignment. MgB_2 tapes were measured at different external field orientation in the split-coil 9T magnet at 4.2K or in the superconducting solenoid with variable temperature insert up to 10T and temperatures 4.2 –22 K. Criterion of 1μV/cm was used for the I_c estimation. The angular dependences $I_c(\alpha)$ and anisotropy factor $k_a = I_{c\text{-par}} / I_{c\text{-perp}}$ (the ratio of critical currents in parallel and perpendicular field direction) as a function of external field were plotted from the measured data. The $\alpha = 90^o$ corresponds to parallel field orientation to the tape surface and $\alpha = 0^o$ to perpendicular one.

3. Results and discussions

3.1. Angular dependences

All MgB_2/metal PIT tapes have a typical angular dependence of critical current with the I_c maximum at parallel external field ($\alpha = 90^o$) and I_c minimum at perpendicular one ($\alpha = 0^o$), see the curve of circles in figure 1. Although only moderate texture is observed by X-ray for PIT tapes (*c*-axes with crystallite size $\approx$ 25.0 nm align perpendicularly to the core/metal interface) [10], the comparable ratio between the I_c maximum to I_c minimum is measured for well textured layer and PIT tape at external field 5 – 5.5 T, see figure 1. The I_c maximum for the external field oriented perpendicularly to the layer surface (parallel to as-grown columnar grains) was measured due to the strongest pinning when the field is aligned paralel to grain boundaries [12]. High texture leads to a narrow peak of critical current for thin MgB_2 layer in comparison to a wide one measured for much less textured MgB_2 tape.

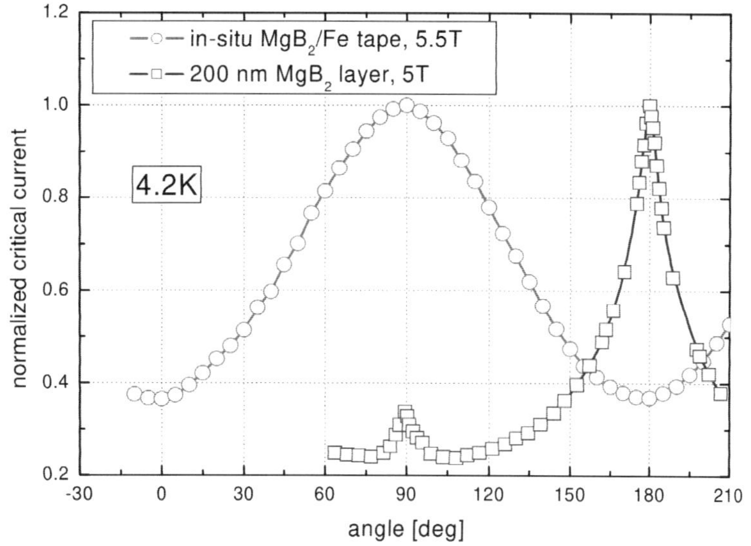

Figure 1. Angular dependences of critical current measured for PIT MgB_2 tape (circles) [8] and thin MgB_2 layer prepared by MBE (squares) [12].

Thin MgB_2 layer shows also second (much smaller) peak at 90°, which is explained by the surface pinning [12]. It is interesting but not yet understood why so differently textured MgB_2 materials show comparable ratio of critical currents in two basic external field directions.

3.2. Anisotropy factor

The anisotropy factor k_a for MgB_2 tapes made by ex-situ, in-situ and MA is plotted in figure 2. The shape of $k_a(\mu_0 H)$ characteristics is similar for all tapes, but the absolute values of k_a are different. The highest k_a is measured for the tapes with MA powder and the smallest for ex-situ one having the powder particle size of 10 μm [6]. Though the resistive measurements of the MA filaments in different field directions show an intrinsic anisotropy value of about 1.5 [13], the transport current anisotropy is much higher and exponentially increasing with field. It can be only partially explained by the texture introduced by rolling. Two k_a dependences in figure 2 plotted for MA/Ti tapes (circles) demonstrate clearly the strong effect of rolling. While $k_a \approx 10$ is measured for the tape deformed by two-axial rolling (TAR, filled circles) at 8.5 T, apparently increased anisotropy (k_a = 70) is observed for the same tape composition and aspect ratio subjected to flat rolling deformation. In the case of MA powder, flat rolling is introducing the texture for already existing MgB_2 grains and also for not reacted Mg particles, both having the hexagonal cell structure. The effect of heat treatment (HT) temperature was also examined. Figure 3 shows not changed $k_a(\mu_0 H)$ characteristics for three identical MA/Fe tapes annealed at different temperatures below and above the melting point of magnesium. The same result was obtained for MA/Ti tapes annealed in the temperature scale 650-800°C. It confirms that the texture of MgB_2 core is controlled only by deformation. But, as already mentioned this texture is really moderate and cannot be alone as a reason of so high k_a values. Another pinning anisotropy influenced by texture (e.g. oriented defects – dislocations) should contribute to the total I_c anisotropy, which has to be verified by a deep structural analysis by TEM. It was already shown that k_a is influenced also by doping (SiC, C) [8, 14], which reduces the grain size and increases the upper critical field through the carbon substitution into boron positions in MgB_2. Changed grain size and/or grain shape is influencing the grain-boundary pinning and also higher H_{c2} is lowering $k_a(\mu_0 H)$. Consequently, both effects are effective ways to reduce high anisotropy factor of PIT MgB_2 tapes.

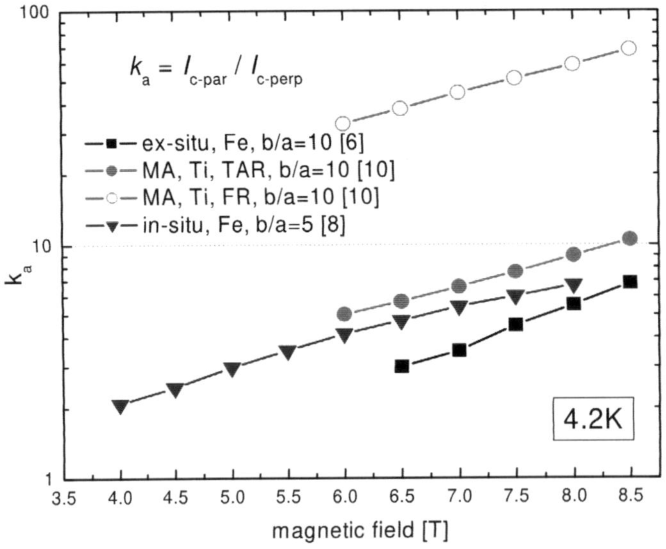

Figure 2. Anisotropy factors measured for ex-situ (squares) [6], in-situ (triangles) [8] and MA (circles) [10] PIT tapes.

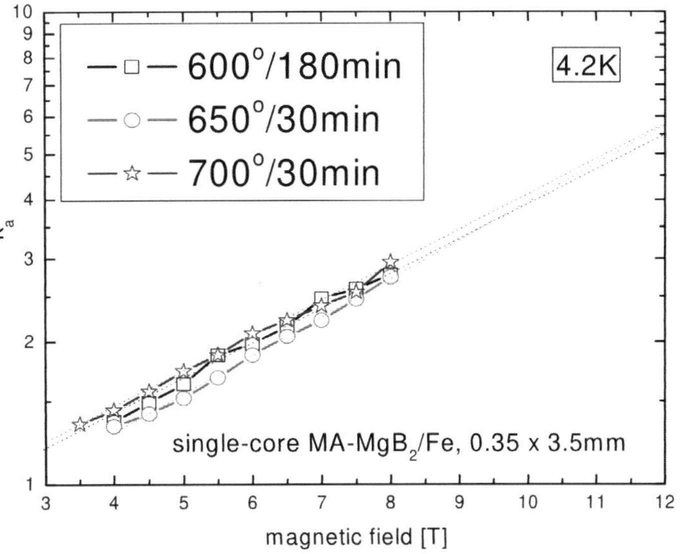

Figure 3. Anisotropy factors measured for MA/Fe tapes annealed at different temperatures.

3.3. The effect of conductor temperature

Anisotropy factors of in-situ MgB_2 tapes sheathed by Fe and Nb were measured at temperature range 4.2 – 22 K. Figure 4 shows the $k_a(\mu_0 H)$ characteristics extrapolated into higher external fields (dashed or dotted lines). One can see the strong effect of conductor temperature on the anisotropy factor k_a. The temperature dependence of k_a at 4T is plotted by figure 5. Though of relatively small tape's aspect ratio b/a = 4.25, the k_a values of Fe sheathed tape are very high especially close to 20 K, which is interesting temperature for a future cryogen-free applications of MgB_2 conductors.

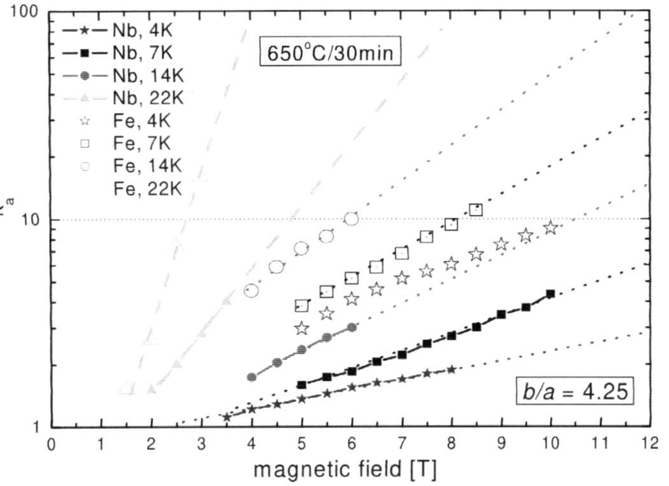

Figure 4. Anisotropy factors measured for two in-situ tapes having Fe and Nb sheath at temperatures 4.2, 7, 14 and 22 K.

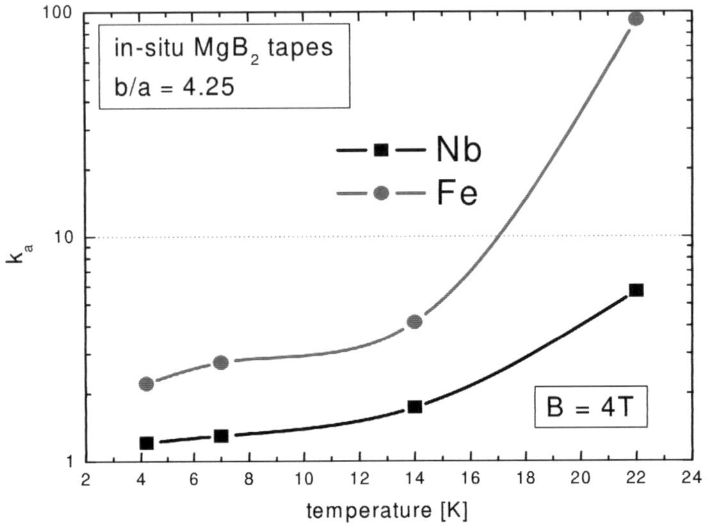

Figure 5. Temperature dependences of k_a for Fe and Nb sheathed in-situ MgB_2 tapes at constant external field 4 T.

Figure 6 compares I_c anisotropy of MgB_2 PIT tape and thin MgB_2 layer with high temperature Bi-2223/Ag tape at liquid helium temperature. Anisotropy factor $k_a > 1$ is advantage for the central part of the coil winding, where the external field is parallel with the flat side of conductor, but the coil current density can be reduced substantially due to the effect of radial field component at the coil edges. In the case of thin layer MgB_2 conductor, $k_a < 1$ will decrease the coil current due to the axial field in the central part. The similar behaviour was observed for flat Nb_3Sn superconductor having columnar grain

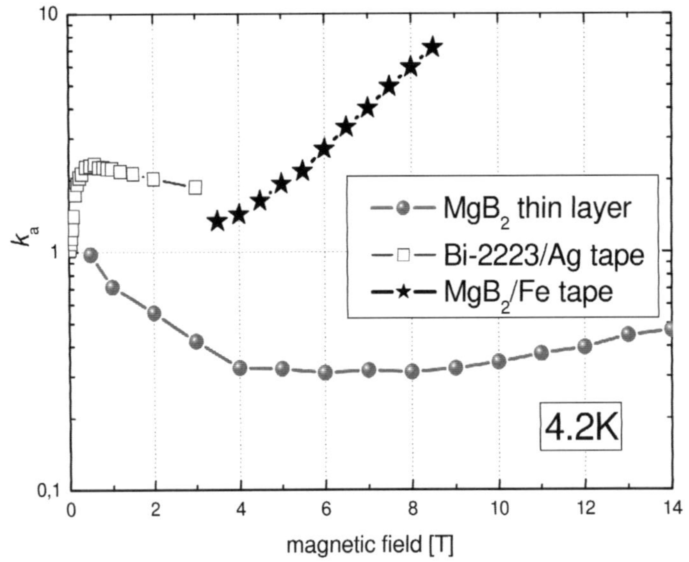

Figure 6. Comparison of I_c anisotropy for MgB_2 PIT tape (stars) and thin MgB_2 layer (circles) [12] with HTS Bi-2223/Ag tape at 4.2K.

structure oriented perpendicularly to the tape surface [15]. Though the I_c anisotropy of high temperature superconductor (HTS) Bi-2223/Ag is very high at liquid nitrogen temperature ($k_a > 100$ at 1 T), it is reduced to $k_a \approx 2$ at 4.2 K. Critical current anisotropy of MgB_2 has to be minimized by the conductor design (shape) and/or by a proper doping to avoid the coil current decrease by the radial component of magnetic field.

4. Conclusions

The I_c anisotropy of differently made MgB_2 tapes has been studied experimentally and compared with the results of thin film made by molecular beam epitaxy [12]. It was shown that observed anisotropy is connected with the texture introduced by rolling deformation and it is increasing with conductor temperature going to 20K. The reason of high current anisotropy factors for the tapes made by PIT technique is not yet fully understood and very high k_a values cannot be explained only by moderate texture of MgB_2. Most probably some another pinning anisotropy influenced by texture (e.g. oriented defects – dislocations) should contribute to so high anisotropy, which has to be verified by TEM analysis. From the point of practical application of MgB_2 it is important to reduce I_c anisotropy especially at temperature close to 20 K by the conductor design (shape) or by a proper doping to avoid the undesired coil current decrease by radial magnetic field.

Acknowledgments

This work was supported by the Science Agency under the contract number APVV-0398-07 and by the EU FP6 contract number NMP3-CT2004-505724. Authors wishing also to acknowledge for the assistance from colleagues: I. Hušek T. Melišek and L. Kopera.

References

[1] Hillman H 1981 Superconductor Materials Science, Metallurgy, Fabrication and Application, Plenm Press, New York, Ed. by S. Foner and B B Schwartz p. 275
[2] Kováč P and Cesnak L 1993 *Sup. Sci. and Technology* **6** 447
[3] Herrmann M, Haessler W, Rodig C, Gruner W, Holzapfel B and Schltz L. 2007 *Appl. Phys. Lett.* **91**, 082507
[4] Haessler W, Herrmann M, Rodig C, Nenkov K, Holzapfel B 2008 *Sup. Sci. Tech.* **21** 062001
[5] Kumakura H, Matsumoto A, Fujii H, Kitaguchi H, Togano K 2002 *Physica C* **382** 93
[6] Lezza P, Gladyshevskii R, Senatore C, Cusanelli G, Flükiger R IEEE 2005 *Trans. on Apll. Superconductivity* Vol. **15** No 2 3196
[7] Kováč P, Hušek I, Melišek T, Dhallé M, Müller M and den Ouen A 2005 *Sup. Sci and Technology* **18** 615
[8] Kováč P, Melišek T and Hušek I 2005 *Sup. Sci. and Technology* **18** L45
[9] Kováč P, Melišek T and Hušek I 2006 *Sup. Sci. and Technology* **19** 470
[10] Kováč P, Hušek I, Melišek T, Haessler W and Herrmann M 2006 *Sup. Sci. and Technology* **19** 998
[11] Kováč P, Hušek I, Dobročka E, Melišek T, Herrmann M and Haessler W 2008 *Sup. Sci. and Technology* **21** 015002
[12] Yamomoto H, Tsukamoto A, Saitoh K, Okada M, Kitaguchi H 2007 *Appl. Phys. Letters* **90** 142516
[13] Haessler W 2008 private communication
[14] Haessler W, Herrmann M, Rodig C, Kairo A, Schubert M, Nenkov K, Fuchs G, Holzapfel B, Kováč P, Melišek T, Abrahamsen A. B and Grivel J. C ASC 2008 contribution 3MB03
[15] Kováč P 2006 *IEEE Trans. on Appl. Superconductivity* **16** 1453

Revisiting the $Y_{1-x}Pr_xBa_2Cu_3O_{7-\delta}$ system- CuO_2 plane disorder induced flux pinning

Anuj Kumar, Arpita Vajpayee, Shiva Kumar, V P S Awana* and H Kishan
Superconductivity & Cryogenics Division, National Physical Laboratory, K.S. Dr. Krishnan Marg, New Delhi-110012, India

*E-mail: awana@mail.nplindia.ernet.in

Abstract. Sample of $Y_{1-x}Pr_xBa_2Cu_3O_7$ with x = 0.0 to 0.80 are synthesized by solid- state reaction route. All the samples crystallized in orthorhombic structure with space group *pmmm*. Slight decrease in orthorhombicity is seen with an increase in Pr content. Superconductivity is destroyed above 50% Pr content. Non-superconducting samples are highly semi conducting. Detailed high field magnetization (*M-H*) measurements showed that the lower critical field (H_{c1}) improves slightly with Pr content of up to 4% and the same later decreases for higher Pr contents. The critical current density calculated by invoking Bean's critical state model improves dramatically until x=0.04 and decrease later for higher Pr contents in comparison to the pristine sample. These results are explained on the basis of Pr^{4f} and O^{2p} hybridization induced disorder in superconducting CuO_2 planes. For low Pr content the mild disorder in superconducting condensate (CuO_2 plane) induces intrinsic pinning and hence resulting in an increased $J_c(H)$ and H_{c1}.

1. Introduction

$REBa_2Cu_3O_7$ (RE = rare earth) is a known high transition temperature (T_c = 90 K) superconducting family with RE = rare earths except Ce, Pr and Tb and is popularly called as RE-123 [1,2]. In case of Ce and Tb the required triple perovskite RE-123 structure is not formed, because the former is in tetravalent state and later reacts actively with Ba to form stable $TbBaO_3$ phase [3,4]. Some of us could induce superconductivity in Tb-123 after replacing Ba by Sr i.e., in $TbSr_2Cu_{2.7}Mo_{0.3}O_{7-\delta}$ [5]. On the other the situation for Pr-123 is different, which crystallizes in required orthorhombic structure with space group *pmmm* but yet not superconducting [5-8]. Also Pr spins in Pr-123, order at unusual high antiferromagnetic order of 17 K [9]. It is known that extended Pr^{4f} orbital hybridizes in perpendicular direction with the adjacent O^{2p} orbitals in superconducting Cu-O_2 planes [10,11]. It is believed that the Pr^{4f} hybridization with adjacent superconducting Cu-O_2 planes introduces disorder and decreases the mobility of charge carriers [10,11]. Also the hybridization of magnetic Pr^{4f} with superconducting Cu-O_2 planes kills the superconductivity by possible pair breaking effects [12]. As test case superconductivity could be restored in insulating Pr doped RE-123 systems by increasing the distance between Pr and Cu-O_2 planes. For example superconductivity could be restored for higher Pr concentration in $Y_{1-x}Pr_x(Ba/Sr)Cu_3O_{7-\delta}$ with successive increase in Ba/Sr substitution [13]. Also there are couple of reports on superconducting $PrBa_2Cu_3O_{7-\delta}$ with an increased c-parameter [14], but yet not reproduced by other groups. What is clear by now is that Pr^{4f} hybridizes with the superconducting Cu-O_2 planes and thereby creates disorder and kills the superconductivity [10-14]. Interestingly, limited disorder being created by low contents of Pr in $Y_{1-x}Pr_xBa_2Cu_3O_{7-\delta}$ could act as a source of intrinsic pinning in superconducting Cu-O_2 planes. This is similar to a situation of intrinsic pinning in $MgB_{2-x}C_x$ by C doping due to disorder in superconducting B-planes [15]. Keeping this in view we studied in the current paper the impact of Pr substitution in $Y_{1-x}Pr_xBa_2Cu_3O_{7-\delta}$. Our studies showed that though the superconducting transition temperature (T_c) decreases monotonically with increase in Pr content the critical current J_c(H) and the lower critical field (Hc_1) increase remarkably for low contents of Pr. We concluded that the limited disorder, being created by low level doping of Pr in RE-123 superconductors, improves tremendously the superconducting performance of the doped systems.

2. Experimental Procedure

The Samples of the series $Y_{1-x}Pr_xBa_2Cu_3O_{7-\delta}$ were synthesized through solid-state reaction. The parent compounds are Y_2O_3, Pr_6O_{11}, $BaCO_3$ and CuO of high purity were mixed in stochiometric ratio and heated at 860°C for 24 hours. Before the final sintering, this exercise was repeated at least three times with intermediate grinding in each stage to ensure thorough mixing of the materials. The resulting powders were grounded, mixed and palletized and heated in air at final sintering temperature of 885°C and further the air sintered samples annealed in Oxygen atmosphere at 600°C for a period of 48 h and subsequently allowed to cool at room temperature over a span of 12 h in the same gas atmosphere. The X-ray diffraction at room temperature was carried out using Rigaku Miniflex-II with Cu K_α radiation. The magnetic measurements were done with the help of SQUID magnetometer (Quantum Design MPMS-XL). The four probe resistivity measurements were carried out by a close cycle refrigerator (CCR) in the temperature range 12-300 K.

3. Results and Discussion

Figure1, depicts the room temperature X-ray diffraction (XRD) patterns for $Y_{1-x}Pr_xBa_2Cu_3O_{7-\delta}$ system with x = 0.30 & 0.80. The observed patterns were fitted using Rietveld analysis and shown in figure1.

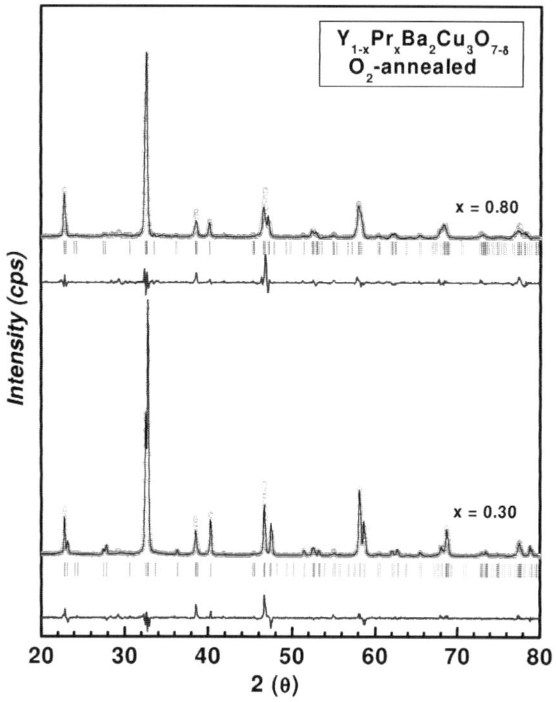

Figure 1. Observed and fitted X-ray diffraction pattern for x = 30% and x = 80%

The samples crystallize in orthorhombic structure with space group *Pmmm*. Figure 1 also depicts that sample with x = 0.80 is less orthorhombic. With the help of Rietveld refinement analysis, the lattice parameters a, b and c of samples were calculated and are listed in table 1. The *c* parameter of the doped samples increases with Pr content (x) for all samples, it indicates that probable substitution of larger ion Pr^{3+}/Pr^{4+} at the site of relatively smaller ion Y^{3+}. These lattice parameter variations with x for the same system are in agreement with reported literature [5-9].

The variation of normalized resistivity with respect to temperature $\rho(T)$ of the system $Y_{1-x}Pr_xBa_2Cu_3O_{7-\delta}$ with x = 0.0, 0.02, 0.04, 0.10, 0.20 and 0.30 is shown in figure 2(a). All the samples exhibit a one-step superconducting transition along with increasing transition width while increasing the doping amount of Pr. The T_c^{onset} of the pristine sample (*YBCO*) is observed at 92.13 K. On increasing the doping amount of Pr, the T_c^{onset} of the system decreases monotonically. The T_c^{onset} values for $Y_{1-x}Pr_xBa_2Cu_3O_{7-\delta}$ are 92.13 K, 91.54 K, 90.68 K, 80.84 K, 77.34 K and 63.21 K for x = 0.0, 0.02, 0.04, 0.10, 0.20 and 0.30 respectively. The residual resistivity value goes on decreasing with Pr concentration. This is because of doping of Pr introduces disorder in superconducting $Cu\text{-}O_2$ planes and decreases the mobility of charge carriers by introducing disorder.

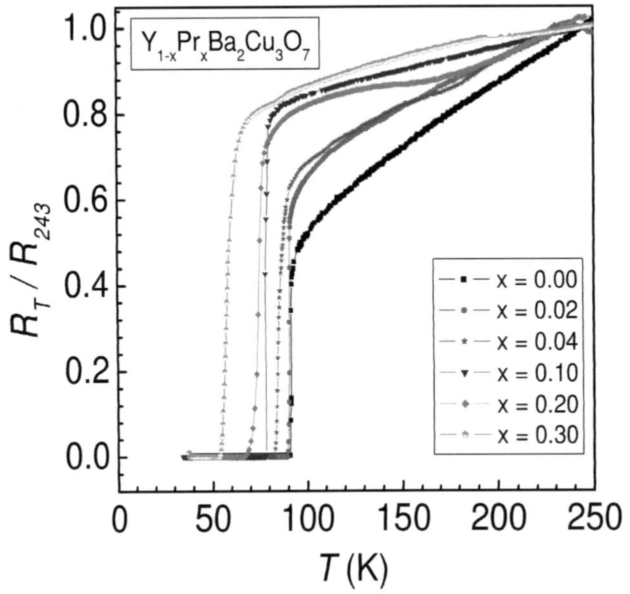

Figure 2(a). Variation of normalized resistivity with respect to temperature $\rho(T)$ for $Y_{1-x}Pr_xBa_2Cu_3O_{7-\delta}$ with x = 0.0, 0.02, 0.04, 0.10, 0.20 and 0.30 samples.

Figure 2(b) depicts the dc susceptibility versus temperature $\chi(T)$ plots in an applied field of 10 Oe in both field-cooled (*FC*) and zero field-cooled (*ZFC*) situations. It is evident from the figure that pure YBCO undergoes a sharper superconducting transition (diamagnetic, T_c^{dia}) at 93.62 K as compared to other doped samples. All the samples exhibit one-step transition from normal state to superconducting state, however the transition width increases a bit by increasing the amount of dopant. The superconducting critical temperature (T_c^{dia}) being seen from $\chi(T)$ measurements is little higher than those T_c^{onset} values revealed by the resistivity measurement.

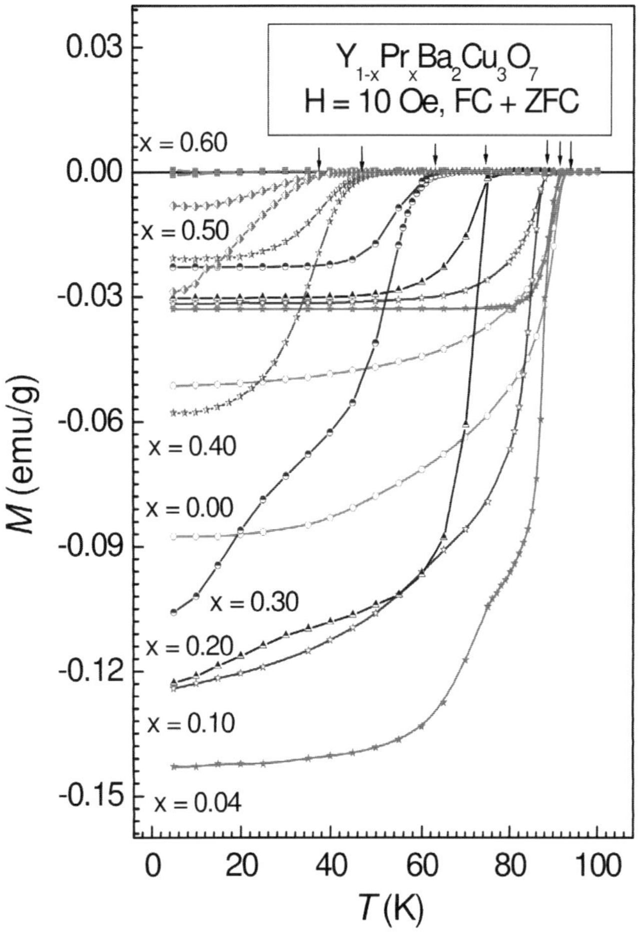

Figure 2(b). Magnetization versus temperature plot for $Y_{1-x}Pr_xBa_2Cu_3O_{7-\delta}$ with x = 0.0, 0.04, 0.10, 0.20, 0.30, 0.40, 0.50 and 0.60 samples in the field-cooled & zero field-cooled situation

The magnetic hysteresis loop for the doped samples $Y_{1-x}Pr_xBa_2Cu_3O_{7-\delta}$ with x = 0.0, 0.02, 0.04 and 0.10 are shown in figure 3 at T = 5 & 20 K. This figure clearly demonstrates that the *M(H)* loop for pure sample is nearly going to saturate at 7 Tesla, while there is a wide opening for other doped samples at the same field. This indicates that irreversibility field values are improved significantly with addition of Pr in parent compound; 2 wt% Pr doped sample is showing the best behaviour among all the doped samples. This indicates that limited disorder being created by low contents of Pr in $Y_{1-x}Pr_xBa_2Cu_3O_7$ could act as a source of intrinsic pinning in superconducting Cu-O_2 planes. The enlarged view of M(H) loop is shown in inset of figure 3. It shows that the lower critical field (H_{c1}) improves slightly with Pr content of up to 4% and the same later decreases for higher Pr contents.

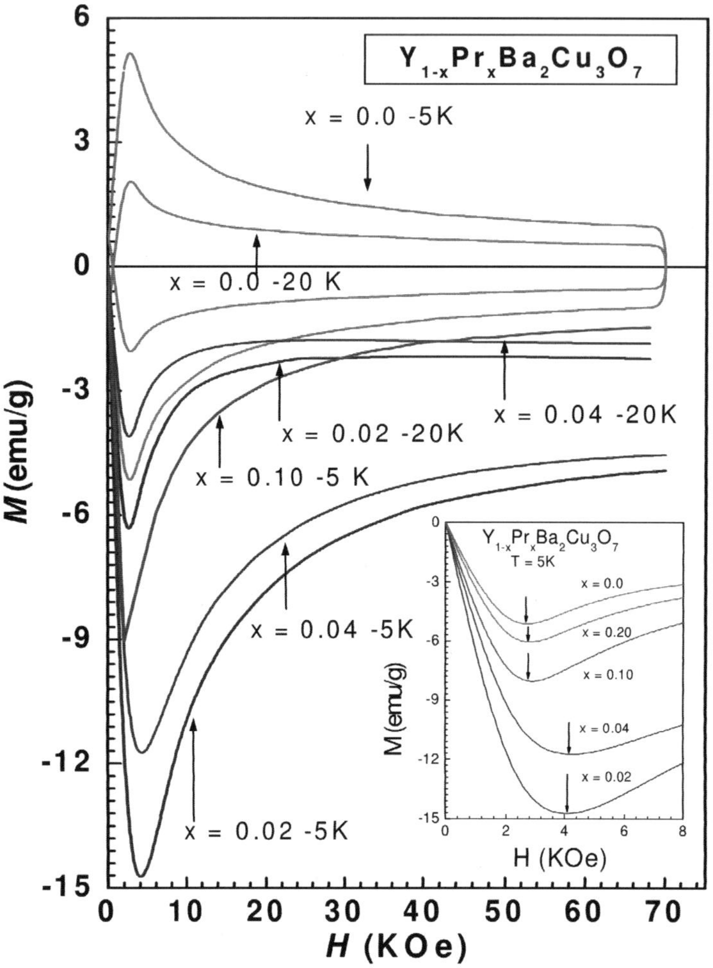

Figure 3. Magnetization loop *M(H)* for $Y_{1-x}Pr_xBa_2Cu_3O_{7-\delta}$ (x=0%, 2%, 4% & 10%) up to 7 Tesla field at 5 & 20 K

The magnetic $J_c(H)$ for all the samples was calculated from the *M(H)* loop at 5 & 20 K. Figure 4 depicts the magnetic J_c vs. *H* for doped samples along with pristine one at 20 & 5 K. At low fields the pristine YBCO sample attains the J_c of the order of 10^3 A/cm^2 at both the temperatures. Significant improvement in $J_c(H)$ values can be seen easily by Pr doping in YBCO; at lower fields the 2% and 4% Pr doped samples attain the J_c value of the order of 10^4 A/cm^2. However for higher doping level the J_c remains more or less same. Among all the doped samples, 2% Pr doped sample gives the best performance. The J_c is 4.5 times higher than pure one at 5 K in 7 Tesla field in case of 2% Pr doped sample and 3.6 times higher at 20 K in 7 Tesla for the same sample. There is also an improvement in J_c values for 4% doped sample but lower than those for 2% doped sample. For 4% Pr doped sample the J_c is 3.7 and 2.7 times higher than the pristine sample at 7 Tesla at 5 & 20 K respectively. Similar results are reported very recently on Pr and Ca doped YBCO samples [16].

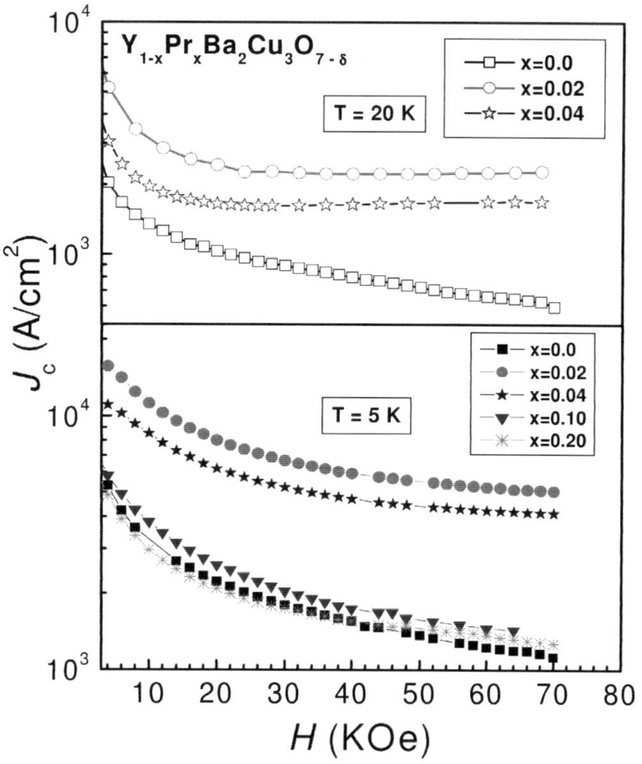

Figure 4. $J_c(H)$ behaviour of $Y_{1-x}Pr_xBa_2Cu_3O_{7-\delta}$ at 5 & 20 K

4. Summary and conclusion

In conclusion, our results strengthen the contention of Pr^{4f} hybridization with Cu-O conduction band. The effect of substituting Pr with different concentrations in YBCO, is investigated to see how the resulting cationic disorder influences the flux pinning and the critical current density $J_c(H)$. The present study brings out the importance of Pr substitution in significantly enhancing $J_c(H)$. Interestingly, the Pr substitution causes disorder in Cu-O_2 planes, although Pr is substituted at the Y site of the YBCO unit cell. Disorder is particularly important in Cu-O_2 planes as they are the seats of both superconductivity and pancake type vortices and therefore the pinning interaction is expected to be pronounced. The Pr substitution helps in marginalizing the adverse effect of grain boundaries either by lowering their resistance or making them (weakly) superconducting by supplying either extra charge carriers or oxygen.

5. Acknowledgement

The authors from *NPL* would like to thank Dr. Vikram Kumar (Director, *NPL*) for constant encouragement in the present work. One of us, Arpita Vajpayee, would also thank *CSIR* for financial support by providing *JRF* fellowship. Authors would also like to thank Prof. E. Takayama-Muromachi from *NIMS* Japan for providing help in carrying out the *SQUID* magnetization measurements.

Table 1. Lattice Parameters of $Y_{1-x}Pr_xBa_2Cu_3O_{7-\delta}$ with x = 0.0, 0.02, 0.04, 0.10, 0.30, 0.40, 0.50, 0.60, 0.70 and 0.80 revealed by Rietveld analysis.

Sample	a (Å)	b (Å)	c (Å)	V (Å^3)	(b-a)/b*100
$YBa_2Cu_3O_{7-\delta}$	3.8229 (2)	3.8864 (3)	11.6660 (13)	173.332 (29)	1.6339
$Y_{0.98}Pr_{0.02}Ba_2Cu_3O_{7-\delta}$	3.8193 (2)	3.8882 (4)	11.6718 (14)	173.332 (30)	1.77203
$Y_{0.96}Pr_{0.04}Ba_2Cu_3O_{7-\delta}$	3.8202 (1)	3.8889 (2)	11.6700 (9)	173.382 (20)	1.76657
$Y_{0.90}Pr_{0.10}Ba_2Cu_3O_{7-\delta}$	3.8281 (2)	3.8910 (5)	11.6738 (18)	173.616 (38)	1.61655
$Y_{0.70}Pr_{0.30}Ba_2Cu_3O_{7-\delta}$	3.8300 (1)	3.8955 (2)	11.6802 (8)	174.267 (18)	1.68143
$Y_{0.60}Pr_{0.40}Ba_2Cu_3O_{7-\delta}$	3.8345 (1)	3.8998 (2)	11.6865 (9)	174.766 (20)	1.67444
$Y_{0.50}Pr_{0.50}Ba_2Cu_3O_{7-\delta}$	3.8380 (1)	3.9030 (2)	11.6892 (9)	175.107 (20)	1.66539
$Y_{0.40}Pr_{0.60}Ba_2Cu_3O_{7-\delta}$	3.8434 (3)	3.9048 (5)	11.6921 (18)	175.480 (39)	1.57242
$Y_{0.30}Pr_{0.70}Ba_2Cu_3O_{7-\delta}$	3.8464 (3)	3.9056 (4)	11.6934 (15)	175.669 (33)	1.51577
$Y_{0.20}Pr_{0.80}Ba_2Cu_3O_{7-\delta}$	3.8582 (6)	3.9028 (10)	11.6979 (40)	176.256 (83)	1.14277

References

[1] Oesterreider H and Smith M 1987 *Mater. Res. Bull.* **22** 1709
[2] Hor P H, Meng R L, Wang Q, Gao L, Huang Z J, Bechtold J, Forster K and Chu C W 1987 *Phys. Rev. Let.* **58** 1891
[3] Yang K N, Lee B W, Maple M B, and Laderman S S 1988 *Appl. Phys. A* **46** 229
[4] Staub U, Soderholm L, Skanthakumar S and Antonio 1995 *Phys. Rev. B* **52** 9736
[5] Awana V P S, Gupta A, Kishan H, Takayama-Muromachi T, Watanabe T, Karppinen M, Yamauchi H, Malik S, Yelon W B, Ganesan and Narlikar A V 2004 *solid State Comm.* **137** 117
[6] Soderholm L, Zhang K, Hinks D G, Beno M A, Jorgensen J D, Segre C U and Schuller I K 2001 *Nature* **328** 604
[7] Kebede A, Jee C S, Schwegler J, Crow J E, Mihalisin T, Myer G H, Salomon R E, Schlottmann P, Kuric M V, Bloom S H and Guertin R P 1989 *Phys. Rev. B* **40** 4453
[8] Malik S K and Tomy C V in: S K Malik and S S Shah (Eds.), Physical and Material Properties of High Temperarue, Superconductor 1994, *NOVA Science, New York* 283
[9] Radousky H B 1992 *J. Mater. Res.* **7** 1917
[10] Fehrenbacher R and Rice T M 1993 *Phys. Rev. Lett.* **70** 3471
[11] Liechtenstein A I and Mazin I I 1995 *Phys. Rev. Lett.* **74** 1000
[12] Malik S K, Tomy C V and Bhargava P 1991 *Phys. Rev. B.* **44** 7042
[13] Awana V P S, Ansari M A, Gupta Anurag, Saxena R B, Kishan H, Rawat R, Ganesan V, Narlikar A V, Buddhikot D and Malik S K 2004 *Physica C* **417** 33
[14] Zou Z, Oka K, Ito T and Nishihara Y 1997 *Jpn. J. Appl. Phys.* **36** L18
[15] Mudgel M, Awana V P S, Kishan H and Bhalla G L 2008 *Solid Stste Commu.* **146** 330
[16] Huhtinen H, *et al*, 2007 *Supercond. Sci. Technol.* **20** S159

Dynamic fatigue behaviour of Ag-doped Bi-2212 textured thin rods

M A Madre[1], Sh Rasekh, J C Diez and A Sotelo

ICMA (UZ-CSIC), Depto. de Ciencia y Tecnología de Materiales y Fluidos, C/María de Luna 3, E-50018, Zaragoza (Spain)

E-mail: amadre@unizar.es

Abstract. The flexural strength of 1 wt.% Ag-doped $Bi_2Sr_2CaCu_2O_{8+\delta}$ thin rods textured by a laser heated floating zone was measured as a function of the environmental conditions (air versus water) at room temperature. Loading rates spanning three orders of magnitude (1, 10 and 100 μm/min) were used to explore their susceptibility to the environmental conditions. These mechanical tests were completed with electrical characterization (critical current at 77K and resistivity from 77 to 300 K) of samples submerged in distilled water for different time lengths (0, 12 and 120h). While $Bi_2Sr_2CaCu_2O_{8+\delta}$ has been shown, in previous works, to be unstable during contact with water molecules, the Ag-doped Bi-2212 textured rods tested in this work are very inert to the water environment, with respect to their mechanical and electrical properties, due to the presence of a narrow (≈150 μm) low textured outer ring formed in the growth process.

1. Introduction

The development of practical applications of bulk high temperature superconductors requires not only the knowledge of their electrical properties, but their mechanical behavior under different environments. Bulk Bi-2212 superconductors have demonstrated that they are suitable for many applications when they are properly processed [1], for example, through the laser floating zone technique [2, 3]. The Bi-2212 bulk materials textured by this or other techniques [4] have very interesting electrical properties that allow developing fault current limiters and current leads [5]. One of the main advantages of this method is that samples can be rapidly grown due to the large thermal gradients present at the solid–liquid interface [6]. A second additional advantage is the absence of crucible, avoiding external contamination of textured samples during their processing.

However, the poor mechanical characteristics of this kind of materials [7], due to their ceramic nature, impose limitations for practical applications. Some attempts to improve their mechanical properties have been performed by means of Ag additions on BSCCO compounds [8-10].

On the other hand, for any practical application, it is necessary to take into account their reactivity to moisture [11, 12], which leads to their chemical degradation and, as a consequence, its effect on their electrical and mechanical properties.

In this work, the environmental susceptibility of Ag-doped Bi-2212 thin rods grown by a laser floating zone method has been measured by means of three-point bending test in air and water, at room temperature. The evolution of the electrical properties, critical current and electrical resistivity, as a function of the immersion time in water, is also reported.

2. Experimental

$Bi_2Sr_2CaCu_2O_{8+\delta}$ + 1 wt.% Ag materials have been prepared from commercial Bi_2O_3 (Panreac, 98+%), PbO (Panreac, 99+%), $SrCO_3$ (Panreac, 98+%), $CaCO_3$ (Panreac, 98.5+%), CuO (Panreac, 97+%) and Ag (Aldrich, 99.9+%) powders. They were weighed in the adequate atomic proportions, and, among the many different synthetic methods [13], the sol-gel method via nitrates has been selected due to the good solubility of $AgNO_3$, as well as the whole metallic nitrates, in dilute nitric acid. Appropriate proportions of citric acid and ethilenglycol were also added to the nitrates solution, followed by heating on a hot plate at 50-80 ºC in order to form the gel by polymerization between citric acid and ethilenglycol when the pH is reduced by decomposition of the metallic nitrates and the HNO_3 added in excess, following the reactions (1) and (2), corresponding to an esterification process (equation 1) and to the polymerisation one (equation 2):

$$\text{R-C}\begin{smallmatrix}\text{=O}\\ \text{OH}\end{smallmatrix} + \text{HO-R'-OH} \longrightarrow \text{H}_2\text{O} + \text{R-C}\begin{smallmatrix}\text{=O}\\ \text{O-R'-OH}\end{smallmatrix} \quad (1)$$

$$\text{n (HO-R-O-R'-OH)} \longrightarrow \text{n}\left[\text{R-O-R'-O}\right]_n + \text{nH}_2\text{O} \quad (2)$$

Further heating leads to the evaporation of the solvent and, finally, to a self combustion of the solid gel. The so obtained powders were thermally treated twice at 750 and 800 ºC for 6 hours, with an intermediate grinding, to avoid grain growth and assure total carbonates decomposition, which is a critical step in order to avoid bubbles formation in the melting process, disturbing the solidification front.

These prereacted powders were then used to prepare cylindrical precursors, 120 mm long and 2-3 mm diameter, approximately, by cold isostatic pressing with an applied pressure of 200 MPa during 1 minute. The obtained cylinders were used as feed in a directional solidification process performed in a laser floating zone melting (LFZ) installation described elsewhere [6]. The textured bars were obtained using a continuous power Nd:YAG laser (λ = 1064 nm), under air, at a growth rate of 30 mm/h and a relative rotation of 18 rpm between the seed and feed. Using these growth conditions and adjusting the laser power input to obtain a molten zone of approximately 1-1.5 times the rod diameter, it is possible to obtain stable growth, which allows the fabrication of very homogeneous bars.

All the studied materials present an incongruent melting and, in consequence, after texturing, it is necessary to perform a thermal treatment in order to form the Bi-2212 superconducting phase [14]. This annealing process is performed under air, and consisted in two steps: 60 h at 850ºC, followed by 12 h at 800ºC and, finally, quenched in air to room temperature.

Table 1. Measured and calculated mechanical properties

Environment	Loading rate (μm/min)	σ_f (MPa)	Weibull parameter, *m*
Air	1	160 ± 7	8.1
Air	10	164 ± 5	11.8
Air	100	169 ± 5	11.5
Water	1	166 ± 6	11.1
Water	10	174 ± 5	14.2
Water	100	183 ± 4	16.8

The flexural strength of the rods was measured by three-point bend tests using a fixture of 10 mm loading spam in a mechanical testing machine (Instron 5565). The tests were performed in air under stroke control at cross-head speeds of 1, 10 and 100 μm/min at room temperature, with and without (only atmospheric moisture) water. The rods tested in water conditions were immersed in distilled water during 1 h before the test. A drop of water, added with a syringe, covered the central section of the rods during the mechanical tests to ensure that the fracture region was immersed in water during the whole test. The flexural strength was computed from the maximum load achieved in the test, according to the strength of materials theory for an elastic beam of circular section. For each condition, 10 samples were tested. The fracture surfaces of the broken specimens were examined in a scanning electron microscope (JEOL JSM 6400).

Low resistance silver contacts were painted on the samples for electrical characterization, before the annealing process. The measurements of the critical current at 77 K, I_c (77 K), and the dependence of the dc resistivity with temperature, $\rho(T)$, were performed on 30 mm long samples using the common four-probe configuration. The current was kept fixed at 1mA in the resistivity measurements, while the critical currents were determined in self-field using the standard 1 μV/cm criterion. All the electrical measurements were performed on samples immersed in water up to 120 h, with intermediate measurements at 0 and 12 h.

3. Results and discussion

The experimental data determined from the flexure tests at different loading rates are displayed in table 1, where the average values of the flexural strength fracture, σ_f, are displayed, together with their corresponding standard errors. The brittle nature of these ceramic materials, together with the high anisotropy of their microstructure is reflected in the Weibull parameter, m, calculated from the data obtained in the bending experiments. An interesting result shown in this table is that the fracture strength values are not so different for each of the environmental conditions and loading rates (from 160 MPa at 1 μm/min under air to 183 MPa at 100μm/min under water). When looking the samples measured under air, the dependence of the mean fracture strength with the loading rate is not so marked ($\Delta \approx 5\%$, from 160 to 169 MPa), indicating that the environmental susceptibility of these textured rods to the moisture is low. On the contrary, a small but more significant dependence with the loading rates can be observed when samples are tested in water ($\Delta \approx 10\%$, from 166 to 183 MPa).

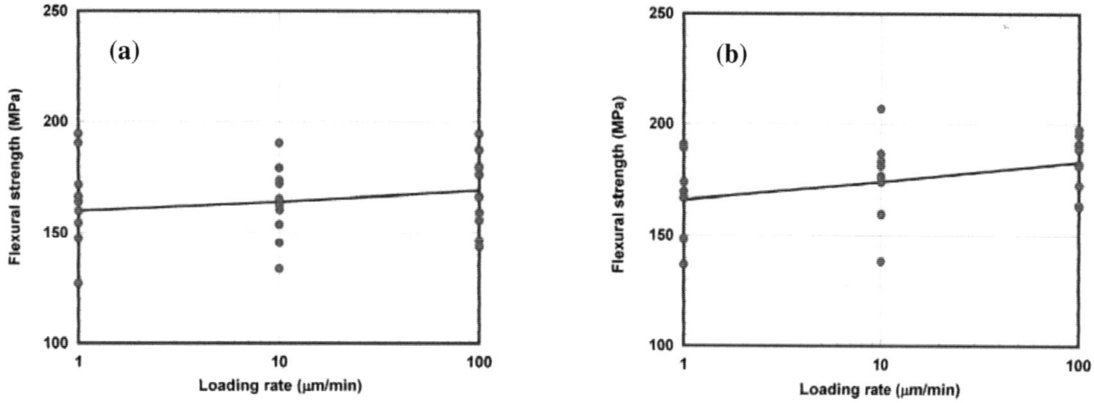

Figure 1. Flexural strength of Ag-doped Bi-2212 materials as a function of the loading rate, measured a) in air; and b) in water. The displayed lines are passing through the mean values for each loading rate.

In order to show this evolution, the individual strength results for each sample and conditions, are displayed in figure 1 (figure 1a for measurements in air and figure 1b, in water). Also, a line is plotted passing through the mean values for each testing speed in order to clarify the strength dependence with the punch displacement speed. This change can be related to the chemical degradation with water [11, 12], which is in agreement with a stress corrosion cracking as a process for the interpretation of the mechanical behavior of this material. The reaction involved in this degradation mechanism takes place in two subsequent step processes, described by the chemical reactions:

$$Bi_2Sr_2CaCu_2O_8 + 3H_2O \rightarrow 2Sr(OH)_2 + Ca(OH)_2 + CuBi_2O_4 + CuO \quad (3)$$

$$Sr(OH)_2, Ca(OH)_2 + CO_2 \rightarrow H_2O + SrCO_3, CaCO_3 \quad (4)$$

The slightly higher mean values for samples measured in water, than those measured in air, can be explained by the formation of a continuous water layer inside the cracks, which avoids the surface contact with the CO_2 necessary to perform the reaction described in equation (4).

The subcritical crack growth is a limitation to fracture strength in ceramics. The stable growth of cracks under applied stresses lower than those for fast fracture (slow crack growth, SCG) is usually studied by means of the power-law function [15]:

$$v = AK_I^n \quad (5)$$

where K_I is the stress intensity factor at the tip of the crack, and A and n are parameters depending on the material and the environment. The n value can be computed from the slope of the lg σ_f versus lg ($d\varepsilon/dt$). The n value obtained in this work for water tested samples is around 47, which is the same order as Al_2O_3 at room temperature in water environment, as it has been previously reported [16]. This result is unexpected when compared with previous works on the influence of moisture in Bi-2212 bulk samples [11, 12]. The water influence is so low that no marked differences in the fractured surfaces have been detected between air and water environment conditions. On the other hand, when samples are measured in air, the n value is increased spectacularly to 82, which is in the order of α-SiC measured at high temperatures [16].

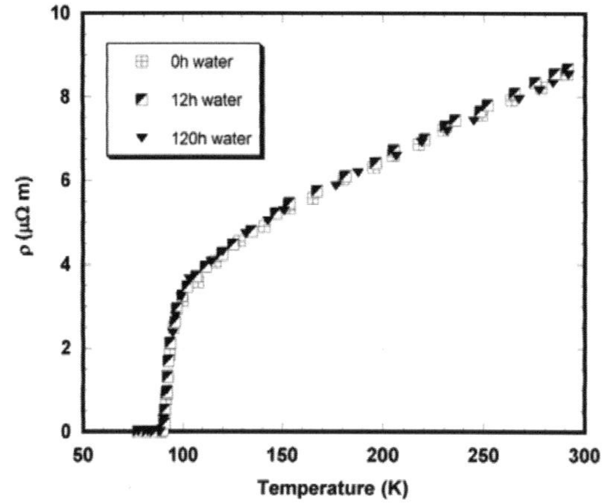

Figure 2. Electrical resistivity of samples as a function of the immersion time in water

As it can be seen in figure 2, the electrical resistivity of the samples remains constant, independently of the water exposition time. This seems to be a surprising result when taking into account the high reactivity of this Bi-2212 compounds with moisture, but it is in agreement with the results obtained in previous works for the pure Bi-2212, when it is processed using the LFZ texturing technique [17]. Moreover, the critical currents have also been measured, and no significant changes have been detected, obtaining values of about 75A at 77K in self-field, corresponding to a value of J_c = 3570 A/cm^2. This good behavior against moisture is related with the special microstructure developed in this material when processed by a LFZ, where a thin outer ring, with low texture and slightly lower grain size than in the inner part, is developed. This ring can be clearly seen in figure 3. As it has been previously reported [17] this zone is acting as a protecting layer against liquid water immersion. This ring really acts like a sealant preventing chemical degradation reactions with moisture, and even in the liquid water conditions, helping to maintain the mechanical and electrical characteristics.

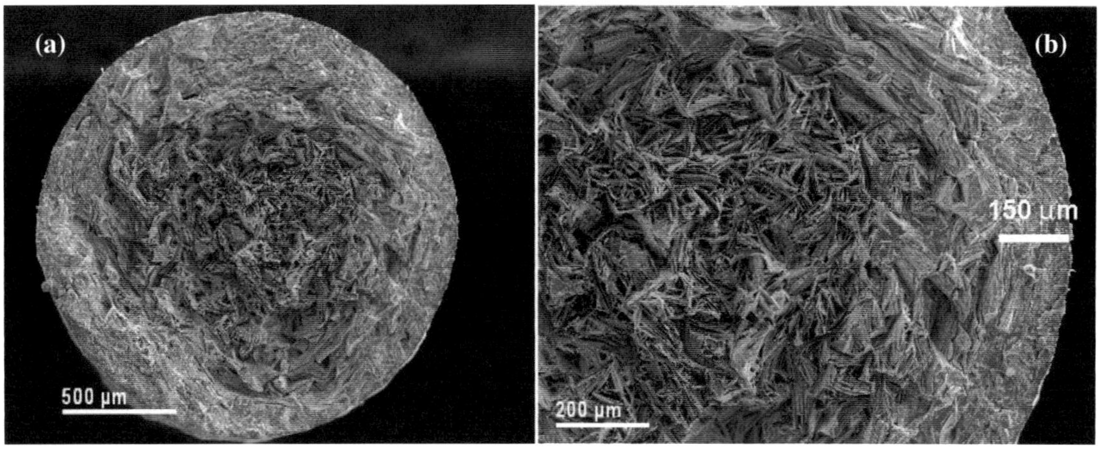

Figure 3. Typical fracture surface of an Ag-doped Bi-2212 textured rod. (a) General view, (b) detail of the outer ring (low texture), and the inner zone (high texture)

4. Conclusions

The environmental susceptibility of 1wt.% Ag-doped Bi-2212 thin rods textured by a laser heated floating zone was studied by means of three-point bending test in air and in water at room temperature. No significant differences have been found for the strength when the flexure tests were performed in air at three different loading rates. Nevertheless, it was found that the strength of the rods tested in water slightly depends on the loading rate, with an n value of 47, while for air tested samples is around 82. The low reactivity of these rods with water has been contrasted with the evolution of the electrical properties. When comparing critical current values at 77K and electrical resistivity curves between 77 and 300 K, no differences were found for samples submerged in water for 0, 12 and 120 h.

While $Bi_2Sr_2CaCu_2O_{8+\delta}$ has been shown, in previous works, to be unstable during contact with water molecules, the thin Ag-doped Bi-2212 textured rods tested in this work, show to be very inert to the environment, with respect to the mechanical and electrical properties, due to the presence of a thin (≈150 µm) low textured outer ring formed in the growth process. The self-coating protection developed in this laser growth process has demonstrated its performance, both in air and in water environment, as it has been reflected in both, mechanical and electrical properties.

Acknowledgments

The authors are indebted the Spanish Ministry for Education and Science (MAT 2005-06279-C03-01) and to the Aragón Regional Government (Research Groups T74 and T12) for financial support. The technical support from C. Gallego, C. Estepa and J. A. Gómez is also acknowledged.

References

[1] Chen M, Donzel L, Lakner M and Paul W 2004 *J. Eur. Ceram. Soc.* **24** 1815-22
[2] Garnier V, Caillard R, Sotelo A and Desgardin G 1999 *Pysica C* **319** 197-208
[3] Costa F M, Silva R F and Vieira J M 2001 *Supercond. Sci. Technol.* **14** 910-20
[4] Diez J C, Angurel L A, Miao H, Fernández J M and de la Fuente G F 1998 *Supercond. Sci. Technol.* **11** 101-6
[5] Ester F J and Peña J I 2007 *Bol. Soc. Esp. Ceram.* V. **46** 240-46

[6] Angurel L A, Diez J C, Martínez E, Peña J I, de la Fuente G F and Navarro R 1998 *Physica C* **302** 39-50
[7] Pastor J Y, Poza P and Llorca J 1999 *J. Am. Ceram. Soc.* **82** 3139-44
[8] Salazar A, Pastor J Y and Llorca J 2003 *Physica C* **385** 404-14
[9] Joo J, Singh J P, Warzynski T, Grow A and Poeppel R B 1994 *Appl. Supercond.* **2** 401-10
[10] Sotelo A, Mora M, Madre M A, Diez J C, Angurel L A and de la Fuente G F 2005 *J. Eur. Ceram. Soc.* **25** 2947-50
[11] Jin S G, Zhu Z Z, Liu L M and Huang Y L 1990 *Solid State Commun.* **74** 1087-90
[12] Lee D, Kondrate Sr R A and Taylor J A 2001 *Physica C* **350** 1-16
[13] Ruiz M T, de la Fuente G F, Badia A, Blasco J, Castro M, Sotelo A, Larrea A, Lera F, Rillo C and Navarro R 1993 *J. Mat. Res.* **8** 1268-76
[14] Mora M, Martínez E, Diez J C, Angurel L A and de la Fuente G F 2000 J. Mat. Res. **15** 614-20
[15] Munz D and Fett T 2001 *Ceramics: Mechanical Properties, Failure Behaviour, Materials Selection,* Springer
[16] Choi S R, Nemeth N N and Gyekenyesi J P 2005 *Fatigue Fract. Eng.* **28** 489-97
[17] Diez J C, Sotelo A, Mora M, Amaveda H and Madre M A 2007 *J. Eur. Ceram. Soc.* **27** 3963-66

Effect of partial substitution of Ca by 4f elements on dissipative processes in Bi:2223 superconductors

D. Marconi[1], **G. Stiufiuc**[1] **and A.V.Pop**[1]

[1] Babes-Bolyai University, Faculty of Physics, 400084 Cluj- Napoca, Romania

dmarconi@phys.ubbcluj.ro

Abstract. (Bi,Pb):2223 superconducting samples were obtained by using solid state reaction method. The effect of partial substitution of Ca by R=Sm, La on the structural and magnetic properties of $(Bi_{1.6}Pb_{0.4})(Sr_{1.8}Ba_{0.2})(Ca_{1-x}R_x)_2Cu_3O_y$ with $0\leq x \leq 0.05$ were investigated by XRD and ac magnetic susceptibility measurements. The intergranular critical current density function of temperature $J_{cJ}(T)$, were obtained and agree with the SNS behavior of grain boundaries.

1. Introduction

Superconductivity of Bi:2223 system may be directly influenced for substitution in the CuO_2 plane itself, or indirectly for substitution in the Ca or Sr positions (affecting the charge transfer to or from CuO_2 planes).

The partial substitution of divalent cation Ca by trivalent cations Y or rare earth was found to induce nanodefects which decreases the hole concentration and affect the normal and superconducting properties. The macroscopic critical current density J_{cJ} is limited by the intergranular vortex pinning force at the grain boundaries (J_{cJ} is much smaller than the critical current density J_{cG} inside the grain). The addition of Sm and Yb on the bulk Bi:2212 lead to increase the critical current density with seven times higher than that of pure sample[1]. Recently was fabricated a Bi:2223 cylindrical bulk current lead (J_c higher than 670 A/cm^2) for use in cryocooler –cooled superconducting magnets [2]. Studies of partial substitution of divalent cation Ca by trivalent cations Y or rare earth (Re) ions in Bi:2223 system are in limited number in literature[3-5].

In this paper we report the effect of the partial substitution for Ca by R=La and Sm up to 5% on the purity of samples and superconducting properties of $(Bi_{1.6}Pb_{0.4})(Sr_{1.8}Ba_{0.2})(Ca_{1-x}R_x)_2Cu_3O_y$, by measurements of the AC magnetic susceptibilities function of temperature.

2. Experimental

Polycrystalline samples with nominal composition $(Bi_{1.6}Pb_{0.4})(Sr_{1.8}Ba_{0.2})(Ca_{1-x}R_x)_2Cu_3O_y$ with R=La, Sm and $0\leq x \leq 0.05$ were prepared by the conventional solid -state reaction of appropriate amounts of the metal oxides and carbonates of 99.99% purity [6].

The partial substitution of Sr by Ba was used to induce the reduction of the modulation period. Appropriate amounts of Bi_2O_3, PbO, $SrCO_3$, BaO, Ca CO_3, La_2O_3, Sm_2O_3 and CuO were mixed in agate mortar and calcinated at 800^0C for 36 hours. The calcinated powder was pressed into pellets and presintered at 845^0C for 200 hours and finally sintered at 850^0C for 150 hours.

The influence of partial substitution of Ca by La, Sm on the structural and magnetic properties of (Bi,Pb):2223 bulk superconducting compound were investigated by XRD and ac magnetic susceptibility measurements. The phase purity was determined by Brucker X-ray diffractometer with Cu-K_{α} radiation.

The real (χ') and imaginary (χ'') parts of the a.c. susceptibility were simultaneously collected with a Lake Shore Model 7000 a.c. susceptometer in the temperature range from 77K to 110K, by using frequencies f and a.c. field amplitudes H_{ac} situated in the ranges from 20 Hz to 1000Hz and from 4 A/m to 800 A/m, respectively.

The temperature T_p for the imaginary part $\chi''(T)$ of the complex magnetic susceptibility performed at various field amplitudes is different influenced by the concentration of 4f element.

3. Results and discussion

Fig.1 shows the X-ray diffraction pattern for x = 0.00 x=0.02 and x=0.05 La and Sm samples. The results have shown that the sample with x=0.00 has a single (2223) phase.

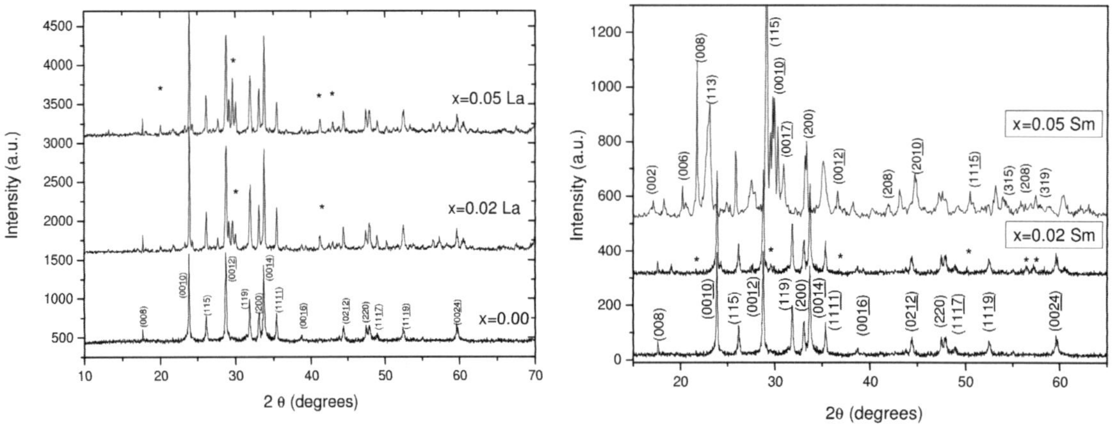

Figure1. Powder XRD pattern for $(Bi_{1.6}Pb_{0.4})(Sr_{1.8}Ba_{0.2})(Ca_{1-x} R_x)_2Cu_3O_y$ with R=La, Sm and 0≤ x ≤0.05, with each peak labeled according to the crystallographic direction that produced it for 2223 phase. Stars marked 2212 phase.

With increasing x up to x=0.05La, peaks corresponding to (2212) phase have increased in number and intensities. In the samples with x≥0.02 Sm the amount of (2212) phase increase from 10%vol. for x=0.02 to 80%vol. for x=0.05 (Table 1). XRD shows that only (2212) phase were found by increasing x up to 0.05 [7,8].

Sample	T_c [K]	Phase content %vol	
		2223	2212
x=0.00	109.0	98	traces-
x=0.02La	107.5	95	5
x=0.05La	106	90	10
x=0.02 Sm	106.5	80	20
x=0.05 Sm	103.5	20	80

Table 1. The phase content of the crystalline in studied samples.

Figure 2 shows the real χ'(T) and imaginary χ"(T) susceptibilities for x=0.00;0.01;0.02 and 0.05 La and Sm samples, by using for a. c. field an amplitude Hac =100 A/m and a frequency f=1000 Hz.

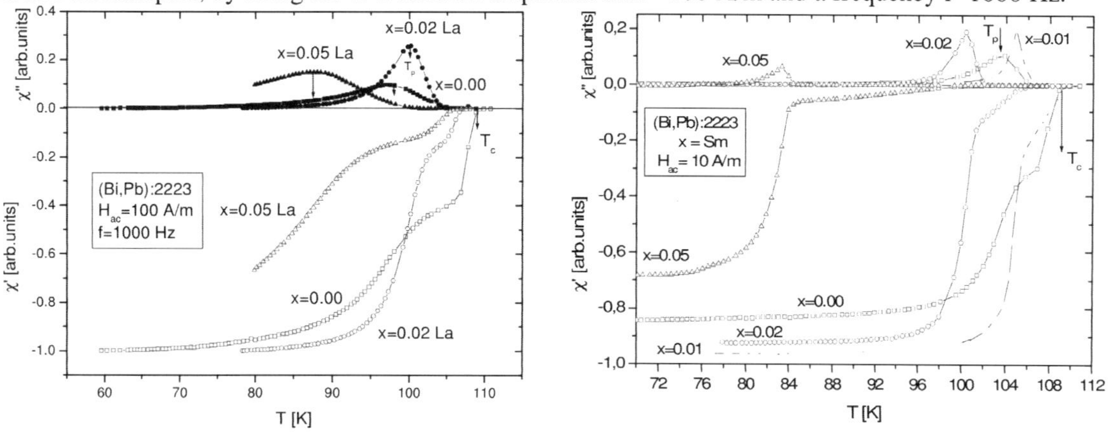

Figure 2. Real and imaginary parts χ' (T) and χ" (T) susceptibilities function of temperature for x=0.00;0.01;0.02 and 0.05 La and Sm samples

With increasing temperature, the real part $\chi'(T)$ shows a two step behavior for La and Sm substitute, characterizing the flux penetration in the intergranular matrix and in the grains respectively. The end of the upper step (the end of the superconductor diamagnetism) corresponds to the intragrain critical temperature. With increasing La and Sm concentration the intergrain drops shift to lower temperatures.

χ''(T) exhibit a single peak at temperature T_p indicating a maximum hysteressis losses due to the motion of the intergranular (Josephson) vortices. With increasing the alternative field amplitude H_{ac}, the χ'' signal shifts to lower temperatures. This result shows that with increasing the ac field amplitude the intergranular coupling decrease. The value of temperature T_p for maximum of χ'' is influenced different by the nature of 4f ions. The increase of T_p with increasing x up to x=0.01 Sm, suggest the increase of intergranular coupling.

To estimate the intergranular critical current density J_{cJ} from the χ'' (T) data we used the magnetic formulas for isotropic superconductors of cylindrical geometry in the Bean critical state model [9].

Figure 3 shows the calculated intergrain critical current densities J_{cJ} function of temperature for La and Sm samples.

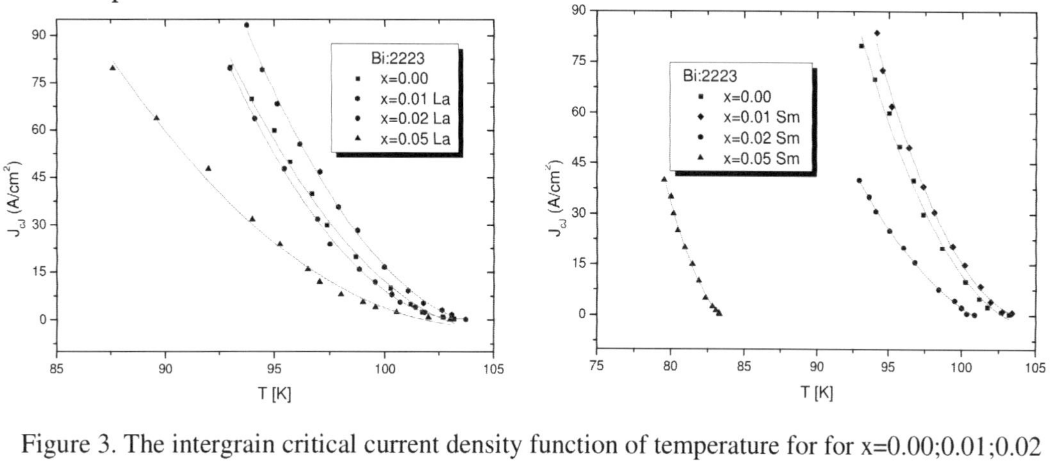

Figure 3. The intergrain critical current density function of temperature for for x=0.00;0.01;0.02 and 0.05 La and Sm samples

Best fit of the data from Fig.3 is described by:

$$J_{cJ}(T)= J_{cJ}(0)\ (\ 1-T/T_{p0}\ )^2 \quad (1)$$

The fitting parameters $J_{cJ}(0)$ and T_{p0} are presented in Table 2.

Sample	J_{cJ} (0) [A/cm^2]	T_{p0} [K]	J_{cJ} (77K) [A/cm^2]
x=0.00	7748	104	522
x=0.01 La	8168	106	613
x=0.02 La	6512	105	463
x=0.05 La	3243	104	220
x=0.01 Sm	7900	105.5	590
x=0.02 Sm	3336	105	243
x=0.05 Sm	11856	85.5	118

Table 2. The fitting parameters for intergranular critical current density J_{cJ}.

At lower temperatures the superconducting grains will be weakly coupled by the proximity effect superconductor-normal–superconductor (SNS) junctions or by Josephson effect in superconductor-insulator–superconductor (SIS) junctions. The dependence for J_{cJ} near T_c described by Eq.1 is expected for superconductor-normal –superconductor (SNS) junctions [10,11].

The critical current density $J_c(x)$ obtained from $\chi''(T)$ data shows a maximum value for $x=x_c$ near 0.01, and suggest that a critical low concentration x_c of 4f element create nanopinnig centers at grain boundaries of monophase Bi:2223 samples, which lead to a maximum value of intergranular critical current density. The $J_{cJ}(0)$=8168 A/cm^2 in 0.01 La sample is larger that $J_{cJ}(0)$=7748A/cm^2 in 0.00 sample. The value for the maximum current density at 77K, J_{cJ}(T=77K), is influenced by the nature of ion which substitute Ca. The temperature dependence of critical current density, $J_{cJ}(T)$, agree with the SNS behaviour of grain boundaries.

The effect of partial substitution of Ca by x=0.02 La, lead to a strong decrease of $J_{cJ}(0)$. This result and the calculated values for J_{cJ} (77K) (Table 2) suggest that the intergranular pinning energy is influenced by the partial substitution of Ca by x=0.02 La.

4. Conclusions

The influence of partial substitution of Ca by La and Sm on the structural and magnetic properties of (Bi,Pb):2223 bulk superconducting compound were studied.

X-ray diffraction measurements shown that the substitution of Ca by La and Sm influenced different the volume fraction of 2223 phase.

The temperature T_p for the imaginary part $\chi''(T)$ of the complex magnetic susceptibility performed at various field amplitudes is different influenced by the concentration of 4f element.

The critical current density Jc(x) obtained from $\chi''(T)$ data shows a maximum value for x=xc near 0.01 for La and Sm substituent and suggest that a critical low concentration xc of 4f element create nanopinnig centers at grain boundaries of monophase Bi:2223 samples, which lead to a maximum value of intergranular critical current density. The value for the maximum current density at 77K, J_{cJ}(T=77K), is influenced by the nature of ion which substitute Ca. The temperature dependence of critical current density, $J_{cJ}(T)$, agree with the SNS behaviour of grain boundaries.

Acknowledgments

The authors are grateful to CNMP-PN II Romania and GAR Romania for the financial support.

References

[1] Prabitha V G, Biju A, Kumar R G, Sarun P M, Aloysius A P and Yamaprasad U S 2005 *Physica C* **433** 28
[2] Ishizuka M and Sakuraba J 2006 *Physica C* **433** 173
[3] Nanda Kishore N, Muralidhar M, Hari Babu V, Pena O, Sergent M and Beniere F 1993 *Physica C* **204** 299
[4] Nanda Kishore N, Satyavathi S, Muralidhar M, Pena O and Hari Babu V 1993 *Physica C* **2235** 1521
[5] Pop A V 2000 *Mod.Phys.Lett. B* **14** 447
[6] Pop A V, Deac I G, Ilonca G, Ciurchea D and Pop V 2000 *Physica* B **284**
[7] Terzioglu C, Yilmazlar M, Ozturk O and Yanmaz E 2005 *Physica C* **423** 119
[8] Ozturk O, Akdogan M, Aydin H, Yilmazlar M, Terzioglu C and Belenli I 2007 *Physica* B **399** 94
[9] Bean C P 1962 *Phys.Rev.Lett.* **8** 250
[10] Ambegaokar V and Halperin B I 1969 *Phys.Rev.Lett.* **22** 1364.
[11] Liu H K, Dou S X, Savvides N , Zhou J P , Tan N X, Bourdillon A J and Sorrell C 1988 *Physica C* **157** 93

Improvement of superconducting properties of MgB_2 by changing the argon ambient pressure and sintering conditions

Burcu Savaskan[1], Kemal Ozturk[1], Sukru Celik[2] and Ekrem Yanmaz[1]
[1] Department of Physics, Faculty of Arts and Sciences, Karadeniz Technical University, 61080 Trabzon, Turkey
[2] Department of Physics, Faculty of Arts and Sciences, Rize University, 53100 Rize, Turkey.

E-mail: **savaskanb@hotmail.com, turkke@hotmail.com**

Abstract. We have investigated various characteristic properties depending on sintering conditions of MgB_2 samples prepared by the standard solid state reaction method. It is inferred from experimental results that the crystallinity of samples were improved when the pressure of the Ar ambient increased. Also, it was found that the sintering temperature above 850 ºC caused extremely high amount of decomposition of the superconductor phase. Finally, it was considered that the sintering process of MgB_2 must be carried out under the pressure of Ar ambient higher than 8 bar to impede the volatility of Mg in the structure of MgB_2. The J_c values of samples systematically enhanced with the increase of sintering time and in particular, the sample sintered for 180 min. exhibited the highest J_c (0) of 4.9×10^3 A cm^{-2} at 30 K. The obtained results demonstrate that the sintering conditions of MgB_2 have a significant influence on $T_{c\ (onset)}$ and J_c, which are directly related to practical applications of MgB_2 based superconductor components.

1. Introduction

The discovery of superconductivity at ~39 K in the magnesium diboride (MgB_2) compound stimulates scientific interest because of it's high critical temperature (T_c) value among the metallic superconductors. It has simple electronic structure, simple binary chemical composition and relatively low fabrication cost [1-4].

Due to the volatility of magnesium and the high melting point of boron, MgB_2 material usually grown in closed systems. In the various studies it is reported that high pressure techniques could be useful to prevent the evaporation of Mg from the compound and to suppress the decomposition of MgB_2 [5]. Any losses of Mg for forming the MgB_2 phase cause a generation of impurities resulting in poor microstructure as well as the superconducting properties such as critical current density [6]. Hence, fabrication of the MgB_2 bulk superconductor sample is generally performed under the inert gas ambient such as Argon, Hydrogen or Nitrogen. Technological applications of superconductors depend primarily on their critical current density property. The various experimental results have demonstrated that the sample of bulk MgB_2 has rather high values of critical current density (J_c) at zero field, but exhibits a rapid decrease of J_c in an applied magnetic field. These problems inflict strong limitations on use of MgB_2 e.g. for energy storage system and superconducting magnets. It is known that the field dependence of J_c are related to the presence of structural defects that can act pinning center and a lack of natural defects in MgB_2 may be responsible for the rapid decline of J_c with increasing field [7]. Many attempts, such as element addition or doping (Cu, Co, Li, Mo, C) [8, 9], nanoparticles (SiC) addition, the introduction of defects by irradiation [10, 11] etc., have been made to improve pinning properties and J_c of the MgB_2 sample. Although the addition or doping of several impurity compounds or elements have been found to be effective in improving the pinning and the critical current properties. Additionally, most likely the actual composition of MgB_2 and the different fabrication and processing conditions are responsible for the different pinning properties.

[1] Corresponding author: Phone: +90 462 325 31 97, Fax: +90 462 325 31 95

Consequently, clarifying the mechanism influencing intrinsic pinning properties of the undoped MgB_2 is very important for practical applications and future investigations.

In the present study, we have investigated the effect of the sintering temperature, sintering time and Ar ambient pressure on the structural and superconducting properties in the MgB_2 samples prepared by a solid state reaction method. We have measured the temperature and field dependence of the magnetization at the different sintering conditions considering the shifting intrinsic pinning behavior of the samples. The experimental result shows that the structural and superconducting properties such as critical current density and flux pinning can be improved with varying of the sintering conditions of the MgB_2 sample.

2. Experimental details

Commercial powder of MgB_2 (Alfa Aesar) with nominal 99% purity was apportioned to 0.6 g, each of which was pressed into a pellet of 13mm in diameter under the pressure of 10 ton. Each pellet was transferred into a stainless steel tube and vacuumed to 10^{-3} bar using a rotary pump at room temperature. Then, vacuum valve was closed and Ar gas valve opened to set Ar gas pressure. After these processes the samples named as the pressure, temperature and time series were performed experimental procedure as below.

For pressure series; after vacuum to 10^{-3} bar using a rotary pump at room temperature the stainless steel tube was put into preheated tube furnace at 1050 °C. In order to investigate the effect of Ar gas pressure, the samples were sintered at 4, 6, 8 and 10 bar in Ar ambient at 1050 °C for 3 h. For temperature series; in order to investigate the effect of sintering temperature, some samples were sintered at 800, 850, 900, 950 and 1000 °C with a 3 h constant sintering time and with 8 bar constant Ar ambient pressure.

For time series; in order to investigate the effect of sintering time, the samples were sintered at 800 °C for 20, 40, 60, 120 and 180 min under constant 8 bar Ar ambient pressure.

The sintering and oxygenation temperatures of the samples were determined from the differential thermal analysis measurement (DTA) with model NETZSCH. The powders and bulk XRD data were collected over a 2θ range from 3°-70°, at a step of 0.02° at room temperature, using a Rigaku D/Max-IIIC X-ray diffractometer with CuK_α radiation. The magnetization properties were measured using a Quantum Design PPMS and VSM system. The M(H) properties of MgB_2 samples dependent on sintering time and temperature were measured up to 0.5 T for the constant temperature with 15 and 30 K while the M(T) properties were measured at 0.1; 0.2; 0.3; 0.4 and 0.5 T under the zero field-cooling regime (ZFC). The measurements were performed by the sweep rate of 5 mT s^{-1}. All the magnetization measurements were made afterwards by the first cooling the sample in zero field and then applying a field to begin the measurement. All samples were rectangular and typical dimensions were approximately 1.3x2.4x4.1 mm^3 respectively for magnetization studies.

3. Results and discussion

Figure 1 shows a DTA curve taken in the temperature ranges 30 °C to 1000 °C on MgB_2 powders in air atmosphere. A wide peak observed at around 250°C with small intensity which is compatible with the beginning of the forming temperature of the MgO phase with joining together magnesium and oxygen in MgB_2 compound. A high endothermic peak occurred at 710 °C because of the oxidation MgB_2 grains in air atmosphere is shown in Fig 1. Finally, the peak seen over the 900 °C is considered as the peritectic temperature of the Mg decomposing in a liquid state [12].

Figure 2 shows the room temperature XRD patterns of the pressure series at 1050 °C for 3 h under 4, 6, 8 and 10 bar Ar gas pressure. The main phase of the all pressure series samples are orthorhombic MgB_4. In the sample sintered at 1050 °C for 4 bar Ar gas only low intensity MgB_4 phase peaks were seen. Figure 2 clearly indicates that portion of the superconductor MgB_2 phase increases when the Ar ambient pressure increases. Due to volatility of Mg at high temperature sintering MgB_2 material decomposed easily and so the increment of the MgB_4 phases as shown in Fig 2 (a) [13].

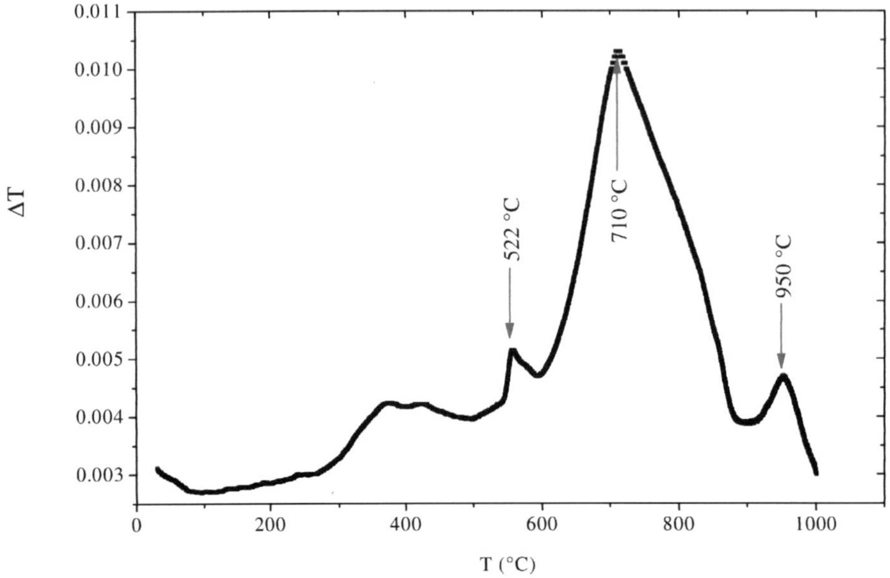

Figure 1. DTA result on MgB_2 powder in air atmosphere from 30°C to 1000 °C.

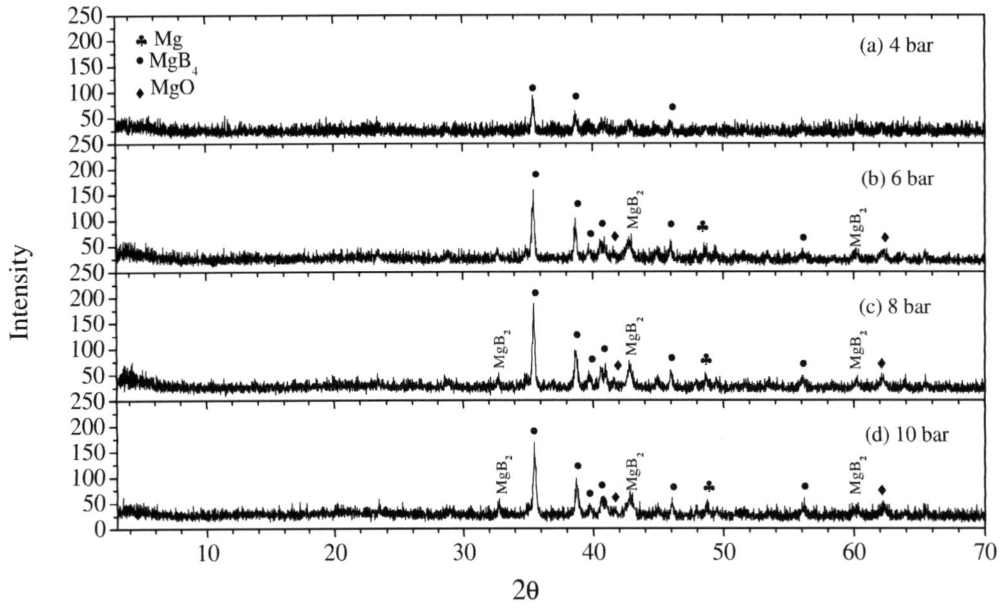

Figure. 2. The X-ray diffaction patterns of bulk MgB_2 sintered at 1050 °C for 3 h under (a) 4 bar , (b) 6 bar, (c) 8 bar and (d) 10 bar Ar gas pressure.

This clarifies that sintering the MgB_2 material at 1050 °C is too high to improve properties. In addition, it was found that during the sintering process the pressure of 4 bar Ar atmosphere is inadequate to prevent skip out of the Mg from sample surface. A little increase of the peak intensities and portion belong to MgB_2 phase with increasing of Ar gas pressure indicate that the skip out of Mg from sample surface can be hindered partially.

Figure 3 shows the typical XRD patterns of MgB_2 bulk samples sintered under 8 bar Ar gas pressure at (a) 800 °C, (b) 900 °C and (c) 1000 °C for 180 min. It is clearly observed that the samples sintered at 800 and 900 °C, show single phase except a small amount of MgO phase. Also, it is revealed that the sample sintered at 1000 °C has MgB_4 and MgO impurity phases in addition to MgB_2 phase. It can be seen from XRD patterns that the MgB_4 and MgO impurity peaks intensity increase as the sintering temperature increases. This case is consistent with the fact that the amount of the MgB_4 phase increases with the Mg-deficiency [13]. So, in this work the increasing amount of MgB_4 induced the Mg-deficient. It was also reported that the obtained MgB_2 sample sintered at 600 °C, revealed poor crystallinity from XRD and unreacted Mg peaks originating from low temperature processing [14, 15]. It is seen in Figure 3 that the peak intensities of MgB_2 decreases when the sintering temperature increased. The reason to that can be attributed to the decomposition of MgB_2 due to volatility of Mg and so the enlargement of the MgB_4 phases. Consequently, the optimum sintering temperature is found to be around 800 °C for forming the bulk MgB_2. In addition, it was seen from the XRD that the sample sintered at 800 °C for (a) 20, (b) 60 and (c) 120 min. sintering times, the dominant peaks were MgB_2 phase and a minor amount of MgO phase found in all the sintering times. Although the sample crystallinity almost the same for all the sintering times.

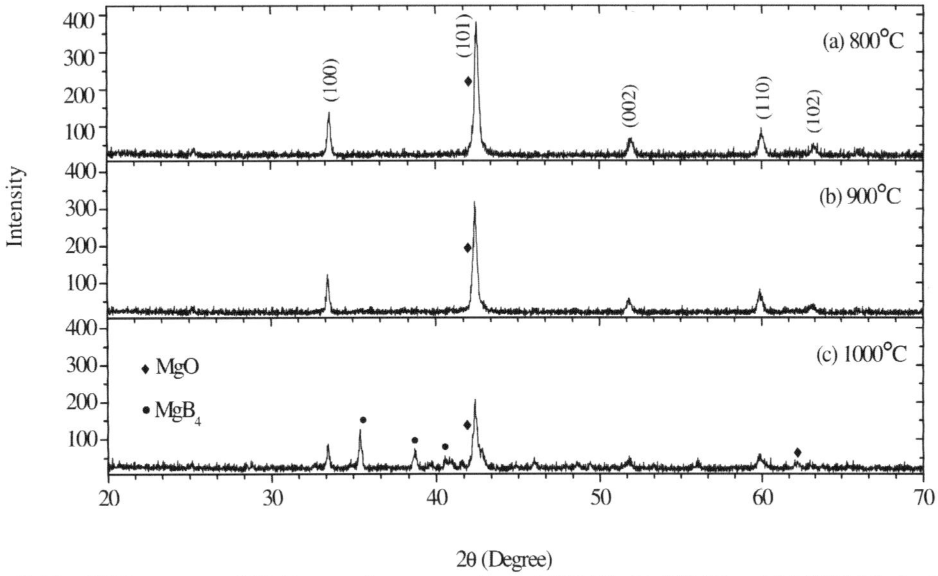

Figure 3. The XRD patterns of MgB_2 samples prepared at (a) 800 °C, (b) 900 °C and (c) 1000 °C for 180 min. under 8 bar Ar gas pressure.

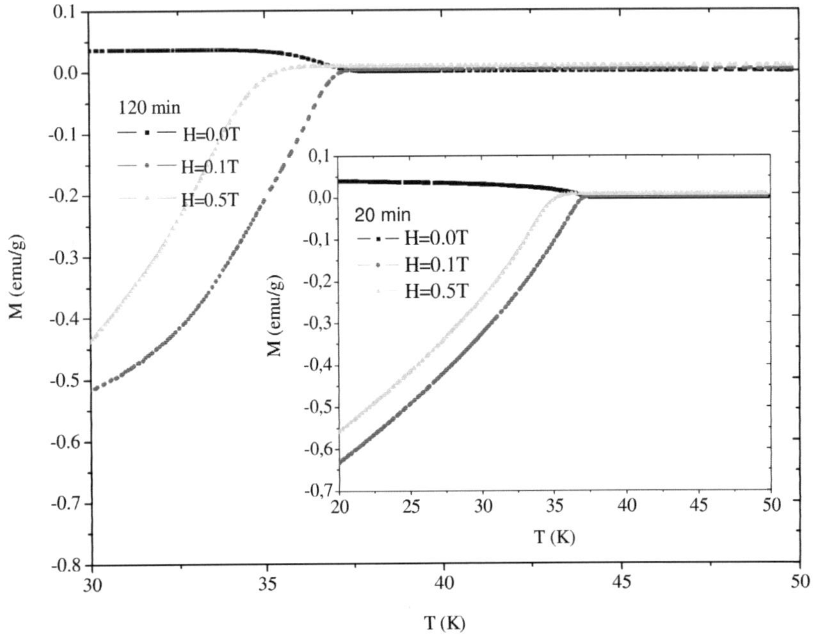

Figure 4. Temperature dependence of magnetization for MgB_2 samples sintered at 800 °C for 20 and 120 min.(see inset figure) under the zero field cooling regime (ZFC)

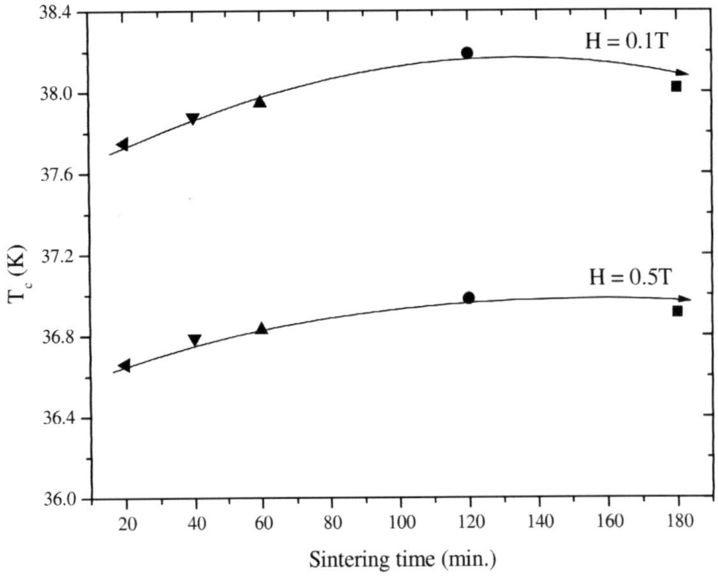

Figure 5. Relationship between superconducting transition temperature $T_{c\ (onset)}$ and sintering time for the sample sintered at 800 °C and Tc measured in the field of 0.1 and 0.5 T under the ZFC regime.

The temperature dependence of magnetization was measured in the field ranged from 0.1 T to 0.5 T in ZFC mode in order to determine the superconducting transition temperature T_c. The superconducting transition temperature $T_{c(onset)}$ for the samples sintered at different temperatures as 800, 850, 900, 950 and 1000 °C were determined using M-T curves to be 38.02, 37.90, 37.87, 37,86 and 37.84 respectively (under 8 bar Ar gas pressure for 3h). It was indicated above that the sintering temperature above 850 °C cause the high amount of decomposition of MgB_2 superconductor phase. Consequently, it is consider that sintering process requires for pressure of Ar atmosphere higher than 8 bar to impede the volatility of Mg for sintering temperatures over 800 °C. Figure 4 shows the temperature dependence of magnetization for MgB_2 samples sintered at 800 °C for 20 and 120 min. (see inset figure) under the ZFC regime. The relationship between superconducting transition temperature $T_{c\ (onset)}$ and sintering time for sample sintered at 800 °C and measured in the field ranged from in 0.1 and 0.5 T was presented in Figure 5. The value of $T_{c\ (onset)}$ ascended with increasing the sintering time between 20-120 min. and descended slightly for 180 min. The lowered $T_{c\ (\mathrm{onset})}$ as 37.75 K at 0.1 T observed for MgB_2 bulks sintered at 800 °C for 20 min. can be explained by poor crystallinity. Because the improvement in superconducting transitions with sintering time coincide well with the improvement of crystallinity as shown Figure 5. It was reported in various studies [13, 16] that the amount of the MgB_4 impurity phase increases because of Mg-deficient due to volatility of the Mg when the sintering time and temperature increases. The decrease in the $T_{c\ (onset)}$ for 180 min. signifies the presence of weak links among grains of the MgB_2.

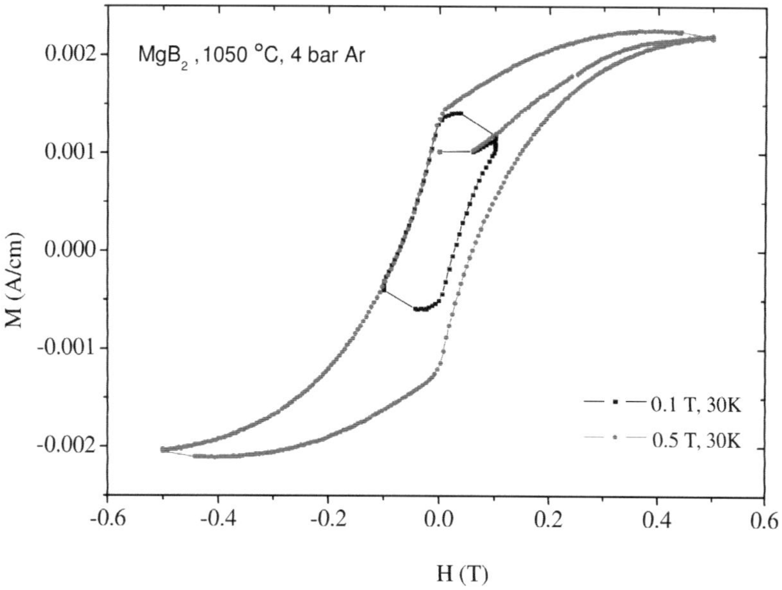

Figure 6. The magnetization hysteresis loop M(H) measured at 30 K of MgB_2 sample sintered at 1050 °C for 3 h under 4 bar Ar gas pressure.

Figure 6 shows the magnetization hysteresis loop M(H) measured at 30 K for the sample sintered at 1050 °C for 3 h under 4 bar Ar gas pressure. As shown in Fig 6 the sample exhibits a typical ferromagnetic behavior. M(H) curves in Figure 6 indicate that during the sintering process the pressure of 4 bar Ar gas is insufficient to prevent skip out of Mg from sample surface and so deterioration of stoichiometry. In addition, it was found that sintering temperature of 1050 °C was too high for MgB_2 superconductor compound. In order to study sintering time effects on the superconducting magnetic properties, we examine the effects of sintering time on the superconducting magnetic hysteresis and critical current density. Figure 7 shows the magnetization hysteresis loops M(H) measured at 30 K for MgB_2 samples sintered at (a) 800, 900 and 1000 °C and (b) sintered at 800 °C for 20, 40, 60, 120 and 180 min. sintering times. It is clearly seen in Figure 7 (a) that the value of magnetization decreased with increasing of sintering temperature and the best condition found to be 800 °C. The curves of Figure 7 (b) clearly indicate that the magnetization values systematically increase with increase of sintering time. Relation between magnetization loops width and the number of flux pinning center in the material were reported in many times [17, 18]. In Figure 7 increasing of magnetization loops width as sintering time increase implies that the amount of the pinning centers improved. The critical current densities J_c values were calculated from the magnetization hysteresis data using the Bean critical state model [19] using the relation: $J_c = 20(M^+ - M^-)/L_1(1-L_1/3L_2)$, where M^+ and M^- are magnetic moment when increasing and decreasing the field, respectively. L_1 and L_2 are sample dimensions perpendicular to the field in cm with $L_1 < L_2$. The magnetic field dependence of critical current densities $J_c(H)$ at 30 K for time series samples were presented in Figure 8. The J_c values are systematically enhanced with an increase of sintering time.

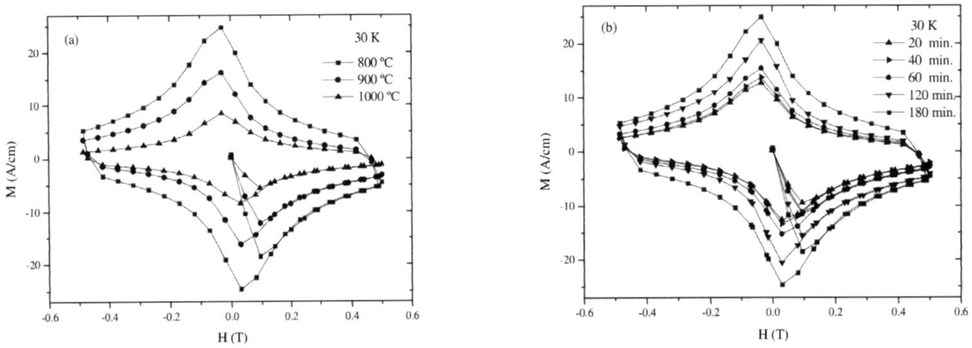

Figure 7. The magnetization hysteresis loops M(H) measured at 30 K for MgB_2 samples (a) sintered at 800, 900 and 1000 °C and (b) sintered at 800 °C for 20, 40, 60, 120 and 180 min. sintering times.

In particular, the sample sintered for 180 min. exhibited the highest J_c value of 4.9×10^3 A cm^{-2} at 30 K. Enhancement of J_c by increasing of sintering time can be explained by the improvement of grain connectivity because of MgB_4 nanoinclusions. As known the small grain size of MgB_4 leads to an increment of effective surface area between MgB_2 and MgB_4 grains that may be increase the quality of grain connection of the MgB_2 polycrystals. Also, the small grain size of MgB_4 sintered 800 °C matches well with the coherence length of MgB_2 (about 12 nm). Consequently, these impurities can act as pinning centres in sample. For all these reasons, we consider that the MgB_4 nanoinclusions are responsible for the increase in $J_c(H)$ for the Mg-deficient samples induced by the convenient sintering time and temperature. This result is consistent with a previous report [20]. In addition, it was found that value of the J_c decreased when the sintering temperature was increased.

In conclusion, we have investigated various characteristic properties depending on sintering conditions of MgB_2 samples synthesis by the solid state reaction method. It was deduced from experimental results that the crystallinity improved when the pressure of the Ar ambient increases. Additionally, it was found that the sintering temperature above 850 °C causes high amount of decomposition of MgB_2 phase. It was thought that the sintering process to make bulk structure have need for pressure of Ar atmosphere higher than 8 bar to impede the volatility of Mg. In general it was found that, the critical transition temperature increases with sintering time and it shows a maximum value as 38.19 K for 120 minutes. The J_c values systematically enhance with decreasing of sintering temperatures and increasing sintering times. Specially, the sample sintered at 800 °C for 180 min. (under 8 bar Ar ambient pressure) exhibited the highest J_c of 1.5×10^4 and 4.9×10^3 A cm^{-2} respectively for 15 and 30 K measuring temperatures. The acquired results demonstrate that the sintering conditions of MgB_2 have a significant influence on $T_{c\ (onset)}$ and J_c, which are directly related to practical applications of MgB_2-based superconductor components. The suitable sintering temperature, sintering time and Ar ambient pressure were determined to be 800 °C, 2–3 h and over 8 bar respectively.

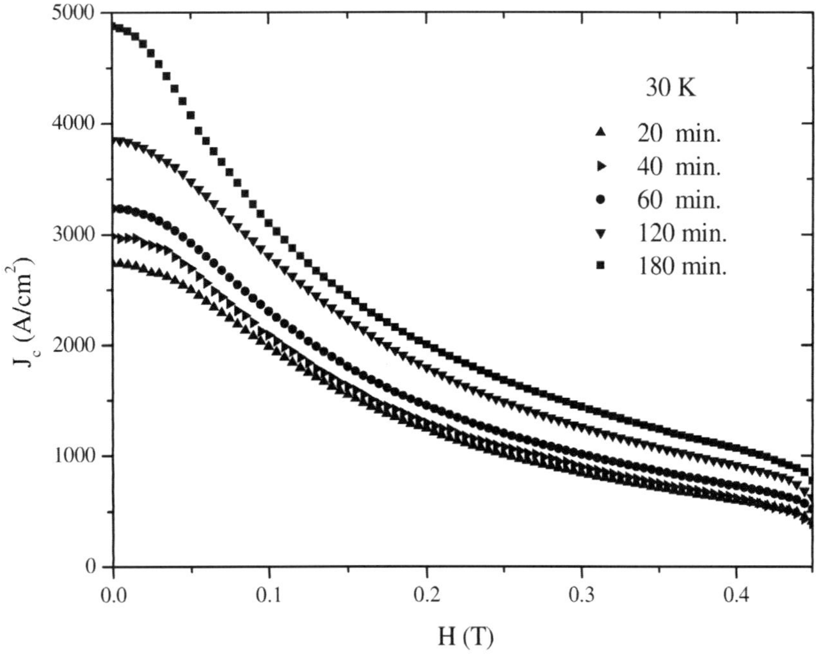

Figure 8. Magnetic field dependence of critical current densities $J_c(H)$ at 30 K for MgB_2 samples sintered at 800 °C for 20, 40, 60, 120 and 180 min.

References

[1] Nagamatsu J, Nakagava N, Muranaka T, Zenitani Y and Akimitsu J 2001 *Nature* **410** 63

[2] Huang X, Mickelson W, Christopher Regan B and Zettl A 2005 *Solid State Commun.* **136**, 278

[3] Du W, Xu H, Zhang H, Xu D, Wang X, Hou X, Wu Y, Jianga F and Qin L 2006 *J. Crystal Growth* **289** 626

[4] Wang S, Zhu Y, Liu Z, Zhou Y, Zhang Q, Chen Z, Lu H and Yang G 2003 *Chin. Phys. Lett.* **20** 1356

[5] Angst M, Puzniak R, Wisniewski A, Jun J, Kazakov S M, Karpinski J, Roos J and Keller H 2002 *Phys. Rev. Lett.* **88** 167004

[6] Yamamoto A, Shimoyama J, Ueda S, Katsura Y, Horii S and Kishio K 2004 *Supercond. Sci. Technol.* **17** 921

[7] Bugoslavsky Y, Perkins G K, Qi X, Colen L F and Caplin A D 2001 *Nature* **410** 563

[8] Hishinuma Y, Kikuchi A, Iijima Y, Yoshida Y, Takeuchi T and Nishimura A 2006 *Supercond. Sci. Technol.* **19** 1269

[9] Yakinci ME, Balci Y, Aksan MA, Adıguzel H I and Gencer A 2002 *J. Superconductivity* **15** 607

[10] Dou S X, Braccini V, Soltanian S, Klie R, Zhu Y, Li S, Wang X L and Larbalestier D 2004 *J. Appl. Phys.* 96 7549

[11] Bugoslavsky Y, Cohen L F, Perkins G K, Polichetti M, Tate T J, Gwilliam R and Caplin A D *Nature* **411** 561

[12] Shen J, Fang M, Zheng Y, Wang H, Lu Y and Xu Z 2003 *Physica C* **386** 663

[13] Bayazit E, Yakinci ME, Balci Y, Aksan M A and Balci Y 2007 *Physica C* **460 610**

[14] Chen S K and Glowacki B A 2006 *Supercond. Sci. Technol* **19** 116

[15] Bayazit E, Altin S, Yakinci ME, Aksan M A and Balci Y 2008 *J. Alloys Compd.* 457 42-46

[16] Zhao Y G, Zhang X P, Qiao P T, Zhang H T, Lia S L, Cao B S, Zhu M H, Han Z H, Wang X L and Gu B L 2001 *Physica C* **366** 1

[17] Feng Y, Zhou L, Wen J G, Koshizuka N, Sulpice A, Tholence J L, Vallier J C and Monceau P 1998 *Physica C* **297** 75

[18] Öztürk K, Çelik Ş, Çevik U and Yanmaz E 2007 *J. Alloys Compd.* **433** 46
[19] Bean C P 1962 *Phys. Rev. Lett.* **8** 250
[20] Xu G J, Pinholt R, Bilde-Sørensen J, Grivel J C, Abrahamsen A B and Andersen N H 2006 *Physica C* **434** 67

Influence of gold diffusion-doped on phase formation, superconducting and microstructure properties of $Bi_{1.8}Pb_{0.35}Sr_{1.9}Ca_{2.1}Cu_3O_y$ superconductors

O Ozturk[1], **M Akdogan**[2], **C Terzioglu**[2,*], **and A Gencer**[3]
[1] Department of Physics, Faculty of Arts and Sciences, Kastamonu University, 37600 Kastamonu-Turkey,
[2] Department of Physics, Faculty of Arts and Sciences, Abant Izzet Baysal University, 14280 Bolu-Turkey,
[3] Department of Physics, Faculty of Arts and Sciences, Ankara University, 06100 Ankara-Turkey.

E-mail: terzioglu_c@ibu.edu.tr

Abstract. We report on low-field magnetic properties of gold diffusion-doped $Bi_{1.8}Pb_{0.35}Sr_{1.9}Ca_{2.1}Cu_3O_y$ superconducting bulk samples by performing ac susceptibility measurements. The undoped samples were prepared by the standard solid-state reaction method. Doping of $Bi_{1.8}Pb_{0.35}Sr_{1.9}Ca_{2.1}Cu_3O_y$ was carried out by means of Au-diffusion during sintering from an evaporated gold film on pellets. To investigate the effect of gold-diffusion and diffusion-annealing duration on transport, magnetic and microstructure properties of the superconducting samples we performed magnetoresistivity, scanning electron microscopy (SEM) and X-ray diffraction (XRD) measurements. The ac susceptibility as a function of temperature measurements were carried out at different values of the ac magnetic field amplitudes (Hac) in the range between 20A/m and 320 A/m for 211 Hz. The imaginary part of ac susceptibility measurements is used to calculate intergranular critical current density $J_c(T_p)$ using the Bean Model. $J_c(T_p)$ is seen to increase from 60 A cm^{-2} to 90 A cm^{-2} with increasing diffusion-annealing time from 10 h to 50 h. The peak temperature, T_p, in the imaginary part of the ac susceptibility is shifted to a lower temperature with decreasing diffusion-annealing duration as well as increasing ac magnetic fields. The force pinning density ($\alpha_j(0)$) increased with increasing diffusion annealing time. The value of T_c in gold-diffused samples, in comparison with the undoped samples, increased from 100 $\pm$ 0.2 K to 104$\pm$0.2 K. It was observed that the value of $T_{c\text{-offset}}$ of the gold-doped samples enhanced with further increasing diffusion-annealing duration. XRD patterns and SEM micrographs are used to obtain information about Bi-2223 phase ratio, lattice parameters and grain size of the samples. Gold doping enhanced the formation high-T_c phase and increased the grain size. The possible reasons for the observed improvements in transport, microstructure and magnetic properties due to Au diffusion and diffusion-annealing time were discussed.

* Cabir Terzioglu, Department of Physics, Faculty of Arts and Sciences, Abant Izzet Baysal University, 14280 Bolu-Turkey.

1. Introduction

The discovery of high temperature superconducting materials has generated great interest of the scientific community for many reasons [1]. Since high-T_c cuprate superconductors are chemically unstable and highly reactive, most materials including the noble metals and oxides react with these superconductors when they come into contact. Bi-based superconductors have attracted the attention of many researchers for its high transition temperature, critical current density and easy synthesis [2-8]. Bi–Sr–Ca–Cu–O system is an important member of the high-T_c superconductor families with an orthorhombic structure and it has been found at least in three phases with atomic ratios in the order of Bi, Sr, Ca, Cu, namely (2201), (2212) and (2223) with zero resistivity transition temperatures of 20 K, 85 K and 110 K, respectively [9]. In these series, Bi-2223 is the most attractive because of having the highest superconducting transition temperature.

As part of a study of doping, the effects of additions of oxides such as SiO_2, Al_2O_3, MgO, ZrO_2 and 3d transition metal oxides on $YBa_2Cu_3O_{7-x}$ (YBCO) [10-11] and $(Bi,Pb)_2Sr_2Ca_2Cu_30_{10-y}$ (BPSCCO) [12] have been investigated. Among the metals, Au and Ag have been found to be non-poisoning to $YBa_2Cu_3O_{7-x}$ [13], while Ag has been found to be the only metal element that is non-poisoning to the BSCCO system [14]. It has been reported that Au doping reduces the superconductivity of Bi(Pb)SrCaCuO [14-15], which is unexpected since Au and Ag are chemically very similar. Dzhafarov et al. [15] investigated the effect of Au diffusion combined with slow and fast cooling on T_c and J_c in Bi-2223 system and found that Au diffusion promotes formation of the high-T_c (Bi-2223) phase and increases significantly J_c. In the resent work [16], the role of diffusion-annealing time on the mechanical properties of gold diffusion-doped Bi-2223 superconductors was investigated. That work indicated that the Au-doping and diffusion-annealing duration increased the Vickers microhardness, Young's modulus, yield strength and fracture toughness of the samples. It was found that the microhardness values decreased with increasing applied load. We also investigated the indentation size effect (ISE) by using different models such as Meyer's law, Hay-Kendall approach and modified proportional specimen resistance model. The Hay-Kendall approach was sufficient for describing the observed ISE [17]. Besides, the gold diffusion in $Bi_{1.8}Pb_{0.35}Sr_{1.9}Ca_{2.1}Cu_3O_y$ has been studied over the temperature range of 500-830$^{\circ}C$ using the technique of successive removal of thin layers and calculation of the lattice parameter c from XRD patterns [18].

In the present work, we reported the results of ac susceptibility, magnetoresistivity, SEM and XRD measurements in gold diffusion-doped Bi-2223 system. The possible reasons for the observed improvements in transport, magnetic and microstructure properties due to Au diffusion and diffusion-annealing duration were discussed.

2. Experimental details

Superconducting $Bi_{1.8}Pb_{0.35}Sr_{1.9}Ca_{2.1}Cu_3O_y$ samples were prepared by the standard solid-state reaction method. The weighed powders of Bi_2O_3 (99.99%), PbO (99.9+%), $SrCO_3$ (99.9+%), $CaCO_3$ (99+%) and CuO (99+%) in the cation ratio of Bi:Pb:Sr:Ca:Cu = 1.8:0.35:1.9:2.1:3 were mixed by grinding machine for 24 h. After milling, the mixed powders were calcined in air at 700, 750 and 800 °C for 24 h. At every calcined temperature, the sample cooled to room temperature to grind. The calcined material was reground and pressed into pellets of 10x4x2 mm3 at 300 MPa. The pellets were sintered in air at 830 °C for 48 h and then cooled down to room temperature. The heating and cooling rates of the temperature were chosen to be 10 and 3 °C min-1, respectively. The Au evaporation (thickness of about 50μm) on one face of the samples was carried out using an AUTO 306 Vacuum Coater (EDWARDS). Doping of $Bi_{1.8}Pb_{0.35}Sr_{1.9}Ca_{2.1}Cu_3O_y$ was carried out by means of Au-diffusion during sintering from an evaporated gold film on pellets. The Au layered superconducting samples were annealed at 830 °C for 10 h (G1), 20 h (G2) and 50 h (G3) (Fig. 1). At the end of this run, Au diffusion was realized through the samples. For comparison, an undoped sample (G0) was also annealed under the same conditions.

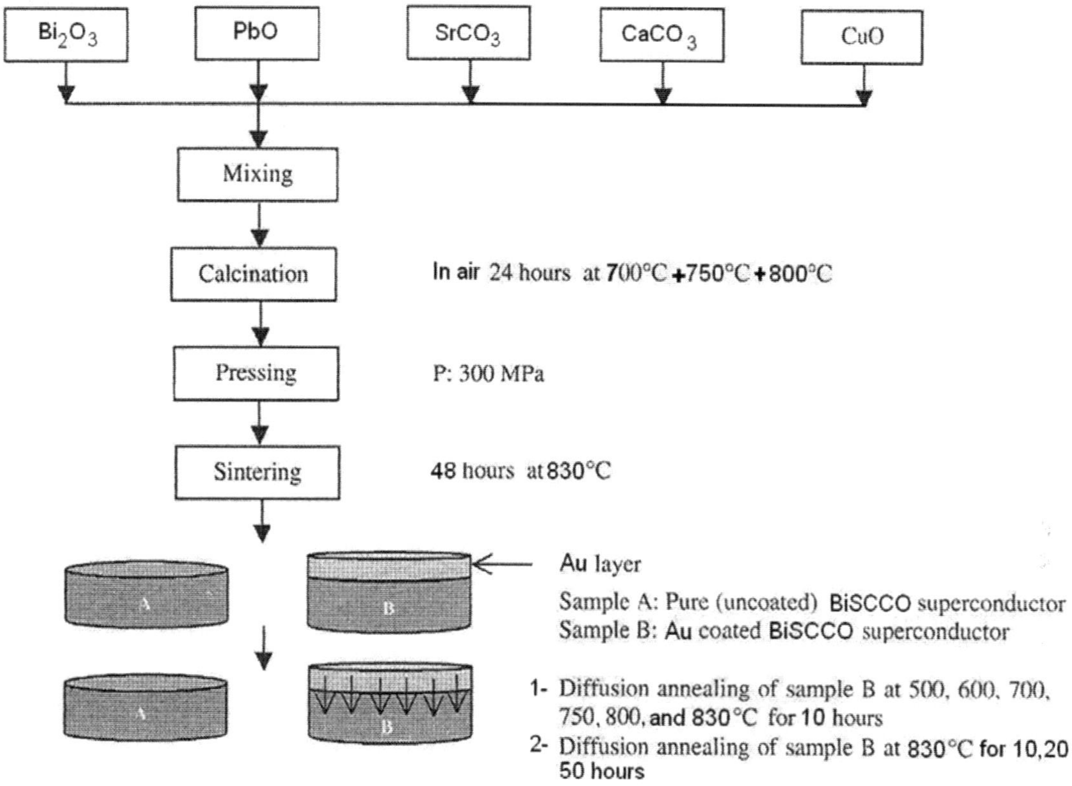

Figure 1. Schematic drawing of the sample preparation.

Ac susceptibility measurements on the samples were performed using a 7130 AC susceptometer of Lake Shore with a closed cycle refrigerator at 211 Hz frequency [19-20]. The ac susceptibility as a function of temperature (40–120 K) at fixed ac magnetic field amplitude (H_{ac}) in a range from 20 to 320 A/m was performed on the samples. The ac magnetic field was applied along the longest dimension of the samples. The measurements of ac susceptibility are commonly used to determine magnetic and superconducting properties of materials. In particular, the ac susceptibility measurement is useful in distinguishing between inter- and intra-grain properties. At high ac fields, two loss peaks can usually be seen in the imaginary part of susceptibility data; a broad peak at low temperature (coupling losses) due to the motion of inter-granular (Josephson) vortices [21] and a narrower peak (intrinsic losses) due to the motion of intra-granular Abrikosov vortices near T_c^{onset} [22]. We observed only one peak (inter-granular) in each of the χ'' vs. T plots at low fields, but at high fields (160 and 320 A/m) one more peak (intragranular) was observed. From the χ' and χ'' vs. T plots, the values of the T_c^{onset} (diamagnetic onset temperature) and T_p (temperature corresponding to χ'' peak) were obtained, respectively. Bean critical-state model [23] is used for determination of inter- and intra-grain critical current densities from ac susceptibility measurements.

The measurement of dc resistivity was performed with the four-probe method on all samples. Both voltage and current contacts were made with silver paint. Magnetoresistivity measurements were made under different dc magnetic field (0, 0.1, 0.2, 0.3 and 0.5 T) in the temperature rage between 60 K and 130 K. The magnetic field was applied parallel to the current direction by an electromagnet. The magnetic field was applied parallel to the current direction. The external dc magnetic field was provided by an electromagnet. The transition temperature T_c was determined as the temperature at which zero resistivity was achieved.

XRD data were taken using a Rigaku D/Max-IIIC diffractometer with CuKα radiation in the range 2Θ = 4°- 60° with a scan speed of 3° /min and a step increment of 0.02° at room temperature. XRD data was taken from powder of the samples. Phase purity and the lattice parameters were obtained from these XRD patterns. The accuracy in determining the lattice parameter c was ± 0.001 Å.
The surface morphologies of the Au diffusion-doped and pure samples were studied by using a Philips XL30 SFEG scanning electron microscope.

3. Results and discussion

The samples produced are named as G0 (undoped sample annealed at 830 $^\circ C$ for 10 h), G1 (gold diffused sample annealed at 830 $^\circ C$ for 10 h), G2 (gold diffused sample annealed at 830 $^\circ C$ for 20 h), and G3 (gold diffused sample annealed at 830 $^\circ C$ for 50 h).

3.1. Ac susceptibility

We have investigated the effect of Au diffusion and diffusion-annealing duration on some physical properties of $Bi_{1.8}Pb_{0.35}Sr_{1.9}Ca_{2.1}Cu_3O_y$ high-T_c superconductors using the low field ac magnetic susceptibility method. The temperature and magnetic field dependence of the complex ac susceptibility, $\chi(T) = \chi'(T) + i\chi''(T)$, of the samples are shown in Fig. 2. χ' and χ'' measure the amount of shielding and the amount of energy loss due to the induced current, respectively. As seen from the real parts of the plots, the onset of critical temperature, T_c^{onset}, appeared at 110 K at different magnetic field values for the all samples. They show field independent behaviors until 102 K. This temperature is the increasing point of the imaginary part. The susceptibility-temperature curves shifted to lower temperature, considerably increased the transition width and also decreased the shielding fraction of the superconducting phase in the samples with increasing the ac field amplitudes and decreasing the diffusion-annealing duration as depicted in Figs. 2 and 3. This similar effect of the diffusion-annealing time and the applied external dc field in transition width is revealed for the dc resistivity measurements in the current study.

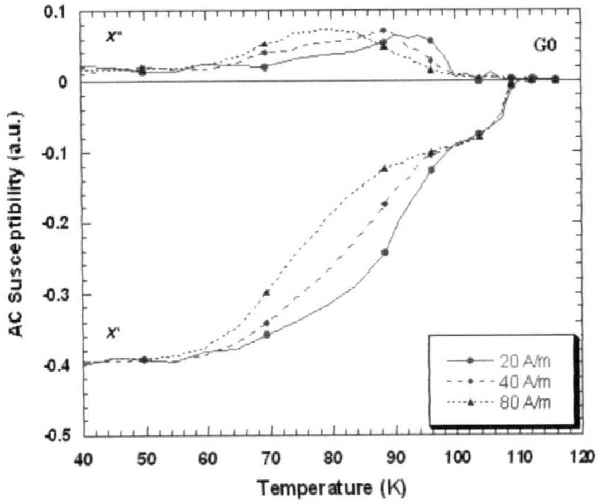

(a)

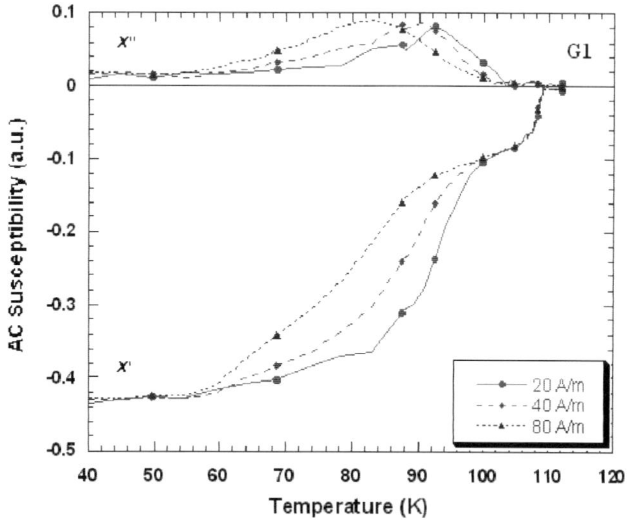
AC Susceptibility (a.u.)
Temperature (K)
X''
X'
G1
20 A/m
40 A/m
80 A/m

(b)

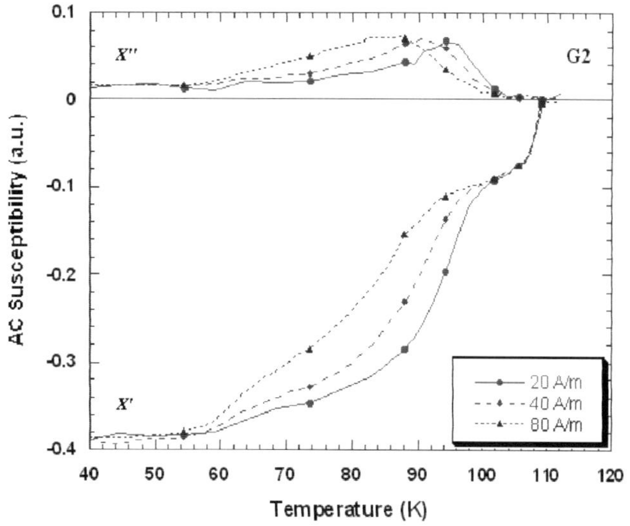
AC Susceptibility (a.u.)
Temperature (K)
X''
X'
G2
20 A/m
40 A/m
80 A/m

(c)

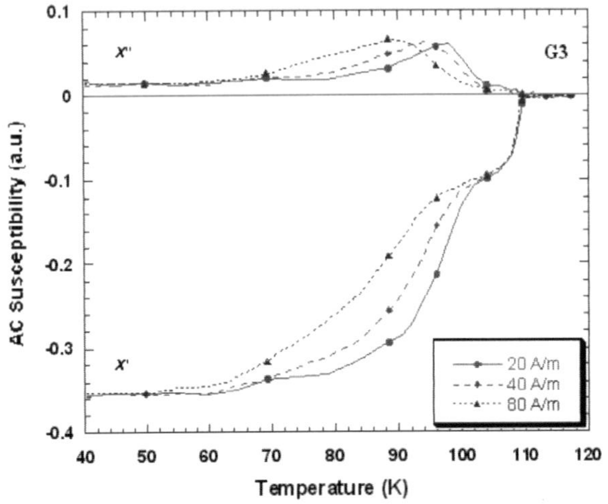

(d)

Figure 2. The temperature dependence of ac susceptibility of the (a) *G0,* (b) *G1*, (c) *G2* and (d) *G3* samples for 20, 40 and 80 A/m.

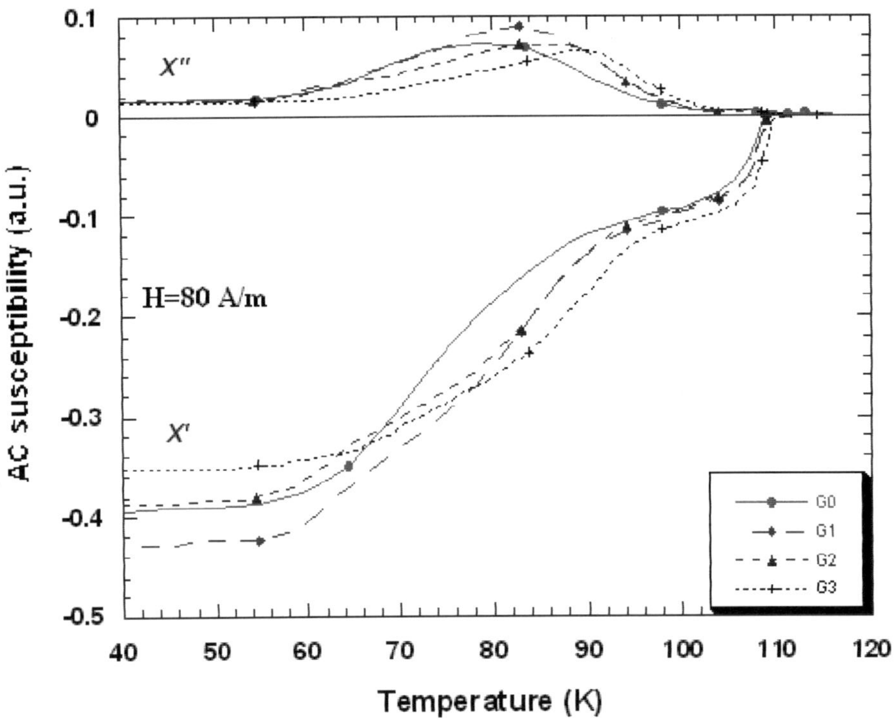

Figure 3. The temperature dependence of ac susceptibility of the *G0, G1,G2* and *G3* samples for 80 A/m.

Table 1. Peak temperature T_p, and applied ac magnetic field amplitude H_{ac} of the samples.

Samples		H_{ac} (A/m) 20	40	80	160	320
G0	T_p(K)	90	88	79	69	64
G1		92	89	83	74	66
G2		95	91	85	76	67
G3		98	94	89	79	70

The maximum peak of χ'' appears at a temperature T_p where the intergranular field penetrates the centre of sample. As seen in Figure 2, T_p values of the samples shifted to lower temperature and the intensity of the imaginary parts of susceptibility slightly increased with increasing the ac field amplitude. The imaginary parts of the G0 sample show peaks at 90, 88, 79, 69 and 64 K for the fields 20, 40, 80, 160 and 320 A/m, respectively. The other samples (G1, G2 and G3) showed the same behaviour and corresponding data are tabulated in Table 1. The shift is related to the pinning force strength. It was found that the overall susceptibility curves of the samples shifted to higher temperatures and increased the shielding fraction of superconducting phase for all ac field amplitudes with increasing the diffusion-annealing duration as shown in Figure 3 for 80 A/m. A similar behavior is seen for the other field values. At large ac field amplitudes two loss peaks are usually found; a broad peak at low temperature and a narrower peak near T_c^{onset}. We did not see any intrinsic loss peaks in our samples for 20, 40 and 80 A/m because of the low value of the applied field. This may be due to small grain size and no flux penetration into the grains [24]. For higher field values namely for 160 and 320 A/m the intragranular (intrinsic loss) peak is clearly observed (Fig.4).

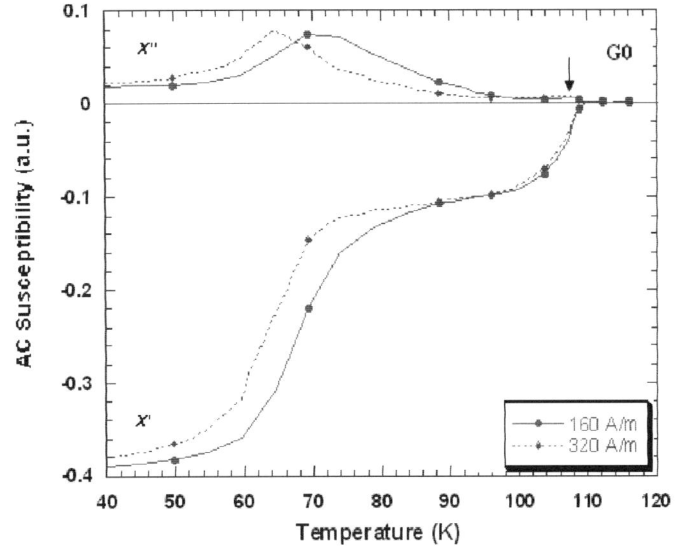

(a)

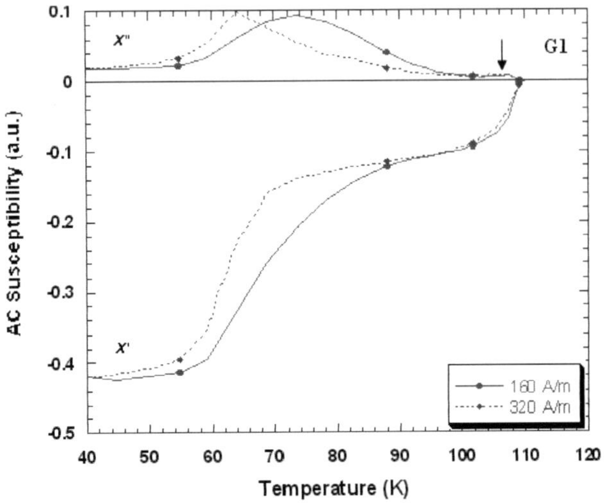
G1
χ"
χ'
160 A/m
320 A/m
AC Susceptibility (a.u.)
Temperature (K)

(b)

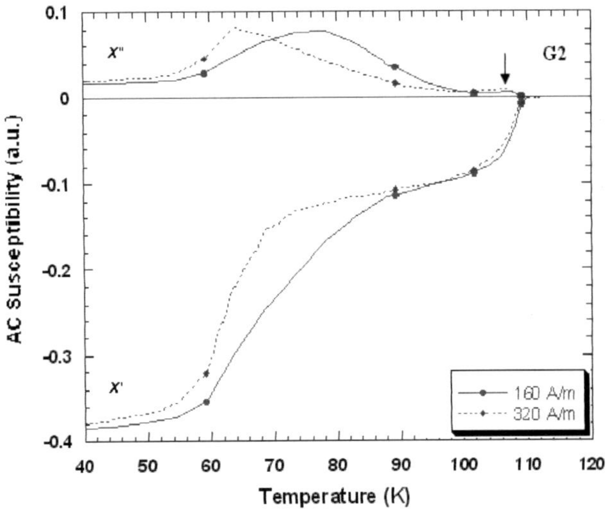
G2
χ"
χ'
160 A/m
320 A/m
AC Susceptibility (a.u.)
Temperature (K)

(c)

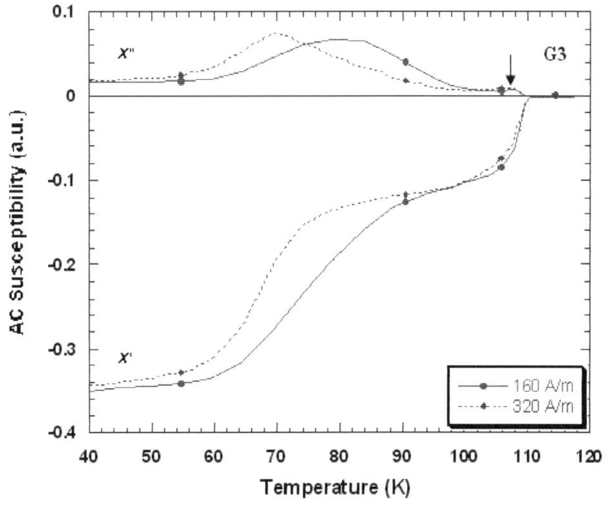

(d)

Figure 4. The temperature dependence of ac susceptibility of the (a) *G0,* (b) *G1*, (c) *G2* and (d) *G3* samples for 160, and 320 A/m.

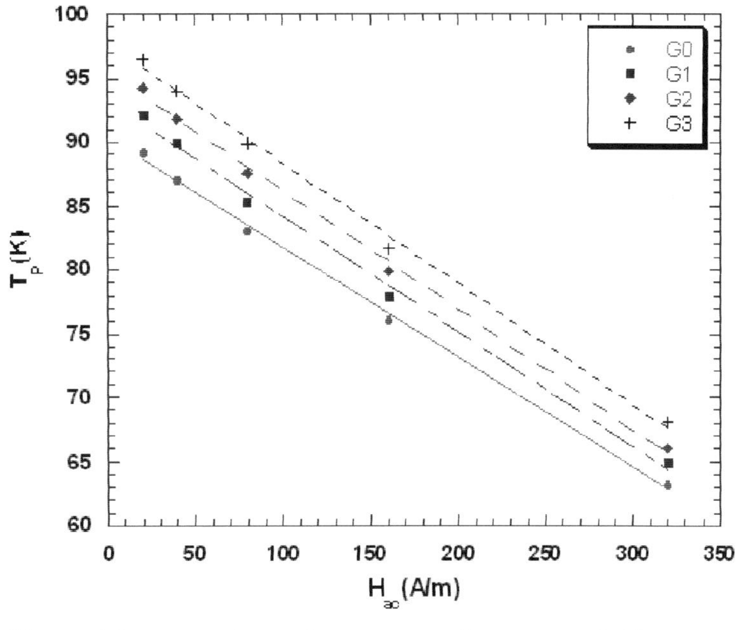

Figure 5. Inter-granular peak temperature T_p versus ac field amplitude for the *G0, G1, G2* and *G3*.

In order to study the effect of Au diffusion and diffusion-annealing time on the intergranular pinning force, the peak temperature dependence as a function of ac magnetic field, H_{ac}, has been investigated. Fig. 5 exhibits graph of the peak temperature vs. ac field amplitudes. As can be clearly seen from the figure, each set of data shows an excellent linear relationship for all the samples. Müller critical state model assumes a magnetic flux independent pinning force densities, and α_J and α_g for inter- and intragranular vortices described by the relation [25]:

$$T_p = T_{p0} - T_{p0}U^{1/2}H_{ac} \tag{1}$$

where U is $\left[\dfrac{\mu_0 \mu_{eff}(0)}{2a\alpha_J(0)}\right]$, a is the length of the samples, $\mu_{eff}(0)$ is the effective permeability of the ceramic, and $\alpha_J(0)$ is the intergranular pinning force density. The slope of each line is proportional to $(\alpha_J(0))^{-1/2}$ and the vertical intercept of each line corresponds to the peak temperature, T_{p0}, at zero ac field amplitude. From a least squares fit of this expression using Eq. (1) to the data T_{p0} and U were extracted for all samples and tabulated in Table 2. Gold doping of the sample (G1) increased T_{p0} from 90 K to 92 K and U decreased from 0.090 and 0.086, in comparison with the undoped sample (G0). As can be seen from the table, the values of T_{p0} increased while U decreased with increasing diffusion-annealing time. A decreasing trend in U means that the values of $\alpha_J(0)$ increase with increasing the diffusion-annealing time, being consistent with our previous transport critical current density measurement such as J_c increased with increasing the diffusion-annealing time [18]. A decreasing trend in pinning force of the samples with decreasing diffusion-annealing time is attributable to grater voids and defects as will be confirmed by SEM results.

Table 2. The extracted values of T_{p0}, U and $J_c^{inter}(T_p)$ of the samples.

Samples	J_c^{inter}(A/cm^2) (at 77 K)	T_{p0} (K)	U (($\alpha_J(0))^{-1/2}$)
G0	35	90.34	0.094194
G1	60	93.17	0.093293
G2	72	95.46	0.089810
G3	90	97.70	0.085977

The temperature dependence of inter-grain and intra-grain critical current density is often determined via the temperature dependence of the inter and intra-grain maximum in the imaginary part, χ''(T) of the ac susceptibility using the Bean model [24,25]. The inter-grain critical current density, $J_c^{inter}(T_p)$, was calculated for our samples using the relation $J_c^{inter}(T_p) = H_{ac}/a$ for the sample having cross section of the rectangular bar shaped, like 2ax2b where a < b. $J_c^{inter}(T_p)$ is the inter-granular current density and H_{ac} is the amplitude of applied field at T_p. The determined $J_c^{inter}(T_p)$ at 77 K for the samples is listed in Table 2. As seen from the table, the value of $J_c^{inter}(T_p)$ increases with increasing diffusion-annealing time. This result is consistent with our previous transport critical current density measurements on the same samples [18]. Moreover, we observed the same evidence for this behavior in ρ-T transport measurements in the samples such as T_c increases with increasing diffusion-annealing time. For intra-grain critical current density, we take the average grain size instead of a. For higher field values namely for 160 and 320 A/m the intra-granular (intrinsic loss) peak is clearly observed (Fig.4). In this work, we are not interested in calculating the intra-grain critical current density values. But it is clear that the intra-grain critical current densities are expected to be higher than the inter-grain critical current densities since intra-grain peak appears at high temperature.

3.2. XRD and SEM analyses

Fig. 6 shows the XRD patterns from the surfaces of (a) undoped Bi1.8Pb0.35Sr1.9Ca2.1Cu3Oy, (b) Au-diffused $Bi_{1.8}Pb_{0.35}Sr_{1.9}Ca_{2.1}Cu_3O_y$ annealed at 830 oC for 10 h, (c) Au-diffused $Bi_{1.8}Pb_{0.35}Sr_{1.9}Ca_{2.1}Cu_3O_y$ annealed at 830 oC for 20 h, and (d) Au-diffused $Bi_{1.8}Pb_{0.35}Sr_{1.9}Ca_{2.1}Cu_3O_y$ annealed at 830 oC for 50 h. Some of the miller indices are indicated in the figure. The calculated lattice parameter c from the (00ℓ) peaks of the XRD data is given in Table 3. The diffusion doping of the sample by gold was exhibited by an increase of the lattice parameter c of the sample (0.19%) for the Au-doped sample (G1) in comparison with that for the undoped sample (G0). The impurity phase (Ca_2PbO_4) and (002)L peaks are observed around 2θ = 17.66o and 5.83o in the G0 sample, respectively. In the G1, G2, G3 samples, (008)H peak is observed around 2θ = 19o while the (008)L peak is appeared in the G0 sample. As can be seen from the table the lattice parameter c increases with increasing diffusion annealing time.

Table 3. Critical temperature T_c^{offset}, lattice parameter *c*, relative volume fraction, and room temperature resistivity of the samples.

Samples	Critical Temperature T_c^{offset} (K)	Lattice Parameter c (Å)	Volume Fraction (%) 2212	2223
G0	100 ± 0.2	36.960	33	67
G1	104 ± 0.2	37.030	23	77
G2	105 ± 0.2	37.190	14	86
G3	106 ± 0.2	37.206	11	89

XRD measurements provide data about lattice parameters, average crystallite size and phase purity for Au diffused Bi-2223 bulk samples. The values of grain sizes can be estimated from XRD measurements by using Scherrer-Warren formula [26], $D = (0.94\lambda / B\cos\theta)$, where D is the crystallite size in nm, λ (15.418 nm) is the wavelength of x-ray in nm, θ is the angle of intensity peak and B is the full width at half maximum (FWHM) of the same intensity peak. With increasing diffusion-annealing time, the intensity of the XRD peaks increases while their positions shift towards smaller angles. This phenomenon indicates that average crystallite size becomes greater as confirmed by SEM investigations. The grain sizes calculated from XRD patterns are about 212, 230, 235and 272 nm for G0, G1, G2 and G3 samples, respectively.

The relative volume fractions of the Bi-2223 and Bi-2212 phases in the samples were calculated from the peak intensities of the XRD results, using the following expressions [27,28];

$$f_{(2223)} = \frac{\sum I_{H(hkl)}}{\sum I_{H(hkl)} + \sum I_{L(hkl)}}, \quad (2)$$

and

$$f_{(2212)} = \frac{\sum I_{L(hkl)}}{\sum I_{H(hkl)} + \sum I_{L(hkl)}} \quad (3)$$

where $I_{H(hkl)}$ and $I_{L(hkl)}$ are the intensities of the (hkl) diffraction lines for Bi-2223 and Bi-2212 phases, respectively (Fig.6).

The relative volume fractions of the phases for all the samples are given in Table 3. As can be seen from the table and from the XRD results one will notice that the volume fraction of Bi-2212 phase decreases while Bi-2223 phase increases with increasing the diffusion-annealing time.

In order to determine the grain sizes and possible precipitation of gold at the grain boundaries the surface morphology of the undoped and Au doped samples was studied by SEM. Fig. 7 represents surface micrographs for (a) the undoped (G0) and (b) gold doped sample annealed at 830 oC for 10 h

(G1). The microstructure of G0 sample is remarkably different from that of G1 sample. Tiny spherical gold particles on the surface of the G1 sample can be seen (Fig. 7b). They are absent on the surface of G0 sample. A broad grain size distribution can be seen for the G1 sample. The average grain sizes calculated using image processing are about 250, 300, 325, 375 nm for the G0, G1, G2 and G3 samples, respectively. These values are consistent with grain sizes calculated from XRD peaks. It is observed that the grain connectivity is worsened gradually with decreasing diffusion- annealing time. The surface of the G1 sample is more uniform with better alignment of grains. Grain size of the gold-doped samples is greater. SEM photograph of G1 shows better crystallinity in comparison with the G0. These results indicate that the surface morphology of the sample is improved with gold diffusion. In the undoped sample flake-type grains were observed and these flake-type grains are believed to be due to the low-T_c phase (Bi-2212) [29,30], being consistent with calculation of the relative volume fraction of the phases in the present study. The flake-type grains are less dominant in sample G1 in comparison with sample G0 while the composition of the needle-type grains grew gradually. These needle-type grains correspond to the high-T_c phase (Bi-2223) [29-31]. The increase of J_c in Bi(Pb)-Sr-Ca-Cu-O by Au diffusion may be caused by the increase of grain size.

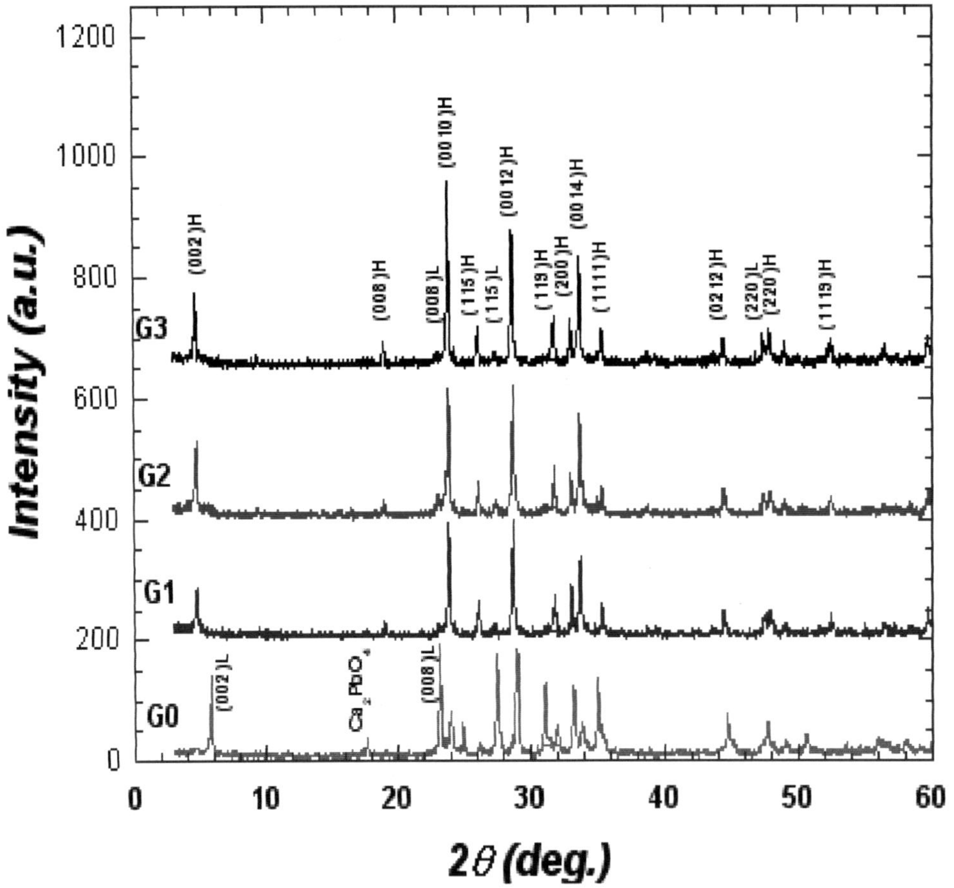

Figure 6. The XRD patterns for the *G0, G1, G2* and *G3*. The peaks indexed L(hkl) and H(hkl) represent the Bi-2212 and Bi-2223 phases, respectively.

(a)

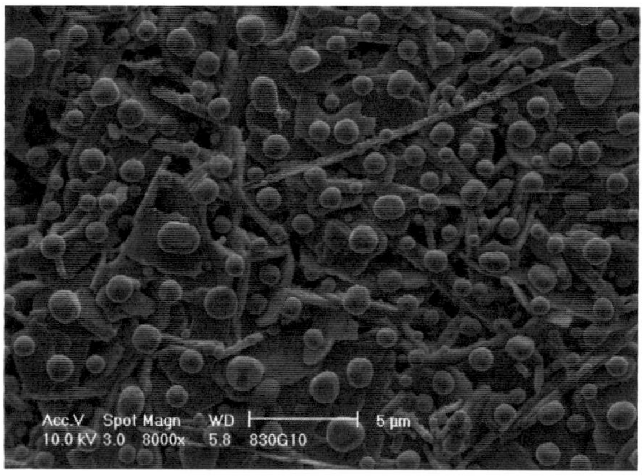

(b)

Figure 7. SEM micrographs of the (a) *G0* and (b) *G1* sample.

3.3. Magnetoresistivity measurements

We have investigated the effect of Au diffusion and diffusion-annealing time on the superconducting properties performing dc resistivity measurements under different dc magnetic fields in the range of zero and 0.5 T. The temperature dependence of magnetoresistivity for all samples (G0, G1, G2 and G3) is shown in Fig. 8. As seen from the figure above 114 K the behaviour of the temperature dependence of the resistivities of the samples is metallic. The room temperature resistivity decreases with increasing the diffusion- annealing time. The zero-resistivity transition temperatures at zero dc fields are 100 K for G0, 104 K for G1, 105 K for G2 and 106 K for G3 sample. It is observed that the zero resistivity transition temperature of the G1 (104 K) is higher than that of G0 sample (100 K), being in agreement with literature [15]. The increased value of T_c in the Au diffusion-doped samples compared with the G0 in our investigations can be interpreted as a result of gold diffusion in inter-grain boundaries. This in turn causes the increase of inter-grains contact surface or decreasing the inter-grain resistivity as can be seen in Fig. 8. It was observed that the T_c of gold-doped samples increases from 104 K to 106 K with increasing diffusion-annealing time. From the above results, it was inferred that the annealing time has the positive effect on the T_c value of the samples. The increase of T_c in Bi(Pb)-Sr-Ca-Cu-O by Au diffusion may also be caused by the increase of grain sizes and their orientations, by improved coupling between Au-diffused superconducting grains, and by the increased number of flux pinning centres due to the presence of Au in the inter-grain regions. In

agreement with literature, XRD patterns obtained from the G0 sample has the characteristic impurity peak due to Ca_2PbO_4 while the gold-doped samples do not have the impurity peak. Presence of this impurity peak in G0 sample is consistent with the experimental finding that the gold-doped samples had lower normal state resistivity and higher zero resisitivity transition temperature (Table 3). It was concluded that formation of Ca2PbO4 will disturb Pb stoichiometry of the sample which in turn will elevate phase segregation from Bi-2223 to Bi-2212. On the other hand, the transition temperature and the lattice parameter c increase with increasing the diffusion-annealing time. This behaviour is due to the change in the Meissner effect as a function of diffusion-annealing time, and is also consistent with the results of the XRD and ac susceptibility measurements.
Fig. 8 displays the temperature dependence of resistivity at magnetic fields of 0, 100, 200, 300 and 500 mT for all the samples. We observed that the broadening of the resistivity transition width increases with increasing external dc magnetic field and with decreasing the diffusion-annealing time, being consistent with our ac susceptibility measurements. It was also observed that, with increasing dc magnetic field up to 0.5 T, T_c^{offset} ranges from 100 K to 67 K for G0, 104 K to 70 K for G1, 105 K to 71 K for G2, and 106 K to 75 K for G3 sample.

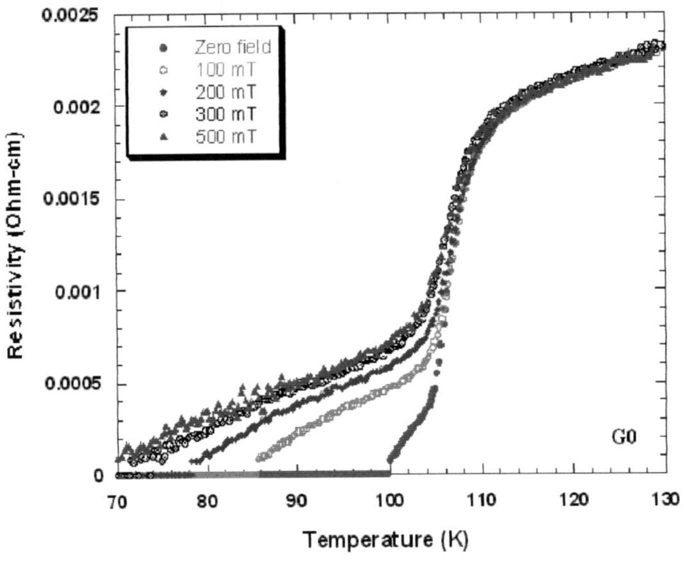

(a)

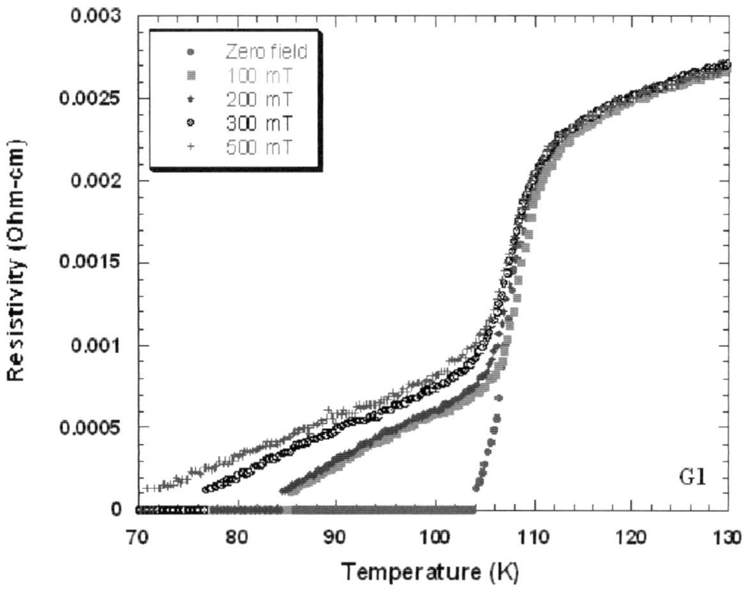
Zero field
100 mT
200 mT
300 mT
500 mT
G1
Resistivity (Ohm-cm)
Temperature (K)
0.003
0.0025
0.002
0.0015
0.001
0.0005
0
70
80
90
100
110
120
130

(b)

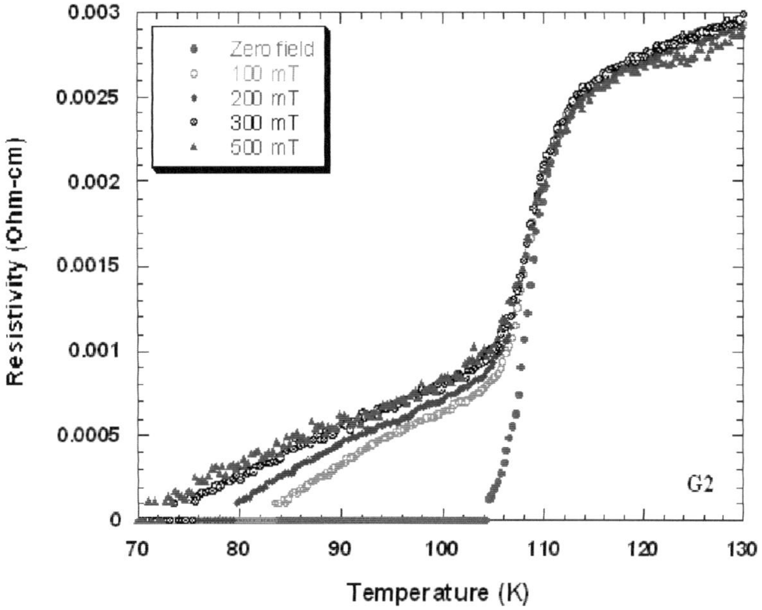
Zero field
100 mT
200 mT
300 mT
500 mT
G2
Resistivity (Ohm-cm)
Temperature (K)
0.003
0.0025
0.002
0.0015
0.001
0.0005
0
70
80
90
100
110
120
130

(c)

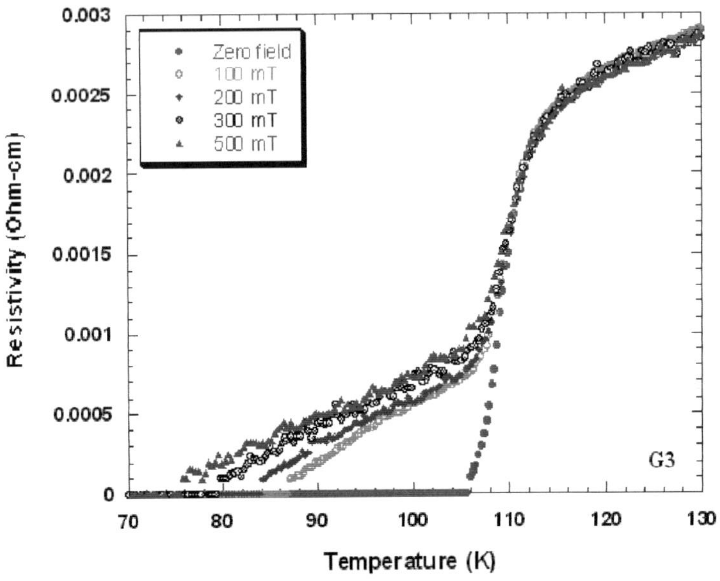

(d)

Figure 8. The temperature dependence of electrical resistivity for the (a) *G0* , (b) *G1,* (c) *G2* and (d) *G3* samples under various applied dc magnetic field up to 0.5 T.

4. Conclusion

In conclusion, $Bi_{1.8}Pb_{0.35}Sr_{1.9}Ca_{2.1}Cu_3O_y$ samples were prepared by solid-state reaction method. The effect of gold diffusion and diffusion-annealing time on the undoped and gold-doped samples was investigated by performing ac susceptibility, magnetoresistivity, XRD and SEM measurements. Our investigations revealed that the gold-diffusion might have improved the coupling of the grains, leading to enhancement of the microstructure and superconducting properties of $Bi_{1.8}Pb_{0.35}Sr_{1.9}Ca_{2.1}Cu_3O_y$ samples. When gold-doped $Bi_{1.8}Pb_{0.35}Sr_{1.9}Ca_{2.1}Cu_3O_y$ samples are compared with the undoped sample, following statements are concluded:

- The peak temperature shifted to higher temperature.
- The inter-granular pinning force density increased.
- The inter-granular critical current density also increased.
- The lattice parameter c increased.
- The relative volume fraction of Bi-2223 phase increased.
- The surface morphology of the microstructure is improved.
- The needle-like grains grew gradually.
- The critical transition temperature increased.
- Room temperature resistivity decreased.

Acknowledgements

This work is supported by the Scientific and Technological Council of Turkey (Project No: 104T325).

References

[1] Müller K H and Pauza A J 1989 *Physica C* **161** 319
[2] Pierre L Schneck J Morin D Toledona J C Primot J Daguet C Savary H 1990 *J. Appl. Phys.* **68** 2296
[3] Nkum R K Datars W R 1992 *Physica C* **190** 465
[4] Nanda Kishore K Muralidhar M and Hari Babu V 1993 *Physica C* **204** 299
[5] Kambe S Guo Y C Dou S X Liu H K Wakahara Y Maeda H Kakimoto K and Yavuz M 1998 *Supercond. Sci. Technol.* **11** 1061
[6] Terzioglu C Yilmazlar M Ozturk O Yanmaz E 2005 *Physica C* **423** 119
[7] Prabitha V G Biju A AbhilashKumar R G Sarun P M Aloysius R P Syamaprasad U 2005 *Physica C* **433** 28
[8] Terzioglu C Ozturk O Kilic A Gencer A and Belenli I 2006 *Physica C* **434** 153
[9] Maeda H Taraka Y Fukutomi M and Asano T 1988 *J. Appl. Phys.* **27**, L209
[10] Dou S X Lin H K Bourdillon A J Zhou J P Tan N X Sun X Y and Sorrell C C 1988 *J.Am.Ceram.Soc.***71** 329
[11] Dou S X Savides N Sun X Y Bourdillon A J Sorrell C C Zhou J P and Easterling K E 1987 *J.Phys.C: Solid State Phys.* **20** 1003
[12] Dou S X Liu H K Guo S J Easterling K E and Mickeel J 1990 *Supercon.Sci.Tech.* **2** 274
[13] Xion G Streitz F H Cieplak M Z Bakhshei A Gourin A and Chien C L 1988 *Phys.Rew.B* **38** 776
[14] Jin S Sherwood R C Tiefel T H Kammlott G W 1998 *Appl. Phys. Lett.* **52** 1628
[15] Dzhafarov T D Altunbaş M Varilci A Küçükömeroğlu T and Nezir S 1996 *Solid State Comm.* **99** 839
[16] Yilmazlar M Ozturk O Gorur O Belenli I and Terzioglu C 2007 *Supercond. Sci. Thecnol.* **20** 365
[17] Nursoy M Yilmazlar M Terzioglu C and Belenli I 2008 *Journal of Alloys and Comp.* **459** 399
[18] Ozturk O Kucukomeroglu T Terzioglu C 2007 *Jurnal of Phys.- Condansed Matter* **19** 346205 (10pp)
[19] Terzioglu C Ozturk O Kilic A Gencer A Belenli I 2006 *Physica C* **434** 153
[20] Ozturk O Akdoğan M Aydın H Yilmazlar M Terzioglu C Belenli I 2007 *Physica B* **399** 94
[21] Müler K H Macfarlane J C Driver R 1989 *Physica C* **69** 158
[22] Müler K H Macfarlane J C Driver R 1991 *Physica C* **203** 178
[23] Bean C P 1964 *Rev. Mod.Phys.* **31** 36
[24] Müler K H Collocott S J Driver R Savvides N 1991 *Supercond. Sci. Thecnol.* **4** 325
[25] Müler K H Collocott S J Driver R 1992 *Physica C* **191** 339
[26] Cullity B D 1978 *Element of X-ray diffraction, Addition – Wesley, Reading, MA*
[27] Chiu C W Meng R L Gao L Huang Z J Chen F and Xue Y Y 1993 *Nature* **365** 323
[28] Halim S A Khawaldeh S A Mohammed S B Azhan H 1999 *Materials Chemistry and Physics* **61** 251
[29] Khalil S M 2001 *Journal of Physics and Chemistry of Solids* **62** 457
[30] Chen Y C Chong K K Meen T H 1991 *Jpn. J. Appl. Phys.* **30** L33
[31] Yoon K H Lee Y B 1991 *J. Mate. Sci.* **26** 5101

Journal of Physics: Conference Series **153** (2009) 012025 doi:10.1088/1742-6596/153/1/012025

Low-field AC susceptibility study of the flux motion in $Bi_2Sr_2CaCu_2Oy$ thin films

A.V.Pop[1*], S.Manolache [1], I.Matei [1] and V.Pop[2]

1. University Babes-Bolyai, Faculty of Physics, University "Babes-Bolyai", 3400 Cluj-Napoca, Romania

2. Technical University Cluj-Napoca, Faculty of Materials Science and Engineering, Department of Physics, Technical University, Cluj-Napoca, Romania 400641, Cluj-Napoca, Romania

[*] Corresponding author: Tel:+4-0264-594315/int.5150; fax:+4-0264-591906

E-mail address:avpop@phys.ubbcluj.ro

Abstract. DC magnetron sputtering technique were used for the synthesis of high quality epitaxial thin films of the superconducting cuprate Bi2Sr2CaCu2Oy (Bi: 2212) thin films. The structural characterization carried by X-ray diffraction indicates that the films are single phase and oriented with their c-axis perpendicular to the substrate surface. The flux motion mechanisms in the vicinity of the irreversibility line have been studied by AC susceptibility technique. The temperature dependencies of real and imaginary parts of AC susceptibility were performed, by using different values of amplitude and frequency. Temperature dependence and low field dependence of the activation energy for flux motion were obtained

1. Introduction

High quality high temperature superconductors (HTS) thin films fabrication and multilayers are important for the fabrication of integrated junction circuits. The Bi2Sr2Can-1CunO2 (n+2)+δ high temperature superconductors (HTS) is very important for fundamental investigations and technical applications. Bi2Sr2Can-1CunO2 (n+2) +δ (BSCCO) has high critical temperatures (Tc~20K for n=1, Tc~85K for n=2 and Tc~110K for n=3) and high critical current densities at low temperature and is used for the fabrication of long lengths HTS tapes.

Superconducting Bi-2212 material suitable for applications at 77 K should have high critical current densities. The films with c-axis aligned to the film plane appear to be preferably suited for the fabrication of sandwich type SNS and SIS Josephson and quasiparticle tunnel junctions [1]. The epitaxial growth of thin films of Bi: 2212 has been realized on substrates like SrTiO3, LaAlO3 and MgO [2-5]. The importance of $Bi_2Sr_2CaCu_2O_{8+d}$ HTSC is being reinvented in the light of its potential use in Terahertz nanotechnology and as a natural SQUID. Bi-2212 has a layered structure, which allows the propagation of EM waves (called Josephson plasma oscillator) [6-8] if transverse magnetic field H_{ab} and longitudinal electric field Ec is applied. Bi-2212 is also used as natural SQUID because there are non-superconducting layers sandwiched between superconducting CuO layers [9].

AC loss is one of the most important parameters in determining the commercial effectiveness of technological applications. In some applications, the reduction of AC loss is more important than other factors. It is well established that the largest contribution to AC losses comes from hysteresis loss, because the motion of flux lines.

The magnetic phase diagram of high-temperature superconductors (HTSC) exhibits an extraordinarily rich variety of phases [10]. Still controversial is the presence of a boundary, called the irreversibility line (IL), which separates a magnetically irreversible state, with a non-linear dissipative behavior, from a reversible region with a linear dissipative property. Some of the suggestions are based on a vortex glass formation [11] or on a flux lattice melting [12].More recent studies investigating the irreversibility line have confirmed the presence of a predicted crossover characteristic value Hcr[13] in the dimensionality of the vortex fluctuations [14] and [15].

Here we present the temperature dependencies of real and imaginary parts of AC susceptibility as a function of the AC driving field, the superimposed DC field and driving frequency for Bi:2212 films.

2. Experimental

Bi:2212 thin films were deposited in situ onto (100) $LaAlO_3$ heated single crystal substrate by using DC magnetron sputtering system [4]. In order to compensate the Bi loss, a bismuth enriched target with chemical composition $Bi_{2.6}\,Sr_{1.97}\,Ca_{0.97}\,Cu_{2.4}\,O_y$ has been synthesized. The optimal parameters used for fabrication of epitaxial quality Bi:2212 thin films were: sputtering gas pressure of 0.9 mbar (with a ratio 1/1 between the argon and oxygen partial pressures), the substrate temperature 790 ^{0}C and d.c. plasma power between 20W and 30W.

X-ray diffraction pattern shows only (00l) reflexions observed for thin films and the reflexions (i00) for $LaAlO_3$ substrate. The FWHM (full width for high maximum) of rocking curve is equal to 0.40^0, and confirmed the high orientation quality of sample.

In order to give an insight into the growth mechanisms of Bi:2212 films, a series of thin films with different thicknesses have been grown.

AC susceptibility measurements were performed by means of a Lakeshore AC susceptometer, equipped with a superconducting coil providing a background DC field of maximum 1 Tesla. The frequencies used may vary from 10 to 1000Hz, while the field amplitude from (0.005-10 Oe)10 to 800 A/m . Superconducting thin film was carefully placed in the centre of one of the secondary coils with its c-axis parallel to the direction of both AC and DC magnetic fields.

3. Results and discussions

Figure 1 shows the real χ 'and imaginary χ "components of AC susceptibility for f=1000Hz and H_{ac}= 0.02 Oe, at different DC values up to 1 Tesla. The real part of susceptibility extrapolates at low temperature to a field independent value , which suggests completing magnetic screening. The increase of B_{dc} leads to a slight broadening of the transition width in χ ' susceptibility transition. The peak in imaginary part of AC susceptibility, χ ",occurs at a temperature T_p (Fig.1) . Many authors assimilate the $T_p(B_{dc},f)$ values measured for different DC fields, $B_{dc} = B_{irr}$, at a given frequency f, as the irreversibility line (IL). In order to avoid nonlinear effects the exciting AC field amplitude H_{ac} was kept below the value 4 A/m.

To analyse the shift of the χ'' intergranular peak we use a linear TAFF model in which vortex displacements are thermally activated and the hopping rate is given by :

$f = f_0 \exp[-U(B_{dc}, T_p)/(k_B T)]$,where f_0 is a typical attempt frequency and $U(B_{dc}, T_p)$ the activation energy. On the basis of this equation we obtain:

$$\mathbf{U(B_{dc}, T_p) = -k_B T_p (\ln f - \ln f_0)} \qquad (1).$$

For a proper interpretation of data , we assume that :

$\mathbf{U(B_{dc}, T_p) = U(B_{dc}) * U(T_p)}$,

These approximations are valid close to T_c and contain the specific dependencies predicted by some models [16,17].

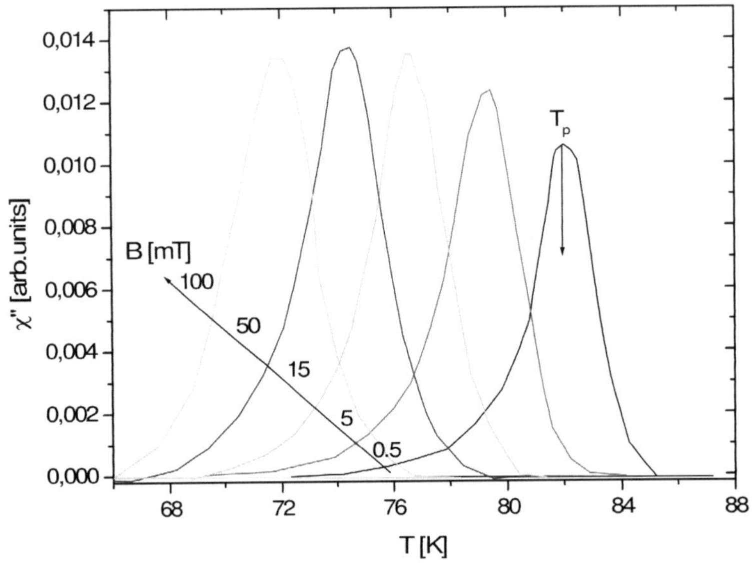

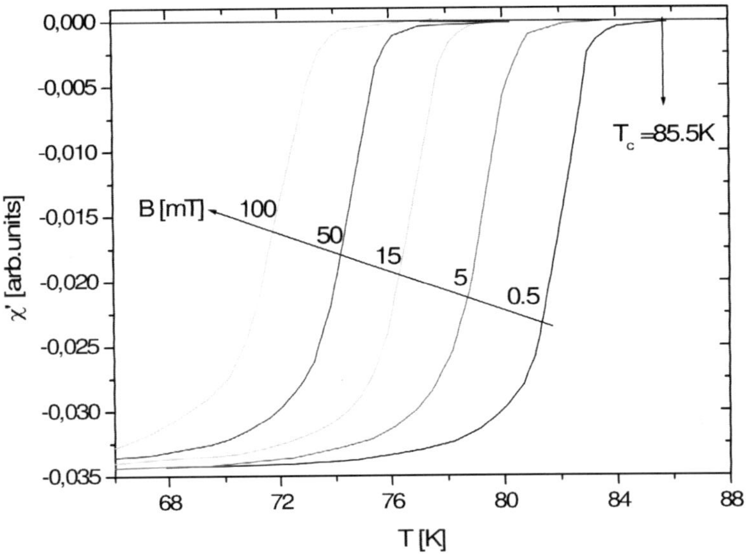

Fig.1.The temperature dependence of AC susceptibilities, χ 'and χ" in Bi: 2212 film for B_{dc} fields up to 1 Tesla, for f=1000Hz and H_{ac}=4A/m.

Fig.2 show the temperature and frequency dependence of imaginary part χ'' of a.c.susceptibility in the presence of a d.c. field H_d = 200A/m superposed over the a.c.field H_{ac}=400 A/m. For a fixed value of B_{dc} the temperature T_p increase with increasing frequency.

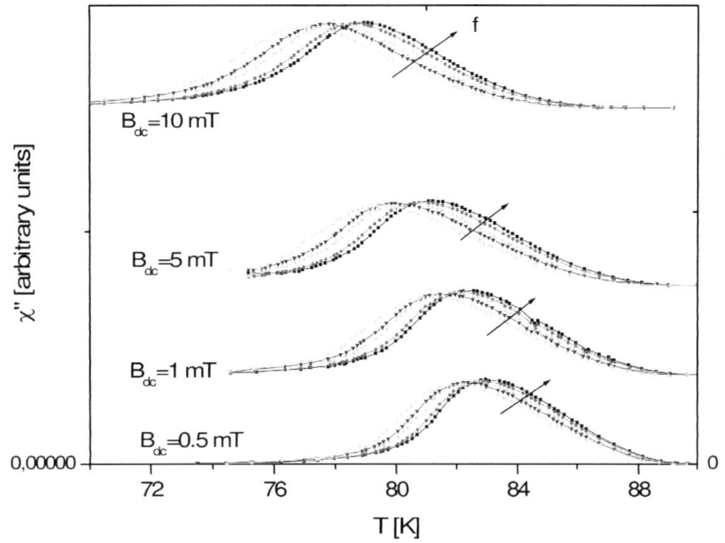

Fig.2. Temperature dependence of the imaginary part χ'' (T) of the complex susceptibility for frequencies f=110, 220,300, 600 and 1000 Hz. The d.c. and a.c magnetic fields were B_{dc} =0.5,1,5 and 10 mT and H_{ac} =10 A/m ,respectively.

Equation (1) indicates that, in linear regime, $\mathbf{U\ (T_p)/\ T_p = [U\ (H_d)]^{-1}\ ln\ (f_0/f)}$ depend on the driving frequency in a logarithmic way. This was checked by representing the Arrhenius plot for $\mathbf{U\ (T_p)/\ T_p}$ versus (lnf), for different B_{dc} values. As shown in fig.3, we obtain a linear dependence for n=1 in $\mathbf{U\ (T_p) = (1\text{-}T_p/T_c)^{\ n}}$.

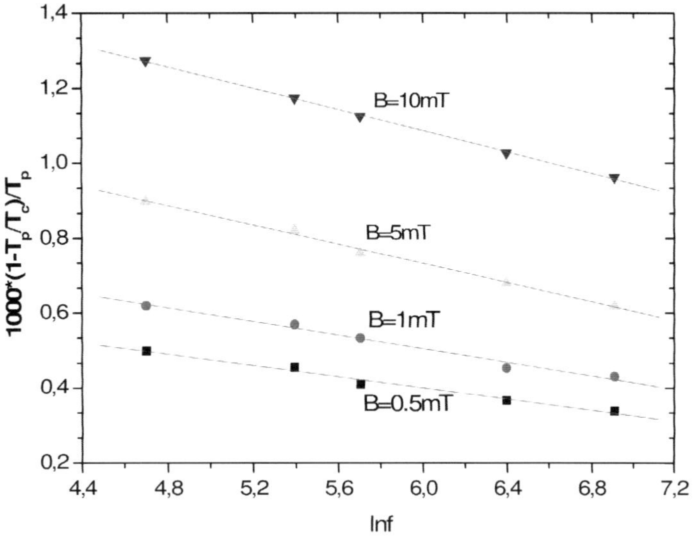

Fig.4. The plot of U $(T_p)/T_p$ = [1-(T_p/T_c)] /T_p versus ln f in different B_{dc} fields.

This result indicates the presence of the ohmic TAFF regime, where the activation barrier is independent of the driving force.

The field dependence of the activation energy was determined by using the slopes of the linear fits determined for each field in fig.4. From the best fit (full line in fig.4) we find that the activation energy is:

$$U = 2920 - 2088 \ln B \text{ [K]}$$

An explanation for the lnB dependence of activation energy may be obtained from the shape of the pinning potential and from the pinning energy distribution. At small magnetic fields, in the individually pinned vortex regime, the motion of individual vortices in the field of unperturbed pinning centers is present. By using for the pinning energies a distribution $f(U) = f_0 \exp[-(U-U_0)/\sigma]$ and assuming that vortices "fill" the strongest pinning sites first, the activation energy can be written as [18, 19]:

$$U(B) = U_0 + \sigma \ln(B_0/B)$$

Where U_0 is the activation energy corresponding to the crossover field B_0 between the single vortex and collective vortex behavior.

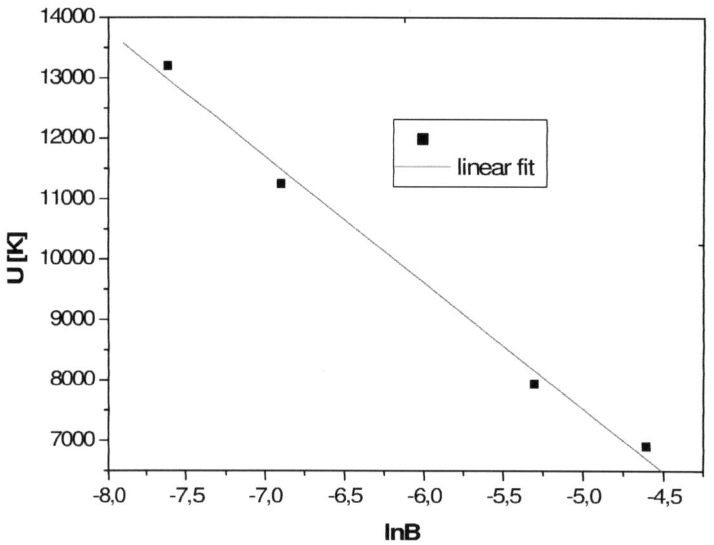

Fig.5.Magnetic field dependence of the low –field activation energy in Bi: 2212 thin film.

4. Conclusions

High quality epitaxial $Bi_2Sr_2CaCu_2Oy$ (Bi: 2212) thin films were obtained by DC magnetron sputtering technique. X-ray diffraction indicates that the films are single Bi: 2212 phase and oriented with their c-axis perpendicular to the substrate surface.

AC susceptibility technique is a good option to study the flux motion mechanisms in the vicinity of the irreversibility line.

The obtained $\mathbf{U(T_p, H_d) = U_0(1-T_p/T_c)*lnB}$ dependence of the pinning barrier is in agreement by the TAFF picture and can be explained by using at small magnetic field, in the individually pinned vortices regime an exponential distribution for pinning energies.

Acknowledgments
The authors are grateful to CNMP-PN II Romania for the financial support.

References

[1] J.Geerk, X.X.Xi, G.Linker, Z.Phys.**B73**, 1327(1988).
[2] P.Wagner,F.Hillmeyer,U.Frey,H.Adrian, T.Steinborn, L.Ranno,A.Elschner, I.Heyvaert, Y.Bruynserade, Physica **C215**, 123(1993).
[3] L.X.You, A.Yurgens,D.Winkler,C.T.Liu, B.Liang,Supercond.Sci.Technol.19,S205(2006)
[4] A.V.Pop,G.Ilonca, M.Pop, JOAM **8**, 480(2006)
[5] T.Onishi, N.Fujiwara,S.Kishida, Physica **C426-431**, 1439(2005).
[6] D. Saeedkia, R.R. Mansour, S.S. Naeini, IEEE Trans. Appl. Supercon. **15** (2005) 3847.
[7] S. Savel'ev, V. Yampol'skii, A. Rakhmanov, F. Noril. Available fromarXiv: hcond-mat/0508722i v1 30 August 2005.
[8] N. Kameda, M. Tokunaga, T. Tamegai, M. Konczykowski, S.Okayasu, Phys. Rev. B **69** 180502 (2004).
[9] Peter J. Lee (Ed.), Engineering Superconductivity, John Wiley and Sons, (2001).
[10] G. Blatter, V. Feigel'man, V.B. Geshkenbein, A.I. Larkin and V.M. Vinokur, Rev. Mod. Phys. 66 (1994), p. 1125.
[11] M.P.A. Fisher, Phys. Rev. Lett. 62 (1989), p. 1415.
[12] A. Houghton, R.A. Pelcovits and A. Sudbo, Phys. Rev. B 40 (1989), p. 6763.
[13] L.I. Glazman and A.E. Koshelev, Phys. Rev. B 43 (1991), p. 2835.
[14] L. Miu, P. Wagner, A. Hadish, F. Hillmer and H. Adrian, Physica C 234 (1994), p. 249.
[15] L. Ghivelder, I. Abrego Castillo, S. Salem-Sugui Jr., R. de Andrade Jr., M. Dhalle, G. Grasso, F. Marti and R. Flukiger, Supercond. Sci. Technol. 12 (1999), p. 566·
[16] V.B.Geshkenbein, M.V.Feigel'man, A.I.Larkin and V.M.Vinokur, Physica C,162-164(1989) p.239.
[17] M.Tinkham, Phys.Rev.Lett.61(1988),p.1658
[18] V.M.Vinokur, P.H.Kes and A.E.Koshelev, Physica C168 (1990)29.
[19] M.Inui,P.B.Littlewood and S.N.Coppersmith, Phys.Rev.Lett.63(1989)2421

Journal of Physics: Conference Series **153** (2009) 012026 doi:10.1088/1742-6596/153/1/012026

Electrothermal performance and growth of single phase in barium doped Bi-2223 high-T_c superconductors

M Anis-ur-Rehman
Applied Thermal Physics Laboratory, Department of Physics, COMSATS Institute of Information Technology, Islamabad 44000, Pakistan.

E-mail:marehman@comsats.edu.pk

Abstract. Solid state reaction method was used to synthesize the high-T_c superconducting material with nominal composition $Bi_{1.6}Pb_{0.4}Sr_{1.6}Ba_{0.4}Ca_2Cu_3O_y$. This paper investigates the associations among macroscopic physical properties and features at atomic level for the prepared samples. The samples were preliminary analyzed by dc electrical resistivity and ac magnetic susceptibility. After that thermal transport, electrothermal conductivity and thermoelectric properties all as a function of temperature (from room down to liquid nitrogen temperature) were measured. X-ray diffraction studies were also done for structural analysis. All the above measurements showed that in the samples, there exists almost a single high- T_c phase with $T_{c,0} \simeq 110\pm1$K. The lattice constants of the material were determined by indexing the diffraction peaks. Samples were investigated for thermal transport properties, i.e. thermal conductivity, thermal diffusivity and heat capacity per unit volume. Thermal transport properties were measured by a newly developed and calibrated Advantageous Transient Plane Source method. Simultaneous measurement of thermal conductivity and thermal diffusivity made it possible to estimate specific heat and the Debye temperature Θ_D. Thermoelectric power (Seebeck Coefficient) and electrothermal conductivity were also measured with a recently developed and attuned apparatus. Using dc electrical resistivity, thermal conductivity and thermoelectric power, the Figure of merit factor was also calculated.

1. Introduction

Detailed investigations of the electrical and thermal transport and ac magnetic susceptibilities in superconductors are important for understanding superconductivity as well as for practical applications. The effect of elements (Pb, Fe, Co, Ni, Zn) doping in Bi-based superconducting materials have been extensively investigated [1-8]. It was reported that the superconducting properties of these materials decrease with increase of the amount of doping, regardless of the nature of the dopants. The suppression of superconductivity was concluded to be due to local disorder induced by the amount of doping. However, the details of the current limiting mechanisms in the Bi-2223 system are not well established. Consequently, it is of interest to try these doping elements in the Bi-2223 system with a different nominal composition, of which we intend to investigate $Bi_{1.6}Pb_{0.4}Sr_{1.6}Ba_{0.4}Ca_2Cu_3O_y$ in order to provide additional observations to contribute further understanding of their role on the

superconductivity of the system. The linear temperature dependence of the electrical resistivity is one of the most important properties of the normal phase kinetics of high-T_c layered cuprates [9].
In superconductors where the dc electrical resistivity diverges to zero below T_c, the thermal conduction is almost a unique measurement to study the transport properties below T_c. The magnitude and temperature dependence of the thermal conductivity are parameters which have an impact on a broad spectrum of devices. In high-T_C superconductors, such information is even more valuable to know how the free carriers and lattice vibrations contribute to the transport of heat. Transient Plane Source (TPS) technique is a well developed and a well known method [10-12] to study the thermal transport properties. The TPS technique is modified and improved for the measurements of thermal transport properties of high-T_c superconductors. The modified arrangement is referred to as the Advantageous Transient Plane Source (ATPS) technique [13, 14]. The modified bridge arrangement is already calibrated with fused quartz, carbon steel and AgCl crystals [13, 15].
Peltier refrigerators use the thermoelectric materials for refrigeration. Peltier thermoelectrics are more reliable than compressor based refrigerators, and are used in situations where reliability is critical like deep space probes. Thermoelectric material applications include refrigeration or electrical power generation. Thermoelectric materials used in the present refrigeration or power generation devices are heavily doped semiconductors. The metals are poor thermoelectric materials with low Seebeck coefficient and large electronic contribution to the thermal conductivity. Insulators have a large Seebeck coefficient and a small contribution to the thermal conductivity, but have too few carriers, which result in a large electrical resistivity. The Figure of merit is the deciding factor for the quality of thermoelectric materials. In order to increase the whole Figure of merit, it is of interest to replace the p-type leg of the Peltier junction by a thermoelectrically passive material with a Figure of merit close to zero [16]. This is why it is interesting to study the Figure of merit of the ceramic superconductors.
One of the important thermomagnetic transport quantities is the electrothermal conductivity and is shown to be one of the powerful probes of high-temperature superconductors. Cryogenic bolometers are sensitive detectors of infrared and millimeter wave radiation and are widely used in laboratory experiments as well as ground-based, airborne, and space-based astronomical observations [17]. In many applications, bolometer performance is limited by a trade off between speed and sensitivity. Superconducting transition-edge bolometer can give a large increase in speed and a significant increase in sensitivity over technologies now in use. This combination of speed with sensitivity should open new applications for superconducting bolometric detectors [18]. Other potent applications for electrothermal conductivity of superconductors is actuators in MEMS technologies, electrothermal rockets etc [19].
The temperature dependence of the dc electrical resistivity, along with low field ac magnetic susceptibility, X-ray diffraction, thermal transport, electrothermal conductivity and thermoelectric power studies and calculations of Figure of merit factor are undertaken in this paper.

2. Experimental
2.1. Preparation and characterization

Precise control over the processing parameters is required to obtain the phase-pure material [20-21]. All specimens were prepared from 99.9% pure powders of Bi_2O_3, PbO, $SrCO_3$, $BaCO_3$, $CaCO_3$ and CuO. The powders were mixed to give nominal composition of $Bi_{1.6}Pb_{0.4}Sr_{1.6}Ba_{0.4}Ca_2Cu_3O_y$ and were thoroughly ground in an agate mortar to give very fine powder. The grind powder was calcined for 21 hours in air at 800°C. A series of pellets was produced in two sizes. Controlled heating and cooling carried out, in air, using a horizontal tube furnace. Poly Vinyl Alcohol (PVA) was used as binder in the samples. Samples were in the shape of cylindrical disks having diameters 13mm and 28mm, and lengths 3mm and 11mm respectively. These samples were sintered at 830°C for the intervals of 24 hours in each sintering step as sintering procedures do affect the properties [22].

The superconducting properties were characterized electrically by using standard four probe method. Contacts were made by high quality silver paste. The temperature was measured by using a calibrated Pt-100 thermometer.

Low field ac susceptibility measurements are very important for the characterization of high-temperature superconductors [23-26]. The sharp decrease in the real part χ' (T) below the critical temperature T_c is a manifestation of diamagnetic shielding. Ac susceptibility of the sample was measured after each sintering step. The low field ac susceptibility properties were studied by the use of mutual inductance bridge method. The measurements were taken from room temperature down to 80K.

X-ray diffractograph (XRD) of sample was taken after the final sintering. The radiation used for XRD was CuK_α and the measurements were made at room temperature. Measurements were done at room temperature since there is no change in the structure of the superconducting materials before and after transition [22, 27].

2.2. Thermal transport properties

Thermal transport measurements, i.e. thermal conductivity, thermal diffusivity and heat capacity per unit volume were performed using the Advantageous Transient Plane Source (ATPS) Technique [13, 15]. Circuit diagram for the method is shown in Figure 1. Simultaneous measurement of thermal conductivity and thermal diffusivity is the foremost advantage of this technique. Heat capacity per unit volume is then calculated using the idea that, if all heat is transported via solid specimen then the thermal conductivity (λ), thermal diffusivity (κ) and heat capacity per unit volume (ρC_p) are expressed by;

$$\kappa = \frac{\lambda}{\rho C_p} \tag{1}$$

A detailed description of this experimental technique can be found elsewhere [10]. The ideal model presupposes that the double spiral sensor, assumed to consist of a set of equally spaced, concentric, and circular line heat sources, is sandwiched in specimens of infinite dimensions. In practice all real specimens do have finite dimensions. However, by restricting the time of the transient, which relates to the thermal penetration depth of the transient heating, a measurement can still be analyzed as if it was performed in an infinite medium. This means that the ideal theoretical model is still valid within a properly selected time window for the evaluation. The scatter in thermal conductivity measurements is about 0.14% and is 0.66% and 0.52% in thermal diffusivity and volumetric heat capacity respectively [13, 15]. Taking into consideration the limitations of the theory of the technique and the experimental sampling errors, the thermal conductivity and thermal diffusivity data contain errors of 4% and 7% respectively. The errors in volumetric heat capacity are around 10% [13-15].

2.3. Thermoelectric power measurements

An easy to use and simple apparatus was designed and developed for thermoelectric power (S) measurements. Circuit diagram along with the sample holder assembly is shown in Figure 2. The sample is subjected to a temperature difference ΔT by using a heating resistor and corresponding voltage difference ΔV across the sample is measured. Thermoelectric power is obtained by taking ratio of the voltage difference to the temperature difference. Chromel-alumel thermocouples are used for measuring the temperature difference, ΔT. The thermocouples are electrically isolated from the sample and thermally connected to the sample. The voltage leads are then silver pasted to the sample in the vicinity of thermocouples to assure that the voltage and temperature gradients are measured at the same locations on the sample for accurate thermoelectric power measurements.

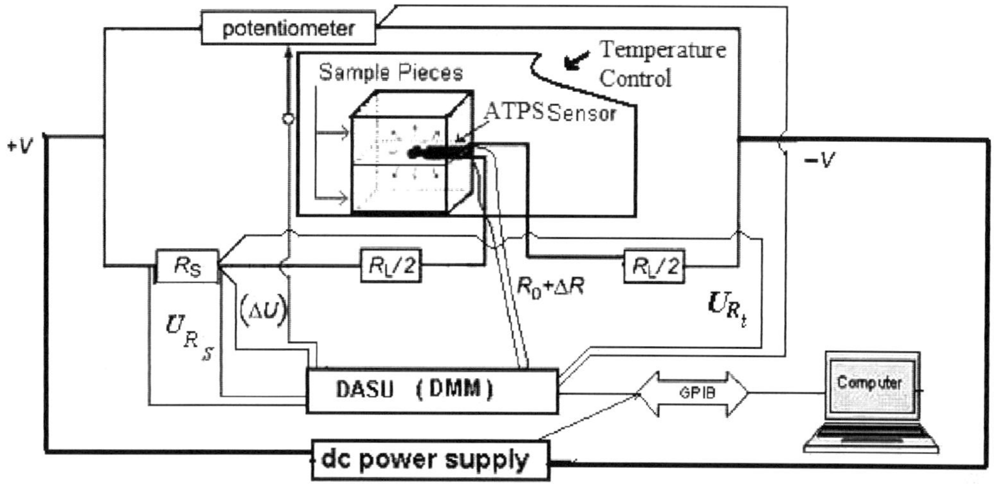

Figure 1. Circuit diagram for the Advantageous Transient Plane Source (ATPS) method.

The next step includes loading the sample assembly into the sample chamber and evacuation of the chamber. The chamber is evacuated to eliminate any water vapour condensation on the sample, which can result in erroneous measurements. Dry nitrogen gas is then filled in the chamber as a conducting media between chamber walls and the sample. This sample chamber is then inserted in liquid nitrogen container for cooling. Data are collected under the computer control. By incorporating multiple measurements in a single run, considerable time is saved by avoiding remounting, and recooling of the samples. In this technique the surface mount resistor (50Ω) was used to heat one end of the sample to establish a measured temperature gradient of approximately 1K.

3. Experimental results and Discussion

3.1. Dc electrical resistivity

Variation of resistivity with change in temperature is recorded after each sintering step and the plots are given in Figure 3. After the final sintering the measured density of the sample was 3.48 gcm^{-3} and $T_{c,0}$ was 110 ± 1K. The added barium (Ba) has increased the $T_{c,0}$. Residual resistivity was 0.19 mΩ-cm and the intrinsic resistivity was 5.9 μΩ-cmK^{-1}. The ratio ρ(273K)/ρ(4.2K) is the residual resistivity ratio (RRR), an important parameter in the design of superconductive applications. In the case of a superconductor, the denominator has to be taken at a temperature slightly above the critical temperature [28]. RRR in our case was in the range 23-33.

3.2. Ac susceptibility

Ac susceptibility measurements were done after each sintering step (Figure 4). Initially two transition phases were present. One of the identified phases is the Bi-2212(low T_c) phase and the other Bi-2223(high T_c) phase. With sintering, the low T_c phase was smoothed out and only phase left is the Bi-2223(high-T_c) phase. Although the resistivity variation with temperature was smoothed out after the third sintering but slight kinks were observed in the susceptibility against temperature plot showing the more sensitivity of the measuring method.

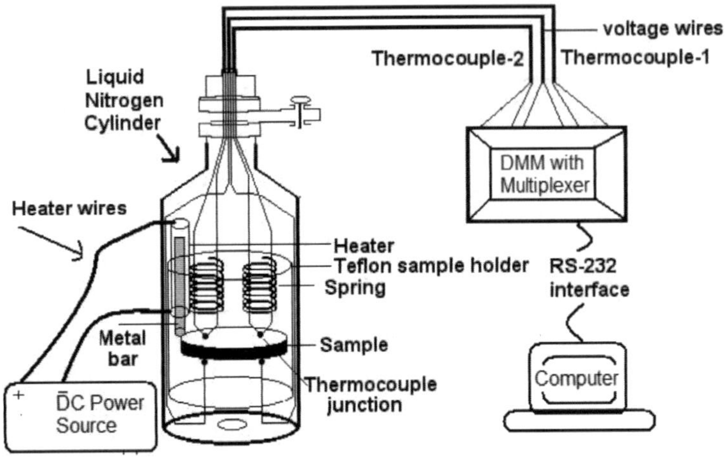

Figure 2. Block diagram of the apparatus developed for thermoelectric power measurements. Scanner card is used with the multimeter for simultaneous measurements at different points as shown. RS-232 is the standard serial interface of the computer.

3.3. X-ray diffraction studies

Almost all the peaks of the XRD pattern were indexed. The only phase is the orthorhombic high-T_c Bi-2223 phase. Lattice parameters were calculated from the (h k l) values of the indexed peaks. The lattice parameters are a = 5.42 (1) Å, b = 5.37 (1) Å and c = 37.12 (8) Å. No peaks were found matching the Bi-2212 low T_c phase. Indexed X-ray diffractograph is shown in Figure 5. The lattice constants agreed with the previous reports [12]. The size of the grains in polycrystalline materials has pronounced effects on many of its properties. Using Scherrer's equation [29];

$$B = \frac{0.9\lambda}{t\cos\theta}, \qquad (2)$$

where;

B = Broadening of diffraction line measured at half its maximum intensity (radians), λ = Radiation source wavelength and

t = Diameter of crystal particle,

particle sizes are determined and the diameter of the crystal particles lies between 172 – 512 Å.

3.4. Thermal properties measurements

After the preliminary characterization of the samples and existence of almost a single phase, large disc-shaped samples (28 mm diameter and 11 mm thickness) were used for the thermal measurements. Figure 6 shows the temperature dependence of the thermal conductivity λ. As the temperature decreases, the conductivity gradually decreases down to near T_c then remarkably increases below T_c. Further decrease in temperature was not possible, due to limitation of the cryostat used, to take the maximum in λ. This temperature dependence agrees with the widely observed behavior of λ for the oxide superconductors [30-34].

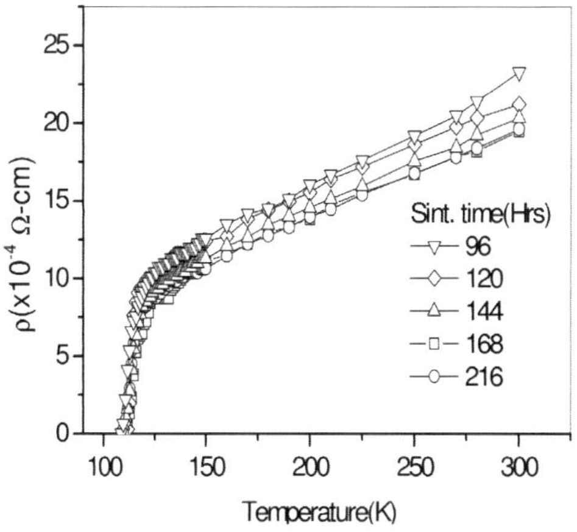

Figure 3. DC electrical resistivity as a function of temperature for sample after each sintering step.

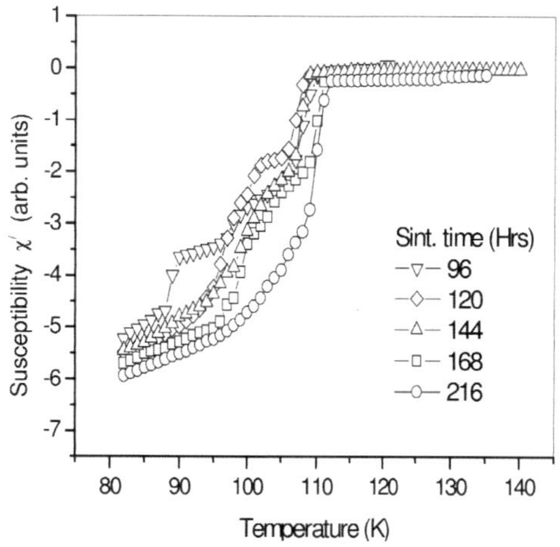

Figure 4. Variation of ac susceptibility (real part) with temperature after each sintering step.

Comparing the results between different laboratories, one notes that the thermal conductivity depends on a particular sample preparation process. The temperature dependence of the conductivity is really similar for all the samples [35, 36]. So the order of magnitude of thermal conductivity (measured by non-steady state method in our case) is comparable to the results obtained by different authors (measured by steady state methods). Figure 6 also shows the variation of measured thermal diffusivity κ with temperature for the sample. Between 294K and T_c, κ increases very gradually with decreasing temperature. The increase of κ becomes very large around T_c and become very steep.

The heat capacity per unit volume, ρC_p, calculated from the thermal conductivity measurements and thermal diffusivity measurements by using equation 1 is shown in Figure 6. ρC_p decreases with decrease in temperature and near T_c a sizeable kink is observed. This jump is mostly due to the improved sharpness of the transition related to the reduction of intergrowth structure by adding Pb [37] and is improved by adding Ba in our case. Since the calculated lattice constants of our sample are similar to Bi-2223 composition so it is assumed that oxygen is 10, and then the composition becomes $Bi_{1.6}Pb_{0.4}Sr_{1.6}Ba_{0.4}Ca_2Cu_3O_{10}$.

Also there is no change in the density of the superconducting sample in the studied temperature range so the value of specific heat C_p is calculated. The absolute value of C_p is 320 $Jmol^{-1}K^{-1}$ at 180K and that is similar to already reported value of a similar composition [37, 38]. Because the phonon contribution is by far dominant than the electronic contribution in the temperature range studied, the specific heat data were fitted by the following Debye formula,

$$C_{p-ph} = 9nR\frac{T^3}{\Theta_D^3}\int_0^{\Theta_D/T}\frac{x^4e^x}{(e^x-1)^2}dx\,, \tag{3}$$

where $C_{p\text{-}ph}$ is molar specific heat, x is the reduced phonon frequency, n (= 19) the number of atoms composing Bi(Ba) 2223 molecules, R the gas constant and Θ_D is the Debye temperature.

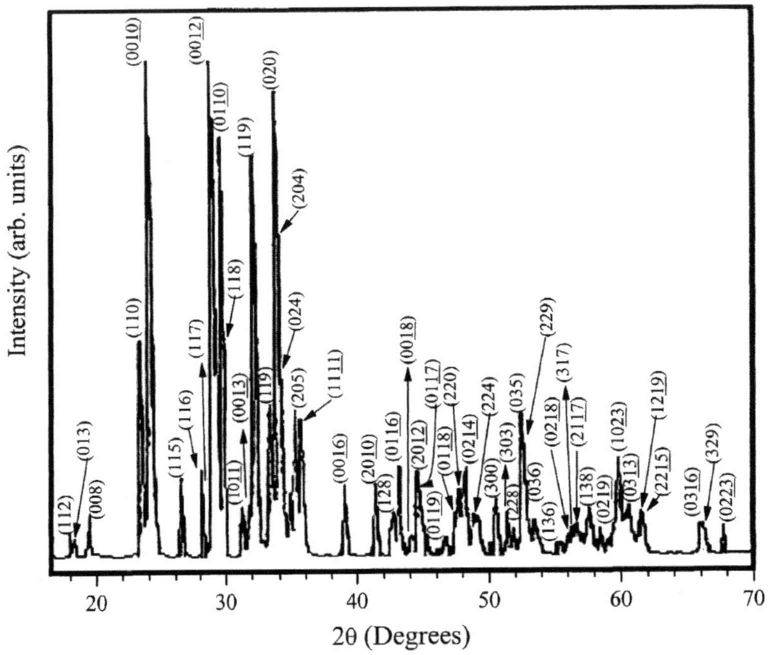

Figure 5. Indexed x-ray diffraction pattern of the sample after the final sintering at room temperature.

Although a single Θ_D fitting fails to give a unified strict fitting over the entire temperature range, but Θ_D = 510 K gives a satisfactory fitting between T = 120 to 230 K as is shown in Figure 7.

3.5. Thermoelectric power

To check the calibration of this new apparatus (Figure 2), thermoelectric power of copper was measured in the temperature range 85-310K. Results of our measurements are shown in Figure 8, indicating an agreement with the already published data [39]. The standard deviation in the data was between 0.01-0.22 μVK^{-1} and the difference between measurements done in this work and the already published [39] data were within 5%.

The thermoelectric power of the high-T_C superconducting sample was measured in temperature range 85-300K. The thermoelectric power (S) reached zero within experimental uncertainty in superconducting state. The thermoelectric power increased with decrease in temperature and after reaching T_c value, thermoelectric power decreased strongly to zero value (Figure 9). At high temperatures the thermoelectric power is almost linear. Thus we can use the Mott expression to determine the Fermi level [39, 40]:

$$S = S_0 - \frac{\pi^2 k_B^{\ 2}}{3|e|E_F} T \qquad (4)$$

where S_0 is a constant. From the slope (-0.03145 $\mu V / K^2$) estimated by a linear extrapolation we have found the Fermi level to be 0.78 eV.

Similar profile for the same kind of superconductors is reported [41-46].

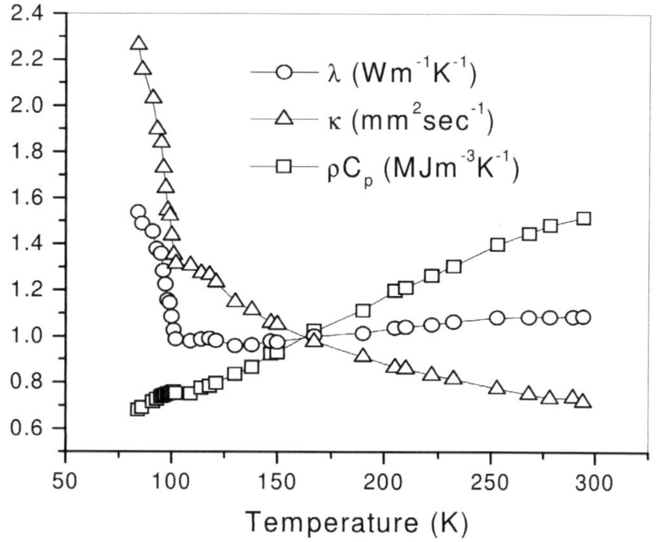

Figure 6. Variation in thermal transport properties with temperature for the sample $Bi_{1.6}Pb_{0.4}Sr_{1.6}Ba_{0.4}Ca_2Cu_3O_y$

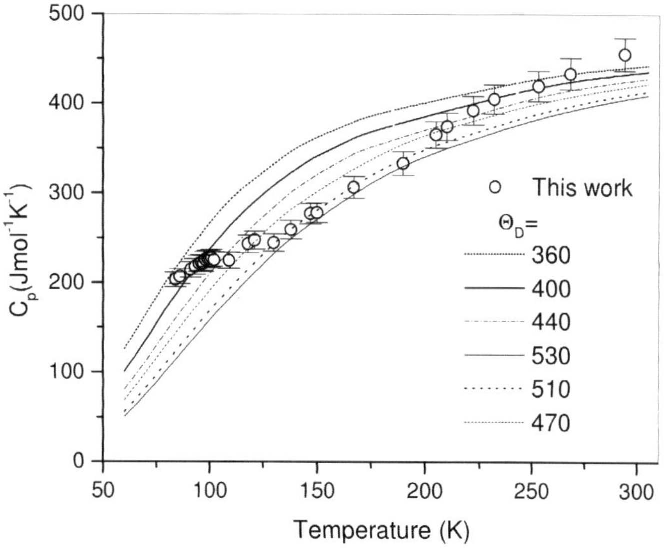

Figure 7. The specific heat estimated from thermal conductivity (λ) and thermal diffusivity (κ). Calculated values for different values of Θ_D are also shown.

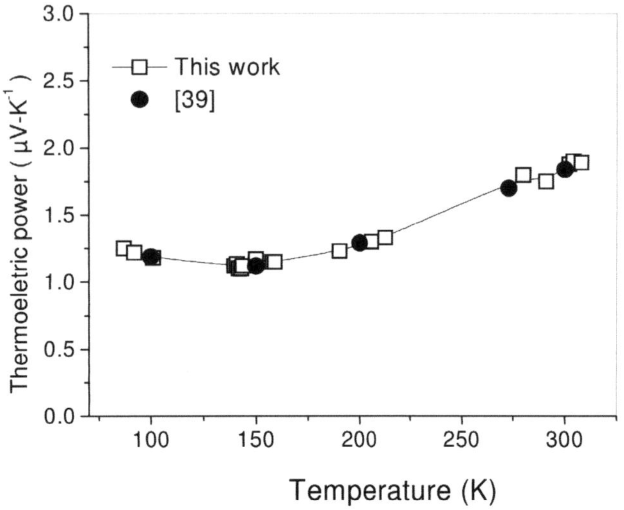

Figure 8. Thermoelectric power of the copper sample with temperature.

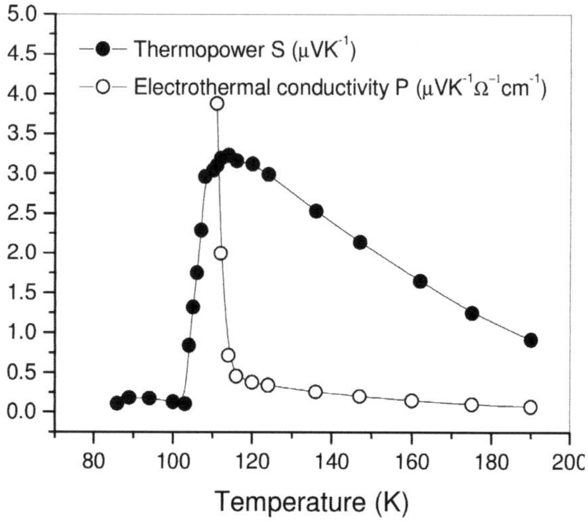

Figure 9. Variation of thermoelectric power (S) and electrothermal conductivity (P) with temperature for the sample.

3.6. Electrothermal conductivity

The electrothermal conductivity (P) is the thermoelectric power divided by the dc electrical resistivity and is given as,

$$P = \frac{S}{\rho}, \qquad (5)$$

Where S is the thermoelectric power and ρ is the dc electrical resistivity.

In the mixed state of a superconductor, the electrothermal conductivity is also defined as the measure of the electrical current density produced by a thermal gradient and is supposed to be independent of the magnetic field. We have utilized the former definition to calculate electrothermal conductivity and are shown in Figure 9.

3.7. Figure of merit

Using the data of electrical resistivity, thermal conductivity and thermoelectric power, Figure of merit factor is calculated and is plotted in Figure 10.

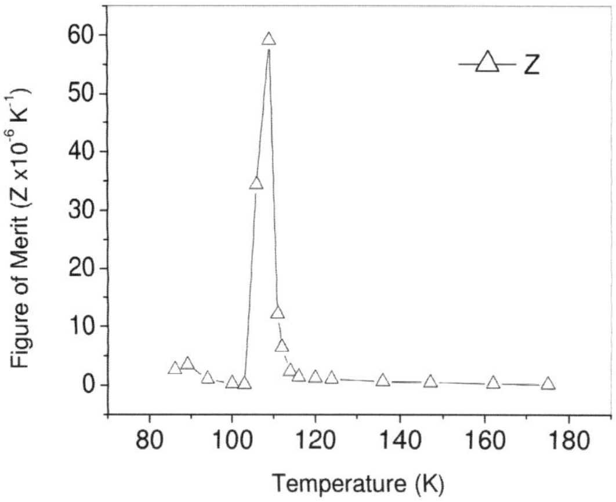

Figure 10. Variation of figure of merit factor (Z) with temperature.

The Figure of merit is calculated from the expression [40],

$$Z(T) = \frac{S^2(T)}{\lambda(T)\rho(T)} \qquad (6)$$

where; Z(T) is the Figure of merit factor, S(T) is the thermoelectric power, λ(T) is the thermal conductivity and ρ (T) is the electrical resistivity.

Near critical temperature the Figure of merit present a remarkable peak for the samples. This peak is due to a quick drop in the electrical resistivity which occurs about 3K before the drop in thermoelectric power. Outside the critical temperature region one can see that the curves of the Figure of merit and thermoelectric power are characterized by a similar behaviour near T_c. Similar trend of the Figure of merit is observed in the Bi-based high-T_c superconductors [40].

4. Summary and conclusions

The samples with nominal composition $Bi_{1.6}Pb_{0.4}Sr_{1.6}Ba_{0.4}Ca_2Cu_3O_Y$ were prepared by a controlled solid state synthesis method. This composition was selected on the basis of experiments conducted by the author with a similar composition [8]. The samples were almost a single phase with Bi-2223 high-T_c phase identified. All the three types of tests i.e. dc electrical resistivity, ac magnetic susceptibility and x-ray diffraction are in agreement with each other. Thermal transport properties include thermal conductivity, thermal diffusivity and heat capacity per unit volume. Simultaneous measurement of thermal conductivity and thermal diffusivity makes it possible to estimate specific heat and the Debye temperature Θ_D. The simultaneous measurement also provides a useful check on the reliability and the consistency of the analyses. Thermal conductivity variation with temperature shows slight decrease initially and then a pronounced increase around T_c. A similar behaviour is observed in all hole-type CuO_2-plane superconductors and in all their structural forms. This effect is due to phonon [47] or quasiparticle scattering [48]. Thermal diffusivity shows a similar trend as that of the thermal conductivity. Heat capacity per unit volume decreases with decrease in temperature. Assuming density of the sample to be constant in the studied temperature range molar specific heat is also calculated. Specific heat jump around T_c is also very prominent. These results indicate a good crystalline structure and an optimal doping. Thermoelectric power was positive in this bismuth-based superconductor. The behavior of thermoelectric power of the sample was approximately linear with temperature as observed in other bismuth-based high-T_C superconductors. The superconducting transition started at 114±1K and after that, thermoelectric power decreased almost to zero value at 103±1K. The known value of the transition temperature of this sample measured from electrical resistivity was 110±1K. Therefore, the difference between thermoelectric transition temperature and resistivity transition temperature were almost in agreement within experimental errors. Electrothermal conductivity increases sharply near transition temperature. A maximum in the Figure of merit, of this ceramic superconductor, is around the superconducting transition temperature. It is then reduced to zero below critical temperature. This system can be useful for application in low-temperature Peltier devices in order to reach temperatures lower than the temperature of liquid nitrogen.

Acknowledgements

The authors would like to acknowledge Higher Education Commission (HEC) Pakistan for the financial support for this research work. Dr. M. Maqsood and Prof. A. Maqsood are appreciated, for their useful suggestions. Mr. Ali Abdullah is acknowledged for lending a hand in preparation of this manuscript.

REFERENCES

[1] Remschnig K, Tarascon J M, Miceli P F and Hull G W 1991 *Phys. Rev.* B **43** 5481
[2] Awana V P S, Agarwal S K, Kumaraswany B V, Singh B P and Narlikar A V 1992 *Supercond. Sci. Technol.* **5** 376
[3] Maeda A, Yabe T, Takebayashi S, Hase M and Uchinokura K 1990 *Phys. Rev.* B **41** 4112
[4] vom Hedt B, Lisseck W, Westerholt K and Bach H 1994 *Phys. Rev.* B **49** 9898
[5] Pop A V, Deltour R, Harabor A, Ciurchea D, Ilonca Gh, Pop V and Todica M 1997 *Supercond. Sci. Technol.* **10** 943
[6] Mori N, Wilson J A and Ozaki H 1992 *Phys. Rev.* B **45** 10633
[7] Kim S H, Kim H S, Lee S H and Kim K H 1992 *Solid State Commun.* **83** 127
[8] Maqsood A, Khaliq M and Maqsood M 1992 *J. Mat. Sci.* **27** 5330
[9] Batlogg B 1990 *High Temperature Superconductivity* (Redwood city, CA: Addison-Wesley)
[10] Gustafsson S E 1991 *Rev. Sci. Instrum.* **62** 797
[11] Maqsood A, Amin N, Maqsood M, Shabbir G, Mahmood A and Gustafsson S E 1994 *Int. J. Energy Res.* **18** 777
[12] Maqsood M, Arshad M, Zafarullah M and Maqsood A 1996 *J. Supercond. Sci. Technol.* **9** 321
[13] Rehman M A and Maqsood A 2002 *J. Phys D: Appl. Phys.* **35** 2040
[14] Maqsood A, Rehman M A, Gumen V and Haq A 2000 *J. Phys D: Appl. Phys.* **33** 2057
[15] Rehman M A, Maqsood A 2003 *Int. J. of Thermophys.* **24** 867
[16] Fee M 1993 *Appl. Phys. Lett.* **62** 1161
[17] Richards P L 1994 *J. Appl. Phys.* **76** 1
[18] Leea A T, Richards L P, Nam S W, Cabrera B and Irwin K D 1996 *Appl. Phys. Lett.* **69** 12
[19] Microsoft Encarta Encyclopedia, Microsoft Corporation, USA, 2003.
[20] Majewski P 2000 *J. Mater. Res.* **15** 4
[21] Balachandran U, Iyer A N, Haldar P, Hoehn J G, Motowidlo L R, Maeda H, Togano K (Eds.), Bi-Based High-T_c Superconductors, Marcel Decker Inc., New York, 1996.
[22] Rehman M A, Maqsood, Ahmad N, Maqsood A and Haq A 1998 *J. Mat. Sci.* **33** 1789
[23] Chen D X, Nogues J and Rao K V 1989 *Cryogenics* **29** 800
[24] Muller K H 1989 *Physica* C **159** 717
[25] Ishida T and Goldfarb R B 1990 *Phys. Rev.* B **41** 8937
[26] Celebi S 1999 *Physica C* **316** 251
[27] Jasiolek G, Gorecka J, Majewski J, Yuan S, Jin S and Liang R 1990 *Supercond. Sci. Technol.* **3** 194
[28] Seeber B 1998 *Handbook of Applied Superconductivity* vol. 1 (Bristol/Philadelphia: Institute of Physics Publishing)
[29] Cullity B D 1967 *Elements of X-ray Diffraction* 3rd ed. (London : Addison-Wesley)
[30] Uher C and Kaiser A B 1987 *Phys. Rev.* B **36** 5680
[31] Peacor S D and Uher C 1989 *Phys. Rev.* B **39** 11559
[32] Mori K, Sasakawa M, Igarashi T, Isikawa Y, Sato K, Noto K and Muto Y 1989 *Physica* C **162** 512
[33] Crommie M F and Zettle A 1990 *Phys. Rev.* B **41** 10978
[34] Cohn J L, Wolf S A and Vanderah T A 1992 *Phys. Rev.* B **45** 511
[35] Ginsberg D M 1992 *High Temperature Superconductivity* (World Scientific Publishing)
[36] Ikebe M, Fujishiro H, Naito T and Noto K 1994 *J. Phys. Soc. Japan.* **63** 3107
[37] Okazaki J, Hasegawa T, Kishio K, Kitazawa K, Kishi A, Ikeda Y, Takano M, Oda K, Kitaguchi H, Takada N and Miura Y 1990 *Phys. Rev.* B **41** 4296
[38] Gordon J E, Prigge S, Collocott S J and Driver R 1991 *Physica* C **185-189** 1351
[39] Barnard R D 1972 *Thermoelectricity in Metals and Alloys* (London : Taylor & Francis)
[40] Bougrine H 1998 Proc. 17th Int.Conf. on Thermoelectrics (IEEE)
[41] Mitra N, Trefny J, Yarar B, Pine G, Sheng Z Z and Hermann A M 1988 *Phys. Rev.* B **38** 7064

[42] Chen G H, Yang G, Yan Y F, Jia S L, Ni Y M, Zheng D N, Yang Q S and Zhou Z X 1989 *Mod. Phy. Lett.* B **3** 1045

[43] Laurent C, Patapi S K, Green S M, Luo L, Politis C, Durczewski K and Ausloos M 1989 *Mod. Phy. Lett* B **3** 241

[44] Lopez A J, Maza J, Yadava Y P, Vidal F, Garcia-Alvarado F, Morán E and Senaris-Rodriguez M A 1991*Supercond. Sci. Technol.* **4** S292

[45] Naqvi S M M R et. al. 1997 Proc. 5th Int. Symp.on Advanced Materials (Islamabad)

[46] Pekala M , Tampieri A, Celotti G, Houssa M and Ausloos M 1996 *Supercond. Sci. Technol.* **9** 644

[47] Tewordt L, Fay D, Wolkhausen Th 1988 *Solid State Commun.* **67** 301

[48] Houssa M and Ausloos M 1994 *Physica* C **235** 1483

Field- and current controlled switching between vortex states in a mesoscopic superconductor

A Rydh, T Golod and V M Krasnov

Department of Physics, Stockholm University, AlbaNova University Center, SE – 106 91 Stockholm, Sweden

E-mail: `arydh@physto.se`

Abstract. We study the controllable manipulation of vortices in a mesoscopic, superconducting "island" of Nb, using an integrated Josephson junction as a field-sensitive vortex detector. The island, divided by a single Josephson junction and suspended by Nb microbridges, was fabricated from a Nb/$Pt_{1-x}Ni_x$/Nb tri-layer using a focused ion beam. We find that the system at select magnetic fields behaves as a vortex memory cell, where current pulses can be used to switch the vortex configuration between metastable states of distinctly different junction critical currents. Non-destructive read-out of a state is then easily done with an intermediate current. Furthermore, we show that the Josephson junction displays a strong magnetoresistive effect at current bias well above the junction critical current but below the onset of flux flow. This enables the junction to be used as a quantitative probe of magnetic field with better than single flux quantum resolution.

1. Introduction

Mesoscopic superconductors are the subject of intense studies with promising potential in devices [1, 2]. Advanced applications typically involve quantum interference through the interplay of several Josephson junctions, but even rather simple systems are rich in physics when vortices are involved. Vortex states in mesoscopic superconductors are determined not only by the applied magnetic field but also by sample boundaries and geometry. This leads to pronounced metastability that could be used in applications such as vortex memory cells. Several methods exist to study vortices in mesoscopic superconductors, including transport, ballistic Hall magnetometry [3], small superconducting-insulator-normal metal tunnel junctions [4, 5] and scanning tunneling microscopy (STM) [6]. Here we use a single Josephson junction embedded into a mesoscopic, superconducting island both as a vortex state detector and as a quantitative, field-sensitive device.

2. Experimental method

Layers of Nb/$Pt_{1-x}Ni_x$/Nb were deposited onto an oxidized Si wafer. The Nb layers were 225 nm and 350 nm thick and the junction layer 20 – 30 nm. Using a focused ion beam (FIB), islands with a central superconductor-ferromagnet-superconductor (SFS) junction and suspended by micro-bridge connectors were cut-out. Details of the sample fabrication can be found in [7]. The results are presented for a sample with 50 at.% Ni, which is sketched in figure 1.

Measurements were carried out in a cryogen-free magnet system with a flowing gas insert reaching 1.6 K base temperature. The magnetic field was applied perpendicular to the island

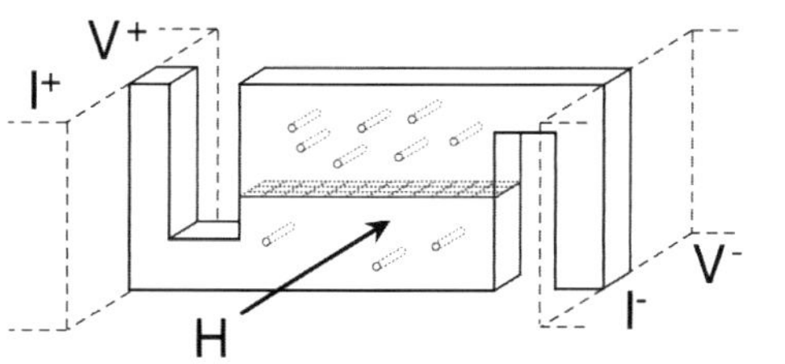

Figure 1. Sketch of the sample. A mesoscopic island of Nb with a center SFS junction was sculptured from a Nb/PtNi/Nb tri-layer. The island width was 1140 nm, total height 495 nm, the lead-to-lead distance 2 μm and the island thickness 230 nm. The magnetic field was applied in parallel with the 20 nm thick junction, driving vortices into the island as illustrated schematically.

surface, yielding maximum geometry-stabilizing effect for vortex trapping, see figure 1. An arbitrary waveform generator was used as a current source. The sample voltage was measured using a 24-bit analog-to-digital converter. For measurements of differential resistance, a set of synchronized lock-in amplifiers built around a field-programmable gate array was used. To avoid hysteresis arising from the magnet inductance while sweeping the magnet, the signal of a Hall sensor was measured simultaneously with the sample and used to determine the magnetic field. Critical currents were determined by finding the highest current with a voltage below the zero-current voltage noise level.

3. Results and discussion

In figure 2, some basic sample characteristics are shown. The bridge critical temperature is $T_c = 8.6$ K, corresponding to the lowest T_c of the two Nb layers. The junction critical current I_c at $T = 2$ K is 1.6 mA, which is slightly lower than the current $I_{ff} \approx 2.5$ mA needed to drive the system into the flux-flow state, see the top right inset of figure 2. This difference between I_c and I_{ff} increases in magnetic field and with increasing temperature, since I_c is more sensitive to magnetic field and temperature than I_{ff}.

The Fraunhofer pattern of figure 2 indicates good junction uniformity with a periodicity $\Delta H \approx 9$ mT. The regular periodicity of the Fraunhofer pattern is, however, limited to measurements at low currents and for moderate changes in magnetic field. Figure 3 illustrates that the Fraunhofer pattern changes depending on how the measurement is performed. At low currents, field sweeps give rise to pronounced hysteresis, as seen in panel (a). When varying the applied magnetic field, the driving force for adding or removing vortices is tuned. In mesoscopic samples, the Meissner screening currents interact with interior vortices, causing metastability. Vortices thus enter at a higher field than they exit so that each state change becomes irreversible.

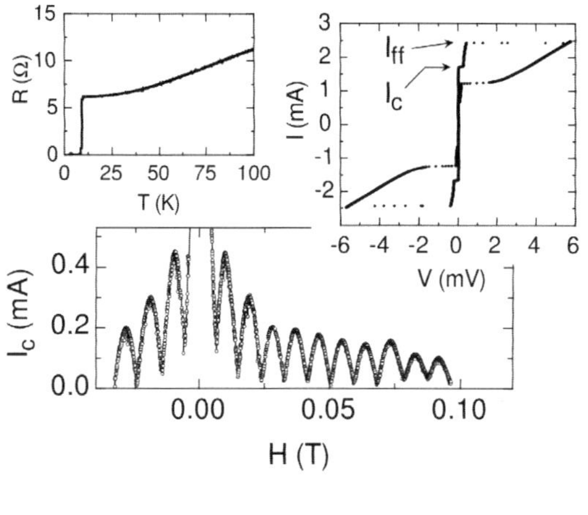

Figure 2. Bottom: Fraunhofer pattern of the junction critical current as a function of magnetic field at $T = 2.0$ K. The periodicity of 9 mT corresponds to one flux quantum in an area of 1.14×0.20 μm^2. Top left: Temperature dependence of the normal state resistance. Top right: Current-Voltage (I-V) characteristics in near zero field. The junction resistance is seen to be about 0.15 Ω, while the flux-flow resistance is several ohms.

Journal of Physics: Conference Series **153** (2009) 012027 doi:10.1088/1742-6596/153/1/012027

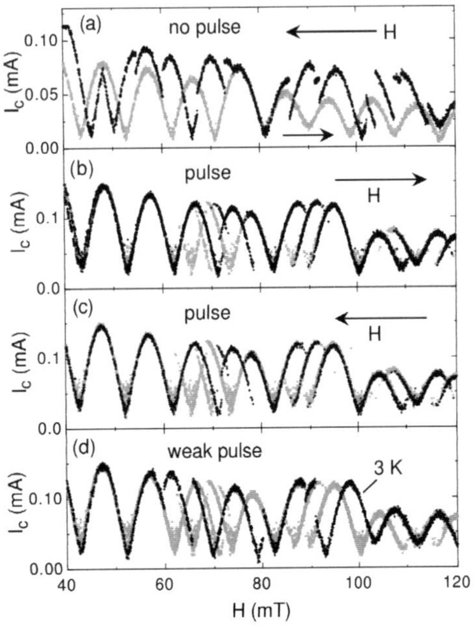

Figure 3. Critical current as a function of magnetic field. **(a)** Sweeping up (gray) and down (black) without pulsing the current at $T = 3$ K. Before sweeping the field down, the system was driven into the flux flow state at $H = 0.4$ T. After equilibration at high current, the obtained vortex state initially remains metastable. However, after the field has been swept down 50 mT to 100 mT, jumps start to occur with an average distance about the same as the Fraunhofer pattern periodicity. **(b)** Increasing magnetic field at $T = 2.3$ K. Before each measurement of I_c, the current was pulsed to 2.4 mA for a few ms, driving the system to the flux flow state. Black curves correspond to a positive pulse while gray corresponds to a negative. **(c)** As in (b) but for decreasing magnetic field. **(d)** $I_c(H)$ at $T = 3$ K (black) compared to the curves of (c) (gray). I_c of the 3 K curve is scaled by 1/1.85.

The system may also be brought between different states by pulsing current through the island. In fact, comparing panels (b) and (c) of figure 3, it is seen that current pulses bring the system back into a pseudo-equilibrium, where information about the field sweep direction is lost. When pulsing the current, the magnetic flux in the junction may take several different values for the same applied magnetic field. Interestingly, positive and negative pulses sometimes favor different, unique states, as is seen by comparing the gray and black curves in the panels at for instance 74 mT. The system at these conditions works as a vortex memory cell.

The observations are consistent with vortices being driven in and out of the mesoscopic island by means of magnetic field and current. When the system is in the Meissner state and the field is increased, the screening currents increase until the Lorentz force for flux entry is exceeded. At this point, the screening currents suddenly decrease and with them the local field in the junction.

It is clear from figure 3 that the vortex configuration can be controlled both by current and magnetic field. However, it is hard to determine the number of flux quanta each jump corresponds to from $I_c(H)$. Fortunately, it appears that the SFS junction displays a fairly strong magnetoresistive effect. By measuring the junction differential resistance dV/dI at a current level well above I_c but below $I_{\rm ff}$, it is possible to obtain a quantitative measure of the magnetic field in the junction. Figure 4 shows dV/dI for increasing and decreasing fields. The Meissner state is clearly discernible and displays a quadratic field dependence. When vortices start to enter the island the screening decreases and the differential resistance drops. In decreasing fields, instead, each expulsion of flux is accompanied by an increase in resistance. The smallest branch spacing is found to be about 3 mT, in good agreement with the value $\Delta H = 3.6$ mT that corresponds to a single flux quantum ϕ_0 in the 1.14×0.495 μm^2 island area. Most jumps, however, involve several vortices.

In figure 5, measurements of dV/dI over a wider field range are shown. In these measurements no current offset was used, to minimize the influence of the Lorentz force. It is seen that the hysteresis disappears at about 1.2 T at 1.6 K, at which point additional flux flow resistance starts to add to the signal. At 6 K the corresponding field is 0.4 T (not shown). Close to T_c irreversible behavior is no longer found, as seen in the inset of figure 5. Rotation of magnetic

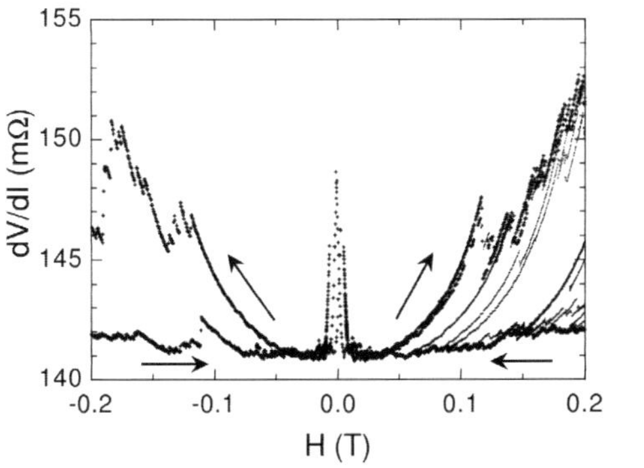

Figure 4. dV/dI as a function of magnetic field at $T = 1.6$ K and $I_{\rm offs} = 1.25$ mA. The arrows indicate the field sweep direction. The branches were obtained by suddenly reversing the field sweep direction. The central peak and small ripples in dV/dI are remaining effects of I_c exceeding or being close to $I_{\rm offs}$.

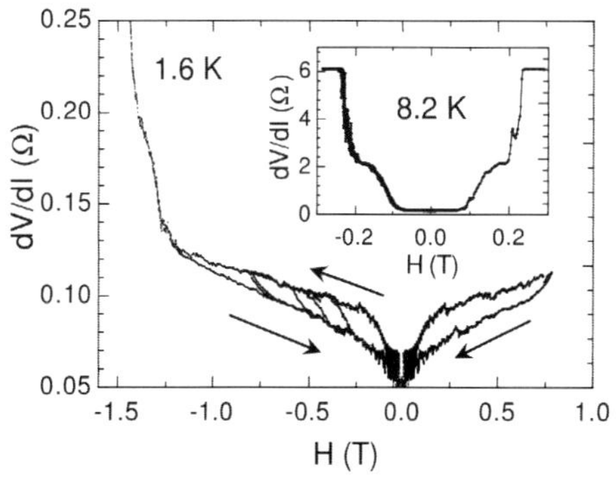

Figure 5. dV/dI as a function of magnetic field at $T = 1.6$ K and $I_{\rm ac} = 0.25$ mA (no offset). The inset shows the corresponding curve at $T = 8.2$ K, illustrating that the hysteresis is strongly temperature dependent as could be expected for vortices but not for magnetic domain rotation.

domains in the $Pt_{1-x}Ni_x$ junction layer could, in principle, cause hysteresis and jumps in the junction local field. The observed temperature dependence, current dependence and behavior of sub-loops, however, all indicate that the current observations are dominated by vortex physics.

4. Conclusion

In conclusion, we have studied the behavior of vortices in a mesoscopic island by using an integrated SFS Josephson junction as a field sensitive detector. We have found that the vortices can be probed both through junction critical current and junction magnetoresistance. The vortices enter and leave the island via a barrier potential, which is overcome either by magnetic field pressure or current drive. At selected magnetic fields, current-programming between persistent vortex states is possible, with non-destructive read-out of distinctly different critical currents.

Acknowledgments

Support from the Knut and Alice Wallenberg Foundation and the Swedish Research Council is gratefully acknowledged.

References

[1] Blamire M G 2006 *Supercond. Sci. Technol.* **19** S132
[2] Van Duzer T 2008 *IEICE Trans. Electron.* **E91-C** 260
[3] Geim A K, Dubonos S V, Lok J G S, Grigorieva I V, Maan J C, Theil Hansen L and Lindelof P E 1997 *Appl. Phys. Lett.* **71** 2379
[4] Kanda A and Ootuka Y 2002 *Microelectronic Engineering* **63** 313
[5] Kanda A, Baelus B J, Peeters F M, Kadowaki K and Ootuka Y 2004 *Phys. Rev. Lett.* **93** 257002
[6] Karapetrov G, Fedor J, Iavarone M, Marshall M T and Divan R 2005 *Appl. Phys. Lett.* **87** 162515
[7] Golod T, Frederiksen H and Krasnov V M submitted to *J. Phys.: Condens. Matter* (proceeding of LT25)

Superconducting Properties of Noncentrosymmetric $Li_2(Pt_{1-x}Pd_x)_3B$ Superconductors

H Takeya[1], S Kasahara[1], M ElMassalami[2], K Hirata[1], K Togano[1]

[1]National Institute for Materials Science, Sengen1-2-1, Tsukuba, Ibaraki 305-0047, Japan
[2]Instituto de Fisica, Universidade Federal do Rio de Janeiro, Caixa Postal 68528, 21945-970 Rio de Janeiro, Brazil

E-mail: takeya.hiroyuki@nims.go.jp

Abstract. Superconductivity has been found in two Li-containing compounds, Li_2Pt_3B and Li_2Pd_3B, which show superconducting transition at temperatures 2.5 K and 7.2 K, respectively, and have the same cubic structure (*P4₃32*) without mirror or inversion symmetry. Because the compounds are noncentrosymmetric, it is interesting to study the symmetries of the superconducting gap function and Cooper pairs. The specific heat of $Li_2(Pt_{1-x}Pd_x)_3B$ [x = 0, 0.5, 1] was measured. The Pd-concentration (x)-dependent electronic specific heat (C_{es}) of $Li_2(Pt_{1-x}Pd_x)_3B$ at low temperatures $\ll T_c$ showed distinctly different features on varying x. The temperature dependence of C_{es} for one end member, Li_2Pd_3B, follows an exponential behavior as in conventional superconductors, while those of $Li_2Pd_{1.5}Pt_{1.5}B$ and Li_2Pt_3B follow a quadratic one, suggesting the existence of line nodes in the superconducting gap. This contrasting result conceivably originates from the difference in the strengths of the spin-orbit coupling of the Pt- and Pd-based compounds.

1. Introduction

The discovery of and research into new metal boride superconductors have been encouraged since the discovery of MgB_2. Some electronic band structure calculations suggest that higher superconducting transitions are possible in materials composed of light elements. For example, LiB_2, BeB_2 [1], and LiBC [2] are the first natural candidates since lighter atoms may help in providing larger phonon frequencies. They have been investigated experimentally and theoretically, but extensive research into these new materials with light elements has not produced positive results. However, some new superconductors have been found in recent years, including $MgCNi_3$ (T_c = 7.6 K) [1], Y_2C_3 (T_c = 18 K) [2], $W_7Re_{13}X$ (X=B, T_c = 7.1 K and X=C, T_c = 7.3 K) [3], $La_3Pd_4Ge_4$ (T_c = 2.75 K) [4], and B-doped diamond (T_c = 4 K) [5].

We have long been interested in multinary borides such as RRh_4B_4, RM_3B, and RNi_2B_2C (R=Y, Lanthanoid Elements, M=Rh, Pd, Pt), which show rich and interesting phenomena arising out of the competition between superconductivity and magnetism at low temperatures. In the recent years, we have also been searching for new superconducting materials containing lithium and boron that has lead to the discovery of superconductivity in Li_2Pt_3B and Li_2Pd_3B at around 2.5 K and 7.2 K, respectively [6,7]. These superconductors crystallize in a cubic structure with a space group of *P4₃32* that lacks both mirror and inversion symmetry-operations. In particular, the Pt and Pd atoms are situated in noncentrosymmeteric sites. Since these compounds are non-magnetic, they may be ideal for studying the influence of this type of broken symmetry, in contrast to a novel heavy fermion (magnetic) superconductor $CePt_3Si$ [8] that lacks a mirror symmetry only along the c-axis. The higher atomic numbers of Pt(Z_{Pt}) and Pd(Z_{Pd}), together with their ratio $(Z_{Pt}/Z_{Pd})^2$~3, emphasize the importance of spin-orbit coupling (SOC), which, as evident, is much stronger for the Pt-based compound than its Pd-based isomorph. According to the results of a penetration depth study by H. Q. Yuan et al. [9] and NMR measurements by M. Nishiyama et al. [10,11], Li_2Pd_3B behaves as a full-gap superconductor, while Li_2Pt_3B has line nodes in the energy gap. In this study, we have investigated the symmetry of the superconducting gap function in $Li_2(Pt_{1-x}Pd_x)_3B$ [x = 0, 0.5, 1] by using specific-heat measurements.

2. Experimental

Samples of $Li_2(Pt_{1-x}Pd_x)_3B$ [x = 0, 0.5, 1] were prepared from an Li lump (Kanto Kagaku Co., above 99%), Pd plate (Rare Metallic Co., 99.95%), Pt wire (Furuya Metal Co., 99.99%), and B flake (Rare Metallic Co., 99.8%). Initially, stoichiometric ratios of Pt/Pd and B were melted together by an arc-furnace in a pure argon atmosphere. Then the ingot of $(Pt/Pd)_3B$ and the lump of Li with a 15% excess were melted together to produce stoichiometric $Li_2(Pt/Pd)_3B$. The internal textures of the reacted materials were observed using a scanning electron microscope (SEM, JEOL JSM-6301F). The microstructural observation was carried out for powdered samples using transmission electron microscopy (TEM, JEOL FX-2000). X-ray powder diffraction (XPD) analyses were performed using a diffractometer equipped with a rotating anode and a graphite monochrometer (JEOL JDX-3500). The composition of Li:Pt/Pd:B was determined by chemical analysis using the induced coupled plasma (ICP) method. The magnetic properties were measured using a SQUID magnetometer (MPMS-XL, Quantum Design Co., Ltd.).

Polycrystalline samples of Li_2Pt_3B, $Li_2Pd_{1.5}Pt_{1.5}B$ and Li_2Pd_3B were cut into rectangular shapes with dimensions of approximately 1.7 x 3.0 x 0.4 mm^3. Their resistivity was measured by a conventional four-probe method using the MPMS-XL system between 1.8 K and room temperature. The lower temperature dependence of the resistivity was implemented using ^{3}He cryostat set-ups operating within the range of 0.4 K to 20 K, up to 50 kOe. The four samples of Li_2Pt_3B (#1, #2), $Li_2Pt_{1.5}Pd_{1.5}B$, and Li_2Pd_3B with weights of 9.1 (#1), -4.4 (#2), 6.5, and 11.1 mg, respectively, were used for the specific-heat measurements using a relaxation-type calorimeter (Oxford Instruments) operated with a ^{3}He refrigerator in the range of 0.4 K to 15 K. The samples were kept in a vacuum or argon gas atmosphere for storage and handling.

3. Results and discussion

The temperature dependence of the resistivity of $Li_2(Pt_{1-x}Pd_x)_3B$ [x = 0, 0.5, and 1] was measured in magnetic fields up to 50 kOe. The upper critical field $H_{c2}(T)$ vs. temperature is shown in Fig. 1, and three of the $H_{c2}(T)$ curves of x = 0, 0.5, and 1 were well fitted by the Werthamer-Hefland-Hohemberg (WHH) formula $H_{c2}(0) = 0.691 \times (dH_{c2} / dT)_{Tc} \times T_c$. Consequently, $H_{c2}(0)$ was estimated to be 19, 25, and 45 kOe for $Li_2(Pt_{1-x}Pd_x)_3B$ [x = 0, 0.5, and 1], respectively. According to the Ginzburg-Landau (G-L) theory, H_{c2} near T_c is given by the following equation:

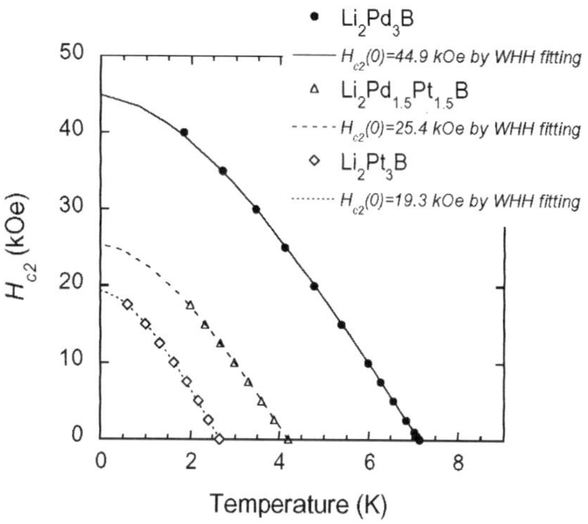

Fig. 1 Upper critical field H_{c2} vs temperature and $H_{c2}(0)$ estimated by the WHH fitting formula for the studied three compositions are shown.

$$H_{c2}(T) = \frac{\Phi_0}{2\pi\xi^2} = \frac{\Phi_0}{2\pi\xi(0)^2}(1 - \frac{T}{T_c}), \qquad (1)$$

where Φ_0 : the fluxoid quantum, ξ: the G-L coherence length, and T_c : the superconducting transition temperature. The coherent length $\xi(0)$ was calculated to be 11.8, 9.9 and 7.6 nm, respectively.

The bulk superconducting transitions were observed at 2.5 K, 3.9 K, and 7.2 K for $Li_2(Pt_{1-x}Pd_x)_3B$ [x = 0, 0.5, and 1, respectively] by the zero field specific-heat measurements. They were completely suppressed at the magnetic fields of 15, 30, and 50 kOe, respectively. These specific-heat temperatures (C) are the sum of an electron contribution (C_e) and a lattice contribution (C_l). C_l is described by $\beta T^3 + \delta T^5 + \varepsilon T^7$ in the harmonic-lattice approximation for an adequate fit. In the normal state, C_e is described by the Sommerfeld term of $\gamma_n T$, where γ_n is proportional to the electron density of states. Estimates of γ_n = 9.6, 9.7, and 9.8 $mJmol^{-1}K^{-2}$ and β= 0.80, 0.81, and 0.90 $mJmol^{-1}K^{-4}$ for x = 0, 0.5, and 1, respectively, were obtained from the normal state data at the magnetic fields described above with fitting at temperatures below 2.5 K, 3.9 K, and 7.2 K, respectively.

The electronic specific heat within the superconducting phase (C_{es}) at low temperatures is calculated by subtracting C_l from C. The $C_{es}/\gamma_n T_c$ of the three compounds is shown in Fig. 2 as a semilogarithmic plot with T_c/T. To evaluate the superconducting gap function in these compounds, C_{es} was analyzed by exponential and quadratic fittings in the low temperature range below $T_C/3$ (T_C/T > 3). The plot (c) of $C_{es}/\gamma_n T_C$ in Li_2Pd_3B follows an exponential, activated process, but the fitting is not good at $C_{es}/\gamma_n T_C$ lower than 0.01. However, any quadratic fitting is impossible. Li_2Pd_3B is estimated to be a conventional BCS superconductor with a fully opened isotropic gap. On the other hand, from the electronic specific heat of Li_2Pt_3B in (a), it is not clear if it follows an exponential or a quadratic behavior in a low temperature range below $T_C/3$. However, the exponential fitting formula is very different from the calculations of the BCS theory, $C_{es}/\gamma_n T_c$ = 8.5 exp (-1.44T_C/T) for 2.5 < T_c/T < 6 [12]. This result suggests that Li_2Pt_3B is an unconventional superconductor, consistent with independent experiments of the penetration depth study [9] and the NMR

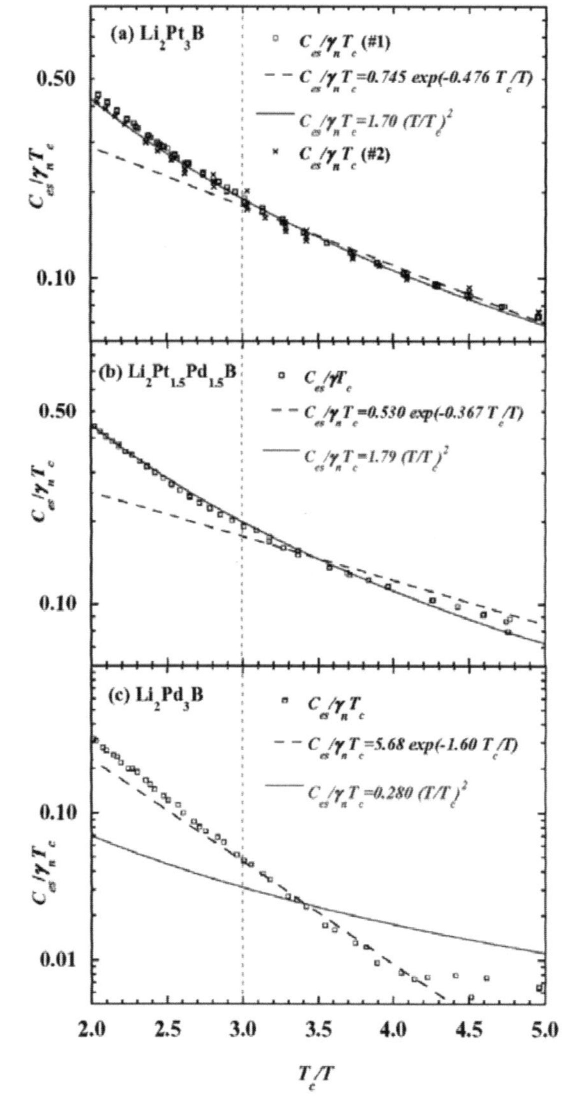

Fig. 2 Reduced electronic specific heat $C_{es}/\gamma T_c$ vs. T_c/T of $Li_2(Pt_{1-x}Pd_x)_3B$ [x=0, 0.5, 1]. The broken and solid lines are the calculated curves fitted by the exponential and quadratic formulas, respectively.

measurements [11]. In the case of $Li_2Pt_{1.5}Pd_{1.5}B$ in plot (b), surprisingly, the 50% Pd-diluted solid-solution still shows the same behavior of the electronic specific heat as that in Li_2Pt_3B.

In the discussion above, C is assumed to be the sum of C_{es} and C_l in the superconducting state. However, in a number of superconductors, notably heavy-fermion compounds and high-T_C cuprates, the total specific heat (C) possibly includes an additional sample-dependent, T-proportional contribution, which is probably associated, at least in part, with residual normal material. Its origin is not entirely clear, however. The specific heat of another Li_2Pt_3B sample (#2) with a different T_C (2.6 K) and transition width was also measured, and the C_{es} was calculated. The low-temperature behaviors of C_{es} in the two samples coincide in the temperature range of $T_c/T > 3$ even though their T_c's are different, which supports the notion that the C_{es} of Li_2Pt_3B shown in Fig. 2 is sample-independent.

In summary, we have investigated the low-temperature specific heat of $Li_2(Pt_{1-x}Pd_x)_3B$ [x = 0, 0.5, 1]. The electronic specific heat (C_{es}) in the superconducting state behaves as a BCS-type conventional superconductor for Li_2Pt_3B but not for $Li_2Pt_{1.5}Pd_{1.5}B$ and Li_2Pd_3B. The physical parameters for the superconductivity in $Li_2(Pd_{1-x}Pt_x)_3B$ [x = 0, 0.5, and 1] are listed in Table 1 below.

Table 1. Normal and superconducting parameters of $Li_2(Pt_{1-x}Pd_x)B$ [x=0, 0.5, and 1].

Compound	T_c [K]	γ [mJ/mol K^2]	θ_D [K]	$\Delta C/\gamma T_c$ [none]	$H_{c2}(0)$ [kOe]	$\xi(0)$ [nm]
Li_2Pt_3B	2.6	9.6	240	0.75	19	11.8
$Li_2Pt_{1.5}Pd_{1.5}B$	3.9	9.7	242	1.1	25	9.9
Li_2Pd_3B	7.2	9.5	228	1.7	45	7.6

This work was supported by the Japan Society for the Promotion of Science and the Brazilian agencies CNPq (485058/2006-5) and Faperj (E-26/171.343/2005).

References

[1] He T, Huang Q, Ramirez A P, Wang Y, Regan K A, Rogado N, Hayward M A, Haas M K, Slusky J S, Inumara K, Zandbergen H W, Ong N P and Cava R J 2001 *Nature* **411** 54
[2] Amano G, Akutagawa S, Muranaka T, Zenitani Y and Akimitsu J 2004 *J. Phys. Soc. Jpn.* **73** 530
[3] Kawashima K, Kawano A, Muranaka T and Akimitsu J 2005 *J. Phys. Soc. Jpn.* **74** 700
[4] Fujii H, Mochiku T, Takeya H and Sato A 2005 *Phys. Rev. B* **72**, 214520
[5] Ekimov E A, Sidorov V A, Bauer E D, Mel'nik N N, Curro N J, Thompson J D and Stishov S M 2004 *Nature* **428** 542
[6] Badica P, Kondo T and Togano K 2005 *J. Phys. Soc. Jpn.* **74** 1014
[7] Togano K, Badica P, Nakamori Y, Orimo S, Takeya H and Hirata K 2004 *Phys. Rev. Lett.* **93** 247004
[8] Bauer E, Hilscher G, Michor H, Paul C, Scheidt E W, Gribanov A, Seropegin Y, Noël H, Sigrist M and Rogl P 2004 *Phys. Rev. Lett.* **92** 027003
[9] H. Q. Yuan, D. F. Agterberg, N. Hayashi, P. Badica, D. Vandervelde, K. Togano, M. Sigrist, M. B. Salamon 2006 *Phys. Rev. Lett.* **97** 017006
[10] Nishiyama M, Inada Y and Zheng G Q 2005 *Phys. Rev. B* **71** 220505
[11] Nishiyama M, Inada Y and Zheng G Q 2007 *Phys. Rev. Lett.* **98** 047002
[12] Gopal E S R 1966 *Sepecific Heat at Low Temperatures* (New York:Plenum)

Transport and ac susceptibility measurements on silver wrapped pellets of $Bi_{1.7}Pb_{0.35}Sr_{1.9}Ca_{2.1}Cu_3O_y$ superconductors

C Terzioglu[1], **D Yegen**[2], **A Varilci**[1], **I Belenli**[1]
[1] Department of Physics, Faculty of Arts and Sciences, Abant Izzet Baysal University, 14280 Bolu-Turkey,
[2] Department of Medical Physics, Dr.A.Y. Ankara Oncology Education & Research Hospital, Ministry of Healthy of Turkey, 06460 Ankara, Turkey.

E-mail: terzioglu_c@ibu.edu.tr

Abstract. The effect of the cooling rates on oxygen content of silver sheathed Bi-2223 bulk superconducting samples was investigated performing dc electrical resistivity, transport critical current density, ac susceptibility and x-ray diffraction (XRD) measurements. The bulk samples with Ag sheathing were annealed under identical conditions and cooled with rates of $25^\circ C/h$, $50^\circ C/h$, $75^\circ C/h$, and $100^\circ C/h$. We estimated the transition temperatures from both dc resistivity and ac susceptibility as a function of temperature measurements. It is observed that T_c and J_c depend on cooling rates of the samples. T_c and J_c decrease with increasing cooling rates. The imaginary part of ac susceptibility measurements is used to calculate inter-granular critical current density using the Bean Model. $J_c(T_p)$ is also seen to decreases with increasing cooling rates. The peak temperature, T_p, in the imaginary part of the ac susceptibility is shifted to a lower temperature with increasing cooling rates as well as increasing ac magnetic fields. The force pinning density decreased with increasing the cooling rates. XRD patterns are given to determine lattice parameter c and obtain information about Bi-2223 and Bi-2212 phases.

1. Introduction

Since the discovery of high-T_c superconductors a large amount of work has been done on Bi(Pb)-Sr-Ca-Cu-O system. It is known that chemical doping and preparation condition play very important roles in high-T_c superconductors. Cooling rates are also important for superconducting properties. This importance is partly due to changing oxygen content during the cooling stage. Thus, it is useful to observe the variation of both the superconducting properties as well as the normal state properties of materials with changing cooling rates in order to understand superconductivity better.

The measurements of ac susceptibility are commonly used to determine magnetic and superconducting properties of materials. In particular, the ac susceptibility measurement is useful in determining the specimen and for distinguishing between inter- and intra-grain properties. At high ac fields, two loss peaks can usually be seen in the imaginary part of susceptibility data; a broad peak at low temperature (coupling losses) due to the motion of inter-granular (Josephson) vortices [1] and a narrower peak

[1] Cabir Terzioglu, Department of Physics, Faculty of Arts and Sciences, Abant Izzet Baysal University, 14280 Bolu-Turkey.

(intrinsic losses) due to the motion of intra-granular Abrikosov vortices near T_c^{onset} [1,2]. These two peaks depend on the sample processing variables as well as the samples composition for the Bi-2223 system [3,4].

The cooling rates reported in the literature range from quenching in liquid nitrogen to controlled cooling as slowly as 1 °C/h. Mainly the rates can be classified as quenching to liquid nitrogen temperature, quenching in icy water, quenching in air, fast cooling, uncontrolled furnace cooling (~150 °C/h) and controlled cooling (1-100 °C/h). Cooling rates following the annealing may have a number of effects on samples. The first effect to be mentioned comes from thermal contraction of the sample and bulk samples, films on substrates and metal-ceramic composites will be differently affected. A second effect is on the oxidization sate of the sample. Another effect worth mentioning is the formation of other phases during cooling stage [5]. Slow cooling rates appear to improve the superconducting properties although they do not affect the phase formation of the Bi-2223 phase provided that the annealing duration is sufficiently long. Fast cooling rates may particularly be more harmful for the composite superconductors where there is an interface between two materials with different thermal expansion coefficients.

In recent years, the cooling rate was found to be an important parameter for optimizing performance of Ag-sheathed Bi-based superconductors [6-8]. It was found that the value of J_c decreases with increasing cooling rate. The effect of cooling rate on transport properties of the samples without silver sheathing has recently been investigated [9]. It was obtained that the values of J_c, T_c, lattice parameter c, and hole concentration were increased with increasing cooling rate.

In this work, we report measurements of electrical resistivity and ac susceptibility as a function of temperature and transport and inter-granular critical current density to investigate the effect of the cooling rates on properties of silver wrapped Bi-2223 bulk superconductors annealed at 830 $^\circ C$ for 48 h and cooled with different cooling rates. Also, XRD analysis is used to characterize microstructures of the sample phases.

2. Experimental details

Production of superconducting powders: Bi_2O_3, PbO, $SrCO_3$, $CaCO_3$ and CuO powders were mixed in appropriate ratios to obtain $Bi_{1.7}Pb_{0.35}Sr_{1.9}Ca_{2.1}Cu_3O_y$ and mixed thoroughly in an agate mortar. The mixture was then calcined in a silver crucible at progressively increasing temperatures up to 820 °C for 12 hours durations. Pellets with rectangular shape were prepared by pressing from the calcined precursor under 1 GPa uni-axial pressure. The pelletized tablets were heat treated in a silver crucible in the furnace all together. Annealing of the samples was performed in two steps at 830 °C for 48 hours durations. The tablets were ground into ceramic powder and re-pelletized before the second annealing. Pellets were completely wrapped in thin silver sheaths and pressed under 1 GPa pressure. These samples were then heat treated at 835 °C for 48 hours. The furnace was heated to the annealing temperature with 100 °C per hour heating rate and 25 °C/h, 50 °C/h, 75 °C/h, and 100 °C/h. cooling rates were applied after the heat treatment. Silver wrapped samples are coded G25 to G100, numbers referring to cooling speeds in units of hour.

XRD data were taken using a Rigaku D/Max-IIIC diffractometer with CuK_α radiation in the range $2\Theta = 4° - 60°$ with a scan speed of 3° /min and a step increment of 0.02° at room temperature. We obtained XRD patterns of the samples after preparation. Silver sheathing was peeled off before XRD examination and patterns were taken from the samples as bulk tablets. Phase purity and the lattice parameters were determined from XRD patterns. The accuracy in determining the lattice parameter c was ±0.001 Å. The relative volume fractions of the Bi-2223 and Bi-2212 phases were determined from the peak intensities of the same particular reflections, using the following expressions [10,11];

$$f_{(2223)} = \frac{\sum I_{H(hkl)}}{\sum I_{H(hkl)} + \sum I_{L(hkl)}} \qquad (1)$$

$$f_{(2212)} = \frac{\sum I_{L(hkl)}}{\sum I_{H(hkl)} + \sum I_{L(hkl)}} \quad (2)$$

Here $I_{H(hkl)}$ and $I_{L(hkl)}$ are the intensities of the (hkl) diffraction lines for Bi-2223 and Bi-2212 phases, respectively (Fig.1).

The measurements of dc resistivity were performed with the four-probe method on all the samples annealed at 830 $^{\circ}C$ for 48 h with different cooling rates (25$^{\circ}C/h$, 50$^{\circ}C/h$, 75$^{\circ}C/h$, and 100$^{\circ}C/h$.). We measured temperature (80-130 K) dependence of resistivity for the samples using a 10 mA dc current in a cryostat. Both voltage and current contacts were made with silver paint. The transition temperature, T_c, was determined as the temperature of zero resistivity. Room temperature resistivities were calculated from room temperature I-V curves. The critical current density J_c was determined from the current-voltage characteristics of the samples at 77 K. The critical current I_c was determined as the onset of a voltage with a criterion of 1 $\mu V/cm$, and the J_c was obtained from I_c and the cross sectional area of the samples.

Ac susceptibility measurements on the samples were made using a home made ac susceptometer and a Lock-in amplifier (Stanford Research SR850) [12]. The ac susceptibility as a function of temperature (77–120 K) at fixed ac magnetic field amplitude (H_{ac}) in a range from 40 to 600 A/m was performed at a frequency of 1 kHz. The ac magnetic field was applied along the longest dimension of the samples. From the χ' and χ'' vs. T plots, the values of the T_c^{onset} (diamagnetic onset temperature) and T_p (temperature corresponding to χ'' peak) were obtained. We could observe only one peak in each of the χ'' vs. T plots. Bean critical-state model [13] is used for determination of inter-granular critical current densities, $J_c(T_p)$, from ac susceptibility measurements. The values of $J_c(T_p)$ in our samples were calculated using the Bean model from the acquired data.

3. Results and discussion

Oxygen stoichiometry can be studied in more detail by thermogravimetric techniques [14]. We measured the masses of each tablet before and after heat treatments in order to record any mass changes. Weight measurements of the pellets revealed a systematic change in masses of the pellets. The change in masses of silver sheathed Bi-2223 bulk samples decreases from -0.36% to -0.46% with increasing cooling rates but the samples without silver sheathing lost more mass [8]. Silver wrapped bulk samples may have experienced mass loss due to oxygen loss through silver sheathing and sublimation of silver at high temperatures [15]. It can be interpreted that there is no silver sheathing so that out diffusion of oxygen is faster and sublimation of Pb and Bi is also easier. Since the out diffusion of oxygen does not happen during cooling, the difference can be attributed to loss of oxygen during annealing, and sublimation Bi and Pb. It should be considered that sample's geometry (volume to surface ratio) plays an important role in sublimation and oxygen intake. Film samples are most prone to sublimation and ease of oxygen intake in this respect.

The XRD patterns from the surface of the G25 and G100 samples are shown in Fig. 1. Some of the Miller indices are indicated in the figure. There is no significant increase in the intensities of particular peaks between the G25 and G100 samples. The lattice parameters for the samples with different cooling rates were determined by using a linear least squares method and were tabulated in Table 1. It is observed that the lattice parameter c decreases with increasing cooling rates. In contrary to this, the lattice parameter c increased with increasing cooling rates in the samples without silver sheathing [9]. The impurity phase (Ca_2PbO_4) is observed around $2\theta = 18^{\circ}$ in both samples in which the intensity of the G25 is higher than that of G100.

The relative volume fraction of the samples calculated using Eqs. (1) and (2) is listed in Table 1. In contrast to the samples without silver sheathing, with increasing cooling rates, the volume fraction of Bi-2223 phase increased and that of Bi-2212 phase decreased.

Table 1. Critical temperature T_c, transport critical current density J_c, lattice parameter c, volume fraction, room temperature resistivity, hole concentration, p of the samples.

Samples	T_c^{offset} (K)	Lattice Parameter c (Å)	J_c^{trans} (A/cm^2)	Hole Concentration p	Volume Fraction (%)		ρ (mΩ.cm) at 300 K
					Bi-2223	Bi-2212	
G25	98$\pm$0.2	37.195	150	0.124	83	17	5.60
G50	91$\pm$0.2	37.187	120	0.114	80	20	7.50
G75	88$\pm$0.2	37.180	50	0.111	77	23	11.20
G100	86$\pm$0.2	37.175	40	0.109	75	25	13.10

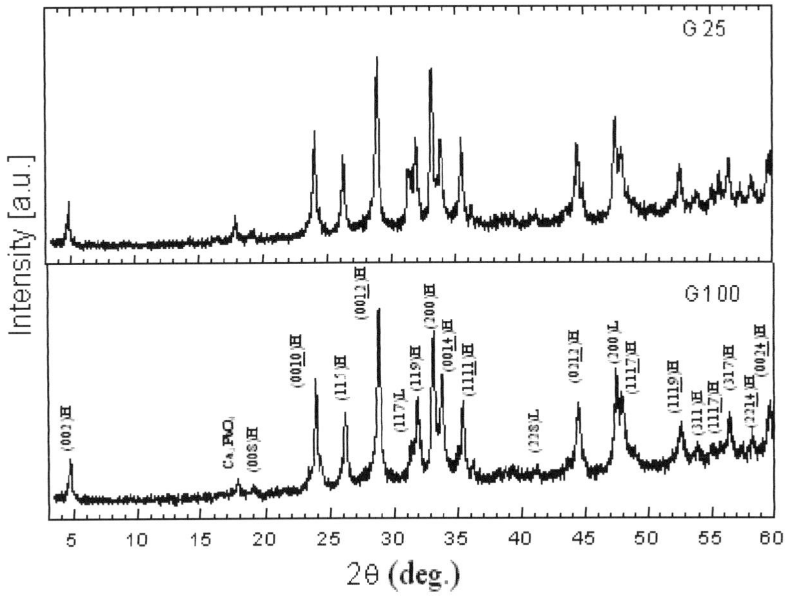

Figure 1. The XRD patterns for the G25 and G100 samples. The peaks indexed (hkl)L and (hkl)H represent the Bi-2212 and Bi-2223 phases respectively.

The electrical resistivity was measured on bar shaped specimens after removal of silver sheathing, using the standard four-probe dc technique, in the temperature range between 80 and 130 K. The temperature dependence of normalized resistivity for all samples annealed at 830 $^\circ C$ for 48 h is shown in Fig. 2. Room temperature resistivities were calculated from room temperature I-V curves (Table 1). Every sample shows linear temperature dependence in normal state, characteristic of Cu-oxide-based (high-T_c) superconductors. The resistivity in the normal state increases with increasing cooling rates. The increase of the normal state resistivity is likely to be due to grain boundaries. Zero-resistivity transition temperatures of the G25, G50, G75, and G100 samples are determined to be 98 K, 91 K, 88 K and 86 K, respectively. The transition temperature decreases with increasing cooling rates. It was observed that the slow cooled sample has lower normal state resistivity, higher zero resistivity transition temperature and longer lattice parameter c. It is observed that the resistivity transition width broadens with increasing the cooling rate, from 25 °C/h to 75 °C/h. The sample cooled with 100 °C/h shows narrower transition but the transition starts at a much lower temperature. As can be seen from the figure, the transition curves from normal to superconducting state indicate double step transition in G50 and G75 samples.

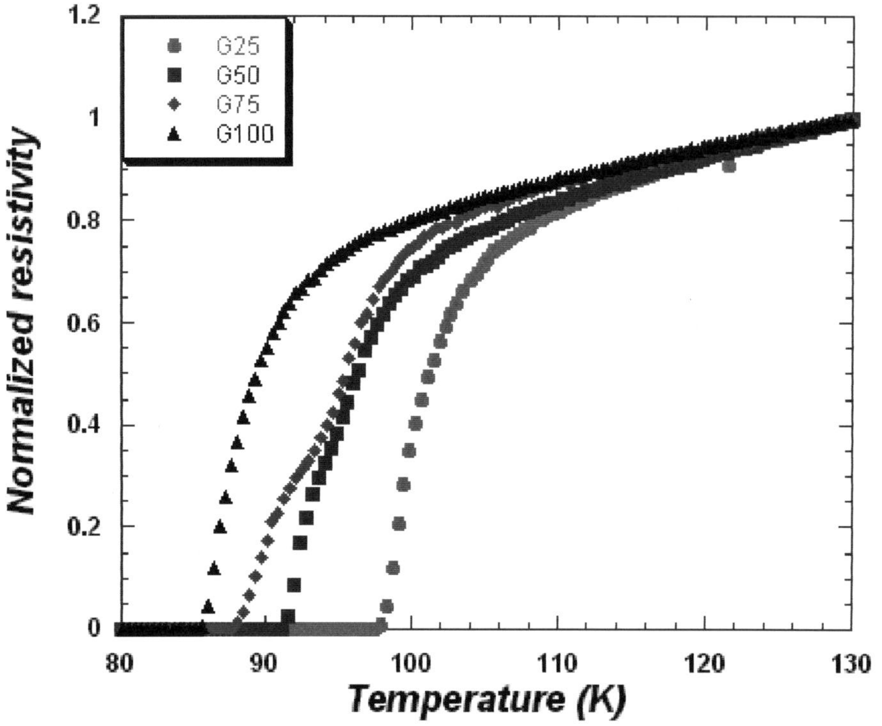

Figure 2. Temperature dependence of normalized resistivity for silver wrapped $Bi_{1.7}Pb_{0.35}Sr_{1.9}Ca_{2.1}Cu_3O_y$ samples annealed at 830 $^{\circ}C$ for 48 h with cooling rate of 25 $^{\circ}C$ /h, 50 $^{\circ}C$ /h, 75 $^{\circ}C$ /h and 100 $^{\circ}C$ /h samples.

The transport critical current densities as a function of cooling rates were measured in liquid nitrogen at self magnetic fields. It was observed that the J_c of the samples decreases from 150 A/cm^2 to 40 A/cm^2 with increasing the cooling rates (Table 1). Low values of critical current density are due to bulk nature of our silver sheathed samples rather than silver sheathed tapes. It was concluded that the enhancement of Jc with decreasing cooling rate is due to increase in the flux pinning strength. Tetenbaum and Maroni reported [16] that slow cooling increases the oxygen content of Bi-2223 phase. It was interpreted that this increase in oxygen content may contribute to the enhanced critical current density observed during the slow cooling for the samples with silver sheathing. In contrast to this, the value of Jc of the samples without silver sheathing increases with increasing cooling rate [9].
The carrier (number of holes) concentration, p, is calculated by using the relation [17]

$$T_c/T_c^{max} = 1 - 82.6\,(p - 0.16)^2 \qquad (3)$$

where T_c^{max} is taken as 110 K for Bi-2223 system. This formula is found to be applicable to several cuprate systems [18,19]. The calculated carrier concentration is listed in Table 1. It is observed that hole carrier concentration decreases from 0.124 to 0.109 with decreasing transition temperature (with increasing cooling rates). It is well known that a parabolic relationship holds between the superconducting transition temperature and hole concentration, being consistent with the present data. The slight increase in T_c with decreasing cooling rate may also be related to the optimization of the hole density. Thus, the cooling rate affects the hole concentration and reduces the carrier density with increasing cooling rates.

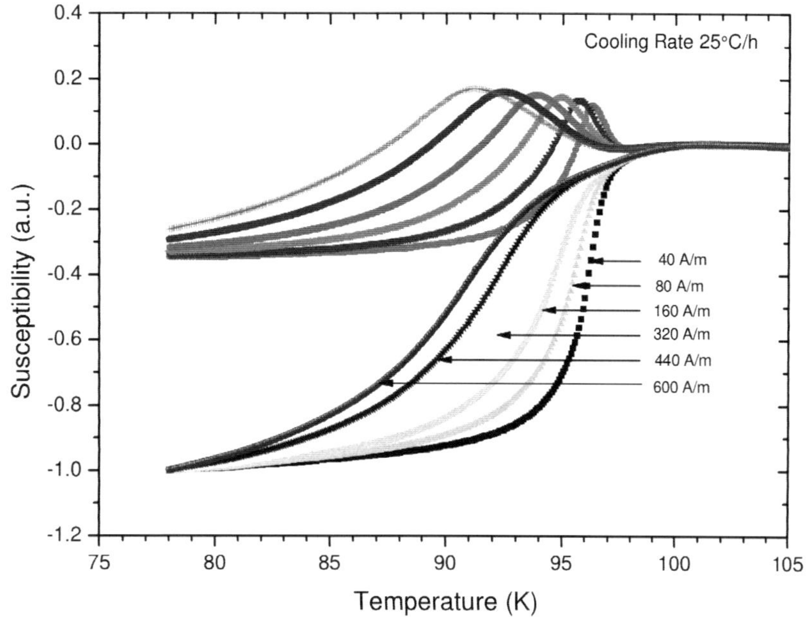

Figure 3. Temperature dependence of the real part and imaginary part of ac susceptibility for silver wrapped $Bi_{1.7}Pb_{0.35}Sr_{1.9}Ca_{2.1}Cu_3O_y$ annealed at 830 $^{\circ}C$ for 48 h with cooling rate of 25 $^{\circ}C$ /h sample at different ac field amplitudes.

We have investigated ac susceptibility as a function of temperature under different ac field amplitudes for the samples. χ' measures the amount of shielding and χ'' the amount of energy loss due to the induced current. Fig. 3 represents some characteristic results of ac susceptibility (χ' and χ'') as a function of temperature under different ac fields for 1 kHz for G25 sample. The maximum of χ'' appears at a temperature T_p where the inter-granular field just penetrates the centre of sample [20]. The intensity of the imaginary parts increases with increasing ac field amplitude. At T_p in the imaginary part, ac field amplitude equals to full flux penetration field, H_p. For sintered high-T_c superconductors temperature dependence of the imaginary part of complex ac susceptibility shows two peaks, T_p and T_g, at high ac fields [21]. The first peak (at high field) appears at a temperature T_g slightly bellow T_c^{onset} due to intra-granular supercurrents, and a second loss peak (at low-field) appears at a temperature T_p lower than T_g due to sample-circulating inter-granular supercurrents. We did not see any intrinsic losses peaks in our samples. The absence of intra-grain peak, T_g, is due to the presence of smaller decoupled grain in the samples [22]. The overall susceptibility curves of the samples are shifted to lower temperature by increasing the ac field amplitude. These results suggest that peak temperatures, T_p, shift to lower temperature with increasing ac field amplitudes (Fig. 3). The shift of T_p with field can be explained in terms of the strength of the pinning force. The effective pinning force density decreases with increasing ac magnetic field causing T_p to shift to lower temperature.

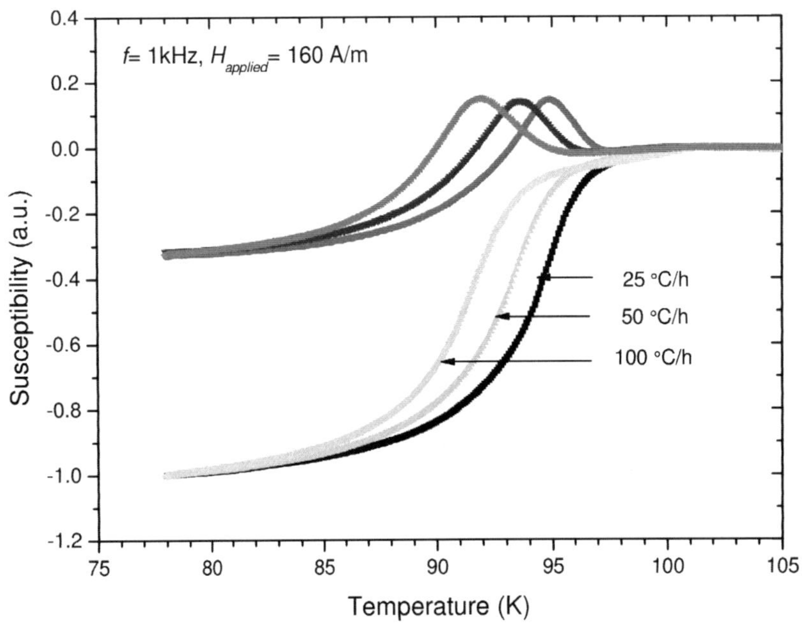

Figure 4. Ac susceptibility ($\chi = \chi' + i\chi''$) versus temperature plots for silver wrapped $Bi_{1.7}Pb_{0.35}Sr_{1.9}Ca_{2.1}Cu_3O_y$ annealed at 830 $^\circ C$ for 48 h with cooling rate of (a) 25 $^\circ C$ /h, (b) 50 $^\circ C$ /h and (c) 100 $^\circ C$ /h samples for H_{ac} = 160 A/m. annealed at 830 $^\circ C$ for 48 h with cooling rate of 25 $^\circ C$ /h sample at different ac field amplitudes.

As can be seen in Fig. 4, the susceptibility-temperature curves shifted to lower temperatures and considerably increased the transition width in χ''-T curves, and also decreased the shielding fraction of the superconducting phase in the samples with increasing the cooling rates. In another word, the susceptibility curves become sharper with decreasing cooling rate, being in agreement with the literature [6]. The sample cooled slowly with 25°C/h (G25) appears to have the strongest connectivity between grains. The fast cooled one (G100) has the weakest links. If we compare these results with resistivity measurements of Fig. 2, sample G25 has better connectivity between grains and sample G100 has the weakest coupling between grains. The data suggest that the faster cooling rates improve the inter-grain connectivity. The ac susceptibility and dc electrical resistivity measurements show the same behaviour such as the transition width increases with increasing the cooling rates. The real and imaginary parts are zero when the sample is in the normal state. χ'' turns positive due to the extent of increasing flux penetration with decreasing temperature below T_c^{onset}. At T_p, full penetration and maximum peak are observed. Below T_p, the intensity of χ'' falls due to decreasing amount of flux penetration. When the temperature reaches to lowest value the whole body starts to shield.

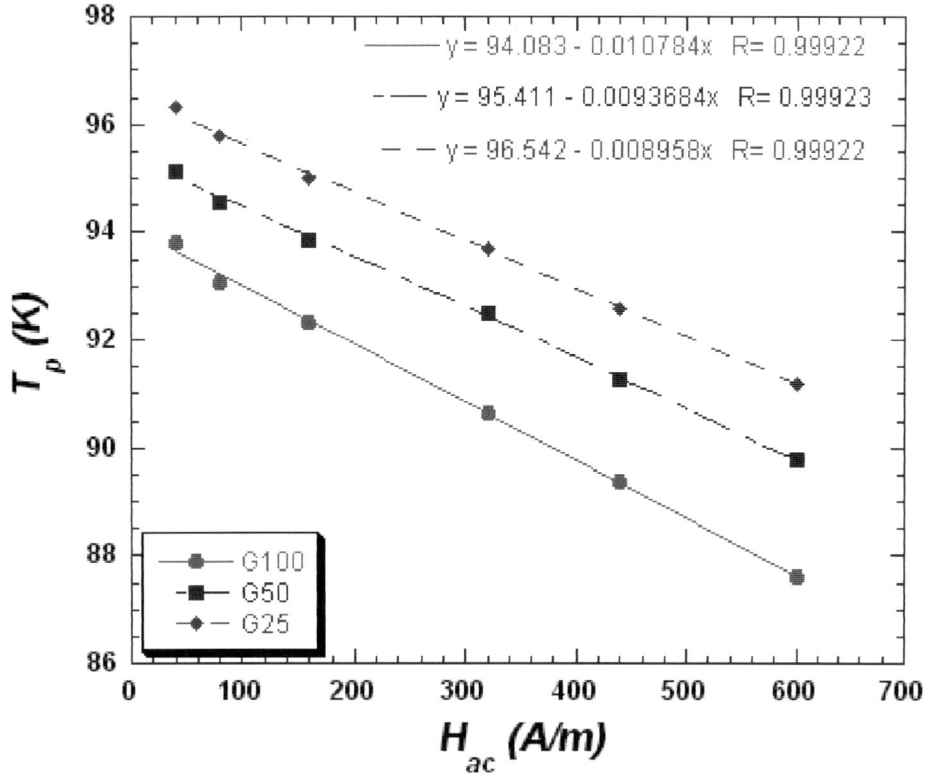

Figure 4. Intergranular peak temperature as a function of ac magnetic field amplitude for silver wrapped $Bi_{1.7}Pb_{0.35}Sr_{1.9}Ca_{2.1}Cu_3O_y$ annealed at 830 $^\circ C$ for 48 h with cooling rate of (a) 25 $^\circ C$ /h , (b) 50$^\circ C$ /h and (c) 100 $^\circ C$ /h samples.

We have studied the peak temperature dependence as a function of ac magnetic field, H_{ac}, in order to investigate the effect of cooling rates on the inter-granular pinning force. As can be seen from Fig. 5 a linear dependence of T_p as a function of H_{ac} is observed for all the samples. Müller critical state model assumes a magnetic flux independent pinning force densities, and α_J and α_g for inter- and intra-granular vortices described by the relation [23]:

$$T_p = T_{p0} - T_{p0} U^{1/2} H_{ac} \qquad (4)$$

where U is $\left[\frac{\mu_0 \mu_{eff}(0)}{2a\alpha_J(0)}\right]$, a is the length of the samples, $\mu_{eff}(0)$ is the effective permeability of the ceramic, and $\alpha_J(0)$ is the inter-granular pinning force density. From a least squares fit of this expression (Eq.(4)) to the data T_{p0} and U were extracted for all samples. The values of T_{p0} decreased from 97 to 94 K while U increased with increasing cooling rates. This means that the values of $\alpha_J(0)$ decreases with increasing cooling rates, being consistent with our transport critical current density measurement such as Jc is decreased with increasing cooling rates. A decreasing trend in pinning force of the samples with increasing cooling rates is attributable to grater voids and defects.

The temperature dependence of inter-granular critical current density, $J_c(T_p)$, is often determined via the temperature dependence of the inter-grain maximum in the imaginary part, χ''(T) of the ac susceptibility using the Bean model [24,25]. $J_c(T_p)$ is calculated for our samples using the relation $J_c(T_p) = H_a/a$ for the sample having cross section of the rectangular bar shaped, like 2ax2b where a < b. $J_c(T_p)$ is the inter-granular current density and Ha is the amplitude of applied field at T_p. The determined $J_c(T_p)$ at 92 K for the samples is determined to be 50 A/cm^2 for G25, 35 A/cm^2 for G50 and 14 A/cm^2 for G100 sample. The value of $J_c(T_p)$ increases with decreasing cooling rate. This result is consistent with present transport critical current density measurements. Moreover, we observed the same evidence for this behavior in ρ-T transport measurements in the samples such as T_c decreases with increasing cooling rates.

4. Conclusion

Changing the cooling rate showed a remarkable effect on the transport properties, lattice parameter c and hole concentration of the silver sheathed bulk samples. The values of J_c, T_c, lattice parameter c and p of the samples were decreased with increasing cooling rate. The temperature dependence of the electrical resistivity is linear in the normal state and room temperature resistivity increased with increasing cooling rate. From ac susceptibility measurements, it is observed that T_p shifts slightly to lower temperature with increasing ac field amplitude and cooling rates. The imaginary part of susceptibility measurement is used to calculate the $J_c(T_p)$ using Bean model and the $\alpha_J(0)$ using Müller critical state model. It is observed that the both $J_c(T_p)$ and $\alpha_J(0)$ were decreased with increasing cooling rates. Thus, the inter-granular properties of the G25 sample are better than that of the G100 sample under low external ac magnetic fields.

Acknowledgements

This work is supported partly by the Scientific and Technological Council of Turkey (Project No: 104T325) and partly the Turkish State Planning Organization (DPT) (Project No: 2004K120200).

References

[1] Calzona V Cimberle M R Ferdeghini C Putti M Siri A S 1989 *Physica C* **157** 425
[2] Müller K H Macfarlane J C Driver R 1991 *Physica C* **203** 178
[3] Pop A V 1999 *Supercond. Sci. Tecnol.* **12** 672
[4] Pop A V Ilonca G Ciurchea D Ye M Geru I I Kantser V G Todic M Delteur R 1996 *J. Alloys Comp.* **116** 241.
[5] Belenli I *D. Phil. Thesis University of Oxford* 1993
[6] Parrell J A Larbalestier D C 1996 *Appl. Phys. Lett.* **69** 2915
[7] Lelovic M Deis T Eror N G Balachanran U Haldar P 1996 *Supercond. Sci. Technol.* **9** 965
[8] Terzioglu C Ozturk O Kilic A Gencer A Belenli I 2006 *Physica C* **434** 153
[9] Ozturk O Yegen D Yilmazlar M Varilci A Terzioglu C 2007 *Physica C* **451** 113
[10] Chiu C W Meng R L Gao L Huang Z J Chen F and Xue Y Y 1993 *Nature* **365** 323
[11] Halim S A Khawaldeh S A Mohammed S B Azhan H 1999 *Materials Chemistry and Physics* **61** 251
[12] Terzioglu C Yegen D Yilmazlar M Görür O Akdogan M Varilci A 2007 *J. Mater. Sci.* **42** 4636
[13] Bean C P 1964 Rev. Mod. Phys. 36 31
[14] Jean F Collin G Andrieux M Blanchard N Forget A Rousseau S and Marucco J F 2003 *Physica C* **345** 384
[15] Belenli I and Turkoglu O 2003 *Supercond. Sci. Technol.* **39** 16
[16] Tetenbaum M Maroni V A 1996 *Physica C* **260** 71
[17] Persland M R Tallon J L Buckley R G Liu R S Floer N E 1991 *Physica C* **176** 95
[18] Terzioglu C Yilmazlar M Ozturk O Yanmaz E 2005 *Physica C* **423** 119
[19] Aksan M A Yakinci M E 2004 *J. Alloys and Compounds* **385** 33

[20] Müller K H Nikolo M Savvides N Driver R Kajimura K Hayakawa H (Eds.) 1991 *Advances in Superconductivity III Springer Verlag Tokyo* p. 119.
[21] Clim J R 1991 *Physica C* **50** 153-155.
[22] Müller K H Collocott S J Driver R and Savvides N 1991 *Supercond. Sci. Technol.* **4** 325-327
[23] Müller K H Matthews D N and Driver R 1992 *Physica C* **191** 339
[24] Babic E Drobac D Horvat J Marohnic Z and Prester M 1989 *J. Less. Comm. Mat.* **151** 89
[25] Gomory F and Lobotka P 1988 *Solid State Commun.* **66** 645

Theory of Nernst Effect in Layered Superconductors

B D Tinh and B Rosenstein

Department of Electrophysics, National Chiao Tung University, Hsinchu 30050, Taiwan, Republic of China

E-mail: tinhdhsp.ep95g@nctu.edu.tw

Abstract. We calculate, using the time-dependent Ginzburg-Landau (TDGL) equation with thermal noise, the transverse thermoelectric conductivity α_{xy}, describing the Nernst effect, in type-II superconductor in the vortex-liquid regime. The method is an elaboration of the Hartree-Fock. An often made in analytical calculations additional assumption that only the lowest Landau level significantly contributes to α_{xy} in the high field limit is lifted by including all the Landau levels. The resulting values in two dimensions ($2D$) are significantly lower than the numerical simulation data of the same model, but are in reasonably good quantitative agreement with experimental data on La_2SrCuO_4 above the irreversibility line (below the irreversibility line at which α_{xy} diverges and theory should be modified by including pinning effects).

1. Introduction

The electric field is induced in a metal under magnetic field by the temperature gradient ∇T perpendicular to the magnetic field $\mathbf{H}$, phenomenon known as Nernst effect [1] (direction of the electric field being perpendicular to both ∇T and $\mathbf{H}$). Recently the Nernst effect in high T_c superconductors attracted attention both theoretically [1, 2, 3, 4, 5, 6] and experimentally [7, 8, 9, 10, 11, 12, 13, 14]. In these materials effect of thermal fluctuations is very strong leading to depinning of Abrikosov vortices created by the magnetic field in type II superconductor below second critical field $H_{c2}(T)$. In the mixed state the Nernst effect is large due to vortex motion, while in the normal state and in the vortex lattice or glass states it is typically smaller. The Nernst effect therefore is a probe of thermal fluctuations phenomena in the vortex matter, but in principle could shed some light on the underlying microscopic mechanism of superconductivity in cuprates.

In low critical temperature superconductors no sign of superconducting fluctuation was reported as the temperature was raised above $T_{c2}(H)$ [15]. In sharp contrast, the appearance of a fluctuation tail above the critical temperature in the Nernst signal was observed in several different high-temperature superconductors [8, 9, 10, 11, 14]. The related Ettignhausen effect was detected as well [7]. At the same time thermal fluctuations in high T_c materials lead to many other remarkable phenomena, most notably vortex lattice melting and thermal depinning well studied both experimentally and theoretically over the last two decades, so that the theory of the Nernst effect should be consistent with the theory of these phenomena.

Theory of the electronic and heat transport (including the Nernst effect) starting from the phenomenological TDGL equation strongly fluctuating superconductors was developed long ago [1, 2]. More recently within the same framework Ussishkin *et al.* [3] calculated perturbatively the low-field the Nernst effect for $T > T_c$ due to contribution of Gaussian fluctuations and obtained results in agreement with microscopic Aslamazov-Larkin [1] calculation. Mukerjee *et al.* [5] numerically simulated two dimensional TDGL equation with Langeven thermal noise for $T < T_c$ and obtained results in reasonable agreement with experimental data on $LaSCO$ at lower temperature, but the thermoelectric conductivity became independent of magnetic field at higher temperatures in contrast to experiment. The simulation of this system, even in 2D, is difficult and it was one of our goals to supplement it with a reliable analytical expression in the region of the vortex liquid, see Fig. 1. Recent understanding of the vortex matter phase diagram is summarized in Fig. 1. There are four phases separated by two transition lines [16]:

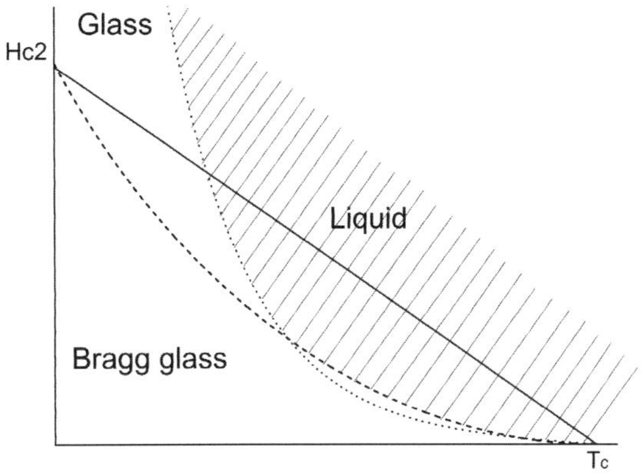

Figure 1. The thermodynamic phase diagram

the first order melting line (sometimes called the order - disorder line at lower temperatures, dashed line in Fig. 1) and the irreversibility (or glass) continuous transition. The melting line separates crystalline phases from a homogeneous phases, while the glass line (dotted line in Fig. 1) separates pinned phases from the unpinned ones. The mean field $H_{c2}(T)$ line (solid line in Fig. 1) in strongly fluctuating superconductors becomes a crossover. Both pinning and crystalline order lead to a strong reduction of the Nernst signal and will not be considered here. Therefore we will concentrate on the vortex liquid phase (dashed area in Fig. 1) and discuss the melting line and disorder only as limits of applicability of the theory. The quantitative theory of the vortex liquid have been developed recently and it was established that the Hartree-Fock approach for the thermodynamic is close to the convergent Borel-Pade one in the wide region of the vortex liquid phase [17].

In this paper we revisit the Hartree-Fock calculation in TDGL originally done in Ref. [2] to obtain explicit expressions for the transverse thermoelectric conductivity α_{xy} in $2D$. Typically only the lowest Landau level contribution was investigated. We extend it to higher Landau levels necessary for exploring the experimentally accessible parameter region and find range of applicability of the results due to approximations made, disorder and crystallization. In this theory the strength of the thermal fluctuations is described by just one dimensionless adjustable parameter η (closely related to the Ginzburg number Gi). The value of the parameter is consistent with the melting line calculated in [18].

2. The Ginzburg-Landau Model in 2D

2.1. Relaxation dynamics and thermal fluctuations

To describe fluctuation of order parameter in thin films or layered superconductors one can start with the Ginzburg-Landau free energy:

$$F = s \int d^2x \frac{\hbar^2}{2m^*}|\mathbf{D}\psi|^2 + a|\psi|^2 + \frac{b'}{2}|\psi|^4, \tag{1}$$

where $\mathbf{A} = (-By, 0)$ describes a constant and practically homogeneous magnetic field (we generally neglect small fluctuations of the magnetic field due to magnetization of order $1/\kappa^2 << 1$ in the region of interest) in Landau gauge and covariant derivative is defined by $\mathbf{D} \equiv \nabla - i(2\pi/\Phi_0)\mathbf{A}$, with $\Phi_0 = hc/e^*$, $e^* = -2e > 0$. For simplicity we assume $a(T) = \alpha(T - T^\Lambda)$, although the temperature dependence can be easily modified to better describe the experimental coherence length. The "mean field" critical temperature T^Λ depends on the ultraviolet (UV) cutoff Λ specified later. It is higher than measured critical temperature due to thermal fluctuations on the mesoscopic scale. The thickness of a layer is s. We apply this model to describe experiments on overdoped $LaSCO$ [12].

Since we are interested in transport phenomena, it is necessary to introduce some kind of dynamics for the order parameter. The simplest is a gauge-invariant version of the "type A" relaxational dynamics,

$$\tau \left(\frac{\partial}{\partial t} + i\frac{e^*}{\hbar}\phi \right) \psi = -\frac{\delta F}{\delta \psi^*} + \zeta, \tag{2}$$

called in the present context TDGL equation. Explicitly the TDGL equation for the superconducting order parameter is

$$\tau \left(\frac{\partial}{\partial t} + i\frac{e^*}{\hbar}\phi \right) \psi = \frac{\hbar^2}{2m^*}\mathbf{D}^2\psi - a\psi - b'|\psi|^2\psi + \zeta, \tag{3}$$

where $\phi(\mathbf{x})$ is the scalar potential describing electric field. To incorporate the thermal fluctuations via Langeven method, the noise term $\zeta(\mathbf{x}, t)$, having Gaussian correlations

$$s\langle \zeta^*(\mathbf{x}, t)\zeta(\mathbf{x}', t')\rangle = 2T\tau\delta(\mathbf{x} - \mathbf{x}')\delta(t - t'), \tag{4}$$

introduced. Here $\delta(\mathbf{x} - \mathbf{x}')$ is the two-dimensional δ function of the in-plane coordinates.

2.2. The heat and the electric transport

We start from the definition of the transport coefficients. Generally the electric and heat transport current densities, $\mathbf{j}^{(e)}$ and $\mathbf{j}^{(h)}$, in metal are related to the applied (sufficiently weak) electric field and the temperature gradient by

$$j_{tr}^{(e)i} = \sigma^{ij}E^j - \alpha^{ij}\nabla^j T, \tag{5}$$

$$j_{tr}^{(h)i} = \widetilde{\alpha}^{ij}E^j - \kappa^{ij}\nabla^j T, \tag{6}$$

where, σ, α, $\widetilde{\alpha}$, and κ are the electrical, the thermoelectric, the electrothermal, and the thermal conductivity components of the conductivity tensor $(i, j = x, y)$. The Onsager relations implies $\widetilde{\alpha} = T\alpha$. The Nernst coefficient (ν_N), under the condition $\mathbf{j}_{tr}^{e} = 0$, is expressed in terms of the above coefficients as

$$\nu_N = \frac{E_y}{(-\nabla T)_x B} = \frac{1}{B}\frac{\alpha_{xy}\sigma_{xx} - \alpha_{xx}\sigma_{xy}}{\sigma_{xx}^2 + \sigma_{xy}^2}. \tag{7}$$

If the system shows no significant Hall effect (only such systems will be considered), then $\sigma_{xy} = 0$ and the expression simplifies:

$$\nu_N = \frac{\alpha_{xy}}{B\sigma_{xx}}. \tag{8}$$

Since near $H_{c2}(T)$ magnetization is very small (of order $1/\kappa^2 \simeq 10^{-4}$ for high T_c materials and in most of the cases the magnetization contributions vanish), we replace the transport currents by the total currents

$$\left\langle \mathbf{j}^h \right\rangle = -\frac{\hbar^2}{2m^*}\left\langle \left(\frac{\partial}{\partial t} - i\frac{e^*}{\hbar}\phi\right)\psi^* \left(\nabla - i\frac{2\pi}{\Phi_0}\mathbf{A}\right)\psi\right\rangle + c.c. \tag{9}$$

These assumptions were extensively discussed in a textbook [1] and [5].

3. The transverse thermoelectric conductivity in the vortex liquid phase

At low temperatures vortex matter organizes itself into a (usually, but not always) hexagonal vortex lattice. When disorder can be effectively neglected (either in very clean materials or when thermal depinning occurs), one can consider transport of the vortex lattice as a whole. Expressions for the electric and the thermal conductivities near $H_{c2}(T)$ neglecting thermal fluctuations were obtained in [2], and according to results the Nernst effect is generally very small compared to one in the vortex liquid. This can be qualitatively understood as a result of rigidity of the lattice. *Below* the melting line the situation in this respect does not change much. Moreover due to unavoidable presence of disorder, the vortex lattice is pinned forming a Bragg glass in most of its domain [16]. However in high T_c superconductors thermal fluctuations are strong enough (especially for high anisotropy and high magnetic fields) to destroy the expectation value of the condensate $\langle\psi\rangle = 0$. We always assume that thermal fluctuations melted away and in addition temperature is high enough to thermally depin the vortex liquid (avoiding the "vortex glass"). As a consequence impurities in the vortex liquid are neutralized.

Due to thermal fluctuations the expectation value of the order parameter in vortex liquid is zero $\langle\psi(\mathbf{x},t)\rangle = 0$. Therefore contribution to the expectation values of physical quantities like the electric and the heat current come exclusively from the correlations. The most important is the quadratic one

$$C(\mathbf{x},t;\mathbf{x}',t') = \langle\psi(\mathbf{x},t)\psi^*(\mathbf{x}',t')\rangle, \tag{10}$$

called the correlation function of the order parameter.

In particular the superfluid density is

$$\langle|\psi(\mathbf{x},t)|^2\rangle = C(\mathbf{x},t;\mathbf{x},t). \tag{11}$$

A simple approximation which captures the most interesting fluctuations effects in the Hartree approximation, in which the cubic term in the GL equation Eq. (3) $b'|\psi|^2\psi$ is replaced by a linear one $b'\left\langle|\psi|^2\right\rangle\psi$

$$\tau\frac{\partial}{\partial t}\psi(\mathbf{x},t) = \left(\frac{\hbar^2}{2m^*}\mathbf{D}^2 - \widetilde{a}\right)\psi(\mathbf{x},t) + \zeta(\mathbf{x},t), \tag{12}$$

leading the "renormalized" value of the coefficient:

$$\widetilde{a} = a + b'\langle|\psi|^2\rangle. \tag{13}$$

The formal solution of this equation is

$$\psi(\mathbf{x},t) = \int d\mathbf{x}'\int dt' G_0(\mathbf{x},t;\mathbf{x}',t')\zeta(\mathbf{x}',t'), \tag{14}$$

where G_0 is the equilibrium Green function which is most easily accomplished by expanding G_0 in terms of the Landau eigenfunctions.

The evaluation of $\langle|\psi(\mathbf{x},t)|^2\rangle$ gives

$$\epsilon_b = \widetilde{\epsilon}_b - \frac{b'T}{2\pi s}\frac{m^*\omega_B}{\hbar\left(\alpha T^\Lambda\right)^2}\sum_{n=0}^{N_f}\frac{1}{\widetilde{\epsilon}_b + 2nb}, \tag{15}$$

where the reduced temperature is defined as $\epsilon = a/\alpha T^\Lambda$, $\epsilon_b = \epsilon + b$, (with similar expression for $\widetilde{\epsilon}$ and $\widetilde{\epsilon}_b$) with $b = B/Hc2(0)$ being the scaled magnetic field, $Hc2(0) = \Phi_0/2\pi\xi^2$ the zero-temperature critical field and $\xi = (\hbar^2/2m^*\alpha T_c)^{1/2}$ the zero-temperature coherence length. The UV cutoff was introduced. It effectively limits the number of Landau levels to $N_f = \frac{\Lambda}{b} - 1$. The "bubble" sum, which diverges logarithmically, can be performed:

$$\frac{b}{\pi}\sum_{n=0}^{N_f}\frac{1}{2nb+\widetilde{\epsilon}_b} = \frac{1}{2\pi}\log\Lambda + u', \tag{16}$$

where the function u' is related by

$$u'(\widetilde{\epsilon}_b, b) = \frac{1}{2\pi}[f_s'(\widetilde{\epsilon}_b/2b) - \log(2b)], \tag{17}$$

to the polygamma function f_s':

$$f_s'(x) = \sum_{n=1}^{\infty}\left[\frac{1}{n+x} - \int_{n-1/2}^{n+1/2}\frac{1}{(y+x)}dy\right] + \left[\frac{1}{x} - \log\left(x+1/2\right)\right]. \tag{18}$$

Thus the critical temperature T_c is significantly renormalized:

$$\epsilon_b^r = \epsilon_b + \frac{b'T}{4\pi s}\frac{m^*\omega_b}{\hbar\left(\alpha T^\Lambda\right)^2}\log\Lambda = \widetilde{\epsilon}_b - \frac{\eta\xi^2 Te^* H_{c2}(0)}{2\hbar c T_c}u'(\widetilde{\epsilon}_b, b), \tag{19}$$

where η is a dimensionless fluctuation parameter

$$\eta = \frac{b'T_c}{\xi^2\left(\alpha T_c\right)^2 s}, \tag{20}$$

introduced in [5]. The relation between η and more often used two dimensional Ginzburg number [1, 17], $Gi_{2D} \equiv \frac{1}{2}\left(8e^2\kappa^2\xi^2T_c/\pi c^2\hbar^2 s\right)^2$, is

$$\eta = 4\sqrt{2Gi_{2D}}\pi^2. \tag{21}$$

Let us assume that the weak electric field $\mathbf{E}$ is along the y axis, generated by the scalar potential $\phi = -E_y y$. The heat current in the vortex liquid phase is given by [2]

$$\left\langle\mathbf{J}^h\right\rangle = -\frac{\hbar^2}{2m^*}\left[\mathbf{D}\left(\mathbf{x}\right)\left(\frac{\partial}{\partial t'} - i\frac{e^*}{\hbar}\phi\left(\mathbf{x}'\right)\right) + \mathbf{D}^*\left(\mathbf{x}'\right)\left(\frac{\partial}{\partial t} + i\frac{e^*}{\hbar}\phi\left(\mathbf{x}\right)\right)\right]C(\mathbf{x},t;\mathbf{x}',t')|_{\mathbf{x}=\mathbf{x}';t=t'}, \tag{22}$$

where

$$C(\mathbf{x},t;\mathbf{x}',t') = \frac{2\tau T}{s}\int_{\mathbf{x}_1,t_1}G(\mathbf{x},t;\mathbf{x}_1,t_1)G^*(\mathbf{x}',t';\mathbf{x}_1,t_1), \tag{23}$$

with G is the Green function of the linearized TDGL equation in the presence of the scalar potential. One finds correction to the Green function to linear order in the electric field

$$G(\mathbf{x},t;\mathbf{x}',t') = G_0(\mathbf{x},t;\mathbf{x}',t') - i\frac{e^*\tau}{\hbar}\int_{\mathbf{x}_1,t_1}\phi(\mathbf{x}_1)G_0(\mathbf{x},t;\mathbf{x}_1,t_1)G_0(\mathbf{x}_1,t_1;\mathbf{x}',t'). \quad (24)$$

In order to determine the transverse thermoelectric conductivity, we need to compute the x component of the heat current to first the electric field. In the chosen gauge, the electrothermal conductivity $\widetilde{\alpha}_{xy} = \frac{\overline{j_x^h}}{E_y}$ (averaged over $\mathbf{x}$) takes a form

$$\widetilde{\alpha}_{xy} = \frac{e^*Tb}{\hbar\pi s}\sum_{n=0}^{N_f}\left[\frac{n+1/2}{2nb+\widetilde{\epsilon}_b} - \frac{n+1}{2(n+1/2)b+\widetilde{\epsilon}_b}\right] = \frac{e^*T(b-\widetilde{\epsilon}_b)}{2\hbar bs}\Big[u'(\widetilde{\epsilon}_b,b) - u'(\widetilde{\epsilon}_b+b,b)\Big], \quad (25)$$

where function u' was defined in Eq. (17). Using the Onsager relation one obtains the transverse thermoelectric conductivity $\alpha_{xy} = \widetilde{\alpha}_{xy}/T$. The vortex liquid energy gap $\widetilde{\epsilon}_b$ as a function of ϵ_b, and substituting the results into Eq. (25). Equations (19) and (25) agree with the calculation of Ullah and Dorsey [2].

4. Comparison with experiment and MC simulation

The experiment results of Y. Wang *et al.* [12] obtained from the Nernst effect and resistivity measurements on an overdoped $LaSCO$ sample with $x = 0.2$ and $T_c = 28K$. The comparison is presented in Fig. 2 (low temperatures in (a) and close to T_c in (b)). The parameters used in the calculation are (see definitions above) $H_{c2}(0) = 45T$ (thus $\xi = 27A^o$) and layer spacing $s = 16A^o$. The fluctuation parameter is $\eta = 0.25$ and provides a reasonable quantitative agreement between theory and experiment. Below irreversibility line where the theory should be modified including both pining and crystalline phase in Fig. 2(a). The deviation develops roughly at the location of the irreversibility line. However, our results are in good quantitative agreement with experiment data for temperature close to T_c in Fig. 2(b), where the numerical simulation gives a nearly constant α_{xy}, while the experiment shows more variation.

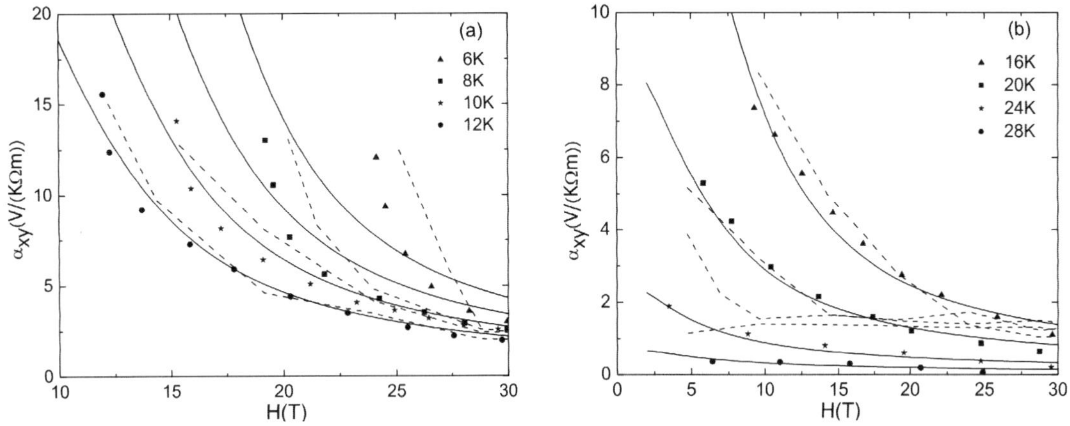

Figure 2. Points are α_{xy} for different temperatures of $LaSCO$ Ref. [12], with x=0.2 (overdoped, $T_c = 28K$). The dashed line is the simulation value of α_{xy} Ref. [5]. The solid line is the theoretical value of α_{xy}, using $H_{c2}(0) = 45T, s = 16A^o, \eta = 0.25$.

5. Conclusion

We obtained, using TDGL equation with thermal noise, explicit expressions for the transverse thermoelectric conductivity α_{xy} in $2D$ including all Landau levels in type-II superconductor in the vortex-liquid regime. The method is the Hartree-Fock. We also obtained the relation between the strength of the thermal fluctuation η and the Ginzburg number Gi. We compared the results to the $2D$ simulation and the experiment results. Our results in $2D$ are significantly lower than the simulation and experiment data below the irreversibility line at which theory should be modified by including both pinning and crystalline effects, but are in reasonably good quantitative agreement with experimental data on La_2SrCuO_4 for temperature close to T_c. In the future work, we will compare with other materials.

Acknowledgments

We are grateful to Y. Wang for proving experiment data, S. Mukerjee for proving simulation data, and Professor A.T. Dorsey for correspondence. This work was supported by NSC of Taiwan.

References

[1] Larkin A and Varlamov A 2005 *Theory of fluctuations in superconductors* (Oxford :Clarendon Press)
[2] Ullah S and Dorsey A T 1990 *Phys. Rev. Lett.* **65** 2066
[3] Ussishkin I, Sondhi S L, and Huse D A 2002 *Phys. Rev. Lett.* **89** 287001
[4] Ussishkin I 2003 *Phys. Rev.* B **68** 024517
[5] Mukerjee S and Huse D A 2004 *Phys. Rev.* B **70** 014506
[6] Tan S and Levin K 2004 *Phys. Rev.* B **69** 064510
[7] Palstra T T M, Batlogg B, Schneemeyer L F, and Waszczak J V 1990 *Phys. Rev. Lett.* **64** 3090
[8] Clayhold J A, Linnen A W, Jr., Chen F, and Chu C W 1994 *Phys. Rev.* B **50** 4252
[9] Hohn C, Galffy M and Freimuth A 1994 *Phys. Rev.* B **50** 15875
[10] Xu Z A, Ong N P, Wang Y, Kakeshita T, and Uchida S 2000 *Nature* (London) **406** 486
[11] Wang Y, Xu Z A, Kakeshita T, Uchida S, Ono S , and Ong N P *Phys. Rev.* B 2001 **64** 224519
[12] Wang Y , Ong N P, Xu Z A , Kakeshita T, Uchida S, Bonn D A, Liang R, and Hardy W N 2002 *Phys. Rev. Lett.* **88** 257003
[13] Capan C, Behnia K, Li Z Z, Raffy H and Marin C 2003 *Phys. Rev.* B **67** 100507
[14] Wang Y, Li L, and Ong N P 2006 *Phys. Rev.* B **73** 024510
[15] Huebener R P 1979 *Magnetic flux structures in supercondutors* (Berlin: Spring-Verlag)
[16] Beidenkopf H, Avraham N, Myasoedov Y, Shtrikman H, Zeldov E, Rosenstein B, Brandt E H and Tamegai T 2005 *Phys. Rev. Let.* **95** 257004
[17] Li D and Rosenstein B 2001 *Phys. Rev.* B **65** 024513
[18] Li D and Rosenstein B 2002 *Phys. Rev.* B **65** 220504

Ammonium Nitrate Melt Technique for the Synthesis of Oxide Compounds

Ugur Topal [1]
Tubitak-UME, P.K.54, 41470 Gebze-KOCAELI/Turkey

E-mail: : ugur.topal@ume.tubitak.gov.tr

Abstract. In this study, we propose a wet chemical method, called ammonium-nitrate-melt-technique, for the synthesis of any type of bulk oxide compounds. This method was successfully applied to the high-T_c superconductors (Y-123, Ru-2212, etc), diluted magnetic semiconductors ($Ti_{1-x}CoO_2$) and hard magnetic barium ferrites ($BaFe_{12}O_{19}$). For instance, we have obtained a single phase M-type barium ferrites with finely granular structure at nanoscale (300-400 nm). Magnetic parameters of these samples are near to the theoretically estimated values. Besides, superconducting transition temperature of ~70 K was observed in ferromagnetism&superconductivity co-existed ruthenates, which is only possible at very high oxygination pressures (600 atm). Consequently, we believe that the method will attract the attention of many material scientists..

1. Introduction

It is well known that solid-state reaction method is generally used for the synthesis of ceramic compounds in which the oxides or carbonates of elements are mixed at appropriate amounts and annealed above the crystallization temperature of desired phase. The solid-state reaction method has some inherent disadvantages such as production of chemically inhomogeneous coarse powders because of high temperature annealing, insufficient mixture of initial oxides and introduction of impurities during ball milling. To avoid these detrimental effects, various chemistry-based synthesis routes have been proposed. These are namely co-precipitation [1,2], glass crystallization [3], hydrothermal synthesis [4,5], sol-gel technique [6], organometallic precursor method [7] micro emulsion synthesis [8,9], etc.

In this study, we present a new chemical method to synthesize oxide compounds called ammonium-nitrate-melt-technique (ANMT). Its principle is based on dissolving the metal oxides in ammonium nitrate melt. As an oxidizing agent or fertilizer, ammonium nitrate combust the materials. Constituent powders may be mixed homogeneously and thus their reaction to form the desired phase starts at lower temperatures compared to the other techniques. We have successfully applied this technique to some oxide compounds which have different application areas, such as; superconducting, magnetic and semiconducting oxides [10-12]. We briefly report the results obtained in these materials to emphasize the contribution of this technique to material science.

---

1 Ugur Topal.

2. Experimental

Required amounts of constituent oxide powders of high purity (99%) are weighed and mixed to form the desired phase. The mixed powder is put into the melted ammonium nitrate kept at 160 °C in a glass beaker. Powder/ammonium nitrate ratio was 1/5 in weight. The solution is kept on the heater until the liquid disappears and a precipitate is formed at the bottom of the beaker. The precipitate is then heated at 450 °C for 5 h. This post-heating is just to eliminate the organic compounds. Then, powders are divided into several parts and independently heat-treated at different temperatures to find the crystallization temperature of desired phase. Calcination time for 1 h is enough to complete the crystallization. It is important to note not to breathe ammonium nitrate vapor. In addition, if one studies with metals instead of metal oxides, combustion process must start at lower temperatures and temperature of the solution must be increased slowly.

3. Results & Discussion

Here, we will report the results obtained in some oxide compounds to show the perfection of the method as an alternative to the other known methods.

3.1. Hard magnetic barium ferrites ($BaFe_{12}O_{19}$)

Powders of $BaCO_3$ and Fe_2O_3, in the ratio required for $BaFe_{12}O_{19}$ were placed in the melt solution of ammonium nitrate and stirred in a magnetic stirrer until a thick liquid was obtained. The solution was heated at 260 °C for 1 day to produce an ash-like reddish precursor. Finally, the precursor was divided into several parts and independently annealed for 1 h at 450, 700, 800, 850, 900, 1000, 1100 and 1200 °C to determine the formation temperature of the $BaFe_{12}O_{19}$ phase.

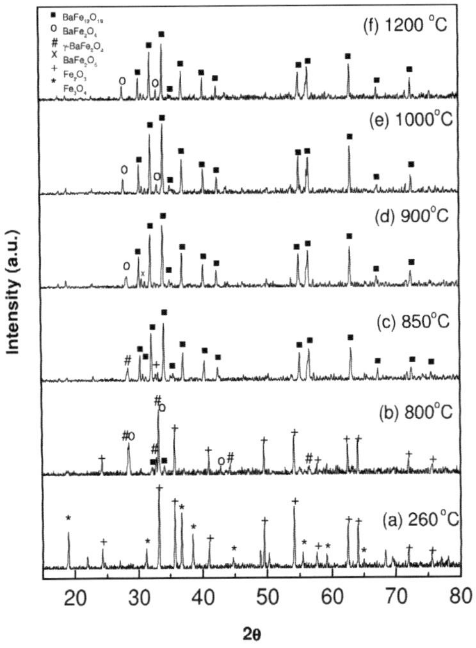

Figure 1. X-Ray graph of barium ferrite calcined at different temperatures.

Figure 2. SEM images of barium ferrite calcined at 1000 °C.

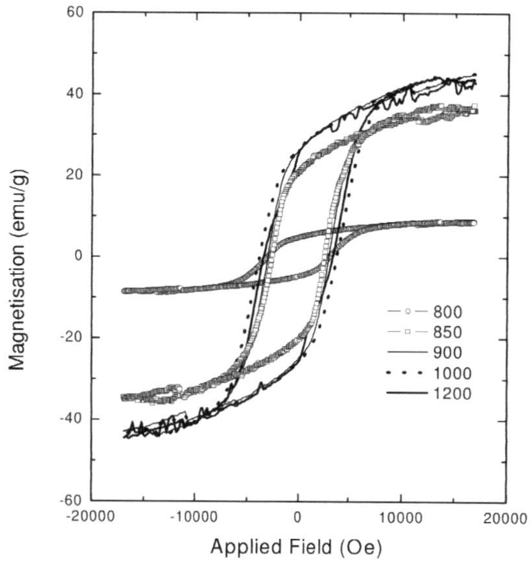

Figure 3. M-H loops of the barium ferrite calcined at different temperatures.

Fine particles of barium ferrite, $BaFe_{12}O_{19}$ have been prepared at temperatures as low as 850 °C (Fig1), considerably lower than that obtained by the typical ceramic process. The powder calcined at 1000 °C for 1 h has grains with a narrow-size distribution in the range 0.3-0.5 μm (Fig.2). These ranges of particle sizes are good for applications in the areas as high-density magnetic recording media and microwave absorber. The structural analysis and magnetic measurements have shown that the particles are single-domain, but when the calcination temperature increases (above 1000 °C) some of the grains start to exhibit a multi-domain behaviour. The magnetic properties of the present samples are better than those obtained by the ceramic method. For instance, coercive field was determined to be 3050 Oe. On the other hand, it was reported to be ~2200 Oe in the samples prepared by traditional methods [13]. Saturation magnetization is as high as 45 emu/g (see Fig.3).

3.2. Magnetic Superconductors: $RuEu_{1.5}Ce_{0.5}Sr_2Cu_2O$

Synthesis procedure is almost same with that of barium ferrite. Only difference comes from the final calcination temperature because the crystallization temperature of ruthenates is higher than the crystallization temperature of barium ferrite. Therefore, we calcined the sample at 1050 °C-1060 °C. Note that calcination temperature is around 1200 °C in the traditional route. In addition, calcination time of 40 h is enough to obtain this ruthenate phase. On the other hand, it was reported to be as long as 100 h in the ceramic route [11].

In this compound, magnetism and superconductivity lives together. Existent magnetism may hinder the superconductivity if the compound is not homogeneously oxidized. Coexistence of superconductivity and magnetism may be achieved by oxygenation at high pressures, such as 600 atm, if one follows the traditional route [14]. On the other hand, it can be obtained by ANMT just under oxygen flow in a tubular furnace. As seen from the Fig.4 and Fig.5, superconducting transition temperature was determined as 50 K from resistivity measurements and near 80 K , which is the highest value reported up to now, from the magnetization measurement.

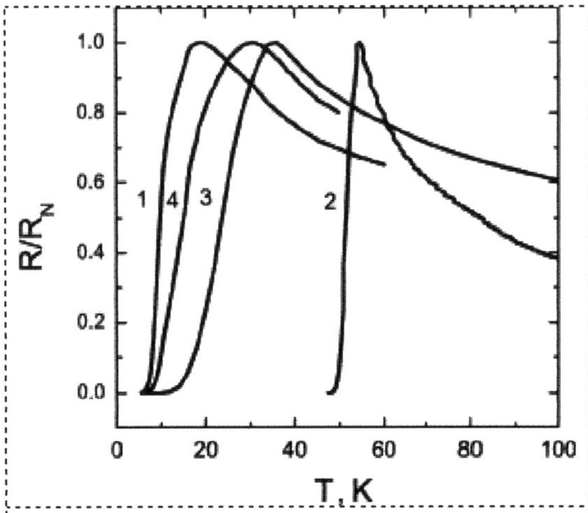

Figure 4. Resistance vs. temperature graph of $RuEu_{1.5}Ce_{0.5}Sr_2Cu_2O$ (2 represents 600 atm oxyg.).

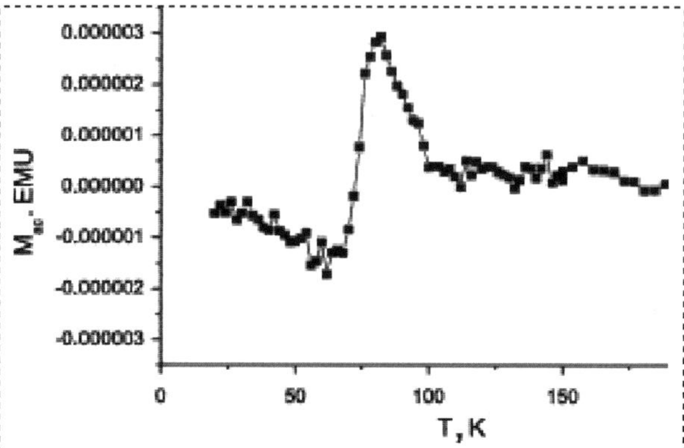

Figure 5. Magnetization vs. temperature graph of $RuEu_{1.5}Ce_{0.5}Sr_2Cu_2O$.

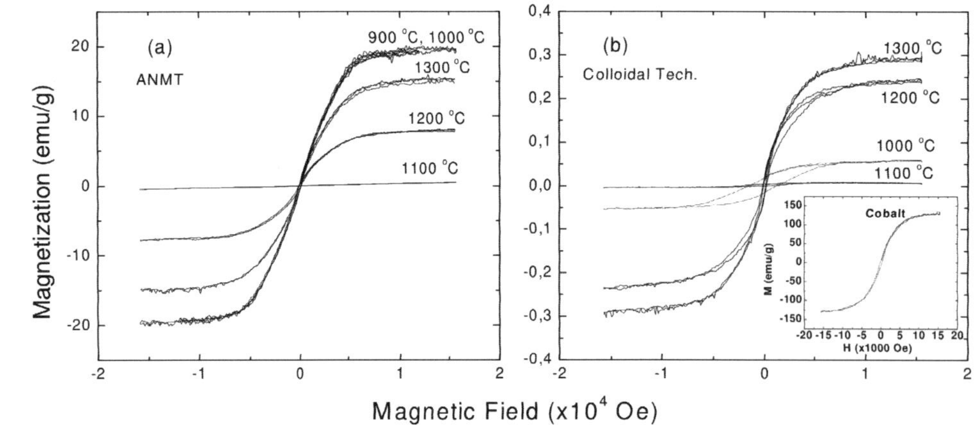

Figure 6. M-H dependence of the materials

3.3. Diluted magnetic semiconductors

Mass storage and recording of information is possible (hard disks, magnetic tapes) by ferromagnetic materials. One of the recent trends in the field of spintronics is to explore magnetic semiconductors with Curie temperature (T_c) higher than room temperature. In this field, $Ti_{1-x}Co_xO_2$ system, known as diluted magnetic semiconductors, supplies all requirements and has attracted much attention.

Synthesis of the $Ti_{1-x}Co_xO_2$ system is quite easy. TiO_2 and cobalt element is mixed at appropriate amounts and the same procedure of ANMT, as explained before, is followed. Calcination temperature may vary according to the desired magnetization from the final product.

Figure 6 shows the M-H curves of $Ti_{1-x}Co_xO_2$ samples obtained by ANMT and a ceramic method with cobalt addition of 25 % at. . As seen, magnetization values of the sample obtained by ANMT are rather high compared to the magnetization of the sample obtained by ceramic route.

4. Conclusion

As a result, advantages of ANMT technique can be summarized as follows:
1. Phases targeted are obtained at low calcination temperatures and in short time
2. As a combustion method, finely granular products can be obtained.
3. As a wet chemical method, final products are rather homogeneous.
4. Magnetic and electrical properties of samples are at optimal level.

5. References

[1] Jacobo S E, Domingo-Pascual C, Rodriguez-Clemente R 1997 *J. Mater. Sci.* **32** 1025
[2] Haneda K, Miyakama C, Kojima H 1974 *J. Am. Ceram. Soc.* **57** 354
[3] Kabo O, Ido T, Yokohama H 1982 *IEEE Trans. Magn.* **18** 1122
[4] Ataie A, Piramoon M R, Harris I R, Ponton C B 1995 *J. Mater. Sci.* **30** 5600
[5] Kumazawa H, Maeda Y, Sada E 1995 *J. Mater. Sci. Lett.* **14** 68
[6] Zhong W, Ding W P, Zhang N 1997 *J. Magn. Magn. Mater.* **168** 196
[7] Licci F, Besagni T 1984 *IEEE Trans. Magn.* **20** 1639
[8] Pillai V, Kumar P, Shah D O 1992 *J. Magn. Magn. Mater.* **116** 299
[9] Pillai V, Kumar P, Multani M S, Shah D O 1993 *Colloids Surf.* A **80** 69
[10] Topal U, Ozkan H, Sozeri H, 2004 *J. Mag. Magn. Mater.* **284** 416
[11] Strukova G K, Zver'kova I I, Dorosinskii L A, Shovkun D V, Zverev V N, Topal U, 2003 *Physica C* **387** 359
[12] Topal U, Ozkan H, Bakan H I, Cankur O, Topal K 2008 *J. Non-crys. Sol.* **354** 1678
[13] Cabanas M V, Calbet J M G, Carvajal J R, Regi M V 1994 *J. Sol. Stat. Chem.* **111** 229
[14] Klamut P W, Dabrowski B, Mini S M, Maxwell M, Kolesnik S, Mais J, Shengelaya A, Khasanov R, Savic I, Keller H, Graber T, Gebhardt J, Viccaro P J, Xiao Y 2001 *Physica C* **364-365** 313

Experimental and numerical study of influence of ferromagnetic cover on critical current of BiSCCO-2223/Ag tape superconductor

M. Vojenčiak[1], J. Šouc[1], F. Gömöry[1]

[1]Institute of Electrical Engineering, Centre of Excellence CENG, Slovak Academy of Sciences, Bratislava, Slovak Republic

elekvoje@savba.sk

Abstract. Samples of commercially available BiSCCO-2223/Ag tape have been partially covered by the ferromagnetic material. Improvement of the self-field critical current up to 15 % has been achieved by this procedure. A critical current of such tape strongly depends on geometric and magnetic properties of both, the superconducting tape as well as the ferromagnetic cover. Numerical simulations, based on the critical state model using commercial finite element method (FEM) code, have been performed. Properties of superconductor are characterized by anisotropic dependence of the critical current density on magnetic field as well as detail geometry of filaments. The ferromagnetic material is characterized by nonlinear magnetization curve. Nonlinear dependences of the critical current on selected parameters are shown in this work. Optimization of the cover parameters using these curves has been made. Samples with various parameters have been manufactured for the confirmation of numerical simulations results. Experimental results are in good qualitative agreement with results obtained by numerical simulations.

1. Introduction

The critical current density of superconductor strongly depends on local magnetic field. Moreover this dependence is nonlinear and anisotropic with respect to the ab-plane of superconductor crystal structure. Magnetic field can be created by both external source (magnet) or superconducting current itself. As was experimentally found [1] and confirmed by numerical simulations [2] it is possible to improve critical current of BiSCCO-2223 tape using ferromagnetic material deposited on the edges of the tape. In a work [2] cross-section of superconductor was simplified to ellipse what was in good agreement with cross-section of investigated tape. In our work we found procedure to calculate critical current of superconductor with arbitrary cross-section also if it is close to ferromagnetic material. Moreover this procedure allows us to analyze the distribution of the magnetic field in detail.

2. Numerical simulation basics

Similar to papers [2,3] we use commercial finite element (FEM) code to find the distribution of current density in superconductor cross-section. We assume tape invariant in longitudinal direction and therefore problem is solved in 2D. Superconductor reaches the critical current when it is filled completely by critical current density (in same direction). Local critical current density is generally different because of different local magnetic field. To assume this phenomenon we described local critical current density by formula:

$$J_c(B_{\parallel}, B_{\perp}) = \frac{J_{c0}}{\left(1 + \frac{\sqrt{k^2 \cdot B_{\parallel}^2 + B_{\perp}^2}}{B_0}\right)^{\beta}} \quad (1)$$

Using proper parameters of this formula FEM code is able to find distribution of current density equivalent to real one. The critical current of such superconductor can be calculated by integration of current density over the whole superconductor cross-section:

$$I_{c,\mathrm{tape}} = \int_{S_{\mathrm{tape}}} J_c(x, y)dS \quad (2)$$

Magnetic field in superconductor is influenced by ferromagnetic material placed close enough to it. In our case the distance between the ferromagnetic nickel and superconductor is less than 50 μm and therefore influence of this material on magnetic field and distribution of current density is evident. For magnetic properties characterization of this material we use formula:

$$\mu_r = \mu_{r\min} + \frac{\mu_{r\max}}{\left(1 + \frac{\sqrt{B_{\parallel}^2 + B_{\perp}^2}}{B_C}\right)^{2\alpha}} \quad (3)$$

which takes into account saturation of magnetic moment, but doesn't assume hysteresis of this material. However, in a case of DC current in superconductor and DC applied magnetic field this simplification is feasible.

3. Particular sample analysis

To estimate proper constants in formulas (1) and (3) it is necessary to measure real properties of both superconductor and ferromagnetic material. In measurement setup material properties itself are affected by finite dimensions of the sample. In a case of superconductor critical current measurement it is affected by magnetic field produced by flowing current itself and in case of magnetic moment measurement it is influenced by demagnetizing field of the sample. To find the parameters in formula (1) the critical current I_c, of the bare superconducting tape sample exposed to the external magnetic field with various orientations was investigated. The constants J_{c0}, k, B_0, β were estimated by comparing the calculated critical current with experimental data.

In this work we use multifilamentary BiSCCO-2223/Ag tape with dimensions 4.2x0.22 mm and self field critical I_{C0} = 100.8 A. Parameters which represent the best fit of measured data are: $J_{C0}=2.46 \cdot 10^8$ A, $k=0.14$, $B_0=0.015$ T, $\beta=0.65$. Average difference of measured and calculated I_C is 1.3 A. Both, measured and calculated dependences of critical current on magnetic field are shown in figure 1

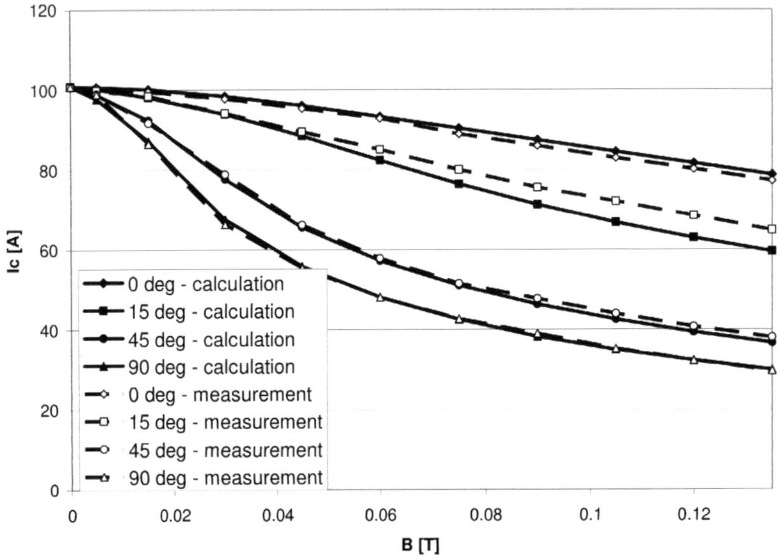

Figure 1. Comparison of dependencies measured (dashed lines) and calculated (solid lines) critical currents on magnetic field with various orientations.

Parameters in formula (2) were estimated using data of magnetic moment measurement in SQUID magnetometer. In this measurement we used Ni layer 10 μm thick deposited on silver substrate with dimensions 4x4 mm. Magnetic moment of this sample was measured in parallel magnetic field to minimize shape effect. We found following parameters as a good approximation of electroplated nickel properties: $\mu_{r\ min}$=1.5, $\mu_{r\ max}$=250, B_C=0.38 T, α=2.2. Real magnetization loop and approximate curve are compared in figure 2.

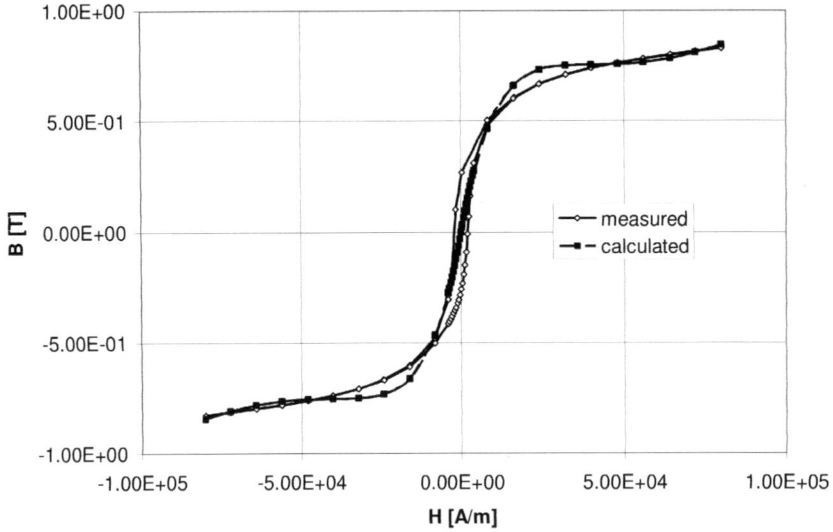

Figure 2. Comparison of measured magnetization loop of electroplated nickel and curve characterizing its properties used in calculations

Using these parameters we are able to predict superconductor-ferromagnetic composite tape properties. As was found in [1] it is possible to improve critical current of BiSCCO-2223/Ag tape by ferromagnetic cover on the edges. However, using this procedure it is also possible to decrease critical current [3]. Therefore it is convenient at first to calculate the influence of this covering for every superconductor parameters and cross-section shape. Main parameters of ferromagnetic cover are width of non-covered area – *d* and thickness of the layer (see insert in figure 3). In a figure 3 dependences of self-field critical current on *d* are shown, parameter of these curves is thickness *t* of Ni layer. From this figure one can see that there is an optimal value of parameter *d*, although dependent on thickness to achieve the highest self-field critical current. Although starting from lower thickness maxima of these curves increases, it saturates reaching thickness of 40 μm.

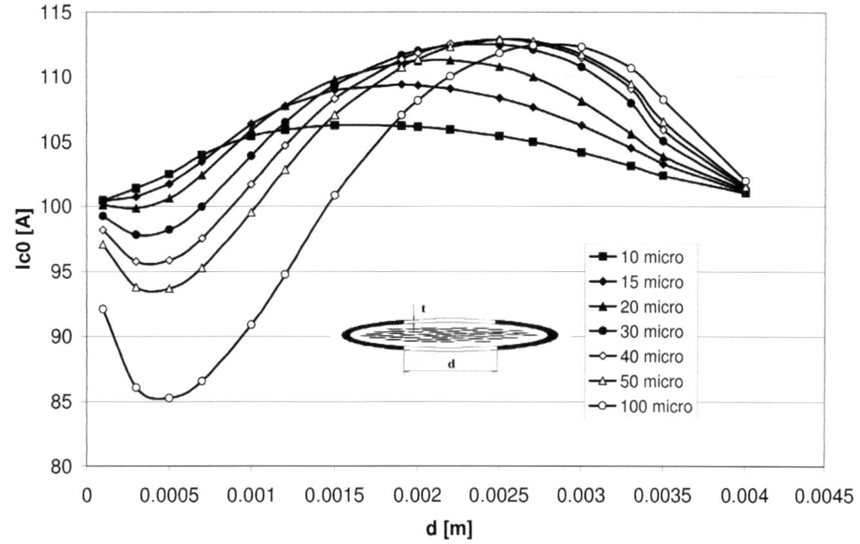

Figure 3. Dependencies of self-field critical current on width of non-covered area – *d*. Parameter is thickness of Ni layer – *t*.

Using these curves we made three samples with various values of both basic parameters. Summary of parameters of these samples and parameters of reference sample are in table 1.

Table 1. Basic properties of prepared samples

Sample	*d* [mm]	*t* [μm]	I_C measured	I_C calculated
A	4.1	0	100.8	100.8
B	1.2	30	108.1	106.4
C	2.4	30	115.7	112.4
D	2.4	15	108.3	108.4

From this comparison is seen good qualitative tendency, but not perfect quantitative agreement. Possible reason is not perfect characterization of the shape of superconductor cross-section. In previous works [2.3] cross-sections of SC were very similar to ellipse. On the other hand, cross-section of the tape investigated in this contribution differs from this assumption (see figure 4).

4. Calculation using image processing

For detail analysis of the effect of ferromagnetic cover we developed procedure, which enable us to take into account arbitrary cross-section of superconductor also with ferromagnetic material in close ambient.

To investigate superconductor-ferromagnetic material composite we use micrograph photo of cross-section. This picture should be modified in sense to increase color contrast between different materials of composite. Resolution of picture was then decreased. We used resolution 393x28 pixel as a compromise between computing time and possibility of detail analysis. In figure 4 there is micrograph of original tape – picture on a top, simplified picture ready for calculations – picture in the middle and simplified picture with nickel cover – picture at a bottom. Last two pictures were directly used as input information for FEM code. Each point in picture represents element with properties of vacuum – white colour, superconductor – red colour or ferromagnetic material – green color.

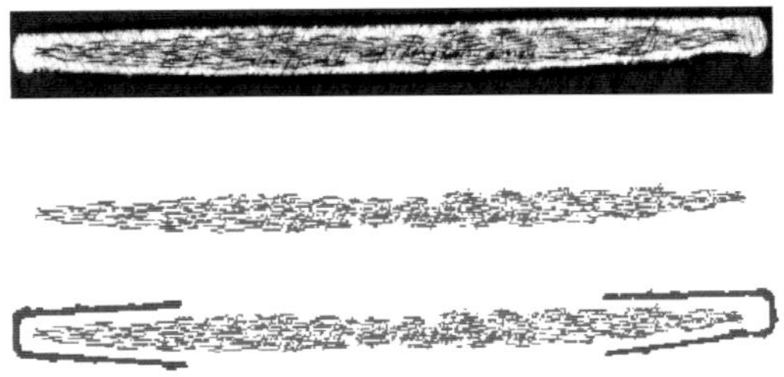

Figure 4. Superconductor cross-section. top – micrograph photo of real tape, middle – simplified picture ready for calculations, bottom – simplified photo with Ni cover on the edges

Using this procedure we are able to calculate also the distribution of magnetic field density in superconductor. Superconductor BiSCCO-2223/Ag has strongly anisotropic properties with respect to the magnetic field. Therefore it is reasonable to investigate parallel and perpendicular component of local magnetic field separately. Orientation of this field is with respect to wide face of the tape. Parallel and perpendicular component of magnetic field was separately discreeted and number of elements with the same level of the magnetic field was divided by total number of elements to quantify influence of magnetic field. We got a curve of the density of elements with certain level of magnetic field by this procedure. As it is visible in a figure 5 a most of elements are exposed to low parallel magnetic field and number of them decreases with increase of level of magnetic field.

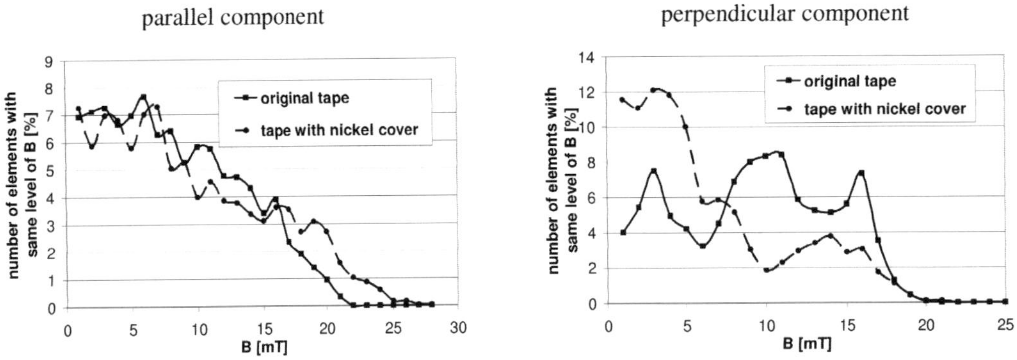

Figure 5. Percentage number of elements with the same level of parallel resp. perpendicular component of magnetic field

Different situation is in a case of perpendicular magnetic field. This curve has a maximum at magnetic field of about 10 mT. By using of the nickel cover the number of elements exposed to the high perpendicular magnetic field strongly decrease

5. Conclusions

Calculations of critical current in case of modifying the BiSCCO-2223/Ag tape covering edges of the tape by ferromagnetic material were performed. Optimization of covering procedure was possible using results of these calculations. This allowed us to improve the critical current ~15%. The procedure, which takes into account the detail geometrical structure of the tape, has been developed. Using this procedure it is possible assume arbitrary cross-section of superconductor and ferromagnetic material. This procedure enables quantitative analysis of magnetic field density. Such comparison of bare tape and tape partly covered by Ni layer has been done.

6. Acknowledgement

This work has been partially supported by the Slovak Research and Development Agency
Contract: APVV-51-045605 and APVV-SK-CN-0015-07

References

[1] Alamgir A K M, Gu C and Han Z 2005 *Physica C* **432** 153
[2] Gömöry F 2006 *Appl. Phy. Lett.* **89** 072506
[3] Gömöry F, Šouc J, Seiler E, Klinčok B, Vojenčiak M, Alamgir A K M, Han Z and Gu C 2007 *IEEE Tr. Appl. Supercond.* **17** 2

Superconductor Ferromagnet bilayers; Experimental investigation of spin polarized transport across the interface

S.K. Wanchoo[1], John J.[2], V.C. Bagwe[2], S. P. Pai[2], A. M. Narsale[3] and R. Pinto[4]

[1]School of Applied Physics & Mathemaics, SMVD University, Katra, India-182 320
[2]*Tata Institute of Fundamental Research, Mumbai-400 005*
[3]*WRIC, University of Mumbai, Mumbai – 400 098*
[4]*Department of Electrical Engineering, IIT Bombay, Mumbai - 400076.*

E-mail: sunilkwanchoo@gmail.com

Abstract. *Deposition of high temperature superconducting thin films by pulsed laser deposition (PLD) started a flurry of activities in the area of rare earth class of oxide manganites. Studies have been carried out by injection of the polarized spins from a ferromagnetic metal, into the superconductor via a thin insulating barrier. However because of the presence of an insulating layer possible role of joule heating in limiting the spin polarized transport across the interface cannot be ruled out. We would like to present some of the results of our work carried out on* $LaAlO_3/La_{0.7}Ca_{0.3}MnO_3/YBa_2Cu_3O_{7-\delta}$ *(LAO/LCMO/YBCO) LCMO on top and* $LaAlO_3//YBa_2Cu_3O_{7-\delta}$ $La_{0.7}Ca_{0.3}MnO_3$ *(LAO/YBCO/LCMO) YBCO on top heterostructures deposited on <100> LAO substrates without any insulating layer separating the two. PLD technique was used to deposit a 500 micron wide line of LCMO followed by deposition of YBCO layer which was subsequently patterned into a 500 micron line directly on top of the LCMO line. Here we report the suppression in the critical current* I_c *of the superconductor due to pair-breaking phenomenon when polarized spins are injected into YBCO in case of LAO/LCMO/YBCO (LCMO on top) sample. Surface and interface characterization of these bilayers was performed by SQUID, AFM, SIMS and four probe techniques.*

1. Introduction:

The coexistence of superconductivity and ferromagnetism is an interesting problem in physics which has attracted the attention of many workers. After the discovery of high temperature superconductors and their realization in thin film form, current interest is focused to look out for materials compatible with them. Deposition of high temperature superconducting thin films by pulsed laser ablation started a flurry of activities in the area of rare earth class of oxide manganites as the same are structurally compatible with the high temperature superconductors. These materials, due to their half-metallic nature, exhibit spin polarization close to unity and thus can act as a reservoirs of polarized spins. These manganites posses a lot of promise for high temperature superconductor based devices involving spin dependent transport of the charge carriers. Devices patterned out of multilayered heterostructures provide considerable potential for applications. The results of the recent studies have put forward the idea that spin polarized electronic transport bringing about suppression of superconductivity[1]. These studies have been motivated by the idea of having three terminal superconducting devices[2] based on polarized quasiparticle injection. It is clear that a study of spin-polarized electron transport from ferromagnetic materials to superconductors is important for the understanding of spin dependent electronic properties and for the realization of possible spin-injection devices.

A number of reports[3,4] involving spin dependent transport in case of low temperature superconductors have been reported in literature. Similar studies in case of high temperature superconductors have also been reported[5,6]. These devices are expected to work at very high speeds. Various studies on suppression of the superconductivity[5-7] by the injection of polarized spins have been phenomenal.

Various devices have been proposed and the experimental data pertaining to these studies has been widely published[6-9] These devices make use of the electron's spin apart from its charge. Almost all these studies are based on trilayer structures of the form, ferromagnet-insulator-superconductor (F-I-S). These studies have been carried out by injecting polarized spins from a ferromagnetic metal into the superconductor via a thin insulating barrier. However because of the presence of an insulating layer possible role of joule heating in limiting the critical currents cannot be ruled out. In this paper we report the synthesis of LAO/LCMO/YBCO (YBCO on top) & LAO/YBCO/LCMO (LCMO on top) heterostructures using a pulsed laser deposition system. We have directly deposited the two layers on each other to avoid any joule heating due to the intentionally deposited insulating thin film separating the two.

2 Experimental Details

The bulk targets of YBCO and LCMO used for the present work have been synthesized via the standard solid-state reaction route. Pulsed laser deposition system was used to grow the LAO/LCMO/YBCO (YBCO on top) & LAO/YBCO/LCMO (LCMO on top) heterostructure on LAO substrates. An excimer laser was used to grow these samples with a wavelength of 248 nm and a repletion rate of 10 Hz. Both the targets were mounted on the target holders which could be rotated using automated motor control. We have used almost identical deposition conditions (after optimizing the same) for the in situ growth of both the two oxide layers. These heterostructures were deposited at a substrate temperature of 800°C. Oxygen partial pressure of 400 mTorr for LCMO and 250 mTorr for YBCO were maintained during their growth respectively. The superconducting transition temperature was established by using a homemade ac susceptibility measurement setup. We have probed the surface morphology of as grown thin films using atomic force microscope. The phase pure nature of the films was established by powder

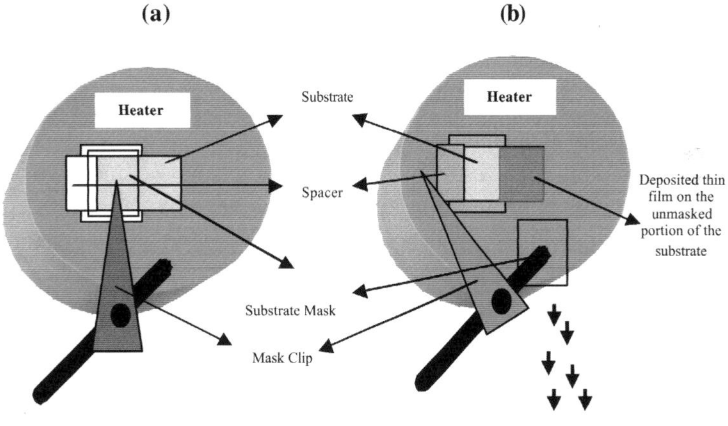

Figure 1. Substrate mounted on the heater with the substrate-mask covering its 50% area before deposition (b) The mask being dropped in situ by disturbing the mask clip with the heater shield (not shown in the sketch here) after the deposition of first layer.

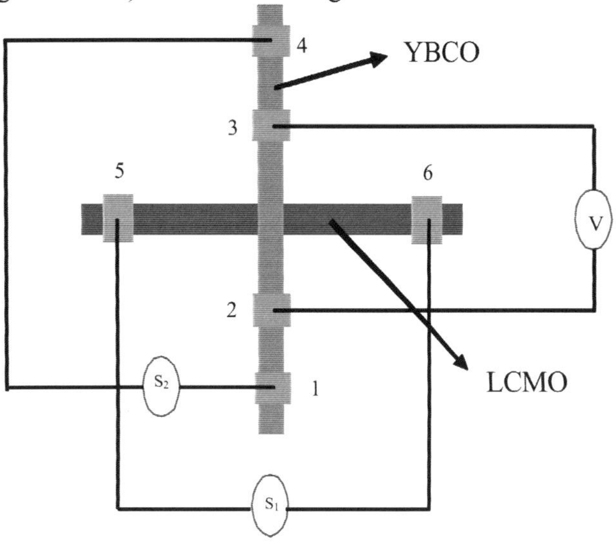

Figure 2. Device structure and electrical contact configurations for measurements of I_c (YBCO) in case of LAO/LCMO/YBCO (YBCO on top) sample. Contacts 1 and 4 were used as current contacts for source S_2 and 2 and 3 were used as voltage contacts for determining I_c of the YBCO line. Contacts 5 and 6 were used as current contacts for source S_1 used to send current through the LCMO line. Note that contacts 5 and 6 are on LCMO layer and the remaining four are on YBCO layer.

x-ray diffraction technique. To carry out the resistivity measurements using a homemade resistivity measurement setup, films were patterned using UV photolithography technique. Contact pads of gold were patterned using the lithography technique after depositing a gold film on the sample by PLD. Another sister sample, which was grown in the same run, was used to carry out the resistivity measurements on the bottom LCMO layer. The interface of the LAO/LCMO/YBCO (YBCO on top) & LAO/YBCO/LCMO (LCMO on top) heterostructures was characterized using secondary ion mass spectroscopy technique (SIMS).

In order to establish the interface quality a depth profiling was carried out using a Cameca 34f SIMS machine. In this measurement O_2^+ was used for ionization at a current of 50nA and at an effective incidence angle of 42°. Finally an analysis zone of 33 micron diameter was used in the center of the erosion zone of 150 micron x 150 micron to carry out the depth profiling.

Further, in order to investigate the effect of the spin polarized current on the critical current of a superconductor another experiment was performed. This device was synthesized by first depositing a LCMO layer (1000 Å thick) in the form of a thin line (500 micron wide) on LAO substrate. This thin line was deposited on a single crystalline (001) oriented LAO substrate. The optimized deposition conditions were used for the growth of the LCMO line. The line was deposited using two other LAO substrates to cover the substrate as a substrate-mask. The substrate-mask was then dropped in situ by a mechanical arrangement (*In situ* substrate mask drop technique) as shown in figure 1. After dropping the substrate-mask, a YBCO layer (2000 Å thick) was deposited in situ on the LAO substrate. Then a 500 micron wide line of YBCO was patterned out of the YBCO layer deposited in situ using the PLD technique. The YBCO layer was patterned using the UV photolithography technique. In this case independent currents, perpendicular to each other, were passed through the LCMO and YBCO layers respectively. Figure 2 shows the connection scheme used for the measurements. S_1 and S_2 denote the two current sources used to send current through the LCMO and YBCO layers respectively. V denotes the voltmeter used to determine the I_c of the YBCO layer. S_1 was connected between 5 and 6 and S_2 was connected between 1 and 4 respectively. The voltmeter V was connected between 2 and 3.

3 Results and Discussion

XRD results of the bilayer grown by pulsed laser deposition technique showed that most of the reflections are along the (001) direction suggesting the epitaxial c axis oriented growth of the both the layers. Figure 3 shows the AFM images of the constituent layers of the LAO/LCMO/YBCO (YBCO on top) sample. The AFM micrograph reveals the granular and morphologic nature of the growth. Figure 4 shows the temperature dependence of resistance for LAO/LCMO/YBCO (YBCO on top) sample. It can be seen from the graph that the metal – insulator transition temperature (T_p) of the LCMO layer is 250K.

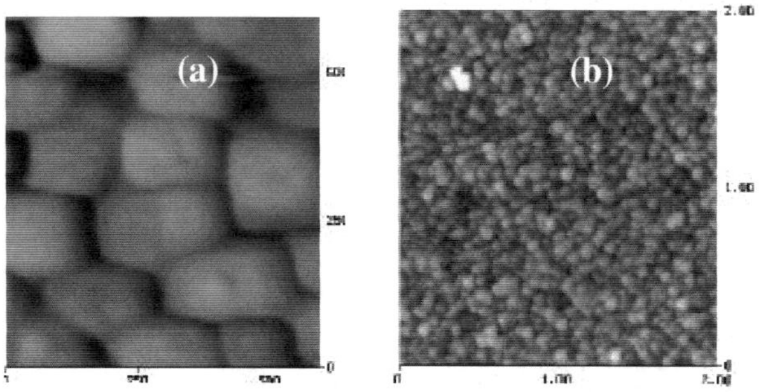

Figure 3. Atomic force microscope micrographs LAO/LCMO/YBCO (YBCO on top) (a) a bilayer and a (b) single LCMO layer.

At 250K LCMO layer shows a clear MI transition while as at 85 K since the resistance of the YBCO microbridge drops to zero and the same conclusively indicates the presence of superconductivity in the microbridge. Hence the four-probe measurements clearly indicate that the LAO/LCMO/YBCO (YBCO on top) sample essentially displays both the properties in a single sample.

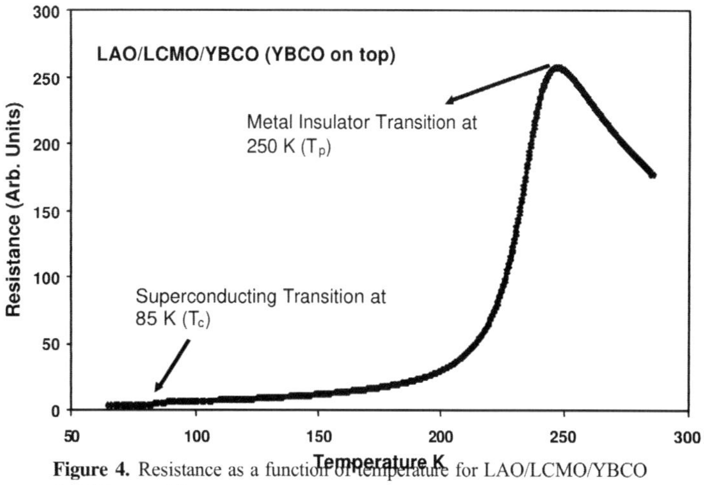

Figure 4. Resistance as a function of temperature for LAO/LCMO/YBCO (YBCO on top) heterostructure.

The four-probe data though does not indicate any change in T_p of LCMO as the same is found to be equal to that of a single layer of LCMO, which was grown for comparison. However the superconducting transition temperature T_c of YBCO does show a decrease from standard 90K to 85 K. In case of LAO/YBCO/LCMO (LCMO on top) sample we do not see any superconducting transition in the fourprobe measurements carried out on the sample. However the top LCMO layer does show a clear metal insulator transition at 250K. The most obvious reason for this seems to be the oxygen deficiency in the bottom YBCO layer. The YBCO layer is not able to get properly oxygenated during the venting of the deposition chamber after the growth of the two layers *in situ* as it is fully covered by the top LCMO layer. In order to probe this top LCMO layer was chemically etched. The sample was then subjected to 3 hrs annealing at 650°C with oxygen flowing in the furnace. Resistivity measurements were then performed on the bottom YBCO layer. The YBCO layer showed a T_c of 88 K which clearly indicates that the film was oxygen deficient.

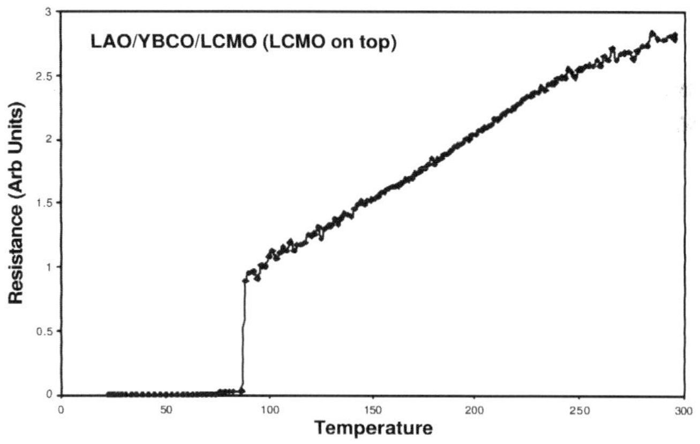

Figure 5. Resistivity measurements on the bottom YBCO layer after the top LCMO layer was et ched and the YBCO layer was subjected to 3h anneal in flowing oxygen.

Figure 6 (a) shows the M Vs H hyteresis loop for LAO/LCMO/YBCO sample was carried out at 40K (well below the T_c of YBCO). The plot clearly demonstrates the excellent display of superconducting ordering in these bilayers. Figure 6 (b) shows the temperature dependence of magnetic moment for the said sample under the FC and ZFC conditions. Once again the plot shows perfect diamagnetic behaviour below 85K (same T_c as was obtained in from the fourprobe measurements). This clearly indicates that the said sample shows excellent diamagnetic behaviour. In order to probe the decrease in T_c we have carried out secondary ion mass spectroscopy (SIMS) studies on the two types of samples to characterize the sharpness of the interface between LCMO and YBCO. Figure shows the SIMS profiles for

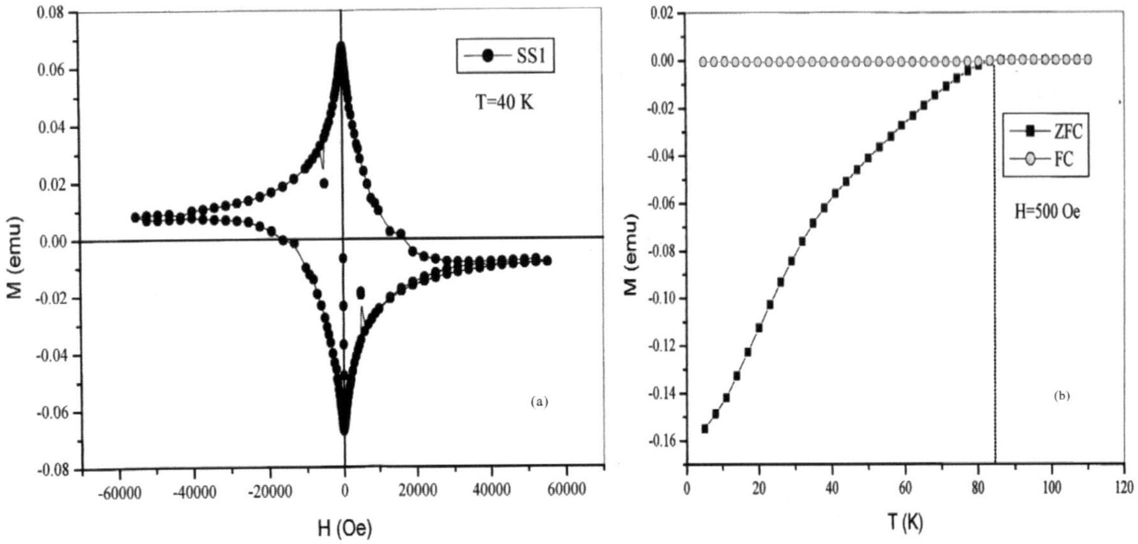

Figure 6. (a) M-H hysteresis loop recorded for L.CMO/YBCO heterostructure 40 K. (b) M vs T curve recorded for LCMO/YBCO heterostructure at a magnetic field of 500 Oe.

LAO/LCMO/YBCO (YBCO on top) & LAO/YBCO/LCMO (LCMO on top) heterostructures. The interface quality depends primarily on the surface roughness of the bottom layer and on the possible inter-diffusion of cations at the interface during the growth of top layer. Atomic force microscopy showed the LCMO roughness (determined partly due to the twinned LAO substrate) to be ~200 Å. We evaluated the cationic inter-diffusion using SIMS. Shown in figure 7 are the

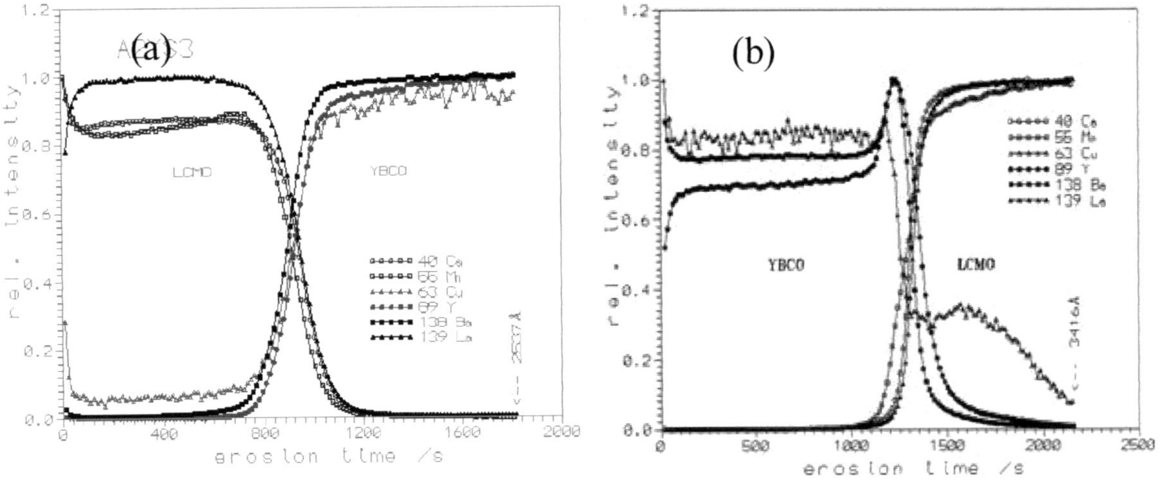

Figure 7. (a) SIMS profiles of a typical LAO/YBCO/LCMO (LCMO on top) sample. The full scale of the x-axis corresponds to 2757 Å. (1820 s in the time axis correspond to an etch crater depth of 2537 Å). (b) LAO/LCMO/YBCO (YBCO on top)sample. The full scale of the x-axis corresponds to 3954 Å. (2150 s in the time axis correspond to an etch crater depth of 3416 Å)

SIMS profiles of (a) LAO/YBCO/LCMO (LCMO on top) interface & (b) LAO/LCMO/YBCO (YBCO on top) interface. If we account ~200 Å due to surface roughness in the ~400 Å inter-diffused layer (SIMS), we see only a nominal inter-diffusion of ~100 Å on each side of the junction, except for diffusion of Cu into the LCMO layer in case of LAO/LCMO/YBCO (YBCO on top) sample. Longer diffusion of Cu into the LCMO layer does not appear to affect the spin carrier density as the T_p was found to be 250 K. This clearly indicates that the minimal diffusion of Cu in case of LAO/LCMO/YBCO (YBCO on top) sample does not affect the properties of LCMO. We believe that the inter-diffused Cu may be sitting at the grain boundaries and does not have any impact on the LCMO properties.

When we probed the LAO/YBCO/LCMO (LCMO on top) sample we could not find any Cu diffusion into the LCMO layer. The interface is sharp and clean in this case however the problem of deoxygenating of the bottom YBCO layer remains a major concern. Therefore it is clear that due to the loss of Cu and also due to a 100 Å inter-diffused layer, the T_c of YBCO shows a decrease of 5 K (in case of LAO/LCMO/YBCO (YBCO on top) sample). When the LCMO is deposited on top we do not see any superconducting transition in the bilayer.

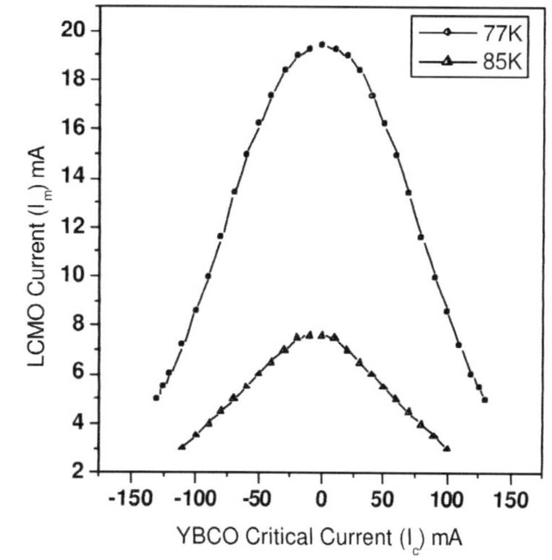

Figure 8. LCMO current I_m vs the YBCO critical current I_c for device shown in figure 2 measured at 77K and 85K respectively.

In case of the device as shown in figure 2 (LAO/LCMO/YBCO (YBCO on top)) initially a sensing current of 10 μA just enough for the four-probe measurements was passed through the YBCO line and the LCMO current was progressively increased. For a particular value of LCMO current (I_m), the YBCO line became normal (lost superconductivity) even when only 10 μA current was flowing through the YBCO line. To investigate this further, the YBCO current (I_c) was also increased in small steps. It was observed that the YBCO line lost superconductivity at progressively lesser LCMO currents. The observed phenomenon was independent of the direction of the injection current (I_m) which is clearly visible in figure 8. Figure shows the plot of I_m vs I_c at 77K and 85K respectively. This result leads to an interesting device application as a "superconducting switch". For relatively large current (less than I_c) flowing through the superconductor (YBCO), even a small current through LCMO (Im) is enough to cause a voltage drop across the superconductor and kill the superconductivity.

4 Conclusion

In conclusion, we have fabricated LAO/LCMO/YBCO (YBCO on top) & LAO/YBCO/LCMO (LCMO on top) heterostructures in situ by PLD on LAO substrates. Measurements showed that T_c of YBCO microbridge on top of LCMO layer is equal to 85 K T_p of the bottom LCMO layer was found to be 250 K. Contribution of ~100 Å due to surface roughness (as seen by AFM) and ~100 Å due to inter-diffusion (as verified by SIMS) does not seem to affect the material properties of LCMO. The LCMO-YBCO heterostructure displays both superconductivity and ferromagnetism simultaneously. On probing the YBCO-LCMO heterostructures we observed that the bilayer shows excellent ferromagnetic properties with a T_p of 250K however there is now superconducting transition which could be observed. On analyzing the heterostructure with SIMS no cationic interdiffucion could be observed. In case of LAO/YBCO/LCMO (LCMO on top) after etching the top LCMO layer and annealing the bottom YBCO layer in flowing oxygen a T_c of 88k could be observed indicating that the YBCO layer was oxygen deficient. An interesting switching experiment was performed and it was observed that the YBCO line lost superconductivity at progressively lesser LCMO currents. In conclusion suppression of superconductivity due to the injection of polarized spins from LCMO into YBCO is successfully demonstrated.

5 Acknowledgements

The authors wish to thank M Neumann-Spallart for useful discussion, and F. Jomard for carrying out SIMS measurements.

References

[1] Y. Gim *et al.* 2001 ***J. Appl. Phys.*** **90,** 4063.

[2] J. Mannhart 1996 ***Semicond. Sci. Technol.*** **9,** 46.

[3] I. Iguchi 1984 ***Physics B*** **126,** 149.

[4] C. C. Chi *et al.* 1979 ***Phys. Rev. B*** **20,** 4465.

[5] I. Iguchi *et al.* 1994 ***Phys. Rev., B*** **50,** 457.

[6] V. A. Vas'ko *et al.* 1997 ***Phys. Rev. Lett.*** **78,** 1134.

[7] Z. W. Dong *et al.* 1997 ***Appl. Phys. Lett.*** **71,** 1718.

[8] S. P. Pai *et al.* 2002 ***Pramana J. of Phys., 58,*** 1147-1151.

[9] Yeh N.C. *et al.* 1999 ***Phys. Rev. B 60,*** 10522 .

The Metal-Insulator Transition in High-T_c Cuprates – An unusual Quantum Transition

W. Winkler
Laboratory for Materials, Wackenbergstr. 84-88, 13156 Berlin, Germany
E-mail: wink-labmat@arcor.de

Abstract. The Hilbert spaces representing the quantum states in the CuO_2 planes are not spanned by crystal symmetry adapted basis states, but by self-consistently renormalized states which form compound systems of Hilbert subspaces. In addition, these self-consistent basis representations are not necessarily stationary in time but behave dynamic in many respects. In particular, the coordinate systems of the Hilbert subspaces form a definite dynamic relative state in space and time, thus the coordinate system itself becomes a variable. This implicates a deterministic space-time relation of quantum states and the quantization of time by an internal time constant, the *eigentime* $\tau_{ei_{DCBF}}$. In undoped and hole doped CuO_2 planes $\tau_{ei_{DCBF}}$ occurs as a conservation quantity, whereas under electron doping a partial fluid exists in which $\tau_{ei_{DCBF}}$ does not occur as quantized quantity. The deterministic space-time behaviour of the Hilbert subspaces represents the emergence of a classical space-time structure in quantum systems. The metal-insulator transition in high-T_c cuprates, usually attributed to an antiferromagnetic Mott transition, results here from an additional splitting into Hilbert subspaces. The transition from the insulator to the conductor is causally related to coordinate transformations from the copper to the oxygen sites. The antiferromagnetism in the CuO_2 planes is not caused by the half-filled valence bands, as usually assumed, but is created by *off-diagonal spin compensations* within filled bands.

1. Introduction

Ever since the experimental investigations of superconducting cuprates, these materials were proven to be unconventional in many respects, in the normal as well as superconducting state. This is primarily concerned within hole-doped cuprates which show, for example, an anomalous low temperature resistivity behaviour $\rho \sim T$, a temperature dependent Hall effect, an optical conductivity $\sigma(\omega) \sim 1/\omega$, non-Korringa copper relaxation rates found in NMR, non-pronounced or unconventional isotopic effects, coexisting spin and superconducting gaps found in neutron scattering, a temperature independent superconducting gap up to T_c distinctly visible in IR measurements, and several other unconventional findings. A comprehensive review can be found in several monographs and review articles [1]–[6] and some particular results have already been discussed in our previous articles [7, 8].

An important phenomenological difference between high-T_c superconducting cuprates and conventional superconductors is the obviously non-s-wave symmetry of the order parameter $\Delta(\alpha)$ in the high-T_c superconductors in dependence on the wave vector direction α. The found π shift, responsible for a half-integer flux quantization of the magnetic flux already existing without an outer magnetic field, is attributed to non-s-wave symmetries [9]. An anisotropic gap $\Delta(\alpha)$ found by angle-resolved photoemission spectroscopy (ARPES), which is reflected by the anisotropy of the momentum dependence of the gap being largest along the $\Gamma - M$ (parallel to the a or b direction) and smallest along $\Gamma - Y$ (diagonal line between them) [10]–[12], can also be attributed to a non-s-wave pairing. However, despite these indications that apparently favour a d-wave symmetry of $\Delta(\alpha)$, the symmetry problem of the superconducting pair state is not yet unambiguously answered. This and other particular problems will appear in a completely different light, if the Topological Resonance (TR) conception [8] is taken as a basis which assumes arrays of parallel one-dimensional electronic states. Under these conditions the symmetry of the gap $\Delta(\alpha)$ has to be related in the first place to one-dimensional electronic states. Hence, all theoretical attempts which are stringently founded on a stationary two-dimensional $d_{x^2-y^2}$ symmetry of the pair-wave function [13]–[16] can not reflect the superconducting state within the CuO_2 planes, if the electronic state is causally founded on one-dimensional electronic states.

The above experimental results were related to hole-doped (p-type) cuprate superconductors. The experimental findings of electron-doped (n-type) superconductors, $Nd_{2-x}Ce_xCuO_{4-\delta}$ (NCCO) and $Pr_{2-x}Ce_xCuO_{4-\delta}$ (PCCO), are different to their hole-doped counterparts in many ways. For example, the highest T_c values in the hole-doped cuprates are more than five times as high as in the highest T_c electron-doped cuprates [17, 18]. The in-plane resistivity is approximately quadratic in temperature, contrary to the pronounced linear behaviour within the hole-doped cuprates. The flat band just below the Fermi-surface (E_F) at $(\pi, 0)$ and $(0, \pi)$ always found in hole-doped cuprates is located at 300 meV below E_F in NCCO [19]. However, the symmetry dependence of the electronic states near the Fermi surface does not seem to essentially be different to the hole-doped case. These different characteristics of the p- and n-type superconductors are causally related to the different influence of the time-quantized phases as it will be discussed in this article.

From the beginnings of theoretical considerations about the electronic states in high-T_c cuprates, the unusual properties in the normal state ($T > T_c$) have questioned the Fermi liquid description as being the right theoretical basis [1]. The occurrence of an antiferromagnetic state in the CuO_2 planes in the hole undoped case has motivated most scientists to attribute the antiferromagnetic couplings to the unusual normal state properties as well as to the existence of a superconducting state. It is assumed that the undoped electronic state of a CuO_2 plane corresponds to an antiferromagnetic Mott insulator and hole doping creates a variety of possible spin frustrations including local spin pair states which may resonate [1]-[6], [13]–[16]. A separation between spin and charge states, which are highly correlated, is explicitly included. This is reflected by holon and spinon states in the RVB conception [1, 2] and the formation of linear spin and charge states, termed as stripes (smectic phases), as well as square structures (nematic phases) [3]. The theoretical postulation of stripes as well as square structures were motivated by extraordinary experimental results which were summarized by the general term *1/8 problem* and the formation of *checkerboard structures*, respectively (see e.g. Refs. [3, 7]). In the theoretical descriptions based on antiferromagnetic couplings, this obvious great variety of electronic states in the CuO_2 planes is comprehended by postulating quantum critical points which lead to quantum critical fluctuations/phases [3], [13]–[16]. In order to differentiate between theories which assume an antiferromagnetic Mott insulator as the fundamental basis and the TR conception, which represents the theoretical basis in this article, some common characteristics of various theories are firstly summarized which are based on the Mott-Hubbard philosophy [1]-[6], [13]–[16]:

(i) The undoped state of the CuO_2 plane represents an antiferromagnetic Mott insulator.

(ii) A hole doped Mott insulator may be conducting. The charge carriers carry also the spin (one-band representation), i.e. the remnant antiferromagnetic couplings are created by the charge carriers themselves.

(iii) Under hole doping, the remnant antiferromagnetic spin interactions are primarily causing the formation of particular quantum phases and the creation of the superconducting phase.

(iv) Various spin-charge separations are possible but they are basically of non-crystalline nature (smectic phases, nematic phases) [3], [13]−[16].

Contrary to these assumptions, which are most popular in describing the electronic states in high-T_c cuprates, we have shown [7, 8] that an antiferromagnetic Mott insulator is not the right frame in order to characterize the electronic states in undoped CuO_2 planes, but a subtle renormalized electronic state exists which was termed as charge and bonding fluctuation state (CBF) [7]. The CBF state includes the antiferromagnetism, however, the antiferromagnetic couplings are not causal for the electronic renormalizations but only a consequence of that. In comparison to the above characteristics of the Mott-Hubbard basis some topics resulting from this new electronic state are summarized:

(a) The basis state in the hole undoped case represents a charge and bonding fluctuation state (CBF) caused by complex electronic renormalizations [7]. The complete quantum state represents a dynamic charge and bonding fluctuation state (DCBF) [8].

(b) A hole doped CBF state may be conducting. The basic principles underlying an insulator-conductor transition are described in this article. The charge carriers do not create the antiferromagnetic spin couplings itself, the antiferromagnetic spin interactions are continously created by lower energetic bands, but the quantum states of the charge carriers and the antiferromagnetic spin states are strongly correlated.

(c) Under hole doping, the antiferromagnetic spin interactions are not causal for the formation of the valence quantum states. Instead quite the reverse applies, the electronic correlations dominate the spin structure. The superconducting state is causally related to electron and phonon states [8].

(d) Spin-charge separations are generally given by (b). The generally existence of topological charge separations which lead to self-consistent crystalline charge orders is deciding. Primarily, the spin structure follows the charge order but not vice versa.

A simple comparison of the topics (i) to (iv) with (a) to (d) shows the different basic assumptions between the Mott-Hubbard model and the theoretical basis underlying the considerations in this article.

In addition to the points selected above, there are further fundamental differences to existing theoretical approaches. The CBF state subdivides the total electronic state into particular electronic substates, corresponding to a subdivision of the unified Hilbert space into Hilbert subspaces. Quantum states which are founded on (i)-(iv) are continuously described within an unified stationary Hilbert space. Contrary to that, an electronic system S which is associated to the Hilbert space $\mathcal{H}$ being decomposed into two subsystems S_1, S_2, and being associated to the Hilbert spaces $\mathcal{H}_1$, $\mathcal{H}_2$, is represented by the compound system $\mathcal{H} = \mathcal{H}_1 \otimes \mathcal{H}_2$, i.e. the tensor product of the Hilbert spaces $\mathcal{H}_1$ and $\mathcal{H}_2$. This separation of the Hilbert space $\mathcal{H}$ into Hilbert subspaces $\mathcal{H}_1$ and $\mathcal{H}_2$ implies two important quantum effects, which are essential within the TR state. Firstly, interferences between states in $\mathcal{H}_1$ and $\mathcal{H}_2$ are prevented, but in return entanglements between quantum states in $\mathcal{H}_1$ and $\mathcal{H}_2$ may be enhanced.

It is of high significance that the Hilbert subspaces themselves can not be considered as being stationary in time but their basis states have to be considered as dynamically variable. Firstly, basis states of particular Hilbert subspaces are continuously exchanged, which particularly may occur in a probabilistic manner but above that there is also a strongly deterministic exchange in space and time. For example, the dynamics of the DCBF state (see(a)) is strongly deterministic in space and time, but the exchange within the **k** and −**k** electron pair states may occur probabilistically in the non-superconducting state (IS quantum hole states).[1] However, these **k**, −**k** pair exchanges are strongly quantized in time in the superconducting state (DS quantum hole states) [8]. The exchange of quantum states is causally related to the exchange of the coordinate systems of Hilbert subspaces. In consequence, the particular coordinate systems of the Hilbert subspaces develop into dynamic variables, either strongly periodic and deterministic in space and time or of probabilistic nature. If the coordinate system occurs as a deterministic space-time variable, internally quantized time scales of the dynamic Hilbert subspaces are created which are termed as *eigentimes* [8].

The goal of this article is to explain the static and dynamic separation of the Hilbert space into Hilbert subspaces and the resulting consequences for the quantum states following from that. It is particularly important to emphasize the conditions for the insulator-conductor transition under the special condition of crystalline topological hole structures and topologically ordered electron structures.

2. The electronic states of the CuO_2 planes

The electronic states of the CuO_2 planes were derived in preceding articles [7, 8]. The crystal symmetry adapted electronic states, usually existing in quantum systems, are renormalized in a way that a charge and bonding fluctuation state (CBF) with the period $2a$, $2b$ is created within the CuO_2 plane, depicted by the +,− signs in Fig. 1. This state implies a general renormalization of the electronic states with the result that the initially equivalent copper sites are separated into two different copper sites, as demonstrated by +,− signs in Fig. 1, which separates the crystal lattice into two sublattices. Only local electronic states associated to one sublattice are involved within every particular quantum state. In the hole undoped case representing half-filled valence states, only quantum valence states that are associated to one sublattice, the lattice corresponding to the − copper sites, are occupied. Therefore, the undisturbed CBF state is characterized by general renormalizations of the electronic states leading to two electronic subspaces with the initial unit cell vectors being divided by half as well as by a charge fluctuation state with the periodicity $2a$, $2b$ along the crystallographic axis a and b, respectively, as depicted in Fig. 1. For all low energetic bands where the Cu-O bonds possess bonding character, the two subspaces associated to the + and − lattice sites still represent two-dimensional electronic states. Especially, band states with nearly exclusively $d_{x^2-y^2}$ orbital symmetry of the Cu orbital parts belong to these two-dimensional low energetic band states. This two-dimensionality of the electronic states is additionally broken for the energetically highest half-filled valence bands where the Cu-O bonds are antibonding. As a result, the two-dimensional quantum states of the energetically highest valence bands are transformed into arrays of one-dimensional quantum states (see Ref. [7]). The one-dimensional electronic states are created along the lines I_x,I'_x and $\bar{I}_x$,$\bar{I}'_x$ as well as I_y,I'_y and $\bar{I}_y$,$\bar{I}'_y$ in Fig. 1, leading to the formation of one-dimensional bands. The particular one-dimensional bands are half-filled for the undoped CuO_2^{2-} plane, leading to the higher populated − copper sites. According to Fig. 1, two arrays of correlated one-dimensional subspaces exist correspondingly to the polarization of the one-dimensional states in $x(a)$ and $y(b)$ direction, respectively.

[1] Although **k**, −**k** pair states generally exist an outer impulse can be transferred to the paired electrons, if the pair state possesses IS quantum hole symmetry (non-superconducting state), as discussed in Ref. [8].

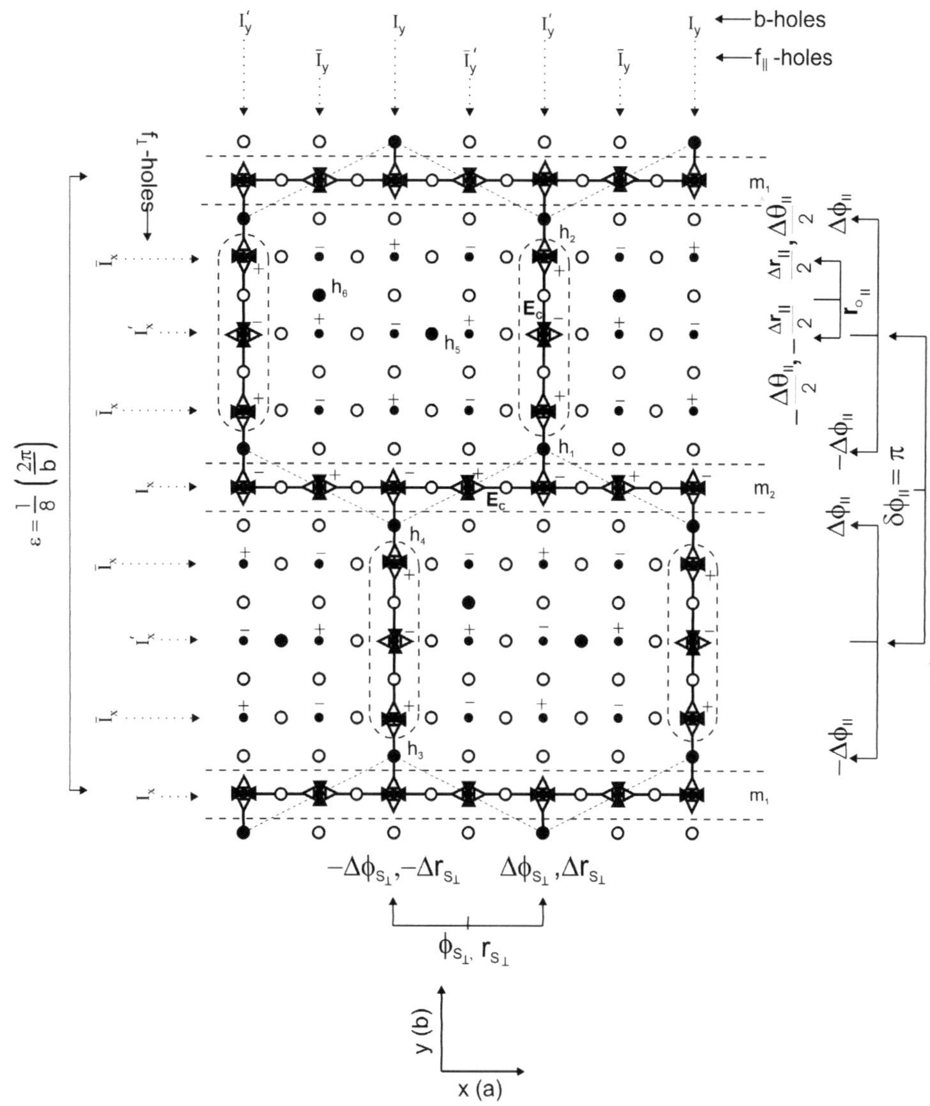

Figure 1. Topological hole configuration of a CuO_2 plane for a nominal hole concentration $n_h = 0.25$ holes/copper, representing an electronic ground state configuration of a static CBF state [7]. The two energetically highest valence bands of the CuO_2 plane are decomposed in two perpendicular arrays of one-dimensional electronic subspaces, along the x coordinate ($I_x, I'_x, \bar{I}_x, \bar{I}'_x$), and the y coordinate ($I_y, I'_y, \bar{I}_y, \bar{I}'_y$). The $+,-$ signs indicate the two different copper sites, corresponding to the renormalized electronic CBF state. The pictogram representation along the bold lines reflects attractive hole - hole interactions with $E_C < 0$ (see Ref. [7]). These holes are termed as b-holes (bonded holes), with n_h(b-holes) $= n_h/2 = 0.125 =$ *1/8* holes/copper (lines I_y, I'_y). The remaining holes are also correlated holes but with $E_C = 0$, and termed as f-holes (free holes). The b-holes and $f_{\|}$-holes are aligned along the *y(b)* coordinate, within the one-dimensional electronic states I_y, I'_y and $\bar{I}_y$, $\bar{I}'_y$, respectively. The $f_{\perp}$-holes belong to equivalent states within I'_x along the *x(b)* coordinate. The phases/coordinates $\frac{\Delta\theta_{||}}{2}/\frac{\Delta\mathbf{r}_{||}}{2}$ and $\frac{-\Delta\theta_{||}}{2}/\frac{-\Delta\mathbf{r}_{||}}{2}$ are topological relative phases/coordinates which differentiate between the + an − copper sublattices. The phase/coordinate $\phi_{S_{||}}/\mathbf{r}_{S_{||}}$ corresponds to the center point phase/coordinate of the entire b-hole structure, $-\Delta\phi_{S_{||}}/-\Delta\mathbf{r}_{S_{||}}$ and $\Delta\phi_{S_{||}}/\Delta\mathbf{r}_{S_{||}}$ are the relative center point phases/coordinates of the b-hole substates along I_y and I'_y, respectively. Equivalent center point phases/coordinates $\phi_{S_\perp}/\mathbf{r}_{S_\perp}$, $-\Delta\phi_{S_\perp}/-\Delta\mathbf{r}_{S_\perp}$ and $\Delta\phi_{S_\perp}/\Delta\mathbf{r}_{S_\perp}$ are defined peperpendicular to the b-hole polarisation direction. The relative phases $-\Delta\phi_{||}$ and $\Delta\phi_{||}$ define the positions of the two b-hole sublattices/substates within every b-hole state along I_y or I'_y having a periodicity of $8b$ and relative shift of $3b$ to each other. This creates the different topological hole states h_1 and h_3 as well as h_2 and h_4, respectively.

Under hole doping, which occurs if electrons are removed from the CuO_2^{2-} plane, the hole density is not widely distributed over the CuO_2 plane but rather localized on particular oxygen atoms. This leads to an additional hole localization energy $E_S < 0$ which is square dependent on the anisotropic local hole density q_h, located predominantly on particular oxygen atoms (Fig. 1). It has been argued [7] that a nominal hole density of $q_h = 1$ is localized on a particular oxygen atom. This means, a one-to-one correspondence exists between the number of doped holes and the number of localized topological hole states. However, these topological hole states are generally not identical with localized quantum states, because the electronic quantum states usually remain delocalized caused by the strong periodicity of the topological hole states. If linear hole - hole arrangements occur, as it is given for the hole pairs (h_1,h_2) or (h_3,h_4) in Fig. 1, an additional hole - hole coupling energy $E_C < 0$ occurs [7]. For non-linear hole - hole arrangements like (h_5,h_6), i.e. a cyclic hole - hole topology [7], the holes are also strongly correlated, but with $E_C = 0$. The localization energy E_S (hole self-energy) and the correlation energy E_C are predominantly based on quadrupolar-polarizations on copper atoms [7]. The different correlations for linear and cyclic hole - hole arrangements create two types of holes, b-holes and f-holes. The linearly arranged b-holes with $E_C < 0$ exist undisturbed within the hole concentration range of $0.125 \leq n_h \leq 0.25$ holes/copper with a definite periodic topological hole order, related to a unite cell of (4a,8b) in Fig. 1, and a constant mean b-hole density of n_h(b-holes) $= 0.125(=1/8)$ holes/copper. Ordered b-hole structures which include b-hole vacancies occur for $n_h < 0.125$ holes/copper. The f-holes with $E_C = 0$ can only exist for $n_h > 0.125$ holes/copper forming strong periodic structures when the hole density of $n_h = 0.25$ holes/copper is reached as given in Fig. 1. The f-holes can exist with different topological hole configurations within every particular b-hole configuration [7, 8].

3. The formation of electron subspaces

The electronic renormalizations lead to crystal symmetry breaking topological charge orders [7]. This becomes obvious by the formation of the CBF state and the occurrence of ordered topological hole structures which are decisively determined by the b-holes. In addition, the symmetry breaking leads to transformations of the coordinate systems of the particular topological charge orders with respect to the fundamental crystal structure. The CBF state which differentiates the copper atoms into two different sites, signed by + and − in Fig. 1, creates two sublattices with the lattice periodicity $2a, 2b$ along the crystal axes, accompanied by shifts of the particular coordinate systems of the two sublattices by $\pm\frac{\Delta \mathbf{r}_{||}}{2}$ with respect to the center point $\mathbf{r}_{o_{||}}$ along the $x(b)$ or $y(b)$ direction (Fig. 1). Similarly, the b-hole sublattice creates topological/spectral phases $-\Delta\phi_{S_{||}}, \pm\Delta\phi_{||}$ and $\Delta\phi_{S_{||}}, \pm\Delta\phi_{||}$ within each particular one-dimensional direction I_y and I'_y in Fig. 1, respectively, and a common center phase $\phi_{S_{||}}$ can be defined for the whole b-hole lattice. All phases are related to the b-hole periodicity of $8b$ in $y(b)$ direction of Fig. 1. The one-dimensional b-hole states along I_y and I'_y can be differentiated by the perpendicular phases $\phi_{S_\perp}, \pm\Delta\phi_{S_\perp}$, which are related to the b-hole periodicity of $4a$ in $x(a)$ direction. Similar topological phases can be defined for periodic f-holes states ($f_{||}$-holes and $f_\perp$-holes), but they are uniquely defined exclusively for $n_h = 0.25$ holes/copper.

The renormalizations of the electronic states which lead to two different copper sites, − and + in Fig. 1, were deduced from MO cluster calculations where every renormalized MO orbital involves predominantly only next neighbour copper sites but one, but not consecutive copper sites. In consequence, the periodic lattice of local electronic states within an infinitely extended CuO_2 plane falls into two sublattices with the lattice periodicity of $2a$, $2b$ along the crystal axes a and b, respectively, correspondingly to the − and + lattice sites in Fig. 1. The two sublattices result from self-consistent electronic renormalizations of the initial crystal symmetry adapted quantum states, which implies a strong association of the quantum states of the two sublattices. Usually, Bloch states are constructed from orthonormalized local electronic states possessing the stationary lattice periodicity. In contrast to that, the above renormalizations leading to symmetry breaking of the electronic states of the CuO_2 plane require the general inclusion of two neighboured lattice sites during the formation of one local state. This means, the local states within a Bloch representation comprehend generally two lattice sites. On the other hand, the local states associated to the − and + copper sites are orthogonal [7]. In consequence, every particular two-dimensional Bloch state can be constructed with respect to the local states of one periodic sublattice − or +. Therefore, two different two-dimensional Bloch states can be defined for a given wave vector $\mathbf{k}$:

$$\overset{[-]}{\varphi}_{\mathbf{k}}(\mathbf{r}_1, -\frac{\Delta\mathbf{r}}{2}) = e^{i(\mathbf{k}\mathbf{r}_1 - \frac{\Delta\theta}{2})}\,\overset{[-]}{u}_{\mathbf{k}}(\mathbf{r}_1 - \frac{\Delta\mathbf{r}}{2}) \in \overset{[-]}{\mathcal{H}}_{-\frac{\Delta\mathbf{r}}{2}} \quad (1)$$

and

$$\overset{[+]}{\varphi}_{\mathbf{k}}(\mathbf{r}_2, +\frac{\Delta\mathbf{r}}{2}) = e^{i(\mathbf{k}\mathbf{r}_2 + \frac{\Delta\theta}{2})}\,\overset{[+]}{u}_{\mathbf{k}}(\mathbf{r}_2 + \frac{\Delta\mathbf{r}}{2}) \in \overset{[+]}{\mathcal{H}}_{+\frac{\Delta\mathbf{r}}{2}} \quad (2)$$

with $\overset{[-]}{u}_{\mathbf{k}}(\mathbf{r}_1 - \frac{\Delta\mathbf{r}}{2})$ and $\overset{[+]}{u}_{\mathbf{k}}(\mathbf{r}_2 + \frac{\Delta\mathbf{r}}{2})$ being periodically local states corresponding to the sublattices − and +, respectively, within a coordinate system that is rotated by 45° with respect to the coordinate system as shown in Fig. 1. The sublattices are assumed to be shifted by $\pm\frac{\Delta\mathbf{r}}{2} = \pm(\frac{a}{2}, 0)$ in $x(a)$ direction or $\pm\frac{\Delta\mathbf{r}}{2} = \pm(0, \frac{b}{2})$ in $x(b)$ direction, within a coordinate system as shown in Fig. 1. Correspondingly, the exponential functions in Eqs. (1) and (2) are phase shifted in relation to $\Delta\theta = \frac{2\pi}{2\Delta\mathbf{r}} \cdot \Delta\mathbf{r} = \pi$, leading to $\frac{\Delta\theta}{2} = \frac{\pi}{2}$ and $-\frac{\Delta\theta}{2} = -\frac{\pi}{2}$. All in all, the original Hilbert space $\mathcal{H}$ being characterized by Bloch states that are not U1 symmetry broken is transformed into a Hilbert space $\mathcal{H}_{ren.}$ with a renormalized basis of U1 symmetry broken states which is separated into two basis representations creating the two subspaces $\overset{[-]}{\mathcal{H}}_{-\frac{\Delta\mathbf{r}}{2}}$ and $\overset{[+]}{\mathcal{H}}_{+\frac{\Delta\mathbf{r}}{2}}$, correspondingly to the two sublattices − and +. The two subspaces $\overset{[-]}{\mathcal{H}}_{-\frac{\Delta\mathbf{r}}{2}}$ and $\overset{[+]}{\mathcal{H}}_{+\frac{\Delta\mathbf{r}}{2}}$ form the orthogonal complement to each other, but in addition, the coordinate systems of the two Hilbert subspaces are different caused by the shift vectors $\pm\frac{\Delta\mathbf{r}}{2}$.

Two electrons can principally be placed within each of the two states described by Eqs. (1) and (2), resulting in a spin-up and a spin-down electron, because the total number of quantum states has to be conserved during the renormalizations. Thus, the dimension of the basis in $\mathcal{H}_{ren.}$ is identical to the the dimension of the initial Hilbert space $\mathcal{H}$. The number of Hilbert spaces is doubled, $\overset{[-]}{\mathcal{H}}_{-\frac{\Delta\mathbf{r}}{2}}$ and $\overset{[+]}{\mathcal{H}}_{+\frac{\Delta\mathbf{r}}{2}}$, but the number of $\mathbf{k}$ states is divided by half caused by the doubling of the lattice constants given by $2a$ and $2b$. An electron occupies either the Bloch state given by Eq. (1) or Eq. (2) making different electronic populations possible. A spin-up and a spin-down electron can simultaneously be placed within the state $\overset{[-]}{\varphi}_{\mathbf{k}}(\mathbf{r}, -\frac{\Delta\mathbf{r}}{2}) \in \overset{[-]}{\mathcal{H}}_{-\frac{\Delta\mathbf{r}}{2}}$ while leaving the state $\overset{[+]}{\varphi}_{\mathbf{k}}(\mathbf{r}, +\frac{\Delta\mathbf{r}}{2}) \in \overset{[+]}{\mathcal{H}}_{+\frac{\Delta\mathbf{r}}{2}}$ empty, or vice versa. In this case spin paired electron states occur at the lattice sites − or +, respectively, i.e. a *diagonal spin compensation/pairing* occurs as it is found in ordinary closed shell states. The "or", however, proves to be as an exclusive "OR", because, either of the two Hilbert subspaces, $\overset{[-]}{\mathcal{H}}_{-\frac{\Delta\mathbf{r}}{2}}$ OR $\overset{[+]}{\mathcal{H}}_{+\frac{\Delta\mathbf{r}}{2}}$, is exclusively populated until one subband is completely filled. This is caused by the energetically favoured imbalanced population of a particular lattice site, without loss of generality the − lattice site in Fig. 1 (see Section 5). Contrary to these local spin paired states, one electron, for example, a spin-down electron, can be placed within the $\overset{[-]}{\varphi}_{\mathbf{k}}(\mathbf{r}, -\frac{\Delta\mathbf{r}}{2})$ state and the spin-up electron in the $\overset{[+]}{\varphi}_{\mathbf{k}}(\mathbf{r}, +\frac{\Delta\mathbf{r}}{2})$ state, or vice versa. If Cu − O bonding states exist, ideally, the "or" is again exclusive. The exclusive OR follows from the fact that under these conditions local unpaired spin populations are favoured for electronic populations of $\mathcal{H}_{ren.}$ higher than half filling (see Section 4). Overall, a spin-down and a spin-up electron belong separately to the lattice sites − and + for every given $\mathbf{k}$ state, i.e. the spins are locally separated so that an *off-diagonal spin compensation/pairing* occurs.

The energetically highest valence states reveal an additional symmetry breaking, the two-dimensional electron states are transformed into arrays of one-dimensional electron states as discussed in Section 2. Each of the two perpendicularly polarized arrays along the $x(a)$ and $y(b)$ coordinate consists of one-dimensional states along $I_i(I_i')$ and $\bar{I}_j(\bar{I}_j')$. This means, the electronic states along a particular one-dimensional state I_i or $\bar{I}_j$ belong to separated subspaces $\mathcal{H}_{I_i}$ or $\mathcal{H}_{\bar{I}_j}$, respectively. Again, every one-dimensional Bloch state can be related to the − or + copper site. Therefore, four different one-dimensional Bloch states can be distinguished for every array corresponding to the shifts of the particular coordinate systems by $\pm\frac{\Delta\mathbf{r}_{||}}{2}$ within I_i and $\bar{I}_j$, according to

$$\overset{[-]}{\varphi}_{\mathbf{k}_{||},I_i}(\mathbf{r}_{1_{||}} - \frac{\Delta\mathbf{r}_{||}}{2}) = e^{i(\mathbf{k}_{||}\mathbf{r}_{1_{||}} - \frac{\Delta\theta_{||}}{2})}\,\overset{[-]}{u}_{\mathbf{k}_{||},I_i}(\mathbf{r}_{1_{||}} - \frac{\Delta\mathbf{r}_{||}}{2}) \in \overset{[-]}{\mathcal{H}}_{I_i,-\frac{\Delta\mathbf{r}_{||}}{2}} \tag{3}$$

$$\overset{[+]}{\varphi}_{\mathbf{k}_{||},I_i}(\mathbf{r}_{2_{||}} + \frac{\Delta\mathbf{r}_{||}}{2}) = e^{i(\mathbf{k}_{||}\mathbf{r}_{2_{||}} + \frac{\Delta\theta_{||}}{2})}\,\overset{[+]}{u}_{\mathbf{k}_{||},I_i}(\mathbf{r}_{2_{||}} + \frac{\Delta\mathbf{r}_{||}}{2}) \in \overset{[+]}{\mathcal{H}}_{I_i,+\frac{\Delta\mathbf{r}_{||}}{2}} \tag{4}$$

$$\overset{[-]}{\varphi}_{\mathbf{k}_{||},\bar{I}_j}(\mathbf{r}_{3_{||}} + \frac{\Delta\mathbf{r}_{||}}{2}) = e^{i(\mathbf{k}_{||}\mathbf{r}_{3_{||}} + \frac{\Delta\theta_{||}}{2})}\,\overset{[-]}{u}_{\mathbf{k}_{||},\bar{I}_j}(\mathbf{r}_{3_{||}} + \frac{\Delta\mathbf{r}_{||}}{2}) \in \overset{[-]}{\mathcal{H}}_{\bar{I}_j,+\frac{\Delta\mathbf{r}_{||}}{2}} \tag{5}$$

$$\overset{[+]}{\varphi}_{\mathbf{k}_{||},\bar{I}_j}(\mathbf{r}_{4_{||}} - \frac{\Delta\mathbf{r}_{||}}{2}) = e^{i(\mathbf{k}_{||}\mathbf{r}_{4_{||}} - \frac{\Delta\theta_{||}}{2})}\,\overset{[+]}{u}_{\mathbf{k}_{||},\bar{I}_j}(\mathbf{r}_{4_{||}} - \frac{\Delta\mathbf{r}_{||}}{2}) \in \overset{[+]}{\mathcal{H}}_{\bar{I}_j,-\frac{\Delta\mathbf{r}_{||}}{2}} \tag{6}$$

with $\mathbf{r}_{i_{||}} = (x, 0)$, $\frac{\Delta\mathbf{r}_{||}}{2} = (\frac{a}{2}, 0)$ or $\mathbf{r}_{i_{||}} = (0, y)$, $\frac{\Delta\mathbf{r}_{||}}{2} = (0, \frac{b}{2})$ for the $x(a)$ or $y(b)$ polarized one-dimensional states in Fig. 1, respectively, and $\frac{\Delta\theta_{||}}{2} = \frac{\pi}{2}$. This means, for every pair of one-dimensional electronic states along I_i and $\bar{I}_{j=i+l}$, with $l = 1$ or $l = -1$ for the neighboured states, it can be distinguished between four separated Hilbert subspaces $\overset{[-]}{\mathcal{H}}_{I_i,-\frac{\Delta\mathbf{r}_{||}}{2}}$, $\overset{[+]}{\mathcal{H}}_{I_i,+\frac{\Delta\mathbf{r}_{||}}{2}}$, $\overset{[-]}{\mathcal{H}}_{\bar{I}_j,+\frac{\Delta\mathbf{r}_{||}}{2}}$ and $\overset{[+]}{\mathcal{H}}_{\bar{I}_j,-\frac{\Delta\mathbf{r}_{||}}{2}}$. Here, $\overset{[-]}{\mathcal{H}}_{I_i,-\frac{\Delta\mathbf{r}_{||}}{2}}$ and $\overset{[+]}{\mathcal{H}}_{I_i,+\frac{\Delta\mathbf{r}_{||}}{2}}$ as well as $\overset{[-]}{\mathcal{H}}_{\bar{I}_j,+\frac{\Delta\mathbf{r}_{||}}{2}}$ and $\overset{[+]}{\mathcal{H}}_{\bar{I}_j,-\frac{\Delta\mathbf{r}_{||}}{2}}$ form the orthogonal complement to each other, respectively. Again, the coordinate systems are different and given by definite relative shifts of $\pm\frac{\Delta\mathbf{r}_{||}}{2}$.

Equivalently to the discussion above, different electronic populations of these particular states are possible. A spin-up and a spin-down electron within I_i or $\bar{I}_j$ can be simultaneously assigned, either to the − OR to the + lattice site, i.e. the two electrons belong either to $\overset{[-]}{\mathcal{H}}_{I_i,-\frac{\Delta\mathbf{r}_{||}}{2}}$ OR to $\overset{[+]}{\mathcal{H}}_{I_i,+\frac{\Delta\mathbf{r}_{||}}{2}}$ and correspondingly either to $\overset{[-]}{\mathcal{H}}_{\bar{I}_j,-\frac{\Delta\mathbf{r}_{||}}{2}}$ OR to $\overset{[+]}{\mathcal{H}}_{\bar{I}_j,+\frac{\Delta\mathbf{r}_{||}}{2}}$. In this case a diagonal spin compensation occurs again. On the other hand, the spin-up and the spin-down electrons may be locally separated so that they belong to the different lattice sites − and +, respectively, i.e. the spin-up electron belongs, for example, to $\overset{[-]}{\mathcal{H}}_{I_i,-\frac{\Delta\mathbf{r}_{||}}{2}}$ and the spin-down electron to $\overset{[+]}{\mathcal{H}}_{I_i,+\frac{\Delta\mathbf{r}_{||}}{2}}$, or vice versa. Correspondingly, the same applies to the $\mathcal{H}_{\bar{I}_j}$ subspaces. This leads again to an off-diagonal spin compensation/pairing.

4. The electronic state of filled bands

If the electron density of a CuO_2^{2-} plane is not reduced, the energetically upper valence bands are half-filled, but the energetic deeper lying bands are filled [7]. In the case of filled bands every − and + copper site is occupied by two electrons, associated to the particular band states of Eqs. (1) and (2), respectively. Therefore, no charge fluctuations occur within these bands; the influence of charge fluctuations is merely given by the charge fluctuations within the upper valence bands (see Section 5). This raises the question, in which way the spin-up and spin-down states are compensated within these filled band states. If the spin-up and spin-down states were compensated within every particular lattice site (diagonal spin compensation), i.e. within the − and + site, respectively, the − and + lattice sites would be completely decoupled/uncorrelated with respect to the spin interactions. This corresponds to the case where a particular $\mathbf{k}$ state belongs to same Hilbert subspace $\overset{[-]}{\mathcal{H}}$ or $\overset{[+]}{\mathcal{H}}$ with its associated spin state functions χ_- and χ_+, respectively, diagonal spin compensation. As already noticed, this is the ordinary case of locally spin paired closed shell states. However, this behaviour can not be expected for states with Cu-O bonding character, because it was deduced [7] that the self-consistent electronic renormalizations, which create the − and + copper sites should lead to a net bonding interaction between these two lattice sites. In particular, the tendency to maintain the total bonding energy for the Cu-O bonding states with $d_{x^2-y^2}$ orbital character as far as possible after renormalization preserves a net bonding interaction between these renormalized states, which belong to low energetic bands. Possibly, quantum state transformations similar as described in the Appendix may occur leading to remnant continuous covalent couplings, which is presently, however, still unproved. All in all, this favours paired spin states where the spins are separately associated to the − and + copper sites, i.e. an antiparallel orientation of the spin states at the − and + copper sites (Fig. 3,(3)). Thus, for every $\mathbf{k}$ state one kind of spins, for example the spin-down states, are favoured to be associated to the local states $\overset{[-]}{u}_{\mathbf{k}}$ and the spin-up states correspondingly to the $\overset{[+]}{u}_{\mathbf{k}}$ states, or (exclusive OR) vice versa, meaning the − and + lattice sites ideally carry only one spin state. Consequently, the spin compensation will occur between the − and + copper sites, corresponding to an off-diagonal spin compensation, but not on the same copper site. This requires the occupation of higher $\mathbf{k}$ states which belong to the second Brillouin zone within every subband. The latter, however, does not lead to particular problems, because diminutive overlap densities between the local states, the distances of the local states along the bond directions are $2a$ and $2b$, within each of the two particular bands result in a very high effective band mass ($m_{\mathbf{k}} \rightarrow \infty$), i.e. the bands representing the electronic states of Fig. 1 will be nearly dispersionless. The two dimensional band states with primarily $d_{x^2-y^2}$ orbital character of the copper parts belong to these low-energetic filled bands. The spin orbit coupling in conjunction with the planar Cu-O bonds of the $d_{x^2-y^2}$ orbital states can cause planar off-diagonal spin pairings resulting in an antiferromagnetically ordered spin state aligned within the CuO_2 plane, as experimentally found (see Section 5). This is a crucial difference to the ad hoc assumptions within other theoretical approaches where the antiferromagnetic spin couplings are generally attributed to half-filled valence bands [1, 3],[5, 6], [13]–[16]. The renormalized half-filled valence bands I_i, $\bar{I}_j$ in Fig. 1, however, prove to have no antiferromagnetic couplings (see Section 5). The same argumentation as given for the $d_{x^2-y^2}$ orbital states is also applicable for the low energetic band states with d_{z^2}, s orbital character, but the stronger localization of the d_{z^2} orbital within the x, y pane leads only to small renormalization influences of the Cu-O bonds and in combination with the additionally existing d_{z^2}, s hybridizations a pronounced two-dimensionally antiferromagnetic ordering of the spins can not be expected.

5. The dynamic quantum states of the undoped CuO_2 plane

In contrast to filled bands, disproportionate populations will occur within the half-filled one-dimensional valence bands I_i and $\bar{I}_j$. By closed shell Hartree-Fock calculations it could be shown that for the case of half filled bands the electrons are associated to one of the two different lattice sites in Fig. 1, leading to a charge and bonding fluctuation state (CBF) [7]. The $-$ copper site in Fig. 1 is assumed to be populated which corresponds simultaneously to a higher electronic charge of this site. In a previous paper it was deduced [7] that a negative electronic correlation energy per copper atom U_{eff} in the order of $-1\,eV$ per copper atom or even higher is generated, if the electronic system is transformed into a state of disproportionate electron densities, favouring the occupation of the $-$ copper site. Consequently, the electrons are driven to populate exclusively one of the two Hilbert subspaces in I_i and $\bar{I}_j$, i.e. per definition the $\overset{[-]}{\mathcal{H}}_{I_i,-\frac{\Delta\mathbf{r}_{||}}{2}}$ and $\overset{[-]}{\mathcal{H}}_{\bar{I}_j,+\frac{\Delta\mathbf{r}_{||}}{2}}$ subspace. Under this condition, the subspaces $\overset{[-]}{\mathcal{H}}_{I_i,-\frac{\Delta\mathbf{r}_{||}}{2}}$ and $\overset{[-]}{\mathcal{H}}_{\bar{I}_j,+\frac{\Delta\mathbf{r}_{||}}{2}}$ comprehend the two spin functions χ_+ and χ_- in their basis representation, realizing a diagonal spin compensation. In this case, the correlation energy U_{eff} has to be compared with half of the antiferromagnetic correlation energy per copper atom $U_{mag} = 4 \cdot U_{\mathbf{S}(-)\mathbf{S}(+)}$ for the case of a filled band, with $U_{\mathbf{S}(-)\mathbf{S}(+)}$ representing half of the antiferomagnetic spin correlation energy between the spin $\mathbf{S}(+)$ at the $+$ copper sites and the opposite spin $\mathbf{S}(-)$ at the $-$ copper site, i.e U_{mag} represents the energy lowering under off-diagonal populations/off-diagonal spin compensations in comparison to diagonal populations/diagonal spin compensations of a filled band. Although the energy $U_{\mathbf{S}(-)\mathbf{S}(+)}$ is not well defined at the moment the highest possible value at all can be estimated, which corresponds to states with $Cu-0$ bonding character. Néel temperatures were measured in the order of 300 K (25.85 meV) for systems with single CuO_2^{2-} planes and experimental values for an antiferromagnetic coupling constant J were deduced to be in the range of 120-165 meV [20][2] (see also Table I in Ref. [3]), for different high-T_c materials. These values lead to the general condition $|U_{eff}| > |U_{mag}|$. Therefore, the tendency to populate the $-$ copper site suppresses the equal population of the $-$ and $+$ copper sites for half-filled valence bands.

[2] The coupling constants J in Ref. [20] are related to a Heisenberg antiferromagnet with $S_{(+)}=S_{(-)}=\frac{1}{2}$, which proves to be too large in relation to the measured Néel temperatures. A Heisenberg antiferromagnet of the here deduced filled band states with off-diagonal spin compensation requires $S_{(+)}=S_{(-)}=1$. However, the experimentally deduced magnetic moments per copper atom are only $\mu = 0.3-0.5\,\mu_B$ [21]. The lowered magnetic moments can be attributed to the annihilation of the spin-up and spin-down spin densities within the oxygen parts of the two spin functions χ_+ and χ_- which are separately associated to the neighboured $\overset{[-]}{\mathrm{Cu}}$ and $\overset{[+]}{\mathrm{Cu}}$ site.

In conclusion, a diagonal spin compensation occurs within the half-filled one-dimensional valence band states in I_i,$\bar{I}_j$, which prevents an antiferromagnetic coupling within these bands (Fig. 3,(1)). This remains also valid for all populations of the bands I_i,$\bar{I}_j$ below half-filling.

The CBF state as given in Fig. 1 is, however, not stable in time, but is driven to turn into its mirror symmetric state $\overline{\text{CBF}}$, caused by the induced lattice deformations that create an additionally positive time dependent potential $V^{deff}(t)$ [8]. The influence of this time dependent positive potential has to be minimized which can be achieved by the solution of the equation of motion

$$i\hbar \frac{dC_i(t)}{dt} = \sum_{j=1,\bar{1}} H_{ij}(t) C_j(t) \tag{7}$$

with the Hamilton matrix elements $H_{ij}(t)$ and the coefficients $C_j(t)$, which are time dependent, where $1 = \text{CBF}$ and $\bar{1} = \overline{\text{CBF}}$. The complete time behaviour can be assigned to time dependent off-diagonal matrix elements $H_{ij}(i \neq j)$ and assuming time stationary diagonal Hamiltonians $H_{ij}(i = j)$ [8]. During the CBF $\leftrightarrow$ $\overline{\text{CBF}}$ transitions U_{eff} is rising but in opposite direction the deformation potential $V^{deff}(t)$ (phonon excitations) will be reduced. Under these conditions an energetic minimum is created which is given by the minimum conditions

$$\frac{dH_{1\bar{1}}}{dt'} = 0 \qquad \text{and} \qquad \frac{dH_{\bar{1}1}}{dt''} = 0. \tag{8}$$

with respect to the time coordinates t',t''. This leads to $t'_{min} = t_o + \tau_{CBF}$ and $t''_{min} = t_o + \tau_{CBF} + \tau_{\overline{CBF}} = t_o + 2\tau_{CBF}$, with t_o the time zero point and τ_{CBF} as well as $\tau_{\overline{CBF}} = \tau_{CBF}$ the lifetimes of the CBF and $\overline{\text{CBF}}$ state, respectively. Identical quantum states and time independent amplitudes will occur for

$$t = n \cdot 2\tau_{CBF} \qquad\qquad (n : integer). \tag{9}$$

In conclusion, the time independent quantum states Ψ_{CBF} and $\Psi_{\overline{CBF}}$, which reflect the total wave function of all the occupied one-dimensional subspaces $\{I_i,\bar{I}_j\}$, form the basis states for a stationary but dynamic quantum state Ψ_{DCBF} which is characterized by an internal time scale, the *eigentime* $\tau_{ei_{DCBF}} = 2\tau_{CBF}$. Hence, the quantum state Ψ_{DCBF} is a dynamic time-quantized but stationary quantum state with an internal time periodicity determined by $\tau_{ei_{DCBF}} = 2\tau_{CBF}$, i.e. Ψ_{DCBF} is strongly deterministic in time within the time period $\tau_{ei_{DCBF}}$.

The transformation of the Ψ_{CBF} state to the $\Psi_{\overline{CBF}}$ state, and vice versa, means that the particular one-dimensional states within I_i and $\bar{I}_j$ of each of the two arrays in Fig. 1 has to be transformed in the same way. This can be reached by the transformation of the Hilbert subspaces according to

$$\overset{[-]}{\mathcal{H}}_{I_i,-\frac{\Delta\mathbf{r}_{||}}{2}} \leftrightarrow \overset{[-]}{\mathcal{H}}_{I_i,+\frac{\Delta\mathbf{r}_{||}}{2}}, \tag{10}$$

$$\updownarrow \qquad\qquad \updownarrow$$

$$\overset{[-]}{\mathcal{H}}_{\bar{I}_j,+\frac{\Delta\mathbf{r}_{||}}{2}} \leftrightarrow \overset{[-]}{\mathcal{H}}_{\bar{I}_j,-\frac{\Delta\mathbf{r}_{||}}{2}} \tag{11}$$

(12)

and

$$\overset{[+]}{\mathcal{H}}_{I_i,+\frac{\Delta\mathbf{r}_{||}}{2}} \leftrightarrow \overset{[+]}{\mathcal{H}}_{I_i,-\frac{\Delta\mathbf{r}_{||}}{2}} \tag{13}$$

$$\updownarrow \qquad \updownarrow$$

$$\overset{[+]}{\mathcal{H}}_{\bar{I}_j,-\frac{\Delta\mathbf{r}_{||}}{2}} \leftrightarrow \overset{[+]}{\mathcal{H}}_{\bar{I}_j,+\frac{\Delta\mathbf{r}_{||}}{2}}\,. \tag{14}$$

Eqs. (10) and (13) as well as Eqs. (11) and (14) describe the transformation within the states I_i and $\bar{I}_j$, respectively. Effectively, the coordinate systems which are related to the periodically local states $\overset{[-]}{u}_k$ and $\overset{[+]}{u}_k$ have been exchanged, represented by the shifts $\pm\frac{\Delta\mathbf{r}_{||}}{2}$. During the transformation of the Hilbert spaces, the electrons which were initially placed at the $-$ lattice site are transferred to the initial $+$ lattice site, i.e. the CBF state is inverted. The electron transformation between the $-$ and $+$ lattice sites can not be described by a simple electron promotion but an exchange of the quantum states is necessary, because the local functions $\overset{[-]}{u}_{\mathbf{k}_{||}}$ and $\overset{[+]}{u}_{\mathbf{k}_{||}}$ depend on the population of the states themselves in a self-consistent manner. Therefore, the ordinary conception of static Hilbert spaces commonly underlying quantum theoretical descriptions has to be replaced by the assumption of Hilbert spaces being periodically transformed in time.

The CBF $\leftrightarrow$ $\overline{\text{CBF}}$ transitions have to occur as fast as possible, because the effective electronic correlation energy U_{eff} possesses its minimum value within the two particular CBF configurations and it is rising proportional to the time period of the existence of an intermediary state between the CBF and $\overline{\text{CBF}}$ state. If it was assumed that the particular one-dimensional states I_i, $\bar{I}_j$ are uncorrelated in time among themselves, the CBF $\leftrightarrow$ $\overline{\text{CBF}}$ transitions would be based on individual transitions within the particular one-dimensional states I_i, $\bar{I}_j$. If at all possible, such uncorrelated transitions would be time consuming leading to a large increase of U_{eff}. Over and above that, the CBF $\leftrightarrow$ $\overline{\text{CBF}}$ transitions would be merely the result of a mean field behaviour where the electrons within the existing CBF state, adopting the lowest U_{eff} value, would tend to prevent the transition to the other $\overline{\text{CBF}}$ state. All in all, the transformations of Eqs. (10)–(14) will be only conceivable, if there is a time correlation between the particular one-dimensional states I_i, $\bar{I}_j$. This can be realized by a time-correlated exchange of paired Hilbert spaces related to I_i and $\bar{I}_j$ according to the vertical arrows in Eqs. (10)–(14). Physically this corresponds to an exchange of the coordinate systems, i.e. an exchange of the $\pm\frac{\Delta\mathbf{r}_{||}}{2}$ coordinates, between the corresponding Hilbert spaces. The dynamically paired Hilbert spaces $\overset{[-]}{\mathcal{H}}_{I_i,-\frac{\Delta\mathbf{r}_{||}}{2}}$ and $\overset{[-]}{\mathcal{H}}_{\bar{I}_j,+\frac{\Delta\mathbf{r}_{||}}{2}}$ as well as $\overset{[+]}{\mathcal{H}}_{I_i,+\frac{\Delta\mathbf{r}_{||}}{2}}$ and $\overset{[+]}{\mathcal{H}}_{\bar{I}_j,-\frac{\Delta\mathbf{r}_{||}}{2}}$ are definitely entangled in space and time in this case. The time-periodic exchange coupling between these entangled Hilbert spaces leads to new dynamic but stationary quantum states with an inherent time scale $\tau_{ei_{DCBF}}$.

For strong correlated identical systems one can define an antisymmetric tensor product $\mathcal{H}_1 \otimes_a \mathcal{H}_1 = \mathcal{H}_1^{\otimes 2}$ for two coupled identical Hilbert spaces $\mathcal{H}_1$ according to the two particle state space $\mathcal{H}_1^{\otimes 2} = \{(\varphi \otimes_a \psi) : \varphi \in \mathcal{H}_1, \psi \in \mathcal{H}_1\}$ with $(\varphi \otimes_a \psi) = \frac{1}{2}(\varphi \otimes \psi - \psi \otimes \varphi)$ [22]. Correspondingly, it results for the one-dimensional states in I_i,$\bar{I}_j$

$$\overset{[-]}{\mathcal{H}}_{I_i} \otimes_a \overset{[-]}{\mathcal{H}}_{\bar{I}_j} \quad \text{and} \quad \overset{[+]}{\mathcal{H}}_{I_i} \otimes_a \overset{[+]}{\mathcal{H}}_{\bar{I}_j}, \tag{15}$$

where the two Hilbert spaces have the same basis representation, but different coordinate systems. The tensor products of Eq. (15) can exist for arbitrary i,j values, i.e. a two-dimensional entanglement of all the one-dimensional states in space and time with respect to their coordinate systems is reached for a given polarization direction. If a Hilbert space $\mathcal{H}_{DCBF}$ is defined as a compound system consisting of subspaces of Eq. (15) and if the particular subspaces of Eqs. (10)−(14) are considered to be basis states, normalized antisymmetric products can be defined according to

$$\overset{[-]}{\mathcal{H}}_{I_i,-\frac{\Delta\mathbf{r}_{||}}{2}} \otimes_a \overset{[-]}{\mathcal{H}}_{\bar{I}_j+\frac{\Delta\mathbf{r}_{||}}{2}} = \frac{1}{\sqrt{2}}(\overset{[-]}{\mathcal{H}}_{I_i,-\frac{\Delta\mathbf{r}_{||}}{2}} \otimes \overset{[-]}{\mathcal{H}}_{\bar{I}_j,+\frac{\Delta\mathbf{r}_{||}}{2}} - \overset{[-]}{\mathcal{H}}_{I_i,+\frac{\Delta\mathbf{r}_{||}}{2}} \otimes \overset{[-]}{\mathcal{H}}_{\bar{I}_j,-\frac{\Delta\mathbf{r}_{||}}{2}}) \tag{16}$$

and

$$\overset{[+]}{\mathcal{H}}_{I_i,+\frac{\Delta\mathbf{r}_{||}}{2}} \otimes_a \overset{[+]}{\mathcal{H}}_{\bar{I}_j,-\frac{\Delta\mathbf{r}_{||}}{2}} = \frac{1}{\sqrt{2}}(\overset{[+]}{\mathcal{H}}_{I_i,+\frac{\Delta\mathbf{r}_{||}}{2}} \otimes \overset{[+]}{\mathcal{H}}_{\bar{I}_j,-\frac{\Delta\mathbf{r}_{||}}{2}} - \overset{[+]}{\mathcal{H}}_{I_i,-\frac{\Delta\mathbf{r}_{||}}{2}} \otimes \overset{[+]}{\mathcal{H}}_{\bar{I}_j,+\frac{\Delta\mathbf{r}_{||}}{2}}). \tag{17}$$

Here, the Hilbert subspaces appear as state vectors, because the antisymmetric product is the result of the exchange of the Hilbert subspaces as a whole, which corresponds to an exchange of the coordinate systems. In the state representation of Eqs. (16) and (17) the time, i.e. the eigentime $\tau_{ei_{DCBF}}$, does not occur, i.e. these antisymmetric products represent time integrated states over the time period $\tau_{ei_{DCBF}}$. This is in opposite to the usually applied principle in quantum mechanics which postulates an objective simultaneity of the states. The eigentime $\tau_{ei_{DCBF}}$ proves to be a real existing and quantized quantity within a stationary DCBF state, but within the pair states given by Eqs. (16) and (17) the eigentime $\tau_{ei_{DCBF}}$ occurs merely as a hidden variable. However, the occurrence of the pair states of Eqs. (16) and (17) themselves is strongly coupled to the existence of the eigentime $\tau_{ei_{DCBF}} = 2 \cdot \tau_{CBF}$. The two product wave functions given by Eqs. (16) and (17) must have the same probability, i.e. they exist for the same lifetime τ_{CBF}, including a complementary deterministic time behaviour within the period $\tau_{ei_{DCBF}}$, and Eqs. (16) and (17) are produced exclusively by exchange processes of quantum states. All pair states represented by Eq. (15) have the same probability to form all possible I_i, $\bar{I}_j$ pairings, where possible pairings at a given time t can perpetually be exchanged without any mass inertia effect, meaning these pair exchanges can occur infinitely fast. So, the complete compound state is given by the Tensor products

$$\overset{[-]}{\mathcal{H}}_{I_1} \otimes_a \overset{[-]}{\mathcal{H}}_{\bar{I}_1} \otimes_a .....\otimes_a \overset{[-]}{\mathcal{H}}_{I_n} \otimes_a \overset{[-]}{\mathcal{H}}_{\bar{I}_n} \qquad \overset{[+]}{\mathcal{H}}_{I_1} \otimes_a \overset{[+]}{\mathcal{H}}_{\bar{I}_1} \otimes_a .....\otimes_a \overset{[+]}{\mathcal{H}}_{I_n} \otimes_a \overset{[+]}{\mathcal{H}}_{\bar{I}_n} \tag{18}$$

for the two polarization directions x and y, respectively. The number n corresponds to the half of the one-dimensional states of a given polarization direction. The state vectors within the antisymmetric tensor product of Eq. (18) result from the simultaneous exchange of all particular Hilbert subspaces $\mathcal{H}_{I_i}$, $\mathcal{H}_{\bar{I}_j}$. There are $n!$ independent possibilities of simultaneous exchanges of paired Hilbert subspaces which have to be included in Eq. (18), i.e. Eq. (18) is restricted to these $n!$ possibilities of simultaneous pair exchanges.

Horizontal transformations of the Hilbert subspaces of Eqs. (10)-(14), depicted by horizontal arrows, correspond to transformations of the individual coordinate systems of the Hilbert subspaces along the particular one-dimensional states where the electrons follow the movements of the coordinate systems. The transformation/exchange of the quantum states can be assumed to be infinitely fast, however, the electron movement between two lattice sites is finite in time, which is given by the mass inertia of the transferred electrons. Hence, the exchange of the Hilbert subspaces occurring with a finite transition time t'_{tr} has to be taken into consideration, leading to the time dependence of the definitely correlated Hilbert subspaces given by:

$$\overset{[-]}{\mathcal{H}}_{I_i}(t') = c_1(t')\cdot \overset{[-]}{\mathcal{H}}_{I_i,-\frac{\Delta\mathbf{r}_{||}}{2}} \oplus c_2(t')\cdot \overset{[-]}{\mathcal{H}}_{I_i,+\frac{\Delta\mathbf{r}_{||}}{2}}, \tag{19}$$

$$\overset{[-]}{\mathcal{H}}_{\bar{I}_j}(t') = c_1(t')\cdot \overset{[-]}{\mathcal{H}}_{\bar{I}_j,+\frac{\Delta\mathbf{r}_{||}}{2}} \oplus c_2(t')\cdot \overset{[-]}{\mathcal{H}}_{\bar{I}_j,-\frac{\Delta\mathbf{r}_{||}}{2}}, \tag{20}$$

$$\overset{[+]}{\mathcal{H}}_{I_i}(t') = c_1(t')\cdot \overset{[+]}{\mathcal{H}}_{I_i,+\frac{\Delta\mathbf{r}_{||}}{2}} \oplus c_2(t')\cdot \overset{[+]}{\mathcal{H}}_{I_i,-\frac{\Delta\mathbf{r}_{||}}{2}}, \tag{21}$$

$$\overset{[+]}{\mathcal{H}}_{\bar{I}_j}(t') = c_1(t')\cdot \overset{[+]}{\mathcal{H}}_{\bar{I}_j,-\frac{\Delta\mathbf{r}_{||}}{2}} \oplus c_2(t')\cdot \overset{[+]}{\mathcal{H}}_{\bar{I}_j,+\frac{\Delta\mathbf{r}_{||}}{2}}, \tag{22}$$

$$\text{with} \quad 1 = c_1(t') + c_2(t'). \tag{23}$$

If a linear time dependence of the expectation values of the states during the exchange process is assumed the coefficients follow a periodic time behaviour

$$c_1(t') = \begin{cases} 1 & & t'_{tr}/2 \le t' < \tau_{CBF} - t'_{tr}/2 \\ \frac{\tau_{CBF}-t'}{t'_{tr}} + \frac{1}{2} & \text{for} & \tau_{CBF} - t'_{tr}/2 \le t' < \tau_{CBF} + t'_{tr}/2 \\ 0 & & \tau_{CBF} + t'_{tr}/2 \le t' < 2\tau_{CBF} - t'_{tr}/2 \\ \frac{t'-(2\tau_{CBF}-t'_{tr})}{t'_{tr}} - \frac{1}{2} & & 2\tau_{CBF} - t'_{tr}/2 \le t' < 2\tau_{CBF} + t'_{tr}/2 \end{cases}, \tag{24}$$

with $t' = t - n \cdot \tau_{ei_{DCBF}} \in [0, \tau_{ei_{DCBF}}]$ and n an integer.

The representation of Eqs. (19)-(22) by a direct sum describing a superposition of quantum states without interferences within a normalized state according to Eq. (23) results from the following arguments: A superposition of quantum states which includes interference terms requires a free accessibility of the quantum states at any time t, reflecting the simultaneity of superimposing quantum states, but additionally, a common Hilbert space for the superimposing states with a unified coordinate system has stringently to be established. Assuming Eqs. (16) and (17) to be basically fulfilled within every time period $\tau_{ei_{DCBF}}$, the time dependence of the Hilbert spaces $\overset{[-]}{\mathcal{H}}_{I_i}(t')..\overset{[+]}{\mathcal{H}}_{\bar{I}_j}(t')$ is exclusively caused by exchanges of the coordinate systems between the one-dimensional states in I_i and $\bar{I}_j$ without any excitation, according to the vertical arrows in Eqs. (10)-(14), i.e. the Hilbert spaces are exchanged. At any time t', an electron with the amplitudes $c_1(t')$ and $c_2(t')$ simultaneously belongs to the two superimposing Hilbert spaces, respectively, but the requirement of identical coordinate systems is never given, which is caused by the time-deterministic and time-quantized exchange of Hilbert subspaces under maintaining of a definite coordinate relative state. As a result, interference terms of the type $c_1(t')c_2(t')\varphi_{\frac{\Delta\mathbf{r}_{||}}{2}}\varphi_{-\frac{\Delta\mathbf{r}_{||}}{2}}$ disappear at any time t', which requires a direct sum of the superimposing quantum states.

The time dependence of the coordinate systems is given by the periodic time dependence of the expectation value of the point of origin $\frac{\delta\mathbf{r}_{||}}{2}(t')$ according to:

$$\frac{\delta\mathbf{r}_{||}}{2}(t')[\overset{[-]}{\mathcal{H}}_{I_i}] = \frac{\Delta\mathbf{r}_{||}}{2}(c_2(t') - c_1(t')) \tag{25}$$

$$\frac{\delta\mathbf{r}_{||}}{2}(t')[\overset{[-]}{\mathcal{H}}_{\bar{I}_j}] = \frac{\Delta\mathbf{r}_{||}}{2}(c_1(t') - c_2(t')) \tag{26}$$

$$\frac{\delta\mathbf{r}_{||}}{2}(t')[\overset{[+]}{\mathcal{H}}_{I_i}] = \frac{\Delta\mathbf{r}_{||}}{2}(c_1(t') - c_2(t')) \tag{27}$$

$$\frac{\delta\mathbf{r}_{||}}{2}(t')[\overset{[+]}{\mathcal{H}}_{\bar{I}_i}] = \frac{\Delta\mathbf{r}_{||}}{2}(c_2(t') - c_1(t')) \tag{28}$$

The individual coordinate shifts of the Hilbert subspaces correlate with phase shifts according to

$$\Delta\theta_{||} = \frac{2\pi}{2a}\cdot\Delta\mathbf{r}_{||} = \mathbf{k}_{||_{ei_{DCBF}}}\cdot\Delta\mathbf{r}_{||} = \frac{2\pi}{\tau_{ei_{DCBF}}}\cdot\tau_{CBF} = \omega_{||_{ei_{DCBF}}}\cdot\tau_{CBF} \tag{29}$$

or

$$\Delta\theta_{||} = \frac{2\pi}{2b}\cdot\Delta\mathbf{r}_{||} = \mathbf{k}_{||_{ei_{DCBF}}}\cdot\Delta\mathbf{r}_{||} = \frac{2\pi}{\tau_{ei_{DCBF}}}\cdot\tau_{CBF} = \omega_{||_{ei_{DCBF}}}\cdot\tau_{CBF} \tag{30}$$

which leads to:

$$\frac{\delta\theta_{||}}{2}(t')[\overset{[-]}{\mathcal{H}}_{I_i}] = \frac{\Delta\theta_{||}}{2}(c_2(t')-c_1(t')) \tag{31}$$

$$\frac{\delta\theta_{||}}{2}(t')[\overset{[-]}{\mathcal{H}}_{\bar{I}_j}] = \frac{\Delta\theta_{||}}{2}(c_1(t')-c_2(t')) \tag{32}$$

$$\frac{\delta\theta_{||}}{2}(t')[\overset{[+]}{\mathcal{H}}_{I_i}] = \frac{\Delta\theta_{||}}{2}(c_1(t')-c_2(t')) \tag{33}$$

$$\frac{\delta\theta_{||}}{2}(t')[\overset{[+]}{\mathcal{H}}_{\bar{I}_i}] = \frac{\Delta\theta_{||}}{2}(c_2(t')-c_1(t')), \tag{34}$$

for the corresponding coordinate center point phase $\frac{\delta\theta_{||}}{2}(t')$. The Bloch states within the time dependent Hilbert spaces given by Eqs. (19)−(22) are defined now with respect to the expectation values of the local periodic function for a given time t' according to

$$\overset{[-]}{\acute{u}}_{\mathbf{k}_{||},I_i}(\mathbf{r}_{||}+\frac{\delta\mathbf{r}_{||}(t')}{2}[\overset{[-]}{\mathcal{H}}_{I_i}]) = c_1(t')\cdot\overset{[-]}{u}_{\mathbf{k}_{||},I_i}(\mathbf{r}_{||}-\frac{\Delta\mathbf{r}_{||}}{2}) \oplus c_2(t')\cdot\overset{[-]}{u}_{\mathbf{k}_{||},I_i}(\mathbf{r}_{||}+\frac{\Delta\mathbf{r}_{||}}{2}) \tag{35}$$

with respect to the time dependent Hilbert space of Eq. (19) and with the expectation value of the coordinate center point $\delta\mathbf{r}_{||}(t')$ as given in Eq. (25). Equivalent representations result for the Eqs. (20)−(22). Exchanged are only the local states $\overset{[-]}{u}_{\mathbf{k}_{||},I_i}(\mathbf{r}_{||}-\frac{\Delta\mathbf{r}_{||}}{2})$ and $\overset{[-]}{u}_{\mathbf{k}_{||},I_i}(\mathbf{r}_{||}+\frac{\Delta\mathbf{r}_{||}}{2})$, but not the exponential function of the Bloch states. Nevertheless, the exponents within the exponential part of the Bloch states are continually symmetry broken in relation to the center coordinate $\mathbf{r}_{o_{||}}$ in Fig. 1. The actual phase will be given by the expectation value $\frac{\delta\theta_{||}}{2}(t')$ corresponding to the effective local state function $\overset{[-]}{\acute{u}}_{\mathbf{k}_{||},I_i}(\mathbf{r}_{||}+\frac{\delta\mathbf{r}_{||}(t')}{2})$. Therefore, the common Bloch state will adopt the general form:

$$\begin{aligned}\overset{[-]}{\varphi}_{\mathbf{k}_{||},I_i}(\mathbf{r}_{||},\frac{\delta\mathbf{r}_{||}(t)}{2}[\overset{[-]}{\mathcal{H}}_{I_i}]) &= e^{i(\mathbf{k}_{||}\mathbf{r}_{||}+\frac{\delta\theta_{||}(t)}{2}[\overset{[-]}{\mathcal{H}}_{I_i}])}\,\overset{[-]}{\acute{u}}_{\mathbf{k}_{||},I_i}(\mathbf{r}_{||}+\frac{\delta\mathbf{r}_{||}(t)}{2}[\overset{[-]}{\mathcal{H}}_{I_i}]) \\ &= e^{i(\mathbf{k}_{||}\mathbf{r}_{||}+\mathbf{k}_{||_{ei_{DCBF}}}\frac{\delta\mathbf{r}_{||}(t)}{2}[\overset{[-]}{\mathcal{H}}_{I_i}])}\,\overset{[-]}{\acute{u}}_{\mathbf{k}_{||},I_i}(\mathbf{r}_{||}+\frac{\delta\mathbf{r}_{||}(t)}{2}[\overset{[-]}{\mathcal{H}}_{I_i}])\end{aligned} \tag{36}$$

where $\frac{\delta\mathbf{r}_{||}(t)}{2}$ is now transformed to a continuous coordinate which is related to the absolute time coordinate t, taking into account that $\overset{[-]}{u}_{\mathbf{k}_{||},I_i}(\mathbf{r}_{||}-\frac{\Delta\mathbf{r}_{||}}{2})$ is equivalent to $\overset{[-]}{u}_{\mathbf{k}_{||},I_i}(\mathbf{r}_{||}+\frac{3\Delta\mathbf{r}}{2})$.

Conventionally, the requirement for the existence of stationary quantum states is determined by the time independence of the expectation values where the expectation value is assumed to be instantaneously given at any time. The basis representation of the corresponding Hilbert space is then required to be time independent, which is the traditional foundation of quantum mechanics. The dynamic Hilbert space $\mathcal{H}_{DCBF}$ is formed by the paired one-dimensional subspaces according to Eqs. (15) and (18). The particular basis states, i.e. the basis of the particular subspaces $\overset{[-]}{\mathcal{H}}_{I_i}(t')$, $\overset{[+]}{\mathcal{H}}_{I_i}(t')$, $\overset{[-]}{\mathcal{H}}_{\bar{I}_j}(t')$ and $\overset{[+]}{\mathcal{H}}_{\bar{I}_j}(t')$, are not stationary in time, as discussed, but behave strongly periodic in time according to Eqs. (19)-(24). The condition for stationarity of the states is now defined by the requirement that the expectation values must be identical for periodically equivalent times. This means, after the time interval $\tau_{ei_{DCBF}} = 2\tau_{CBF}$ the basis representation of all the subspaces $\{\overset{[-]}{\mathcal{H}}_{I_i}(t'), \overset{[+]}{\mathcal{H}}_{I_i}(t'), \overset{[-]}{\mathcal{H}}_{\bar{I}_j}(t'), \overset{[+]}{\mathcal{H}}_{\bar{I}_j}(t')\}$ must be identically recreated. Hence, for every electron belonging to the dynamic DCBF state the eigentime $\tau_{ei_{DCBF}}$ occurs as a conservation quantity. If the conservation of $\tau_{ei_{DCBF}}$ is violated, i.e. if the strong time-periodicity of the Hilbert subspaces according to Eqs. (19)-(24) is not given, the two summands in Eq. (16) and Eq. (17) are differently weighted, i.e. pair states according to Eq. (16) and Eq. (17) do not exist. The two dimensional coupling in time of the particular one-dimensional subspaces is then broken up and, in consequence, the DCBF state can not exist.

A time dependence of the Hilbert subspaces means that the periodically local states $\overset{[-]}{\acute{u}}_{\mathbf{k}_{||}}$ and $\overset{[+]}{\acute{u}}_{\mathbf{k}_{||}}$ are time dependent corresponding to the time dependent amplitudes $c_1(t'), c_2(t')$ according to Eq. (35). Hence, the two one-dimensional subspaces in I_i and $\bar{I}_j$ are strongly time-correlated multi-electronic pair states with respect to their time dependent phases $\delta\theta_{||}(t')$ of Eqs. (31) - (34), but not yet in relation to the their particular $\mathbf{k}_{||}$ states.

The strongly time-periodic phase $\delta\theta_{||}(t')$ prevents that this phase can appear as a free variable. The phase $\delta\theta_{||}(t')$ is the only existing phase under half-filling, i.e. every accidental electronic transition from an occupied local site $\overset{[-]}{u}_{\mathbf{k}_{||}}$ to free states of the local site $\overset{[+]}{u}_{\mathbf{k}_{||}}$ is connected with changes of the phase $\Delta\theta_{||}$ by $\pm\pi$ (Eqs. (3)-(6)). In consequence, every local electronic transition which is not time dependent according to the amplitudes $c_1(t'), c_2(t')$ is forbidden. In addition, any single elctronic excitation of $\mathbf{k}_{||}$ states within the filled bands of the subspaces $\overset{[-]}{\mathcal{H}}_{I_i}$ or $\overset{[-]}{\mathcal{H}}_{\bar{I}_j}$ also requires electronic transitions to local states $\overset{[+]}{u}_{\mathbf{k}_{||}}$, because, the closed shell electronic state representing a diagonal spin pairing within $\overset{[-]}{\mathcal{H}}_{I_i}$ or $\overset{[-]}{\mathcal{H}}_{\bar{I}_j}$ is removed in favour to open shell states, which include off-diagonal spin pairings. Moreover, the electronic excitations are strongly local due to the large effective band mass $m_{\mathbf{k}_{||}} \to \infty$. Even if delocalized $\mathbf{k}_{||}$ excitations were assumed, the accompanied modulation of the electron density would include the population of local states $\overset{[+]}{u}_{\mathbf{k}_{||}}$, because the local states $\overset{[-]}{u}(\mathbf{R}_i)$ at the lattice sites $\mathbf{R}_i$ are completely filled. In either case, this requires local transitions from a filled local state $\overset{[-]}{u}_{\mathbf{k}_{||}}$ to an empty local state $\overset{[+]}{u}_{\mathbf{k}_{||}}$. In this case, the time quantization of the states given by $\tau_{ei_{DCBF}}$ would be violated again.

In consequence, a hole undoped CuO_2^{2-} plane, representing half-filled valence bands, is an insulator as it has always experimentally been found.

6. Separated worlds – unified worlds

Symmetry broken Bloch states according to Eq. (36) implicate a commutator based on the space and impulse operator $\delta\mathbf{r}_{||}(t)$ and $\frac{\hbar}{i}\frac{\partial}{\partial(\delta\mathbf{r}_{||})}$ which describes the dynamics of the coordinate system given by

$$[\delta\mathbf{r}_{||}(t), \frac{\hbar}{i}\frac{\partial}{\partial(\delta\mathbf{r}_{||})}] = i\hbar, \tag{37}$$

implying the uncertainty relation

$$\Delta(\delta\mathbf{r}_{||}(t))_{\mathbf{k}_{||_{ei_{DCBF}}}} \cdot \Delta p_{\mathbf{k}_{||_{ei_{DCBF}}}} \geq \frac{\hbar}{2}, \tag{38}$$

with $\Delta p_{\mathbf{k}_{||_{ei_{DCBF}}}}$ representing the impulse uncertainty of a dynamic coordinate system, with $\Delta p_{\mathbf{k}_{||_{ei_{DCBF}}}} = 0$ for an infinitely extended CuO_2 plane. This noncommutative behaviour seems to be very surprising at first glance, because $\delta\mathbf{r}_{||}(t')$ is deterministic in time which should lead to a classical relation between space coordinate and impulse, i.e. one would expect $\delta\mathbf{r}_{||}(t)$ and $\frac{\hbar}{i}\frac{\partial}{\partial(\delta\mathbf{r}_{||})}$ to commute. The phenomenon can be explained by the cyclic behaviour of the absolute time t, $t = t' + n \cdot \tau_{||ei_{DCBF}}$ ($\tau_{||ei_{DCBF}} \equiv \tau_{ei_{DCBF}}$), which leads to identical quantum states. The wave function in Eq. (36) is also cyclic with respect to the space coordinate, $\delta\mathbf{r}(t) = \delta\mathbf{r}(t') + m \cdot (2\Delta\mathbf{r}_{||})$. The numbers m and n are not necessarily equivalent, but can adopt completely different values. Therefore, at a given time t the space coordinate $\delta\mathbf{r}_{||}(t)$ can have all values $\delta\mathbf{r}_{||}(t')$ modulo $2\Delta\mathbf{r}_{||}$. The relative coordinate systems of the particular Hilbert spaces $\overset{[-]}{\mathcal{H}}_{I_i}$, $,\overset{[-]}{\mathcal{H}}_{\bar{I}_j}$, $\overset{[+]}{\mathcal{H}}_{I_i}$,$\overset{[+]}{\mathcal{H}}_{\bar{I}_j}$ are definitely entangled in space and time, but the absolute reference coordinate system given by $\mathbf{r}_{o_{||}}$ is uncertain for multiples of the time quantization $\tau_{||ei_{DCBF}}$ and the space quantization $2\Delta\mathbf{r}_{||}$, respectively. In brief, the total common quantum state encompasses a classical space-time behaviour. In view of this extraordinary characteristic, it is only justified to compare the quantum properties for times t modulo $\tau_{||_{ei_{DCBF}}}$ and for space coordinates $\delta\mathbf{r}_{||}$ modulo $2\Delta\mathbf{r}_{||}$ or one considers quantum states that are integrated states over the time period $\tau_{||_{ei_{DCBF}}}$ as it is implicitly done in Eqs. (16) and (17). The latter requirement reflects the fact that the complete symmetry of the stationary quantum state is only defined after the time period $\tau_{||_{ei_{DCBF}}}$. A very important result is that at any time t the total quantum state is self-consistently separated in particular branches which are assigned to four entangled worlds, i.e. entangled by the unified time function Eq. (24), represented by four distinguishable Hilbert spaces $\overset{[-]}{\mathcal{H}}_{I_i}$, $,\overset{[-]}{\mathcal{H}}_{\bar{I}_j}$, $\overset{[+]}{\mathcal{H}}_{I_i}$,$\overset{[+]}{\mathcal{H}}_{\bar{I}_j}$. The state vectors of the complete compound system representing entangled states in Eq. (18) do not explicitly include the eigentime $\tau_{||_{ei_{DCBF}}}$, i.e. these state vectors describe the whole world time independent. The internal time $\tau_{||_{ei_{DCBF}}}$ of these state vectors exists merely as a hidden variable.

The eigentimes introduced in this paper describe periodic movements of the coordinate systems of the quantum states. This may also occur in ordinary Bloch states which adapt the crystal symmetry, i.e. states without broken U1 symmetry. A given eigentime τ_{ei} would be related to the coordinate transformation from a given lattice site to the next one. This, however, coincides with the general uncertainty of the space coordinate which is given by $\mathbf{r}$ modulo $\Delta\mathbf{r}$, where $\Delta\mathbf{r}$ corresponds to one lattice spacing. If the coordinate system is moved from lattice site i to lattice site $i+1$ the eigentime τ_{ei} can be simultaneously defined with respect to the lattice site i or to the lattice site $i+1$, or to any other lattice site $i+k$. In consequence, the eigentime τ_{ei} is completely indefinite. Hence, it is not appropriate to define the eigentime as a variable due to its arbitrariness. If eigentimes possibly exist they will merely occur as hidden indefinite variables which can not be detected. In consequence, the definition of non-symmetry broken Bloch states within a stationary Hilbert space when assuming an unified coordinate system is justified. The energy of the Bloch states is decisively determined by time independent transition matrix elements $H_{i,i\pm1}$ between the local site i and the neighboured local sites $i \pm 1$. This means, an electron belongs simultaneously to the lattice site i and to the lattice sites $i \pm 1$. This simultaneity, or better the simultaneous accessibility of the states i and $i \pm 1$ is not queried, if one assumes indefinite eigentimes in ordinary Bloch states, i.e. the existence of dynamic but probabilistically varying coordinate systems. The complete uncertainty of the coordinate systems guarantees a simultaneous accessibility of different lattice sites at any time t. Although the description of all Bloch states can be done within a common and time stationary Hilbert space every particular Bloch state could thus also possess its own coordinate system, i.e. in reality the whole quantum system could represent a *hidden* dynamic many-worlds quantum state. Therefore, the high-T_c materials may offer the chance to get more clarity about the many-worlds problematic [23]-[26].

7. Dynamic quantum states of the CuO_2 plane under hole doping

The individual orthonormalized local states within the periodic functions $\overset{[-]}{u}_{\mathbf{k}_{||}}$ and $\overset{[+]}{u}_{\mathbf{k}_{||}}$ of the valence states comprehend three atomic wave functions, a copper orbital state being $s, d_{x^2-x^2}, d_{z^2}$ hybridized and two oxygen orbitals $p_{||}$ that correspond to the oxygen atoms on the left and right hand side of the included copper atom along a one-dimensional state in I_i or $\bar{I}_j$. This is visualized by an assumed probability distribution $u^2_{o_{||}}$ in Fig. 2(a). The shifts $\pm\frac{\Delta\mathbf{r}_{||}}{2}$ of the particular coordinate systems within the Bloch states of the Eqs. (3)-(6) are related to the positions of the copper atoms. This reflects the fact that the left and right positioned oxygen atoms contribute to the same amount to a given local state, and the periodicity of the local states is determined by the copper positions of the corresponding copper sublattices $-$ and $+$, respectively. In more common terms, the probability to find an electron at a given oxygen site is the same for all oxygen atoms at any time t, i.e. the left and right oxygen atoms are indistinguishable with respect to their populations. As a result, the coordinate center point is identical with the center of mass of the local state, i.e. given by the copper site within a $O_{||} - Cu - O_{||}$ bonding-unit. This remains valid as long as the electronic states are independent/uncorrelated Bloch states, which is given for the hole undoped uncorrelated half-filled one-dimensional valence bands within the subspaces $\overset{[-]}{\mathcal{H}}_{I_i}, \overset{[-]}{\mathcal{H}}_{\bar{I}_j}$.

The situation will distinctly be changed if ordered topological hole states are formed, caused by reduced electronic populations, i.e. far from half-filling. An undisturbed topological hole ordering as created by the b-holes in Fig. 1 occurs for hole concentrations of $0.125 \leq n_h \leq 0.25$ holes/copper. The f-holes, occurring exclusively for $n_h > 0.125$ holes/copper, also tend to ordered states, as given in Fig. 1, but a clustering or an increased periodicity has to be expected which is inverse proportional to $n_h (< 0.25)$ holes/copper. In the subsequent considerations, infinitely extended ordered topological hole structures are considered, i.e. hole structures with periodic boundary conditions. Such extended strongly ordered topological hole states, clearly given for the b-holes within the above concentration range, are created by coherent interferences of $\mathbf{k}_{||}, -\mathbf{k}_{||}$ paired Bloch states according to [8]:

$$\begin{aligned}
\overset{[-]}{\varphi}_{\substack{\mathbf{k}_{||},-\Delta\phi_{S_{||}},\\ \Delta\phi_{||},-\Delta\phi_{S_\perp}}}(\mathbf{r}_{1_{||}}+\Delta\mathbf{r}_{\Delta\phi_{||}}) &= \sum_l c_l e^{i[(\mathbf{k}_{||}+l\mathbf{q}_{||})\mathbf{r}_{1_{||}}+\phi_{S_{||}}-\Delta\phi_{S_{||}}+\Delta\phi_{||}+\phi_{S_\perp}-\Delta\phi_{S_\perp}]} \\
&\cdot\ e^{i\Delta\theta_{\Delta\phi_{||}}}\overset{[-]}{u}_{\mathbf{k}_{||}+l\mathbf{q}_{||},-\Delta\phi_{S_{||}},\Delta\phi_{||},-\Delta\phi_{S_\perp}}(\mathbf{r}_{1_{||}}+\Delta\mathbf{r}_{\Delta\phi_{||}}) \\
&= \sum_l c_l \overset{[-]^l}{\varphi}_{\mathbf{k}_{||},-\Delta\phi_{S_{||}},\Delta\phi_{||},-\Delta\phi_{S_\perp}}(\mathbf{r}_{1_{||}}+\Delta\mathbf{r}_{\Delta\phi_{||}}) \\
&\in \overset{[-]}{\mathcal{H}}_{I_i,\Delta\mathbf{r}_{\Delta\phi_{||}}} \qquad (39)
\end{aligned}$$

$$\begin{aligned}
\overset{[-]}{\varphi}_{\substack{-\mathbf{k}_{||},-\Delta\phi_{S_{||}},\\ -\Delta\phi_{||},-\Delta\phi_{S_\perp}}}(\mathbf{r}_{2_{||}}+\Delta\mathbf{r}_{-\Delta\phi_{||}}) &= \sum_l c_l e^{i[(-\mathbf{k}_{||}-l\mathbf{q}_{||})\mathbf{r}_{2_{||}}+\phi_{S_{||}}-\Delta\phi_{S_{||}}-\Delta\phi_{||}+\phi_{S_\perp}-\Delta\phi_{S_\perp}]} \\
&\cdot\ e^{i\Delta\theta_{-\Delta\phi_{||}}}\overset{[-]}{u}_{-\mathbf{k}_{||}-l\mathbf{q}_{||},-\Delta\phi_{S_{||}},-\Delta\phi_{||},-\Delta\phi_{S_\perp}}(\mathbf{r}_{2_{||}}+\Delta\mathbf{r}_{-\Delta\phi_{||}}) \\
&= \sum_l c_l \overset{[-]^{-l}}{\varphi}_{-\mathbf{k}_{||},-\Delta\phi_{S_{||}},-\Delta\phi_{||},-\Delta\phi_{S_\perp}}(\mathbf{r}_{2_{||}}+\Delta\mathbf{r}_{-\Delta\phi_{||}}) \\
&\in \overset{[-]}{\mathcal{H}}_{I_i,\Delta\mathbf{r}_{-\Delta\phi_{||}}} \qquad (40)
\end{aligned}$$

$$\begin{aligned}
\overset{[-]}{\varphi}_{\substack{\mathbf{k}_{||},\Delta\phi_{S_{||}},\\ \Delta\phi_{||},\Delta\phi_{S_\perp}}}(\mathbf{r}_{3_{||}}+\Delta\mathbf{r}_{\Delta\phi_{||}}) &= \sum_l c_l e^{i[(\mathbf{k}_{||}+l\mathbf{q}_{||})\mathbf{r}_{3_{||}}+\phi_{S_{||}}+\Delta\phi_{S_{||}}+\Delta\phi_{||}+\phi_{S_\perp}+\Delta\phi_{S_\perp}]} \\
&\cdot\ e^{i\Delta\theta_{\Delta\phi_{||}}}\overset{[-]}{u}_{\mathbf{k}_{||}+l\mathbf{q}_{||},\Delta\phi_{S_{||}},\Delta\phi_{||},\Delta\phi_{S_\perp}}(\mathbf{r}_{3_{||}}+\Delta\mathbf{r}_{\Delta\phi_{||}}) \\
&= \sum_l c_l \overset{[-]^l}{\varphi}_{\mathbf{k}_{||},\Delta\phi_{S_{||}},\Delta\phi_{||},\Delta\phi_{S_\perp}}(\mathbf{r}_{3_{||}}+\Delta\mathbf{r}_{\Delta\phi_{||}}) \\
&\in \overset{[-]}{\mathcal{H}}_{I'_i,\Delta\mathbf{r}_{\Delta\phi_{||}}} \qquad (41)
\end{aligned}$$

$$\begin{aligned}
\overset{[-]}{\varphi}_{\substack{-\mathbf{k}_{||},\Delta\phi_{S_{||}},\\ -\Delta\phi_{||},\Delta\phi_{S_\perp}}}(\mathbf{r}_{4_{||}}+\Delta\mathbf{r}_{-\Delta\phi_{||}}) &= \sum_l c_l e^{i[(-\mathbf{k}_{||}-l\mathbf{q}_{||})\mathbf{r}_{4_{||}}+\phi_{S_{||}}+\Delta\phi_{S_{||}}-\Delta\phi_{||}+\phi_{S_\perp}+\Delta\phi_{S_\perp}]} \\
&\cdot\ e^{i\Delta\theta_{-\Delta\phi_{||}}}\overset{[-]}{u}_{-\mathbf{k}_{||}-l\mathbf{q}_{||},\Delta\phi_{S_{||}},-\Delta\phi_{||},\Delta\phi_{S_\perp}}(\mathbf{r}_{4_{||}}+\Delta\mathbf{r}_{-\Delta\phi_{||}}) \\
&= \sum_l c_l \overset{[-]^{-l}}{\varphi}_{-\mathbf{k}_{||},\Delta\phi_{S_{||}},-\Delta\phi_{||},\Delta\phi_{S_\perp}}(\mathbf{r}_{4_{||}}+\Delta\mathbf{r}_{-\Delta\phi_{||}}) \\
&\in \overset{[-]}{\mathcal{H}}_{I'_i,\Delta\mathbf{r}_{-\Delta\phi_{||}}} \qquad (42)
\end{aligned}$$

For the + lattice sites equivalent states result correspondingly to:

$$\overset{[+]}{\varphi}_{\substack{\mathbf{k}_{||},-\Delta\phi_{S_{||}},\\ \Delta\phi_{||},-\Delta\phi_{S_\perp}}}(\mathbf{r}_{5_{||}}+\Delta\mathbf{r}_{\Delta\phi_{||}}) \in \overset{[+]}{\mathcal{H}}_{I_i,\Delta\mathbf{r}_{\Delta\phi_{||}}} \qquad (43)$$

$$\overset{[+]}{\varphi}_{\substack{-\mathbf{k}_{||},-\Delta\phi_{S_{||}},\\ -\Delta\phi_{||},-\Delta\phi_{S_\perp}}}(\mathbf{r}_{6_{||}}+\Delta\mathbf{r}_{-\Delta\phi_{||}}) \in \overset{[+]}{\mathcal{H}}_{I_i,\Delta\mathbf{r}_{-\Delta\phi_{||}}} \qquad (44)$$

$$\overset{[+]}{\varphi}_{\substack{\mathbf{k}_{||},\Delta\phi_{S_{||}},\\ \Delta\phi_{||},\Delta\phi_{S_{\perp}}}}(\mathbf{r}_{7_{||}}+\Delta\mathbf{r}_{\Delta\phi_{||}}) \in \overset{[+]}{\mathcal{H}}_{I'_i,\Delta\mathbf{r}_{\Delta\phi_{||}}} \tag{45}$$

$$\overset{[+]}{\varphi}_{\substack{-\mathbf{k}_{||},\Delta\phi_{S_{||}},\\ -\Delta\phi_{||},\Delta\phi_{S_{\perp}}}}(\mathbf{r}_{8_{||}}+\Delta\mathbf{r}_{-\Delta\phi_{||}}) \in \overset{[+]}{\mathcal{H}}_{I'_i,\Delta\mathbf{r}_{-\Delta\phi_{||}}} \tag{46}$$

All these states are related to the b-holes in Fig. 1. Here $||$ means parallel to the polarization direction of the b-hole structure, i.e. along the one-dimensional electronic states, and $\perp$ perpendicular to that. Every periodic b-hole structure along a particular line I_y or I'_y in Fig. 1 can be considered as a superposition of two phase shifted periodic b-hole substructures, each having a periodicity of $8b$ but a relative shift to each other of $3b$. This leads to the four coherent states given by Eqs. (39)-(42), as well as Eqs. (43)-(46) [8]. The phases $\pm\Delta\phi_{||}$ are spectral phases of U1 symmetry broken one-dimensional Bloch states, parallel to the b-hole polarization direction along I_y, I'_y and $\pm\Delta\phi_{S_\perp}$ are the relative center phases of the one-dimensional Bloch states perpendicular to the b-hole polarization direction. The absolute assignment between the $\mathbf{k}_{||}$ wave vectors and the sign of the phases $\pm\Delta\phi_{||}$ is arbitrarily chosen. For the b-hole state in Fig. 1 one obtains: $\mathbf{k}_{||} = (0, k_y)$, $\mathbf{q}_{||} = (0, q_y)$ and $\mathbf{r}_{||} = (0, y)$. The coherence conditions are defined both by the spectral phases $\pm\Delta\phi_{||}$, the relative center point phases $\pm\Delta\phi_{S_{||}}$ as well as the absolute center point phase $\phi_{S_{||}}$, defining the relative and absolute local positions of the particular b-hole states along the y direction in Fig. 1, and particularly by the fundamental excitation wave vector $\mathbf{q}_{||}$. If the b-holes are polarized in $y(b)$ direction $\mathbf{q}_{||} = (0, \frac{2\pi}{8b})$ is given and for $x(a)$ directed b-hole states $\mathbf{q}_{||} = (\frac{2\pi}{8a}, 0)$ results.

A coherent superposition of Bloch states according to Eqs. (39)-(46) leads to electronic density modulations along the periodically local states $\overset{[-]}{u}_{k_{||}}$ and $\overset{[+]}{u}_{k_{||}}$. If the local states were continuously given by the local state function $u_{o_{||}}$ in Fig. 2(a), the hole density would symmetrically be distributed at the given copper site, i.e the two neighboured oxygen atoms would equivalently be involved. This would represent quantum states corresponding to the hole undoped case. However, the topological holes are exclusively assigned to a particular oxygen site as depicted in Fig. 1 [7], i.e. a topological hole state is assigned to the left or right oxygen atom of a given bonding-unit $\mathrm{O}_{||} - \mathrm{Cu} - \mathrm{O}_{||}$. Therefore, a coherent superposition of Bloch states with the fundamental excitation vector of $\mathbf{q}_{||} = (0, \frac{2\pi}{8b})$ is a necessary but not a sufficient condition for representation of the topological b-hole structure. A splitting of the local state $u_{o_{||}}$ into two parts which are separately associated to the left and right oxygen atom has additionally to occur. Then, the local state $u_{o_{||}}$ is represented by $u_{o_{||}} = (u_{R,\frac{\Delta\mathbf{r}_{||}}{2}} + u_{L,-\frac{\Delta\mathbf{r}_{||}}{2}})$ with the local wave functions $u_{R,\frac{\Delta\mathbf{r}_{||}}{2}}$ and $u_{L,-\frac{\Delta\mathbf{r}_{||}}{2}}$ which are associated to the left and right oxygen site, respectively. Once again for the sake of clarity, the local function $u_{o_{||}}$ is not changed but it is merely split into two parts $u_{R,\frac{\Delta\mathbf{r}_{||}}{2}}$ and $u_{L,-\frac{\Delta\mathbf{r}_{||}}{2}}$ which are associated to different coordinate systems given by the shifts $\frac{\Delta\mathbf{r}_{||}}{2}$ and $-\frac{\Delta\mathbf{r}_{||}}{2}$, respectively. There is not a real renormalization of the quantum state $u_{o_{||}}$. Hence, one has currently to start from $u_{o_{||}}$ if the electron density distribution is discussed, as it is done below. The separation of the local states $u_{o_{||}}$ into the particular states $u_{R,\frac{\Delta\mathbf{r}_{||}}{2}}$ and $u_{L,-\frac{\Delta\mathbf{r}_{||}}{2}}$ without a real renormalization of $u_{o_{||}}$ means that the transformation of the local state function requires no additional energy, i.e. the transformation reflects merely a symmetrically splitting of the existing coordinate system into two subsystems which are relatively shifted to each other by $\Delta\mathbf{r}_{||}$. Such coordinate transformations reflect merely a topological polarization of the local state functions. This fact will be very significant for

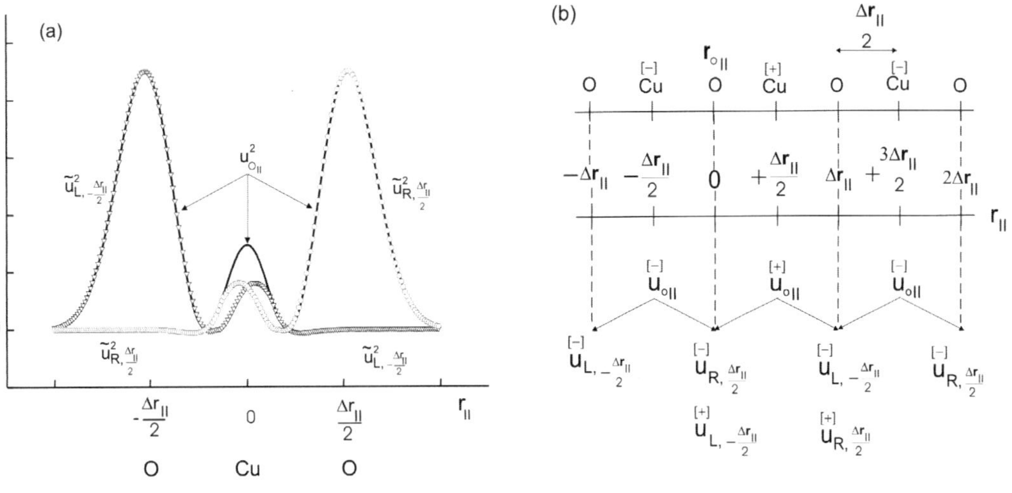

Figure 2. (a) Visualization of local wave functions (density distributions) at a given copper site. In the hole undoped case, the local function $u_{o_{||}}$ is centered at the copper site including two neighboured oxygen orbitals which are not differentiated (full line). If periodic topological hole states exist, the local function $u_{o_{||}}$ is split up into two oxygen centered particular local states $u_{L,-\frac{\Delta r_{||}}{2}}$ and $u_{R,\frac{\Delta r_{||}}{2}}$ (triangles, circles). (b) The local state function $u_{o_{||}}$ is split into the local states $u_{L,-\frac{\Delta r_{||}}{2}}$ and $u_{R,\frac{\Delta r_{||}}{2}}$ which implicates a coordinate transformation of $-\frac{\Delta r_{||}}{2}$ and $\frac{\Delta r_{||}}{2}$, respectively. This leads to a transformation of the coordinate systems of the particular Hilbert subspaces which were initially centered at the copper sites to new coordinate systems, now, being centered at the oxygen sites (vertical dashed lines).

various unusual properties of high-T_c materials. Principally the same considerations as given for the b-hole system can be applied to a periodically ordered f-hole system. However, broken periodicity of the topological hole states have to be included for $n_h < 0.25$ holes/copper.

The splitting of the periodically local functions $\overset{[-]}{u}_{k_{||}}$ and $\overset{[+]}{u}_{k_{||}}$ into two sets of periodic wave functions which are locally separated by $\mathbf{\Delta r}_{||} = a$ or $\mathbf{\Delta r}_{||} = b$ makes it possible that only one oxygen atom left or right to a copper atom in a $\mathrm{O}_{||} - \mathrm{Cu} - \mathrm{O}_{||}$ bonding-unit is predominantly involved in the spectral superposition of the Bloch states according to Eqs. (39)-(46). Only under these conditions the periodic hole structures along the particular one-dimensional states I_i, $\bar{I}_j$ in Fig. 1 can be realized by a constructive interference of Bloch states.

The periodic lattices, defining the periodically local states $\overset{[-]}{u}_{k_{||}}$ and $\overset{[+]}{u}_{k_{||}}$, which were initially defined by the copper sites, are now associated to the neighbouring oxygen sites that belong either to the − or + copper sites , respectively, as depicted in Fig. 2(b). These transformations define the periodically local states of Eqs. (39)-(46). The periodically local states of Eqs. (39)-(46) are causally related to the spectral phases $\Delta\phi_{||}$ and $-\Delta\phi_{||}$ which result from the formation of coherent Bloch states of $\mathbf{k}, -\mathbf{k}$ paired electron states, which create the topological hole orders of Fig. 1 [8]. Hence, the redefined coordinate systems in Eqs. (39)-(46) being related to the phases $\Delta\phi_{||}$ or $-\Delta\phi_{||}$ correspond to new relative states of the particular coordinates and phases according to:

$$\mathbf{\Delta r}_{-\Delta\phi_{||}} = 0 \qquad (\mathrm{modulo}\, 2\mathbf{\Delta r}_{||}) \tag{47}$$

$$\mathbf{\Delta r}_{\Delta\phi_{||}} = -\mathbf{\Delta r}_{||} \qquad (\mathrm{modulo}\, 2\mathbf{\Delta r}_{||}) \tag{48}$$

and

$$\Delta\theta_{-\Delta\phi_{||}} = 0 \qquad \text{(modulo } 2\pi) \tag{49}$$

$$\Delta\theta_{\Delta\phi_{||}} = -\pi \qquad \text{(modulo } 2\pi) \tag{50}$$

with respect to the original coordinate system given by $\mathbf{r}_{o_{||}}$, visualized by the vertical dashed lines in Fig. 2(b). Equations (47)-(50), originating from the shifts of the particular coordinate systems, represent new coordinate systems for the transformed states within the DCBF state. Under these conditions, the initial Hilbert spaces are transformed as follows:

$$\overset{[-]}{\mathcal{H}}_{I_i,-\frac{\Delta\mathbf{r}_{||}}{2}} \quad \begin{matrix}\nearrow \\ \searrow\end{matrix} \quad \begin{matrix} \overset{[-]}{\widetilde{\mathcal{H}}}_{I_i,-\Delta\mathbf{r}_{||}} \\ \overset{[-]}{\widetilde{\mathcal{H}}}_{I_i,0} \end{matrix} \tag{51}$$

$$\begin{matrix} \uparrow & & \uparrow \\ \mathrm{U} & & \mathrm{U} \\ \downarrow & & \downarrow \end{matrix} \tag{52}$$

$$\overset{[+]}{\mathcal{H}}_{I_i,+\frac{\Delta\mathbf{r}_{||}}{2}} \quad \begin{matrix}\nearrow \\ \searrow\end{matrix} \quad \begin{matrix} \overset{[+]}{\widetilde{\mathcal{H}}}_{I_i,0} \\ \overset{[+]}{\widetilde{\mathcal{H}}}_{I_i,+\Delta\mathbf{r}_{||}\equiv-\Delta\mathbf{r}_{||}} \end{matrix} \tag{53}$$

and equivalently for I_i'. If the b-hole state is assigned to the one-dimensional states $\bar{I}_i$, it results

$$\overset{[-]}{\mathcal{H}}_{\bar{I}_i,+\frac{\Delta\mathbf{r}_{||}}{2}} \quad \begin{matrix}\nearrow \\ \searrow\end{matrix} \quad \begin{matrix} \overset{[-]}{\widetilde{\mathcal{H}}}_{\bar{I}_i,+\Delta\mathbf{r}_{||}\equiv-\Delta\mathbf{r}_{||}} \\ \overset{[-]}{\widetilde{\mathcal{H}}}_{\bar{I}_i,0} \end{matrix} \tag{54}$$

$$\begin{matrix} \uparrow & & \uparrow \\ \mathrm{U} & & \mathrm{U} \\ \downarrow & & \downarrow \end{matrix} \tag{55}$$

$$\overset{[+]}{\mathcal{H}}_{\bar{I}_i,-\frac{\Delta\mathbf{r}_{||}}{2}} \quad \begin{matrix}\nearrow \\ \searrow\end{matrix} \quad \begin{matrix} \overset{[+]}{\widetilde{\mathcal{H}}}_{\bar{I}_i,0} \\ \overset{[+]}{\widetilde{\mathcal{H}}}_{\bar{I}_i,-\Delta\mathbf{r}_{||}} \end{matrix} \tag{56}$$

However, the dynamics of the DCBF state is further on related to center coordinates with respect to the copper sites, because the time dependence of the split Hilbert spaces is still related to $u_{o_{||}}$ (see below). Therefore, the time dependence of the coefficients $c_1(t')$, $c_2(t')$ of the DCBF state is conserved and only the relative coordinate systems of the split Hilbert subspaces are different, i.e $\tau_{ei_{DCBF}}$ is unchanged and still represents a conservation quantity for all transformed Hilbert subspaces in Eqs. (51)-(56).

The dynamics within the DCBF state is based on the exchange of the the relative coordinates $\Delta\mathbf{r}_{\Delta\phi_{||}}$ and $\Delta\mathbf{r}_{-\Delta\phi_{||}}$ as well as the relative phases $\Delta\theta_{\Delta\phi_{||}}$ and $\Delta\theta_{-\Delta\phi_{||}}$ which must be strongly symmetric in time. Every exchange has to obey to a strong symmetric time function as it is given in Eq. (24), in order to realize the pair symmetry within Eqs. (16) and (17). In addition, only a symmetric splitting of the initial coordinates on the left side of Eqs. (51),(53),(54) and (56) by $\pm\frac{\Delta\mathbf{r}_{||}}{2}$, leading to the Hilbert subspaces on the right side, may conserve the relative state quantization in space of the particular Hilbert subspaces. In this case, effectively a horizontal exchange of the local state functions

$\overset{[-]}{u}_{R,\frac{\Delta\mathbf{r}_{||}}{2}}$ and $\overset{[+]}{u}_{R,\frac{\Delta\mathbf{r}_{||}}{2}}$ as well as $\overset{[-]}{u}_{L,-\frac{\Delta\mathbf{r}_{||}}{2}}$ and $\overset{[+]}{u}_{L,-\frac{\Delta\mathbf{r}_{||}}{2}}$ of neighboured oxygen sites in I_y and $\bar{I}_y$ in Fig. 1 occurs, which corresponds to a global shift of the center point coordinate $\mathbf{r}_{o_{||}}$ by $\Delta\mathbf{r}_{||} = b$. Altogether, the transformed coordinate systems of the particular Hilbert subspaces are exclusively located at the oxygen sites (Fig. 2b). Therefore, the center points of the coordinate systems of the local function $u_{o_{||}}$ and the particular local functions $u_{R,\frac{\Delta\mathbf{r}_{||}}{2}}$, $u_{L,-\frac{\Delta\mathbf{r}_{||}}{2}}$ are localized at the copper sites and oxygen sites, respectively, but nowhere else.

In conclusion, the total $\mathbf{k}_{||}$ space within the one-dimensional bands I_y, $\bar{I}_y$, already differentiated with respect to the $-$ and $+$ copper sites, is additionally split up into two subspaces having coordinate systems which are relatively shifted to each other by $\Delta\mathbf{r}_{||}$ according to Eqs. (51), (53), (54) and (56). This results in new basis representations of the particular Hilbert subspaces which possess local states that are associated to one of the two oxygen sites within a $O_{||} - Cu - O_{||}$ bonding-unit. Uncorrelated one-electron states, usually given for half-filled bands, continuously belong to the non-differentiated local states $u_{o_{||}}$, which are associated to the copper sites. Periodic topological hole orders as given in Fig. 1 are connected with $\mathbf{k}_{||}, -\mathbf{k}_{||}$ pair states [8]. The coherent pair excitations lead to the localized hole densities at particular oxygen atoms and in consequence to the separated states $u_{R,\frac{\Delta\mathbf{r}_{||}}{2}}$ and $u_{L,-\frac{\Delta\mathbf{r}_{||}}{2}}$. Nevertheless, one has further on to start from $u_{o_{||}} = u_{R,\frac{\Delta\mathbf{r}_{||}}{2}} + u_{L,-\frac{\Delta\mathbf{r}_{||}}{2}}$ if the electron density distribution is considered, because $u_{o_{||}}$ is not renormalized as mentioned above. The electron density $u^2_{o_{||}}$ is separated into three parts $u^2_{R,\frac{\Delta\mathbf{r}_{||}}{2}}$, $2u_{R,\frac{\Delta\mathbf{r}_{||}}{2}} \cdot u_{L,-\frac{\Delta\mathbf{r}_{||}}{2}}$ and $u^2_{L,-\frac{\Delta\mathbf{r}_{||}}{2}}$. The diagonal parts $u^2_{R,\frac{\Delta\mathbf{r}_{||}}{2}}$ and $u^2_{L,-\frac{\Delta\mathbf{r}_{||}}{2}}$ are definitively associated to the coordinate systems which result from the splitting and the shifts $\frac{\Delta\mathbf{r}_{||}}{2}$ and $-\frac{\Delta\mathbf{r}_{||}}{2}$ of the initial coordinate system, respectively. The off-diagonal term $2u_{R,\frac{\Delta\mathbf{r}_{||}}{2}} \cdot u_{L,-\frac{\Delta\mathbf{r}_{||}}{2}}$ can not exist, because there is not an unambiguously defined coordinate system. Only off-diagonal states $u_{R,\frac{\Delta\mathbf{r}_{||}}{2}} \cdot u_{L,\frac{\Delta\mathbf{r}_{||}}{2}}$ or $u_{R,-\frac{\Delta\mathbf{r}_{||}}{2}} \cdot u_{L,-\frac{\Delta\mathbf{r}_{||}}{2}}$ can really exist. If the local state $u_{R,\frac{\Delta\mathbf{r}_{||}}{2}}$ is populated the local state $u_{L,\frac{\Delta\mathbf{r}_{||}}{2}}$ is also particularly involved, and if the local state $u_{L,-\frac{\Delta\mathbf{r}_{||}}{2}}$ is populated the state $u_{R,-\frac{\Delta\mathbf{r}_{||}}{2}}$ is particularly involved, too. In a dynamic picture this means, the off-diagonal terms $2u_{R,\frac{\Delta\mathbf{r}_{||}}{2}} \cdot u_{L,-\frac{\Delta\mathbf{r}_{||}}{2}}$ are dynamically associated/transformed to the product of the local state functions $u_{R,\frac{\Delta\mathbf{r}_{||}}{2}} \cdot u_{L,\frac{\Delta\mathbf{r}_{||}}{2}}$ or $u_{R,-\frac{\Delta\mathbf{r}_{||}}{2}} \cdot u_{L,-\frac{\Delta\mathbf{r}_{||}}{2}}$. This implicates that these off-diagonal density terms may be exchanged between the two local state functions $u_{R,\frac{\Delta\mathbf{r}_{||}}{2}}$ and $u_{L,-\frac{\Delta\mathbf{r}_{||}}{2}}$ in a probabilistic manner. The phases $\Delta\theta_{\Delta\phi_{||}}$ and $\Delta\theta_{-\Delta\phi_{||}}$ in Eqs. (39)-(46) which are related to the $\mathbf{k}_{||}$ states within the Bloch functions, however, are not probabilistic but strongly associated to the coordinate system of $u_{R,\frac{\Delta\mathbf{r}_{||}}{2}}$ or $u_{L,-\frac{\Delta\mathbf{r}_{||}}{2}}$. The additional probabilistic exchange coupling between the local state functions $u_{R,\frac{\Delta\mathbf{r}_{||}}{2}}$ and $u_{L,-\frac{\Delta\mathbf{r}_{||}}{2}}$ guarantees the conservation of an electronic coupling between these two local functions, which is a precondition for conductivity. These exchanges of the coordinate systems may be infinitely fast, because no mass transfer occurs, i.e. there is only an exchange of information. Each of the local functions $u_{R,\frac{\Delta\mathbf{r}_{||}}{2}}$ and $u_{L,-\frac{\Delta\mathbf{r}_{||}}{2}}$, which obey to a strongly deterministic time function with respect to the dynamics of their coordinate systems according to Eq. (24), additionally include a probabilistic off-diagonal part, i.e. $u_{R,\frac{\Delta\mathbf{r}_{||}}{2}}$ creates $u_{L,\frac{\Delta\mathbf{r}_{||}}{2}}$ parts and $u_{L,-\frac{\Delta\mathbf{r}_{||}}{2}}$ creates $u_{R,-\frac{\Delta\mathbf{r}_{||}}{2}}$ parts with the coordinate systems being conserved, respectively. The probability to find $u_{R,\frac{\Delta\mathbf{r}_{||}}{2}} \cdot u_{L,\frac{\Delta\mathbf{r}_{||}}{2}}$

or $u_{R,-\frac{\Delta \mathbf{r}_{||}}{2}} \cdot u_{L,-\frac{\Delta \mathbf{r}_{||}}{2}}$ will depend on the relative strength of polarization caused by the local states $u_{R,\frac{\Delta \mathbf{r}_{||}}{2}}$ and $u_{L,-\frac{\Delta \mathbf{r}_{||}}{2}}$, i.e. it will be proportional to the relative amplitudes of $u^2(\mathbf{r})_{R,\frac{\Delta \mathbf{r}_{||}}{2}}$ and $u^2(\mathbf{r})_{L,-\frac{\Delta \mathbf{r}_{||}}{2}}$ at a given position $\mathbf{r}$. Hence, the total effective local density associated to the right and left oxygen atoms can be represented by

$$\begin{aligned} \tilde{u}^2_{R,\frac{\Delta \mathbf{r}_{||}}{2}}(\mathbf{r}) &= u^2_{R,\frac{\Delta \mathbf{r}_{||}}{2}}(\mathbf{r}) + 2 \cdot u_{R,\frac{\Delta \mathbf{r}_{||}}{2}}(\mathbf{r}) \cdot u_{L,\frac{\Delta \mathbf{r}_{||}}{2}}(\mathbf{r}) \\ &\quad \cdot \left(\frac{u^2_{R,\frac{\Delta \mathbf{r}_{||}}{2}}(\mathbf{r})}{u^2_{R,\frac{\Delta \mathbf{r}_{||}}{2}}(\mathbf{r}) + u^2_{L,-\frac{\Delta \mathbf{r}_{||}}{2}}(\mathbf{r})} \right), \end{aligned} \tag{57}$$

$$\begin{aligned} \tilde{u}^2_{L,-\frac{\Delta \mathbf{r}_{||}}{2}}(\mathbf{r}) &= u^2_{L,-\frac{\Delta \mathbf{r}_{||}}{2}}(\mathbf{r}) + 2 \cdot u_{R,-\frac{\Delta \mathbf{r}_{||}}{2}}(\mathbf{r}) \cdot u_{L,-\frac{\Delta \mathbf{r}_{||}}{2}}(\mathbf{r}) \\ &\quad \cdot \left(\frac{u^2_{L,-\frac{\Delta \mathbf{r}_{||}}{2}}(\mathbf{r})}{u^2_{R,\frac{\Delta \mathbf{r}_{||}}{2}}(\mathbf{r}) + u^2_{L,-\frac{\Delta \mathbf{r}_{||}}{2}}(\mathbf{r})} \right), \end{aligned} \tag{58}$$

with

$$u^2_{o_{||}}(\mathbf{r}) = \tilde{u}^2_{o_{||}}(\mathbf{r}) = \tilde{u}^2_{R,\frac{\Delta \mathbf{r}_{||}}{2}}(\mathbf{r}) + \tilde{u}^2_{L,-\frac{\Delta \mathbf{r}_{||}}{2}}(\mathbf{r}), \tag{59}$$

described within a common coordinate system, which is graphically depicted in Fig. 2(a).

The physical background of Eqs. (57) and (58) implicate two properties which seem to be contradictory:

(a) The formation of particular Hilbert subspaces which belong to different coordinate systems.

(b) Simultaneously, the unification of the two Hilbert subspaces by common/exchanged off-diagonal densities.

This dualism of the electronic states is only comprehensible if we consider the coordinate systems as dynamically variable. A more detailed examination is given in the Appendix with the result that the local state function $\tilde{u}_{o_{||}}(\mathbf{r})$ is given by the direct sum

$$\tilde{u}_{o_{||}}(\mathbf{r}) = \tilde{u}_{R,\frac{\Delta \mathbf{r}_{||}}{2}}(\mathbf{r}) \oplus \tilde{u}_{L,-\frac{\Delta \mathbf{r}_{||}}{2}}(\mathbf{r}) \tag{60}$$

of particular local functions $\tilde{u}_{R,\frac{\Delta \mathbf{r}_{||}}{2}}(\mathbf{r})$ and $\tilde{u}_{L,-\frac{\Delta \mathbf{r}_{||}}{2}}(\mathbf{r})$ reflecting local states that are associated to the two coordinate systems, respectively, and which are on their part a direct sum of local states again being associated to the two coordinate systems (Eqs.(A.12) and (A.13)). The Hilbert space axiomatic is furthermore defined within the initial Hilbert spaces represented by the left side of Eqs. (51), (53), (54) and (56), but on the condition of varying coordinate systems. All in all the quantum states are dynamically separated in terms of relative states of the coordinate systems. This behaviour can be termed as an incomplete separation of one Hilbert space into several Hilbert subspaces, depicted by the tilde on the Hilbert space symbol on the left side of Eqs. (51), (53), (54) and (56).

The symmetric splitting of the Hilbert spaces according to Eqs. (51), (53), (54) and (56) does not influence the space-time quantization of the DCBF state, as discussed above. This means, at every time t' the two particular basis representations defined by $u_{o_{||}}$ and $\tilde{u}_{R,\frac{\Delta \mathbf{r}_{||}}{2}}$,

$\tilde{u}_{L,-\frac{\Delta\mathbf{r}_{||}}{2}}$, which belong to the different coordinate systems given by $\mathbf{r}_{o_{||}}$ and $\frac{\Delta\mathbf{r}_{||}}{2}, -\frac{\Delta\mathbf{r}_{||}}{2}$, may exist simultaneously with a certain probability. Hence, these two representations may superimpose according to

$$u_{||}(t') = \alpha_{1_{||}}(t')u_{o_{||}} \oplus \alpha_{2_{||}}(t')(\tilde{u}_{R,\frac{\Delta\mathbf{r}_{||}}{2}} \oplus \tilde{u}_{L,-\frac{\Delta\mathbf{r}_{||}}{2}}) \quad (61)$$

with real coefficients $\alpha_{1_{||}}(t')$, $\alpha_{2_{||}}(t')$ and $\alpha_{1_{||}}(t') + \alpha_{2_{||}}(t') = 1$. Again the direct sum occurs, because, any interference term must disappear based on the different coordinate systems which forme a definite coordinate relative state. The quantum state of Eq. (61) is the most common quantum state including the completely uncorrelated local state for $\alpha_{1_{||}} = 1$ and $\alpha_{2_{||}} = 0$ and the strongly correlated/entangled partial local states for $\alpha_{1_{||}} = 0$ and $\alpha_{2_{||}} = 1$. Equation (61) with $\alpha_{1_{||}} \neq 0$ and $\alpha_{2_{||}} \neq 1$ may be important, for example, if correlated electron states are induced even under half-filling of the valence bands.

The fact that the transformed Hilbert subspaces according to the right sides of Eqs. (51), (53), (54) and (56) create common coordinate systems is of decisive importance. According to Fig. 2(b), the transformed local states $\overset{[-]}{u}_{L,-\frac{\Delta\mathbf{r}_{||}}{2}}$ and $\overset{[+]}{u}_{R,\frac{\Delta\mathbf{r}_{||}}{2}}$ as well as $\overset{[+]}{u}_{L,-\frac{\Delta\mathbf{r}_{||}}{2}}$ and $\overset{[-]}{u}_{R,\frac{\Delta\mathbf{r}_{||}}{2}}$ belong to common coordinate systems. The local states $\overset{[+]}{u}_{L,-\frac{\Delta\mathbf{r}_{||}}{2}}$ and $\overset{[-]}{u}_{R,\frac{\Delta\mathbf{r}_{||}}{2}}$ as well as $\overset{[-]}{u}_{L,-\frac{\Delta\mathbf{r}_{||}}{2}}$ and $\overset{[+]}{u}_{R,\frac{\Delta\mathbf{r}_{||}}{2}}$ are quasi-stationary related to each other within the absolute coordinate systems $\mathbf{r}_{o_{||}} - \Delta\mathbf{r}_{||}$ (modulo $2\Delta\mathbf{r}_{||}$) or $\mathbf{r}_{o_{||}}$ (modulo $2\Delta\mathbf{r}_{||}$), respectively. These local functions are synchronously changed in dependence on $c_1(t')$, $c_2(t')$, with respect to a common coordinate system. Therefore, electronic transitions between the local states $\overset{[+]}{u}_{L,-\frac{\Delta\mathbf{r}_{||}}{2}}$ and $\overset{[-]}{u}_{R,\frac{\Delta\mathbf{r}_{||}}{2}}$ as well as $\overset{[-]}{u}_{L,-\frac{\Delta\mathbf{r}_{||}}{2}}$ and $\overset{[+]}{u}_{R,\frac{\Delta\mathbf{r}_{||}}{2}}$ can occur independently on $c_1(t')$, $c_2(t')$ and the time-quantization $\tau_{ei_{DCBF}}$. A probabilistic electronic transition from $\overset{[-]}{u}_{R,\frac{\Delta\mathbf{r}_{||}}{2}}$ to $\overset{[+]}{u}_{L,-\frac{\Delta\mathbf{r}_{||}}{2}}$ as well as $\overset{[-]}{u}_{L,-\frac{\Delta\mathbf{r}_{||}}{2}}$ to $\overset{[+]}{u}_{R,\frac{\Delta\mathbf{r}_{||}}{2}}$ or vice versa is not connected with a jump of $\pm\Delta\theta_{||}$. As a result, the states $\overset{[-]}{u}_{R,\frac{\Delta\mathbf{r}_{||}}{2}}$ and $\overset{[+]}{u}_{L,-\frac{\Delta\mathbf{r}_{||}}{2}}$ as well as $\overset{[-]}{u}_{L,-\frac{\Delta\mathbf{r}_{||}}{2}}$ and $\overset{[+]}{u}_{R,\frac{\Delta\mathbf{r}_{||}}{2}}$ can freely interfere at any time t. In this case, the Hilbert subspaces $\overset{[-]}{\widetilde{\mathcal{H}}}_{I_i,0}$ and $\overset{[+]}{\widetilde{\mathcal{H}}}_{I_i,0}$ as well as $\overset{[-]}{\widetilde{\mathcal{H}}}_{I_i,-\Delta\mathbf{r}_{||}}$ and $\overset{[+]}{\widetilde{\mathcal{H}}}_{I_i,-\Delta\mathbf{r}_{||}}$ in Eqs. (51) and (53) form a common Hilbert space, respectively, which is the linear span of the corresponding subspaces, $\mathrm{lin}(\overset{[-]}{\widetilde{\mathcal{H}}}_{I_i,0} \cup \overset{[+]}{\widetilde{\mathcal{H}}}_{I_i,0})$ and $\mathrm{lin}(\overset{[-]}{\widetilde{\mathcal{H}}}_{I_i,-\Delta\mathbf{r}_{||}} \cup \overset{[+]}{\widetilde{\mathcal{H}}}_{I_i,-\Delta\mathbf{r}_{||}})$ (Eq. 52). Equivalent linear spans, $\mathrm{lin}(\overset{[-]}{\widetilde{\mathcal{H}}}_{\bar{I}_j,0} \cup \overset{[+]}{\widetilde{\mathcal{H}}}_{\bar{I}_j,0})$ and $\mathrm{lin}(\overset{[-]}{\widetilde{\mathcal{H}}}_{\bar{I}_j,-\Delta\mathbf{r}_{||}} \cup \overset{[+]}{\widetilde{\mathcal{H}}}_{\bar{I}_j,-\Delta\mathbf{r}_{||}})$ (Eq. 55), result for the one-dimensional states along $\bar{I}_j$ according to Eqs. (54) and (56). As a result, at any time t an electron can be transferred from the occupied local site $\overset{[-]}{u}_{L,-\frac{\Delta\mathbf{r}_{||}}{2}}$ to the empty local site $\overset{[+]}{u}_{R,\frac{\Delta\mathbf{r}_{||}}{2}}$ or from the occupied local site $\overset{[-]}{u}_{R,\frac{\Delta\mathbf{r}_{||}}{2}}$ to the empty local site $\overset{[+]}{u}_{L,-\frac{\Delta\mathbf{r}_{||}}{2}}$, under conservation of the phase $\delta\theta_{||}(t')$. In terms of the copper-oxygen bonding that means, an electron initially giving a contribution to the $\overset{[-]}{\text{Cu}}$–O bonding is transferred to the $\overset{[+]}{\text{Cu}}$–O bonding. In this way the charge center point of the electronic system is changed. In consequence, an electron transfer along the $\mathbf{r}_{||}$ coordinate occurs, i.e. the electronic system becomes conducting. The formation of common coordinate systems of particular Hilbert subspaces and the existence of the off-diagonal densities in Eqs. (57) and (58) form the basis for dynamic quantum states throughout all atomic sites along the coordinate $\mathbf{r}_{||}$. The deterministic quantized time-behaviour of the quantum states given by $\delta\theta_{||}(t')$, being initially a barrier for conductivity in the hole undoped case, is bypassed, even under the continuous existence of strongly ordered topological hole states.

Under hole doping the electronic states are basically $\mathbf{k}_{||}, -\mathbf{k}_{||}$ pair states, if ordered topological hole structures as given in Fig. 1 exist, as it was previously elaborated [8]. Only pair excitations are possible within the $\mathbf{k}_{||}$ space which leave the center of mass point and charge center point unchanged. Hence, a directed electron movement can not be created within the $\mathbf{k}_{||}$ space, but will only result from unrestricted superpositions of the local states $\overset{[-]}{u}_{L,-\frac{\Delta\mathbf{r}_{||}}{2}}$ and $\overset{[+]}{u}_{R,\frac{\Delta\mathbf{r}_{||}}{2}}$ as well as $\overset{[+]}{u}_{L,-\frac{\Delta\mathbf{r}_{||}}{2}}$ and $\overset{[-]}{u}_{R,\frac{\Delta\mathbf{r}_{||}}{2}}$. An impulse that is transferred to the electron pair states will only lead to a finite effective mass $m_{||_{eff}} < \infty$ of the paired charge carriers, if an electronic transition between the local states $\overset{[-]}{u}_{L,-\frac{\Delta\mathbf{r}_{||}}{2}}$ and $\overset{[+]}{u}_{R,\frac{\Delta\mathbf{r}_{||}}{2}}$ as well as $\overset{[-]}{u}_{R,\frac{\Delta\mathbf{r}_{||}}{2}}$ and $\overset{[+]}{u}_{L,-\frac{\Delta\mathbf{r}_{||}}{2}}$ takes place. Principally the same considerations can be applied to a periodically ordered f-hole system. Singular topological f-holes or strongly clustered topological f-hole states, which can exist for low hole doping $0.125 < n_h \ll 0.25$ holes/copper [7], however, can form non-conducting regions within the CuO_2 plane. The same will occur for the b-holes under the condition $n_h \ll 0.125$ holes/copper.

In conventional superconductors, the conductivity is attributed to the phase of the superconducting condensate which appears as a free variable. The center phase $\phi_{S_{||}}$ of the topological b-hole state in Fig. 1 is strongly coupled to the center coordinate $\mathbf{r}_{o_{||}}$ of the CBF state. Therefore, $\phi_{S_{||}}$ is strongly determined by the center coordinate $\mathbf{r}_{o_{||}}$ with an uncertainty given by modulo $2\Delta\mathbf{r}_{||}$, reflecting the periodicity of the CBF state. If the electronic system is transferred from the CBF to the $\overline{\text{CBF}}$ state the phase $\phi_{S_{||}}$ is also transferred, by $\delta\phi_{S_{||}} = \frac{2\pi}{8}$ (OR (exclusive) by $\delta\phi_{S_{||}} = -\frac{2\pi}{8}$) with respect to the b-hole state in order to occupy an equivalent next neighboured topological hole state, if the b-hole state is moved in $y(b)$ direction of Fig. 1 [8]. In addition, a phase transformation according to $\phi_{S_{||}} \to \phi_{S_{||}} + \delta\phi_{S_{||}}$ during the lifetime τ_{CBF} of a CBF configuration is generally not possible, since it would be connected with a jump of $\pm\Delta\theta_{||}$ which is forbidden due to the deterministic time behaviour of $\delta\theta_{||}(t')$. In other words, such probabilistic transitions would violate the strongly deterministic space-time separation of the Hilbert spaces. Hence, the variation $\delta\phi_{S_{||}}(t')$ will be time-dependent obeying the same functionality as given for $\delta\mathbf{r}_{||}(t')$ and $\delta\theta_{||}(t')$, i.e. $\delta\phi_{S_{||}}(t')$ is strongly deterministic in space and time, too. Thus, the phase $\phi_{S_{||}}(t')$ can not be a free variable of the b-hole condensate, and neither the phases $\pm\Delta\phi_{S_{||}}(t')$ nor $\pm\Delta\phi_{||}(t')$ which are directly related to $\phi_{S_{||}}(t')$.

The phases $\phi_{S_{||}}$,$\pm\Delta\phi_{S_{||}}$ and $\pm\Delta\phi_{||}$ in Fig. 1 which characterize the b-hole state are strongly periodic in space with a periodicity of $8b$ with respect to the I_y or I'_y direction. In the full TR state, introduced in Ref. [8], the topological b-hole state is transferred in $y(b)$ direction (see Fig. 1) by 8 subsequent steps of one lattice spacing, if we consider the non-superconducting state, i.e. a state with IS quantum hole symmetry (see Ref. [8]). After 8 phase steps $\delta\phi_{S_{||}} = \frac{2\pi}{8}$ (OR $\delta\phi_{S_{||}} = -\frac{2\pi}{8}$) the initial b-hole state is recreated. The strong coupling of $\delta\phi_{S_{||}}(t')$ to the time function Eq. (24) of the DCBF state requires that a particular topological b-hole state as given in Fig. 1 has exclusively a lifetime of τ_{CBF} in the non-superconducting state, although the uncertainty $2n \cdot \Delta\mathbf{r}_{||}$ (n integer) of the DCBF state would principally permit jumps of $\pm 2n \cdot \delta\phi_{S_{||}} = \pm 2n \cdot \frac{2\pi}{8}$ during the time τ_{CBF}. This conclusion, i.e. the stationarity of the b-hole state during the lifetime

τ_{CBF}, was justified by the correlation energy $E_C < 0$ for the b-holes [8]. However, there are even more fundamental reasons. Every transformation to a different b-hole state along y coordinate changes the hole topology. Among others, this leads to variations of the quadrupolar polarizations on the copper atoms (see Ref.[7]) along the perpendicular one-dimensional states I_x, $\bar{I}_x$ of Fig. 1. This is connected with bonding transformations $\overset{[-]}{\text{Cu}}\text{–O} \leftrightarrow \overset{[+]}{\text{Cu}}\text{–O}$ (see Fig.5(c) in Ref. [7]). As deduced here, such transformations are only allowed, if a basis representation given by $\overset{[-]}{u}_{L,\frac{-\Delta\mathbf{r}_{||}}{2}}$ and $\overset{[+]}{u}_{R,\frac{\Delta\mathbf{r}_{||}}{2}}$ exists, because the deterministic time behaviour according to Eq. (24) is violated otherwise. However, most one-dimensional subspaces I_x, $\bar{I}_x$ carry no holes, so the basis states of these hole free one-dimensional subspaces have only $u_{o_{||}}$ character.[3] Therefore, additional polarizations which would inevitably occur, if the b-hole state is topologically transformed in $y(b)$ direction of Fig. 1 are forbidden, i.e. the b-hole state must be stationary during the time τ_{CBF}. Hence, the topological b-hole state can only be periodically changed in steps of τ_{CBF} correspondingly to the CBF $\leftrightarrow$ $\overline{\text{CBF}}$ transitions with an characteristic inherent time function similar as given by Eq. (24). This defines a stationary periodicity of the dynamic b-hole state corresponding to the *eigentime* $\tau_{ei_{||}} = 8 \cdot \tau_{CBF}$. The dynamic b-hole state is absolutely stationary with respect to this eigentime, if a strong periodic b-hole structure is assumed and if all eight equivalent b-hole states along the b direction in Fig. 1 have the same probability. However, even if the b-hole state were not stationary with respect to the eigentime, i.e. if different eigentimes would exist, a promotion into states with other eigentimes $\tau'_{ei_{||}}$ can only happen if the existing periodic cycle is finished. Only under these requirements the stationarity of the Bloch states in Eqs. (39)-(46) is realized, if the phase $\phi_{S_{||}}(t)$ is time dependent. Therefore, $\phi_{S_{||}}(t)$ is strongly periodic in time during a given cycle defined by $\tau_{ei_{||}}$. This prevents that $\phi_{S_{||}}(t)$ and the stationary related phases $\pm\Delta\phi_{S_{||}}$ and $\pm\Delta\phi_{||}$ can exist as free variables in the hole doped (p-type) cuprates.

The f-hole states are strongly coupled to the b-hole state with their own intrinsic eigentimes, if the f-hole structures are strongly periodic in space. Under these conditions the same argumentation holds true for the f-hole states as discussed for the b-holes. However, different f-hole configurations are possible during the lifetime τ_{CBF} [8]. The one-dimensional states along I'_x in Fig. 1 where $f_{\perp}$-holes are placed implicate a basis representation of $\overset{[-]}{u}_{L,-\frac{\Delta\mathbf{r}_{||}}{2}}$, $\overset{[+]}{u}_{R,\frac{\Delta\mathbf{r}_{||}}{2}}$ with the possibility of $\overset{[-]}{\text{Cu}}\text{–O} \leftrightarrow \overset{[+]}{\text{Cu}}\text{–O}$ transitions. In consequence, transformations between different topological f-hole states are possible during the life time τ_{CBF}, if the transformations occur within the one-dimensional bands $\bar{I}_y$, $\bar{I}'_y$, I'_x in Fig. 1 where f-holes already exist. Altogether, the phases within a strong periodic b-hole condensate can generally not occur as free variable. The f-hole states possess a certain variability of their phases which is, however, distinctly restricted.

In consequence, the conductivity in undisturbed largely extended periodic hole states as given in Fig. 1 is exclusively realized by the phase conserving local transitions $\overset{[-]}{u}_{L,-\frac{\Delta\mathbf{r}_{||}}{2}} \leftrightarrow \overset{[+]}{u}_{R,\frac{\Delta\mathbf{r}_{||}}{2}}$ and $\overset{[-]}{u}_{R,\frac{\Delta\mathbf{r}_{||}}{2}} \leftrightarrow \overset{[+]}{u}_{L,-\frac{\Delta\mathbf{r}_{||}}{2}}$ with respect to the b-holes, and also for the f-holes the restricted variability of the phases will largely constrain the conductivity on these local transitions.[4]

[3] This is only valid in the non-superconducting state having IS quantum hole symmetry and if polarization influences do not exist which may possibly induce $u_{o_{||}} \leftrightarrow u_{L,-\frac{\Delta r_{||}}{2}}, u_{R,\frac{\Delta r_{||}}{2}}$ transitions.

[4] If the topological hole states in Fig. 1 are clustered, as has to be expected for $n_k \ll 0.25$ holes/copper with respect to f-holes and for $n_k \ll 0.125$ holes/copper with regard to b-holes, electrons may be promoted or may tunnel between the clusters including not well-defined periodic phase shifts. The eigentime $\tau_{ei,||}$

8. The quantum state dynamics of the CuO_2 plane under electron doping

The dynamics of the quantum states under electron doping into the initially undoped CuO_2^{2-} plane proves to be very different to the quantum dynamic behaviour in hole doped CuO_2^{2-} planes. This can be found in the n-type Ce based compounds NCCO and PCCO. The states which are associated to the − copper sites within the undoped CuO_2^{2-} plane belonging to $\overset{[-]}{\mathcal{H}}_{I_i,-\frac{\Delta \mathbf{r}_{||}}{2}}$ or $\overset{[-]}{\mathcal{H}}_{\bar{I}_j,+\frac{\Delta \mathbf{r}_{||}}{2}}$ are filled, so that every additional electron has to be injected into states at the + copper site, which belong to $\overset{[+]}{\mathcal{H}}_{I_i,+\frac{\Delta \mathbf{r}_{||}}{2}}$ or $\overset{[+]}{\mathcal{H}}_{\bar{I}_j,-\frac{\Delta \mathbf{r}_{||}}{2}}$, as depicted in Figs. 3,(2) and 3,(2'). This is connected with a reduced strength of the charge fluctuation potential within the existing CBF state. Every additional electron being injected at the + copper site compensates the charge fluctuation influence of one electron within the DCBF state which was already localized at the − copper site. If the number of additionally injected electrons at the + copper site is n_e, $2 \cdot n_e$ electrons will not any longer contribute to the exchange energy of the DCBF state. Or more precisely, the $2 \cdot n_e$ electrons will either contribute to off-diagonal populations as depicted in Fig. 3,(2) or to pairwise diagonal populations given in Fig. 3,(2'), both representing time independent states. Therefore, these electrons are free, i.e. they are not quantum coupled to the fluctuations of the DCBF state. The charge fluctuations of the DCBF state occur merely as a mean field influence for these electrons, but they are not quantum entangled to the dynamics of the quantum states within the DCBF state. The quantum states populated by these electrons can be considered as static with the result that the eigentime $\tau_{ei_{DCBF}}$ is not a conservation quantity for these states, because it does not exist for those.

For the moment we assume that these $2 \cdot n_e$ electrons represent one-electron states,

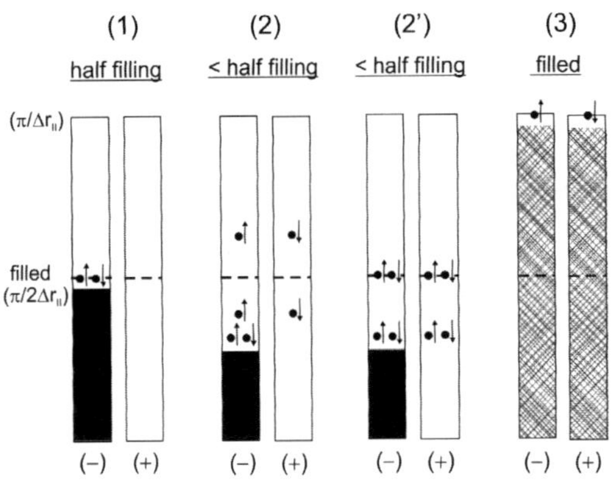

Figure 3. Population of the two different copper sites (+) and (−) in Fig. 1. (1) For half filled bands, only the (−) copper sites are populated forming diagonal spin pairings, within the antibonding valence bands. (2),(2') For populations above half filling, the (+) and (−) lattice sites are simultaneously populated where off-diagonal (2) or diagonal (2') spin pairings may occur. (3) For filled bands, the (+) and (−) lattice sites are simultaneously populated where ideally off-diagonal spin pairings occur in the case of bonding states, which is the case for lower energetically filled bands. Filled areas indicate populated states.

i.e. not paired electron states. At any time t an electron which belongs, for example, to the − copper site can be promoted to the + copper site or vice versa, i.e. these transitions are not coupled to $\tau_{ei_{DCBF}}$, as it is given for half-filled or hole doped valence bands. In consequence, these states which are independent on $\tau_{ei_{DCBF}}$ may freely interfere, i.e. they may belong simultaneously in a quantum probabilistic sense to the Hilbert spaces $\overset{[-]}{\mathcal{H}}_{I_i,+\frac{\Delta \mathbf{r}_{||}}{2}}$ and $\overset{[+]}{\mathcal{H}}_{I_i,+\frac{\Delta \mathbf{r}_{||}}{2}}$ as well as $\overset{[-]}{\mathcal{H}}_{I_i,-\frac{\Delta \mathbf{r}_{||}}{2}}$ and $\overset{[+]}{\mathcal{H}}_{I_i,-\frac{\Delta \mathbf{r}_{||}}{2}}$ or to $\overset{[-]}{\mathcal{H}}_{\bar{I}_j,+\frac{\Delta \mathbf{r}_{||}}{2}}$ and $\overset{[+]}{\mathcal{H}}_{\bar{I}_j,+\frac{\Delta \mathbf{r}_{||}}{2}}$ as well as

and the local phase $\phi_{S_{||}}$ as well as their definite interrelation become uncertain. In the same way this uncertainty will be important within tunnel junctions at crystal boundaries, for example, with respect to Josephson pair tunnelling.

$\overset{[-]}{\mathcal{H}}_{\bar{I}_j,-\frac{\Delta\mathbf{r}_{||}}{2}}$ and $\overset{[+]}{\mathcal{H}}_{\bar{I}_j,-\frac{\Delta\mathbf{r}_{||}}{2}}$, under condition of probabilistic varying coordinate systems. As a result, these $2 \cdot n_e$ electrons may probabilistically occupy states within Hilbert spaces which are the linear spans $\mathrm{lin}(\overset{[-]}{\mathcal{H}}_{I_i,+\frac{\Delta\mathbf{r}_{||}}{2}} \cup \overset{[+]}{\mathcal{H}}_{I_i,+\frac{\Delta\mathbf{r}_{||}}{2}})$, $\mathrm{lin}(\overset{[-]}{\mathcal{H}}_{I_i,-\frac{\Delta\mathbf{r}_{||}}{2}} \cup \overset{[+]}{\mathcal{H}}_{I_i,-\frac{\Delta\mathbf{r}_{||}}{2}})$ or $\mathrm{lin}(\overset{[-]}{\mathcal{H}}_{\bar{I}_j,+\frac{\Delta\mathbf{r}_{||}}{2}} \cup \overset{[+]}{\mathcal{H}}_{\bar{I}_j,+\frac{\Delta\mathbf{r}_{||}}{2}})$, $\mathrm{lin}(\overset{[-]}{\mathcal{H}}_{\bar{I}_j,-\frac{\Delta\mathbf{r}_{||}}{2}} \cup \overset{[+]}{\mathcal{H}}_{\bar{I}_j,-\frac{\Delta\mathbf{r}_{||}}{2}})$. The uniform band states, given for half-filled and hole doped bands, are split up into energetic lower band states which belong further on separately to the Hilbert spaces on the left side of Eqs. (51),(53),(54) and (56) and energetic upper band states that belong to the corresponding static linear spans of these Hilbert spaces. The energetically lower band states are filled with $n' = n_{\frac{\pi}{2}} - n_e$ electrons and the energetically upper band states with $n'' = 2n_e$ electrons, with $n_{\frac{\pi}{2}}$ being the number of electrons for half-filled bands. The n' electrons are lowered in energy based on their additional negative correlation energy within the DCBF state. The n'' electrons have lost the DCBF correlation energy and are probably additionally shifted to higher energies resulting from the continually existing dynamic charge fluctuations which create a time dependent potential that generates a non-stationary electron dynamics.

If the degree of electron doping is increased, one has to expect that ordered topological electron structures may also be formed, similar as it was given under hole doping. The driving force for such topologically structured electron states should again be quadrupolar-polarization induced correlations, similar as described for hole doped states [7]. Again, this leads to $\mathbf{k}_{||}, -\mathbf{k}_{||}$ electron pair states and transformations of the coordinate systems of the Hilbert spaces according to the right sites of Eqs. (51) to (56). At present, the topological electron structure for n-type cuprates is not yet known which prevents a deeper elaboration of the quantum states. Nevertheless, an ordered topological electron structure may again lead to a periodic transformation of a topologically ordered electron state in phase steps of $\delta\phi_{S_{||}}$. This again will lead to an eigentime $\tau_{ei_{||}} = n \cdot \tau_{CBF}$ of this dynamic state, with $n \cdot \Delta\mathbf{r}_{||}$ being the periodicity of the topological electron structure along its polarization direction. The lifetime τ_{CBF} remains the determining time constant, because an ordered topological electron structure will also be strongly coupled to the CBF superstructure $(-,+)$ caused by the same electronic polarizations (quadrupolar-polarizations), as it is given for hole doped states. One can say, an ordered topological electron structure will topologically triggered in time by τ_{CBF}. The main difference to the hole doped case is again given by the fact that the electron pair states $\mathbf{k}_{||}, -\mathbf{k}_{||}$ creating a topological electron structure are not quantum-coupled to the DCBF state. These electron pair states belong at all times probabilistically to the linear spans $\mathrm{lin}(\overset{[-]}{\mathcal{H}}_{I_i,+\frac{\Delta\mathbf{r}_{||}}{2}} \cup \overset{[+]}{\mathcal{H}}_{I_i,+\frac{\Delta\mathbf{r}_{||}}{2}})$, $\mathrm{lin}(\overset{[-]}{\mathcal{H}}_{I_i,-\frac{\Delta\mathbf{r}_{||}}{2}} \cup \overset{[+]}{\mathcal{H}}_{I_i,-\frac{\Delta\mathbf{r}_{||}}{2}})$ or $\mathrm{lin}(\overset{[-]}{\mathcal{H}}_{\bar{I}_j,+\frac{\Delta\mathbf{r}_{||}}{2}} \cup \overset{[+]}{\mathcal{H}}_{\bar{I}_j,+\frac{\Delta\mathbf{r}_{||}}{2}})$, $\mathrm{lin}(\overset{[-]}{\mathcal{H}}_{\bar{I}_j,-\frac{\Delta\mathbf{r}_{||}}{2}} \cup \overset{[+]}{\mathcal{H}}_{\bar{I}_j,-\frac{\Delta\mathbf{r}_{||}}{2}})$, as discussed for single electron states. As a result, $\tau_{ei_{DCBF}}$ does not exist for these states and, in consequence, the phase $\theta_{||}$ is uncertain for these quantum states. Hence, the electron doped quantum pair states can be described as follows: There will be a strong deterministic space-time relation of the ordered topological electron state, which is given by the eigentime $\tau_{ei_{||}} = n \cdot \tau_{CBF}$ and the fixed phase steps $\delta\phi_{S_{||}}$ including the periodicity of $n \cdot \delta\phi_{S_{||}}$. The deterministic space-time relation reflects the coherent transformation of all Hilbert subspaces in space and time. However, the expectation values of the phases $< \theta_{||} >$, $< \phi_{S_{||}} >$ are free variables at any time t, based on the possibility of probabilistic superpositions of states which belong initially to different Hilbert subspaces which are associated to the $(-)$ and $(+)$ lattice sites, respectively. Therefore, at any time t the strongly time-periodic oscillations of the topological electron structure, given by $\tau_{ei_{||}}$ and $n \cdot \delta\phi_{S_{||}}$, can be superimposed by shifts of the expectation values of the particular phases which imply charge density variations.

In conclusion, the phase of the condensate can occur as free variable, with respect to the energetically higher n'' electrons. The corresponding phases $\theta_{||}$ and $\phi_{S_{||}}$ of the energetically deeper n' electrons are further on time-quantized, at least for the b-holes, and can not behave as a free variable. Altogether, the electron dynamics and especially the conductivity in hole doped and electron doped high-T_c materials will be distinctly different, as pointed out in the introduction.

9. Experimental proofs

Direct experimental proofs of the here deduced extraordinary characteristics of the quantum states in space and time require well-directed experiments which have yet to be done. Nevertheless, there is a broad experimental support for the here deduced quantum states as already discussed in our previous articles [7, 8]. More detailed experimental comparisons usually require a broader discussion, which will be done in separated articles.

A rather definite proof of the here deduced quantum behaviour can be given by experimental methods which are sensitive to the effective mass of the charge carriers, because, the effective mass $m_{||}$ [5] along the particular one-dimensional states I_i, $\bar{I}_j$ will be infinite, $m_{||} = \infty$, if the local state function $u_{o_{||}}$ exclusively exists but $m_{||}$ will be finite, if $u_{o_{||}}$ is split up into the coordinate shifted relative states states $u_{L,-\frac{\Delta r_{||}}{2}}$ and $u_{R,\frac{\Delta r_{||}}{2}}$, according to Fig. 2. If at no time holes exist within the particular one-dimensional states I_i or $\bar{I}_j$, the local states within these one-dimensional states will have exclusively $u_{o_{||}}$ character,[3] but the local states $u_{o_{||}}$ will be split up into the states $u_{L,-\frac{\Delta r_{||}}{2}}$ and $u_{R,\frac{\Delta r_{||}}{2}}$, if an ordered topological hole structure exists. Even if topological holes are only temporarily existent within the particular states I_i or $\bar{I}_j$, contributions of the split local states $u_{L,-\frac{\Delta r_{||}}{2}}$, $u_{R,\frac{\Delta r_{||}}{2}}$ have to be assumed.[6]

A decisive proof can be given by Hall effect measurements, because the b-holes cannot create a Hall current, if the b-hole state is permanently polarized along one polarization direction, e.g. $y(b)$ direction in Fig. 1, and if no f-holes exist which is given for $n_h \leq 0.125$ holes/copper. In this case, the states within I_x, $\bar{I}_x$ in Fig. 1 never carry holes, i.e. $m_{||} = \infty$ for these states. This means, the effective mass $m_\perp$ for perpendicular electronic state transformations with respect to the initial electronic states which belong to I_y, $\bar{I}_y$ is infinite, $m_\perp = \infty$. Therefore, a Hall current can not exist. If additionally f-holes exist ($f_{||}$-holes + $f_\perp$-holes), the effective mass $m_{||}$ along the states I_x, $\bar{I}_x$ which carry $f_\perp$-holes will be permanently or temporarily finite, i.e. a Hall current can be formed. Even this is found by Hall effect measurements of T. Noda et al. on $La_{2-x-y}Nd_ySr_xCuO_4$ with $x = 0.10$, 0.12, 0.13, 0.14 and $y = 0.4$, 0.6 [27]. Below a characteristic temperature T_o the Hall coefficient R_H changes significantly in dependence on x. For $x = 0.1$, 0.12 R_H drops to zero when lowering the temperature, whereas for $x = 0.13$, 0.15 R_H remains finite for $T \rightarrow 0$. This reflects the different behaviour of the b-hole and f-hole system. Below a temperature T_o the hole states are assumed to be "pinned" [27]. In the theory derived here this means that there is only one polarization direction for the topological b-hole state ($k_B T_o$ represents an activation barrier for switching the polarization direction). For $x = 0.1$, 0.12 ($n_h =$ 0.1, 0.12 holes/copper) only b-holes exist, i.e. $m_\perp = \infty$, whereas for x = 0.13, 0.15 ($n_h =$ 0.13, 0.15 holes/copper) b-holes and f-holes exist meaning $m_\perp$ will become finite, even if the b-hole polarization direction is stationary.

If the b-hole structure fluctuates between the two polarization directions $x(a)$ and $y(b)$

[5] The effective mass $m_{||}$ is related to transitions between local states $\overset{[-]}{u}$ and $\overset{[+]}{u}$, but not identical with the effective band mass $m_{\mathbf{k}_{||}}$ which must be set infinite for $-\mathbf{k}_{||}$,$\mathbf{k}_{||}$ pair states.

[6] This, however, includes the problem of quantum relaxations, which is generally very complex.

(full dynamic TR state [8]), the one dimensional states I_x, $\bar{I}_x$ as well as I_y, $\bar{I}_y$ carry temporarily hole states, leading to the formation of a Hall current, even if exclusively b-hole states given for $n_h \leq 0.125$ holes/copper exist. This is impressively reflected by the experiments of S. Arumugan et al. [28]. It was shown that a pressure induced structural transition from the LTT (low temperature tetragonal) to the LTO (low temperature orthogonal) phase in $La_{1.48}Nd_{0.4}Sr_{0.12}CuO_4$ ($n_h \simeq 0.12$ holes/copper) [see also the neutron scattering experiments for LTT↔LTO transitions of M. Fujita et al.[29]] leads to a recovering of the Hall coefficient R_H that is different from zero. This was interpreted as "de-pinning". In the TR state theory, the pinning and de-pinning effect is not caused by a magnetic proximity effect but describes the quantum transition from the constrained TR state, where only one b-hole polarization direction exists, to the full TR state, where the b-hole state is temporarily polarized in x(a) and y(b) direction.

10. Conclusions

The comprehensive analysis of the electronic states within the CuO_2 planes [7] has shown that the quantum states in the CuO_2 planes can not be assumed any longer to be adapted to the crystal symmetry. Self-consistent renormalizations of the electronic states lead to an electronic superstructure, the CBF state, which subdivides the initial crystal lattice into two sublattices. Inevitably, the initial Hilbert space $\mathcal{H}$ is renormalized ($\mathcal{H}_{ren.}$) and $\mathcal{H}_{ren.}$ is transformed into the Hilbert subspaces $\overset{[-]}{\mathcal{H}}_{-\frac{\Delta\mathbf{r}}{2}}$ and $\overset{[+]}{\mathcal{H}}_{+\frac{\Delta\mathbf{r}}{2}}$. The two subspaces form the orthogonal complement to each other, but beyond that they represent a definite relative state with respect to their particular coordinate systems given by the relative coordinate shifts $\pm\frac{\Delta\mathbf{r}}{2}$. Regarding the half-filled valence bands, the two dimensional Hilbert subspaces are additionally transformed into arrays of one-dimensional Hilbert subspaces $\{\overset{[-]}{\mathcal{H}}_{I_i,-\frac{\Delta\mathbf{r}_{||}}{2}}, \overset{[+]}{\mathcal{H}}_{I_i,+\frac{\Delta\mathbf{r}_{||}}{2}}, \overset{[-]}{\mathcal{H}}_{\bar{I}_j,+\frac{\Delta\mathbf{r}_{||}}{2}}, \overset{[+]}{\mathcal{H}}_{\bar{I}_j,-\frac{\Delta\mathbf{r}_{||}}{2}}\}$. Again, the subspaces $\overset{[-]}{\mathcal{H}}_{I_i,-\frac{\Delta\mathbf{r}_{||}}{2}}$ and $\overset{[+]}{\mathcal{H}}_{I_i,+\frac{\Delta\mathbf{r}_{||}}{2}}$ as well as $\overset{[-]}{\mathcal{H}}_{\bar{I}_j,+\frac{\Delta\mathbf{r}_{||}}{2}}$ and $\overset{[+]}{\mathcal{H}}_{\bar{I}_j,-\frac{\Delta\mathbf{r}_{||}}{2}}$ form the orthogonal complement to each other, and their individual coordinate systems are definitely related to each other by the relative shifts $\pm\frac{\Delta\mathbf{r}_{||}}{2}$ of the coordinate systems of the two Hilbert subspaces along I_i and $\bar{I}_j$ in Fig. 1, respectively. Altogether, the set of one-dimensional subspaces $\{\overset{[-]}{\mathcal{H}}_{I_i,-\frac{\Delta\mathbf{r}_{||}}{2}}, \overset{[+]}{\mathcal{H}}_{I_i,+\frac{\Delta\mathbf{r}_{||}}{2}}, \overset{[-]}{\mathcal{H}}_{\bar{I}_j,+\frac{\Delta\mathbf{r}_{||}}{2}}, \overset{[+]}{\mathcal{H}}_{\bar{I}_j,-\frac{\Delta\mathbf{r}_{||}}{2}}\}$ define a definite relative state with respect to their coordinate systems. For half-filled valence bands and under hole doping, the exceptional property is that these Hilbert subspaces are deterministically time dependent, i.e. their coordinate systems are periodic functions of time along the space coordinate $\mathbf{r}_{||}$ within a space interval of $2\Delta\mathbf{r}_{||}$, with $\Delta\mathbf{r}_{||}$ being the lattice constant in a or b direction. In this case, the stationarity of a basis state is not defined as a time independent state which is instantaneously given, as usually supposed in quantum mechanics, but is defined for periodically equivalent internal times t_{int} according to

$$t_{int} = t' + n \cdot \tau_{ei_{DCBF}} \qquad (n\,\text{integer}). \tag{62}$$

Here, $\tau_{ei_{DCBF}}$ defines the oscillation time (eigentime) for periodic oscillations of the coordinate systems of the Hilbert subspaces. The time $t'\epsilon[0, \tau_{ei_{DCBF}}]$ is a definite internal time variable which describes the periodic time dependence of the coordinate systems of the Hilbert subspaces (Eq. (24)). The integer number n occurs as a probabilistic variable describing the uncertainty of the absolute position of the common center point coordinate of the Hilbert subspaces $\{\overset{[-]}{\mathcal{H}}_{I_i,-\frac{\Delta\mathbf{r}_{||}}{2}}, \overset{[+]}{\mathcal{H}}_{I_i,+\frac{\Delta\mathbf{r}_{||}}{2}}, \overset{[-]}{\mathcal{H}}_{\bar{I}_j,+\frac{\Delta\mathbf{r}_{||}}{2}}, \overset{[+]}{\mathcal{H}}_{\bar{I}_j,-\frac{\Delta\mathbf{r}_{||}}{2}}\}$ in time, which is quantized by $\tau_{ei_{DCBF}}$. Two possible quantum state interpretations are valid:

(a) Time integrated states over t' between 0 and $\tau_{ei_{DCBF}}$.

(b) Time instantaneous states with $t' = const.$ according to Eq. (62).

In both cases the time scale t_{int} has to be considered as quantized in time by $\tau_{ei_{DCBF}}$. In case (a) the eigentime $\tau_{ei_{DCBF}}$ appears to be a hidden variable, because the integrated quantum state is equivalent for all times $n \cdot \tau_{ei_{DCBF}}$, i.e. the quantum state is stationary. However, one has to bear in mind that the time scale is not continuous but quantized by $\tau_{ei_{DCBF}}$. In (b) the time t' occurs explicitly, but equivalent quantum states are only given for multiples of $\tau_{ei_{DCBF}}$, i.e. the stationarity is again defined for multiples of $\tau_{ei_{DCBF}}$. The full symmetry of the whole quantum state is, however, only defined for case (a), because the antisymmetric pair states according to Eqs. (16)-(18) exist only as integrated states over the full time period $\tau_{ei_{DCBF}}$. If crystal symmetry adapted quantum states are given, individual dynamic coordinate systems of particular quantum states may exist but the corresponding eigentimes are completely uncertain. This is equivalent with a disappearing of the definite time variable t'. This corresponds to $\tau_{ei} = 0$, i.e. the internal time variable t_{int} disappears, as usually supposed in quantum mechanics, leading to time stationary quantum states which are instantaneously defined and unchanged at all times.

If additional electrons n_e are injected into the two times negatively charged CuO_2^{2-} plane, i.e. electron-doped (n-type) cuprates, the initially unpopulated Hilbert subspaces $\overset{[+]}{\mathcal{H}}_{I_i,+\frac{\Delta\mathbf{r}_{||}}{2}}$ and/or $\overset{[+]}{\mathcal{H}}_{\bar{I}_j,-\frac{\Delta\mathbf{r}_{||}}{2}}$ will be populated. As a result, $2n_e$ electronic states in $\overset{[-]}{\mathcal{H}}_{I_i,-\frac{\Delta\mathbf{r}_{||}}{2}}$ and $\overset{[+]}{\mathcal{H}}_{I_i,+\frac{\Delta\mathbf{r}_{||}}{2}}$ remain static, i.e. they are not quantum entangled to the DCBF state. The same occurs for the one-dimensional states along $\bar{I}_j$, if electrons are additionally injected into $\overset{[+]}{\mathcal{H}}_{\bar{I}_j}$. Therefore, the eigentime $\tau_{ei_{DCBF}}$ does not exist for these quantum states, so that t_{int} becomes zero for these quantum states. At any time, states within the linear spans $\mathrm{lin}(\overset{[-]}{\mathcal{H}}_{I_i,+\frac{\Delta\mathbf{r}_{||}}{2}} \cup \overset{[+]}{\mathcal{H}}_{I_i,+\frac{\Delta\mathbf{r}_{||}}{2}})$, $\mathrm{lin}(\overset{[-]}{\mathcal{H}}_{I_i,-\frac{\Delta\mathbf{r}_{||}}{2}} \cup \overset{[+]}{\mathcal{H}}_{I_i,-\frac{\Delta\mathbf{r}_{||}}{2}})$ or $\mathrm{lin}(\overset{[-]}{\mathcal{H}}_{\bar{I}_j,+\frac{\Delta\mathbf{r}_{||}}{2}} \cup \overset{[+]}{\mathcal{H}}_{\bar{I}_j,+\frac{\Delta\mathbf{r}_{||}}{2}})$, $\mathrm{lin}(\overset{[-]}{\mathcal{H}}_{\bar{I}_j,-\frac{\Delta\mathbf{r}_{||}}{2}} \cup \overset{[+]}{\mathcal{H}}_{\bar{I}_j,-\frac{\Delta\mathbf{r}_{||}}{2}})$ may be occupied. In this respect more customary single or paired quantum states occur for these $2n_e$ electrons. However, the dynamic DCBF state is existent further on which generates some exceptional dynamic conditions, for these $2n_e$ electrons as well.

If the intrinsic eigentime $\tau_{ei_{DCBF}}$ exists for the quantum states, free electronic transitions between different Hilbert subspaces are forbidden at any time t, as long as the Hilbert subspaces belong to different coordinate systems. Any electron transition between Hilbert subspaces that does not obey the strongly deterministic time behaviour according to Eq. (24) breaks/destroys the quantum state itself, because $\tau_{ei_{DCBF}}$ is no longer a conservation quantity. This is the reason way the undoped twice negatively charged CuO_2^{2-} plane behaves like an insulator. These conditions would also be valid under hole doping, if the Hilbert subspaces $\{\overset{[-]}{\mathcal{H}}_{I_i,-\frac{\Delta\mathbf{r}_{||}}{2}}, \overset{[+]}{\mathcal{H}}_{I_i,+\frac{\Delta\mathbf{r}_{||}}{2}}, \overset{[-]}{\mathcal{H}}_{\bar{I}_j,+\frac{\Delta\mathbf{r}_{||}}{2}}, \overset{[+]}{\mathcal{H}}_{\bar{I}_j,-\frac{\Delta\mathbf{r}_{||}}{2}}\}$ remained unchanged. However, the highly correlated $-\mathbf{k}_{||}, \mathbf{k}_{||}$ electron pair states under hole doping lead to an additional splitting of every particular Hilbert subspace into two further subspaces according to Eqs. (51),(53),(54) and (56). This creates Hilbert subspaces according to $\overset{[-]}{\widetilde{\mathcal{H}}}_{I_i,0}$ and $\overset{[+]}{\widetilde{\mathcal{H}}}_{I_i,0}$ as well as $\overset{[-]}{\widetilde{\mathcal{H}}}_{I_i,-\Delta\mathbf{r}_{||}}$ and $\overset{[+]}{\widetilde{\mathcal{H}}}_{I_i,-\Delta\mathbf{r}_{||}}$ (modulo $2\Delta\mathbf{r}_{||}$), and along $\bar{I}_j$ equivalent transformations occur. Now, the common coordinate systems in $\overset{[-]}{\widetilde{\mathcal{H}}}_{I_i,0}$ and

$\overset{[+]}{\widetilde{\mathcal{H}}}_{I_i,0}$ as well as $\overset{[-]}{\widetilde{\mathcal{H}}}_{I_i,-\Delta\mathbf{r}_{||}}$ and $\overset{[+]}{\widetilde{\mathcal{H}}}_{I_i,-\Delta\mathbf{r}_{||}}$ enables accidental electronic transitions between the Hilbert spaces $\overset{[-]}{\widetilde{\mathcal{H}}}_{I_i,0}$ and $\overset{[+]}{\widetilde{\mathcal{H}}}_{I_i,0}$ as well as $\overset{[-]}{\widetilde{\mathcal{H}}}_{I_i,-\Delta\mathbf{r}_{||}}$ and $\overset{[+]}{\widetilde{\mathcal{H}}}_{I_i,-\Delta\mathbf{r}_{||}}$ at any time t without breaking $\tau_{ei_{DCBF}}$. As a result, electronic transformations along the one-dimensional states I_i, $\bar{I}_j$ will be possible and, in consequence, conductivity can occur. In conclusion, the insulator-conductor transition in high-T_c cuprates is the result of an additional splitting of the Hilbert subspace being accompanied with coordinate transformations where the copper site centered Hilbert subspaces have been transformed into oxygen site centered Hilbert subspaces. In electron doped (n-type) cuprates, $\tau_{ei_{DCBF}}$ is not a conservation quantity for the $2n_e$ conduction electrons and, therefore, electronic transitions between particular Hilbert subspaces may always occur. The dynamic influence of the DCBF state on the $2n_e$ conduction electrons is merely a mean field effect, the electron doped quantum states are not time-quantization by $\tau_{ei_{DCBF}}$. However, isolating states can also be formed if the doped holes or electrons localize, which may be possible for small hole or electron doping, if clustered topological electron structures occur.

In summary, the TR conception of the quantum states in high-T_c cuprates, recently derived [8], enables a complete new interpretation of the electron dynamics in the CuO_2 planes of these materials. Separated Hilbert subspaces form a highly correlated compound system including a definite/deterministic relative state of their coordinate systems in space and time. The coordinate systems themselves occur as variables. A particular splitting of the Hilbert subspaces accompanied with coordinate transformations are responsible for the insulator-conductor transition under hole doping. Mostly unusual is the quantization of time by an internal time constant $\tau_{ei_{DCBF}}$ termed as eigentime, which occurs for undoped (half-filled) and hole doped cuprates as a conservation quantity but for electron doping a separated electron fluid appears in which $\tau_{ei_{DCBF}}$ is not a conservation quantity. The time quantization of the quantum states implicates a deterministic space-time behaviour of quantum states, meaning that the quantum realm encompasses classical behaviour. This result motivates the reflection about the general interpretation of quantum mechanics.

Appendix: Probabilistic superposition of Hilbert spaces

The separation of Hilbert spaces in Hilbert subspaces has been elaborated as a striking characteristic of the quantum states within the CuO_2 planes. This is causally connected with the formation of definite relative coordinates of the separated Hilbert spaces, i.e. a definite relative state of the individual coordinate systems, which may be dynamical. Most unusually, a time-deterministic dynamic relative state of the coordinate systems was concluded as described in Sec. 5. Over and above that time independent, i.e. probabilistic dynamic relative states of the coordinate systems, have to be taken into consideration as well, as concluded in Sec. 7.

Under half filling, separated Hilbert spaces occur which are related to the copper sites. Subsequently, these particular Hilbert spaces are defined by the Hilbert space $\mathcal{H}_o$.

$$\mathcal{H}_o \equiv \overset{[-]}{\mathcal{H}}_{I_i,-\frac{\Delta\mathbf{r}_{||}}{2}}, \overset{[-]}{\mathcal{H}}_{I_i,+\frac{\Delta\mathbf{r}_{||}}{2}}, \overset{[+]}{\mathcal{H}}_{I_i,-\frac{\Delta\mathbf{r}_{||}}{2}}, \overset{[+]}{\mathcal{H}}_{I_i,+\frac{\Delta\mathbf{r}_{||}}{2}} \tag{A.1}$$

or

$$\mathcal{H}_o \equiv \overset{[-]}{\mathcal{H}}_{\bar{I}_i,-\frac{\Delta\mathbf{r}_{||}}{2}}, \overset{[-]}{\mathcal{H}}_{\bar{I}_i,+\frac{\Delta\mathbf{r}_{||}}{2}}, \overset{[+]}{\mathcal{H}}_{\bar{I}_i,-\frac{\Delta\mathbf{r}_{||}}{2}}, \overset{[+]}{\mathcal{H}}_{\bar{I}_i,+\frac{\Delta\mathbf{r}_{||}}{2}} \tag{A.2}$$

and equivalently for I'_i and $\bar{I}'_i$. If the electron density is reduced, the formation of topological hole states is accompanied by a splitting of the local wave function $u_{o_{||}}$ into left and right local wave functions $u_{L,-\frac{\Delta\mathbf{r}_{||}}{2}}$ and $u_{L,\frac{\Delta\mathbf{r}_{||}}{2}}$ as discussed in Sec. 7, where the latter are associated with the oxygen sites contrary to the $u_{o_{||}}$ function which is assigned to the copper sites. This leads to the splitting of the Hilbert space $\mathcal{H}_o$ into separated subspaces $\mathcal{H}_{o,+\frac{\Delta\mathbf{r}_{||}}{2}}$ and $\mathcal{H}_{o,-\frac{\Delta\mathbf{r}_{||}}{2}}$ which are related to the Hilbert spaces in Eqs. (39) - (46) according to

$$\begin{aligned}\mathcal{H}_{o,} \to \mathcal{H}_{o,+\frac{\Delta\mathbf{r}_{||}}{2}}, \mathcal{H}_{o,-\frac{\Delta\mathbf{r}_{||}}{2}} \equiv & [\overset{[-]}{\mathcal{H}}_{I_i,\Delta\mathbf{r}_{-\Delta\phi_{||}}}, \overset{[-]}{\mathcal{H}}_{I_i,\Delta\mathbf{r}_{\Delta\phi_{||}}}], \\ & [\overset{[+]}{\mathcal{H}}_{I_i,\Delta\mathbf{r}_{-\Delta\phi_{||}}}, \overset{[+]}{\mathcal{H}}_{I_i,\Delta\mathbf{r}_{\Delta\phi_{||}}}], \\ & [\overset{[-]}{\mathcal{H}}_{\bar{I}_i,\Delta\mathbf{r}_{-\Delta\phi_{||}}}, \overset{[-]}{\mathcal{H}}_{\bar{I}_i,\Delta\mathbf{r}_{\Delta\phi_{||}}}], \\ & [\overset{[+]}{\mathcal{H}}_{\bar{I}_i,\Delta\mathbf{r}_{-\Delta\phi_{||}}}, \overset{[+]}{\mathcal{H}}_{\bar{I}_i,\Delta\mathbf{r}_{\Delta\phi_{||}}}],\end{aligned} \tag{A.3}$$

and equivalently for I'_i and $\bar{I}'_i$, where the assignment between the Hilbert spaces on the left and right side of Eq. (A.3) depends on the considered states in Eqs. (A.1) and (A.2). The subsequent local states are defined in dependence on the given coordinate system according to

$$u_{R,\frac{\Delta\mathbf{r}_{||}}{2}}(\mathbf{r}), u_{L,\frac{\Delta\mathbf{r}_{||}}{2}}(\mathbf{r}) \;\; \epsilon \;\; \mathcal{H}_{o,+\frac{\Delta\mathbf{r}_{||}}{2}} \tag{A.4}$$

$$u_{R,-\frac{\Delta\mathbf{r}_{||}}{2}}(\mathbf{r}), u_{L,-\frac{\Delta\mathbf{r}_{||}}{2}}(\mathbf{r}) \;\; \epsilon \;\; \mathcal{H}_{o,-\frac{\Delta\mathbf{r}_{||}}{2}} \tag{A.5}$$

There is no time ordering for a dynamic association of the states u_L and u_R in Eqs. (A.4) and (A.5) to the two Hilbert subspaces which allows a probabilistic superposition of these two quantum states within a common or separated Hilbert subspace, i.e. if the coordinate system is preserved or changed into the complementary coordinate system. Hence, the subsequent states, respective Hilbert spaces, may occur simultaneously in a quantum probabilistic sense.

$$u'_{R,\frac{\Delta\mathbf{r}_{||}}{2}}(\mathbf{r}) = u_{R,\frac{\Delta\mathbf{r}_{||}}{2}}(\mathbf{r}) + f_{R,\frac{\Delta\mathbf{r}_{||}}{2}}(\mathbf{r}) u_{L,\frac{\Delta\mathbf{r}_{||}}{2}}(\mathbf{r}) \, \epsilon \, \mathcal{H}_{o,+\frac{\Delta\mathbf{r}_{||}}{2}} \tag{A.6}$$

$$u''_{L,-\frac{\Delta\mathbf{r}_{||}}{2}}(\mathbf{r}) = f_{R,\frac{\Delta\mathbf{r}_{||}}{2}}(\mathbf{r}) u_{L,-\frac{\Delta\mathbf{r}_{||}}{2}}(\mathbf{r}) \, \epsilon \, \mathcal{H}_{o,-\frac{\Delta\mathbf{r}_{||}}{2}} \tag{A.7}$$

with

$$f_{R,\frac{\Delta\mathbf{r}_{||}}{2}}(\mathbf{r}) = \left(\frac{u^2_{R,\frac{\Delta\mathbf{r}_{||}}{2}}(\mathbf{r})}{u^2_{R,\frac{\Delta\mathbf{r}_{||}}{2}}(\mathbf{r}) + u^2_{L,-\frac{\Delta\mathbf{r}_{||}}{2}}(\mathbf{r})} \right) \tag{A.8}$$

and

$$u'_{L,-\frac{\Delta\mathbf{r}_{||}}{2}}(\mathbf{r}) = u_{L,-\frac{\Delta\mathbf{r}_{||}}{2}}(\mathbf{r}) + f_{L,-\frac{\Delta\mathbf{r}_{||}}{2}}(\mathbf{r}) u_{R,-\frac{\Delta\mathbf{r}_{||}}{2}}(\mathbf{r}) \, \epsilon \, \mathcal{H}_{o,-\frac{\Delta\mathbf{r}_{||}}{2}} \tag{A.9}$$

$$u''_{R,\frac{\Delta\mathbf{r}_{||}}{2}}(\mathbf{r}) = f_{L,-\frac{\Delta\mathbf{r}_{||}}{2}}(\mathbf{r}) u_{R,\frac{\Delta\mathbf{r}_{||}}{2}}(\mathbf{r}) \, \epsilon \, \mathcal{H}_{o,+\frac{\Delta\mathbf{r}_{||}}{2}} \tag{A.10}$$

with

$$f_{L,-\frac{\Delta\mathbf{r}_{||}}{2}}(\mathbf{r}) = \left(\frac{u^2_{L,-\frac{\Delta\mathbf{r}_{||}}{2}}(\mathbf{r})}{u^2_{R,\frac{\Delta\mathbf{r}_{||}}{2}}(\mathbf{r}) + u^2_{L,-\frac{\Delta\mathbf{r}_{||}}{2}}(\mathbf{r})} \right). \tag{A.11}$$

From this, two particular quantum states can be defined according to

$$\begin{aligned} \tilde{u}_{R,\frac{\Delta \mathbf{r}_{||}}{2}}(\mathbf{r}) &= \beta_{\frac{\Delta \mathbf{r}_{||}}{2}} \cdot u'_{R,\frac{\Delta \mathbf{r}_{||}}{2}}(\mathbf{r}). \oplus \beta_{-\frac{\Delta \mathbf{r}_{||}}{2}} \cdot u''_{L,-\frac{\Delta \mathbf{r}_{||}}{2}}(\mathbf{r}) \\ &\epsilon\ \widetilde{\mathcal{H}}_{o,+\frac{\Delta \mathbf{r}_{||}}{2}} \end{aligned} \tag{A.12}$$

with $\beta_{\frac{\Delta \mathbf{r}_{||}}{2}} = 1$ and $\beta_{-\frac{\Delta \mathbf{r}_{||}}{2}} = -1$ and

$$\begin{aligned} \tilde{u}_{L,-\frac{\Delta \mathbf{r}_{||}}{2}}(\mathbf{r}) &= \beta_{-\frac{\Delta \mathbf{r}_{||}}{2}} \cdot u'_{L,-\frac{\Delta \mathbf{r}_{||}}{2}}(\mathbf{r}) \oplus \beta_{\frac{\Delta \mathbf{r}_{||}}{2}} \cdot u''_{L,\frac{\Delta \mathbf{r}_{||}}{2}}(\mathbf{r}) \\ &\epsilon\ \widetilde{\mathcal{H}}_{o,-\frac{\Delta \mathbf{r}_{||}}{2}} \end{aligned} \tag{A.13}$$

with $\beta_{\frac{\Delta \mathbf{r}_{||}}{2}} = -1$ and $\beta_{-\frac{\Delta \mathbf{r}_{||}}{2}} = 1$, respectively. Then, the electron density distribution is given by

$$\begin{aligned} \tilde{u}^2_{R,\frac{\Delta \mathbf{r}_{||}}{2}}(\mathbf{r}) &= u^2_{R,\frac{\Delta \mathbf{r}_{||}}{2}}(\mathbf{r}) + 2 \cdot u_{R,\frac{\Delta \mathbf{r}_{||}}{2}}(\mathbf{r}) \cdot u_{L,\frac{\Delta \mathbf{r}_{||}}{2}}(\mathbf{r}) \\ &\cdot \left(\frac{u^2_{R,\frac{\Delta \mathbf{r}_{||}}{2}}(\mathbf{r})}{u^2_{R,\frac{\Delta \mathbf{r}_{||}}{2}}(\mathbf{r}) + u^2_{L,-\frac{\Delta \mathbf{r}_{||}}{2}}(\mathbf{r})} \right). \end{aligned} \tag{A.14}$$

and

$$\begin{aligned} \tilde{u}^2_{L,-\frac{\Delta \mathbf{r}_{||}}{2}}(\mathbf{r}) &= u^2_{L,-\frac{\Delta \mathbf{r}_{||}}{2}}(\mathbf{r}) + 2 \cdot u_{L,-\frac{\Delta \mathbf{r}_{||}}{2}}(\mathbf{r}) \cdot u_{R,-\frac{\Delta \mathbf{r}_{||}}{2}}(\mathbf{r}) \\ &\cdot \left(\frac{u^2_{L,-\frac{\Delta \mathbf{r}_{||}}{2}}(\mathbf{r})}{u^2_{R,\frac{\Delta \mathbf{r}_{||}}{2}}(\mathbf{r}) + u^2_{L,-\frac{\Delta \mathbf{r}_{||}}{2}}(\mathbf{r})} \right), \end{aligned} \tag{A.15}$$

respectively, which leads to the common state function

$$\begin{aligned} \tilde{u}_{o_{||}}(\mathbf{r}) &= \tilde{u}_{R,\frac{\Delta \mathbf{r}_{||}}{2}}(\mathbf{r}) \oplus \tilde{u}_{L,-\frac{\Delta \mathbf{r}_{||}}{2}}(\mathbf{r}) \\ &= \left(u'_{R,\frac{\Delta \mathbf{r}_{||}}{2}}(\mathbf{r}) \ominus u''_{L,-\frac{\Delta \mathbf{r}_{||}}{2}}(\mathbf{r}) \right) \oplus \left(u'_{L,-\frac{\Delta \mathbf{r}_{||}}{2}}(\mathbf{r}) \ominus u''_{R,+\frac{\Delta \mathbf{r}_{||}}{2}}(\mathbf{r}) \right). \end{aligned} \tag{A.16}$$

For the electron density distribution results

$$u^2_{o_{||}}(\mathbf{r}) = \tilde{u}^2_{o_{||}}(\mathbf{r}) = \tilde{u}^2_{R,\frac{\Delta \mathbf{r}_{||}}{2}}(\mathbf{r}) + \tilde{u}^2_{L,-\frac{\Delta \mathbf{r}_{||}}{2}}(\mathbf{r}), \tag{A.17}$$

which is in accordance with prior discussions in Sec. 7 (Eq. (59)). Eqs. (A.6) and (A.9) reflect superpositions of quantum states within the same Hilbert space, which include interference terms, whereas Eqs. (A.12) and (A.13) are dynamic superpositions of different Hilbert spaces preventing interferences as the consequence of different coordinate systems. This interplay between dynamic and static coordinate systems makes in the first place the quantum states represented by Eqs. (A.12) and (A.13) possible, as discussed in Sec. 7. The norm of Eqs. (A.12) and (A.13) is 1/2 which means that only one electron is nominally associated with particular oxygen site, with the consequence that maximally one topological hole may be placed at a given oxygen site, as it has previously already been deduced [7].

References

[1] Anderson P W 1997 *Theory of Superconductivity in the High-T_c Cuprates* (Princeton University Press, Princeton)
[2] Ogata M and Fugkuyama H 2008 *Rep. Prog. Phys.* **71** 036501
[3] Kivelson S A, Bindloss I P, Fradkin E, Oganesyan V, Tranquada J M, Kapitulnik A and Howald C 2003 *Rev. Mod. Phys.* **75** 1201
[4] Basov D N and Timusk T 2005 *Rev. Mod. Phys.* **77** 721
[5] Edegger B, Muthukumar V N and Gros C 2007 *Adv. in Phys.* **56** 927
[6] Lee P A 2008 Rep. Prog. Phys. **71** 012501
[7] Winkler W and Winkler K 2006 *Physica* C **450** 1
[8] Winkler W and Winkler K 2007 *Physica* C **457** 1
[9] Tsuei C C and Kirtley J R 2000 *Rev. Mod. Phys.* **72** 969
[10] Shen Z X and Dessau D S 1995 *Phys. Rep.* **253** 1
[11] Ding H, Norman M R, Mochiku T, Kadowaki K and Giapinzakis J 1996 *Nature (London)* **382** 51
[12] Ma J, Quitmann C, Kelley R J, Berger H, Margaritondo G and Onellion M 1995 *Science* **267** 862
[13] Sachdev S 2000 *Science* **288** 475
[14] Sachdev S 2003 *Rev. Mod. Phys.* **75** 913
[15] Emery V J, Kivelson S A and Zachar O 1997 *Phys. Rev.* B **56** 6120
[16] Kivelson S A, Fradkin E and Emery V J 1998 *Nature (London)* **393** 550
[17] Tokura Y, Takagi H and Uchida S 1989 *Nature (London)* **337** 345
Takagi H, Uchida S, and Tokura Y 1989 *Phys. Rev. Lett.* **62** 1197
[18] Tsuei C C, Gupta A and Koren G 1989 *Physica C* **161** 415
Peng J L, Maiser E, Venkatesan T, Greene R L, and Czjzek G 1997 *Phys. Rev.* B **55** R6145
[19] Damascelli A, Hussain Z and Shen Z W 2003 *Rev. Mod. Phys.* **75** 473
[20] Keimer B, Belk N, Birgeneau R J, Cassanho A, Chen C Y, Greven M, Kastner M A, Aharony A, Endoh Y, Erwin R W and Shirane G 1992 *Phys. Rev.* B **46** 14034
Matsuda M, Yamada K, Kakurai K, Kadowaki H, Thurston T R, Endoh Y, Hidaka Y, Birgeneau R J, Kastner M A, Gehring P M, Moudden A H and Shirane G 1990 *Phys. Rev.* B **42** 10098
Vaknin D, Sinha S K, Stassis C, Miller L L, and Johnston D C 1990 *Phys. Rev.* B **41** 1926
Greven M, Birgeneau R J, Endoh Y, Kastner M A, Matsuda M and Shirane G 1995 *Z. Phys.* B **96** 465
Shamoto S, Sato M, Tranquada J M, Sternlieb B J, and Shirane G 1993 *Phys. Rev.* B **48** 13817
Thurston T R, Matsuda M, Kakurai K, Yamada K, Endoh Y, Birgeneau R J, Gehring P M, Hidaka Y, Kastner M A, Murakami T and Shirane G 1990 *Phys. Rev. Lett.* **65** 263
[21] Uemura Y J, Kossler W J, Yu X H, Kempton J R, Schöne H E, Opie D, Stronach C E, Johnston D C, Alvarez M S and Goshorn D P 1987 *Phys. Rev. Lett.* **59** 1045
Nachumi B, Fudamoto Y, Keren A, Kojima K M, Larkin M, Luke G M, Merrin J, Tschernyshyov O, Uemura Y J, Ichikawa N, Goto M, Takagi H, Uchida S, Crawford M K, McCarron E M, MacLaughlin D E and Heffner R H 1998 *Phys. Rev.* B **58** 8760
[22] Blanchard P and Brüning E 2003 *Mathematical Mehtods in Physics: Distributions, Hilbert Space Operators, and Variational Methods* (Birkhäuser Boston) pp 227-234
[23] Everett H 1957 *Rev. Mod. Phys.* **29** 454
[24] DeWitt B S 1970 *Physics Today* **23** 155
[25] DeWitt B S in: 1971 *Foundations of Quantum Mechanics* (Academic Press Inc., New York)
[26] DeWitt B S and Graham N (Eds.) 1973 *The Many-Worlds Interpretation of Quantum Mechanics* (Princeton University Press, Princeton, N.J.)
[27] Noda T, Eisaki H and Uchida S *Science* 286 (1999) 265
[28] Arumugam S, Môri N, Takeshita N, Takashima H, Noda T, Eisaki H and Uchida S 2002 *Phys. Rev. Lett.* **88** 247001
[29] Fujita M, Goka H, Yamada K and Matsuda M 2002 *Phys. Rev. Lett.* **88** 167008

Magnetic torque transferring study for bulk High-T_c superconductors and permanent magnets

A. Wongsatanawarid[1]**, A. Suzuki, H. Seki and M. Murakami**

Superconducting Materials Laboratory, Shibaura Institute of Technology, 3-7-5 Toyosu, Koto-ku, Tokyo 135-8548 Japan

Abstract. The torque transferring mechanism taking place in a superconducting mixer design has been studied. Several coupling magnetic arrangements were investigated for more details in the engineering design. A bulk superconductor sample was used to study the torque forces for various cooling gaps, and the twist angle dependence was also monitored for the rotational stiffness in stability. The experimental data with four permanent magnet configurations have been studied in the present work. The maximum torque forces are summarized for usage of engineering design with various gaps. The torque/gap characteristics for four configurations were also measured for the optimisation of the torque at a designed operating gap.

1. Introduction

One of the attractive properties of bulk RE-Ba-Cu-O (RE: rare earth elements) superconductors is that they can levitate a permanent magnet, and can firmly hold it in midair by virtue of flux pinning effect. Such levitation effect can lead to the applications to avoid the contact between materials and perfectly frictionless mechanism, for example, superconducting bearings, superconducting flywheel energy storage system, contact-less conveying system and levitated vehicles [1]. Superconducting materials can also provide torque forces when multi-pole magnetic fields were trapped inside the bulk superconductors. A mixing process without contamination can be performed with a superconducting mixer [2]. Such a device is attractive for medical applications where an ultra clean environment is required. In this study, we try to maximize the torque force between a bulk superconductor and permanent magnet in various configurations of magnetic poles. Furthermore, we studied the torque force as a function of gap in all the configurations.

The understanding of torque transferring mechanism is a goal of this study, which will lead us to develop a robust non-contact superconducting mixer that our research group have tackled in the previous studies [2, 3]. The new design of non-contact torque transferring mechanism was innovated as shown in Figure 1.

Bulk superconductors used in the present study were commercially available Dy-Ba-Cu-O bulk superconductors (Nippon steel) produced by a typical top-seeded melt-growth process. The chemical composition of the sample was $DyBa_2Cu_3O_{6.9}$ 75.0 wt%, $Dy_2BaCuO_{5.0}$ 24.5 wt% and Pt 0.5 wt%. The sample dimensions were 30 mm in diameter and 5 mm in thickness.

[1] Corresponding author. Tel. +81-3-5859-8117, Fax. +81-3-5859-81010, E-mail: lot407@yahoo.com

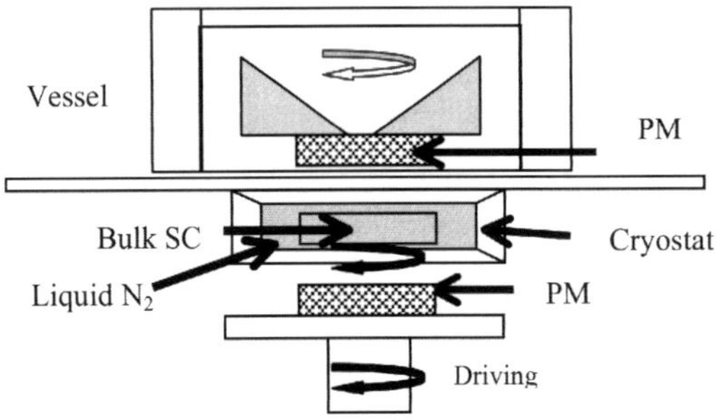

Figure 1. Schematic illustration for the basic idea of a non-contact superconducting mixer

The experimental study will verify the suitable configuration that will yield the optimum torque transferring characteristics. The operating gap dependence is also taken into account. However, uncomplicated arrangement of magnetic poles will be the first candidate to avoid the additive cost in practical applications.

2. Analytical model

Magnetic torque transferring characteristics have already been studied in various superconductor/permanent magnet configurations [4-6]. Tixador *et al.* [4] have studied the superconducting coupling device which is based on transverse force acting between multiple sets of superconductor/permanent magnet couples. However, they still need a rotational support to hold the coupling set of permanent magnets. Thus perfectly non-contact torque transferring was not achieved in their work.

Type-II superconductors can trap magnetic fields and thus behave like a permanent magnet in that the quantized fluxoids were pinned at normal conducting particles in a superconductor matrix. Physically, the pinning force which is induced by approaching a pole of permanent magnet toward a superconductor can repulse magnetic field, which leads to levitation and then free rotation in mid air. Therefore, we induced the pinning force of the superconductor with the different magnetic poles on the same face, which generated the torque coupling between them in a perfectly non-contact mode. However, we learned that the force between a permanent magnet and a bulk superconductor must be balanced in a vertical direction, because the repelling force of the superconductor against a permanent magnet was much higher than the attractive force at the designed gap.

In a radial direction, the total pinning force of an activated volume with the differential of magnetic flux in both radial and angular directions at various gaps have been studied. In this work we also tried to maximize the torque force by optimising the configuration of permanent magnets. The superconductor/permanent magnets interactions are considered based on Maxwell's equation and Ampere's law. Once we simplify by the magneto-quasi-static reduction thus we obtained,

$$\nabla \times B = \mu_0 J \tag{1}$$

In the torque coupling mechanism, however, the torque forces acting between a superconductor and permanent magnets will include the transverse force. To yield the maximum torque, the trapped pinning forces have to be induced as large as possible. Therefore, the transverse force between a superconductor and permanent magnet can be expressed as follows:

$$F_r = \int_{v_{sc}} \mu_0 M(H) \frac{dH}{dr} dv \tag{2}$$

Then we applied the classical Bean model approximation to the magnetization in the case of full penetration field. We obtained thus

$$F_r \approx \mu_0 J_c a \frac{dH}{dr} V_{act} \tag{3}$$

Where H is the field created by the permanent magnets and V_{act} is the superconducting volume, which develops magnetization. In alternating magnetic poles excitation, the single domain grain size is considered large, so that a is equal to the elementary magnet width. Then the torque force will be a function of r, leading to a fundamental equation:

$$d\tau = dr \times F_r \tag{4}$$

As for the definition of rotational stiffness, k, is the derivative of the applied torque respect to the angle of distortion, θ, in particular when considering the stability limit of the mechanism.

$$k = \left(\frac{\partial \tau_r}{\partial \theta} \right) \tag{5}$$

Following the standard of International Electrotechnical Commission (IEC) code 61788-9/CDV, the measurements for bulk high temperature superconductors of magnetic field trapped was described in details [7]. Thus we followed the procedure of the standard, the J_c value will be calculated by the empirical equation as expressed below.

$$B_z = C\left((z+D)\ln\frac{R+\sqrt{R^2+(z+D)^2}}{z+D} - z\ln\frac{R+\sqrt{R^2+z^2}}{z} \right) \tag{6}$$

where B_z = the trapped flux density
z = a gap between a superconductor and a magnetic sensor
D = the height of a superconductor
R = a radius of a superconductor
C = the constant related to $\mu_0 J_c$

The numerical evaluation in comparison to the experimental results is also necessary in the engineering design. Once the simulation method is established, one can estimate the torque forces for large devices in practical applications scale.

3. The experimental set up

We used commercially available Dy-Ba-Cu-O samples (Nippon Steel) produced by the top-seeded melt growth process, the detail of which is described in ref [1]. The chemical composition of the sample was $DyBa_2Cu_3O_{6.9}$ 75.0 wt%, $Dy_2BaCuO_{5.0}$ 24.5 wt% and Pt 0.5 wt%. The samples were single-domain with dimensions of 30 mm diameter and 5 mm thickness in which *c*-axis was aligned parallel to the axial direction.

The magnetic field trapped by the sample have been investigated to estimate the J_c value by following the International Electrotechnical Commision standard number 61788-9/CDV. For instance, the sample was cooled in the presence of the external magnetic field at 77 K. The excited magnetic field was applied by Nd-Fe-B permanent magnet with the surface flux density of 0.45 T and remnant of 1.1 T. The magnets are concentrically set to the sample to apply the magnetic field parallel to the *c*-axis of the sample. The magnetic field trapped by the sample was measured after the removal of external field. As suggest in the standard, the trapped flux density values are those measured after at least 15 minutes of removal of external field [7]. We have designed an experimental set up which can be used for the trapped field measurements in a static field of a permanent magnet. The upper and lower stand can move in both ascending and descending modes smoothly controlled by a thread-levelling fixture. After cooled to a superconducting state in a field-cooled condition as explained above, a set of permanent magnets were removed from the sample. In this process, the field generated from the permanent magnet was trapped by the superconductor. Then, the field distribution trapped by the sample was measured by scanning a Hall probe sensor at every 1 mm pitch on the squared scanning area on the sample surface as the same in ref. [8].

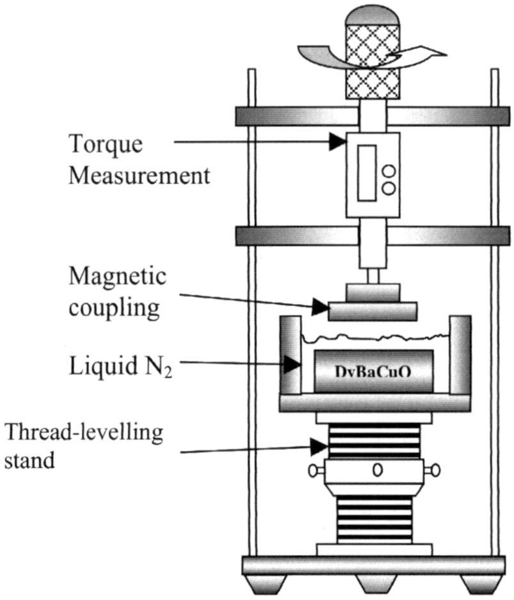

Figure 2. An illustration of torque measuring set up

The torque coupling measurements can also be performed using the configuration shown in Fig. 2. Here the upper magnet part is fixed to a torque measuring instrument aligned concentrically with the frame rods. The exciting set of permanent magnets will also be rigidly fixed with the torque measuring instrument that allows only the angular direction movement. The experiments were studied for four different configurations of Nd-Fe-B magnets such as, two poles conventional magnetic coupling, 2-poles bar shape, 2 poles square and 4 poles square permanent magnets, which are schematically shown in Fig. 3. The specification details of the components are summarized in Table 1.

Table1. Specification details of components

Component	Material	Dimensions (mm)	Surface flux density (T)
Superconductor	Dy-Ba-Cu-O	$\varnothing 30 \times 5^t$	N/A
2 poles coupling	Nd-Fe-B	$\varnothing 40 \times 10^t$	0.466
2 poles bar	Nd-Fe-B	40x20x10	0.457
2 poles square	Nd-Fe-B	40x40x10	0.394
4 poles square	Nd-Fe-B	40x40x10	0.472

The relationship between the torque force and the gap has also been studied by varying the gap in the field-cooling mode at every 1 mm. The torque forces were measured with a digital programmable torque-measuring instrument with a resolution of 0.001 Nm. The data have been recorded at every 10-degree angle of twist.

4. Results and Discussion

Figure 4 shows the trapped field distribution of Dy-Ba-Cu-O sample, which shows the peak trapped field value of 0.323 T. Following the empirical equation expressed in IEC standard as shown in eq.(6), the average J_c value over the volume of the sample was calculated to be 4.36 kA/cm^2 at 77 K by 0.45 T of external field $H//c$-axis.

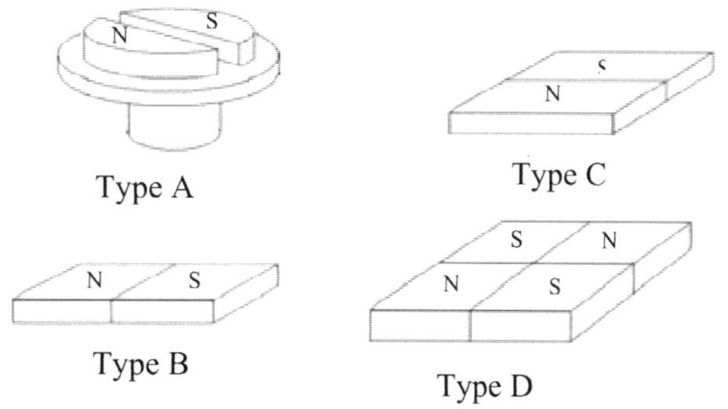

Figure 3. Configurations of permanent magnets for magnetic coupling used in the present study: (A) conventional magnetic coupling, (B) 2 poles bar, (C) 2 poles square, and (D) 4 poles square.

A single peak and symmetric field distribution at the height of 1 mm over the sample confirmed that the sample is a single domain without any appreciable defects [8]. Furthermore, we also measured the trapped magnetic field distribution of the same sample when excited by 4 poles permanent magnets and the result is shown in Fig. 5, which shows that the sample can also trap the magnetic flux in different poles symmetrically.

The torque forces were also measured at various gaps in every 1 mm step in the field-cooling mode until the measured torque values were reduced below 5 N·cm. The results are plotted in Figs. 6 (a) to (d) for type A, B, C, and D configurations, respectively. The torque force decreases gradually with increasing the cooling gap. The maximum torque yields at different angle up to the poles configuration. The type A, B and C configurations, which have 2 magnetic poles on the coupling surface, exhibited the maximum torque force at about 60 degree twist. In contrast, the type D configuration of 4 magnetic poles yields the maximum torque at about 45 degree twist. The different of the maximum torque values are an effect of the arrangement of magnetic coupling configuration, which can useful for an engineering design of further mechanism.

The maximum torque forces for all the configurations are summarized in Fig. 7. The results clearly show that the torque forces decreased with increasing the cooling gap. The torque force achieved at 1mm cooling gap was the highest for the type D configuration, however the force decreased more rapidly with gap than the other configurations. It is interesting to note that the gap dependences of the torque forces for types A, B and C configurations are similar, which suggests that these gap dependences reflect the characteristics of 2 poles configuration.

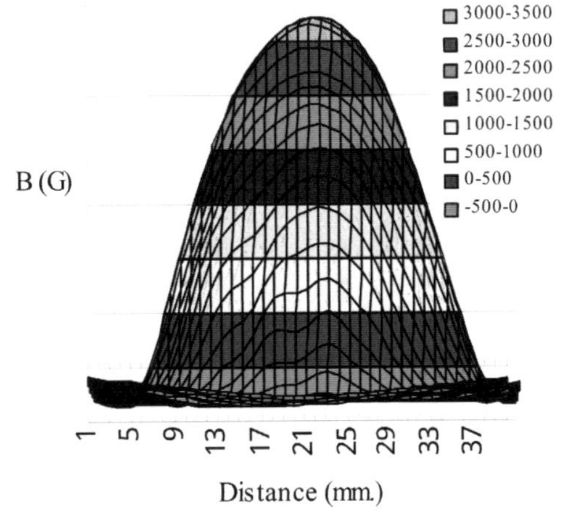

Figure 4. Trapped field distribution of Dy-Ba-Cu-O measured at 1 mm above the sample excited by a single pole magnet.

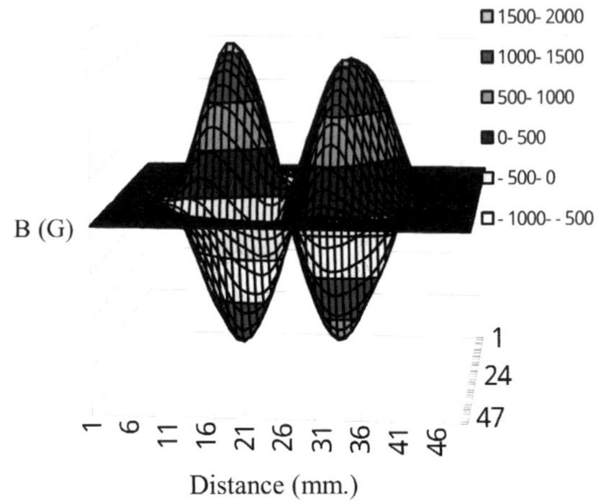

Figure 5. Trapped field distribution of Dy-Ba-Cu-O measured at 1 mm. above the sample excited by 4 poles magnet

In addition, the results of a conventional coupling configuration type A showed that its torque force is similar to that obtained in the 2 poles bar like shape configuration (type B). Since the conventional coupling configuration has the wider poles pitch, however, unstable rotation has occurred during the agitating process. A similar situation was observed in the experimental mixing process in that the levitated component centrifugally dropped and touched the ground.

In contrast, the 4 poles square shape never experienced such a drop in the levitating component in the mechanism. The higher in the poles pitch was the smaller the torque force, which was also found in the work of Zheng *et al.* [5]. Furthermore, we achieved higher torque force in a quite similar coupling type as compared to their work, presumably due to the fact that the present superconductor exhibited stronger flux pinning.

5. Summary

We have studied the torque transferring mechanism between a bulk superconductor and permanent magnets with four useful configurations. The designed operating gap and torque coupling force have to compromise with the permanent magnets array. The 4 poles configuration produced the highest torque of about 32 N·cm at 1mm gap, however the force decreased significantly with increasing the gap. At the larger gaps, the stronger coupling force can be obtained from 2 poles square configuration. It was also found that the configuration those have more magnetic poles on the surface exhibited the maximum torque at lower twist angles, however, the torque force reduces more rapidly than the lesser magnetic poles array. The qualitative comparison of the results led us to the conclusion that one can design a superconducting mixer that can stir the liquid level of 30ℓ at 5 cm operating gap.

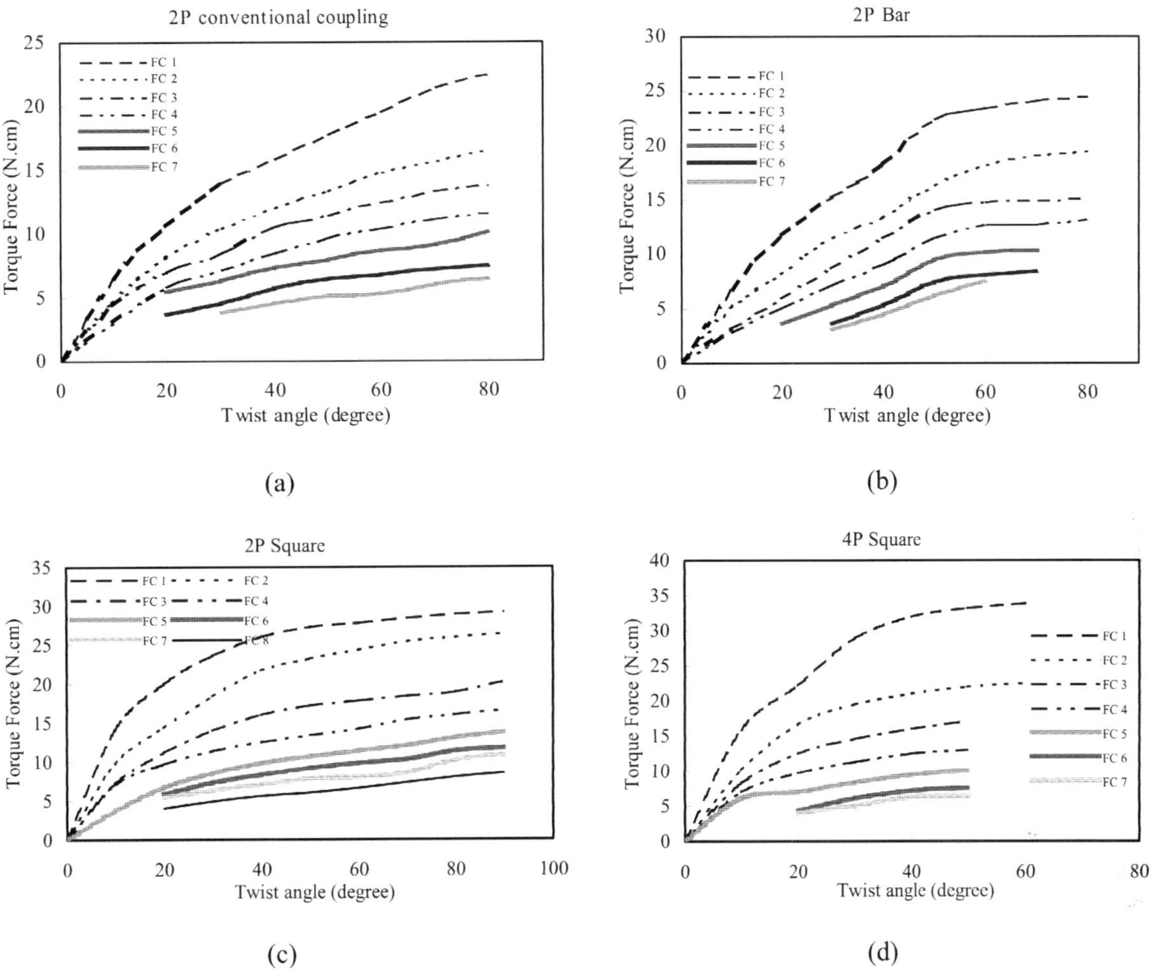

Figure 6 Plots of torque-gap relation against twist angles for four configurations of experiments: (a) 2 poles conventional coupling, (b) 2 poles bar, (c) 2 poles square, and (d) 4 poles square. Here FC1-8 denote the field cooling gap of 1 to 8 mm.

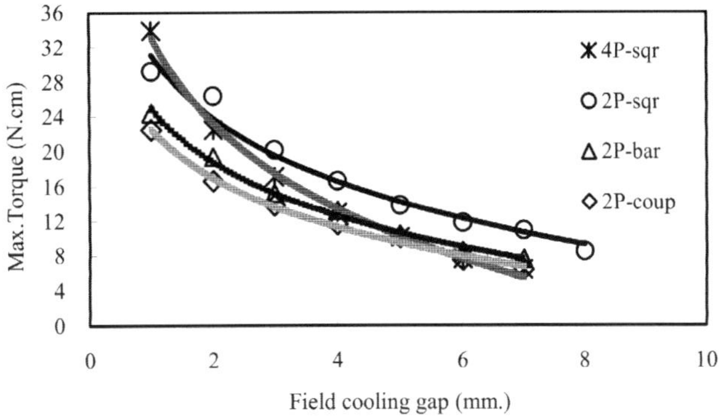

Figure 7. Maximum torque plotted against the cooling gap for four types of magnetic couplings.

6. References

[1] Murakami M 2007 *Int. J. Appl. Ceram. Technol.* 4[3] 225-241
[2] Koyama F, Akiyama S and Murakami M 2006 *Supercond. Sci. Techonol.* 19. S572-S574
[3] Akiyama S and Murakami M 2006 *European Patent.* EP 1731217A1
[4] Tixador P, Hiebel P, Hotier E, Chuad X and Beaugnon E 1995 *Cryogenics.* 35, 567-572
[5] Zheng P , Haik Y, Kilani M and Chen C 2002 *J. of Magnetism. And Magnetics. Mat.* 214, 292-302
[6] Yonnet J.P, Hemmerlin S, Rulliere C and Lemarquand G. 1993, *IEEE Trans Magn.* MAG-29 2932-2934.
[7] *International Electrotechnical Standard*, IEC 61788-9/CDV
[8] Nariki S, Murakami M 2002, *Physica C* 378-381, 759-763.

Reel-to-Reel Continuous Deposition of $Ce_xZr_{1-x}O_2$ Single Buffer Layer for YBCO coated conductors

Jie Xiong[1], Bowan Tao, Wenfeng Qin, Xiao Feng, Xiaoke Song, Fei Zhang, Yanrong Li

State Key Laboratory of Electronic Thin Film and Integrated Devices, University of Electronic Science and Technology of China, Chengdu, 610054, Republic of China

E-mail: bearbear622@163.com

Abstract. In this paper, a study regarding the epitaxial growth of $Ce_xZr_{1-x}O_2$ film on biaxially textured Ni-5at.%W substrate and its use as a single buffer layer of a YBCO coated conductors was reported. Films of Ce–Zr mixed oxide were prepared by direct-current (d.c.) reactive magnetron sputtering with the two sputtering guns arranged symmetrically with respect to the substrate. In sputtering process, d.c. power of Zr was fixed in 200 W while that of Ce was varied with 30 W, 50 W, 75 W, and 100 W, respectively. It was confirmed that the composition of the films could be controlled with modulating power of Ce target. All samples exhibited good epitaxial growth with c-axis perpendicular to the substrate surface. Atomic force microscope revealed a continuous, dense, and crack-free surface morphology for $Ce_{0.32}Zr_{0.68}O_2$ thin films, which provided themselves as the good single buffer to the $YBa_2Cu_3O_{7-\delta}$ (YBCO) coated conductors. High quality $Ce_{0.32}Zr_{0.68}O_2$ buffer layers up to 100-m length could be fabricated with production speed of about 1.2m/h. X-ray scans have been performed as a function of length to determine the crystallographic consistency of the epitaxial $Ce_{0.32}Zr_{0.68}O_2$ over length.

1 . Introduction

Due to its lower ac loses, better in-field performance and the lower processing costs over first-generation high-temperature superconductor (HTS) tapes, second-generation $YBa_2Cu_3O_{7-\delta}$(YBCO) coated conductors have attracted the attention of the researchers worldwide[1-3]. However, it is well known that the initial issue for the successful development

[1] Corresponding author Tel.: +86-28-83202550; fax:+86-28-83202569.

of coated conductors for high temperature superconducting (HTS) wire applications is to chemically isolate the HTS film from the metal substrate, preventing diffusion of Ni into the superconductor during high-temperature processing, and also provide an epitaxial template for growth of the HTS coated conductors. Cerium dioxide CeO_2 and yttria stabilized Zirconium dioxide (YSZ) films have been widely used as a 'buffer layer' for high superconductor [4]. In the case of YBCO coated conductor, CeO_2 film is used for seed layer as well as cap layer, while YSZ film plays a major role of diffusion barrier [5-6]. To our knowledge, these multi-layer buffers make the fabrication process complicated and expensive. Many efforts on the use of a single buffer layer in a coated conductors have been done with yttria stabilized zirconia[7], Y_2O_3[8-9], $La_2Zr_2O_7$[10], $La_{0.7}Sr_{0.3}MnO_3$[11], $LaMnO_3$[12-13], and $SrTiO_3$[14]. As a promising candidate of buffer layer material for YBCO coated conductors, CeO_2 has been prepared by many groups due to its excellent chemical compatibility with Ni-based alloy substrates as well as good lattice match with YBCO or ReBCO. However, CeO_2 has seldom been used as a single buffer layer due to the existence of a critical thickness of around 50 nm [15-16] beyond which cracks are to generate and micro-cracks to propagate. Consequently, the diffusion of Ni towards YBCO is inevitable, which would greatly reduce the critical current densities. Since the crack formation in cerium oxide layers mainly depends on the stresses introduced by the lattice mismatch with Ni alloy substrate and on the arrangement of the oxygen vacancies in the lattice, it is possible to reduce these stresses by optimizing the doped-Zr in the Ceria layer, thus obtaining very thick and crack-free $Ce_xZr_{1-x}O_2$ (CZO) buffer layers.

It was reported that Y_2O_3 not only could stabilize high temperature cubic zirconia phase but CeO_2 also could stabilize cubic or tetragonal zirconia phases [17-18]. In this study, using reactive sputtering system with two metallic target of simple chemical composition such as Ce and Zr, the preparation and properties of CZO films with adjustable lattice constants for better lattice matching and chemical compatibility are investigated as a single buffer layer for YBCO coated conductors. The growth of subsequent YBCO films by direct-current (d.c.) sputtering is also investigated.

2. Experimental

The short segments (5-15mm), which were cut from a 10m long Ni-5at.%W tape, were manufactured and supplied by evico GmbH. The metal alloy tape had thickness of 70μm and width of 10mm. Prior to deposition the chamber was pumped down to a base pressure below 1×10^{-4}Pa.

Films of Ce–Zr mixed oxide with thickness around 300nm were deposited by d.c. reactive magnetron sputtering with the two sputtering guns arranged symmetrically with respect to the substrate. Two disk type Ce metal (99.99%) and Zr metal (99.95%) target were used. In sputtering process, d.c. power of Zr was fixed in 50 W while that of Ce was varied with 30 W, 50 W, 75W, and 100 W, respectively. It was confirmed that the composition of the films could be controlled with controlling power of Ce target. The samples were heated to 750°C at forming gas pressure (96%Ar+4%H_2) of 0.5 Pa, and water vapour for reaction with deposited Cerium and Zirconium atoms was introduced inside the chamber near the NiW substrates to the pressure of about 10^{-3} Pa. When all the parameters for deposition were in equilibrium, the NiW tape was moved at the speed of 1.2 m/h by the reel-to-reel moving system.

To assess the quality of the CZO single buffer layer, YBCO films were deposited by d.c. sputtering on short segments of the coated substrates. Details of the experimental conditions were reported elsewhere [19-20]. Typical YBCO film thicknesses were 500 nm.

X-ray diffraction analysis (θ-2θ, ω-scan and Φ-scan) was used to characterize the phase, structure, texture and lattice parameter of the buffer layers. A Bede D1 X-ray diffractometer with CuKα radiation was used in the present studies. An atomic force microscopy (AFM) and a SEIKO SPA300HV microscope were used to analysis the surface morphology of the samples. The critical current density (J_c) was measured using J_c-scan Leipzig system.

3. Results and discussion

The compositions of CZO films varied with increasing d.c. power to Ce target were analyzed by EDS, the values for x were 0.32, 0.5, 0.6 and 0.75 when d.c. power to Ce target were 30, 50, 75, and 100W, respectively. The epitaxial nature of CZO on NiW tape was investigated using XRD. Figure 1 shows the XRD θ-2θ scans for CZO films with various compositions. For each of the samples, only the (002) CZO peak is observed, indicating that the films are preferentially aligned. Also, the θ-2θ scans clearly illustrate a shift in the (002) peak position as a function of the atomic fraction (x) of Ce in CZO films. Note that the position of the (002) Ni peak (from the substrate) is invariant. The shift in CZO peak positions toward lower diffraction angles indicates an increase in the lattice parameter as the Ce content of the films increases. A linear dependence of the lattice parameter with x is observed shown in Fig. 2. Due to big lattice mismatch more than 50% between CZO and NiW, a 45° rotation of CZO with respect to the NiW was observed. The lattice mismatch between CZO and NiW varies from 7.8% ($Ce_xZr_{1-x}O_2$ with x=0.75) to 4.9% ($Ce_xZr_{1-x}O_2$ with x=0.32). Considering the effect from lattice match, $Ce_{0.32}Zr_{0.68}O_2$ was used as the buffer layer for the subsequent YBCO coated conductors.

For implementation of $Ce_{0.32}Zr_{0.68}O_2$ single buffer layers into YBCO wire manufacturing, it is important to grow dense and crack-free layers. AFM micrographs of the surface morphology for 300 nm thick $Ce_{0.32}Zr_{0.68}O_2$ films fabricated on biaxially textured Ni-alloy substrates are presented in Figs. 3. The sample exhibits a continuous, crack-free, and dense surface morphology, average roughness around 3.3nm, which should best serve the goal of obtaining high-quality YBCO coatings.

One-meter-long biaxially textured $Ce_{0.32}Zr_{0.68}O_2$ buffer layers were fabricated with production speed of about 1.2m/h. XRD ω- and Φ-scan measurements have been performed in order to examine the uniformity of the $Ce_{0.32}Zr_{0.68}O_2$ buffer layer texture over length. Figure 4 plots the XRD ω-rocking curve FWHM distributions for the $Ce_{0.32}Zr_{0.68}O_2$ (002) peak reflections taken along a 1 m length of NiW alloy tape. Uniform c-axis alignment of the $Ce_{0.32}Zr_{0.68}O_2$ buffer layers can be seen from this figure. Similarly, excellent uniformity for the in-plane texture of the $Ce_{0.32}Zr_{0.68}O_2$ layer is evident from the plot (Fig. 5) of XRD Φ-scan FWHM distribution recorded from the $Ce_{0.32}Zr_{0.68}O_2$ (111) reflection along the tape length. A 1-m $Ce_{0.32}Zr_{0.68}O_2$ film was obtained with Δω of 3-3.5^o and ΔΦ of 5-6°. Further, we also observed the texture was uniform along the entire length of the tape.

After optimizing the fabrication process for $Ce_{0.32}Zr_{0.68}O_2$ films on NiW substrate, we deposited the 500nm thick YBCO film by d.c. sputtering. A typical XRD θ-2θ scan of the YBCO film is shown in Fig.6. The YBCO film exhibits excellent c-axis oriented growth. Weak $NiWO_4$ and NiO impurity peaks are observed at 19° and 37°, respectively, implying minor oxidation of the metal substrate interface. Electrical transport and superconducting property measurements were conducted on the same sample. The self-field Jc measured with Jc-scan Leipzig system about $0.75MA/cm^2$ was obtained at temperature of 77K.

4. Conclusion

Co-sputtered CZO films were obtained by reactive sputtering of Ce and Zr metal targets simultaneously. It has been shown that CZO film with proper composition could used as single buffer layer in the fabrication of YBCO coated conductor. AFM observations demonstrated that CZO buffer layers with thickness up to 300nm were crack-free and dense. XRD ω- and Φ-scans of these buffer layers showed good and uniform out-of-plane and in-plane texture over lengths. YBCO coated conductors have also been prepared by d.c. sputtering on 300nm-thick CZO simple layers, indicating excellent c-axis oriented growth, and a critical current, Jc, of about $0.75MA/cm^2$ at 77K and self-field was obtained.

Reference

[1] Norton D P, Goyal A, Budai J D, Christen D K, Kroeger D M, Specht E D, He Q, Saffian B, Paranthaman M, Klabunde C E, Lee D F, Sales B C and List FA 1996 *Science* **274** 755

[2] Goyal A, Norton D P, Budai J D, Paranthaman M, Specht E D, Kroeger D M, Christen D K, He Q, Saffian B, List F A, Lee D F, Martin P M, Klabunde C E, Hartfield E and Sikka V K 1996 *Appl. Phys. Lett* **69** 1795

[3] Xiong J, Tao B W, Qin W F, Tang J L, Han X and Li Y R 2008 *Supercond. Sci. Technol.* **21** 025016

[4] Vratskikh V, Drozdov Y and Talanov V, 1997 *Supercond. Sci. Technol.* **10** 766

[5] Goyal A, Norton D P, Budai J D, Paranthaman M, Specht E D, Kroeger D M, Christen D K, He Q, Saffian B, List F A, Lee D F, Martin P M, Klabunde C E, Hartfield E and Sikka V K 1996 *Appl. Phys. Lett.* **69** 1795

[6] Lee D F, Paranthaman M, Mathis J E, Goyal A, Kroeger D M, Specht E D, Williams R K, List F A, Martin P M, Park C, Norton D P and Christen D K 1999 *Jpn. J.Appl. Phys.* **38** L178

[7] Park C, Norton D P, Lee D F, Verebelyi D T, Goyal A, Christen D K and Budai J D 2000 *Physica C* **341** 2481

[8] Shi D Q, Ko R K, Song K J, Chung J K, Ha H S, Kim H S, Moon S H, Yoo S I and Park C 2005 *Supercond. Sci. Technol.* **18** 561

[9] Kim H S, Park C, Ko R K, Shi D Q, Chung J K, Ha H S, Park Y M, Song K J and Youm D J 2005 *Physica C* **426** 926

[10] Sathyamurthy S, Paranthaman M P, Zhai H Y, Kang S K, Christen H M, Cantoni C, Goyal A and Martin P M 2003 *IEEE Trans. Appl. Supercon*d. 13 2658

[11] Aytug T, Paranthaman M, Kang B_W, Sathyamurthy S, Goyal A and Christen D K 2001 *Appl. Phys. Lett.* **79** 2205

[12] Paranthaman M P, Aytug T, Kang S, Feenstra R, Budai J D, Christen D K, Arendt P N, Stan L, Groves J R, De Paula R F, Foltyn S R and Holesinger T G 2003 *IEEE Trans.Appl. Supercond.* **13** 2481

[13] Aytug T, Paranthaman M, Kang S, Zhai H Y, Leonard K J, Vallet C E, Sathyamurthy S, Christen H M, Goyal A and Christen D K 2003 *IEEE Trans. Appl. Supercond.* **13** 2661

[14] Dawley J T, Ong R J and Clem P G 2002 *J Mater. Res.* **17** 1678

[15] Paranthaman M, Goyal A, List F A, Specht E D, Lee D F, Martin P M, He Q, Christen D K, Norton D P, Budai J D and Kroeger D M 1997 *Physica C* **275** 266

[16] Xiong J, Chen Y, Qiu Y, Tao B_W, Qin W_F, Cui X_M, Tang J_L and Li Y R *2006 Supercond. Sci. Technol.* **19** 1068

[17] Rossignol S, Madier Y and Duprez D, 1999 *Catal. Today* **50** 261

[18] Sammes N M, Tompsett G A and Cai Z, 1999 *Solid State Ionics* **121** 121

[19] Xiong J, Qin W F, Cui X M, Tao B W, Tang J L and Li Y R 2007 *J. Crys. Growth* **300** 364

[20] Xiong J, Qin W F, Cui X M, Tao B W, Tang J L and Li Y R 2006 *Physica C* **442** 124

Figure captions:

Fig.1 XRD θ-2θ scans of CZO films deposited with different d.c. power of Ce (a) $Ce_{0.32}Zr_{0.68}O_2$ film (Ce power of 30 W) (b) $Ce_{0.5}Zr_{0.5}O_2$ film (Ce power of 50 W) (c) $Ce_{0.6}Zr_{0.4}O_2$ film (Ce power of 75 W) (d) $Ce_{0.75}Zr_{0.25}O_2$ film (Ce power of 100 W)

Fig.2 Lattice parameters calculated from the XRD data as functions of atomic fraction x of Ce in CZO

Fig.3 Typical AFM micrograph for $Ce_{0.32}Zr_{0.68}O_2$ film

Fig.4 XRD ω-rocking curve FWHM distributions of the $Ce_{0.32}Zr_{0.68}O_2$ (002) reflections as a function of the 1m length tapes.

Fig.5 XRD Φ-rocking curve FWHM distributions of the $Ce_{0.32}Zr_{0.68}O_2$ (111) reflections as a function of the 1m length tapes.

Fig.6 Typical XRD θ-2θ pattern of the YBCO films on $Ce_{0.32}Zr_{0.68}O_2$/NiW substrate

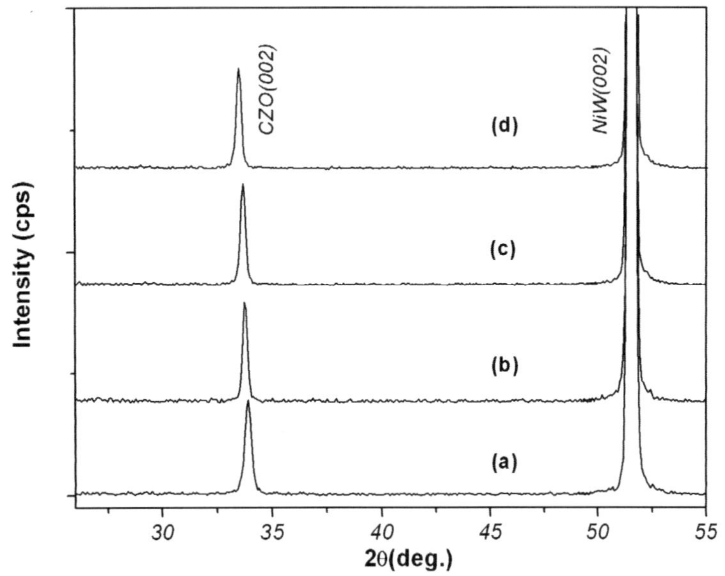

Fig.1 Jie Xiong et al.

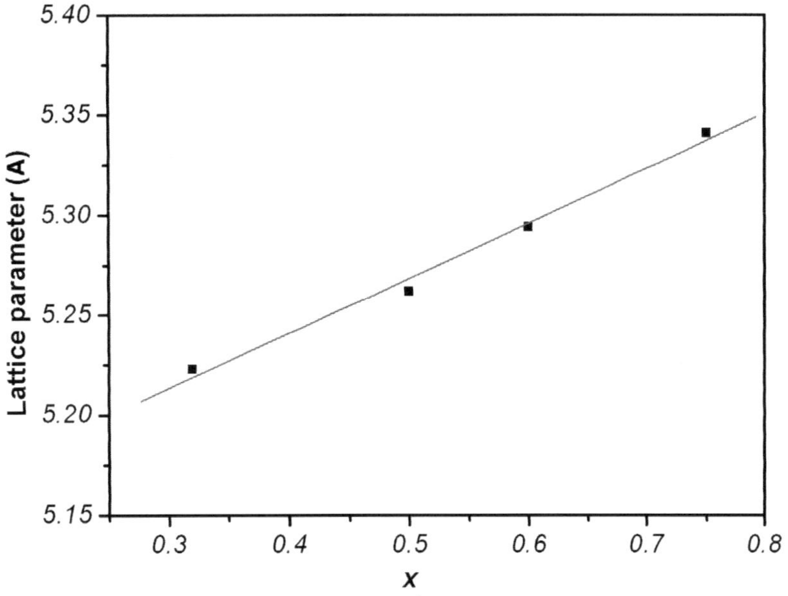

Fig.2 Jie Xiong et al.

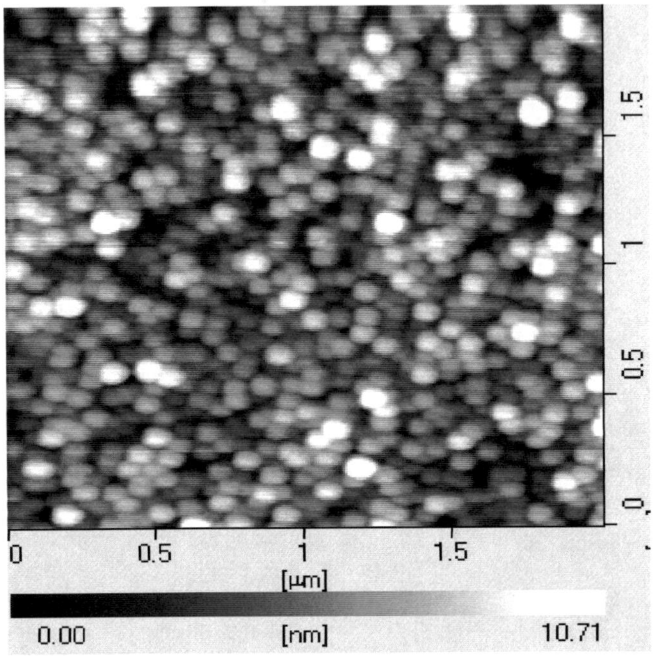

Fig.3 Jie Xiong et al.

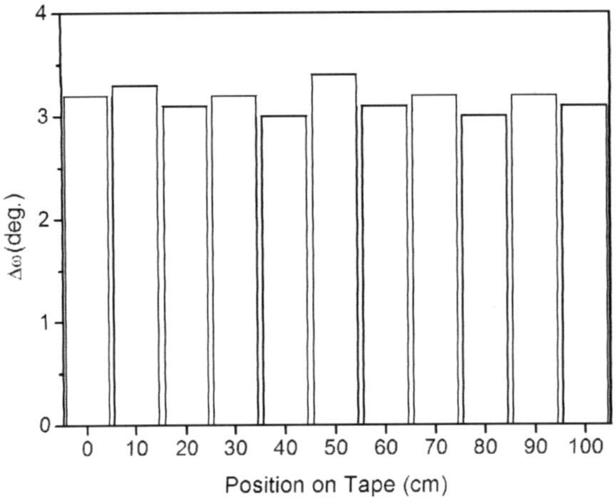

Fig.4 Jie Xiong et al.

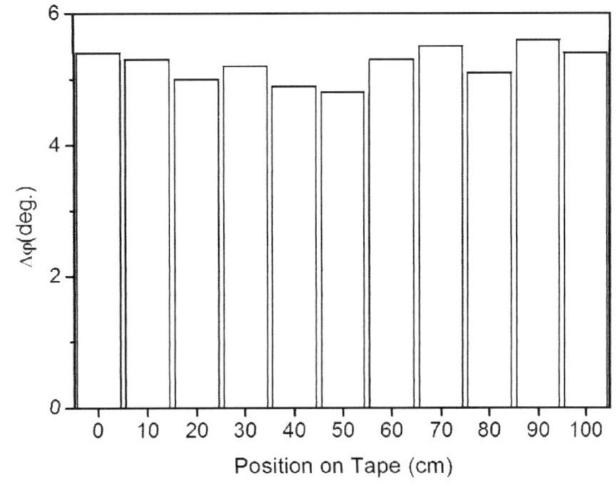

Fig.5 Jie Xiong et al.

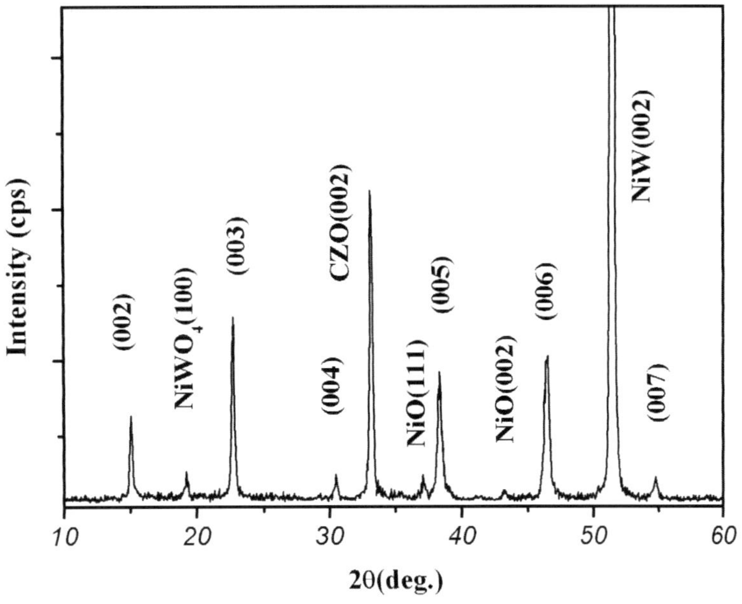

Fig.6 Jie Xiong et al.

Influence of Pr Substitution on Acoustic Gruneisen Parameter and Lattice Vibrational Anharmonicity Contribution in Ceramic Superconductors

A K Yahya

School of Physics and Materials Studies, Faculty of Applied Sciences, Universiti Teknologi MARA, 40450 Shah Alam, Selangor, Malaysia

E-mail: ahmad191@salam.uitm.edu.my

Abstract. This paper reports the effect of Pr substitution on lattice anharmonicity contribution to temperature dependent longitudinal and shear ultrasonic velocity data in $(Er_{1-x}Pr_x)Ba_2Cu_3O_{6.9}$ (x = 0.0, 0.1 and 0.2) and $TlSr_2(Ca_{1-y}Pr_y)Cu_2O_7$ (y = 0.5 and 1.0) superconductors. Pr was observed to influence the deviation of the elastic behaviour of both samples from the lattice anharmonicity curves at higher temperatures. The calculated effective Gruneisen parameter for the longitudinal mode (γ_{eff}^L) for both compounds was found to increase with Pr content, indicating increasing anharmonicity contribution. In addition, both series also showed that γ_{eff}^L is larger than the effective Gruneisen parameter for the shear mode (γ_{eff}^S) throughout the substitution. The influence of Pr substitution on γ_{eff} and the BCS electron-phonon coupling constant (λ) as well as its possible relationship to superconductivity are discussed.

1. Introduction

Ultrasonic velocity measurement is an interesting technique which is sensitive to bulk changes in elastic properties and lattice instabilities taking place in a material. Previous reports on high-temperature superconductors revealed several anomalies in the temperature dependent ultrasonic velocity profiles where a large change in velocity was normally reported at high temperatures especially above 200 K [1-3]. The percentage of velocity change was observed to be affected by oxygen content and chemical substitution. A few reports have shown that sound velocity in high-temperature superconductors is dominated by lattice anharmonicity at lower temperatures but deviated at higher temperatures [4-5]. However, it is not known if chemical substitution can affect the lattice anharmonicity contribution to elastic properties of high-temperature superconductors.

A comparison between sound velocity curve and vibrational anharmonicity curve is interesting as it allows for the lattice anharmonicity contribution to be measured. A means of assessing the anharmonic contribution to the temperature dependence of the velocity of sound for normal solids was given by Nava et. al.[6]. In this model, the change in sound velocity is given by

$$\frac{\Delta v}{v} = -\frac{TC_v}{2\rho v^2}\gamma_{eff}^2 \tag{1}$$

where v is the sound velocity, ρ is the mass density, C_v is the specific heat per unit volume and γ_{eff} is the effective Gruneisen parameter. The specific heat (C_v) in the equation above is given by

$$C_v = \left(\frac{n}{V}\right)(3N_A k_B)\left[3\left(T/\theta_D\right)^3 \int_0^{\theta_D/T} \frac{x^4 e^x}{\left(e^x - 1\right)^2} dx\right] \qquad (2)$$

where N_A is the Avogadro's number, k_B is the Boltzmann constant, n is the mole number and V is the volume of the sample. The change in velocity ($\Delta v/v$) can be fitted to this model using equation (1) if values of $\Delta v/v$ at two different temperatures along with Debye Temperature (θ_D) of a material are known. Anharmonic contribution to lattice vibration can be estimated from the computed value of the effective Gruneisen parameter (γ_{eff}) which can be calculated from the equations above. Nava et. al. successfully used the model on ultrasonic longitudinal and shear velocities of $REBa_2Cu_3O_{7-\delta}$ (RE=Y, Gd) [6].

In addition, it would be interesting to see if lattice anharmonicity contribution affects superconductivity. BCS electron-phonon coupling constant (λ) in the weak coupling limit can be computed and compared to the calculated values of Gruneisen parameter (γ_{eff}) to see if a relationship exists. This would be of particular importance as the mediating role of lattice vibrations in Cooper pair formation has regained attention [7].

This paper reports the effect of Pr substitution on lattice anharmonicity contribution to longitudinal and shear ultrasonic wave velocity in $(Er_{1-x}Pr_x)Ba_2Cu_3O_{6.9}$ (x = 0.0, 0.1 and 0.2) and $TlSr_2(Ca_{1-y}Pr_y)Cu_2O_7$ (y = 0.5 and 1.0) superconductors. The possible relationship between the acoustic Gruneisen parameter and electron-phonon coupling constant (λ) will be discussed.

2. Experimental details

2.1 Sample preparation

The $(Er_{1-x}Pr_x)Ba_2Cu_3O_{6.9}$ (x = 0, 0.1 and 0.2) samples were prepared by mixing appropriate amounts of high purity (≥99.99 %) Er_2O_3, Pr_6O_{11}, $BaCO_3$ and CuO powders [8]. The materials were ground in a mortar and then calcined in air at around 900 °C for 48 hours with several intermittent grindings and oven cooled. The powders were then pressed into pellets of 13 mm diameter and around 3 mm thickness. The pellets were heated at the same temperature for 24 hours and cooled to room temperature at 40 °C per hour. The $TlSr_2(Ca_{1-y}Pr_y)Cu_2O_7$ (y = 0.5 and 1.0) samples were prepared using high purity (≥ 99.99 %) powders of Tl_2O_3, Pr_6O_{11}, $SrCO_3$, CaO and CuO using conventional solid state and the precursor method [9]. Final sintering was around 1000 °C in flowing O_2 for 3-5 minutes followed by oven cooling to room temperature.

2.2 Characterization

Electrical resistance (d.c.) measurements were carried out using the four-point-probe technique with silver paint contacts. The samples were also examined by X-ray powder diffraction with CuK_α radiation using Siemens D 5000 diffractometer. The ultrasonic velocity was measured using a Matec 7700 system which utilizes the pulsed-echo-overlap technique. The sample was bonded to a quartz transducer using Nonaq stopcock grease. The longitudinal and shear velocity measurements were performed at 10 MHz in an Oxford Instrument liquid nitrogen cryostat model DN 1711 and the temperature was varied at a rate of about 1 K/min. during warming and cooling. Details on the ultrasonic technique and sample preparation procedure can be found in our earlier publications [8,9]. Ultrasonic velocity data collected during sample heating was used for computations of the effective Gruneisen parameters and construction of anharmonicity curves.

3. Results and discussion

Powder X-ray diffraction analysis showed all $(Er_{1-x}Pr_x)Ba_2Cu_3O_{6.9}$ (x = 0.0, 0.1 and 0.2) samples to be single phased 123 structures [8]. Temperature dependent electrical resistance measurements for $(Er_{1-x}Pr_x)Ba_2Cu_3O_{6.9}$ with x = 0-0.2 exhibit metallic normal state behaviour with $T_{c\ zero}$ of 92 K, 85 K and 73 K, respectively. The oxygen content of the superconducting $ErBa_2Cu_3O_{6.9}$ sample was estimated based on lattice parameter values [8] using results of lattice constant versus oxygen content variation of Maletta H. et. al.[10]. The Pr substituted samples were also estimated to have an oxygen content of $O_{6.9}$ since low Pr substitution was reported not to affect the oxygen content appreciably [8]. X-ray diffraction analysis for $TlSr_2(Ca_{1-y}Pr_y)Cu_2O_7$ (y = 0.5 and 1.0) showed dominant 1212 phase with minor 1201 phase [9]. $TlSr_2(Ca_{0.5}Pr_{0.5})Cu_2O_7$ was superconducting with $T_{c\ zero}$ of 77 K. Increasing Pr substitution in $TlSr_2PrCu_2O_7$ caused the sample to be insulating. Table 1 gives the zero-resistance transition temperature ($T_{c\ zero}$), density, longitudinal velocity (v_l), shear velocity (v_s), Debye temperature (θ_D) measured at 80 K, effective Gruneisen parameters (γ_{eff}) and BCS electron-phonon coupling constant λ for both series. A discussion of ultrasonic propagation in polycrystalline ceramics is given in ref. [11].

Table 1. Zero-resistance transition temperature ($T_{c\ zero}$), density, longitudinal velocity (v_l), shear velocity (v_s), Debye temperature (θ_D) measured at 80 K, effective Gruneisen parameters (γ_{eff}) and BCS electron-phonon coupling constant (λ) for a) $ErBa_2Cu_3O_{6.9}$, b) $(Er_{0.9}Pr_{0.1})Ba_2Cu_3O_{6.9}$, c) $(Er_{0.8}Pr_{0.2})Ba_2Cu_3O_{6.9}$, d) $TlSr_2Ca_{0.5}Pr_{0.5}Cu_2O_7$ and e) $TlSr_2PrCu_2O_7$. The longitudinal and shear velocity values are corrected for porosity as in ref. [11].

Sample	$T_{c\ zero}$ (K)	Density (g/cm^3) ± 0.02	v_l (kms^{-1}) ± 0.02	v_s (kms^{-1}) ± 0.02	θ_D (K) ± 5	γ_{eff}^L ±0.1	γ_{eff}^S ± 0.1	λ ± 0.01
a) $ErBa_2Cu_3O_{6.9}$	92	5.62	4.44	2.71	375	8.0	5.3	0.65
b) $(Er_{0.9}Pr_{0.1})Ba_2Cu_3O_{6.9}$	85	5.86	4.60	2.79	387	8.6	5.5	0.61
c) $(Er_{0.8}Pr_{0.2})Ba_2Cu_3O_{6.9}$	77	5.75	4.66	3.12	428	9.3	6.6	0.54
d) $TlSr_2Ca_{0.5}Pr_{0.5}Cu_2O_7$	77	4.54	3.49	2.03	282	7.4	4.6	0.70
e) $TlSr_2PrCu_2O_7$	-	4.96	3.50	2.03	279	8.6	4.0	-

Ultrasonic velocity measurements on the samples revealed different elastic response with Pr substitution. Equation (1) has been fitted to the low-temperature experimental data for both modes assuming γ_{eff} is temperature independent using θ_D values given in Table 1. Figure 1 shows temperature dependencies of longitudinal and shear wave velocities of superconducting $ErBa_2Cu_3O_{6.9}$. For superconducting $ErBa_2Cu_3O_{6.9}$, the ultrasonic velocity for both modes (figure 1(a)) is dominated by vibrational anharmonicity below 160 K. Above 160 K, ultrasonic velocity for both modes deviates very much from the calculated anharmonicity curves. For $(Er_{0.9}Pr_{0.1})Ba_2Cu_3O_{6.9}$, the ultrasonic velocity for the longitudinal (figure 2(a)) and shear modes (figure 3(a)) are also dominated by vibrational anharmonicity below 160 K, above which, both velocities start to deviate from the anharmonicity curves. On the other hand, for $Er_{0.8}Pr_{0.2}Ba_2Cu_3O_{6.9}$, the longitudinal velocity deviates from anharmonicity curve starting at around 180 K (figure 2(b)). However, the shear velocity starts to deviate from the vibrational anharmonicity curve at around 160 K (figure 3(b)). The Pr substitution also caused the percentage change in velocity do decrease for both modes.

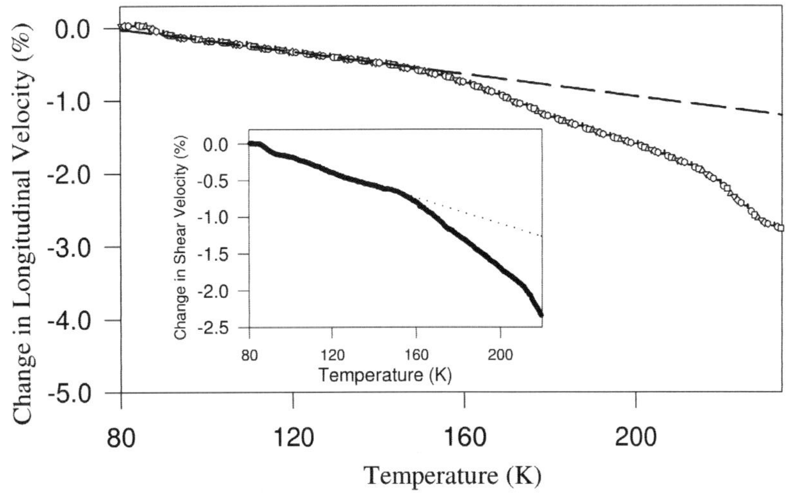

Figure 1. Temperature dependencies of longitudinal wave velocities of superconducting $ErBa_2Cu_3O_{6.9}$. Inset shows the shear wave velocity versus temperature for the same sample. Fits of equation (1) to the experimental curves are shown by the broken lines.

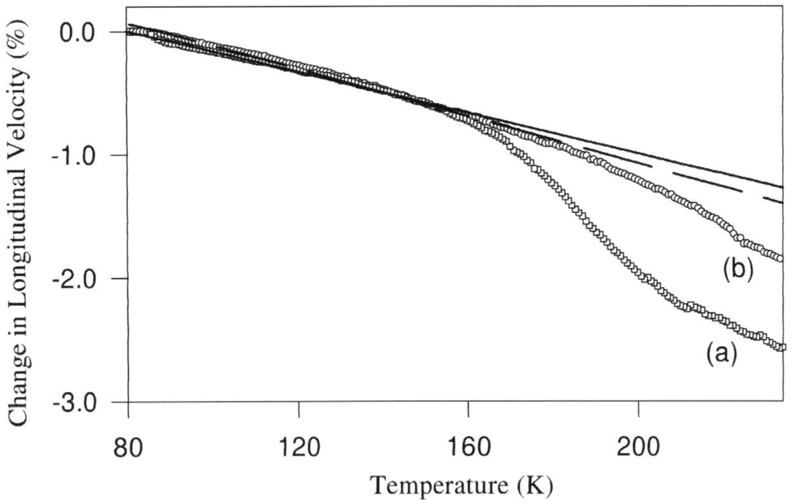

Figure 2. Temperature dependencies of longitudinal wave velocities of (a) $(Er_{0.9}Pr_{0.1})Ba_2Cu_3O_{6.9}$ and (b) $(Er_{0.8}Pr_{0.2})Ba_2Cu_3O_{6.9}$. Fits of equation (1) to curve (a) and curve (b) are shown by the solid and broken lines, respectively.

For superconducting $TlSr_2(Ca_{0.5}Pr_{0.5})Cu_2O_7$ the curves based on the anharmonicity model given by equation (1) are in good agreement with the experimental data for the longitudinal mode (figure 4(a)). Increasing Pr substitution in $TlSr_2PrCu_2O_7$ (figure 4(b)) caused the velocity curve to deviate above 180 K.

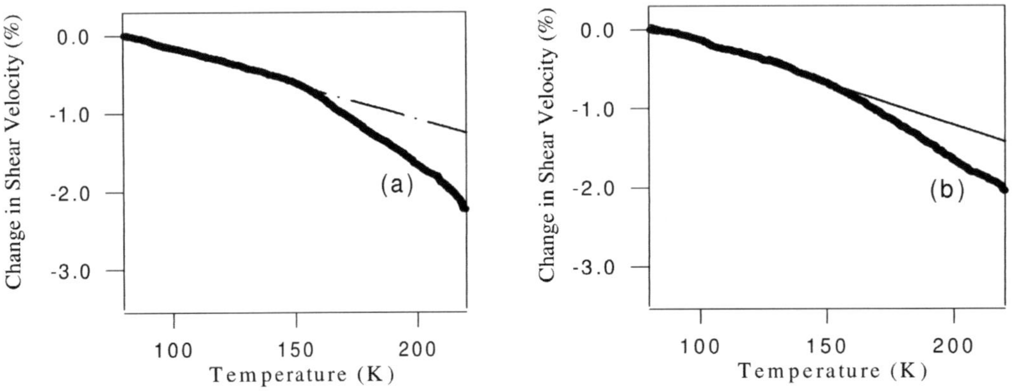

Figure 3. Temperature dependencies of shear wave velocities of (a) $(Er_{0.9}Pr_{0.1})Ba_2Cu_3O_{6.9}$ and (b) $(Er_{0.8}Pr_{0.2})Ba_2Cu_3O_{6.9}$. Fits of equation (1) to curve (a) and curve (b) are shown by the broken and solid lines, respectively.

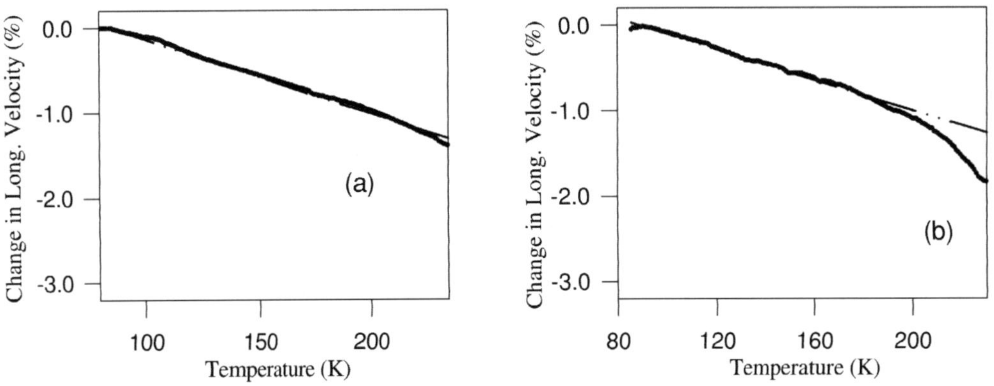

Figure 4. Fits of equation (1) (broken lines) to the temperature dependencies of the longitudinal wave velocity of (a) $TlSr_2(Ca_{0.5}Pr_{0.5})Cu_2O_7$ and (b) $TlSr_2PrCu_2O_7$. Fits of equation (1) to curve (a) and curve (b) are shown by the broken lines.

From the fitting of the anharmonicity curve to the velocity data at low temperatures, the Gruinesen parameter was computed. For superconducting $ErBa_2Cu_3O_{6.9}$, the values of γ_{eff}^L is substantially larger than γ_{eff}^S for all samples (Table 1). This indicates that in these samples, the longitudinal anharmonicity acoustic mode is more dominant than the shear mode anharmonicity. A similar observation was reported in Y123 [4] and Gd1113 [5]. In addition, for $(Er_{1-x}Pr_x)Ba_2Cu_3O_{6.9}$ it was observed that both γ_{eff}^L and γ_{eff}^S increased with Pr content. For $TlSr_2(Ca_{0.5}Pr_{0.5})Cu_2O_7$ and $TlSr_2PrCu_2O_7$ samples, γ_{eff}^L was also observed to be substantially larger than γ_{eff}^S (Table 1). This indicates that the contribution of the longitudinal mode anharmonicity is larger than the shear mode. However, a comparison between the superconducting $TlSr_2(Ca_{0.5}Pr_{0.5})Cu_2O_7$ and non-superconducting $TlSr_2PrCu_2O_7$ showed that the superconducting sample has a larger γ_{eff}^S.

A discussion on the role of lattice anharmonicity in superconductivity of $TlSr_2(Ca_{1-y}Pr_y)Cu_2O_7$ (y = 0.5 and 1.0) is not clear as Pr substitution caused γ_{eff}^L to increase but at the same time γ_{eff}^S also decreases. In addition, given the possible valence of Pr as +3 or +4, simple valence calculation shows that the average Cu valence, which is an indication of hole concentration, also decreased with the substitution. For $(Er_{1-x}Pr_x)Ba_2Cu_3O_{6.9}$, given the slow T_c suppression rate, the valence of Pr is suggested to be 3+. Since both Er and Pr have a common valence of +3, there is no change in the average valence of Cu throughout the substitution. As such it can be suggested that the increase in both γ_{eff}^L and γ_{eff}^S has probably caused the decrease in T_c. In addition, for the $(Er_{1-x}Pr_x)Ba_2Cu_3O_{6.9}$ further analysis involving electron-phonon coupling constant (λ) can be made. Standard BCS theory [12] in the weak coupling limit suggests T_c to be dependent on electron-phonon coupling constant (λ), in addition to the Debye temperature, θ_D i.e. $T_c = 1.13\theta_D e^{-1/\lambda}$. Our results showed that the decrease in T_c following an increase in θ_D as Pr is increased is also accompanied by a decrease in λ (Table 1). Interestingly, the decrease in λ with Pr substitution is in turn accompanied by an increase in anharmonicity which is indicated by γ_{eff} (Table 1). Since γ_{eff} is evaluated in the low temperature regions, it can be suggested that an increase in anharmonicity contribution may have contributed to the suppression of T_c. This is supported by a previous suggestion that in phonon-mediated superconductors, the presence of lattice anharmonicity does not enhance T_c [13]. As such, it is suggested from this work that an increase in lattice anharmonicity contribution at low temperatures due to Pr substitution may be detrimental to electron-phonon coupling in $(Er_{1-x}Pr_x)Ba_2Cu_3O_{6.9}$ superconductors.

4. Conclusion

In conclusion, effective Gruneisen parameters (γ_{eff}) have been computed from ultrasonic velocity data for $(Er_{1-x}Pr_x)Ba_2Cu_3O_{6.9}$ (x = 0, 0.1 and 0.2) and $TlSr_2(Ca_{1-y}Pr_y)Cu_2O_7$ (y = 0.5 and 1.0) ceramics to evaluate lattice vibrational anharmonicity contribution to their elastic behaviours. Pr substitution was found to strongly influence the deviation of longitudinal and shear mode elastic response from the calculated vibrational anharmonicity curves above 160 K. For both series γ_{eff}^L increased with Pr content and γ_{eff}^L were also observed to be substantially larger than γ_{eff}^S. This indicates that not only the contribution of the longitudinal acoustic mode anharmonicity increased with Pr but it is also larger than the contribution for the shear mode. However, for $(Er_{1-x}Pr_x)Ba_2Cu_3O_{6.9}$ (x = 0, 0.1 and 0.2) the increase in γ_{eff} with Pr is also accompanied by the decrease in electron-phonon coupling constant, λ which suggests that an increase in lattice anharmonicity contribution may be detrimental to formation of Cooper pairs.

References

[1] Fukami T, Kondo R, Kobayashi T, Kawatani K, Ichikawa F and Anomie T 1999 *Physica C* **241** 336
[2] Cankurtaran M and Saunders G A 1992 *Supercond. Sci. Technol.* **5** 529
[3] Yahya A K, Hamid N A, Abd-Shukor R and Imad H 2004 *Ceramics International* **30** 1597
[4] Cankurtaran M, Saunders G A, Wang Q and Celik H 1995 *Supercond. Sci. and Technol.* **8** 811
[5] Yahya A K, Yusoff M H and Abd-Shukor R 2002 *J. Solid State Sci. and Technol. Letters* **9** 210
[6] Nava R, Saint-Paul M, Noel H, Levet J C, Potel M and Gougeon P 1990 *Physica C* **167** 59
[7] Shimada D, Tsuda N 2002 *Physica C* **371** 52
[8] Yahya A K and Abd-Shukor R 1998 *Supercond. Sci. Technol.* **11** 173
[9] Yahya A K, Koh A K and Abd-Shukor R 1999 *Physics Letters* **259** 295
[10] Maletta H 1990 *Physica C* **169** 371
[11] Levy M, Xu M F, Sarma B K and Sun K J 1992 Ultrasonic propagation in sintered high-T_c superconductor *Physical Acoustics (XX) Ultrasonics of High-Tc and Other Unconventional Superconductors*, ed M Levy (London: Academic Press) pp 237-298
[12] Bardeen J, Cooper L N and Schrieffer J R 1957 *Phys. Rev.* **108** 1175
[13] Freericks J K, Jarrell M and Mahan G D 1996 *Phys. Rev. Letters* **77** 4588

Journal of Physics: Conference Series **153** (2009) 012038 doi:10.1088/1742-6596/153/1/012038

The influence of cooling rates on microstructure and mechanical properties of $Bi_{1,6}Pb_{0,4}Sr_2Ca_2Cu_3O_y$ superconductors

H A Cetinkara[1], M Yilmazlar[2], O Ozturk[3], M Nursoy[4], C Terzioglu[5,*]
1 Department of Physics, Faculty of Science and Arts, Mustafa Kemal University, 31034 Antakya , Hatay, Turkey,
2 Faculty of Education, Sakarya University, 54300 Hendek, Sakarya, Turkey,
3 Department of Physics, Faculty of Science and Arts, Kastamonu University, 37100 Kastamonu, Turkey,
4 Department of Mechanical Engineering, Faculty of Engineering and Architecture, Abant Izzet Baysal University, 14280 Bolu, Turkey,
5 Department of Physics, Faculty of Science and Arts, Abant Izzet Baysal University, 14280 Bolu, Turkey.

E-mail: terzioglu_c@ibu.edu.tr

Abstract. We investigated the effect of cooling rates on the microstructure and mechanical properties of $Bi_{1,6}Pb_{0,4}Sr_2Ca_2Cu_3O_y$ superconductors prepared by standard solid state reaction methods. The samples were annealed under identical condition and cooled with different cooling rates. The investigations consisted of Vickers microhardness, SEM and XRD measurements. XRD examination of the samples showed that high percentage of Bi-2212 phase was observed and low-T_c phase increased with increasing the cooling rates. From SEM analysis, flake-like grains were more pronounced with increasing cooling rates. The indentation load versus diagonal length of the samples under different indentation loads in the range of 0.245-2.940 N were presented. We calculated Vickers hardness, Young's modulus, yield strength, fracture toughness values. These mechanical properties of the samples were found to be load and cooling rate dependent. In addition, we calculated the load independent microhardness, Young's modulus, and yield strength and fracture toughness of the samples using different models. The possible reasons for the observed changes in microstructure and mechanical properties of the samples due to cooling rates were discussed.

1. Introduction

BSCCO superconductor is one of the most interesting nominate material for technological applications [1]. Despite the potential advantages of this material compared with other superconductors (a lesser

* Cabir Terzioglu, Department of Physics, Faculty of Arts and Sciences, Abant Izzet Baysal University, 14280 Bolu-Turkey.

tendency to react with water or carbon dioxide, no need for special annealing in oxygen), it is often limited by its poor mechanical performance.

Variations in superconducting characteristics of pure and doped BiPbSrCaCuO reported by different research groups are often ascribed to different sintering conditions [2-4]. To find optimum preparation conditions for pure samples several studies have investigated the influence of cooling effect such as lattice parameters, critical temperature and volume fraction [5,6].

The mechanical properties of bulk BSCCO superconductor, as reported in the literature, can be summarized as low elastic modulus [7,8] low yield stress [9] and high degree of brittleness. However, Vickers microhardness indention testing is a useful tool to probe the mechanical properties of these brittle materials because the requirement for large specimens is relaxed. Vickers indention measurement of BSCCO superconductors have been reported by Jie Luo et al [10]. Veernder et al. reported the results of Vickers microhardness of the $Bi_{2-x}Pb_xSr_2Ca_2Cu_3O_y$ superconductor at different loads [11].

It is generally known that cooling rates plays very important role as well as chemical doping in high-T_c superconductor materials. On the other hand, cooling rates are also very important for improving superconducting, microstructure and mechanical properties. It is useful to observe the variation both the microstructure and mechanical properties.

In this paper, we report the influence of the cooling rate on the microstructure and mechanical properties of bulk BiPbSrCaCuO samples. We prepared three different samples, via cooling in liquid nitrogen (fast quench), air (quench) and furnace (unquenched).

2. Experimental details

The samples were prepared with the help of the standard solid-state reaction method. The starting materials consisted of mixture of oxides (Bi_2O_3, Pb_3O_4, and CuO) and carbonates ($SrCO_3$, $CaCO_3$) with a nominal composition $Bi_{1.6}Pb_{0.4}Sr_2Ca_2Cu_3O_y$. The reaction mixture was calcinated at 800°C for 24 h. This sample cooled to room temperature to grind. The resulting powders were then pressed into pellets at the pressure of 300 MPa. The pellets were sintered in air 832°C for 120 h and then cooled. Samples were prepared way to different cooling. First group sample heated and cooled in furnace (NC) (the heating and cooling rates of the temperature were chosen to be 5°C/min. and 3°C/min., respectively). Second group sample heated and cooled quench (SC). Third group sample heated and cooled rapid quench (FC) (in liquid nitrogen). The temperature dependence of the resistivity of the samples was measured by the standard four probe technique. X-ray diffraction (XRD) analysis was carried out by a rigaku D/ Max-IIIC diffractometer using high intensity Cu Kα radiation. The lattice parameters were determined from the d values of XRD peaks by a standard least-square.

Hardness measurements of the BSCOO samples were performed on polished surfaces with a digital microhardness tester at room temperature. A Vickers pyramidal indenter with different loads (0.245, 0.490, 0.980, 1.960, and 2.940 N) and a single loading time of 10s were applied and diagonals of indentation were measured with an accuracy of $\pm 0.1\mu m$. Indentations were made at different parts of the samples surface such that the distance between any two indentations was no less than two times the diagonal of the indentation mark to avoid surface effects due to neighbouring indentation. An average of 5 readings at different locations of specimen surfaces was taken to obtain reasonable mean values for each load.

3. Results and discussion

The T_c^{offset} of transition temperature values are 103 K, 101K and 91K for the FC, SC and NC samples, respectively. The transition temperature increases with increasing the cooling rate. This result is good agreement with previous result [6].

In order to investigate the effect of cooling rates on the mechanical properties, we measured the diagonal length as a function of test load. Conventional Vickers microhardness measurement consist of applying a load F on the material via geometrically defined indenter and after the indenter is removed, measuring the characteristic dimension d of the resultant impression. The Vickers microhardness (apparent) values of different applied loads were calculated using the equation [12]

$$H_v = 1854,4(F/d^2) \quad \text{(GPa)} \qquad (1)$$

where F is the applied load in Newton (N) and d is the diagonal length of indentation mark in micrometer (μm). The calculated load dependent microhardness values for different applied loads are summarized in Table 1. Figure 1 shows the variation of microhardness as a function of applied load for the NC, SC and FC samples. The variation of microhardness with load has similar shape irrespective of the cooling rates, although the numerical values are different. We have observed that the microhardness values increased with rapid quench cooling (FC). The reason for this ascribed to rapid quench cooling filling the intergrain space, resulting in better grain growth and causing larger grains, as a result leading to remarkable increase in the mechanical resistance of the samples. It was observed that Vickers hardness increased with increasing amount of doping in Bi-2223 superconductors [13-16]. It is also obvious from the figure that the Vickers microhardness values are load dependent for all samples; the calculated microhardness value decreases non linearly as the applied load decreased until 2N, then it tends to attain saturation. As reported by Khalil [14], this behaviour can be explained as follows; (a) at larger indention loads, the Vickers hardness registered smaller values, this observation may be due to the presence of weak grain boundaries of the superconducting ceramics; (b) at smaller indention loads, the Vickers hardness recorded higher values, this is ascribed to the fact that the measured hardness values were more indicative of the monocrystalline state without interference from grain boundaries. This non-linear behaviour has also been observed in the literature for Bi–Pb–Sr–Ca–Cu–O samples [13,14,17] and is known as the indentation size effect (ISE) [18–21].

Table 1. The load dependent values of H_v, E, Y, K_{IC} and B_i for the samples.

Sample	F (N)	H_v(GPa)	E (GPa)	Y (GPa)	K_{1c} (Pa.m$^{-1/2)}$	B_i (m$^{1/2}$)10^{+5}
NC	0.2450	0.4573	37.4819	0.1524	492.1	9.2920
	0.4900	0.4282	35.0973	0.1427	476.2	8.9915
	0.9800	0.3785	31.0219	0.1262	447.7	8.4534
	1.9600	0.3511	28.7735	0.1170	431.2	8.1413
	2.9400	0.3313	27.1520	0.1104	418.9	7.9085
SC	0.2450	0.4789	39.2523	0.1596	510.1	9.3876
	0.4900	0.4481	36.7278	0.1494	493.5	9.0807
	0.9800	0.3757	30.7937	0.1252	451.8	8.3148
	1.9600	0.3557	29.1544	0.1186	439.7	8.0905
	2.9400	0.3423	28.0561	0.1141	431.3	7.9366
FC	0.2450	0.5100	41.8014	0.1700	551.6	9.2450
	0.4900	0.4736	38.8215	0.1579	531.6	8.9094
	0.9800	0.3824	31.3407	0.1275	477.7	8.0051
	1.9600	0.3588	29.4047	0.1196	462.7	7.7539
	2.9400	0.3536	28.9802	0.1179	459.3	7.6978

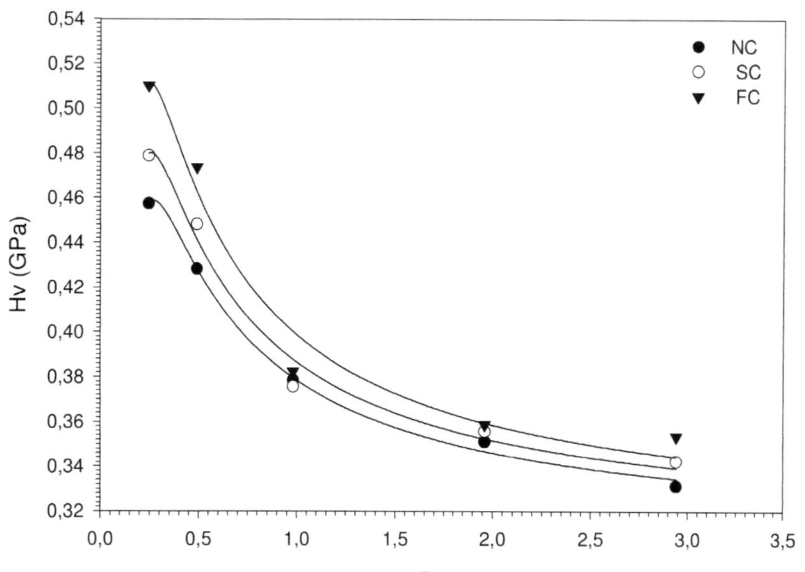

Figure 1. The variation of microhardness with load for the samples.

To account for this effect, several relationships between the applied load and the resulting indentation size have been suggested [22–25]. This effect can be explained by two different methods [12]. The first method assumes that the indentation contains an elastic portion. The elastic part of the deformation is relaxed upon loading. This can be accounted for by adding an elastic component, d_e, to the measured plastic indentation semidiagonal, d_p. Thus, a true hardness, H_0, is defined from [12, 26]

$$H_0 = 1854,4\left(\frac{F}{(d_p + d_e)^2}\right) \qquad \text{(GPa)} \qquad (2)$$

Therefore, Eq. (2) indicates that measured indentation diagonals should be linear with the square root of the applied load and the slope of such a curve is proportional to $(H_0)^{-1/2}$ and the vertical intercept of this graph is proportional to the elastic part of the indentation semi diagonal, d_e. The load dependence of the indentation diagonals for the NC, SC and FC samples was reanalyzed as d_p versus $F^{1/2}$ plots, as in Fig. 2. It was obvious that such plots are linear with the estimated linear regression coefficients (LRC) always better than 99.9%, implying that Eq. (2) provides a satisfactory description to calculate the true hardness of the indentation data for the samples. The extracted values of H_0, d_e and LRC are listed in Table 2. As can be seen from the table, it was observed that the values of H_0 of the samples decreased slightly but that of d_e increased. Quin et al. [27] investigated the variation of Vickers microhardness as a function of indentation load for a variety of ceramic materials. They observed that such a hardness–load curve shows a distinct transition to a plateau of the constant hardness level and concluded that the transition in such curves corresponds to the intrinsic hardness value of the material. In this study, this plateau is reached at 2 N applied load for the samples. As can be seen from Table 2, the true microhardness value of the NC sample (0.795 GPa) is higher than the hardness results (see Table 1) in the plateau (saturated) region (H_v = 0.331 and 0.351 GPa). This behaviour is observed in other samples (SC and FC) in this work. This result indicates that the true hardness of the sample is higher than that of the traditionally calculated ones. The calculation of the H_0 values for the investigated samples also indicated the magnitude of d_e to be 8.8 for NC and 8.9–9.4 μm for SC and

FC samples. These extracted magnitudes of d_e infer that the amount of relaxation of the diagonal length is significant with respect to the measured diagonals at low indentation loads [28] and hence the ISE is pronounced for the low load range. This method has been applied to YBaCuO and BiPbSrCaCuO [12,26,28] materials but consistency was not very good.

Table 2. Best-fit results of experimental data corresponding to Eq. (2).

Samples	H_0(GPa)	d_e(μm)	LRC	H_v (GPa) in plateau region
NC	0.783	8.875	0.99952	0.331–0.351
SC	0.787	8.911	0.99963	0.342–0.356
FC	0.795	9.451	0.99933	0.354–0.359

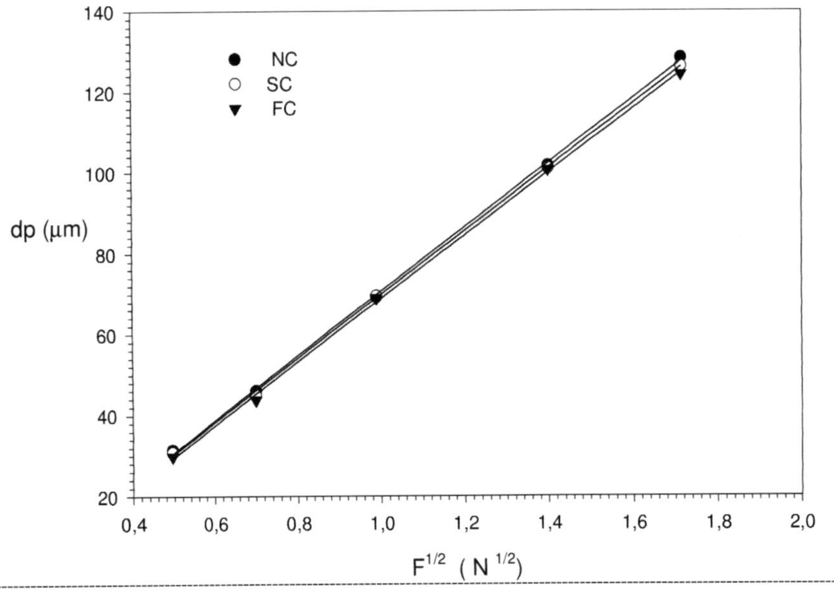

Figure 2. Plots of diagonal length versus square root of applied loads for the samples.

The second method considers energy dissipative processes during the indentation rather than elastic processes. In this model, a true microhardness can be defined by subtracting a dissipative part, F_0, from the applied load [12]

$$H_0 = 1854{,}4\left(\frac{F - F_0}{d^2}\right) \quad \text{(GPa)} \qquad (3)$$

Fig. 3 shows applied load as a function of the square of the diagonal length for the samples. Each set of data shows an excellent linear relationship (LRC > 0.999). The slope of each line corresponds to the load independent hardness constant, H_0, and the intercept of each line represents the sample resistance pressure, F_0. The extracted values of F_0, H_0 and LRC are listed in Table 3. As can be seen from this table, the values F_0 of the samples decreased and the values H_0 of the samples increased with decreasing the cooling rates. The LRC of each sample is very high, implying that Eq. (3) provides a satisfactory description of the indentation data for the samples. The energy dissipation model is also supported in our work by Fig. 4, which shows a typical impression of the NC sample for a load of 0.98 N.

Table 3. Best-fit results of experimental data corresponding to Eq. (3).

Samples	H_0(GPa)	F_0(N)	LRC
NC	0.174	0.117	0.99939
SC	0.179	0.108	0.99953
FC	0.184	0.107	0.99967

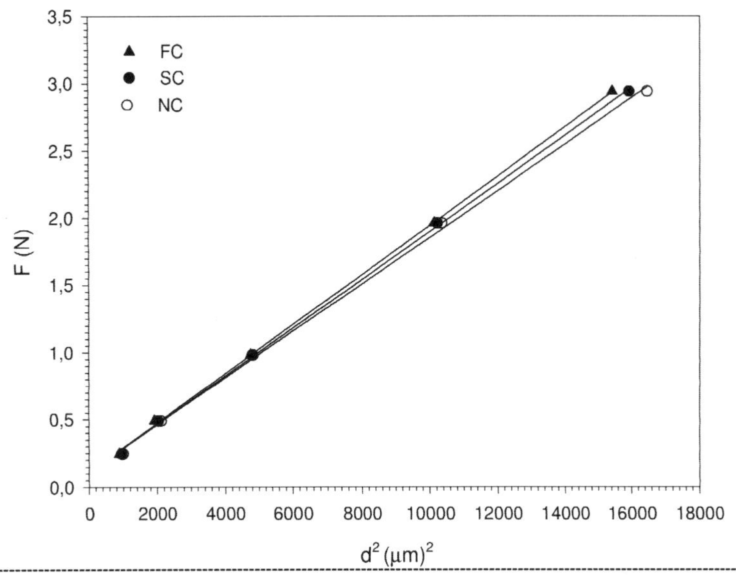

Figure 3. Graph of the applied load against the square of the diagonal length for the samples.

On the other hand, it is observed that the diagonal length is strongly dependent on the applied load from the experimental observations. This observation is governed by

$$\frac{F}{d} = H_0 d + \gamma \qquad (4)$$

proposed in [13,22,29,30]. Fig. 5 shows the values of F/d against the diagonal length of indentation, d, for the samples. Each set of data shows an excellent linear relationship. The slope of each line corresponds to the true hardness, H_0, and the intercept of each line represents the surface energy, γ. The extracted values of H_0, γ and LRC are listed in Table 4. It is observed that the values of H_0 and γ of all samples increased with decreasing cooling rates. This observation is ascribed to the dissipation of the energy of cracks at the interfaces [12] and a similar behaviour in ceramics was also reported in indentation studies [13,30,31].

Table 4. Best-fit results of experimental data corresponding to Eq. (4).

Samples	H_0(GPa)	$\gamma x10^{-3}$ ($N\mu m^{-1}$)	LRC
NC	1.554	3.231	0.99976
SC	1.584	3.315	0.99980
FC	1.594	3.640	0.99960

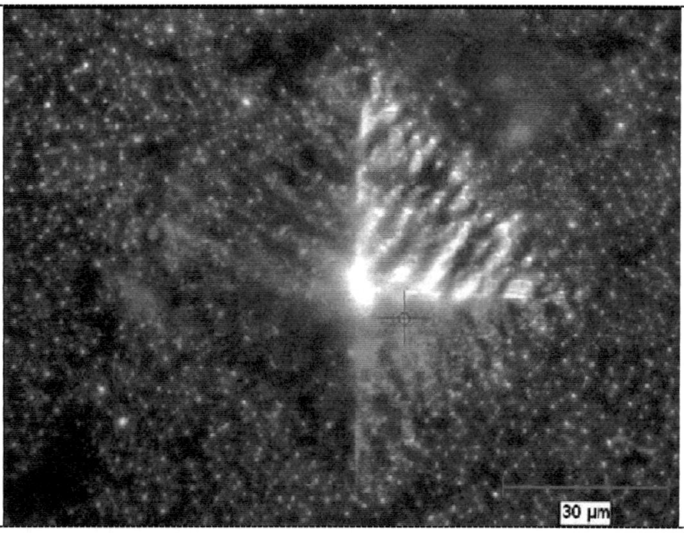

Figure 4. The typical indentation for an applied load of 0.98 N for the NC sample.

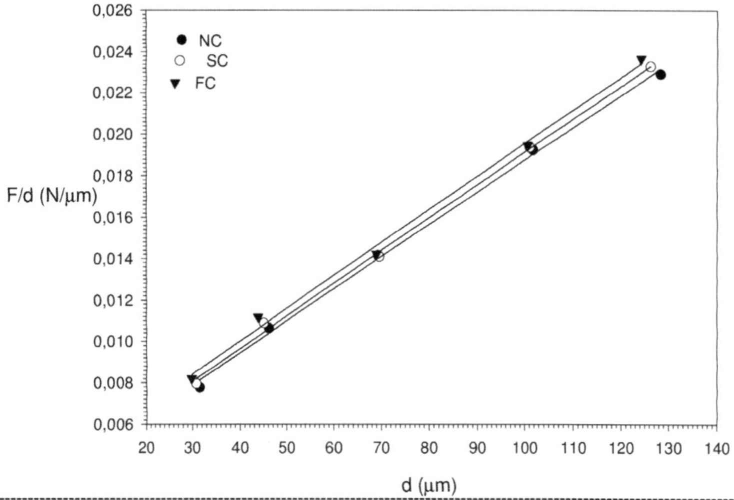

Figure 5. Plots of F/d versus d for the samples.

In most materials, the elastic modulus (Young's modulus), E, is related to the Vickers microhardness (apparent) by the relation [11]

$$E = 81.9635H_v \quad (5)$$

and yield strength Y is related to the hardness by the relation [32, 33]

$$Y \approx H_v/3 \quad (6)$$

The values of load dependent E and Y were calculated for each load using Eqs. (5) and (6), and are summarized in Table 5. As seen in this table, the load dependent E and Y decrease significantly with increasing cooling rates and loads. This behaviour is due to crack initiation and improvement of microhardness. Similar changes in the yield strength and elastic modulus were reported in the literature [11,34].

Table 5. The load independent values of Hv, E, Y, KIC and Bi for the samples.

Samples	H_0 (GPa)	E (GPa)	Y (GPa)	K_{1c} (Pam$^{-1/2)}$	$B_i(m^{1/2})$
NC	0.1540	12.6224	0.4620	285.6	53.92
SC	0.1580	12.9502	0.4740	293.0	53.93
FC	0.1590	13.0322	0.4770	308.0	51.62

It is useful to mention the fracture toughness, K_{IC}, as it is one of the main mechanical properties of superconducting samples. The fracture toughness is an important parameter for the selection of materials for applications. Owing to the nature of intrinsic brittleness, microindentation may result in microfracture around the impressed region on the surface of the samples [35]. As microfracture occurs mainly during the loading a portion of the energy, which is used to create the indentation deformation, will be dissipated by the crack formation. Owing to the definition of K_{IC} as the critical stress intensity factor, it is directly related to γ of the crack faces [36, 37]

$$K_{1C} = \sqrt{2E\gamma} \quad (7)$$

where E is the load dependent Young's modulus. The values of load dependent K_{IC} were calculated using Eq. (7) and are summarized in Table 1. From this table, it is observed that K_{IC} increases significantly with increasing loads. A similar change in the fracture toughness was reported in the literature [31,34]. Owing to this relation, an increase in K_{IC} corresponds to an increase in the average surface energy as proposed from the hardness calculations. From the values of K_{IC} and microhardness, one can calculate the brittleness index, Bi, using the relation [38]

$$B_i = \frac{H_v}{K_{1C}} \quad (8)$$

The calculated values of load dependent Bi for each load is listed in Table 1. From the table, it was observed that the brittleness index is load dependent up to 2 N and then remains close to a saturation value at higher applied loads for all samples. Note that the apparent microhardness, Young's modulus, yield strength, fracture toughness and brittleness index of the samples in the present work indicate strong dependency on applied load. In addition, we focused on the load independent values of Young's modulus, yield strength, fracture toughness and brittleness index of the samples. Instead of using apparent microhardness (H_v) values to calculate load dependent Y , E, K_{IC} and Bi in Eqs. (5)–(8) for each load, one can calculate E, Y , K_{IC} and Bi using true microhardness (load independent, H_0)

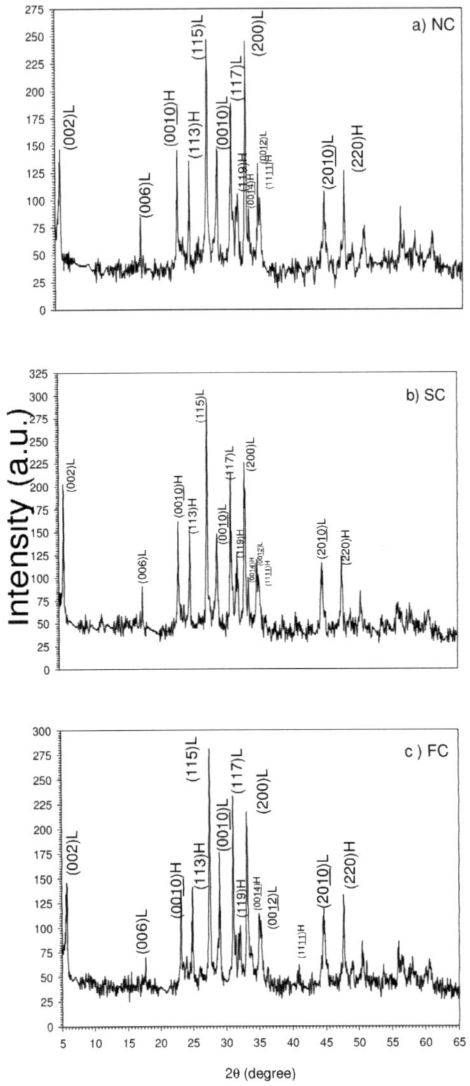

Figure 6. XRD patterns for the samples a) NC b) SC and c) FC.

calculated by Eq. (4) for each sample. The obtained load independent Y, E, K_{IC} and Bi for the samples are tabulated in Table 5. From the table, it is observed that the load independent values of E, Y, K_{IC} increase, whereas that of Bi decrease. Comparing Tables 1 and 5, one can conclude that the load independent values are lower than those of the load dependent values. This is in agreement with the literature [12,39]. The above results revealed that by influence of cooling rates, it is possible to control the mechanical properties of the samples.

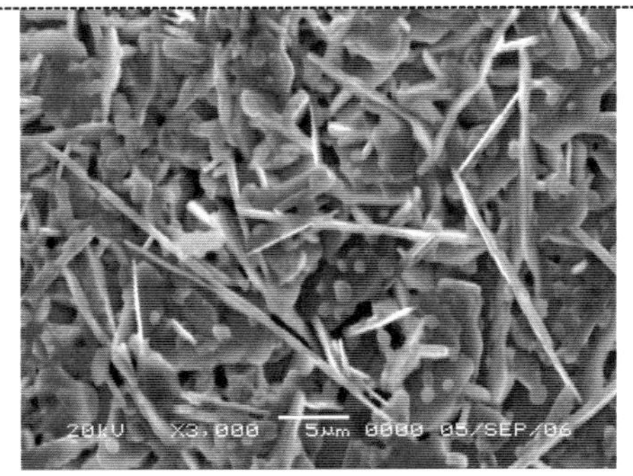

a) NC

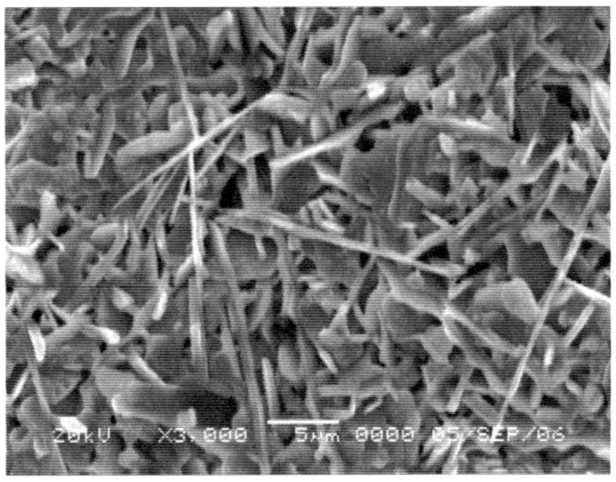

b) SC

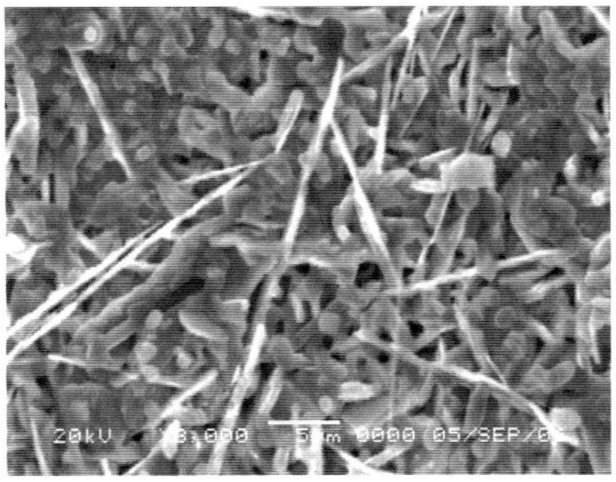

c) FC

Figure 7. SEM photos for the samples.

The XRD patterns from surface of the samples are shown in Fig. 6. Miller indices are indicated in the figure. (hkl)H and (hkl)L mean high-Tc phase (Bi-2223) and low-Tc phase (Bi-2212), respectively. We used the linear least squares method to calculate the lattice parameters of the samples. All of the samples are almost low phases. Comparison of XRD patterns did not reveal any significant shift at position of the peaks.

The structure of surface morphology of the samples was studied by SEM in order to determine the grain sizes. Fig. 7 shows surface micrographs for the NC, SC and FC samples. Grain size was small and granularity was apparent with sub-micron grains. Together with the XRD results, we deduce that the annealing temperature was lower than optimum and as a result grain connectivity was not sufficient.

4. Conclusion

This research represents the experimental results of cooling rate effect on Bi1,6Pb0,4Sr2Ca2Cu3Oy superconductors prepared by standard solid state reaction methods. The phase evaluation, microstructure, and mechanical properties of the samples have been reported. XRD examination of the samples showed that the Bi-2212 phase increased while the Bi-2223 phase decreased with increasing the cooling rates. From SEM analysis, flake-like grains were more pronounced with increasing cooling rates. The indentation load versus diagonal length of the samples under different indentation loads in the range of 0.245-2.940 N were presented. We calculated Vickers hardness, Young's modulus, yield strength, fracture toughness values. These mechanical properties of the samples were found to be load and cooling rate dependent. Faster cooling rate degrades the denser microstructure of the surface morphology and mechanical properties. These degradations are due to a increase in voids, impurity phase and resistance to crack propagation and to a decrease in the contact area among the superconducting grains.

Acknowledgements

This work is supported partly by the Scientific and Technological Council of Turkey (Project No: 104T325) and partly the Turkish State Planning Organization (DPT) (Project No: 2004K120200).

References

[1] Tamperi A Masini R Dimesso L Guicciardi S and Cristina Malpezzi M 1993 *Jpn. J. Appl. Phys.* **32** 4490
[2] Halim S A Khawaldeh S B Mohamed H 1999 *Azhan Materials Chemistry and Physics* **61** 251-259
[3] Ilonca G Pop A V Yang T R Jurcut T Lunga C Stiufiuca G Stiufiuca R Panfilescu I A 2001 *International Journal of Inorganic Materials* **3** 763–767
[4] Yilmazlar M Ozturk O Aydin H Akdogan M and Terzioglu C 2007 *Chinese J. Phys.* **Vol. 45** No. 2-I
[5] Ozturk O Yegen D Yılmazlar M Varilci A Terzioglu C 2007 *Physica C* **451** 113-117
[6] Terzioglu C Ozturk O Kılıç A Gencer A Belenli I 2006 *Physica C* **434** 153-156
[7] Chang F G Ford P J Saunders G A Li J Almond D P Chapman B 1993 *Supercond. Sci. Tech.* **6** 484
[8] Muralidhar M Nanda Kishore K Ramana Y V and Babuu V H 1992 *Mat. Sci. Eng. B* **13** 215
[9] Lo W Campbell A M Luo J and Stevens R 1995 *J. Amter. Res.* **10** 568
[10] Luo J Stevens R Lo W and Campbell A M 1995 *J. Mater. Sci.* **30** 3050
[11] Veerender C Dumke V R and Nagabhooshanam M 1994 *Phys. Stat. Sol. (a)* **144** 199
[12] Leenders A Ullrich M and Freyhardt H C 1997 *Physica C* **279** 173
[13] Yilmazlar M Cetinkara H A Nursoy M Ozturk O and Terzioğlu C 2006 *Physica C* **442** 101
[14] Khalil S M 2001 *J. Phys. Chem. Solids.* **62** 457
[15] Bruneel E Degrieck J Van Diriessche I and Hoste S 2002 *Physica C* **372-376** 1063
[16] Yilmazlar M Ozturk O Gorur O and Terzioğlu C 2007 *Supercond. Sci. Technol.* **20** 365-371
[17] Murakami A Katagari K Noto K Kasaba K Sohoji Y Muralidhar M Sakai N and Murakami M

2002 *Physica C* **378-381** 794
[18] Gong J Wu J and Guan Z 1999 *Matter. Lett.* **38** 197
[19] Sangwai K and Surowska B 2003 *Mater. Res. İnnov.* **7** 91
[20] Tickoo R Tandon R P Bamzai K K and Kotru P N 2003 *Mater. Chem. Phys.* **42** 446
[21] Elmustafa A A and Stone D S 2003 *J. Mech. Phys. Solid* **51** 357
[22] Frohlinch F Grau P and Grellmann W 1997 *Phys. Status Solidi* **42** 79
[23] Li H and Bradt R C 1993 *J. Mater. Sci.* **22** 917
[24] Hays C and Kendall E G 1973 *Metallography* **6** 275
[25] Gong J Wu J and Guan Z 1999 *J. Eur. Ceram. Soc.* **19** 2625
[26] Li Z Ghosh A and Kobayashi A S 1989 *J. Am. Soc.* **72** 904
[27] Quinn J B and Quinn G D 1997 *J. Mater. Sci.* **32** 4331
[28] Dutta A K Narasaiah N Chattopadhyaya A B and Ray K K 2001 *Ceram. Int.* **27** 407
[29] Hirao K and Tomozawa M J 1997 *Am. Ceram. Soc.* **70** 497
[30] Bernhardt E O 1941 *Z. Metall.* **33** 135
[31] Khalil S M 2005 *Smart Mater. Struct.* **14** 804
[32] McClintock F A and Argon A S 1996 *Mechanical Behaviour of Materials (Reading, MA: Addison-Wesley)* p 455
[33] Tabor D 1951 *The Hardness of Metals (Oxford: Clarendon)*
[34] Goyal A Funkenbusch P D Kroeger D M and Burns S J 1992 *J. Appl. Phys.* **71** 1363
[35] Lawn B R and Wilshaw T R 1975 *J. Mater. Sci.* **10** 1049
[36] Farber B Y Sidorov N S Kulakov V I Lunin A Y Izotov A N Emel'chenko G A Bobrov V S Fomenko L S Natsik V D and Lubenets S V 1991 *Superconductivity* **4** 2296
[37] Farber B Y Sidorov N S Kulakov V I Lunin A Y Izotov A N Emel'chenko G A Bobrov V S Fomenko L S Natsik V D and Lubenets S V 2006 *Compounds* **415** 300–6
[38] Nihara K Movena R and Hasselman D P H 1982 *J. Mater. Sci.***1** 13
[39] Uzun O Kolemen U Celebi S and Guclu N 2005 *J. Eur. Ceram. Soc.* **25** 969

Effect of Pr-doping on magnetic and transports in $YBa_2Cu_3O_{7-\delta}$ single crystal superconductor

J C Zhang[1,4], Z P Qin[1], G Jin[2], M Q Chen[1], X Yao[3], C B Cai[1], S X Cao[1]
[1] Department of Physics, Shanghai University, Shanghai 200444, China
[2] Department of Physics, Zhoukou Nomal University, Zhoukou 466000, China
[3] Department of Physics, Shanghai Jiaotong University, Shanghai 200030, China

Abstract. In cuprate oxide YBCO superconductor, Pr-doping is an interesting project for a long time. The reason is that many experiments have shown that of all the rare earth elements, only Ce, Pr and Tb are exceptive, does not result in superconducting samples while with the same crystal structure as other Y123 superconductors. Here, $CeBa_2Cu_3O_{7-\delta}$ and $TbBa_2Cu_3O_{7-\delta}$ isn't crystalline stable, while $Y_{1-x}Pr_xBa_2Cu_3O_{7-\delta}$(YPr123 systems) is easily formed in single-phase structure for all Pr content x=0.0-1.0. Therefore, to understand the question why $PrBa_2Cu_3O_{7-\delta}$(Pr123) is not superconductor is possibly essential to help to determine the nature of the superconductivity. The typical general results for Pr substitution are that there are a monotonic decrease in T_c and eventually a metal-insulator transition with increasing Pr concentration and that there are also ion size effect which shows that in $R_{1-x}Pr_xBa_2Cu_3O_{7-\delta}$ (R=Rare Earth), Pr suppressed superconductivity more seriously as the rare earth's radius increase. Theoretically, some models, such as magnetic pair breaking, hole filling, carrier localization, and so on, have been proposed to explain the phenomena above. Magnetic and pinning force behavior is a kind of important characteristics in both conventional superconductor and layered cuprate superconductor. Studies of flux behavior in superconductor are of interest from the viewpoints of both fundamental research and practical application. In order to explain the anomalous behavior of Pr123, in the present work, the magnetization and flux pinning results at various temperatures are reported for $Y_{0.93}Pr_{0.07}Ba_2Cu_3O_{7-\delta}$ (YPBCO) single crystals. The samples were prepared by top-seeded solution growth method. The magnetic field dependence of critical current density J_c, pinning force density F_p and the phase diagram are also given for YPBCO single crystals. The results show that the value of the ratio of the maximum pinning force field *vs* irreversibility field is at around 0.333. Flux pinning mechanisms is also discussed in different magnetic field by the scaling function as well as the ratio of the pinning force field at H_{on} and H_{peak}. The results are compared with doping free YBCO sample, which probably explains that Pr suppresses the superconductivity.

1. Introduction

There are two reasons for many interests on the flux pinning mechanism in high temperature superconductors. First, the study complex photograph, disorder structure and anisotropies may be a good angle of view for the study of high-temperature superconductor (HTSC) mechanism[1-6]. Second, it is also of great interest from technological aspect of high T_c superconductors. The critical current density J_c is one of important parameters in the application of superconductors[7,8]. As we all know, PBCO is not a superconductor. Pr has a large size in ion radius and own big spin magnetic moment comparing with other rare-earth element[9]. Chemical substitute is an important method to

[4] jczhang@staff.shu.edu.cn

increase pinning density. This method does not break molecular structure and can be used for discussion of the relationship between the density of doping and critical temperature. The size of non-superconductor area, caused by doping Pr, is closely near the coherence length, so it is an effective pinning, which has been proved by experiments. The magnetization curves at different temperatures have been measured in $Y_{0.928}Pr_{0.072}Ba_2Cu_3O_{7-\delta}$ single crystals, which were annealed for 72h or 172h at 500 ℃. The magnetic field dependence of critical current density J_c and pinning force density F_p are obtained according theoretical model. Peak effect is shown. We give preliminary explanation from flux pinning mechanisms.

2. Sample Preparation and Experiment

The samples used in the experimental are grown by the Top-seeded Solution Growth Method. The raw materials with a weight ratio of Ba:Cu=5:3 were synthesized from $BaSO_4$, Y_2O_3 and CuO powder of 99.9% purity using solid state reaction. The speed of single crystal growth is 0.02*mm/h* with 120*r/h* of rotating rate. Inductively Coupled Plasma Technique shows that the value of x is 0.072. The raw crystals were further annealed for 72h (sample P-72h) or for 172h (sample P-172h) at 500 ℃. Magnetization measurements were performed using Quantum Design PPMS-9 system with a constant sweep rate of 100 *Oe/sec*. Before each measurement we set temperature to 180 ℃ to release the captured flux.

3. Results and Discussion

3.1. Magnetic measurement

Figure 1 shows a typical measurement of DC susceptibility (FC) of all samples, with the external magnetic field perpendicular to their c axis. The superconducting transition temperature, at which the magnetic moment decrease 10% of the whole transition procession, is about 87K for P-72h and 86.5K for P-172. There is a second peak around 79.5K for P-72h.

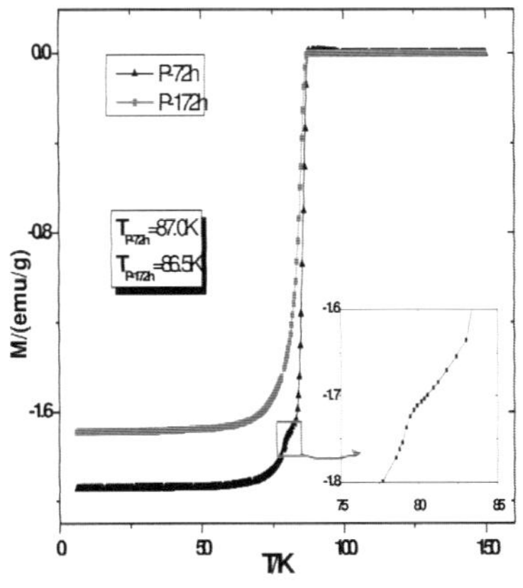

Figure 1. The DC susceptibility for annealed 72h and 172h samples measured. The applied field H=100 Oe, with H//c axis.

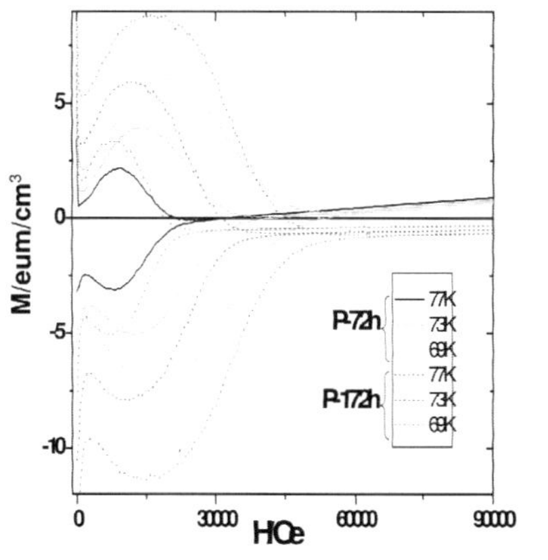

Figure 2. Magnetic Moment, M, as a function of the applied field, H, at 77K, 73K and 69K where the pronounced peaks can be observed. Solid lines for P-72h and dotted lines for P-172h.

For YBCO, the strength of superconductivity usually increases with the increase of the annealing time, but a different result has been obtained in our experiments. Why does this happen? We think that one reason is the non-uniform distribution of Pr ions in microstructures during the growth procedure. We got that the value of x is 0.072, but this value is the average density of Pr. Such micro-phase separation results in the different T_c superconductivity areas with size of dozens of Angstrom. The second peak in P-72h reveals that there exists micro-phase separation. Therefore, areas with a doping density less than 0.072 should exist. Since critical temperature decrease with increasing doping density of Pr, higher onset transition temperature should correspond to a lower doping density. Thus we can speculate that the ions of Pr distribute more uniformly in microstructures after 172h annealing. This can be proved by the disappearing of the second peak for P-172h. In other word, the values of x for the different T_c superconductivity areas converged to 0.072 for P-172. The interface between the different T_c superconductivity areas can be the effect pinning center, which can be shown in a later paper.

Figure 2 shows the magnetization hysteresis loops for P-72h and P-172h at 77K, 73K and 69K. We can observe the peak effect in the two samples. For superconductors, the value of M is usually zero when the magnetic field reaches the upper critical field H_{c2}. But figure 2 shows a different behavior: when the magnetic field increase above the irreversible field, M changes the sign for P-72h. This shows that the sample shows paramagnetism behavior above the irreversible field. Tan Mingqiu *et al.* [9] attributed this new feature to which the large magnetic moment from the spin polarization of Pr ions. Similar features can also be observed in other magnetic doping, for example, the Gd-doping in YBCO was reported by Feng Yong *et al.*[10] We usually consider two main sources for magnetic moment. One is the ring current; the other is the electron spin polarization. When the sample is in superconducting state, the ring current is more important than the electron spin polarization, so that we can omit the latter. When H is above H_{irr}, flux line lattices start to melt, and then distribute disorderly. Under such circumstances, the effect of electron spin polarization should be considered. The different regions with the size of dozens of Angstrom and non-uniform distributions of Pr ions can be in favor of the emergence of paramagnetic. The ions of Pr distribute more uniformly in microstructures after a 172h annealing of raw single crystal. Simultaneously, the spin orientation of Pr ion is more disordered.

3.2. The Peak Effect

We can calculate J_c for different temperatures by the Bean model. Then, the pinning force density F_p is given by $\boldsymbol{F_p}=\boldsymbol{J_c}\times\boldsymbol{B}$. Using the above two equations, we get (J_{on}, H_{on}), the onset of critical current increase with magnetic field, (J_{peak}, H_{peak}), the maximum of the peak effect, (F_{peak}, H_{fpeak}), the maximum of the pinning force, and the irreversibility field H_{irr} which correspond to the branch point in the magnetization-hysteresis-loops. The temperature scaling of the current $f(H/H_{peak})= J_c(H,T)/J_{peak}$ is shown in figure 3 for two samples at different temperatures. According to H.Küpfor *et al.* [11], the temperature scaling of the current is a well-known property in twinned and twin-free crystals. In the higher temperature region, the pinning interaction of twins and other large scale defect suffer less from thermal oscillations of the vortices than point-like defects or the oxygen vacancy clusters. And also, twin-free crystals always show a broadening of the peak at lower temperatures because the effective defect concentration increases with decreasing temperature. We can find that the curves for P-172h are more dispersible than for P-72h in figure 3. Because of the non-uniform distribution of Pr ions in microstructures, the interface between the areas different T_c superconductivity can be effective pinning center. After the annealing of a longer time, the density of effective pinning center will decrease. Compared with P-72h in figure 3, we find the peak effect for P-172h is obviously restrained. First, the value of J_{peak} /J_{peak} is about 0.6 for P-72h, and P-172h is 0.75. In other word, the ratio of critical current gain for P-72h which is caused by the peak effect is higher than for P-172h. Second, the peak of P-172h is broader than that of P-72h. The reason is that the interface between the different T_c superconductivity areas can be the effective pinning center for P-72h, which is useful to the appearance of peak effect. We conclude that the peak effect is related to micro-phase separation introduced by Pr doping.

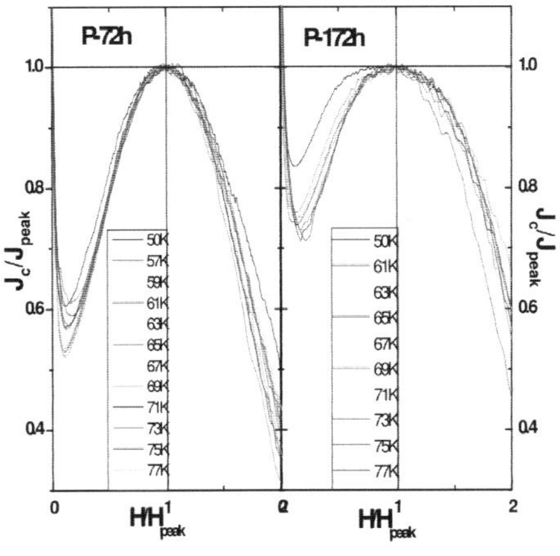

Figure 3. Normalized current density J_c/J_{peak} *vs* normalized applied magnetic field H/H_{peak} at different temperatures for P-72h and P-172h.

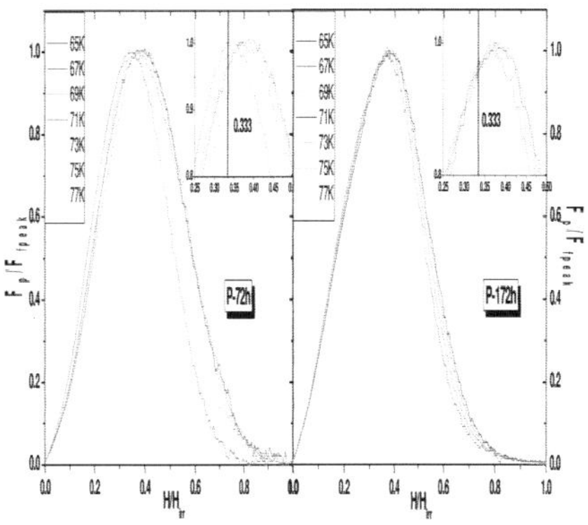

Figure 4. Normalized pinning force density F_p/F_{fpeak} *vs* normalized applied magnetic field $h=H/H_{irr}$ at different temperatures for P-72h and P-172h. The insets show magnifications of the maximum, where the vertical line with h=0.333 is only an eyeguide.

There are many reports about the peak effect of YBCO. Wen *et al.*[12] proposed a model concerning the non-uniform distribution of upper critical field among the pinning sites to explain the fishtail effect. According to their model, in the high temperature region, small size normal cores are the main pinning centers. Therefore, the pinning force density should be proportional to $h(1-h)^2$, as predicted by Dew-Hughes model[13], yielding a maximum at h=0.333, with $h=H/H_{c2}$. For a simple estimation, we will use the irreversible field H_{irr} determined in the experiment as H_{c2}. In figure 4, we present the correlation between F_p/F_{fpeak} and $h=H/H_{irr}$ from 65K to 77K. It is evident that all these curves show a maximum around h=0.333, which is in good agreement with the theoretical predication for small size normal core pinning. We observe carefully that the maximum is above 0.333. Which we can explain using the slightly higher H_{c2} than H_{irr} in the vicinity of the critical temperature. But we must note here that we can only judge the main flux pinning mechanism using Dew-Hughes model. The flux pinning mechanism in HTS is very complex, which may change with the field. Hence we observe that the H_{irr}/F_{fpeak}-h curves of P-72h are more broad than that of P-172h. The reason may be that the non-uniform distribution of Pr ions in microstructures can induce the increase of mean free path pinning. This explanation agrees with the former.

4. Conclusions

The magnetization curves at different temperatures have been measured in $Y_{0.928}Pr_{0.072}Ba_2Cu_3O_{7-\delta}$ single crystals, which were annealed for 72h or 172h at 500 ℃. We find that the peak effect is related to micro-phase separation introduced by Pr dopants. Comparing P-72h with P-172h, we come to a conclusion that the interface between the different T_c superconductivity areas can be the effective pinning center, which enables the appearance of peak effect. Flux pinning mechanisms are discussed in different magnetic field using the scaling function. At low temperature, the single vortex pinning takes an important role. The value of the ratio of the maximum pinning force field *vs* irreversibility field is near around 0.333. This strongly suggests that the effect of the small normal core pinning is the main reason for the fishtail effect.

Acknowledgments
This work is supported by NSFC(No.10774097,10504019), Program for Changjiang Scholars and Innovative Research Team in University(No:IRT0739), The Key-project of the Science & Technology Committee of Shanghai Municipality (No.07JC14022), Special Research Fundation for the Doctoral Discipline of University (Grant No. 20050280004).

References

[1] Park A R, Chio S M, Dender D C, Lynn J W and Ling X S 2003 Phys. Rew. Lett. 91 167003.
[2] Yeshurun Y, Bontemps N, Burlachkov L and Kapitulnik A 1994 Phys. Rew. B 49 1548.
[3] Grzegorz H and Nagi A D S 1996 Phys. Rew. B 54 15463.
[4] Neier-Hirmer R, Kupfer H and Scheurer H, 1985 Phys. Rew. B 31 183.
[5] Brandt E H 1983 Phys. Rew. Lett. 50 1599.
[6] Takuya H and Katsukuni Y 2002 Phys. C 383 48.
[7] Diosa J E, Vargas R A and Mellander B E 1997 J Phys: Condens. Matter 9, 4621.
[8] Kokkaliaris S, Zhukov A A, Groot P A J D, Wolf T, Gagnon R and Taillefer L 2000 Phys. Rew. B 61 3655.
[9] Tan M Q, Tao X M, Zhang Q R and Xu Y B 1995 Progress in Physics (in chinese) 15 173.
[10] Feng Y, Zhou L, Yang W M and Zhang C P 2000 Acta Physica Sinica (in chinese) 49 146 .
[11] Küpfor H, Wolf T, Lessing C, Zhukov A A, Lancon X, Meier-hirmer R, Schauer W and Wühl H 1998 Phys. Rev. B 58 2886.
[12] Wen H H and Zhao Z X 1996 Appl. Phys. Lett 68 856.
[13] Hughes D D 1974 Philos. Mag. 30 293.

The structure and magnetic properties of $Fe_{1.2}In_{1.87}Se_4$ single crystals

J.Hasani Barbaran, G.G.Guseinov, Y.I.Aliyev, G.M.Agamirzoeva
Institute of Physics Azerbaijan
National Academy of Science, AZ-1143 Baku, pr.Cavid.33, Azerbaijan
E-mail: g.guseinov@rambler.ru

Abstract: $Fe_{1.2}In_{1.87}Se_4$ single crystals have been fabricated by partial substitution of In atoms with Fe atoms in In_2Se_3 in vacuum with heat treatment up to 1270°K and then cooling to room temperature. Based on X-ray diffraction studies and crystallographic calculations it has been shown to crystallize into rhombohedral lattice with $ZnIn_2S_4$ type three-packet (layer) structure and with lattice parameters a=4.08, c=39.96Å, Z=3 and space group $R\bar{3}m$. The dependence of the magnetic susceptibility of produced crystals with temperature has been investigated; it shows that a transition from ordered paramagnetic to disordered spin-glass state takes place below 19K.

1. Introduction

A number of AB_2X_4-type natural and artificial crystals are raw materials to fabricate heat-resistant materials, some metals and jewels, due to their physico-chemical properties. We note that in the given crystals there has been also found high ion mobility, ferroelectric properties and existence of commensurate phase, but in crystals with paramagnetic and magnetic cations, phase transitions of magneto-ordering type have been observed. As AB_2X_4-type compounds have the ability to crystallize in various polytype phases, they can be used as model objects when investigating various chemical phenomena. In this case, the search and investigation of general physico-chemical properties of AB_2X_4-type chalcogenides can be considered as one of the main trends of material science. Taking into consideration the interest in magnetic semiconductors with layered and layered-chained crystal structure as used in developing modern electronics and microelectronics, authors of works [1, 2] obtained the compound of $Fe_{0.5}In_{1.5}S_3$=$Fe_{0.74}In_{2.16}S_4$ composition (by EDX analysis); there has been carried out investigations of magnetic properties, Mössbauer and neutron diffraction. Also on the basis of auto-diffractometric experiments its crystal structure has been studied. As a result it can be seen that $Fe_{0.74}In_{2.16}S_4$ compound has been crystallized in a cubic lattice with a lattice parameter of 10.69 Å, a space group of Fd3m, Z=8, and a crystal structure refered to as transformed spinel. Analysis of Mössbauer and neutron diffraction shows that Fe atoms in the compound are in two valence states, and so crystallochemical formula of composition must be in the form $Fe_{0.74}In_{2.16}S_4$.
Following the above-mentioned work, this paper deals with synthesis result, structure and investigation of magnetic properties of $Fe_{1.2}In_{1.87}Se_4$ compound single crystals obtained by partial substitution of In atoms in In_2Se_3 with Fe atoms.

2. Experimental Details

By partial subsititution of In atoms with Fe atoms in In_2Se_3 semiconducting compound and by conserving total valence balance, we have synthesised $Fe_{1.2}In_{1.87}Se_4$ and $Fe_{0.74}In_{2.16}Se_4$ compounds (NB this paper deals only with the investigation on the first of these crystals).

Compound synthesis has been made in high-quality, transparent quartz ampoules. The elements used for synthesis have chemical purity 99.998%. Fe, In and Se constituents are ground into small pieces and placed in quartz ampoule in mass appropriate to weight of 3g. The ampoules are then evacuated up to 10^{-2} Pa and then sealed. The ampoule is then placed in one-zone furnace, heated up to 920°K with rate ~100°K/h and kept at this temperature for 1 hour with regular vibration. Then the temperature of furnace is increased to 1270°K with same rate and kept at this temperature for 1.5 hour with vibration. Next, the furnace temperature is reduced to T=870 K at the same rate, and then annealed at this temperature for 24 hours, and then the furnace is turned off and cooled down to room temperature. At the end of the process, the ampoule is taken out of the furnace and visual observation of synthesis product is carried out by the microscope. The sample appears to be a crystalline mass with layered morphology and easily to be split into small thin layers which shows the perfection of its exfoliation.

The prepared layered crystals have been studied by Roentgen unit of powder auto-diffractometer on CuKα-radiation. Hence the lattice constants, the symmetry and some crystallochemical parameters have been determined. Magnetic properties of $Fe_{1.2}In_{1.87}Se_4$ sample have been measured with SQUID magnetometer in magnetic fields up to 50 kOe and temperature range 2 K≤T≤400 K.

3. Results and discussion

3.1. X-ray diffraction studies

After growing $Fe_{1.2}In_{1.87}Se_4$ crystals, undeformed parallelepiped samples with ~0.1x0.3mm sizes have been extracted for X-ray studies. In order to determine the crystal quality by photometric methods, Laue diffraction patterns have been obtained, which indicate normal perfectness of grown single crystals. The Roentgen diffraction shows that the crystal lattice of the samples under investigation have hexagonal (trigonal) symmetry. The diffractogram obtained and the crystallographic data calculated on its basis are given in Fig.1.

As a result of crystallographic calculations of investigated samples, it can be seen that they have crystallized in rhombohedral lattice. The rhombohedral lattice constants in hexagonal structure are: a=4.08 Å, c=39.69 Å, sp.gr. $R\bar{3}m$, Z=3, V=544.8 Å^3, $V_{lat}/\Sigma Se$=45 Å, R_{roent}=5.48 g/sm^3.

It is known that lattice constants obtained for $Fe_{1.2}In_{1.87}Se_4$ are also characteristic of a number of complex chalcogenides with three-packet structure. Therefore structural model of the investigated samples must not be different from the models of semiconducting phases such as $ZnIn_2S_4$ and $FeIn_2Se_4$. As already known, stoichiometric compounds such as $ZnIn_2S_4$ [3] and $FeIn_2Se_4$ [4] have been crystallized in polytype structures of similar behaviour. The compound we have investigated is different from above-mentioned analogues by virtue of the metal atom relation. This, by maintaining the general crystallochemical character of structure between them, leads to certain differences in metal atom distribution in octahedron and tetrahedral vacancies. These differences were seen first in the diffraction patterns (which is their characteristic).

The diffraction pattern taken from $FeIn_2Se_4$ [4] differs distinctly from the diffraction pattern of $Fe_{1.2}In_{1.87}Se_4$ in the intensity of interference lines, which must be due to the relation of In atoms to Fe atoms. The scattering factor of In atoms compared with Fe ones is about twice as many as Fe atoms and so this must be due to the comparative decrease in In atoms and increase in Fe atom content. We note that structural calculations have confirmed the above-mentioned facts.

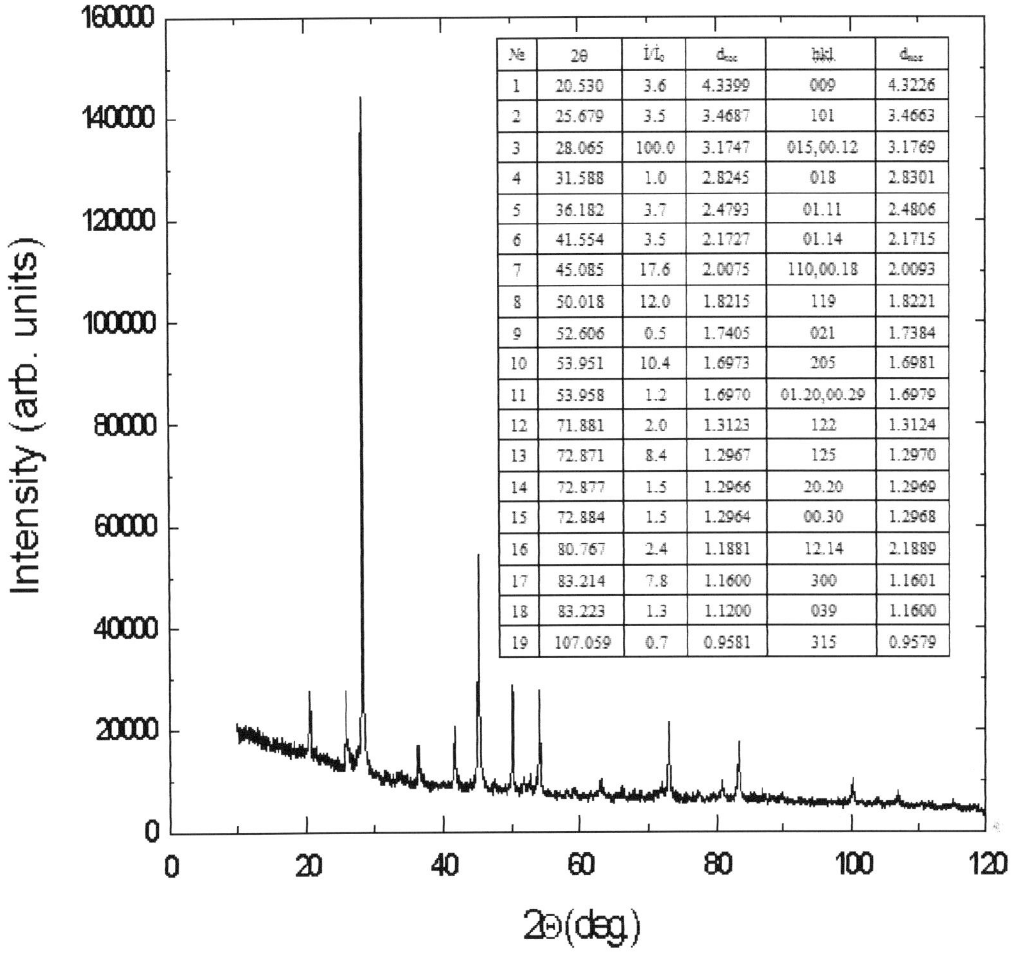

№	2θ	I/I_0	d[illegible]	hkl	d[illegible]
1	20.530	3.6	4.3399	009	4.3226
2	25.679	3.5	3.4687	101	3.4663
3	28.065	100.0	3.1747	015,00.12	3.1769
4	31.588	1.0	2.8245	018	2.8301
5	36.182	3.7	2.4793	01.11	2.4806
6	41.554	3.5	2.1727	01.14	2.1715
7	45.085	17.6	2.0075	110,00.18	2.0093
8	50.018	12.0	1.8215	119	1.8221
9	52.606	0.5	1.7405	021	1.7384
10	53.951	10.4	1.6973	205	1.6981
11	53.958	1.2	1.6970	01.20,00.29	1.6979
12	71.881	2.0	1.3123	122	1.3124
13	72.871	8.4	1.2967	125	1.2970
14	72.877	1.5	1.2966	20.20	1.2969
15	72.884	1.5	1.2964	00.30	1.2968
16	80.767	2.4	1.1881	12.14	2.1889
17	83.214	7.8	1.1600	300	1.1601
18	83.223	1.3	1.1200	039	1.1600
19	107.059	0.7	0.9581	315	0.9579

Fig.1. The X-ray diffractions and crystallographic data of $Fe_{1.2}In_{1.87}Se_2$

Interatomic vector in one-dimensional section-00*l* and electron density ρ(001) show that there has been certain difference in metal distribution in octahedron and tetrahedral positions of these structures. In Fig.2. there has been given structural model for c/3 layer (packet) of triplex polytype phase.

Analysis of these investigations shows that the statistical distribution of atoms in the structure and replication of periodicity in the third layer of triplex structure is the cause of polytype phase occurrence as a result of tetrahedral layer displacement against each other for ½ edge of tetrahedrons.

Atom coordinates in structure are in below-mentioned order: In octahedron layer (0.6Fe+0.4In) in position 3(a), coordinates (000), in tetrahedral vacancies ($Fe_{0.3}$+$In_{0.7}$) in position 6(c), coordinates 00.25; 6 atoms of Se in position 6(c), coordinates 00.12, other 6 atoms of Se also in position 6(c), coordinates 00.30.

So the structure of investigated crystals presents isomorphic structure with $ZnIn_2S_4$ and $FeIn_2Se_4$ triplex structure where sequence of one layer corresponds to the sequence of tetrahedron-octahedron-tetrahedron-vacancy (TOTV).

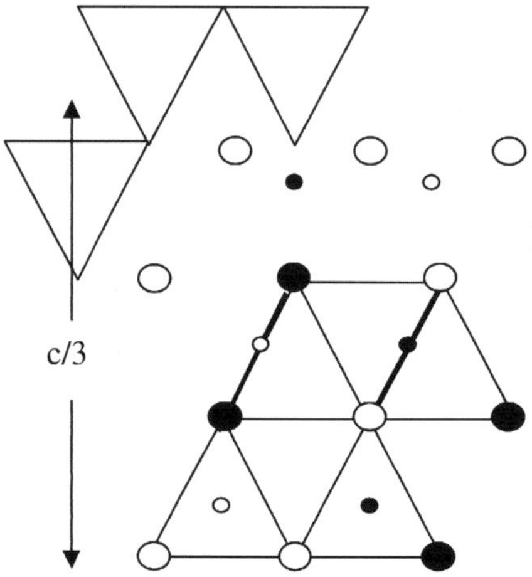

Figure.2. The model crystal structure of $Fe_{1.2}In_{1.87}Se_4$

3.2. Magnetic measurements

Fig.3 shows the temperature dependence of magnetic susceptibility ZFC and FC of $Fe_{1.2}In_{1.87}Se_4$ sample. Regimes correspond to preliminary cooling of the sample down to 2 K
(or 4 K) either in zero field ZFC or at $H \neq 0$ FC as usual.

It can be seen that at high temperatures, the magnetic susceptibility obeys the Curie-Weiss law. The paramagnetic Curie temperature as determined from the curve of reverse magnetic susceptibility is negative θ=-205K which indicates the dominance of the antiferromagnetic exchange interaction. Effective magnetic (paramagnetic) moment H_{ef} is 5.00 μ_B/Fe, which is slightly bigger than the moment for Fe^{2+} (Fe^{2+}, H_{ef}=4.9 μ_B). From the insert in Fig.3, it can be seen that at low temperatures (below 19°K on magnetic susceptibility curves) magnetic irreversibility is observed, i.e. magnetic susceptibility in ZFC and FC regimes are different. According to Fig.3 in ZFC regime, the observed maximum value is shifted to the side of low temperatures with the increase of field. It is known that similar irreversibility is one of the signatures of spin-glass (S-G) disorder [5].

The frustration parameter is determined from the paramagnetic Curie temperature relation to junction structure (S-G) T_f is equal to f=10.7, which indicates a bigger degree of frustration[6, 7].

It is known that frustration and disorder are necessary conditions for spin-glass behaviour [5]. In Fig.4 we show the dependence of $Fe_{1.2}In_{1.87}Se_4$ magnetization on magnetic field at T=4 and 400 K in regime FC.

At T=400 K, the sample magnetization increases linearly with the field, which is characteristic of the paramagnetic state. For T=4K (and below) we observe shift hysteresis (insert in Fig.4). For spin glasses (S-G), one of the magnetic properties is the display of macroscopic anisotropy after the sample is first cooled down to $T<T_f$ (freezing temperature of S-G) in external magnetic field (regime FC). The presence of macroscopic anisotropy in S-G has been determined unambiguously in experiments on investigation of transverse susceptibility, electron paramagnetic resonance and rotational moment, and has been also revealed in occurrence of M_{FC} magnetization hysteresis. In this case, in rather low temperature ranges $T<<T_f$ there has been observed displacement of hysteresis cycles that is the sign of unidirectional anisotropy occurrence [8]. Experimental observations of the macroscopic anisotropy of S-G shows it to be due to correlations between local exchange fields and anisotropy fields which are brought about by cooling sample in the field to $T<T_f$.

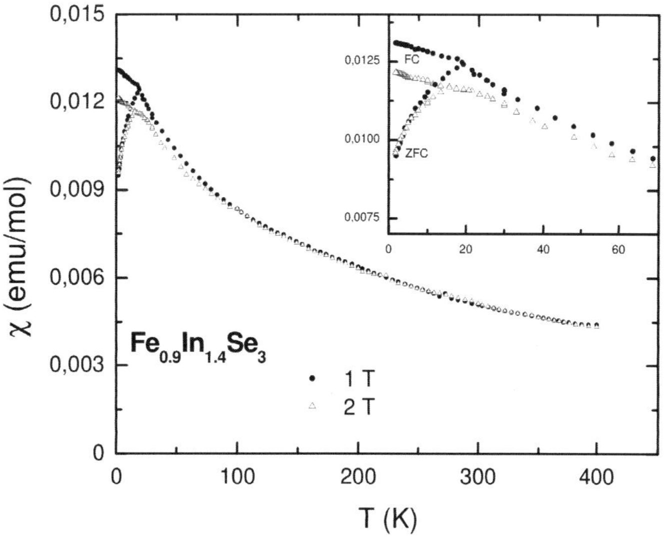

Figure.3.The magnetic susceptibility of $Fe_{1.2}In_{1.87}Se_4$ crystals vs. temperature at various magnetic field in the ZFC and FC regimes. In the insert: the low-temperature range of magnetic susceptibility corresponding to the magnetic field values H=1T and H=2T in the ZFC and FC regimes.

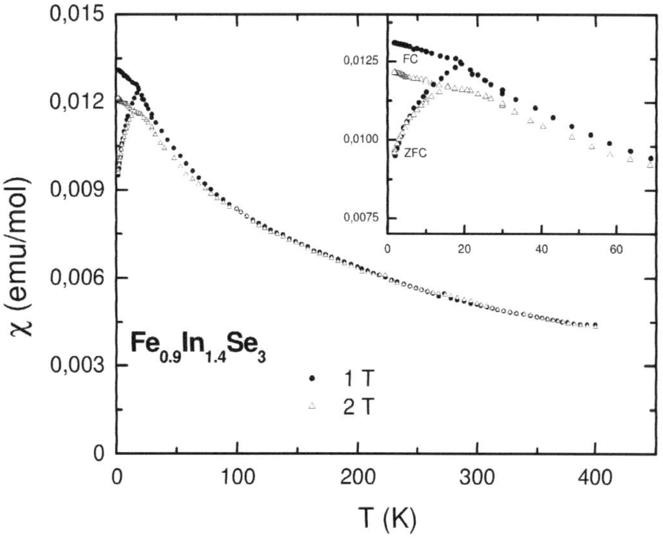

Figure.4. The magnetization of $Fe_{1.2}In_{1.87}Se_4$ crystals vs. magnetic field in the T=4÷400°K temperature range. In the insert: the hysteresis at T=4°K in the FC regime.

(3000 to 20000 Oe). The paramagnetic behavior, being linear, has been fitted to a straight line.

4. Conclusion

In this work $Fe_{1,2}In_{1,87}Se_4$ single crystals have been prepared with isomorphic structure with $ZnIn_2S_4$ and $FeIn_2Se_4$ triplex structure where sequence of one layer corresponds to the sequence of tetrahedron-octahedron-tetrahedron-vacancy. while the lattice is rhombohedral with a=4.08, c=39.96 Å, Z=3 and space group $R\bar{3}m$.

We also suggest, based upon the obtained experimental results and observed magnetic irreversibility (χZFC$\neq\chi$FC) and the displacement of hysteresis cycles at low temperatures and the high degree of frustration (f=10.7), that a spin-glass state is possible in $Fe_{1.2}In_{1.87}Se_4$ crystals below T_f=19K .

References

[1] Krimmel A et al. 2005 *J.Phys. Condens Matter.* **17** 3611
[2] Peissner M et al. 2006 *Hepertine Interact* **169** 1305
[3] Donika F G et al. 1971 *Kristallography* **16** 235 (Ras.)
[4] Reil S et al. 1998 *J.of Alloys and Compounds* **270** 83
[5] Bruder K et al. 1986 *Rev.Mod.Phys.* **58** 801
[6] Goya G F et al. 2002 *J. Of Solid State Chemistry* **164** 326
[7] Ramikez A P 2001 *Hanbook of Magnetic Materials* ed Schow K H J (Amsterdam: Elsevier) **13** 423
[8] Levy P M et al. 1982 *J.Appl.Phys.* **53** 2168

Kinetics of first stage of formation magnetite nanoparticles prepared by chemical precipitation method

R A Ali-zade[1]
Institute of Physics, Azerbaijan National Academy of Sciences,
H.Javid str.33, Baku Az-1143, Azerbaijan

E-mail: alizade@lan.ab.az

Abstract. First stage of formation magnetite nanoparticles is described by kinetics equation. Solution of kinetics equation is Gauss function. Mean value and standard deviation of Gauss function are $\bar{x} = \frac{Kt}{4r_0}$, $\sigma = \sqrt{\frac{Dt}{2}\frac{1}{r_0}}$. Critical size, determinant of collapse and growth of nanoparticles are determined. It's showed nanoparticles less critical size may be is collapsed, bigger – grow. The dependence of the size and speed of the growth size on time are determined.

1. Introduction

Investigation of growth mechanism of nanoparticles is present large scientific and practical interest. As, nanoparticles with given size and characteristics are required in nanotechnology. Nanoparticles growth mechanism determines distribution function of nanoparticles on size, physical-chemical properties of nanoparticles medium and etc. Because of, known of growth mechanism give possibility control of preparation of nanoparticles and to obtain nanoparticles with given parameters (mean diameter, standard deviation, coefficient polydispersity and other) and characteristics (magnetic moment).

Nanoparticles growth mechanism is enough complex process and depended from many conditions (temperature, viscosity, concentration of medium and etc.). Conditions determinant of nanoparticle growth are changed in the dependence on method preparation of nanoparticles. Mechanisms of formation of nanoparticles, preparation by chemical precipitation, laser photolysis, and condensation metal atoms method are different [1-5].

Two stages of nanoparticles growth are suggested in preparation of nanoparticles by chemical precipitation method [1-3,7-12]. In the first stage, nanoparticles growth to occur due to sequent jointing of atoms or molecules to nucleus. In the second stage to occur coagulation of formatted nanoparticles. In the literature, in base second stage of nanoparticle (water drop, aerosol, gel and etc.) growth is described [1-3,7-12]. Second stage of nanoparticles growth is described by diffusion equation and it is solved at different boundary condiyions. Authors of works [1-3,7-11] obtained results, matched with experiment. Moreover, is having without coagulation growth model of nanoparticles for description second stage of growth [10]. This model is different tradition diffusion model and also gives asymmetric distribution function nanoparticles on size.

---

[1] Rasim A.Ali-zasde.

At presented work, we described first stage of formation nanoparticles prepared by chemical precipitation method with kinetc equation. It has been showed that solution of equation is Gauss function and mean value and standard deviation of function are $\bar{x} = \frac{Kt}{4r_0}$, $\sigma = \sqrt{\frac{Dt}{2}\frac{1}{r_0}}$. Critical size, determinant of collapse and growth of nanoparticles are determined. It's showed nanoparticles less critical size may be is collapsed, bigger – grow. The dependence of the size and speed of the growth size on time are determined.

2. Kinetics of the first stage of nanoparticles growth.

One of wide used magnetite nanoparticles preparation method is chemical precipitation of two 2 and 3 valent iron salts by alkali [13-15]. We determined also two stages of formation of nanoparticles, prepared by this method [1-3,9,16,17]. Stages of nanoparticles formation has been determined on the base of analysis of distribution function nanoparticles on size [16,17]. Analysis showed, distribution function is asymmetric and is flat to the big sizes. It's has been showed initial part of distribution may be described by normal function, final- by logarithmic normal function [16,17]. It's known, these function described different mechanisms of nanoparticles growth. In the first stage, nanoparticles are formatted from saturated solution. Growth of nanoparticles occurs due to consecutive jointing of molecules to the nucleus. In the second stage, nanoparticles have enough big size and growth from saturated solution is impossible. In this stage, coagulation of nanoparticles dominates. Transition diameter of between these two stages has been determined. Mean and transition diameters for magnetite nanoparticles samples I-VI are: (9.5; 9.3), (7.7; 7.5), (7.3; 7.2), (5.4; 5.3), (10.2; 10.1), (4.8; 3.4) nm, rspectively [16,17].

In the first stage, in addition to nanoparticles growth its melting is possible, with small probability. It's related with no stability of nanoparticles (superfluous of surfaces energy) [18,19].

Thus, take into account above indicating, the first stage formation of nanoparticles may be described by follow differential equation:

$$\frac{\partial\rho(r,t)}{\partial t} = D\frac{\partial^2\rho(r,t)}{\partial r^2} - K\frac{\partial\rho(r,t)}{\partial r}$$

Change of density state on time may be occurs consecutive jointing molecules to the nucleus $\left(K\frac{\partial\rho(r,t)}{\partial r}\right)$, and melting of nanoparticles big size $\left(D\frac{\partial^2\rho(r,t)}{\partial r^2}\right)$. Introduce non dimensional parameter $x=r/r_{max}$, where r_{max} is maximum diameter of nanoparticles in the first stage (or transition diameter of between different growth mechanisms). Kinetic equation of the first stage of nanoparticles growth is written in non dimensional value as:

$$\frac{\partial\rho(x,t)}{\partial t} = D\frac{\partial^2\rho(x,t)}{r_0^2\partial x^2} - K\frac{\partial\rho(x,t)}{r_0\partial x} \tag{1}$$

Take into account above indicating, function of density state must satisfy following boundary conditions:

$$\rho(x,t)\big|_{t=0,x=0} = 1;\quad \rho(x,t)\big|_{t=0,x=1} = 0;\quad \rho(x,t)\big|_{t\to\infty,x=0} = 0;\quad \rho(x,t)\big|_{t\to\infty,x=1} = 1$$

Solution this equation will take as:

$$\rho(x,t) = e^{mx+nt}\cdot p(x,t)$$

where $m = \frac{K}{x_0 2\frac{D}{x_0^2}} = \frac{Kx_0}{2D}$, $n = -\frac{K^2}{x_0^2 4\frac{D}{x_0^2}} = -\frac{K^2}{4D}$

Then equation (1) transformed to:

$$\frac{\partial p(x,t)}{\partial t}=\frac{D}{r_0^2}\frac{\partial^2 p(x,t)}{\partial x^2}$$

Solution of last equation will take as multiplication of two functions $p(x,t)=R(x)\cdot S(t)$ [20]. We obtain after division both sides of last equation to $R(x)\cdot S(t)$:

$$\begin{cases}\dfrac{r_0^2}{D}\dfrac{1}{S(t)}\dfrac{\partial S(t)}{\partial t}=-\lambda^2\\ \dfrac{1}{R(x)}\dfrac{\partial^2 R(x)}{\partial x^2}=-\lambda^2\end{cases}$$

Solutions of these equations are:

$$S(t)=e^{-\frac{D}{r_0^2}\lambda^2 t},\quad R(x)=e^{i\lambda^2 x}$$

Thus, we obtain one quotient solution of equation (1)

$$\rho_\lambda(x,t)=e^{-(\frac{D}{r_0^2}\lambda^2-n)t}\,e^{(m+i\lambda^2)x}$$

We obtain full solution of equation by integration of quotient solution on λ

$$\rho(x,t)=\int A(\lambda)\rho_\lambda(x,t)d\lambda$$

We obtain $A(\lambda)=\frac{1}{\pi}$ use boundary conditions. After integration we obtain

$$\rho(x,t)=\frac{1}{\sqrt{2\pi\sigma}}e^{-\frac{(x-\bar{x})^2}{2\sigma^2}}$$

As see $\rho(x,t)$ is normal function with mean value $\bar{x}=\frac{Kt}{4r_0}$, and standard deviate $\sigma=\sqrt{\frac{Dt}{2}\frac{1}{r_0}}$, which depended on time.

3. Determination of coefficients *K* and *D*

Value of K has been calculated by formula in [21, 22].

$$K=N_A\sigma_{AB}^2\left[8\pi RT\left(\frac{1}{M_A}+\frac{1}{M_B}\right)\right]^{1/2}e^{-E_a/RT}$$

where N_A is Avogadro number, σ_{AB} - effective section of collision of molecule *A* and *B* ($\sigma_{AB}=\pi r_A^2+\pi r_B^2$), M_A and M_B - molecular mass (M_A-molecular mass of magnetic molecule is equal to 231.54, M_B - molecular mass of formatted nanoparticles, $M_B=n\cdot M_A$, *n* – number of molecules), E_a-activation energy (≈2-5eV) [23]. Evaluation of geometrical size of stable configuration of nanoparticles showed that at increasing number of molecules up 1 to 10, value of r_B is changed at interval $r_A(1\div 2.8)$. Thus, evaluation of mean value of *K* with above indicated formula is: 1.58 10^{-8} м/c.

Value *D* is determined from determination of distribution parameter [24-26].

$$\sigma^2=\bar{x}^2-\overline{x^2}$$

Where $\bar{x}$ and $\overline{x^2}$ are determined as:

$$\overline{x^m}=\int x^m\cdot\rho(x,t)\cdot dx$$

Result of integration substituted to determination of σ and we obtain.

$$\left(\sqrt{D}\right)^2+\left(\sqrt{D}\right)^2\frac{r_0}{\sqrt{t}}-\sqrt{D}\frac{K^2t}{2}+K\frac{r_0^2}{2\sqrt{t}}=0$$

Solution of cubic equation, don't take into account term with ($<10^{-18}$) is [27]:

$$D=\frac{r_0^2}{t}\left(0.167\frac{Kt}{r_0}\right)^{2/3},\qquad \sigma=\frac{1}{\sqrt{2}}\left(0.167\frac{Kt}{r_0}\right)^{1/3}$$

All figures have been plotted by ORIGIN-7 on the base of data obtained by porogram written by FORTAN algorithmic lqanguage.

In the figures 1, 2 the curves $\overline{x}(t)$, $\sigma(t)$ are presented for samples magnetic nanoparticles.

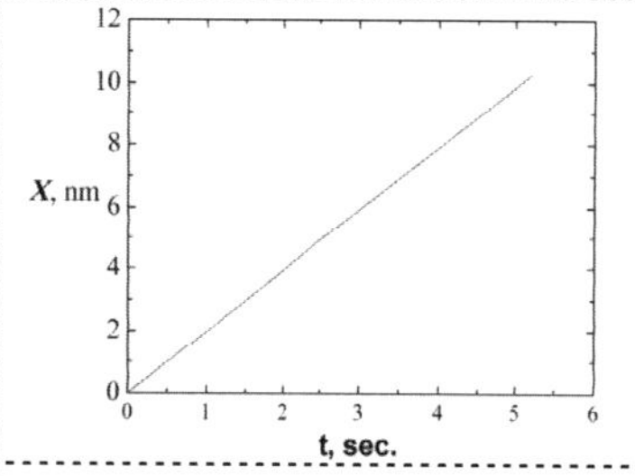

Figure 1. The dependence of mean value of distribution on time of nanoparticles formation.

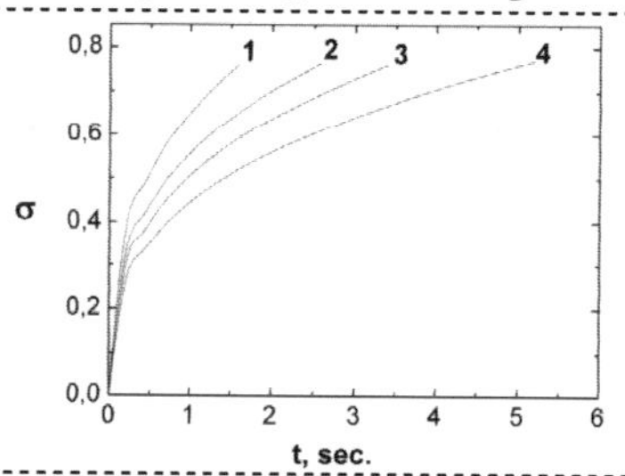

Figure 2. The dependence of deviation of distribution on time of nanoparticles formation.

As see figures 1, 2 mean diameter and standard deviation are increased with increasing time. There value is agreed to value obtained by histogram distribution nanoparticles [16, 17]. In the figure 3 are presented the dependence of coefficient diffusion on time for nanoparticles samples VI, III and VI. As see coefficient of diffusion is decreased on time. This is related with increasing nanoparticle size on time, when diffusion is decreased. This is agreeing with result of work [28].

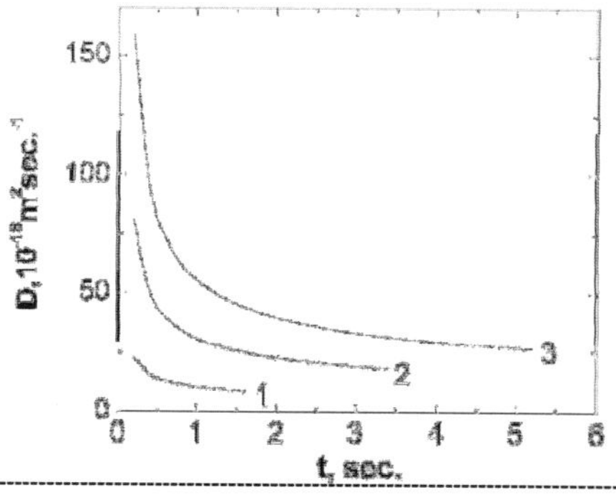

Figure 3. The dependence of coefficient of diffusion on time of nanoparticles formation.

4. Results and discussion

In the figure 4 are presented curves of distributions nanoparticles on size. Curves 1, 2 and 3 obtained at different time of growth. As see for all cases distribution is gauss. Mean value and standard deviation of distribution are increased with increased time. This distribution not changed to transient diameter of nanoparticles (figure 4).

Growth and collapse of nanoparticles are determined from relations $\frac{\partial\rho(x,t)}{\partial t}<0$ and $\frac{\partial\rho(x,t)}{\partial t}>0$.

This obtained, that at $x>\frac{Kt}{4r_0}$ growth, and at $0<x<\frac{Kt}{4r_0}$ collapse of nanoparticles is possible.

Growth of nanoparticles may be occurs by jointing molecules of collapsed nanoparticles. Collapse of nanoparticles is related superfluous energy of nanoparticles [18,19].

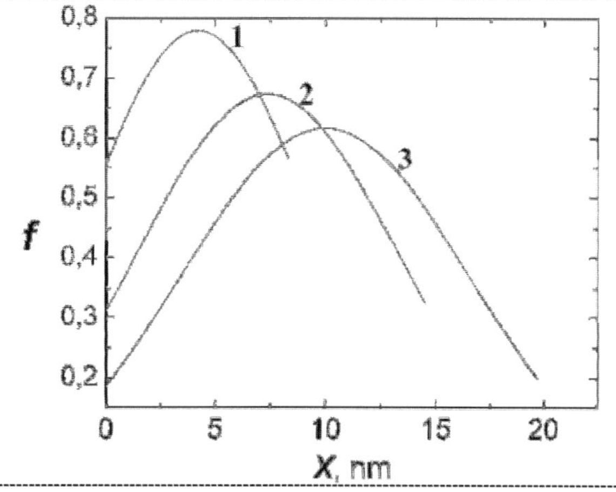

Figure 4. Distribution nanoparticles on size determined at different time of nanoparticles formation (1- 2sec., 2- 3.5sec., 3- 4.75sec.)

Stationary state of nanoparticles formatting is determined from condition $\frac{\partial\rho(x,t)}{\partial t}=0$. Stabilise time is equal $t=\frac{4r_0}{K}(x+0.52)$. Evaluations of time showed, that it approximately are equal to time necessary for formatting transition diameter for every sample of nanopapticles.

As see all expressions depended on transition diameter of nanoparticles.

The dependence of nanoparticles size on time formation of nanoparticles may be obtain from accumulation of mass at formatting of nanoparticles

$$M(t)=\int \rho_0 4\pi x^2 \rho(x,t)dx$$

where ρ_0 - is density of magnetite (5,24q/sm^3).

Expressions for size and rate of speed of the growth size of nanoparticles on time are:

$$r(t)=r_0\left(3\bar{x}^2+\sqrt{\frac{2}{\pi}}\bar{x}\sigma+\sigma^2\right)^{1/3}$$

$$v(t)=\frac{1}{3}K\frac{1}{\left(3\bar{x}^2+\sqrt{\frac{2}{\pi}}\bar{x}\sigma+\sigma^2\right)^{2/3}}\left(\frac{3}{2}\bar{x}+\sqrt{\frac{2}{\pi}}\frac{\sigma}{4}+\frac{0,059}{3\sigma}\left(\sqrt{\frac{2}{\pi}}\cdot\frac{\bar{x}}{\sigma}+2\right)\right)$$

In the figures 5, 6 are presented size and speed of the growth size of nanoparticles on time. As see size of nanoparticles is increased with increasing time and at the end reach transient diameter. Initial time, speed of the growth size is big and it is decreased by time. This may be explained that concentration of jointing molecule around nanoparicles (number of jointing molecules) is decreased at increasing time.

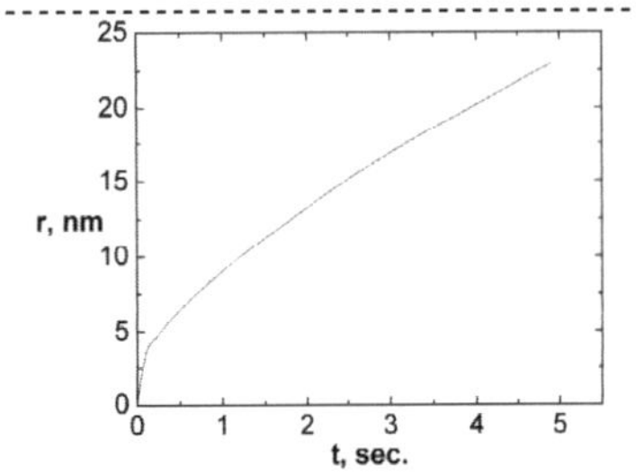

Figure 5. The dependence of nanoparticle size on time of nanoparticles formation.

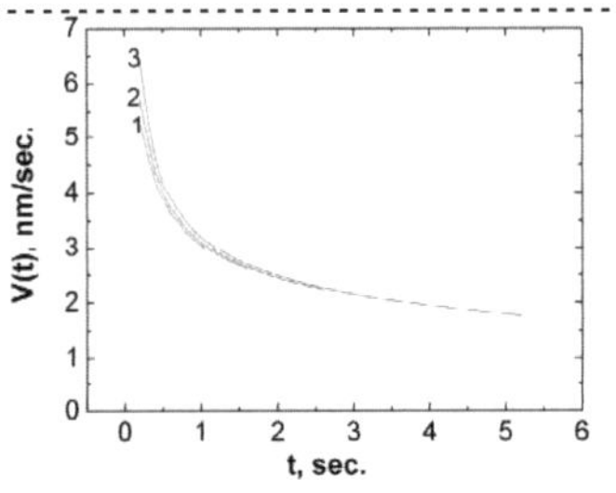

Figure 6. The dependence of nanoparticle rate size on time of nanoparticles formation (curves 1,2 and 3 correspond to samples VI, III and V).

The dependence of speed of the growth size of nanoparticles on size may be explained as nanoparticles size depended on physical-chemical properties of medium (viscosity, temperature and etc.). This dependent must be taking into account in the determination of coefficients *K* and *D*.
Analysis of result showed that, at formation nanoparticles size is change by law $t^{1/3}$, although mean diameter of nanoparticles - by *t*. This may be explained as nanoparticles mean diameter is determinate by averaging of nanoparticles set.

5. Conclusion

In the fist stage of nanoparticle formation, distribution nanoparticles on size are described by Gauss function. Parameters of this function depended on time. Growth and collapse of nanoparticles depended on nanoparticle size. Size and speed of the growth size of nanoparticles depended on time. Size of nanoparticles is increased with increasing time and at the end reach transient diameter. Initial time, speed of the growth size is big and it is decreased by time.

5. References

[1] Todes O M 1946 *Russian J. Phys.Chem.* **10** 629
[2] Liftsich I M, Slezov V V 1958 *Russian J.Expr.Teor.Phys.* **35** 479
[3] Frank S H 1960 *J. Appl. Phys.* **30** 1518
[4] Emelianov A, Eremin A, Jander H, Wagner H Gg. 2003 *Z.Phys.Chem.* **217** 1361
[5] Starke R, Kock B, Roth P 2003 *Shock Waves.* **12** 351
[6] Stepanov A L 2003 *Pisma JTF* **28** 58
[7] Sarian S, Weart H W 1966 *J. Appl. Phys.* **37** 1675
[8] Lushnikiov A A 1978 *Russian J.Phys.Atmp.Ocean.* **14** 1946
[9] Granqvist C G, Buhrman R A, 1976 *J. de Physique.* **47** 2200
[10] Söderlund J. , Kiss L B, Niklasson G A , Granqvist C G 1998 *Phys. Rev.Lettr.* **80** 2386
[11] Shigeta M et al, 2007 *J. Phys. D: Appl. Phys.* **40** 2407
[12] Morochov I D, Trusov L I, Lapovok V N 1984 *Fizicjesjiye yavleniya v ultradispersnich sredakh* (Moscow: Energoatomizdat) p 224
[13] Turkin, S.I., Lukin, Yu.V., Markvicheva, E.A., et al., 1989 USSR Inventor's Certificate no. 1486515, *Byull. Izo-bret..* **22** 100
[14] Turkin, S.I., Lukin, Yu.V., Markvicheva, E.A., et al., 1990 USSR Inventor's Certificate no. 1567623, *Byull. Izo-bret.* **20** 105

[15] Buryakov A N, Grichkova I A, Zubov V P and etc. 1991 USSR Inventor's Certificate no.1628478 USSR, BI **6** 194
[16] Ali-zade R 2008 *Inorganic Materials* **44** 165
[17] Ali-zade R A 2008 *Inorganic Materials* **44** 1105
[18] Karpenko S V, Temrokov A I 2004 *Russian J.Tech.Phys.* **74** 115
[19] Koshkin V M, Slezov V V 2004 *Russian J.Pisma v JETF.* **30** 38
[20] Tikhonov A N, Samarski A A 1977 *Uravneniya Matematicheskoi Fiziki* (Moskva: Nauka) p 288
[21] Gerasimov Ya I 1973 *Kurs Fizicheskoi Khimii* (Moscow: Khimiya) p 624
[22] Eyring H, Lin S H , Lin S M 1980 *Basic Chemical Kinetics* (New York-Chichester-Brisbane-Toronto: John Willey & Sons.) p 528
[23] Hoare M R, Pal P 1971 *Advances in Physics* **20** 161
[24] Hudson D J 1964 *Statistics. Lecture on Elementary Statistics and Probability* (Geneva) p 206
[25] Feller W 1970 *An Introduction to probability theory ad its applications.* (New York-Chichester-Brisbane-Toronto: John Willey & Sons.) p 528
[26] Pugachev V S 1979 *Theoria veroyatnostei i matemativheskaya statistika* (Moscow: Nauka) p 496
[27] Korn G A, Korn T M 1968 *Mathematical Handbook for Scientists and Engineers. Definitions. Theorems and Formulas for Refernce and Review.* (Moscow: Mir) p 832
[28] Kolchov E M, Aqashin A V, Gordeyev L S 2000 *Russian J.Phys.Khimii.* **74** 888

Localized magnetic polaritons in antiferromagnetic superlattice with impurity

R T Askerbeyli (Tagiyeva)[1] and B Tanatar[1]
[1] Physics Department, Bilkent University, Ankara, Turkey

E-mail: rtagiyeva@ yahoo.com

Abstract. The dispersion relation for magnetic polaritons localized at the antiferromagnetic impurity film in the antiferromagnetic superlattice (antiferromagnetic/nonmagnetic or antiferromagnetic/antiferromagnetic) are derived in the effective –medium approximation and calculations are performed for the properties of long-wavelength electromagnetic modes. In such systems one finds both surface polaritons which are localized near the surface and guided modes where excitations have a standing-wave –like character and the impurity region acts as a waveguide, because of the magnetic polaritons propagating freely over the impurity layer and dampen in the perpendicular direction on either sides of this region. We assume an external magnetic field parallel to the magnetization and the film interfaces. The dispersion curves and frequency region of the existence of the surface-guided modes of the magnetic polaritons localized at the impurity layer of the antiferromagnetic superlattice are investigated for two values of the external magnetic field: H=0 and H=0.3T.

1.Introduction

The properties of the collective excitations such as polaritons (coupled-mode excitations originating from dipole-active elementary excitations such as phonons, plasmons, magnons, interacting with photons) which propagate in the various magnetic (ferromagnetic, antiferromagnetic) superstrurtures (superlattices, thin film) have been the subject of increasing interest in recent years [1-8]. A number of different anferromagnetic-based superstructures and properties of the bulk and surface excitations propagating in such systems have been considered theoretically in the literature [9-12]. Superlattices constructed from antiferromagnetic materials are of interest to communications and signal processing technologies for devices that work at wavelengths in the infrared [13,14]. In this paper we derive the general dispersion relation for magnetic polaritons, which appear at the impurity layer of the superlattice composed of alternating antiferromagnetic or antiferromagnetic and non-magnetic layers. This problem is considered within the framework of a macroscopic theory in the Voigt configuration in the presence of an external magnetic field.

2. Theory

It is known that the effective-medium approximation can be applied in the regions of high frequency dispersion, where wavelength is much greater than the superlattice period L (L=a+b) and the wavevector k appearing in the dispersion relation is small compared to L^{-1}. In this case the antiferromagnetic superlattice behaves like an anisotropic bulk medium, because of antiferromagnetic resonance frequencies are in the far infrared region.

The effective medium is described by the effective-medium permeability tensor with the following components [3]:

$$\langle\mu_{xx}\rangle = \frac{(a+b)^2\mu_1^{(1)}\mu_1^{(2)} + ab[(\mu_1^{(1)}-\mu_1^{(2)})^2-(\mu_2^{(1)}-\mu_2^{(2)})^2]}{(a+b)(a\mu_1^{(2)}+b\mu_1^{(1)})} \quad (1)$$

$$\langle\mu_2\rangle = \frac{a\mu_2^{(1)}\mu_1^{(2)}+b\mu_{2x}^{(2)}\mu_1^{(1)}}{(a+b)(a\mu_1^{(2)}+b\mu_1^{(1)})}, \quad (2)$$

$$\langle\mu_{yy}\rangle = \frac{(a+b)\mu_1^{(1)}\mu_1^{(2)}}{a\mu_1^{(2)}+b\mu_1^{(1)}}, \quad (3)$$

For the TE mode, the relevant component of the dielectric tensor in the effective-medium description is ε_{xx}^{SL}:

$$\varepsilon_{xx}^{SL} = \frac{a\varepsilon_1+b\varepsilon_2}{a+b}, \quad (4)$$

where $\varepsilon_i(i=1,2)$ is the dielectric constant of the i-th component of the superlattice.

Here $\mu_1^{(j)}$ and $\mu_2^{(j)}$ are non-vanishing components of the frequency-dependent magnetic permeability tensor:

$$\hat{\mu}^{(i)}(\omega) = \begin{bmatrix} \mu_1^{(i)} & i\mu_2^{(i)} & 0 \\ -i\mu_2^{(i)} & \mu_1^{(i)} & 0 \\ 0 & 0 & 1 \end{bmatrix}, \quad (5)$$

and for an antiferromagnet the elements of $\hat{\mu}^{(i)}(\omega)$ are

$$\mu_1(\omega) = 1 + \frac{\Omega_a\Omega_m}{\Omega_1^2-\omega_+^2} + \frac{\Omega_a\Omega_m}{\Omega_1^2-\omega_-^2}, \quad (6)$$

$$\mu_2(\omega) = \frac{\Omega_a\Omega_m}{\Omega_1^2-\omega_+^2} - \frac{\Omega_a\Omega_m}{\Omega_1^2-\omega_-^2}, \quad (7)$$

where $\Omega_m = \gamma 4\pi M_0$ (M_0 is the sublattice magnetization), $\Omega_0 = \gamma H_0$ and $\omega_\pm = \omega \pm \Omega_0$. The antiferromagnetic resonance frequency in zero applied field Ω_1 is given by the anisotropy H_{an} and the exchange H_{ex} fields as

$$\Omega_1 = \gamma[H_{an}(2H_{ex}+H_{an})]^{\frac{1}{2}}, \quad (8)$$

Using the Maxwell's equations we derive the dispersion relation for localized magnetic polaritons. The dynamic magnetic field satisfies the following equation:

$$\nabla^2\vec{h} - \vec{\nabla}(\vec{\nabla}\vec{h}) - \frac{\varepsilon}{c^2}\frac{\partial^2}{\partial t^2}(\vec{h}+4\pi\vec{m}) = 0, \quad (9)$$

The general solution of equation (9) can be written in the form:

$$\boxed{h^{(sl)} = (\langle H_x^{(sl)}\rangle, \langle H_y^{(sl)}\rangle)e^{\beta y}e^{i(kx-\omega t)}} \quad (10)$$

for the superlattice and

$$h^{(0)} = [(H_{1x}^{(0)}, H_{1y}^{(0)})e^{\alpha_0 y} + (H_{2x}^{(0)}, H_{2y}^{(0)})e^{-\alpha_0 y}]e^{i(kx - \omega t)} \qquad (11)$$

for the impurity film occupying the region 0<y<d.

Here $\langle H_{x,y}^{(sl)} \rangle, \langle B_{x,y}^{(sl)} \rangle$ are the average values of $H_{x,y}^{(sl)}, B_{x,y}^{(sl)}$ in the superlattice. The parameter α_0 is defined as $\alpha_0^2 = k^2 - \omega^2 / c^2 \mu_v^{(0)}$. Here $\mu_v^{(0)} = \mu_1^{(0)} - (\mu_2^{(0)2} / \mu_1^{(0)})$ is the effective magnetic permeability of the antiferromagnetic impurity layer in the Voigt geometry. The parameter α_0 can now be either purely real (this case corresponds to the surface modes) or purely imaginary (for guided modes). The parameter β ($\beta^2 = -k_y^2$) is the decay parameter of the magnetic polaritons in the antiferromagnetic superlattice and $(\mathrm{Re}\,\beta)^{-1} > 0$ is the penetration depth of the magnetic polaritons into the antiferromagnetic superlattice. The expression for the parameter β can be obtained from the dispersion relation for bulk magnetic polaritons in the effective medium [3]:

$$\begin{aligned} \beta^2 &= k^2[f_a^2 + f_b^2 + f_a f_b \frac{\mu_v^1 \mu_1^1 + \mu_v^2 \mu_1^2 + 2\mu_2^1 \mu_2^2}{\mu_1^1 \mu_1^2}] - \frac{\omega^2}{c^2}(f_a e_1 + f_b e_2)(f_a \mu_v^1 + f_b \mu_2^2), \\ f_a &= \frac{a}{a+b}, f_b = \frac{b}{a+b}. \end{aligned} \qquad (12)$$

Here L=a+b is the superlattice period.

In order to derive the dispersion relation for magnetic polaritons which are localized at the impurity film we apply the electromagnetic boundary conditions to the left (y=0) and right (y=d) surfaces of the impurity layer, namely, the continuity of the tangential component of the magnetic field $\vec{h}$ and normal component of $\vec{b}$. After some algebra, we obtain:

$$\begin{aligned} &\frac{1}{A}\{\mu_1 \alpha_0 ch(\alpha_0 d)[(k\langle \mu_2^{(sl)} \rangle + \beta \langle \mu_{yy}^{(sl)} \rangle)(\varepsilon_{SL} \frac{\omega^2}{c^2} \langle \mu_{yy}^{(sl)} \rangle - k^2) - (k\langle \mu_2^{(sl)} \rangle - \beta \langle \mu_{yy}^{(sl)} \rangle)(\varepsilon_{sl} \frac{\omega^2}{c^2} \langle \mu_{yy}^{(sl)} \rangle - k^2)] + \\ &sh(\alpha_0 d)[(k\langle \mu_2^{(sl)} \rangle + \beta \langle \mu_{yy}^{(sl)} \rangle)\{k\mu_2^{(0)}(\varepsilon_{sl} \frac{\omega^2}{c^2} \langle \mu_{yy}^{(sl)} \rangle - k^2) - (k\langle \mu_2^{(sl)} \rangle - \beta \langle \mu_{yy}^{(sl)} \rangle)(\varepsilon_0 \frac{\omega^2}{c^2} \mu_1^{(0)} - k^2)\} + \\ &(\varepsilon_{sl} \frac{\omega^2}{c^2} \langle \mu_{yy}^{(sl)} \rangle - k^2)\{k\mu_2^{(0)}(k\langle \mu_2^{(sl)} \rangle - \beta \langle \mu_{yy}^{(sl)} \rangle) - \mu_v^{(0)} \mu_1^{(0)} (\varepsilon_{sl} \frac{\omega^2}{c^2} \langle \mu_{yy}^{(sl)} \rangle - k^2)\}]\} = 0, \\ &A = (\varepsilon_{sl} \frac{\omega^2}{c^2} \langle \mu_{yy}^{(sl)} \rangle - k^2)^2 (\varepsilon_0 \frac{\omega^2}{c^2} \mu_1^{(0)} - k^2). \end{aligned} \qquad (13)$$

Equation (13) together with equation (12) determines the frequencies of localized the magnetic polaritons, which appear in the antiferromagnetic superlattice with antiferromagnetic impurity film in the effective-medium description. This equation is the general dispersion relation for surface-guided localized magnetic polaritons propagating parallel to the impurity film and can be applied to both ferromagnetic and antiferromagnetic systems. Only those solutions of equation (13) for which the condition $\beta > 0$ are fulfilled describe physical localized magnetic polaritons.

3. Numerical calculations

For numerical calculations we have introduced the following dimensionless parameters. We discuss our analytical results for antiferromagnetic superlattice SL(MnF_2/ZnF_2) with antiferromagnetic impurity film FeF_2 . Here we assume $f_a = f_b = 0.5$ and the thickness of impurity film d=0.0005cm. In figure1 we present the magnetic polariton spectra through a plot of the reduced frequency $\omega^* = \omega / \Omega_m^{(MnF_2)}$ against the wavevector $k^* = ck / \Omega_m^{(MnF_2)}$ in zero external magnetic field $H_0 = 0$. The different symbols (open and solid squares, triangles, dots) denote the surface (SM) and guided (GM)

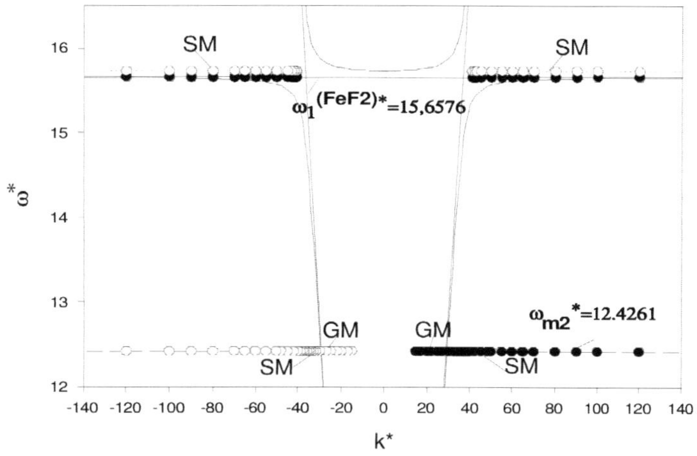

Figure 1. Dispersion relation for bulk and surface (guided) polaritons in FeF_2 antiferromagnetic film in zero field.

modes spectral branches. We observe two frequency regions which represent the bulk excitations and four branches of the localized magnetic polariton modes (three surface and one guided) for each direction of k^* . Two high-frequency surface modes exist between the bulk bands and start at $\omega_1^{(FeF_2)^*} = 15.6576$ (where $\omega_1^{(FeF_2)}$ is the antiferromagnetic resonance frequency in $H_0 = 0$ and determine by equation (8)). In the MnF_2 –resonance region the low-frequency branch starts as guided mode for small value of the wavevector but with increasing $|k^*|$ transforms to the surface mode and tend to $\omega_{m2}^{(MnF_2)^*} = 12.4261$, where $\omega_{m2}^{(MnF_2)}$ is the magnitostatic frequency in MnF_2, as well as in the semi-infinite antiferromagnetic superlattice.

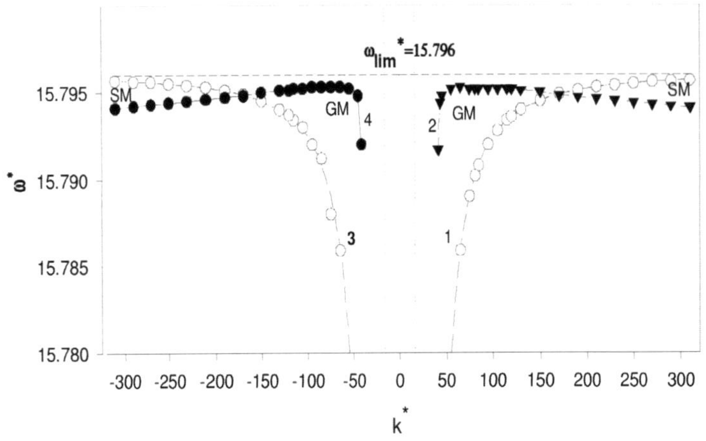

Figure2. High-frequency bulk and surface (guided) modes of magnetic polaritons in antiferromagnetic film FeF_2 with applied field H=0.3 T.

The frequency of the surface (guided) mode does not depend on the sign of the wavevector, i.e. the surface modes are reciprocal.

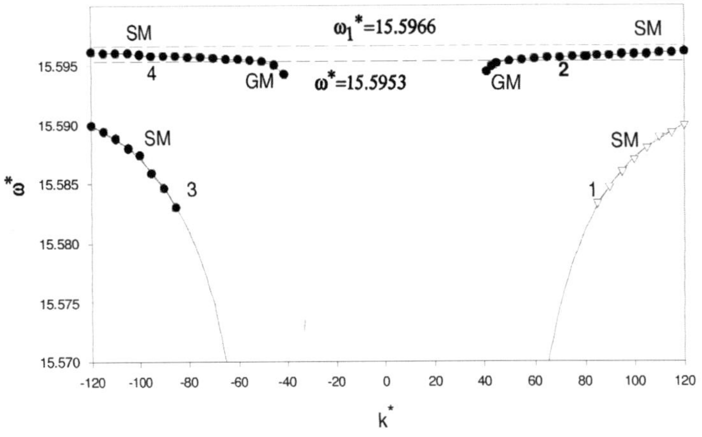

Figure 3. The dispersion relation for low-frequency bulk and surface (guided) modes of polaritons in FeF_2 antiferromagnetic film at H=0.3 T.

Figures 2-5 show the surface - guided modes for the same structure with H=0.3T. In contrast to the above, there are now three bulk bands and in the presence of the external magnetic field the new localized modes appear making the spectra more complex.

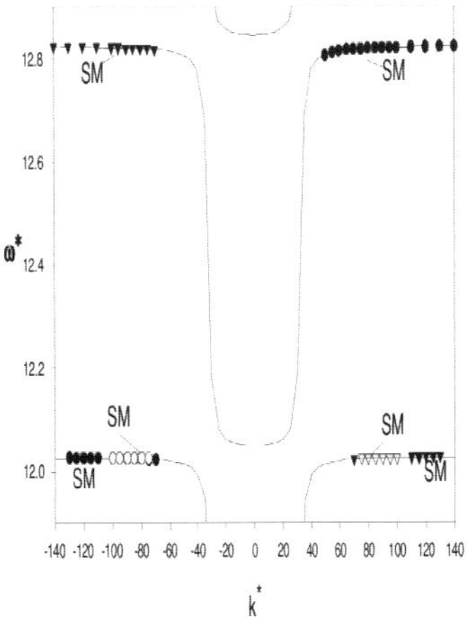

Figure 4. Dispersion relation for bulk and surface (guided) modes in a superlattices $SL(MnF_2/ZnF_2)$.

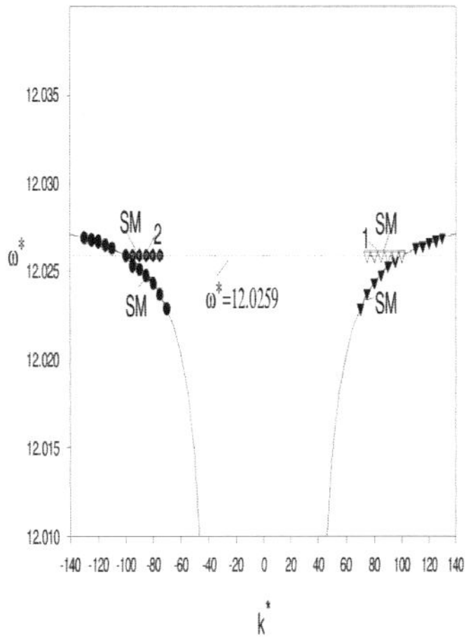

Figure 5. Same as in figure 4, but in other scale

As one can see, there are two types of localized modes: those, which separate from and lies below a bulk band (denoted as 1 and 3 on figures 2,3), these branches are pure surface modes and the other one, which start as guided modes and then transforms to surface modes at $|k^*| \to \infty$ (branches 2 and 4 on figures). The frequencies $\omega_{\lim}^* = 15.796$ and $\omega_l^* = 15.5966$ on figures 2 and 3 are the limiting frequencies for high and low bulk branches, respectively. Generally, we observe seven surface - guided modes of localized magnetic polaritons for both directions of the wavevector, respectively: four curves in the FeF2- resonance region (see figures 2,3) and three surface modes in the MnF_2 – resonance region (see figures 4,5). It is essential that the surface mode denoted as 1 and 2 on the figure 5 exists for the restricted values of the frequency ω^* and the wavevector k^* (k^* =75-100).

4. Conclusions

In summary, we have derived the general effective-medium expression for the surface -guided magnetic polaritons, which propagate in the antiferromagnetic superlattice with antiferromagnetic impurity film and investigate the influence of the external magnetic field on the energy of localized magnetic polaritons . It is essential that the spectrum of magnetic polaritons in the presence of an external magnetic field is weakly non-reciprocal, in contrast to ferromagnetic superlattice. In the system under consideration one finds both surface polaritons which are localized near the surface and guided modes where excitations have a standing-wave –like character and the impurity region acts as a waveguide, because of the magnetic polaritons propagating freely over the impurity layer and dampen in the perpendicular direction on either side of this region. We hope that our theoretical predictions will motivation further experimental work.

Acknowledgements

This study was supported by TUBITAK-BIDEP Postdoc Grant 2218 B.T. Is supported by TUBITAK and TUBA.

5. References

[1] Camley R E 1999*J. Magn.Magn.Mat.* **136** 583
[2] Abraha K and Tilley D 1996 *Surf.Sci.Rep.* **24**
[3] Raj N and Tilley D 1987 *Rhys.Rev.B* **36** 7003
[4] Papanicolaou N 1999 *J.Phys.: Condens.Matter*. 59 (printed in the UK)
[5] Schuller Ivan K, Kim S and Leighton C 1999 *J.Magn.Magn.Mat.* **200** 571
[6] Barnas J 1987 *Solid State Commun.* **61,** 405
[7] Agranovich V M and Mills D L 1982 *Surface Polaritons* North-Holland Amsterdam
[8] Stamps R and Camley R 1989 *Rhys.Rev.B* **40** 596
[9] Almeida N S and Tilley D R 1990 *Solid State Commun.* ***73*** 23
[10] Camley R E ,Cottam M G and Tilley D R 1982 *Solid State Commun.* ***81*** 571
[11] Guimaraes E and Albuquerque E 2002 *Solid State Commun.**122*** 623
[12] Tagiyeva R, Saglam M and Boyacioglu B 2007 *J. Phys. and Chemistry of Solids* **68** 1296
[13] Abraha K, Brown D E, Dimenlow T, Parker T J and Tilley D R 1994 *Rhys.Rev.B* **50** 6808
[14] Brown D E, Dimenlow T , Parker T J, Abraha K and Tilley D R 1994 *Rhys.Rev.B* **49** 12266

Rashba Spin-Orbital Splitting in Kane Type Quantum Anti-Wires

A M Babayev[1,2], S Cakmaktepe[3] and D Turkoz Altug[1]

[1]Department of Physics, University of Suleyman Demirel, Isparta 32260, Turkey
[2]Institute of Physics, Azerbaijan Academy of Sciences, 370143, Baku, Azerbaijan
[3]Department of Physics, Kilis 7 Aralik University, Kilis, Turkey

E-mail: arifb2004@yahoo.co.uk, cakmaktepe2001@yahoo.com

Abstract. In the present study, using three-band Kane's model including the conduction band, light and spin-orbital hole bands, the electronic states of a semiconductor quantum anti-wire is studied in presence of Rashba spin-orbital interaction and external magnetic field and compared with those of a quantum anti-wire of the same size. It's calculated the radii, height, coupling strength and magnetic field dependence of Rashba splitting for carriers respectively. It has been found for the InSb quantum anti-wire that Rashba splitting of electrons are decreased with the increasing of radius.

1. Introduction

The study of semiconductor nanocrystals in recent years has been of explosive interest both experimental and theoretical points [1]. The interest concentrates on size quantization in solids in those objects. The electron energy spectrum consists of a set of discrete levels for an ideal quantum dot. So the semiconductors become very attractive for possible applications in micro and nano opto-electronics [2]. The electron spin plays an important role in the quantum nanostructure design. Spin orbital quantum interaction can arise in quantum dots by various mechanisms related to electron confinement and are generally introduced into Hamiltonian via Rashba [3] and Dresselhaus terms [4]. Dresselhauss showed that in bulk semiconductors of zinc blend symmetry the Γ_6 band (conduction band of III-V compounds) is characterized by an anisotropic spin splitting proportional to k^3 for small k values. As demonstrated by a Dyakonov and Perel [5] the splitting leads to electron spin relaxation. Rashba interaction was used firstly to controllably rotate the electron spin by Datta and Das [6]. The earliest theoretical study in the spin splitting of conduction band in semiconductor heterostructures was done by Okava and Uemura [7]. Lassnig was the first to recognize the essential point that the spin inversion asymmetry splitting in the conduction band is related to the gradient of the potential including the valance offsets at the interfaces [8]. Pfeffer and Zawadski used the complete five level k.p approach to the band structure of III-V compounds, which allowed to treat bulk inversion asymmetry and structure inversion asymmetry mechanism of spin-splitting within the same model [9]. Analytical solutions to the problem of the Rashba spin-orbit coupling were done in semiconductor quantum dots taking into account the parabolic band model [10,11]. Hashimzade *et al* were studied theoretically the electronic states of electrons in Kane type semiconductor quantum disk in presence of Rashba spin orbital interaction [12]. Kudryashov were calculated the Rashba spin orbit interaction, wave functions and energy spectrum in a two dimensional quantum dot [13]. In the study of [14], Babayev *et al* calculated the energy spectrum of a quantum anti-wire taking into account the real band

[1] Corresponding Author: arifb@fef.sdu.tr

structure of Kane type materials. Babayev and collaborates calculated the electron eigenstates in a quantum anti-wire in presence of Rashba spin-orbital splitting using parabolic band model [15]. However, the experimental advantages of using narrow-gap semiconductors for the low-dimensional systems make it necessary to account for the real band structure of these materials. In the present study, it's aimed to take into account the coupling of the conduction and valence bands for considering the non-parabolicity of the electron dispersion in narrow and medium gap semiconductors. It's calculated the total spin-splitting energy in an InSb quantum anti-wire with hard walls both without and with an applied constant axial magnetic field. The system used in this study might simply be a very long cylindrical cavity of radius r_0 etched out of a bulk composite semiconductor material. It's shown an example of an anti-wire in Figure 1. Every cavity in Figure 1 corresponds to an anti-wire. The whole figure is called as photonic crystals which are now envisaged as an essential building of future photonic devices [16]. We describe the confinement of electrons in quantum anti-wire by a separable potential $V(\mathrm{r})=V(z)+V(x, y)$, where $V(x, y)$ is the confinement potential in the xy plane and $V(z)$ is the potential in the z direction. The confining potential in the xy plane is assumed to be symmetric, $V(x, y)= V(\rho)$. Here we will mainly consider a hard-wall confining potential in the xy plane, $V(\rho)=0$ for $\rho > r_0$ and $V(\rho)=\infty$ for $\rho = r_0$, r_0 being the radius of the quantum anti-wire. The perpendicular confinement $V(z)$ is assumed strong enough that only the lowest z-sub-band is occupied. Here $V(z)$ is assumed to be asymmetric.

Figure 1. Anti-wire example.

In the three bands Kane's Hamiltonian; the valence and conduction band interaction is taken into account via the only matrix element P (so called Kane's parameter). The system of Kane equations including the non-dispersional heavy hole bands have the form [17,18,19].

$$-\varepsilon\varphi_1-\frac{Pk_+}{\sqrt{2}}\varphi_3+\sqrt{\frac{2}{3}}Pk_z\varphi_4+\frac{Pk_-}{\sqrt{6}}\varphi_5-\frac{Pk_z}{\sqrt{3}}\varphi_7+\frac{Pk_-}{\sqrt{3}}\varphi_8=0 \tag{1}$$

$$-\varepsilon\varphi_2-\frac{Pk_+}{\sqrt{6}}\varphi_4+\sqrt{\frac{2}{3}}Pk_z\varphi_5+\frac{Pk_-}{\sqrt{2}}\varphi_6-\frac{Pk_+}{\sqrt{3}}\varphi_7-\frac{Pk_z}{\sqrt{3}}\varphi_8=0 \tag{2}$$

$$\frac{Pk_-}{\sqrt{2}}\varphi_1-(\varepsilon+\varepsilon_g)\varphi_3=0 \tag{3}$$

$$\sqrt{\frac{2}{3}}Pk_z\varphi_1-\frac{Pk_-}{\sqrt{6}}\varphi_2-(\varepsilon+\varepsilon_g)\varphi_4=0 \tag{4}$$

$$\frac{Pk_+}{\sqrt{6}}\varphi_1+\sqrt{\frac{2}{3}}Pk_z\varphi_2-(\varepsilon+\varepsilon_g)\varphi_5=0 \tag{5}$$

$$\frac{Pk_+}{\sqrt{2}}\varphi_2-(\varepsilon+\varepsilon_g)\varphi_6=0 \tag{6}$$

$$-\frac{Pk_z}{\sqrt{3}}\varphi_1-\frac{Pk_-}{\sqrt{3}}\varphi_2-(\Delta+\varepsilon+\varepsilon_g)\varphi_7=0 \tag{7}$$

$$\frac{Pk_+}{\sqrt{3}}\varphi_1-\frac{Pk_z}{\sqrt{3}}\varphi_2-(\Delta+\varepsilon+\varepsilon_g)\varphi_8=0 \tag{8}$$

Here P is Kane parameter, E_g is the band gap energy, Δ is the value of spin-orbital splitting and $k_\pm=k_x\pm ik_y$, $\vec{k}=-i\vec{\nabla}$. The energy origin has been chosen at the conduction-band minima.

2. Zero Magnetic Field

In the first part, the solution of Kane equations is investigated in presence of Rashba spin-orbital splitting in zero magnetic field.

Firstly, if it's substituted the expression (3-8) into formulas (1) and (2), it's obtained;

$$\begin{aligned}&\left(-\varepsilon+\frac{P^2}{2}\frac{k_+k_-}{\varepsilon+\varepsilon_g}+\frac{2}{3}\frac{P^2k_z^2}{\varepsilon+\varepsilon_g}+\frac{P^2}{6}\frac{k_+k_-}{\varepsilon+\varepsilon_g}+\frac{P^2k_z^2}{3}\frac{1}{\Delta+\varepsilon+\varepsilon_g}+\frac{P^2k_-k_+}{3(\Delta+\varepsilon+\varepsilon_g)}\right)\varphi_1\\&+\left(-\frac{P^2k_zk_-}{3}\frac{1}{\varepsilon+\varepsilon_g}+\frac{P^2k_-k_z}{3(\varepsilon+\varepsilon_g)}+\frac{P^2k_zk_-}{3}\frac{1}{\Delta+\varepsilon+\varepsilon_g}-\frac{P^2k_-k_z}{3(\Delta+\varepsilon+\varepsilon_g)}\right)\varphi_2=0\end{aligned} \tag{9}$$

Applying the perturbation approach it' presented the Hamiltonian H as;

$$H=H_0+H_R \tag{10}$$

where the Hamiltonian H_0 describes Kane Hamiltonian, and the Hamiltonian H_R describes the effect of a weak Rashba spin-orbital interaction. H_0 and H_R are;

$$H_0=\begin{bmatrix}H_{11}-\varepsilon & 0\\ 0 & H_{22}-\varepsilon\end{bmatrix},\quad H_R=\begin{bmatrix}0 & -iRk_-\\ iRk_+ & 0\end{bmatrix} \tag{11}$$

where

$$H_{11}=\frac{P^2}{2}\frac{k_+k_-}{\varepsilon+\varepsilon_g}+\frac{2}{3}\frac{P^2k_z^2}{\varepsilon+\varepsilon_g}+\frac{P^2}{6}\frac{k_+k_-}{\varepsilon+\varepsilon_g}+\frac{P^2k_z^2}{3}\frac{1}{\Delta+\varepsilon+\varepsilon_g}+\frac{P^2k_-k_+}{3(\Delta+\varepsilon+\varepsilon_g)} \tag{12}$$

$$H_{22}=\frac{P^2}{6}\frac{k_+k_-}{\varepsilon+\varepsilon_g}+\frac{2}{3}\frac{P^2k_z^2}{\varepsilon+\varepsilon_g}+\frac{P^2}{2}\frac{k_-k_+}{\varepsilon+\varepsilon_g}+\frac{P^2k_-k_+}{3(\Delta+\varepsilon+\varepsilon_g)}+\frac{P^2k_z^2}{3}\frac{1}{\Delta+\varepsilon+\varepsilon_g} \tag{13}$$

if the above Equations are used, equation (10) can be written as;

$$\begin{bmatrix}H_{11}-\varepsilon & -iRk_-\\ iRk_+ & H_{22}-\varepsilon\end{bmatrix}\begin{bmatrix}\varphi_1\\ \varphi_2\end{bmatrix}=0 \tag{14}$$

for the lowest subband we obtain the following set of two equations for the envelope functions of $\varphi_1(r)$ and $\varphi_2(r)$:

$$\left(-\varepsilon+\frac{P^2}{2}\frac{k_+k_-}{\varepsilon+\varepsilon_g}+\frac{2}{3}\frac{P^2k_z^2}{\varepsilon+\varepsilon_g}+\frac{P^2}{6}\frac{k_+k_-}{\varepsilon+\varepsilon_g}+\frac{P^2\widehat{k}_z^2}{3}\frac{1}{\Delta+\varepsilon+\varepsilon_g}+\frac{P^2k_-k_+}{3(\Delta+\varepsilon+\varepsilon_g)}\right)\varphi_1-iRk_-\varphi_2=0 \tag{15}$$

$$\left(-\varepsilon+\frac{P^2}{6}\frac{k_+k_-}{\varepsilon+\varepsilon_g}+\frac{2}{3}\frac{P^2k_z^2}{\varepsilon+\varepsilon_g}+\frac{P^2}{2}\frac{k_-k_+}{\varepsilon+\varepsilon_g}+\frac{P^2k_-k_+}{3(\Delta+\varepsilon+\varepsilon_g)}+\frac{P^2\widehat{k}_z^2}{3}\frac{1}{\Delta+\varepsilon+\varepsilon_g}\right)\varphi_2+iRk_+\varphi_1=0 \tag{16}$$

is obtained, and operators $\widehat{k}_\pm$ in cylindrical coordinates take the form;

$$k_\pm=-ie^{\pm i\varphi}\left(\frac{d}{d\rho}\pm\frac{i}{\rho}\frac{d}{d\varphi}\right) \tag{17}$$

We consider a quantum anti-wire modeled as a quantum cylindrical cavity of radius r_0 etched out of a bulk composite semiconductor material and the height *d*. Because of the symmetry of the problem it is convenient to use cylindrical coordinates (ρ, φ).

The eigenfunctions of the total momentum operator with a half integer eigenvalue m are of the following form;

$$\begin{pmatrix} \varphi_1 \\ \varphi_2 \end{pmatrix} = \begin{pmatrix} e^{im\varphi + ik_z z} f(\rho) \\ e^{i(m+1)\varphi + ik_z z} g(\rho) \end{pmatrix} \tag{18}$$

where m=0, ±1, ±2, ±3, ... are the magnetic quantum numbers through the z-axis. The radial function $f(\rho)$ and $g(\rho)$ is found to satisfy the following differential equation;

$$\{-\varepsilon + \frac{P^2}{3}(\frac{2}{\varepsilon+\varepsilon_g} + \frac{1}{\Delta+\varepsilon+\varepsilon_g})(-\frac{d^2}{d\rho^2} - \frac{1}{\rho}\frac{d}{d\rho} + \frac{(m+1)^2}{\rho^2} + k_z^2\} f(\rho) + R(\frac{d}{d\rho} - \frac{m}{\rho}) g(\rho)) = 0 \tag{19}$$

$$\{-\varepsilon + \frac{P^2}{3}(\frac{2}{\varepsilon+\varepsilon_g} + \frac{1}{\Delta+\varepsilon+\varepsilon_g})(-\frac{d^2}{d\rho^2} - \frac{1}{\rho}\frac{d}{d\rho} + \frac{(m+1)^2}{\rho^2} + k_z^2\} g(\rho) + R(\frac{d}{d\rho} - \frac{m}{\rho}) f(\rho)) = 0 \tag{20}$$

The perturbation theory can be applied by accepting the Rashba spin-orbital interaction term as weak perturbation term. Then the wave functions are taken as Bessel function;

$$\begin{bmatrix} f(\rho) \\ g(\rho) \end{bmatrix} = \begin{bmatrix} d_1 Y_m(\nu\rho) \\ d_2 Y_{m+1}(\nu\rho) \end{bmatrix} \tag{21}$$

If it's used the wave functions in equation (21) in the equations (19) and (20) respectively, the following differential equations are obtained;

$$x^2 \frac{d^2}{dx^2} Y_m(x) + x \frac{d}{dx} Y_m(x) + x^2 (\frac{\varepsilon_0}{\nu^2} + \frac{d_2}{d_1}\frac{R_0}{\nu}) Y_m(x) - m^2 Y_m(x) = 0 \tag{22}$$

and

$$x^2 \frac{d^2}{dx^2} Y_{m+1}(x) + x \frac{d}{dx} Y_{m+1}(x) + x^2 (\frac{\varepsilon_0}{\nu^2} + \frac{d_1}{d_2}\frac{R_0}{\nu}) Y_{m+1}(x) - (m+1)^2 Y_{m+1}(x) = 0 \tag{23}$$

where;

$$\varepsilon_0 = \frac{3\varepsilon}{P^2(\frac{2}{\varepsilon+\varepsilon_g} + \frac{1}{\varepsilon+\varepsilon_g+\Delta})} - k_z^2 \qquad R_0 = \frac{3R}{P^2(\frac{2}{\varepsilon+\varepsilon_g} + \frac{1}{\varepsilon+\varepsilon_g+\Delta})}$$

Let's accept that; the solutions of the equation (22) and equation (23) are;

$$\frac{\varepsilon_0}{\nu^2} + \frac{d_2}{d_1}\frac{R_0}{\nu} = 1 \text{ and } \frac{\varepsilon_0}{\nu^2} + \frac{d_1}{d_2}\frac{R_0}{\nu} = 1 \tag{24}$$

If the two expressions in the equation (24) are multiplied by each other, then it's obtained;

$$(\frac{R_0}{\nu})^2 = (1 - \frac{\varepsilon_0}{\nu^2})^2 \tag{25}$$

where

$$\nu^2 \pm R_0 \nu - \varepsilon_0 = 0 \tag{26}$$

If the equation (26) is solved for the *v* then

$$\nu_{1,2}^{\pm} = \pm \frac{R_0}{2} \pm \sqrt{\frac{R_0^2}{4} + \varepsilon_0} \tag{27}$$

is found. In addition to this, if the expressions in equation (24) is subtracted each other;

$$\frac{R_0}{\nu}\left(\frac{d_2}{d_1}-\frac{d_1}{d_2}\right)=0 \tag{28}$$

is found. It's clearly seen in equation (28) that $\frac{R_0}{\nu}\neq 0$, so $\left(\frac{d_2}{d_1}-\frac{d_1}{d_2}\right)=0$. This implies two possibilities which are $d_1=d_2$ and $d_1=-d_2$. Hence there are two degenerate solutions for $\upsilon_\pm$ which are combined in the general solution;

$$\begin{bmatrix} f(\rho) \\ g(\rho) \end{bmatrix} = A\begin{bmatrix} d_1^{+}Y_m(\nu_+\rho) \\ d_2^{+}Y_{m+1}(\nu_+\rho) \end{bmatrix} + B\begin{bmatrix} d_1^{-}Y_m(\nu_-\rho) \\ d_2^{-}Y_{m+1}(\nu_-\rho) \end{bmatrix} \tag{29}$$

where

$$\nu_\pm = \pm\frac{R_0}{2}+\sqrt{\frac{R_0^2}{4}+\varepsilon_0}$$

It's implied the Drichlet boundary condition as $f(\rho=r_0)=g(\rho=r_0)=0$ for ν linear combination of solution (29).

$$\begin{bmatrix} d_1^{+}Y_m(\nu_+\rho) & d_1^{-}Y_m(\nu_-\rho) \\ d_2^{+}Y_{m+1}(\nu_+\rho) & d_2^{-}Y_{m+1}(\nu_-\rho) \end{bmatrix}\begin{bmatrix} A \\ B \end{bmatrix}=0 \tag{30}$$

It gives us the following exact equation for energy spectrum of the quantum anti-wire with spin orbital interacting

$$\frac{d_1^{+}}{d_2^{+}}\frac{d_1^{-}}{d_2^{-}}Y_m(\nu_+r_0)Y_{m+1}(\nu_-r_0)-Y_m(\nu_-r_0)Y_{m+1}(\nu_+r_0)=0 \tag{31}$$

For value of $\nu_\pm$ $\frac{d_1^{+}}{d_2^{+}}\frac{d_2^{-}}{d_1^{-}}=-1$ equation (31) simplifies to

$$Y_m(\nu_+r_0)Y_{m+1}(\nu_-r_0)+Y_m(\nu_-r_0)Y_{m+1}(\nu_+r_0)=0 \tag{32}$$

Not that for the quantum wire with the finite height d $k_z=\frac{\pi}{d}n,\ n=1,2,3,...$ and the minimal value for k_z must be taken as $k_z=\frac{\pi}{d}$.

The n th solution of equation (32), is shown the n th energy level of the electron which have a momentum of m+1. Equation (32) is invariant under the change m→-m reflecting the Kramer's degeneracy. According to the standard analysis the spin-orbit coupling splits all the l≠0 states into two Kramers doublets with m=l+1/2 and m=l-1/2, while l=0 states naturally remain Kramer's doublets. We have analyzed Equation (32) numerically, labeling the energy eigenstates as (m,n) where n is a non-negative integer such that $E_{m,n}<E_{m,n+1}$ at R=0. In Figure 2, it's shown the radii dependence of Rashba spin-orbit splitting in quantum anti-wire including both the parabolic and Kane models using the band structure parameters for InSb; $m=0.014m_0$, E_g=0.24 eV, R=2.5 10^{-9} eV.cm [12]. As it's seen from the Figure 3, the value of the Rashba spin-splitting is decreasing with the increasing of the radius for both models. According to this Figure 2, with increase of r_0, the electron energy levels in both cases are close to each other. At rather small sizes of r_0, the variance electron dispersion laws becomes more and more important and, therefore, the curves for $\Delta E/E_g(R)$ ΔE=E(J=1/2,m=1)-E(J=3/2,m=0) separate from each other.

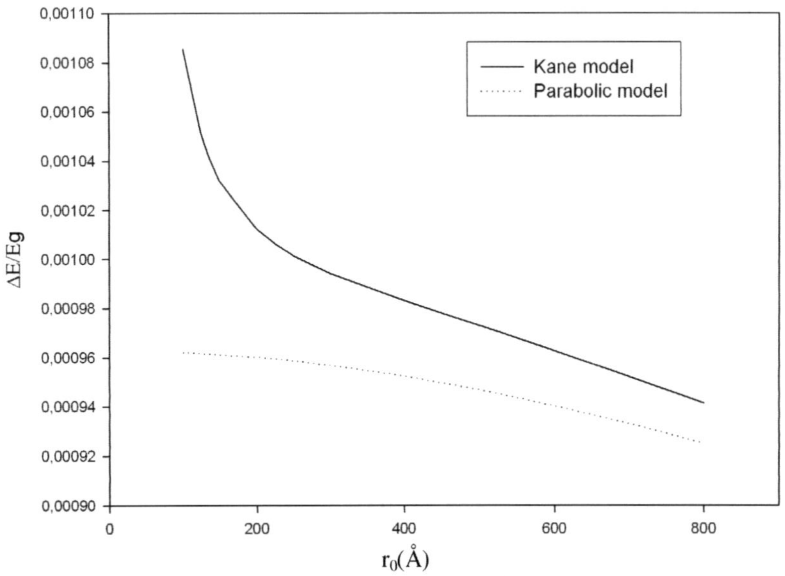

Figure 2. The change of Rashba spin-orbit splitting in InSb quantum anti-wire (B=0).

In Figure 3 and 4, the radius and height dependence of electron energies in InSb quantum anti-wire, are shown where Rashba parameter is constant again comparing the parabolic band model and Kane model. It's seen from the Figure 2 and 3 that the energies are decreasing with the increasing of radius and height. It can be seen from the Figure 4 that the curves for the parabolic band model and Kane model becomes closer in the early height values.

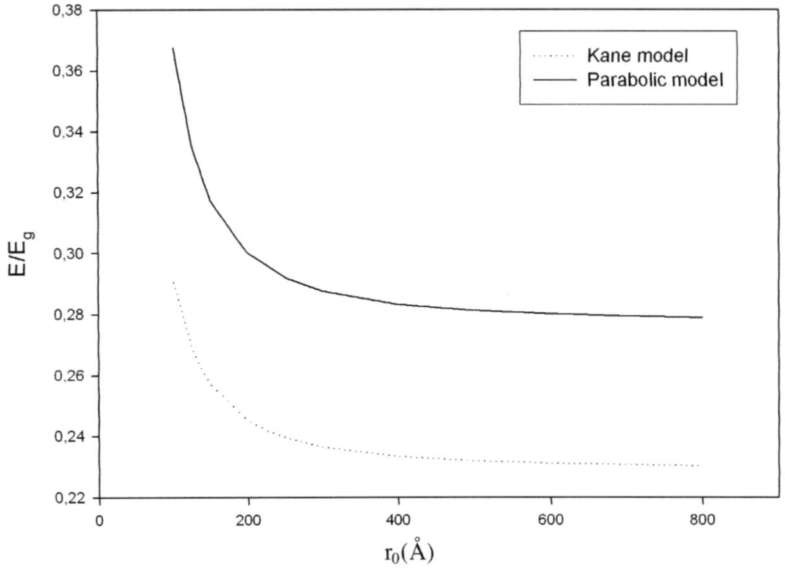

Figure 3. The radii dependence of electron energies in InSb quantum anti-wire (B=0).

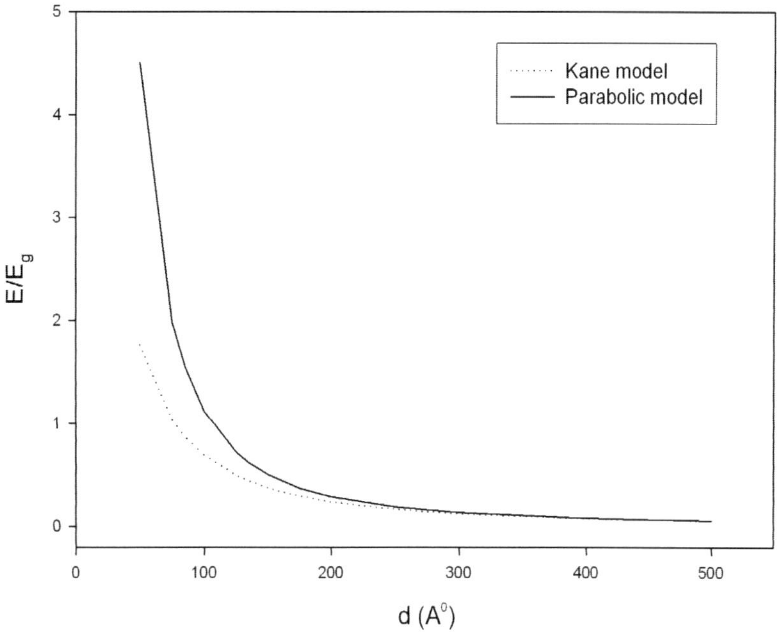

Figure 4. The hight dependence of electron energies in InSb quantum anti-wire (B=0).

The evolution of the first few energy levels with the Rashba parameter plotted in Figure 5. The curves are labeled by quantum numbers (n, m).

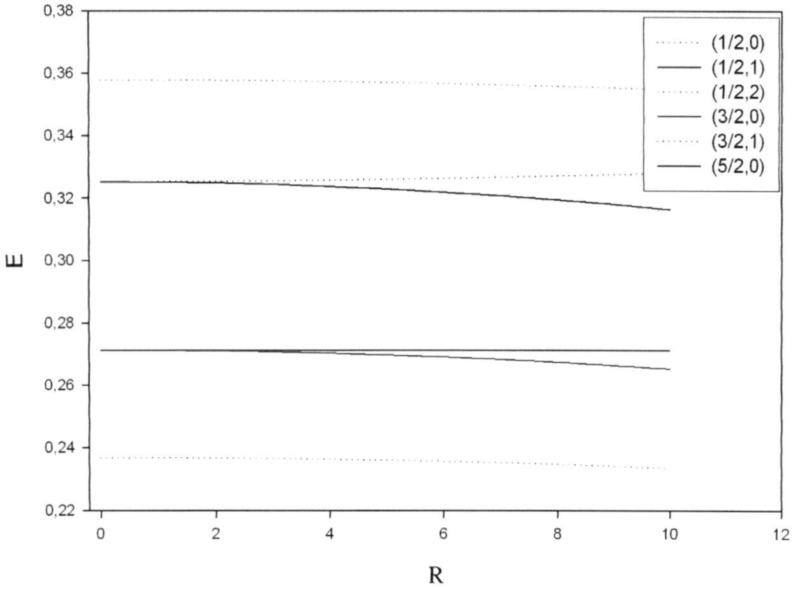

Figure 5. The Rashba parameter change of electron energies in InSb quantum anti-wire (B=0).

3. Applied Magnetic Field

In the second case the problem is solved under an applied magnetic field. For a uniform magnetic field H, directed along the z-axis, the vector potential may be chosen in the form;

$$\vec{A} = \left(-\frac{Hy}{2}, \frac{Hx}{2}, 0 \right) \tag{33}$$

the operators $k_\pm$ in Eqs. (15) and (16) become cylindrical coordinates and take the form

$$k_\pm = i \exp(\pm i\varphi) \left(\frac{\partial}{\partial \rho} \pm \frac{i}{\rho} \frac{\partial}{\partial \varphi} \pm \frac{\rho}{2\alpha^2} \right) \tag{34}$$

where $\alpha = \sqrt{\frac{\hbar c}{eH}}$ is the magnetic length. Since the system is cylindrically symmetric, the wave function can be represented as [19];

$$\begin{pmatrix} \varphi_1 \\ \varphi_2 \end{pmatrix} = \begin{pmatrix} \exp(im\varphi + ik_z z) \exp\left(-\frac{x}{2} \right) x^{|m|/2} Y(x) \\ \exp(im\varphi + ik_z z) \exp\left(-\frac{x}{2} \right) x^{|m|+1/2} Y(x) \end{pmatrix} \tag{35}$$

Where; $x = \frac{\rho^2}{2\alpha^2}$.

Substituting expressions (3)-(8) into formulas (1) and (2) and using the formulas of (34) and (35), we obtain second order differential equations for the radial functions;

$$x \frac{d^2Y(x)}{dx^2} + (-x + 1 + |m|) \frac{dY(x)}{dx} - \frac{1}{2}\left(1 + |m| + m + h\alpha^2 - \alpha^2 k_t^2\right) Y(x) + \frac{\alpha\beta}{\sqrt{2}} x^{\frac{1+|m+1|-|m|}{2}} \left(\frac{dZ(x)}{dx} + \frac{1}{2}\frac{1}{x}\left(1 + |m+1| + m\right) Z(x) \right) = 0 \tag{36}$$

$$x \frac{d^2Z(x)}{dx^2} + (-x + 1 + |m|) \frac{dZ(x)}{dx} - \frac{1}{2}\left(1 + |m| + m + h\alpha^2 - \alpha^2 k_t^2\right) Z(x) + \frac{\alpha\beta}{\sqrt{2}} x^{\frac{1-|m+1|-|m|}{2}} \left(\frac{dY(x)}{dx} + \frac{1}{2}\frac{1}{x}\left(1 + |m| - m - 2x\right) Y(x) \right) = 0 \tag{37}$$

In Equations (37) and (38) the following notations are used;

$$h = \frac{\lambda_H \left(\frac{1}{E + E_g} - \frac{1}{E + E_g + \Delta} \right)}{\left(\frac{2}{E + E_g} + \frac{1}{E + E_g + \Delta} \right)} \tag{38}$$

$$k_t^2 = \frac{3E}{P^2 \left(\frac{2}{E + E_g} + \frac{1}{E + E_g + \Delta} \right)} - k_z^2 \tag{39}$$

$$\beta = \frac{3R}{P^2 \left(\frac{2}{E + E_g} + \frac{1}{E + E_g + \Delta} \right)} \tag{40}$$

Let us introduce a_1 and a_2 parameters which are related to the equation in the following form:

$$x \frac{d^2Y(x)}{dx^2} + (-x + 1 + |m|) \frac{dY(x)}{dx} - a_1 Y(x) = 0 \tag{41}$$

$$x \frac{d^2Z(x)}{dx^2} + (-x + 1 + |m| + 1) \frac{dZ(x)}{dx} - a_2 Z(x) = 0 \tag{42}$$

The equations are canonical form of Kummer's equations for the confluent hypergeometric function [20]. Equation (36) and (37) can be written as follows taking into account the equations of (41) and (42).

$$a_1 - \frac{1}{2}\left(1 + |m| + m + h\alpha^2 - \alpha^2 k_t^2\right) Y(x) + \frac{\alpha\beta}{\sqrt{2}} x^{\frac{1+|m+1|-|m|}{2}} \left(\frac{dZ(x)}{dx} + \frac{1}{2}\frac{1}{x}\left(1 + |m+1| + m\right) Z(x) \right) = 0 \tag{43}$$

$$a_2 - \frac{1}{2}\left(1+|m|+m+h\alpha^2-\alpha^2 k_t^2\right)Z(x)+\frac{\alpha\beta}{\sqrt{2}}x^{\frac{1-|m+1|-|m|}{2}}\left(\frac{dY(x)}{dx}+\frac{1}{2}\frac{1}{x}\left(1+|m|-m-2x\right)Y(x)\right)=0 \quad (44)$$

The solutions of Equations (43) and (44) that is bounded is $Y(x)=d_1U(a_1,b_1,x)$ and $Z(x)=d_2U(a_2,b_2,x)$, where

$$a_1 = \frac{m+|m|}{2} - \varepsilon_0, \quad b=|m|+1 \quad (45)$$

$$a_2 = \frac{m+1+|m+1|}{2} - 1 - \varepsilon_0, \quad b=|m+1|+1 \quad (46)$$

$$\varepsilon_{0\pm} = \frac{1}{2}(-2+\frac{1}{2}\alpha^2\beta_R^2+\alpha^2k_t^2 \pm \sqrt{1-4s+4s^2+\frac{1}{4}\alpha^4\beta_R^4+\alpha^4\beta_R^2k_t^2}\ ; \ h\alpha^2=-2s \quad (47)$$

The wave function can be written;

$$R^+(x)=c_1d_{1+}U\left(\frac{m+|m|}{2}-1-\varepsilon_{0+},b_1,x\right)+c_2d_{2+}U\left(\frac{m+1+|m+1|}{2}-1-\varepsilon_{0+},b_2,x\right) \quad (48)$$

$$R^-(x)=c_1d-U\left(\frac{m+|m|}{2}-1-\varepsilon_{0-},b_1,x\right)+c_2d_{2-}U\left(\frac{m+1+|m+1|}{2}-1-\varepsilon_{0-},b_2,x\right) \quad (49)$$

In a nano anti-wire with an infinite potential barrier, the wave function must vanish at the quantum anti-wire's surface giving the dispersion equation for the electron quantum size levels:

$$\frac{d_{1+}}{d_{1-}}\frac{d_{2-}}{d_{2+}}\left(U\left(\frac{m+|m|}{2}-1-\varepsilon_{0+},b_1,x\right)+U\left(\frac{m+1+|m+1|}{2}-1-\varepsilon_{0-},b_2,x\right)-U\left(\frac{m+|m|}{2}-1-\varepsilon_{0-},b_1,x\right)+U\left(\frac{m+1+|m+1|}{2}-1-\varepsilon_{0+},b_2,x\right)\right) \quad (50)$$

The above equation provides all information about the energy spectrum of electrons. The energy is a complicated function of the anti-wire parameters ρ, d and the electron angular momentum. Equation (50) can be useful for analyzing the influence of nonparabolicity on the energy spectrum of electrons near a quantum anti-wire.

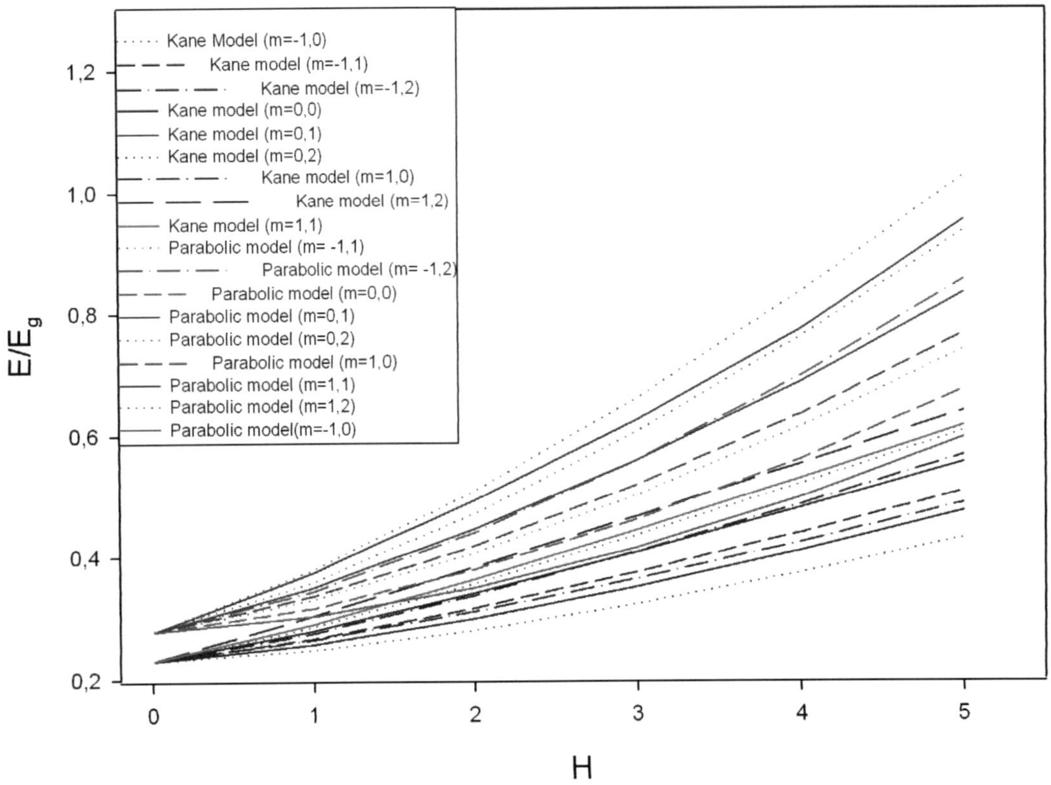

Figure 6. Energy as a function of magnetic field for different (m,n) states.

It's seen the magnetic field dependence of the sub-bands for different n and different m values for electrons in InSb type quantum anti-wire with a constant radius 200 Å and the height 20 Å. It can be from the Figure 6 that seen that the lowest sub-band corresponds to the state having ($m=-1$, $n=0$) which differs from the same graph in the quantum-wire problem [11]. As it's seen from Figure 6, the electron energies in Kane model is smaller than that of parabolic band model.

4. Conclusion

It's presented the analytical solutions of Kane equations for a quantum anti-wire in presence of Rashba spin-orbit interaction with and without an external magnetic field. The calculations are based on the eight-band **k·p** theory, which takes into account the nonparabolicity of the electron energy levels. A nonzero magnetic field breaks the time-inversion symmetry and removes the corresponding degeneracy. It's shown that Rashba splitting is increasing with the increasing of the radius of the quantum anti-wire in Kane model.

Acknowledments
We would like to acknowledge the financial support (Project no: 107T263) by the Turkish Scientific and Technical Research Institute (TUBİTAK).

References

[1] Chakraborty T 1999 *Quantum Dots* (Amsterdam: Elsevier)
[2] Bimberg D, Grundman M, Ledentsov M 2001 *Quantum Dot Herosutructures* (New York:Wiley)
[3] Rashba E I 1960 *Sov. Phys. Solid State* **2** 1109

[4] Dresselhaus G 1955 *Phys. Rev.* **100** 580
[5] Dyakonov M I and Perel V I 1971 *Zh. Eksp. Teor. Fiz.* **60** 1954
[6] Data S and Das B 1990 *Appl. Phys. Lett.* **56** 665.
[7] Ohkawa F and Uemura Y 1974 *J. Phys. Soc. Jpn.* **37** 1325
[8] Lassnig R 1985 *Phys. Rev. B* **31** 8076
[9] Pfeffer P and Zawadzki W 2004 *Semiconductor Science Technology* **19** R1-R17
[10] Tsitsishvili E, Lozano G S and Gogolin A O 2004 *Phys. Rev. B* **70** 115316
[11] Bulgakov E N, Sadreev A F 2001 *Pis'ma Zhetf* **73** 10
[12] Hashimzade F M, Babayev A M, Çakmak S, Çakmaktepe S 2004 *Physica E: "Lowdimensional Systems and Nanostructures* **25/1** 78-85
[13] Kudryashov V V 2007 *Math-Ph.* **2** 0708.3904
[14] Babayev A M, Çakmak S, Artunc E, Çakmaktepe Ş 2005 *Physica E: "Low-dimensional Systems and Nanostructures* **28/1** 37-42
[15] Babayev A M, Çakmaktepe Ş, Turkoz D. A., to be published
[16] Lourtioz J M 2008 *Photonic Crystals* (Berlin Heidelberg: Springer Verlag) p.507
[17] Askerov B M 1970 *Kinematc Effects in Semiconductors* (Leningrad: Nauka)
[18] Kane E O 1957 *J. Phys. Chem. Solids* **1** 249
[19] Hashimzade F M, Babayev A M, Mehdiyev B M 2006 *Phys. Rev. B* 73 245321
[20] Abramovich M, Stegun A 1965 *Handbook of Mathematical Functions* (New York: Dover) p.504

Journal of Physics: Conference Series **153** (2009) 012044 doi:10.1088/1742-6596/153/1/012044

Magnetism in wide band gap semiconductors implanted with non-magnetic ions

R P Borges[1], M M Cruz[1,3], R C da Silva[4,5], F Bern[3], P Venezuela[2], M Dionízio Moreira[2], A T Costa[2], M Godinho[1,3]

[1]Centro de Física da Matéria Condensada - Universidade de Lisboa, Campo Grande, Ed. C8, 1749-016 Lisboa, Portugal
[2]Instituto de Física, Universidade Federal Fluminense, 24210-346 Niterói, RJ, Brazil
[3]Dep. de Física, Faculdade de Ciências, Universidade de Lisboa, Ed. C8, Campo Grande, 1749-016 Lisboa, Portugal
[4]Laboratório de Feixe de Iões, Dep. Física, Instituto Tecnológico e Nuclear, Estrada Nacional 10, 2686-953 Sacavém, Portugal
[5]Centro de Física Nuclear da Universidade de Lisboa, Av. Prof. Gama Pinto 2, 1649-003 Lisboa, Portugal

E-mail: rpborges@fc.ul.pt

Abstract. Single crystals of ZnO, TiO_2 and $LaAlO_3$ have been implanted with Ar with 100 keV and different fluencies. The Ar implanted crystals showed a week ferromagnetic-like signal between 10 K and 400 K. Hysteresis curves obtained at room temperature allowed confirming the ferromagnetic behaviour of the implanted samples. Spin polarised first principles density functional calculations were performed in the case of ZnO considering Zn interstitials and O vacancies. No net magnetic polarisation was found for O vacancies, but in the case of Zn vacancies a magnetic moment of $1\mu_B$ was obtained.

1. Introduction

Diluted magnetic semiconductors (DMS), in which ferromagnetism arises from the doping of a non-magnetic semiconductor with a transition metal element, was first predicted by T. Dietl in oxide systems for the case of Mn-doped ZnO [1]. The large amount of experimental and theoretical work that followed Dietl's calculation has been far from conclusive with a wide spread of results published by different groups. One of the most discussed issues in these systems is the formation of secondary phases that could be responsible for many of the reports of ferromagnetism in magnetic semiconductors [2,3]. However, the presence of ferromagnetic ordering in a variety of systems that do not include magnetic elements [4,5], together with band structure calculations showing that cation vacancies may be associated with a magnetic moment [6,7], has highlighted the possible contribution of structural defects to the observed magnetic signal. In ZnO, ferromagnetism has been observed in doped systems with structural defects such as thin films [8] and implanted crystals [9], and in non-doped ZnO nanoparticles capped with different organic molecules [10].

In this work the implantation of Ar into ZnO, TiO_2 and $LaAlO_3$ single crystals was carried out in order to investigate the role of the induced defects in the magnetic properties of the material.

2. Experimental

Single crystals of ZnO, TiO_2 and $LaAlO_3$ were implanted at room temperature with Ar ions of energy 100 keV using different fluencies as shown in Table 1

Table 1. Implantation fluencies of Ar ions in different single crystals

ZnO	TiO_2	$LaAlO_3$
1×10^{16}		
1×10^{17}	1×10^{17}	1×10^{17}
2×10^{17}	2×10^{17}	

The implantation procedure with these high fluencies induces disorder in the implanted region. To recover the crystalline structure thermal annealing treatments were carried out at temperatures related with the recovery stages of each lattice. The magnetic behaviour of the virgin crystals and implanted samples was analyzed through magnetization measurements performed with a Squid magnetometer. Rutherford Backscattering Spectroscopy (RBS) and RBS-C were used to investigate the implantation profiles and the lattice induced disorder. Particle induced X-ray emission (PIXE) was also used to check for the presence of impurities.

3. Results and Discussion

For all samples a ferromagnetic component was detected after implantation. In Figure 1 the results obtained for the magnetisation as a function of applied field measured at room temperature for the virgin crystals are presented. The magnetic susceptibilities (M/H) of the virgin crystals, determined at different temperatures, are found to agree within 3% with reported values [11, 12].

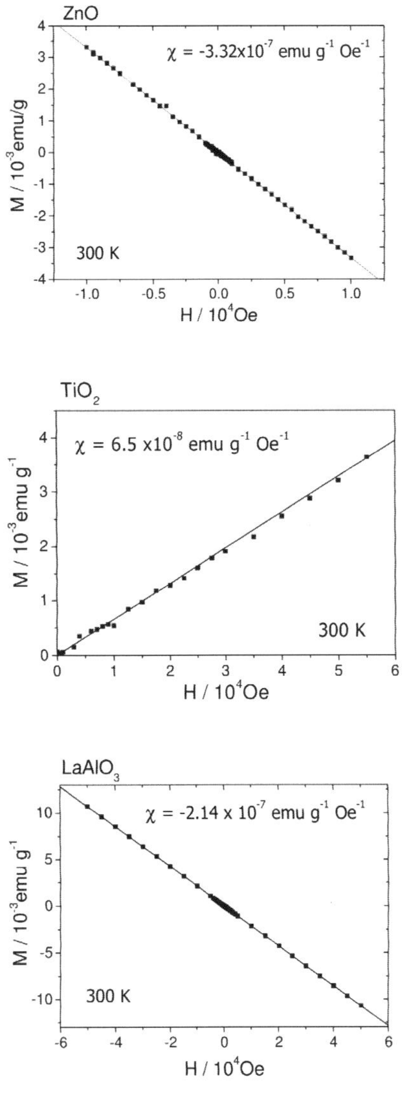

Figure 1. Room temperature magnetisation of oxide non implanted single cristals .

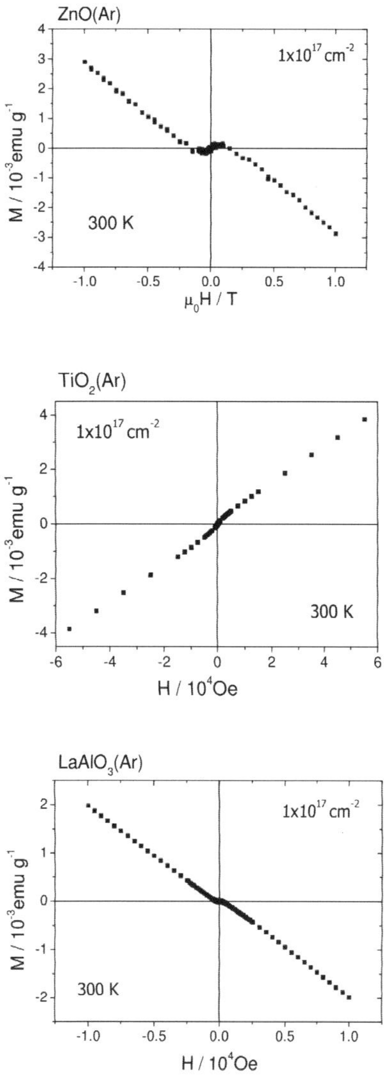

Figure 2. Hysterisis curves for the three oxides after implantation with argon with a fluence of $1\times10^{17}\,cm^{-2}$

In Figure 2 hysteresis curves are presented for the 1x 10^{17} cm^{-2} implanted samples at 300 K. The results evidence the presence of ferromagnetic behaviour at low fields and, as major contribution, the diamagnetic (paramagnetic in the case of TiO_2) signal attributed to the non-implanted volume. For the samples implanted with different fluencies the results are similar. In order to isolate the magnetic contribution of the implanted region, the diamagnetic (paramagnetic) signal of the non-implanted part was subtracted from the raw measurements. Since the implanted region is only a very small fraction of the crystal volume (~0.02%), the total mass of the crystals were used to calculate this contribution. All the experimental magnetisation results presented below already include this correction.

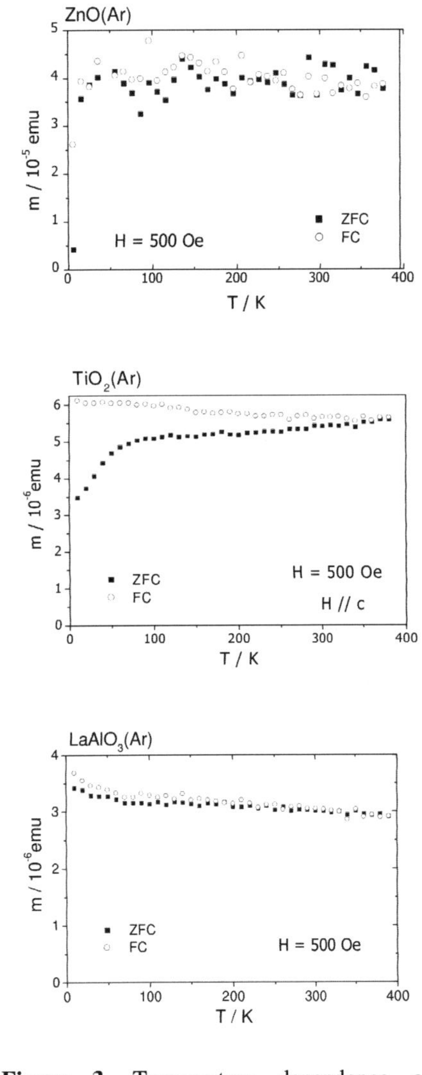

Figure 3. Temperature dependence of the magnetization for the three implanted oxides .

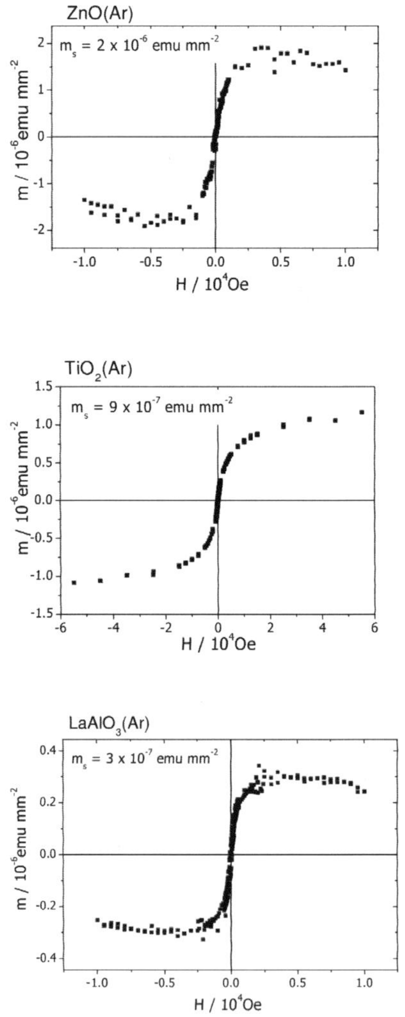

Figure 4. Magnetic moment hysteresis curves at 300 K for the samples implanted with 1×10^{17} Ar cm^{-2}.

In Figure 3 the temperature dependence of magnetisation for the three implanted samples is shown. The curves display the common feature of an approximately constant magnetisation value up to 380 K, suggesting a Curie temperature well above room temperature. The hysteresis loops measured for the same samples at 300 K (Figure 4) confirm the existence of ferromagnetism at this temperature.

In order to check for the existence of impurities particle induced X-ray emission (PIXE) was performed on the samples. In the case of TiO_2, no impurities have been detected. For the implanted and virgin ZnO crystals emission peaks for various transition metal impurities were observed with a concentration smaller than 0.04 at. % and an estimated contribution to the total magnetic moment between 5 and 50 times smaller than the measured values.

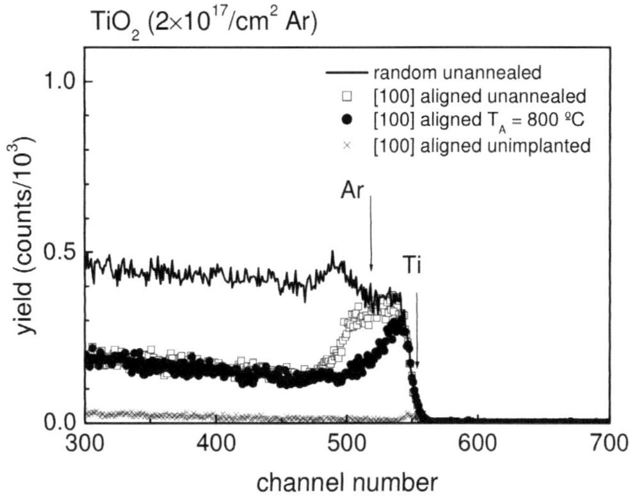

Figure 6. RBS-C spectra for TiO2 implanted with argon.

To recover the lattice structure and modify the concentration of defects, the following heat treatments were carried out: for ZnO two consecutive annealings in air at 400°C and 500°C for 6 h and for TiO_2 one annealing at 800°C for 1 h in argon atmosphere.

The RBS-C spectra shown in Figure 6 show that prior to the annealing the implanted region is highly damaged in both cases and that the damage induced after implantation is significantly reduced with the thermal treatments, although some disorder still remains This information is obtained comparing the yield for incident ions along one of the crystalline directions of the crystals with the values of the same yield for the non implanted sample. The high yield observed in the RBS-C, in the implanted depth spectra, after implantation almost equals the yield obtained for a non aligned direction indicating a highly disordered region. After annealing this yield decreases in a large depth.

The magnetization results after the heat treatments are presented in Figure 7 for Ar implanted ZnO and TiO_2. The results show that the process of recrystallization and lattice recovery induced by the annealing is accompanied by a reduction or disappearance of the ferromagnetic signal, respectively for TiO_2 and ZnO in the case of the thermal treatments referred.

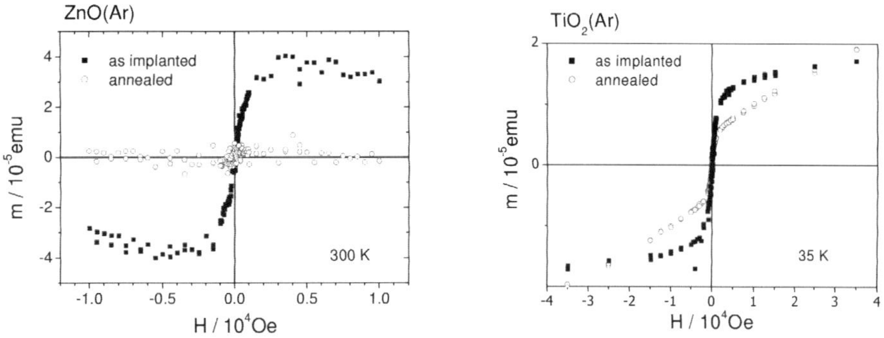

Figure 7. Comparison between hysteresis curves obtained, before and after the thermal treatment, for samples implanted with 1×10^{17} Ar cm^{-2}

Given the low impurity concentration, the formation of aggregates with nanosize dimensions induced by the implantation procedure could be put forward to explain the observed magnetic signal in ZnO. However, such nanosized particles would have superparamagnetic behaviour and no evidence of such behaviour can be found neither for the virgin or the implanted crystals (Figure 3).

In order to understand further the mechanism behind the appearance of the magnetic moment, band structure calculations using density functional theory within the generalised gradient approximation were performed for ZnO. A supercell of 128 atoms and periodic boundary conditions were used. Two types of defects, Zn vacancies and oxygen vacancies, were considered the calculated values of the associated moments being 1.05 and 0 μ_B respectively. For the case of Zn vacancies the magnetic moment results from the polarisation of the oxygen p bands giving rise to defect states just above the valence band as shown in Figure 8. The defect state bands cross the Fermi energy resulting in a metallic solution and are fully polarised.

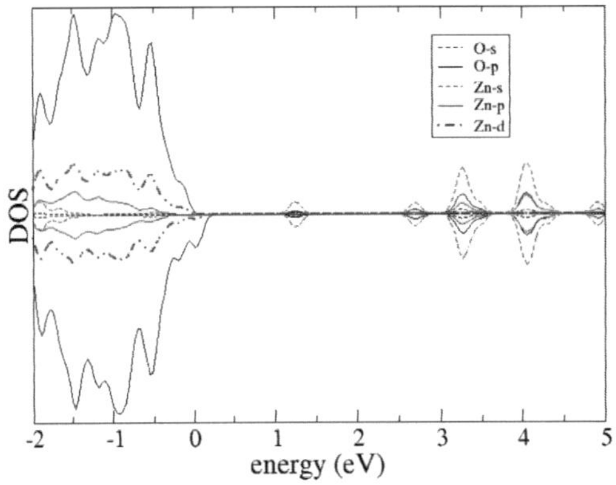

Figure 8. Partial densities of states for the case of a Zn vacancy. The zero of energy is at Fermi energy.

3. Conclusions

In this work we have shown that oxide semiconductor single crystals implanted with non magnetic Ar ions develop a ferromagnetic signal at room temperature. The results show that the obtained magnetic behaviour can not be assigned to impurities or to the formation of secondary phases or aggregates and is ascribed to the lattice defects induced by the implantation procedure. Band structure calculations performed for ZnO show that point defects can give rise to polarised bands inside the semiconducting gap, pointing to a possible mechanism for the long range ordering of the magnetic moments. Further experiments and calculations are under progress.

Aknowledgements

This work was financed by the Portuguese Foundation for Science and Tecnology (FCT) through project PTDC/FIS/66262/2006. R. P. Borges thanks the FCT for support through grant SFRH/BPD/28221/2006.

References

[1] Dietl T, Ohno H, Matsukura F, Cibert J and Ferrand D 2000 *Science* **287** 1019

[2] Pinto J.V., Cruz M.M., da Silva R.C., Franco N., Casaca A., Alves E., and Godinho M., 2007 *Eur. Phys: J. B* **55** 253

[3] Borges R P, Pinto J V, da Silva R C, Gonc¸alves A P, Cruz M M and Godinho M 2007 *J. Magn. Magn. Mater.* **316** e191

[4] Coey J M D, Venkatesan M, Stamenov P, Fitzgerald C B and Dorneles L S 2005 *Phys. Rev.* B **72** 024450
[5] Young D P, Hall D, Torelli M E, Fisk Z, Sarrao J L, Thompson J D, Ott H R, Oseroff S B, Goodrich R G and
Zysler R 1999 *Nature* **397** 412
[6] Elfimov I S, Yunoki S and Sawatzky G 2002 *Phys. Rev. Lett.* **89** 216403
[7] Chaitania Das P and Sanvito S 2005 *Phys. Rev. Lett.* **94** 217205
[8] Hong N H, Sakai J and Briz´e V 2007 *J. Phys.: Condens. Matter* **19** 036219
[9] Borges R P et al 2007 J. *Phys.: Condens. Matter* **19** 476207
[10] Garcia M A *et al* 2007 *Nano Lett.* **7** 1489
[11] *CRC Handbook of Chemistry and Physics* 1999 80th edn (Boca Raton, FL: CRC Press) p 4-134
[12] Janisch R, Gopal P and Spaldin N A 2005 *J. Phys.: Condens. Matter* **17** R657

Journal of Physics: Conference Series **153** (2009) 012045 doi:10.1088/1742-6596/153/1/012045

Strain and interface effects on the magnetic and transport properties of $La_{0.7}Ca_{0.3}MnO_3$/CaO multilayers

A Casaca[1,2,6], R P Borges[1], P Ferreira[3], A Saraiva[1], M A Rosa[1], R C da Silva[4], W C Nunes[1], S Magalhães[4] and M Godinho[1,5]

[1] Centro de Física da Matéria Condensada, Universidade de Lisboa, Campo Grande, Ed. C8, P-1649-003 Lisboa, Portugal
[2] Área Científica de Física, Instituto Superior de Engenharia de Lisboa, P-1959-007 Lisboa, Portugal
[3] Departamento de Engenharia Cerâmica e do Vidro, CICECO, Universidade de Aveiro, Campus universitário de Santiago, P-3810-193 Aveiro, Portugal
[4] Departamento de Física, Instituto Tecnológico e Nuclear, P-2686-953 Sacavém, Portugal
5 Departamento de Física, Faculdade de Ciências, Universidade de Lisboa, Campo Grande, Ed. C8, P-1749-016 Lisboa, Portugal

Abstract. $La_{0.7}Ca_{0.3}MnO_3$/CaO (LCMO/CaO) multilayers were deposited on (100) $LaAlO_3$ substrates by pulsed injection metal organic chemical vapour deposition. The samples were characterised by X-ray diffraction, transport and magnetization measurements. An anomalous behaviour of the electrical resistance under applied magnetic field and an exchange bias effect for the magnetization are observed for the multilayers with thinner LCMO layers. These results suggest the existence of a short range antiferromagnetic coupling at the layer interfaces.

1. Introduction

The discovery of colossal magnetoresistance (CMR) in the manganite compounds [1,2] generated an enormous interest on their fundamental properties and on their potential applications in the spintronics industry. In particular, much attention has been devoted to the magnetic and magnetotransport properties of thin film and heterostructures, in view of their relevance for device fabrication. In these materials, the lattice mismatches with the substrate induce tensile or compressive strains that strongly influence magnetic and transport properties such as the metal to insulator transition temperature or the magnetoresistance. As an example, it is now well established that the metal to insulator transition temperature decreases with increasing biaxial distortions, since these distortions increase the Jahn-teller effect and, therefore, increase electron localization [3,4].

[6] Corresponding author. Email address: amcasaca@deea.isel.ipl.pt

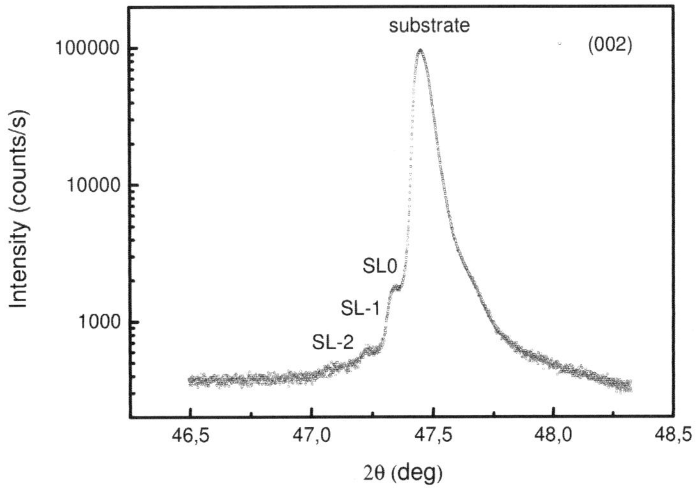

Figure 1. X-ray θ-2θ diffraction scan around the substrate (LAO) (002) peak for the 20/20 sample: the small height of the satellite peaks adjacent to the main peak (SL0) indicates a heterostructure growth with significant interface roughness.

Several types of heterostructures and superlattices have been fabricated, conjugating the ferromagnetic manganese oxides with different kinds of spacers: magnetic conductors, non magnetic conductors or non magnetic insulators [5,6]. A large enhancement of the magnetoresistance at low temperatures was found, for instance, in $La_{0.67}Ca_{0.33}MnO_3/SrTiO_3$ multilayers [5] and $La_{0.67}Ca_{0.3}MnO_{33}/SrRuO_3$ multilayers [6]. A possible explanation for this interesting effect is the creation of magnetic nonuniformities near the interfaces due to disorder, functioning as additional spin-dependent scattering centres.
The manganite perovskite $La_{0.7}Ca_{0.3}MnO_3$ (LCMO) and the insulator CaO coexist naturally in the Ruddlesden-Popper (RP) phases. RP phases $(AO)(ABO3)_n$ consist of perovskite slabs of n layers of corner-sharing BO_6 octahedra separated from neighbouring blocks by an AO rock-salt layer. Thus, LCMO and CaO are good candidates for the engineering of multilayers and other heterostructures, keeping in mind, nevertheless, that the rather large (≈20%) lattice mismatch between LCMO and CaO is likely to impose strong strains on the LCMO layers.
In this work we present magnetization and magnetoresistance measurements performed on a $La_{0.7}Ca_{0.3}MnO_3$ film, on a $La_{0.7}Ca_{0.3}MnO_3/CaO/La_{0.7}Ca_{0.3}MnO_3$ trilayer and on $[(La_{0.7}Ca_{0.3}MnO_3)/(CaO)]_{\times N}$ multilayers.

Table 1. Sample description.

Type	**Designation**	**Layer thicknesses**	**T_p (K)**	**T_c (K)**
film	—	LAO/LCMO(120 nm)	257	240
trilayer	—	LAO/LCMO(120nm)/CaO(30nm)/LCMO(40 nm)	225	205
multilayer	10/10	LAO/[LCMO(10nm)/CaO(10nm)]×12/LCMO(20 nm)	<55	—
multilayer	20/20	LAO/[LCMO(20nm)/CaO(20nm)]×12/LCMO(20 nm)	58	—
multilayer	30/30	LAO/[LCMO(50nm)/CaO(50nm)]×2/LCMO(20 nm)	63	—

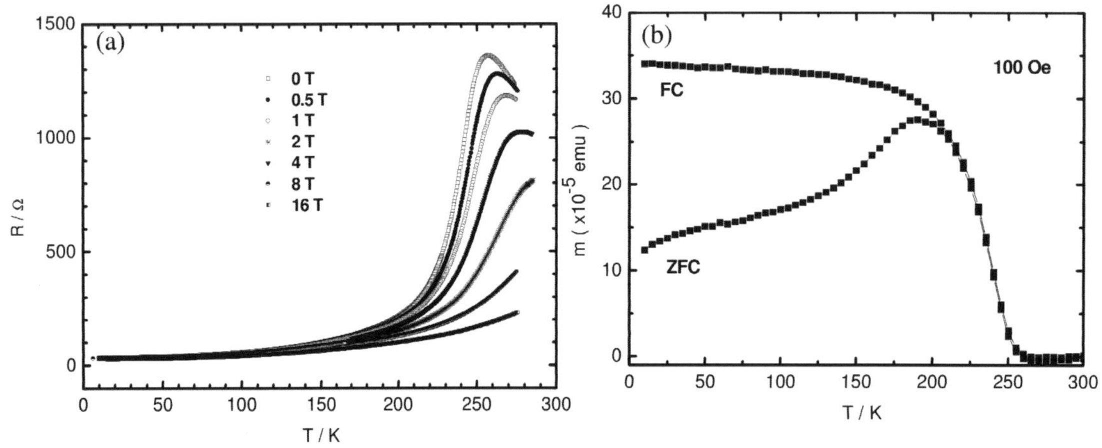

Figure 2. (a) Temperature dependence of the resistance for applied magnetic fields up to 16 T for the LCMO film. **(b)** Temperature dependence of the magnetization for the same LCMO film: zero field cooled (**ZFC**) and field cooled under a 100 Oe field (**FC**).

2. Experimental

The samples were grown by pulsed injection metal organic chemical vapour deposition on (100) single crystalline $LaAlO_3$ (LAO) substrates. In this technique, the metalorganic precursors are injected to the reactor in a liquid solution and the number of droplets for each deposition is accurately controlled. Substrate temperature was kept at 700°C and the droplets were evaporated at a temperature of 200°C. The injection rate was 2 Hz and the injector opening time was 2 ms. The reactor pressure was kept at a constant pressure of 5 torr with Ar flow rate of 900 ml/min and O_2 flow rate of 600 ml/min. All the samples were slowly cooled down to room temperature in situ in 1 bar of O_2 after deposition. In the multilayers, an additional LCMO thin layer (20 nm thick) was deposited on top, to avoid the degradation of the last CaO layer. To control the layers' thickness, the deposition rate was established by Rutherford Backscattering Spectroscopy: for LCMO a deposition rate of 0.1 nm per droplet was obtained.

The structural characterization of the samples was performed by X-ray diffraction with a Bruker diffractometer using the Cu $K_{\alpha 1}$ line. Symmetric (002) 2θ-ω scans were collected with a point detector positioned 290 mm away from the sample and operating with an acquiring time of 2 s per 2θ step (0.001°). Magnetization measurements were performed on a 5 T Quantum Design SQUID magnetometer with the field applied parallel to the samples surface. The electrical resistance was determined by the standard dc four-probe method, with indium contacts directly attached to the sample surface. The measurements were performed inside an OXFORD cryostat, in magnetic fields up to 16 T, with the field parallel to the current. Sample properties are summarized in Table 1, where T_p refers to the zero field peak temperature in the resistance curves and T_c is the Curie temperature.

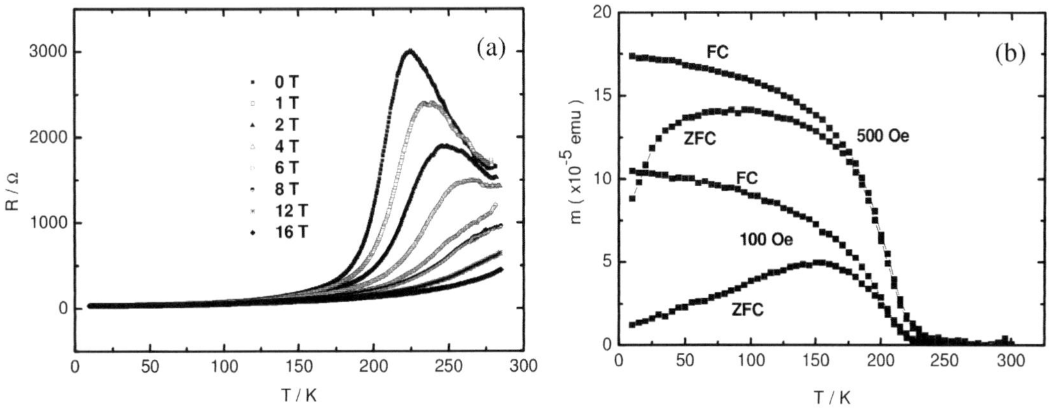

Figure 3. **(a)** Temperature dependence of the resistance for applied magnetic fields up to 16 T for the LCMO/CaO/LCMO trilayer. **(b)** Temperature dependence of the magnetization for the same LCMO/CaO/LCMO trilayer : zero field cooled and field cooled curves under 100 Oe and 500 Oe fields.

3. Results and discussion

Figure 1 shows a 2θ-ω scan for sample 20/20 around the LAO substrate (002) reciprocal lattice point. The main peak is centered at $2\theta \approx 47.45°$ which is the experimental Bragg angle obtained for the LAO (002) substrate. Keissig fringes can be observed but the small height of the satellite peaks and the fact that the 2θ angular separation between SL0 and SL-1 is slightly different from the separation between SL-1 and SL-2, implies the existence of a small thickness period fluctuation or intermixing and, consequently, reveals a significant interface roughness.

The temperature dependence of the electrical resistance at several magnetic fields for the LCMO film is shown in figure 2a: the film presents the usual behaviour found in good quality $La_{1-x}Ca_xMnO_3$ films at optimal doping ($x \approx 0.3$), with the peak temperature (T_p) at zero field equal to 257 K; as the applied magnetic field is increased, the peak in the resistance shifts to higher temperatures, as a result of the increased alignment of the Mn^{3+} and Mn^{4+} ion moments that reduces the electron scattering. Figure 2b shows the temperature dependence of the magnetization for the same LCMO film, in agreement with the expected behaviour and indicating a Curie temperature of 240 K.

Figure 3a shows the temperature dependence of the electrical resistance at several magnetic fields for the LCMO/CaO/LCMO trilayer: the curves present essentially the same behaviour of the film but with the peaks appearing at lower temperatures: the zero field peak (T_p) occurs at 225 K. This temperature is close to the Curie temperature T_c = 205 K derived from magnetization curves, a value that is smaller than that obtained for the single layer. The difference between the values of T_p and T_c, indicates some degree of structural disorder in the LCMO layers [7]. The shift of the metal to insulator transition temperature to lower temperatures can be ascribed to the additional tensile strain at the LCMO/CaO interfaces due to the lattice mismatch between CaO and LCMO: the larger unit cell parameters of CaO induce an in-plane tensile stress leading to a smaller *c* lattice parameter in the LCMO layers. It is known that tensile strain in LCMO [4] has the same effect of compressive strain: it increases the Jahn-Teller (JT) lattice distortion, with the consequent increase in the e_g levels splitting and enhancement of electron localization; consequently the metal to insulator transition temperature decreases. Furthermore we should consider the possibility of Ca diffusion into

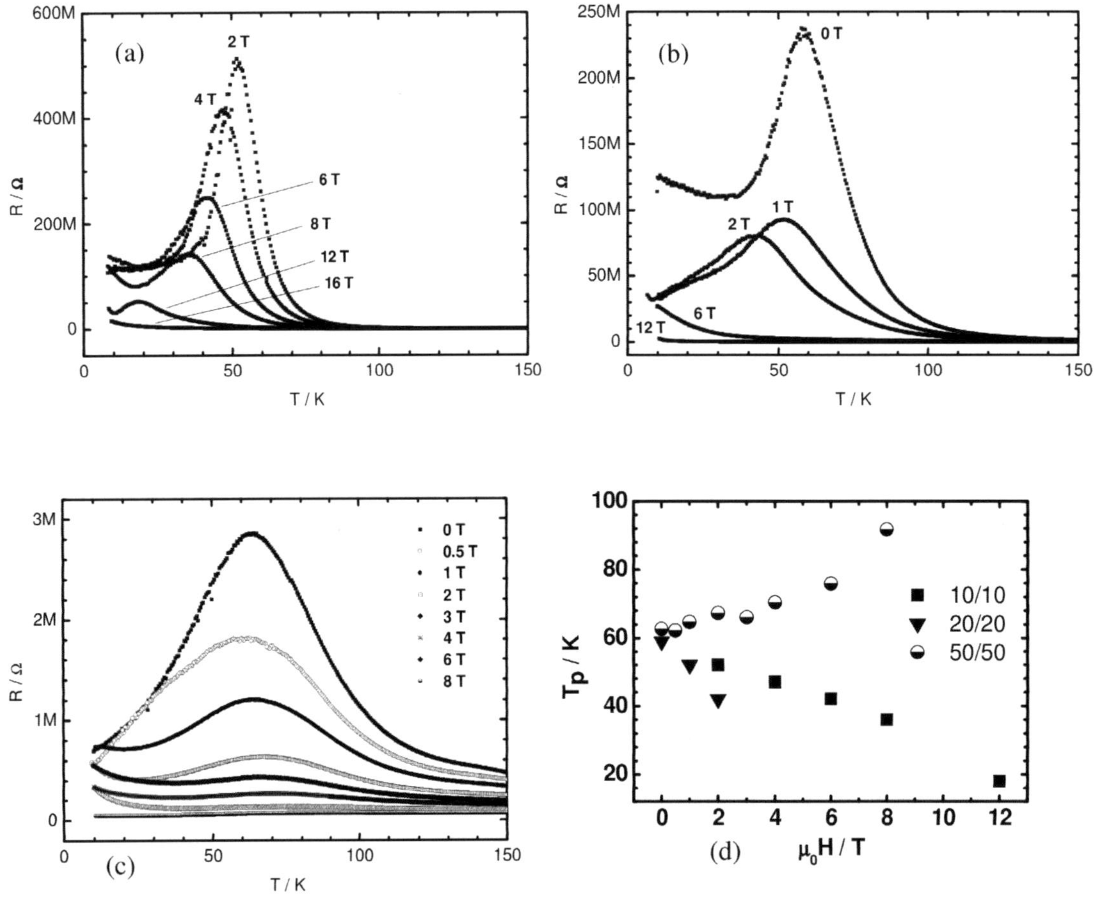

Figure 4. Temperature dependence of the resistance under various applied magnetic fields: **(a)** for the 10/10 multilayer; **(b)** for the 20/20 multilayer and **(c)** for the 50/50 multilayer. **(d)** Resistance peak temperature as a function of the magnetic field for the three samples: for the 10/10 and 20/20 samples, the peaks presents an odd magnetic field dependence, shifting to lower temperatures as the field increases.

the ferromagnetic layers, giving rise to non-optimal hole doping and a lowering of the critical temperature [8] in a volume fraction of the ferromagnetic layers.

In figure 4, the temperature dependence of the electrical resistance is plotted for several magnetic fields for the three multilayers, hereby designated according to their LCMO/CaO thicknesses by (10/10), (20/20) and (50/50) as shown in table 1. For the 10/10 sample, the resistance at zero field is extremely large below 100 K and can not be measured with our experimental setup. Nevertheless, we can estimate that T_p(0T) is below 55 K, since the magnetoresistance curve obtained at this temperature displays a H^2 dependence, usually observed above the metal-insulator transition in these compounds [9]. For samples 20/20 and 50/50, the zero field resistance peak occurs at 58 K and 63 K, respectively. The overall reduction of the metal to insulator transition temperature can be the result of several combined effects: the strain effects, already discussed for the trilayer; a reduction in the e_g band width due to the reduced dimensionality and the existence of phases with suppressed ferromagnetism (or with antiferromagnetic ordering). As to the high resistance values, particularly high for the 10/10 and 20/20 multilayers, they can indicate magnetic barriers between misaligned ferromagnetic regions resulting in increased spin dependent scattering [10]. The most striking feature in the curves of figure 4 is the odd magnetic field dependence exhibited by the resistance peak for the samples 10/10 and

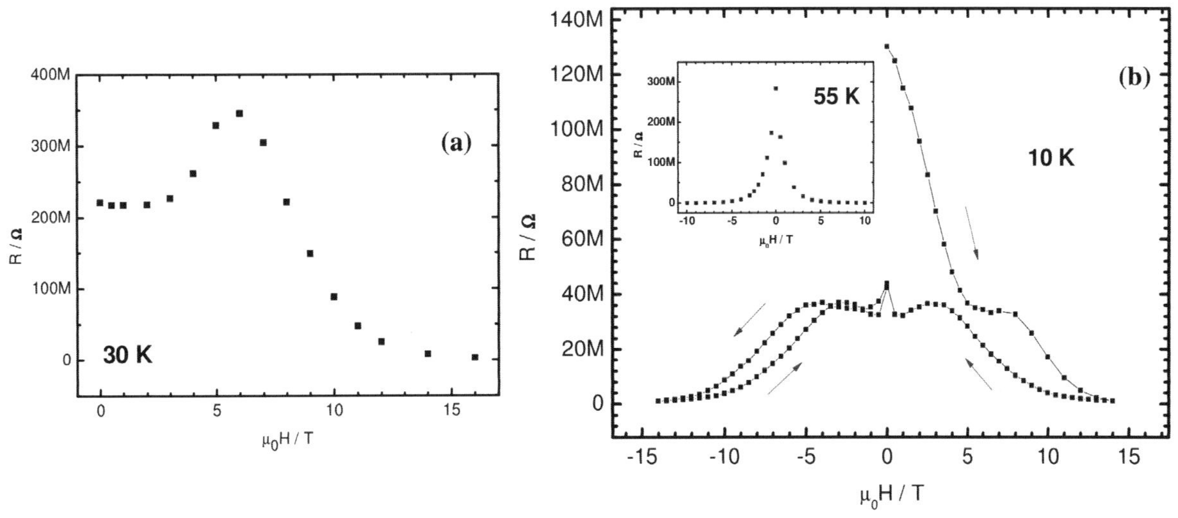

Figure 5. **(a)** Magnetic field dependence of the resistance for the 10/10 multilayer at T = 30 K, showing a maximum at a 6 T field. **(b)** Magnetic field dependence of the resistance for the 20/20 multilayer at T = 10 K, evidencing hysteretic behaviour; the arrows indicate the field sweep sequence (0 T → 14 T → -14 T →0 T). Inset: at T = 55 K, no hysteresis is observed.

20/20: as the field increases, the height of the peak decreases as expected, but the peak temperature (T_p) decreases. This unusual magnetic field dependence of T_p has never been observed, to our knowledge, in manganite systems and cannot be understood in the framework of a simple double exchange mechanism. For the multilayer with the thicker layers (50/50), the resistance peak temperature exhibits the usual behaviour, shifting to higher temperatures with increasing field (cf. figure 4d).

In order to clarify these observations, the resistance of samples 10/10 and 20/20 was measured as a function of the magnetic field at several fixed temperatures. Figure 5a shows that, for sample 10/10, at a constant temperature of 30 K the resistance at first increases with applied magnetic field, up to a field of about 6 T and then decreases consistently up to the maximum scanned field of 16 T. Figure 5b shows that for sample 20/20, at T = 10 K and after the initial curve, a hysteretic resistance develops with a resistance maximum between 3 T and 6 T. For a temperature of 55 K, near the zero field resistance peak, the hysteresis is no longer observable and the resistance decreases monotonically with applied field (cf. inset to figure 5). The main features exhibited by the R(H) curves in our samples are similar to those found in Gd/Co multilayers, which were interpreted in the framework of a field induced spin-flop transition [11], as a consequence of the existence of an antiferromagnetic coupling in the interface region between alternating Co and $Gd_{0.7}Co_{0.3}$ layers. Furthermore, strong hysteretic effects on the magnetoresistance were observed in systems where a canted antiferromagnetic state is transformed into a metastable ferromagnetic state by application of large magnetic fields [8].

In order to get further insight on the samples' properties, we performed magnetization measurements of the multilayers. The magnetization curves as a function of temperature displayed in figure 6 for the samples 10/10 and 20/20 are reminiscent of those for systems with some type of magnetic frustration such as spin glasses or cluster glasses. Figure 7a shows the magnetization hysteresis loops for sample 10/10 at 5 K, obtained after field cooling in a 10 kOe field: the loop presents an evident shift towards the negative values of applied field. This shift is the so-called exchange bias effect and can be evaluated through the exchange bias field $H_{EB} = -(H_1+H_2)/2$, H_1 and H_2 being, respectively, the negative and positive coercive fields (cf. figure 7a). This effect has also been observed in sample

20/20, whereas it is negligible for sample 50/50. Figure 7b shows the temperature dependence of the

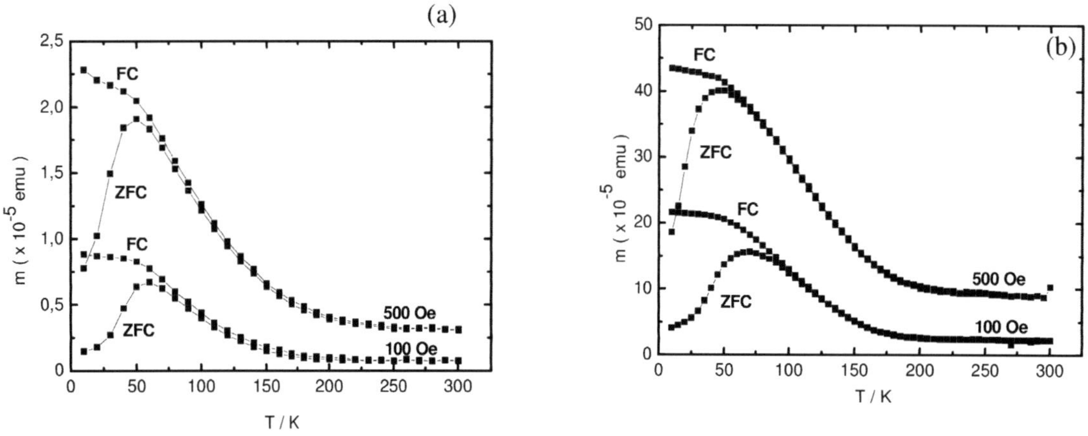

Figure 6. Temperature dependence of the magnetization for samples: **(a)** 10/10 and **(b)** 20/20.

exchange field H_{EB} for the three samples (all data obtained after field cooling in a 10kOe field). Two features are evident from the plot: the exchange bias increases in magnitude and persists to higher temperatures with decreasing layer thickness. These features point to the existence of a strong interfacial effect that becomes less significant as the volume of the ferromagnetic layers increases. This result correlates with the unusual resistance behaviour under applied magnetic field observed for the 10/10 and 20/20 samples. The exchange bias effect is usually attributed to the existence of adjacent ferromagnetic and antiferromagnetic layers, with $T_N < T_C$: if the ferromagnetic and the antiferromagnetic layers are cooled in an applied field, the exchange coupling between the layers makes it energetically favourable for the ferromagnetic layer to maintain the magnetization direction acquired while cooling; for temperatures well below the Néel temperature (T_N) a shift in the hysteresis loop will occur. The exchange bias effect has also been observed in samples with a spin glass phase surrounding a ferromagnetic region [12,13], when the freezing temperature is $T_g < T_c$. In order to investigate the spin glass hypothesis we measured the time dependence of the resistance under zero applied field for sample 20/20, at 10 K and 5 K after zero field cooling from room temperature for both measurements. No relaxation of the resistance is observable for a time scan of 50000 s; therefore, in the absence of any other direct evidence, we must rule out the spin glass scenario.

In the multilayers considered in the present work the relative importance of the exchange bias effect on the overall sample properties increases as the thickness of the LCMO layers decreases. As suggested by the X-ray analysis, the interfaces in the multilayers may present considerable roughness, and thus, we expect an important degree of mixing between the two materials. This could result in the formation of Ca rich phases of the $La_{1-x}Ca_xMnO_3$ compound ($x > 0.5$) that are antiferromagnetic [8]. The possibility of formation of antiferromagnetic Ruddlesden-Popper phases may also be considered; several authors have shown that RP phases can appear at the interface between different perovskite structures as way of accommodating lattice mismatches through the formation of planar defects [14]. These considerations are schematically illustrated on figure 8.

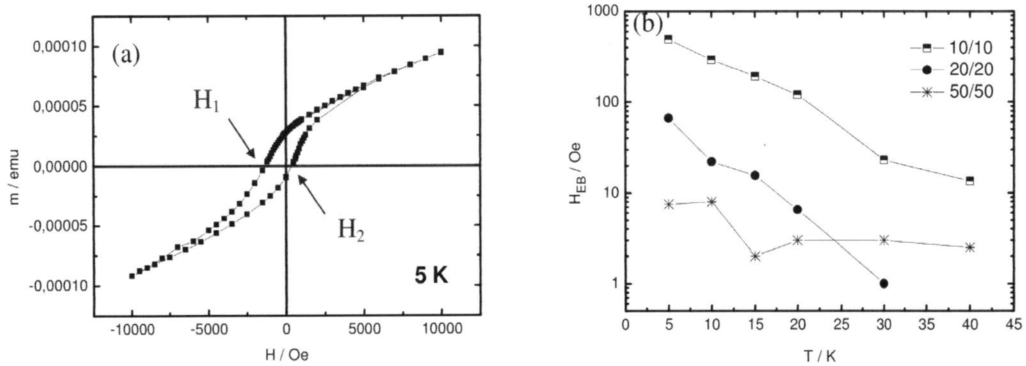

Figure 7. (a) Magnetization hysteresis loop for sample 10/10 at 5 K, obtained after field cooling in a 10 kOe field: a shift towards the negative values of the field is evident, signal of the existence of an exchange bias effect; the arrows indicate the coercive fields, H_1 and H_2. **(b)** Temperature dependence of the exchange bias field $H_{EB} = -(H_1+H_2)/2$, for samples 10/10, 20/20 and 50/50.

Addressing now the field dependence of the resistance displayed in figure 5, we propose a preliminary interpretation: at a fixed temperature, by applying a magnetic field the ferromagnetic metallic state associated with the LCMO phases is promoted. However, on the adjacent antiferromagnetic phases, whose existence is inferred from the exchange bias effect, the antiferromagnetic order starts to be disrupted as the field increases and so, the magnetic disorder increases in the percolation path of the conducting electrons, resulting in an electrical resistance enhancement. At a large enough field, the canted antiferromagnetic state is transformed into a metastable ferromagnetic state, with the magnetic moments aligned with the applied field; with the application of even larger fields, more magnetic moments align with the applied field and the resistance of the sample decreases. This explanation is in agreement with the magnetic field dependence of the resistance for sample 10/10 shown in figure 5a, where a change in behaviour is observed for a 6 T field, which we ascribe to a spin-flop transition as the one observed in the Co/Gd multilayers [11]. A transition from an antiferromagnetic to a ferrimagnetic state and then to a ferromagnetic state has also been theoretically proposed by Rößler and Bogdanov [15], in antiferromagnetic multilayers with uniaxial anisotropy. The possible existence of Ruddlesden-Popper (RP) phases in our multilayers could provide the strong anisotropy required to observe a transition similar to the one predicted. At this moment, it is not clear if such a mechanism can account for the odd temperature dependence of the resistance peak in samples 10/10 and 20/20, but it is clear that such behaviour is correlated to the predominance of interfacial effects associated with the exchange bias mechanism. Further experimental and theoretical work is necessary in these and similar systems in order to clarify these issues.

4. Conclusions

We have grown a $La_{0.7}Ca_{0.3}MnO_3$ film, a $La_{0.7}Ca_{0.3}MnO_3/CaO/La_{0.7}Ca_{0.3}MnO_3$ trilayer and $[(La_{0.7}Ca_{0.3}MnO_3)/(CaO)]_{\times N}$ multilayers by pulsed injection metal organic chemical vapour deposition on $LaAlO_3$ substrates. The film presents magnetic and transport properties in agreement with those found in the literature. The trilayer exhibits a Curie temperature and a metal to insulator transition temperature lower than the single layer, a behaviour ascribed both to strain effects and to possible Ca diffusion into the ferromagnetic layers. The multilayers show very large electrical resistance values and metal to insulator transition temperatures close to 60 K; under applied magnetic field the resistance presents an anomalous behaviour for the multilayers with the thinner layers, suggesting interfacial effects and not readily understood in the framework of the double exchange mechanism.

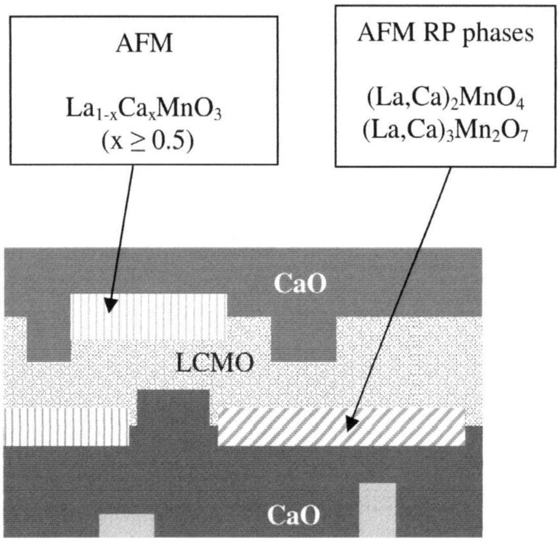

Figure 8. Schematic view of the interface structure of the LCMO/CaO multilayers, showing the possible antiferromagnetic phases.

The multilayers also exhibit an exchange bias effect for the magnetization that increases in magnitude and persists to higher temperatures with decreasing layer thickness, again suggesting a link to interfacial features. We interpret these results as a consequence of the interfaces' roughness, where antiferromagnetic Ca rich or Ruddlesden-Popper phases may have formed. The coexistence of ferromagnetic and antiferromagnetic phases can account for the observed large resistance values resulting from the increased spin dependent scattering. We also propose that the existence of a field induced spin-flop transition from a canted antiferromagnetic state into a metastable ferromagnetic state may account for the isothermal field dependence of the resistance.

Acknowledgments
Authors wish to acknowledge the portuguese FCT for financial support under contract no. PPCDT/FIS/60153/2004. Authors are also indebted to A P Gonçalves for valuable help with the X-ray measurements.

References

[1] von Helmolt R, Wecker J, Holzapfel B, Schultz L and Samul S 1993 *Phys. Rev. Lett.* **71** 2331
[2] Jin S, Tiefel T H, McCormack M, Fastnacht R A, Ramesh R and Chen L 1994 *Science* **264** 413
[3] Lu Y, Klein J, Herbstritt F, Philipp J B, Marx A and Gross R 2006 *Phys. Rev. B* **73** 184406
[4] Zhang H D, Li M, An Y K, Gao J, Hu F X, Wang Y and Jia Q J 2007 *J. Phys. D: Appl. Phys.* **40** 2723
[5] Kwon C, Kim K C, Robson M C, Gu J Y, Rajeswari M, Venkatesan T and Ramesh R 1997 *J. Appl. Phys.* **81** 4950
[6] Gong G Q, Guptha A, Xiao G, Lecoeur P and McGuire T R 1996 *Phys. Rev.* B **54** 3742
[7] Steren L B, Sirena M and Guimpel J 2000 *J. Magn. Magn. Mater.* **211** 28
[8] Coey J M D, Viret M and von Molnár S 1999 *Advances in Physics* **48** 167
[9] Venimadhav A, Hegde M S, Prasad V and Subramanyam S V 2000 *J. Phys. D: Appl. Phys.* **33** 2921
[10] Coey J M D, Viret M, Ranno L and Ounadjela K 1995 *Phys. Rev. Lett.* **75** 3910

[11] Colino J, Andrés J P, Riveiro J M, Martínez J L, Prieto C and Sacedón J L 1999 *Phys. Rev. B* **60** 6678

[12] Kodama R H, Berkowitz A E, McNiff E J, Jr and Foner S 1996 *Phys. Rev. Lett.* **77** 394

[13] Zhu M H, Zhao Y G, Cai W, Wu X S, Gao S N, Wang K, Luo L B, Huang H S and Lu L 2007 *Phys. Rev. B* **75** 134424

[14] Wu J S, Jia C L, Urban K, Hao J H and Xi X X 2002 *J. Cryst. Growth* **234** 603; Jia C L, Rodriguez Contreras J, Schubert J, Lentzen M, Poppe U, Kohlstedt H, Urban K and Waser R 2003 *J. Cryst. Growth* **247** 381; Qin Y L, Jia C L, Urban K, Hao J H and Xi X X 2003 *J. Cryst. Growth* **252** 279

[15] Rößler U K and Bogdanov A N arXiv:cond-mat/0605493v1

Magnetic and magnetocaloric properties of monovalent substituted $La_{0.65}Ca_{0.3}M_{0.05}MnO_3$ (M=Na, Ag, K) perovskite manganites

W. Cheikh-Rouhou Koubaa[a], **M. Koubaa**[a], **A. Cheikhrouhou**[a,b]

[(a)]Laboratoire de Physique des Matériaux, Faculté des Sciences de Sfax, B. P. 1171, 3000 Sfax Tunisia, [(b)]Institut NEEL, CNRS, B.P.166, 38042 Grenoble cedex9, France

abdcheikhrouhou@yahoo.fr

Abstract. The effects of monovalent doping in the A-site on the structural, magnetic and magnetocaloric properties in $La_{0.65}Ca_{0.3}M_{0.05}MnO_3$ (M=Na, Ag, K) powder perovskite manganites have been investigated. Our samples have been synthesized using the conventional solid state reaction at high temperature. Rietveld refinements of the X-ray diffraction patterns at room temperature show that all our powder samples are single phase and crystallize in the orthorhombic structure with Pbnm space group. Magnetization measurements versus temperature in a magnetic applied field of 50mT indicate that our investigated samples display a paramagnetic-ferromagnetic transition with decreasing temperature. The Curie temperature T_C is found to be 295K, 254K and 285K for M=Na, Ag and K respectively. From the measured magnetization data of $La_{0.65}Ca_{0.3}M_{0.05}MnO_3$ (M=Na, Ag, K) samples as a function of magnetic applied field, the associated magnetic entropy change close to their respective Curie temperature T_C and the relative cooling power RCP have been determined. The maximum magnetic entropy change, $\left|\Delta S_m^{max}\right|$, in a magnetic field change of 5T is found to be 3, 5.16 and 3.54 Jkg/K for M=Na, Ag and K respectively. The $La_{0.65}Ca_{0.3}K_{0.05}MnO_3$ sample exhibits the largest RCP value of 295J/kg.

1. Introduction

The magnetoresistive oxides $RE_{1-x}M_xMnO_3$ (RE=La, Pr, … and M=Ca, Sr, Ba, …) provide a rich variety of structural, magnetic and transport properties depending on the doping concentration (Mn^{3+}/Mn^{4+} ratio), the average size of A cation-site $\langle r_A \rangle$ which controls the effective electron bandwidth W and the mismatch effect [1,2]. Recently, novel properties have been observed in perovskite-type ferromagnetic manganese oxides related to magnetocaloric effect (MCE) [3,4]. The MCE originates from the heating or the cooling of magnetic material due to the application of magnetic field. Increasing the strength of the applied magnetic field in a ferromagnetic material induces the alignment of the magnetic moments with the direction of the magnetic field which leads to a decrease of the spin entropy. This process is accompanied by a rise of the lattice entropy when the field is applied adiabatically. However, if we remove the magnetic applied field, the spin system tends to randomize which increases the spin entropy, reduces the lattice one and consequently lowers the temperature of the system. Pecharsky and Gschneidner [5-6] discovered a giant MCE in the pseudo-binary alloy $Gd_5(Si_xGe_{1-x})_4$ in the temperature range 50-280K. More recently, giant MCE at about 300K has been measured in $MnFeP(O_{0.45}As_{0.55})$ [7]. An overview of the magnetocaloric properties in perovskite manganese oxides have been provided by Phan and Yu [8]. In the present work, we synthesized $La_{0.65}Ca_{0.3}M_{0.05}MnO_3$ (M=Na, Ag and K) powder perovskite manganites by the conventional solid state reaction at high temperature and studied the crystallographic, magnetic and magnetocaloric effects due to the substitution of 5% of calcium by monovalent element Na, K and Ag.

2. Experimental techniques

Polycrystalline samples of $La_{0.65}Ca_{0.3}M_{0.05}MnO_3$ (M=Na, K and Ag) were synthesized using the solid state reaction method at high temperature by mixing La_2O_3, $CaCO_3$, MnO_2 and A_2CO_3 (A=Na, Ag and K) up to 99.9% high purity in the desired proportion. The starting materials were intimately mixed in an agate mortar and then heated in air up to 1000°C for 60h. The obtained powders were then pressed into pellets (of about 1mm thickness) and sintered at 1100°C in air for 60h with intermediate regrinding and repelling. Finally, these pellets were rapidly quenched to room temperature in air in order to freeze the structure at the annealed temperature. Phase purity, homogeneity and cell dimensions were determined by powder X-ray diffraction at room temperature. Structural analysis was carried out using the standard Rielveld method [9]. Magnetization measurements versus temperature in the range 20–350K and versus magnetic applied field up to 7T were carried out using a vibrating sample magnetometer. MCE were deduced from the magnetization measurements versus magnetic applied field up to 7T at several temperatures.

3. Results and discussions

X-Ray powder diffraction (XRD) patterns at room temperature reveal that all our synthesized samples crystallize in the orthorhombic structure with Pbnm space group. For all samples, the structural data are summarized in Table 1.

Table1. Refined structural parameters of $La_{0.65}Ca_{0.3}M_{0.05}MnO_3$ samples at room temperature

	$La_{0.65}Ca_{0.35}MnO_3$	M = Na	M = Ag	M = K
Space group	Pbnm	Pbnm	Pbnm	Pbnm
a (Å)	5.469(8)	5.449(1)	5,455(8)	5.447(9)
b (Å)	5.432(5)	5.483(3)	5,474(1)	5.462(7)
c (Å)	7.740(6)	7.688(4)	7,703(2)	7.739(4)
V (Å^3)	230.01	229.72	230,06	230.32
T_C (K)	248	295	254	285
$<r_A>$ (Å)	1.146	1.149	1.154	1.165
σ^2	0.36x10^{-3}	0.37x10^{-3}	1.16x10^{-3}	6.57x10^{-3}

In this table the average ionic radius $<r_A>$ and the mismatch size σ^2 in the A-site are also listed. The unit cell volume of $La_{0.65}Ca_{0.35}MnO_3$ (LCMO) sample is found to be 230.01Å^3, which is close to that reported by Zhang et al. [10]. The average crystallite size can be estimated using Scherer's formula: $\langle S\rangle = \frac{K\lambda}{\beta_{1/2}\cos\theta}$, where λ, $\beta_{1/2}$ and K are the X-ray wave length, the full width at half maximum and the Scherer crystal shape constant respectively [11]. The average crystallite size, ranging between 21nm and 35nm, influences the magnetic and magnetocaloric properties as will be discussed below.

Magnetization measurements versus temperature in the range 20-350K in a magnetic applied field of 50mT showed that all our substituted samples exhibit a sharp transition from paramagnetic to ferromagnetic state with decreasing temperature (Fig. 1a). The Curie temperature T_C, determined from the position of the inflexion point in the M (T) curve, is found to be 248K for the parent compound LCMO and 295K, 254K and 285K for M=Na, Ag and K respectively. In order to confirm the ferromagnetic behavior of our samples at low temperatures, we performed magnetization measurements versus magnetic applied field up to 7T at several temperatures. A typical evolution of the magnetization versus magnetic applied field for LCMO sample is plotted in Fig. 1b. Below T_C, the magnetization M increases sharply with magnetic applied field for H<0.5T and then saturates above 1T. The saturation magnetization shifts to higher values with decreasing temperature. We plot in Fig. 1c the Arott curves $M^2(\mu_0H/M)$ for LCMO sample. As can be seen, Arrott plots above T_C show a linear behavior which indicates that a second order magnetic transition occurs. The magnetic entropy change, $|\Delta S_M|$, has been deduced from isothermal magnetization measurements. Based on Maxwell's relations, $|\Delta S_M|$, can be evaluated using the following equation $|\Delta S_M| = \sum_i \frac{M_i - M_{i+1}}{T_{i+1} - T_i}\Delta H_i$ where M_i and M_{i+1} are the experimental values of magnetization measured at temperatures T_i and T_{i+1} respectively, under magnetic applied field H_i [12]. Fig. 1d shows the temperature dependence of the magnetic entropy change, $|\Delta S_M|$, of LCMO sample.

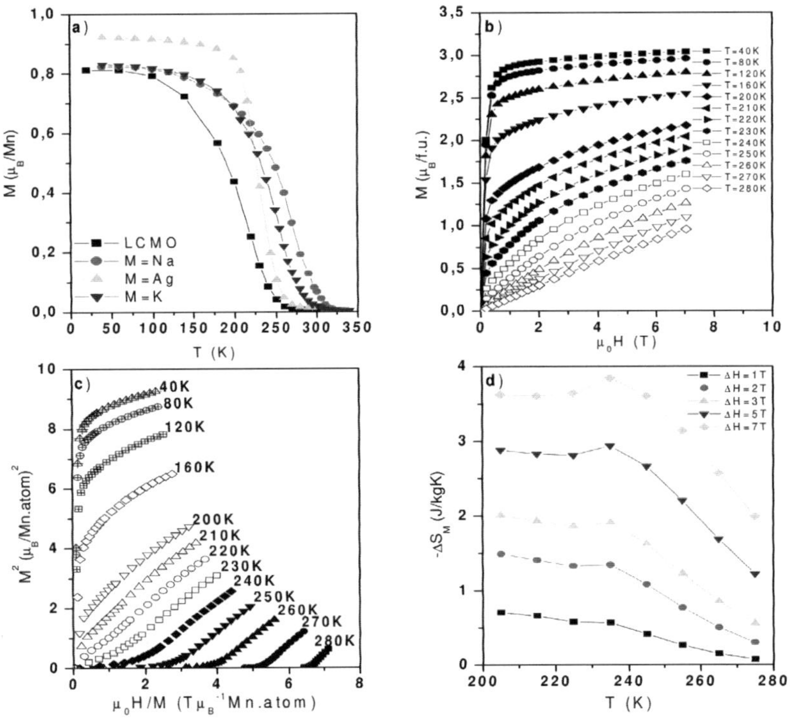

Fig. 1. a) M(T) at 50mT for $La_{0.65}Ca_{0.3}M_{0.05}MnO_3$ samples, b) M(H) at several temperatures, c) Arott curves $M^2(\mu_0H/M)$ and d) $|\Delta S_M(T)|$ at several magnetic applied field for LCMO sample

We can observe that with decreasing temperature, the magnitude of $|\Delta S_M|$ increases progressively up to 235K and then remains roughly constant at low temperatures. The same behaviour has been observed by Pekala et al. [13] on $La_{0.7}Ca_{0.3}MnO_3$ sample prepared by sol-gel method and has been interpreted in terms of nano-crystalline grain size. Figures 2a-c show the temperature dependence of the magnetic entropy change under several magnetic applied fields for $La_{0.65}Ca_{0.3}M_{0.05}MnO_3$ samples

(M=Na, Ag, K). All our samples exhibit large MCE effect around the magnetic transition temperature and the magnitude of $|\Delta S_M|$ increases with increasing the magnetic applied field. The maximum magnetic entropy change observed, $|\Delta S_M^{Max}|$, is found to be 3 $Jkg^{-1}K^{-1}$, 5.16$Jkg^{-1}K^{-1}$ and 3.54$Jkg^{-1}K^{-1}$ in a magnetic field change of 5T for M=Na, Ag and K respectively. The relative cooling power (RCP) is evaluated as $RCP = -\Delta S_M(T,H) \times \delta T_{FWHM}$ where δT_{FWHM} is the full-width at half-maximum of $|\Delta S_M|$ versus temperature [14]. For our samples, the RCP values are respectively 250.3J/kg, 279J/kg and 295J/kg at 5T for M=Na, Ag and K. The magnitude of the RCP for $La_{0.65}Ca_{0.3}K_{0.05}MnO_3$ specimen is about 72% of that of pure Gd. The higher RCP value is observed in $La_{0.845}Sr_{0.155}MnO_3$ [15] which reaches 670J/kg under 7T around 235K making difficult its application in magnetic refrigeration near room temperature. Moreover $La_{0.7}Ca_{0.25}Sr_{0.05}MnO_3$ [15] and $La_{0.7}Ca_{0.2}Sr_{0.1}MnO_3$ [16] compounds are characterized by an RCP value of 462 J/kg (5T) and 374 J/kg (5T) respectively around room temperature which are comparable to our results. According to Landau Theory, $|\Delta S_M|$ can be evaluated through the following formula: $\Delta S_M = -\frac{1}{2}\frac{\partial A}{\partial T}M^2 - \frac{1}{4}\frac{\partial B}{\partial T}M^4$ where A and B coefficients are deduced from the linear fit of Arrott plots. Amaral et al. [17] have reported that the calculated value agrees well with the experimental data for $La_{0.7}Sr_{0.3}MnO_3$ compound, which is known to be a typical system governed by double exchange interaction and is the less affected by other instabilities. We plot in Fig.2d the experimental and calculated curves of $|\Delta S_M|$ as a function of temperature in a magnetic field change of 5T for $La_{0.65}Ca_{0.3}K_{0.05}MnO_3$ sample.

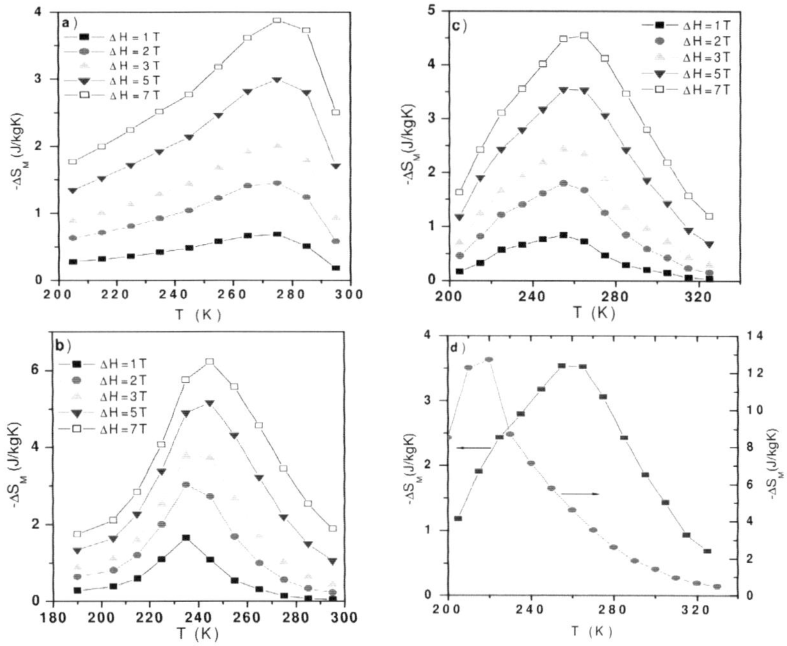

Fig. 2 $|\Delta S_M(T)|$ at several magnetic applied field for a) M=Na, b) M=Ag and c) M=K. d) Experimental and calculated $|\Delta S_M|$ for $La_{0.65}Ca_{0.3}K_{0.05}MnO_3$ sample under 5T.

A shift of 30K to lower values is observed in the calculated curve while the magnitude is four time lager compared to the experimental one. This discrepancy has been also observed by Yang et al. [18]

where the authors compared the theoretical and calculated magnetocaloric effect of electron-doped manganites. This behavior has been explained in terms of Jahn-Teller distortion.

4. Conclusions

We investigated the effect of monovalent doping on the physical properties of $La_{0.65}Ca_{0.3}M_{0.05}MnO_3$ (M=Na, Ag, K) powder samples. The structural study shows that all our synthesized samples crystallize in the orthorhombic structure with Pbnm space group. A large magnetocaloric effect around room temperature is observed in our samples. $La_{0.65}Ca_{0.3}K_{0.05}MnO_3$ exhibits the highest RCP value of 295J/kg in a magnetic applied field change of 5T.

5. Acknowledgments

This study has been supported by the Tunisian Ministry of Higher Education, Scientific Research and Technology.

References

[1] Radaelli P G, Marezio M, Hwang H Y and Cheong S W 1996 *J. Solid State Chem.* **122** 444
[2] Rodriguez-Martinez L M and Attfield J P 1996 *Phys. Rev.* B **54** R15622
[3] Guo Z B, Du Y W and Feng D 1997 *Phys. Rev. Lett.* **78** 1142
[4] Guo Z B and Du Y W *Appl. Phys. Lett.* **70** 904
[5] Pecharsky V K and Gschneidner Jr. K A 1997 *J. Magn. Magn. Mater.* **167** L179-L184
[6] Pecharsky V K and Gschneidner Jr. K A 1997 *Phys. Rev. Lett.* **78** 4494-4497
[7] Tegus G, Brück E, Buschow K H J and de Boer F R 2002 *Nature* **415** 150-152
[8] Phan M H and Yu S C 2007 *J. Magn. Magn. Mater.* **308** 325
[9] Wiles B and Young R A 1981 *J. Appl. Crystallogr.* **14** 149
[10] Zhang J, Li N, Feng M, Pan B C and Li H 2008 *J. Alloys Comp.* (in press)
[11] Taylor A 1961 *X-ray Metallography* (New York- Wiely)
[12] McMichael R D, Ritter J J and Shull R D 1993 *J. Appl. Phys.* **73** 6946
[13] Pekala M, Drozd V, Fagnard J F, Ph. Vanderbemden and M. Ausloos (in press)
[14] Gschneidner K A and Pecharsky V K 2000 *Annu. Rev. Mater.* Sci. **30** 387
[15] Szewczyk A, Gutowska M, Dabrowski B, Plackowski T, Danilova N P and Gaidukov Y P 2005 *Phys. Rev.* B **71** 224432
[16] Phan M H, Yu S C and Hur N H 2005 *Appl. Phys. Lett.* **86** 072504
[17] Amaral J S, Reis M S, Amaral V S, Mendonça T M, Araùjo J P, Sà M A, Tavares P B and Vieira J M *J. Magn. Magn. Mater.* **290-291** 686-689
[18] Yang J, Lee Y P and Li Y *J. App. Phys.* **102** 033913

Development of Halbach magnet for portable NMR device

N. Doğan[1,*], R. Topkaya[1], H. Subaşı[1], Y.Yerli[1], B.Rameev[1,2]

[1] *Gebze Institute of Technology, 41400 Gebze/Kocaeli, Turkey*
[2] *Kazan Physical-Technical Institute, 420029 Kazan, Russian Federation*

E-mail: ndogan@gyte.edu.tr

Abstract: Nuclear magnetic resonance (NMR) has enormous potential for various applications in industry as the on-line or at-line test/control device of process environments. Advantage of NMR is its non-destructive nature, because it does not require the measurement probe to have a contact with the tested media. Despite of the recent progress in this direction, application of NMR in industry is still very limited. This is related to the technical and analytical complications of NMR as a method, and high cost of NMR analyzers available at the market. However in many applications, NMR is a very useful technique to test various products and to monitor quantitatively industrial processes. Fortunately usually there is no need in a high-field superconducting magnets to obtain the high-resolution spectra with the detailed information on chemical shifts and coupling-constant. NMR analyzers are designed to obtain the relaxation parameters by measuring the NMR spectra in the time domain rather than in frequency domain. Therefore it is possible to use small magnetic field (and low frequency of 2-60 MHz) in NMR systems, based on permanent magnet technology, which are specially designed for specific at-line and on-line process applications. In this work we present the permanent magnet system developed to use in the portative NMR devices. We discuss the experimental parameters of the designed Halbach magnet system and compare them with results of theoretical modelling.

Keyword: NMR analyzer, Halbach magnet

1. Introduction

Nuclear magnetic resonance (NMR) is a general technique which is widely used in various areas of basic research, in industry for testing broad range of products and to control manufacturing processes [1], in diagnostic medical imaging [2] and analytical chemical spectroscopy [3]. Usually, for good spectroscopic resolution a superconducting magnet is used to produce the high magnetic fields (9-20 T) with homogeneity of the sample volume as high as possible ($\Delta B_0/B_0 < 10^{-7}$) [4]. Inhomogeneous magnetic fields had been avoided in the previous 50 years because of the understanding that chemical-shifts cannot be resolved in a polarization field B_0 that is inhomogeneous across the sample (see Ref. [5]). Today, however, the ideas of multi-dimensional NMR [3,6] are extensively explored with mobile sensors and *inhomogeneous fields* (see, for instance, Ref. [7]). In multidimensional NMR techniques the correlations of the parameters other than the resonance frequencies or space coordinates in images are utilized to increase resolution and to make the NMR data more informative. Now it is commonly accepted that NMR in inhomogeneous fields inside and outside the magnet is very useful instrument of characterization and research. Especially the mobile

* Corresponding author. Address: Gebze Institute of Technology, Department of Physics, 41400 Gebze, Kocaeli, Turkey. Tel.: +90(262)605 1341, Fax:+90(262)653 8490. E-mail: ndogan@gyte.edu.tr.

NMR devices equipped with light-weight magnets with homogeneous fields and simple field profiles have very high potential to use in various applications. To realize these devices the Halbach magnet design is considered as the most prospective magnet system (see Ref. [7-9]). This design is a quasi-cylinder magnet constructed from many magnet blocks in such a way that the magnetic field is largely homogeneous inside the cylinder, transverse to the cylinder axis, and zero outside [10]. Nearly homogeneous fields for applications in the one-sided NMR are conventionally generated by the classical C-shaped magnet geometry[11].

In last years, low-field NMR devices (NMR analyzers) have found very wide applications in different industrial and agricultural fields, such as food products, medicine and petrochemical industry. Widespread application fields of low-field NMR include determining moisture, oil, and hydrogen and flour amount. For instance, manufacturers of eatable oils prefer to buy raw products according to the oil content in a substance. However, most of the NMR analyzers commercially available today are not portable. There are table-top versions to be used at the laboratory conditions (at-line NMR devices). Besides, some NMR devices are specially designed heavy (not portable) constructions to use in the specific on-line processes (on-line devices). On the other hand, some prototypes of the portable devices have already been developed but not found yet a wide application area. The main problem in developing portable NMR devices is to produce light-weighting magnet system that could provide a magnetic field with desirable homogeneity rather than to create a high magnetic field. Therefore, the compact and light magnet designs for homogeneous magnetic field, such as Halbach design, have attracted a special attention of researchers [7-16].

Ideal Halbach magnet is formed from the bar magnets with equal magnetization that are oriented and positioned with analytic equations [10]. This design has the following advantages which are important for NMR applications [10]:

- High homogeneity of the static magnetic field
- Optimal use of the material magnetization
- Transverse field direction (allows using the solenoidal coils for NMR)
- Very small stray fields
- Simple, compact and robust construction consisting of identical magnets
- Easy and cheap production process

Several researchers have recently applied Halbach magnets to build NMR magnets for using in a tabletop NMR [17], a mobile NMR for measuring drilled rock [12, 18] and pre-polarized samples in the earth-field NMR experiments [19]. A Halbach magnet generates a highly homogeneous static magnetic flux density near its center and a moderately homogeneous static magnetic flux density near its top; it can also be built to be very compact. It has been demonstrated that the field homogeneity of the Halbach magnet is high enough to be used in various versions of portable devices (see Refs. [10-14, 20]).

The above mentioned motivated us to construct and to study the magnetic properties of most simple version of Halbach magnet design with aim to check its potential for application in the mobile NMR devices.

2. Theory of Halbach magnet construction

Theory predicts that the circular structure with magnetization direction rotating as shown in Fig. 1 will produce homogeneous magnetic field. However, it is hardly possible to realize practically such rotation through a continuous media, while it is easy to assemble a discrete structure having individually polarized parts. Therefore, the real Halbach design consists of the permanent magnets with equal magnetization that are oriented and positioned with analytic equations [7,18] (see below).

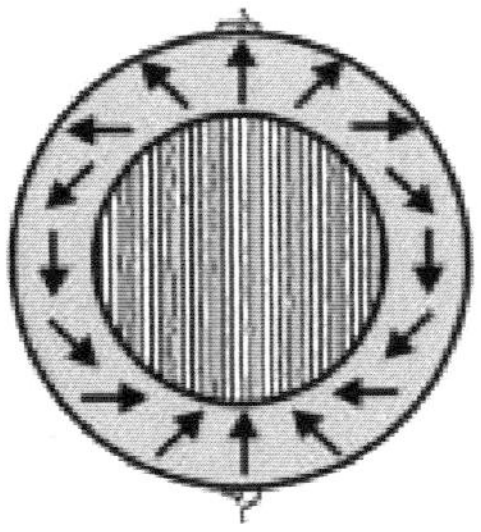

Figure 1. Ideal Halbach magnet, producing strong, homogeneous field inside and low stray field outside.

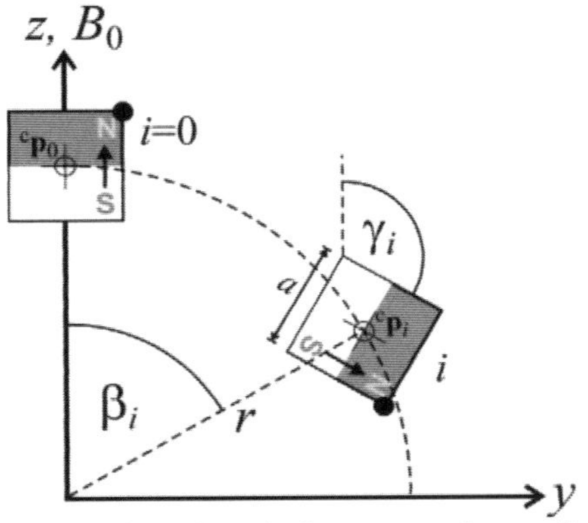

Figure 2. Infinitely long array for perfect magnetic homogeneity [10].

For the calculation of the exact geometry of the magnet arrangement, two parameters are predetermined: the radius of the ring, r, and the number of magnets, n. Each magnet is positioned such, that its center (vector ${}^{c}p_i$) has the same distance r from the origin. The coordinates of each magnet center are then represented with matrix representation:

$$ {}^{c}p_i = \begin{pmatrix} {}^{c}y_i \\ {}^{c}z_i \end{pmatrix} = r \begin{pmatrix} \sin\beta_i \\ \cos\beta_i \end{pmatrix} \qquad \text{where} \quad \beta_i = i\alpha \quad \text{with } i = 0,1,\ldots,n-1. $$

Here, β_i is angle between the magnet centre and direction of the magnetic field B_0. Orientation of each magnet relative to the z axis is determined with $\gamma_i = 2\beta_i$ angle. Once the n magnets are spatially arranged (calculating $n/8$ magnets or one octant is sufficient due to symmetry), their size, a, is usually scaled such that the most dense arrangement results. This gives the following coordinates of each corner of the ith bar-magnet:

$$^{1}p_i = \begin{pmatrix} ^{1}y_i \\ ^{1}z_i \end{pmatrix} = {}^{c}p_i + \frac{a}{\sqrt{2}}\begin{pmatrix} \cos\xi_i \\ \sin\xi_i \end{pmatrix}$$

$$^{2}p_i = \begin{pmatrix} ^{2}y_i \\ ^{2}z_i \end{pmatrix} = {}^{c}p_i + \frac{a}{\sqrt{2}}\begin{pmatrix} -\sin\xi_i \\ \cos\xi_i \end{pmatrix}$$

$$^{3}p_i = \begin{pmatrix} ^{3}y_i \\ ^{3}z_i \end{pmatrix} = {}^{c}p_i + \frac{a}{\sqrt{2}}\begin{pmatrix} -\cos\xi_i \\ -\sin\xi_i \end{pmatrix}$$

$$^{4}p_i = \begin{pmatrix} ^{4}y_i \\ ^{4}z_i \end{pmatrix} = {}^{c}p_i + \frac{a}{\sqrt{2}}\begin{pmatrix} \sin\xi_i \\ -\cos\xi_i \end{pmatrix}$$

$$\text{with } \xi_i = \frac{\pi}{4} - 2\beta_i$$

These matrix representations give coordinates of corners of each magnet.

Here each corner is represented by a vector $^{j}p_i$, where the index i identifies the magnet and j denotes the corner, numbered in the sense of quadrants as shown in Fig. 3.

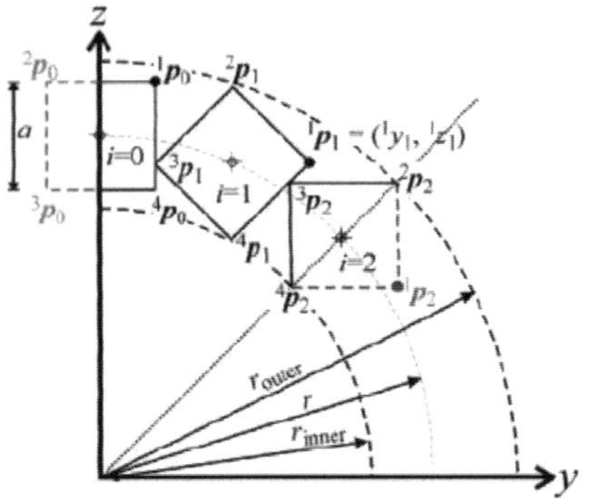

Figure 3. Symbolic representation of magnet coordinates [9].

The length of a magnet's side, a, is found by solving

$$^{4}p_{(n/8)-1} + \lambda\,(^{1}p_{(n/8)-1} - {}^{4}p_{(n/8)-1}) = {}^{3}p_{n/8}\ ,$$

which gives $a = 2r\Xi(\alpha)$ with:

Hence, the area, A, occupied by the n magnets is

$$A = na^2 = 4nr^2\Xi(\alpha)^2 .$$

The inner radius (accessible for experiments) is then given by

$$r_{inner} = r(1-\sqrt{2}\Xi(\alpha)),$$

and the outer radius is

$$r_{outer} = r(1+\sqrt{2}\Xi(\alpha)) .$$

Homogeneity of the magnetic field depends on the number bar magnets used in the Halbach magnet design. Each magnet should have the same magnetization and sizes to produce homogeneous magnetic field for NMR devices. A best compromise between homogeneity and field strength is obtained at high magnet numbers. However, n=4 will also results in a good homogeneity but not maximal magnetic field. To hold the magnets in the correct orientation and fixed position, the supporting structure withstanding mutual magnetic forces of these magnets is needed (see Ref. [10,18]).

3. Experimental results and their discussion.

In this work we have constructed the simplest Halbach magnet design having 4 permanent magnets with square cross section (Fig. 4). It uses the minimal number of permanent magnets providing at the same time enough high homogeneity of the magnetic field of the target volume. Magnetic field depends on the distance between the sample and the magnets as well as on the magnetization of the magnets.

At room temperature NdFeB (Neodymium-Iron-Boron) type permanent magnets have very suitable magnetic properties comparing with other permanent magnet compounds. Therefore, we use grade N45 of NdFeB magnets and the dimensions of these permanent blocks were 50×50×50 mm^3. Magnetic properties of the magnets were as follows: remanence B_r=1320-1380 mT, coercivity H_c=923 kA/m and maximum energy product $(BH)_{max}$=43-46 MGOe.

Position of the magnets is very important to obtain good homogeneity at the centre of the structure. Four magnet blocks were positioned in aluminium frame with diameter of 18 cm and depth of 5 cm (Fig.5). In order to fix the magnets, 5×5 cm aluminium block with 3.4 cm hole was placed in the middle of the magnets. Aluminium disks with thickness of 5 mm were used to cover construction from the bottom and the top. Four brass screws were used to fix the overall structure consisting of the frame, central blocks, top and bottom disk and permanent magnet blocks (Figs. 4-5). The weight of overall construction was about 6 kg.

The static magnetic flux density profile of the magnet was measured using a Hall-probe gaussmeter (Hirst GM05 Gaussmeter with TP002 Hall Probe). Measurements with aid of Hall-effect gaussmeter using a home-made 2D position fixture revealed rather high magnetic field (nearly 5.60 kG) in the center of the magnet construction. Dependence of the magnetic field in the central lateral plane showed good homogeneity for the radial direction perpendicular to magnetic lines (the horizontal direction in Fig. 4): the magnetic field deviation is very close the measurement error of the gaussmeter of about % 0.5. However, the homogeneity in the radial direction parallel to magnetic lines (it is vertical in Fig. 4) reveal rather large magnetic inhomogeneity through the hole (>1%)..

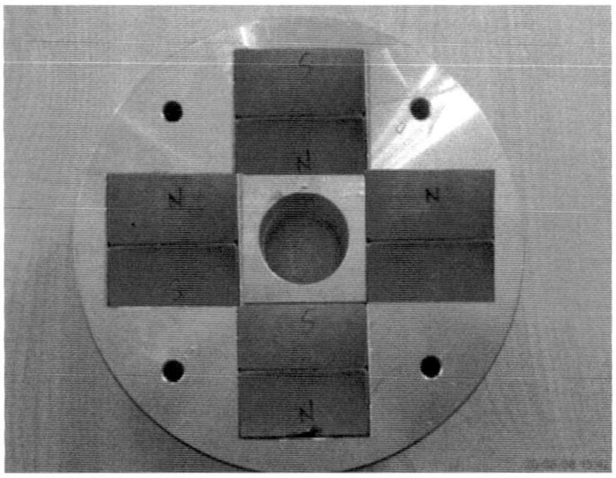

Fig. 4. Permanent magnets installed in the aluminium frame.

Fig. 5. Magnet construction as a whole.

To compare the experimentally measured magnetic fields with theoretically expected profiles FEMLAB software packet has been used. The calculated magnetic field profiles in the directions perpendicular (x-axis) and parallel (y-axis) to the magnetic lines are shown in Fig. 6 and Fig. 7, respectively. The theoretical calculations have revealed the magnetic field changes along x of nearly the same magnitude as in experiment, while the magnetic inhomogeneity along the field lines (y-axis) is expected to be much smaller than observed in our Hall-probe measurements. This discrepancy is attributed to the tolerances of the construction and in the magnetic parameters of the permanent magnets used. Thus the results of our theoretical modelling shows that an improvement of the construction is still possible and it will be the subject of further work.

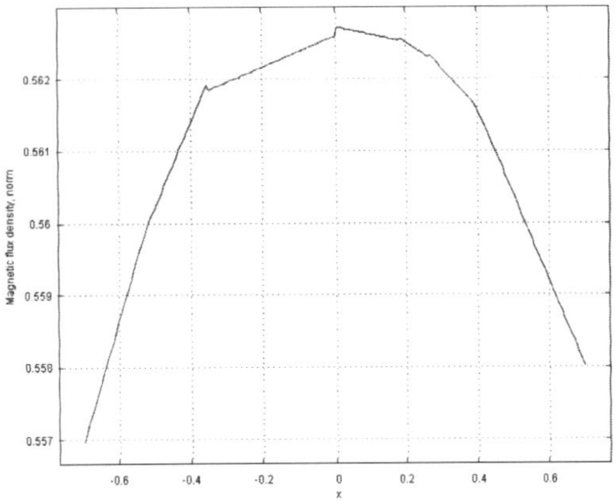

Fig. 6. The calculated magnetic field profile the central z-plane of the construction in the direction ***perpendicular*** to the magnetic lines.

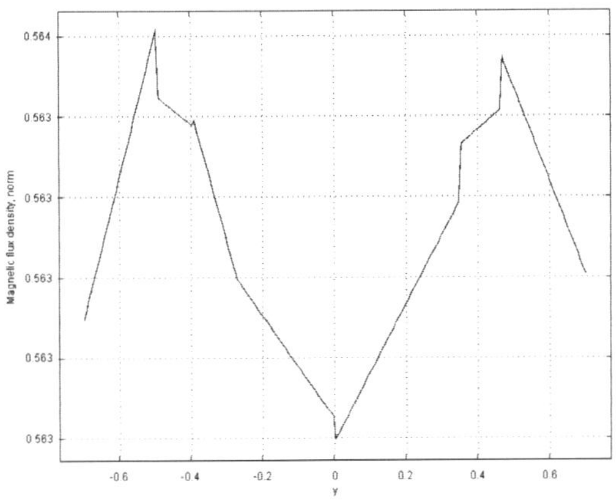

Fig. 7. The calculated magnetic field profile the central z-plane of the construction in the direction ***parallel*** to the magnetic lines.

4. Conclusions

Halbach magnet design has a number of advantages for using in the portable NMR analyser as a system to produce homogeneous magnetic field. Our results reveal that even a simple construction consisting of four permanent magnets provide magnetic parameters which are suitable for possible applications. There is some potential to improve homogeneity using the shim magnets in the bore region (see Refs. [21,22]). Their orientation, size and vertical position can be adjusted to control the field homogeneity or to make field gradients if needed. Depending on the application planned higher number of individual magnets or another approach, where differently shaped individual magnets are employed, may be used to increase the horizontal field strength or homogeneity. Besides, doubling of magnet construction can be considered as another way to improve the field homogeneity along the z-axis. Further research is planned to improve the magnetic field homogeneity of the Halbach construction to use in portable NMR devices.

Acknowledgements.
This work was supported by the grant No. 106T321 of TÜBİTAK (The Scientific and Technological Research Council of Turkey) and by the grant No. 00062.STZ.2007-1 of Ministry of Industry and Trade of Turkey. Authors also acknowledge the support under NATO SfP grant No. 982836.

References.
[1] Nordon A, McGill C A and Littlejohn D, 2001 *Analyst* **126** 260
[2] Haacke E M, Brown R W, Thompson M R and Venkatesan R 1999 *Physical Principles and Sequence Design* (New York: Wiley)
[3] Ernst R R, Bodenhausen G and Wokaun A 1987 *Principles of Nuclear Magnetic Resonance in One and Two Dimensions* (Oxford: Clarendon Press)
[4] Blümich B 2000 *NMR Imaging of Materials* (Oxford: Clarendon Press)
[5] Meriles C A, Sakellariou D, Heise H, Moulé A J and Pines A 2001 *Science* **293** 82
[6] Aue W P, Bartholdi E and Ernst R R 1976 *Journal Chem. Phys.* **64** 2229
[7] Blümich B, Perlo J and Casanova F 2008 *Progress in Nuclear Magnetic Resonance Spectroscopy* **52** 197
[8] Halbach K 1980 *Nuclear Instrum. Methods* **169** 1
[9] Halbach K 1979 *IEEE Trans. Nucl. Sci. NS*-**26** 3882
[10] Raich H and Blümler P 2004 *Concepts Magnetic Resonance B* **23** 16

[11] Perlo J, Casanova F, Blumich B, 2007 *Science* **315** 110
[12] Anferova S, Anferov V, Arnold J, Talnishnikh E, Voda M A, Kupferschla¨ger K, Blumler P, Clauser C, Blumich B, 2007 *Magnetic Resonance Imaging* **25** 474
[13] Leupold H A, Potenziani II E and Abele M G 1988 *Journal Applied Physics* **64** 5994
[14] Leupold H A, McLane G F 1994 *Journal Applied Physics* **76** 6253
[15] Podol'skii A 2002 *IEEE Trans. Magn.* **38** 2
[16] Xu X N, Lu D W, Yuan Q, Han Y S and Jin X 2004 *Journal Applied Physics* **95** 6302
[17] Moresi G and Magin R 2003 *Concept Magnetic Resonance* **B19** 35
[18] Anferova S, Anferov V, Rata DG, Blumich B, Arnold J, Clauser C, et al. 2004 *Concept Magnetic Resonance* **B23** 26
[19] Appelt S 2005 NMR goes mobile in the Earth's magnetic field. *Book of abstracts:* 5th colloquium on mobile NMR. p. L03.
[20] Perlo J, Demas V, Casanova F, Meriles C A, Reimer J, Pines A, et al. 2005 *Science* **308** 1279
[21] Bertora F, Trequattrini A, Abele M G and Rusinek H 1993 *Journal Appl Physics* **73** 68641102
[22] Chang W H, et al. 2006 *Magnetic Resonance Imaging* **24** 1095

Observation of room temperature ferromagnetism in ZnTe:Cr films grown onto glass substrate by thermal evaporation method

D Soundararajan[1], D Mangalaraj[1,2], D Nataraj[1], L Dorosinskii[3, 6] and J Santoyo-Salazar[4,5]
[1]Thin Film Laboratory, Department of Physics, Bharathiar University, Coimbatore – 641 046, India
[2]Department of Nanoscience and Technology, Bharathiar University, Coimbatore – 641 046, India
[3]National Institute of Metrology (TUBITAK - UME), P.K. 54, 41470, Gebze – Kocaeli, Turkey, e-mail: dorosins@ume.tubitak.gov.tr
[4]Universidad Nacional Autonoma de Mexico, Instituto de Investigaciones en Materiales, Mexico D.F. 04510, Mexico

Abstract: ZnTe and ZnTe:Cr films were prepared onto glass substrates using thermal evaporation method. Structural properties of the prepared samples were analyzed using X-ray diffractometer, and the presence of ZnCrTe phase was identified along with poor crystallinity. Composition analysis was done using XPS and the Cr content in the film was found to be 0.05 atomic percent. Transmittance spectra were recorded using UV-Vis spectrophotometer. The valence state of Cr in ZnTe:Cr film is determined to be +2 using electron spin resonance (ESR) spectroscopy. Magnetic moment data as a function of magnetic field were recorded using Superconducting Quantum Interference Device (SQUID) magnetometer at temperatures 5, 77 and 300 K. The results showed minority ferromagnetic behavior even at room temperature. Magnetic domains were observed using Magnetic Force Microscopy and the average value of domain size is 3.7 nm.

1. Introduction

Diluted magnetic semiconductors (DMS) are currently a very active area of research due to their technological potential for use in spintronic devices and their interesting magnetic and transport properties [1]. The discovery of the ferromagnetism in (In,Mn)As [2] and (Ga,Mn)As [3] promoted diluted magnetic semiconductors to be fundamental materials for spintronics because of a compatibility with semiconductors used in present electronics. However, their Curie temperatures (T_C) are not high enough for practical applications. From application point of view, it is necessary to realize room-temperature ferromagnetism in DMS systems. Besides III–V compound semiconductors, II–VI compound semiconductors have also been used as DMS by doping with transition metal impurities such as Fe, Mn and Co. The 3d transition metals doped into II-IV compound semiconductors with zinc-blend type, such as (Zn, Mn)Se, (Cd, Mn)Te or (Zn, Mn)Te, have a high potential as magneto optical and semiconductor-spintronics materials, because the solubility of the 3d-TMs in II-VI compound semiconductors is much higher than that in the III-V compound semiconductors. Although

[6] Corresponding author

a lot of fundamental studies have been carried out in this type of DMS, little application has been realized. The reason is that spin glass or antiferromangetism was observed in II-VI compound semiconductors when doped with transition metal impurities [4]. However it has been recently reported that highly p doped ZnTe films showed ferromagnetism at the temperature of 3 K [5]. This is due to the mediation of the ferromagnetic ordering between magnetic ions by the added hole carriers [6]. On the contrary, when transition metal impurity such as Cr is doped into a II-VI compound semiconductor, net ferromagnetic exchange interaction has been predicted without p- or n-type doping [7]. The reason is that in the case Cr doping into II-VI semiconductors, replacement of cation position (Zn, Cd) itself acts as a source of hole as well as a source of magnetic impurity, unlike the case of Fe or Mn doping into II-VI semiconductors where a separate hole dopant (N) is also required in addition to the magnetic impurity dopants (Fe, Co, and Mn) [8]. Recently, room temperature ferromagnetic behavior was observed when Cr is heavily doped into ZnTe films and (Zn,Cr)Te is prooved to be a real DMS. Many experimental and theoretical results are available on the origin of observed ferromagnetic behavior [9-13]. In general, ZnTe:Cr films were prepared by Molecular Beam Epitaxy (MBE) technique [14, 15]. In this paper, we present a detailed study of the magnetic properties of Cr doped ZnTe films prepared using conventional thermal evaporation method. Microstructure and optical transmittance spectra of undoped ZnTe and Cr doped ZnTe films have been compared. Electron Spin Resonance spectra were measured for the Cr doped ZnTe film at room temperature to identify the valence state of Cr in ZnTe lattice. As a continuation of these studies, magnetic moment versus magnetic field data were recorded at temperatures of 5, 77 and 300 K to observe ferromagnetic hysteresis loops. Magnetic domains were observed using Magnetic Force Microscopy. A discussion of the obtained results is given.

2. Experimental Details

2.1 Film Preparation

Appropriate weights of high purity Zn (Alfa Aeser), Te (Alfa Aeser) and Cr (Alfa Aeser) metals were taken together in a quartz ampoule for the preparation of $Zn_{1-x}Cr_xTe$ alloys with x = 0, and 0.05. The quartz ampoule was vacuum sealed at a vacuum of 10^{-5} Torr. This vacuum sealed quartz ampoule was kept inside a rotating furnace at 1100°C for 24 hours. The temperature was slowly increased step by step to the required value which was maintained for 24 hours. The temperature of the zone was then slowly decreased in steps to room temperature. The molten alloy was taken out and made into fine powder by grinding in a mortar and then used for further characterizations. Thus prepared ZnTe:Cr powder was thermally evaporated from a tungsten dimple source onto well cleaned glass substrates under a vacuum of 4 x 10^{-5} mbar. The substrates were kept at room temperature throughout the experiment.

2.2 Characterization techniques

The crystalline structure of the as-grown ZnTe and ZnTe:Cr films was analyzed using X-ray diffractometer [Model – Shimadzu (XRD-6000), λ = 1.5406 Å] operated in the 2θ scan range of 20° to 80°. In the present study, XPS measurements were carried out using an Escalab. 220iXL system (VG Scientific Inc.) with a monochromatic Mg Kα X-ray source (1284.6 eV). The relative atomic percentage of Zn, Te and Cr was obtained using XPS peak fit software. Optical transmittance spectra of the films were recorded using a UV-VIS-NIR (CARY 2390) spectrophotometer at room temperature. Electron Spin Resonance (ESR) spectroscopy was used to record the first derivative absorption spectra using an E-112 Varian ESR Spectrometer at X-band frequency at room temperature. ZnTe:Cr films grown onto glass substrate were cut into 5 mm x 5 mm size and placed into the microwave cavity. The microwave frequency was kept constant at 9 GHz, while the magnetic field was varied between 1 and 8 kG. The ESR was measured with a X-band (9.0 GHz) spectrometer in a 100 kHz field modulation. Magnetic properties were recorded using a Superconducting Quantum Interference Device (SQUID) magnetometer at various temperatures viz 5, 77 and 300 K. Ex-situ topography and surface magnetic domain pattern study was performed at room temperature by means of Scanning Probe Microscope (SPM) (JEOL JSPM-4210) to scan the particle size and magnetic domains at room temperature conditions. Topography of films was scanned by AFM in Tapping™ mode, and magnetic domains were analyzed by lift mode. A Magnetic cantilever tip NSC14/Co–Cr, Mikro-masch, Co. was used with resonant frequency of 160 kHz. The magnetic domains were observed in lift mode MFM with interaction in the height of 25 nm and output of 0.025 Amp/V. The 2D and 3D images, profiles, and domain measurements were processed with the software win SPM DPS, JEOL Ltd.

3. Results and discussion

3.1 Structural analysis

The XRD patterns of ZnTe and Cr doped ZnTe films in the 2Θ scan range of 20° to 80° are shown in figure 1. The XRD pattern of the ZnTe film shows a main peak along (111) crystallographic orientation indicating cubic zinc-blende structure as observed by Akram et al for ZnTe films grown onto glass substrate using thermal evaporation method [16]. On the other hand, the diffraction pattern of ZnTe:Cr film shows amorphous background along with a main peak along (200) crystallographic direction indicating ZnCrTe cubic zinc blende structure. This peak is similar to a peak observed for ZnCrTe film on GaAs (100) substrate prepared by using MBE technique as observed by Shinji Kuroda et al [17, 18]. In the doped sample, in addition to the peak along (200) crystallographic direction indicating ZnCrTe cubic zinc-blende phase, peaks corresponding to Cr related phases such as Cr, Cr2O3 and CrTe are also observed in the amorphous background [19, 20]. The calculated lattice constant values of ZnTe and ZnTe:Cr films are 5.96 Å and 6.08 Å respectively. The change in lattice constant is an indication that Cr is substitutionally incorporated into the ZnTe lattice.

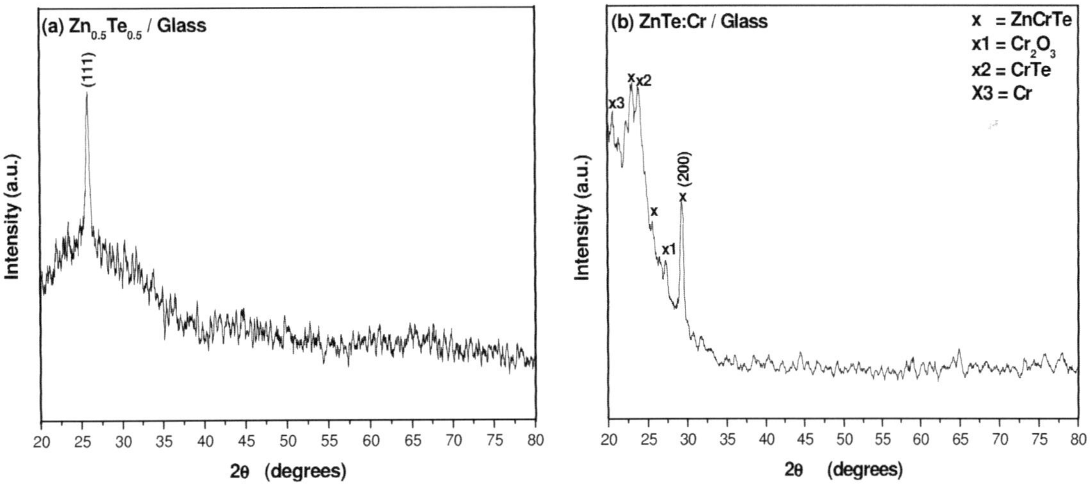

Figure 1. X-ray diffraction patterns of (a) as-deposited ZnTe and (b) Cr doped ZnTe films grown onto glass substrates. Inset to figure 1b shows expanded view of the same result in the 2θ scan range of 20° to 30°.

3.2 Composition analysis

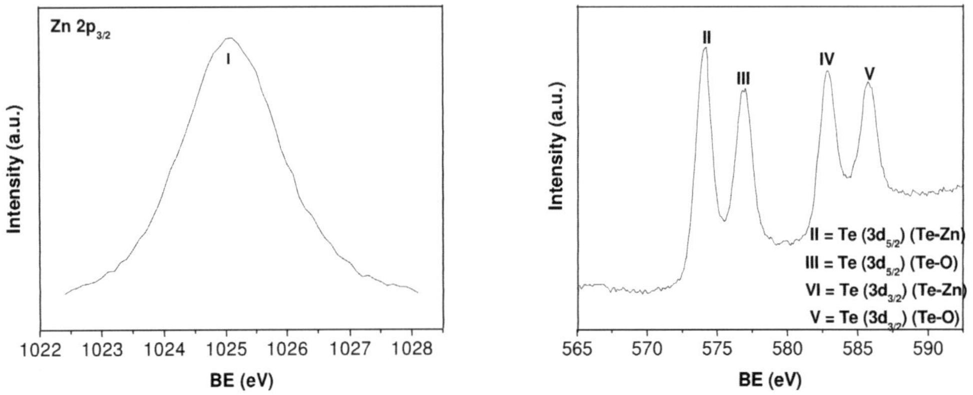

Figure 2. High resolution XPS spectra of $Zn_{0.5}Te_{0.5}$ film.

The microscopic surface composition of the prepared ZnTe and ZnTe:Cr films was investigated by X-ray Photoelectron Spectroscopy (XPS). Figures 2 and 3 show typical high resolution XPS spectra of the $Zn_{0.5}Te_{0.5}$, and $Zn_{0.45}Cr_{0.05}Te_{0.5}$ films. The most probable assignments to the origin of the components are presented in tables 1 and 2 [21]. Relative percentage of the components as evaluated from the high resolution spectra are given in table 3. Figure 2 exhibits the characteristic Zn $2p_{3/2}$ and Te $3d_{5/2}$ peaks at the binding energies of 1025.08 eV and 574.2 eV respectively. One can see that the Te peak splits into two peaks, Te $3d_{5/2}$ and Te $3d_{3/2}$ at binding energies of 574.2 eV and 582.9 eV respectively. The corresponding core level spectra of Te $3d_{5/2}$ and Te $3d_{3/2}$ are split into two peaks at binding energies of 576.9 eV and 585.7 eV due to the Te – O bonding in TeO_2. The stoichiometry of the as-grown ZnTe films was determined by the $2p_{3/2}$ peak for Zn and $3d_{5/2}$ and $3d_{3/2}$ peaks for Te. This shows that the films were slightly rich in Zn (Zn: Te = 52.8: 47.17). Figure 3 shows the characteristic Zn $2p_{3/2}$, Te $3d_{5/2}$ and Cr $3d_{5/2}$ peaks at their corresponding binding energies of 1021.8 eV, 572.9 eV and 574.3. Moreover, the presence of TeO_2 and antiferromangetic Cr_2O_3 components was observed at their corresponding binding energies of 576.5 eV and 575.6 eV. The stoichiometry of the ZnTe:Cr films was determined by the $2p_{3/2}$ peak for Zn, 3d peaks for Te and 3d peaks for Cr. Estimated values of the relative atomic percentage of Zn, Te and Cr in the $Zn_{1-x}Cr_xTe$ (x = 0, 0.05) films are given in table 3.

Table 1. Assignments to the component peaks of the high resolution XPS scans of the $Zn_{0.5}Te_{0.5}$ film.

Peaks	Component	Binding energy (eV)	Assignment
1	I (Zn $3d_{5/2}$)	1025.08	Zn - Te
2	II (Te $3d_{5/2}$)	574.2	Te – Zn
3	IV (Te $3d_{5/2}$)	576.9	Te – O_2
4	VI (Te $3d_{3/2}$)	582.9	Te – Zn
5	VIII (Te $3d_{3/2}$)	585.7	Te – O_2

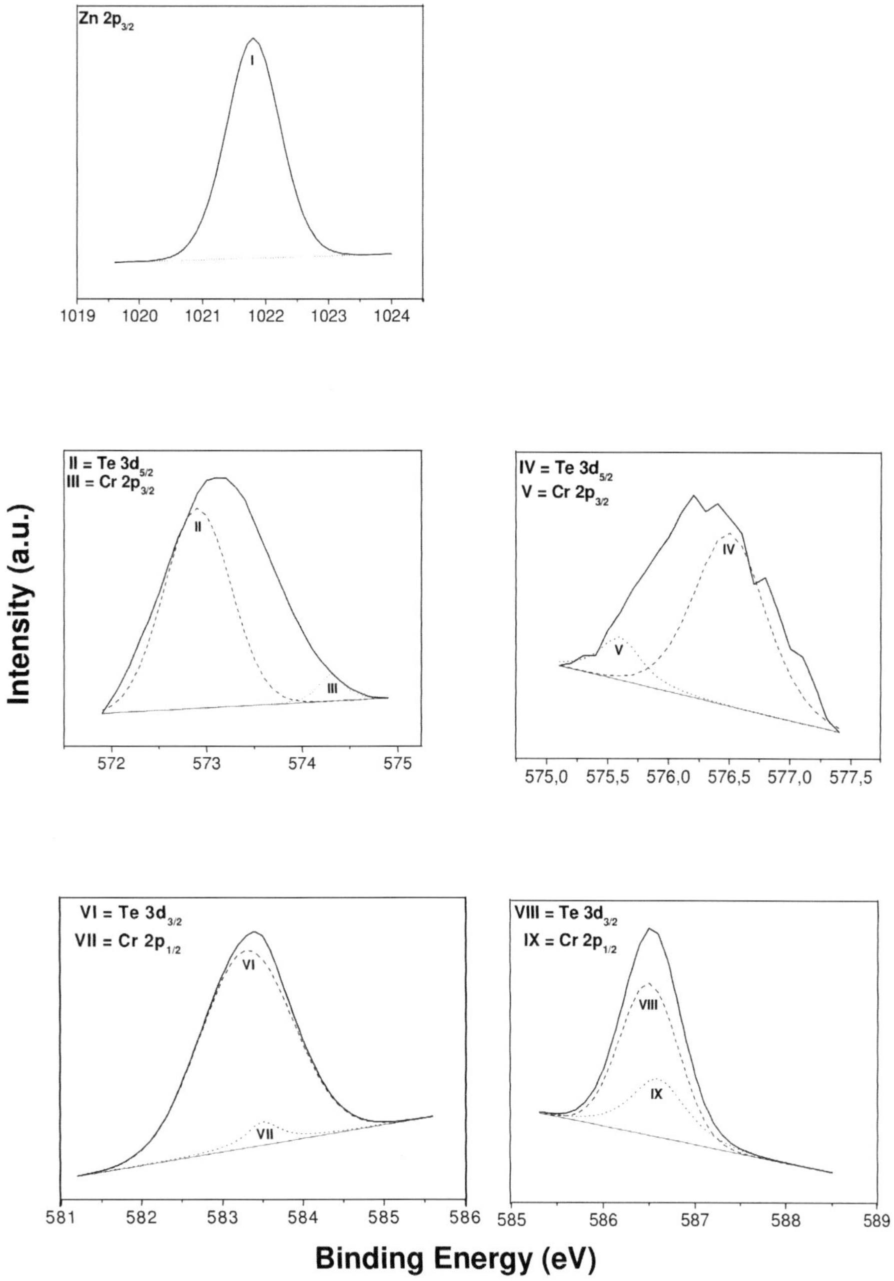

Figure 3. High resolution XPS spectra of $Zn_{0.45}Cr_{0.05}Te_{0.5}$ film.

Table 2. Assignments to the component peaks of the high resolution XPS scans of the $Zn_{1-x}Cr_xTe$ (x = 0.0, 0.05) films.

Peaks	Component	Binding Energy (eV)	Assignment
1	I (Zn $3d_{5/2}$)	1021.8	Zn - Te
	II (Te $3d_{5/2}$)	572.9	Te – Zn
2	III (Cr $3d_{3/2}$)	574.3	Cr - Te
	IV (Te $3d_{5/2}$)	576.5	Te – O_2
3	V (Cr $3d_{3/2}$)	575.6	Cr_2 – O_3
	VI (Te $3d_{3/2}$)	583.3	Te – Zn
4	VII (Cr $2p_{1/2}$)	583.5	Cr – Te
	VIII (Te $3d_{3/2}$)	586.5	Te – O_2
5	IX (Cr $2p_{1/2}$)	586.6	Cr_2 – O_3

Table 3. Relative percentage of the components of Zn $2p_{3/2}$, Te $3d_{5/2}$, Te $3d_{3/2}$, Cr $3d_{3/2}$ and Cr $3d_{1/2}$ peaks as evaluated from the high resolution XPS spectra of the $Zn_{1-x}Cr_xTe$ (x = 0, 0.05) films.

Samples	Relative percentage of the components (%)								
	Peak 1	Peak 2		Peak 3		Peak 4		Peak 5	
	(Zn $2p_{3/2}$)	(Te $3d_{5/2}$)	(Cr$3d_{3/2}$)	(Te $3d_{5/2}$)	(Cr $3d_{3/2}$)	(Te $3d_{3/2}$)	(Cr $2p_{1/2}$)	(Te $3d_{3/2}$)	(Cr$2p_{1/2}$)
$Zn_{0.5}Te_{0.5}$	52.81	16.99	-	11.4	-	10.44	-	8.34	-
$Zn_{0.45}Cr_{0.05}Te_{0.5}$	44.998	28.049	1.5902	1.3	0.21	15.05	0.9	5.63	2.2683

3.3. Thickness Measurement

Thickness of as-grown $Zn_{0.5}Te_{0.5}$ and $Zn_{0.45}Cr_{0.05}Te_{0.5}$ films was estimated using cross sectional analysis as shown in figure 4. The calculated thicknesses are 146 nm and 148 nm.

3.4. Optical transmittance

Transmission studies were carried out in the wavelength range of 300 – 1100 nm to investigate the optical absorption properties of the as-grown $Zn_{1-x}Cr_xTe$ (x = 0, 0.05) films. The results are plotted in figure 5. ZnTe and ZnTe:Cr films exhibit a difference in the optical transmittance. This difference is attributed to structural, compositional and crystallinity changes in the films taking place when Cr is heavily doped into the ZnTe lattice. Fundamental absorption edge was found to shift towards lower wavelength region (from 750 nm to 620 nm) when Cr concentration in as grown $Zn_{1-x}Cr_xTe$ (x = 0, 0.05) film increases. The cause of the observed blue shift shall be due to the decrease of Zn concentration and increase of Cr concentration in the films.

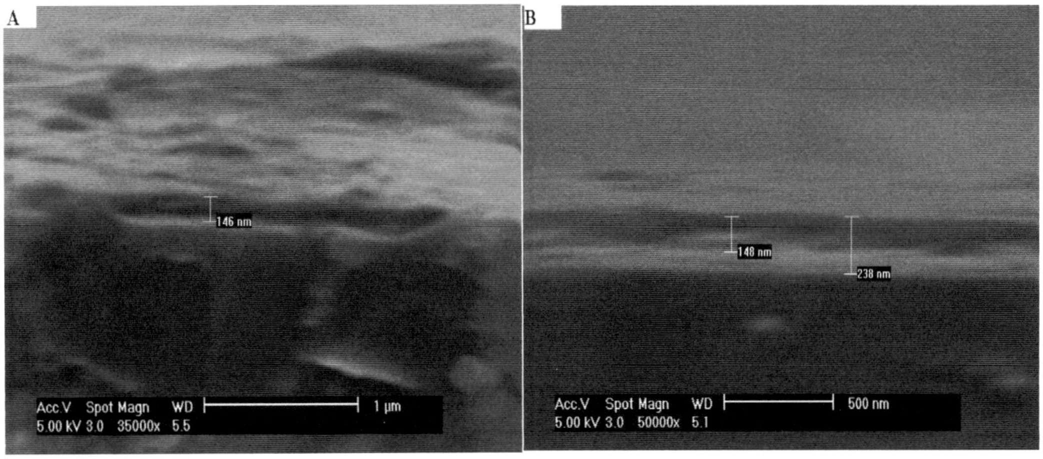

Figure 4. Cross sectional SEM images of as-grown $Zn_{1-x}Cr_xTe$ (x = 0.0, 0.05) films grown onto glass substrates. The calculated values are 146 nm and 148 nm for $Zn_{0.5}Te_{0.5}$ and $Zn_{0.45}Cr_{0.05}Te_{0.5}$ films respectively.

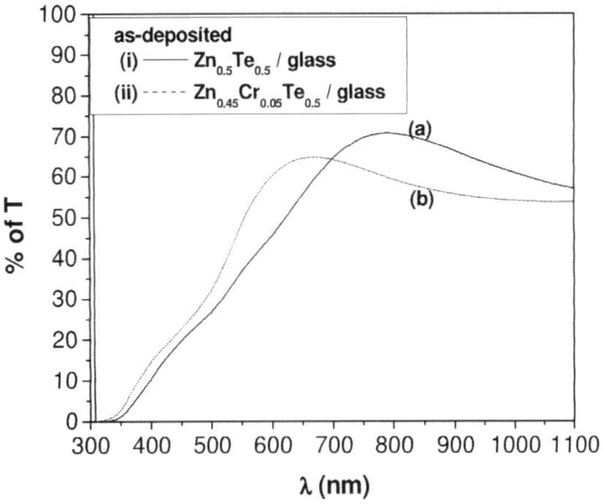

Figure 5. Transmittance spectra of as-grown $Zn_{1-x}Cr_xTe$ (x = 0.0, 0.05) films grown onto glass substrates.

3.5. Electron Spin Resonance (ESR) analysis

The advantage of ESR spectroscopy is its great sensitivity to the microscopic environment of the paramagnetic centre. Besides, it is a tool to know the valence state of Cr in the film so that the origin of magnetic moment in the Cr doped ZnTe film can be inferred. The electronic configuration of free Cr atom is 1s2 2s2 2p6 3s2 3p6 3d5 4s1. Chromium enters the ZnTe lattice substitutionally for the divalent cation and is expected to be in the Cr2+ valence state. Figure 6 displays the room temperature ESR profiles measured without sample and measured for the ZnTe:Cr film grown onto glass substrate (figure 6 b). The spectrum of ZnTe:Cr film shows a broad signal which consists of two sharp peaks and a minute unresolved hump at 85 mT. The large sharp signal at B = 163 mT is due to the presence of impurities in the glass substrate [22]. The second small sharp signal at B = 335.3 mT is due to DPPH (2, 2-Diphenyl-1-Picrylhydrazyl) free radical, a similar peak is also observed for the ESR spectra without sample. It is confirmed that an unresolved minute peak is observed without hyperfine

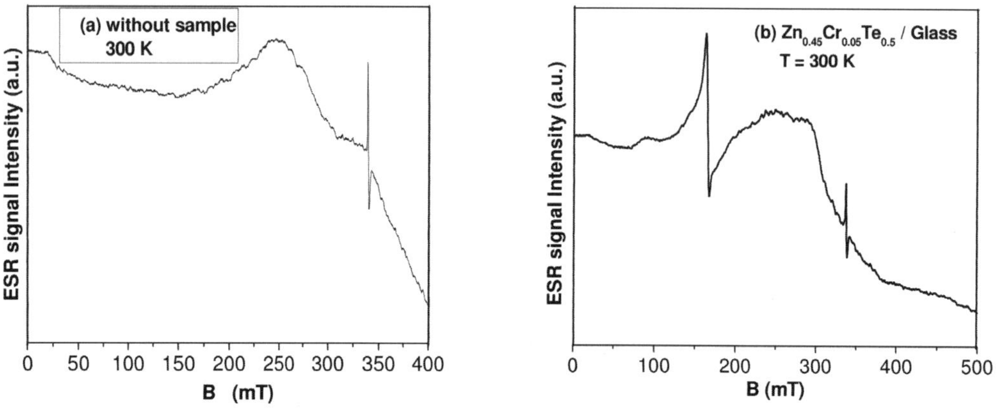

Figure 6. ESR spectra measured (a) without sample and (b) for $Zn_{0.45}Cr_{0.05}Te_{0.5}$ film grown onto glass substrate.

splitting when comparing the ESR signals recorded at room temperature in the presence and absence of $Zn_{0.45}Cr_{0.05}Te_{0.5}$ film. The ESR spectrum of the I = 0 chromium isotopes can be analyzed using the following spin Hamiltonian:

$$hv = g\beta B$$

Where, h ---- Planck's constant
v ---- Microwave Frequency (9 GHz)
β ---- Bohr Magneton ($9.27402 * 10^{-24}$ JT^{-1})
B ---- Magnetic Field (T)
and g ---- gyromagnetic tensor

We derive $g_{||}$ and $g_{\perp}$ values from the observed spectra. The calculated value of '$g_{||}$' in the Zeeman interaction term is 7.900 at B = 85 mT, which is close to 7.92 of '$g_{||}$' value for Cr^{2+} ion in ZnTe crystal as reported elsewhere [23]. $g_{\perp}$ could not be estimated as there is no signal detected from the recorded room temperature ESR spectra. As the calculated $g_{||}$ is close to the earlier reported value, it is assigned to isolated Cr^{2+} ions substitutionally incorporated in the zinc lattice site with the valence state of +2. Therefore, any ferromagnetic property must be due to the Cr^{2+} state of chromium.

3.6. Magnetic Moment measurements

Magnetic moment as a function of applied magnetic field was measured at temperatures of 5, 77 and 300 K. The results are plotted in figure 7. Magnetic moment data were recorded by varying the external magnetic field from 0 to 20 000 Oe. The expanded view of hysteresis loops in the low field region is shown in the inset of figure 7. The results show hysteresis loops with different magnitudes. At T = 5 K, a ferromagnetic like hysteresis curve was observed. It should be noted that the ferromagnetic characteristic survived even at T = 300 K. In general, for magnetic materials, a better signal is expected at lower temperatures. At T = 5 K, the magnetic property of the film shows the coexistence of ferromagnetic and paramagnetic phase. Ferromagnetic hysteresis curve is observed in the low field region (0 to 3000 Oe). But, paramagnetic behavior is observed in the high field region (3000 to 20000 Oe). The paramagnetic behavior, being linear, has been fitted to a straight line with an intercept (about 9.1×10^{-4} emu). Since the intercepts are identical for positive and negative field regions, the expectation of paramagnetism at high field is confirmed [24]. It is believed that the presence of isolated Cr ions shall acts as a source of paramagnetic in addition to the observed ferromagnetic like behavior. At high temperatures (77 and 300 K) ferromagnetic and diamagnetic

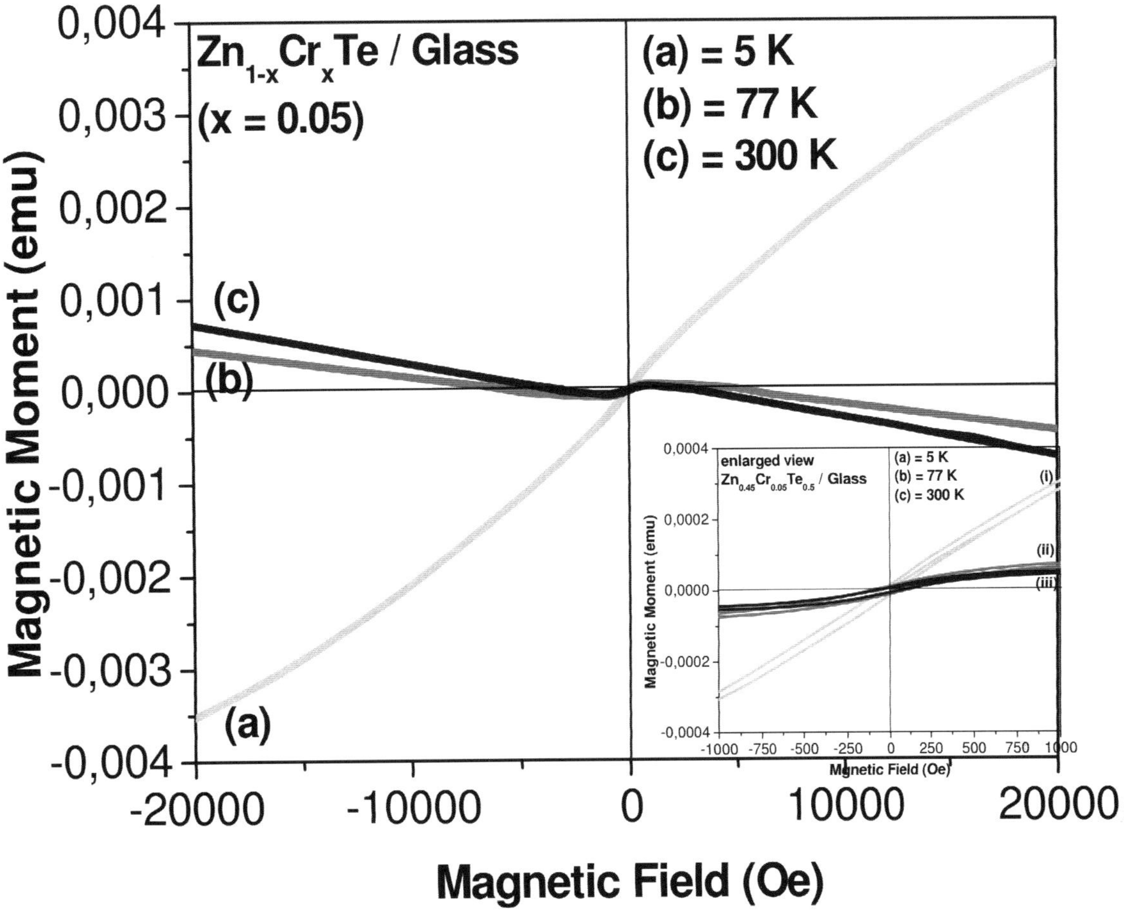

Figure 7. Magnetic moment versus magnetic field plot of the $Zn_{0.45}Cr_{0.05}Te_{0.5}$ film at temperatures 5, 77 and 300 K (top of figure 5). Inset to figure shows the expanded view of same plot in the low field region (bottom of figure 5). An obvious hysteresis loop is observed in the lower field region of M vs H plot confirming a ferromagnetic ordering induced by the Cr doping in ZnTe.

behaviour was observed respectively at low (0 to 3000 Oe) and high (3000 to 20000 Oe) fields. When temperature increased, the contribution of the paramagnetic phase rapidly vanishes. The diamagnetic contribution is temperature independent and becomes dominant at high fields following the significant reduction in paramagnetic component. In this sample, magnetization measured at all temperatures exhibits hysteresis in the low field region. The hysteresis loop observed even at room temperature shall originate from the presence of the ZnCrTe zinc-blende phase and also from CrTe precipitates. The diffraction pattern of Cr doped ZnTe film shows poor crystalline nature with a small peak due to the ZnCrTe zinc-blend phase. In addition to it, other Cr related precipitates such as Cr, Cr_2O_3 and CrTe were observed. The Cr and Cr_2O_3 phases are antiferromagnetic stable state of chromium oxide at ambient conditions. Its metastable phase is chromium dioxide (CrO_2) which is a half metallic ferromagnet and it becomes Cr_2O_3 at ambient conditions [25]. Hence the observed ferromagnetic behavior is assumed to arise from the ZnCrTe zinc-blende phase and from CrTe precipitates. It is already reported that the cubic zinc-blende crystal structure of ZnCrTe and CrTe are half metallic ferromagnetic with Curie temperature above room temperature and it plays a key role in the ferromagnetic behavior of ZnTe:Cr films [26-29]. It is therefore concluded that the presence of zinc-blende ZnCrTe phase along with CrTe precipitates shall be the reason for the observed ferromagnetic behavior in the low field region at all the measured temperatures 5, 77 and 300 K.

3.7. Magnetic domains

MFM is becoming one of the most used techniques for magnetic domain imaging. Prior to MFM analysis, the film was exposed to minute argon gas blow and consequently topographic observation was made using AFM. Then the magnetic domains were observed using MFM. Figure 8(a-b) shows surface topography and magnetic domain images of $Zn_{0.45}Cr_{0.05}Te_{0.5}$ film grown onto glass substrate. The average size of particles over the film surface was 45 nm. The contact between the nanoparticles produced small islands where the magnetic interactions were observed in lift mode. The stripes inside the islands contain magnetic domains which have different positions due to the magnetic force. Figure 8b indicates the tendency of the material to form mono domains or bigger domains if the material would be exposed to thermal treatment. The marked region indicated in figure 8b was taken in order to analyze the magnetic domains in the islands as shown in figure 9a. The magnetic domains were measured in an area of 30 x 30 nm as shown in figure 9b. Rectangular shape of domains was seen with an average size of 3.70 nm as shown in figure 9c.

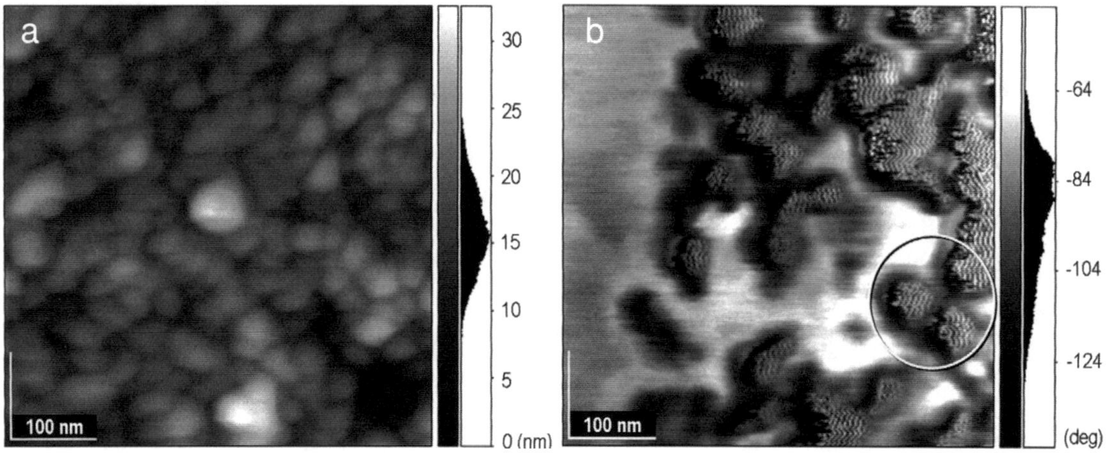

Figure 8. Topography and MFM analysis in $Zn_{0.45}Cr_{0.05}Te_{0.5}$ film grown onto glass substrate. (a) Distribution of nanoparticles in the film in the area of 500 x 500 nm and (b) Magnetic domains in form of islands.

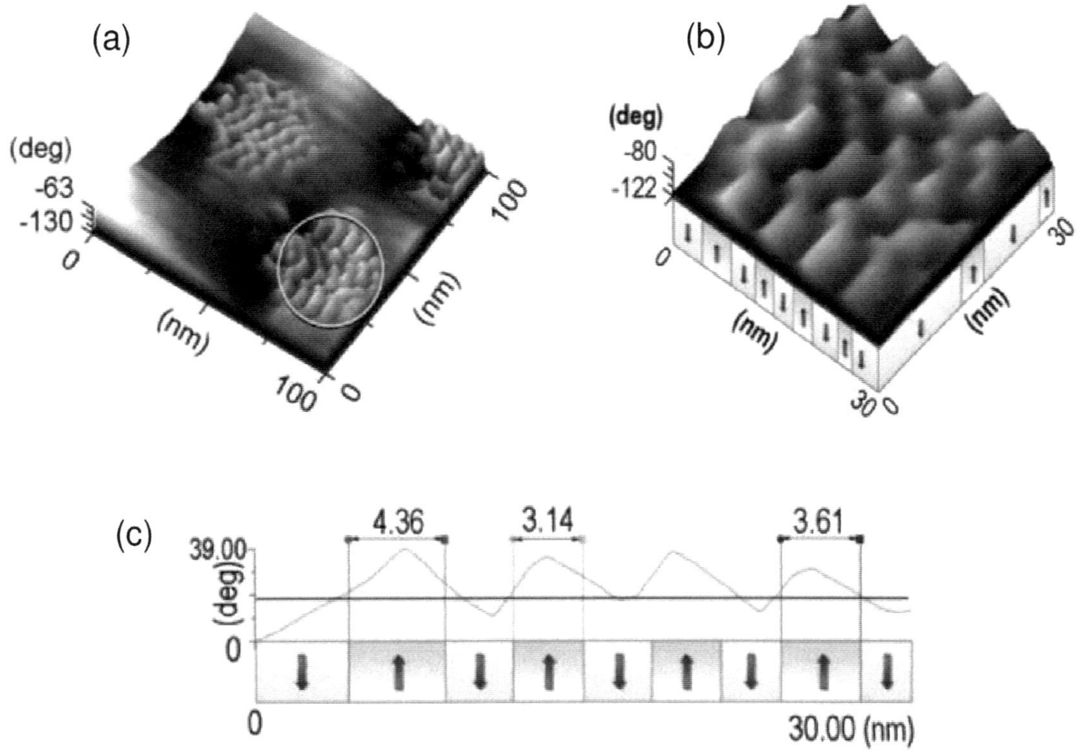

Figure 9. Magnetic multi domains of $Zn_{0.45}Cr_{0.05}Te_{0.5}$ film grown onto glass substrate. (a) domains in the surface of area of 100 x 100 nm; (b) selected area of 30 x 30 nm. The domain scheme below the 3D image shows bright zones corresponding to positive charge (↑) and dark zones corresponding to negative charge (↓). (c) Magnetic domains profile in cross section of the image of (b). The average size of magnetic domains is 3.7 nm.

4. Conclusion

An attempt was made to study the room temperature ferromagnetism in Cr doped ZnTe films grown onto glass substrates prepared by thermal evaporation method. Diffraction pattern of Cr doped ZnTe film showed a main peak along (200) crystallographic direction indicating ZnCrTe zinc blende phase along with the presence of small peaks related to Cr precipitates such as Cr, Cr_2O_3 and CrTe. From ESR spectra, the valence state of Cr ions in $Zn_{0.45}Cr_{0.05}Te_{0.5}$ film was found to be +2. Magnetic moment data measured at T = 5, 77 and 300 K show ferromagnetic minority behavior in the lower field region. Magnetic multi domains were observed using MFM and the average domain size was found to be 3.7 nm. It is concluded that the presence of ZnCrTe zinc blende phase as well as CrTe related precipitates should be the reason for the observed hysteresis curve and the ferromagnetic domains.

References

[1] Prinz G 1998 *Science* **282** 1660; Ohno H 1998 *Science* **281** 951; Pearton S J, et al. 2003 *Mat. Sci. Eng.* **R40** 137.
[2] Ohno H, Shen A, Matsukura F, Oiwa A, Endo A, Katsumoto S and Iye Y 1996 *Appl.Phys.Lett.* **69** 363.

[3] Munekata H, Ohno H, Ruf R R, Gambino R J and Chang L L 1991 *J. Cryst. Growth* **111** 1011.
[4] Buyanova I A, Rudko G Yu, Chen W M, Toropov A A, Sorokin S V, Ivanov S V and Kop'ev P S 2003 *Appl. Phys. Lett.* **82** 1700.
[5] Ferrand D, Cibert J, Wasiela A, Bourggnon C, Tatarecko S, Fishman G, Andrearczyk T, Jaroszynski J, Jolesnik S, Dietl T, Barbara B and Dufeu D, 2001 *Phys.Rev. B* **63** 85201.
[6] Park S E, Lee H J, Cho Y C and Jeong S Y, 2002 *App. Phys.Lett.* **80** 4187.
[7] Tetsuya Fukushima, Kazunori Sato, Hiroshi Katayama-Yoshida and Dederichs P H 2004 *Jpn. J. Appl. Phys.* **43** L1416.
[8] Guo X G, Cao J C, Chen X S and Lu W 2006 *Solid state communications* **138** 275.
[9] Saito H, Yamagata S and Ando K 2004 *J. Appl. Phys.* **95** 7175.
[10] Ozaki N, Nishizawa N, Kuroda S and Takita K 2004 *J. Phys. Condens. Matter* **16** 5773.
[11] Saito H, Zayets V, Yamagata S and Ando K 2003 *Phys. Rev. Lett.* **90** 207202.
[12] Saito H, Zayets V, Yamagata S and Ando K 2003 *J. Appl. Phys.* **93** 6796.
[13] Saito H, Yamagata S and Ando K 2004 *J. Appl. Phys.* **95** 7175.
[14] Saito H, Zayets V, Yamagata S and Ando K 2003 *J. Appl. Phys.* **93** 6796.
[15] Wang Q, Sun Q and Jena P 2005 *J. Appl. Phys.* **97** 043904.
[16] Akram K S, Ali Z and Maqsood A 1999 *Appl. Surf. Sci.* **143** 39.
[17] Ozaki N, Nishzawa N, Marcet S, Kuroda S, Eryu O and Takita K 2006 *Phys. Rev. Lett.* **97** 037201.
[18] Kuroda S, Ozaki N, Nishizawa N, Kumekawa T, Marcet S and Takita K 2005 *Science and Technology of Advanced Materials* **6** 558.
[19] JCPDS, X-ray powder diffraction file, Joint Committee for Powder Diffraction Standards.
[20] Ye H, Zhang Q, Saito S, Jeyadevan B, Tohji K and Tsunoda M 2003, *J. Appl. Phys.* **93** 6856.
[21] Handbook of the Elements and Native Oxides.
[22] Diaconu M, Schmidt H, Poppl A, Bottcher R, Hoentsch J, Rahm A, Hochmuth H, Lorenz M and Grundmann M 2005 *Superlatices and Microstructures* **38** 413.
[23] Vallin J T, Slack G A, Roberts S and Hughes A E 1970 *J. Phys. Rev. B* **2** 4313.
[24] Jayakumar O D, Salunke H G, Kadam R M, Mohapatra M, Yaswant G and Kulshreshtha S K 2006 *Nanotechnology* **17** 1278.
[25] Stanoi D et al. 2005 *J.Mat.Sci.Eng B.* **118** 74–78.
[26] Xie W H and Liu B G 2004 *J. Appl. Phys.* **96** 3559.
[27] Xie W H, Xu Y Q, Liu B G 2003 *Phys. Rev. Lett.* **91** 037204.
[28] Galanakis I and Phivos Mavropoulos 2003 *Phys. Rev. B* **67** 104417.
[29] Sreenivasan M G, Hou X J, Teo K L, Jalil M B A, Liew T and Chong T C 2006 *Thin Solid Films* **505** 133–36.

Production, characterization and magnetic properties of Bismuth-substituted Yttrium Iron Garnet layers prepared by sol–gel process

M. EROL[1], Y. OZTURK[2], I. AVGIN[2] and E. CELIK[1]
[1] *Metallurgical and Materials Engineering Department, Dokuz Eylul University, 35160 Izmir, Turkey*
[2] *Electrical and Electronics Engineering Department, Ege University, 35100 Izmir, Turkey*
E-mail: mustafa.erol@ogr.deu.edu.tr

Abstract. In this study, $(BiY_2)Fe_5O_{12}$ (Bi:YIG) magneto-optic layers were deposited on Si(100) and glass substrates using a sol-gel process. Bi, Y and Fe based-precursors were dissolved using methanol and glacial asetic acid as a solvent and chelating agent respectively. The turbidity, viscosity, gel point and pH values of the prepared solutions were determined using turbidimeter, rheometer and pHmeter. The prepared xerogels were dried and used for DTA/TG analysis to determine heat treatment regime in thin film production. Based on DTA/TG results, the deposited films were annealed at 800°C for 2 hours in air. Inasmuch as the magnetic and morphological properties are important for applications such as superconductive material vortex investigations, the produced films were investigated in details. The structural properties of the coatings were characterized by using XRD, SEM-EDS and AFM. The magnetic properties of the films were evaluated by the substitution of Bi and process conditions using VSM. The obtained results are presented.
Keywords: Thin films, Bi:YIG, Sol-gel, Magneto-optic properties

1. Introduction

With the development of microwave devices and magneto-optical (MO) integrated devices; the research of yttrium iron garnet (YIG) thin films is becoming more and more important because of properties that can be applied in both microwave and MO devices [1]. Artificial media providing the enhancement of magneto-optical MO effects attract much interest, due to the fact that they may advance functional capabilities of known optical devices such as isolators, magneto-optical memory systems, and magneto-optical spatial light modulators. Large angles of rotation, wide working spectral ranges, and nanosized or microsized dimensions of new MO media are essential properties demanded by applications [2]. To realize these applications, high-performance wide-band and temperature-stabilized optical isolators are essential, which in turn require Faraday rotators capable of ensuring a Faraday rotation stable against wavelength and temperature variation. Nevertheless, the Faraday rotation (θ_F) coefficient of YIG (approximately only 220 deg/cm at 1300 nm) is too small for practical use [3].

In order to improve the MO properties, the ions with +3 balance and suitable ionic radii to yttrium such as the Bi ion or Ce ions are usually substituted with yttrium in YIG films, which can increase the Faraday angle efficiently. Faraday rotation of Bi-substituted magnetic garnet reveals strong wavelength and temperature dependencies [3]. Produced YIG thin films serve as served as Faraday rotator for wide-band and temperature-stabilized optical isolators by the following reasons: (1) the Bi^{3+} and Y^{3+} ions offer the opposite contributes to the wavelength and temperature characteristic of Faraday rotation (2) doping Bi^{3+} ions on dodecahedral sites of garnet structure can enhance greatly the Faraday rotation and increases in proportion to Bi^{3+} ions concentration without increasing absorption loss.

In recent researches YIG or Bi-YIG films have been prepared by a variety of methods such as radiofrequency sputtering, liquid phase epitaxy, pyrolysis, sol–gel process, and pulsed laser ablation deposition on several substrates such as glass, Si(100) wafer or gadolinium gallium garnet (GGG). As mentioned, several techniques can be employed to make YIG based materials such as RF magnetron sputtering, sol-gel and pulse laser deposition technique [4-9]. Of these techniques, the sol-gel processing has a number of advantages. To illustrate this, it is possible to synthesize quite good polycrystalline ferrites with the sol-gel method. The sol–gel process offers considerable advantages such as better mixing of the starting materials and excellent chemical homogeneity in the final product. Moreover, the molecular level mixing aids the structure evolution lowering the crystallization temperature [10] and the sol-gel layer can be deposited desired thickness in one step because this thickness depends only on precursor's concentration. The available Bi-YIG material research has mainly on single thin film crystals. Here a polycrystalline Bi-YIG, studied infrequently, have been produced using the sol-gel method and eventually its magneto-optical properties are presented.

2. Experimental Procedures

The solutions were prepared by dissolving Bi-, Y- and Fe-based alkoxide based precursors in glacial asetic acid as a chelating agent and methanol as solvent respectively. The transparent solutions were formed through the hydrolysis of Bi, Y and Fe alkoxides after stirring in an ultrasonic cleaner at room temperature for 30 minutes. The obtained transparent solutions can remain stable approximately for a month. Four solutions, as listed in Table 1, were prepared and differ from each other depending on their concentrations since we aimed a research on the effect of concentration to film thicknesses and magneto-optical properties in this study. The solutions were called as BM1, BM2, BM3 and BM4 (see Table 1 for more details).

Table 1. Descriptions of solutions

Solutions	**BM1**	**BM2**	**BM3**	**BM4**
Fe-alkoxide (g)	0,7063	0,3532	0,2648	0,1765
Y- alkoxide (g)	0,308	0,154	0,1155	0,077
Bi- alkoxide (g)	0,152	0,076	0,057	0,038
Glacial asetic acid (ml)	10	10	10	10
Methanol (ml)	2	2	2	2
C.R	1	2	3	4

C.R. Concentration Rate. If the concentration of BM1 is the unit concentration. The rate of other solution to unit solution is concentration rate.

To determine solution characteristics which affect thin film structure, turbidity, pH values and rheological properties of the prepared solutions were respectively measured using turbidimeter, pH meter and rheometer machines before coating processes. Turbidity properties of the solutions were measured using standard solutions for coating process by TB 1 Velp Scientifica Model turbidimeter according to ISO 7027 nepheometric method. The sample was placed in a vessel with a dimension of Ǿ25 mm and height of 50 mm. Formazin has been recognized throughout the world as a primary standard. Thus formazin solution was used to calibrate the turbidity in the range of 0 and 1000 ntu (nephelometric turbidity unit). Having prepared transparent solutions, we measured pH values of the solutions to check their acidic and basis characteristics using a standard pH meter with Mettler Tolede electrode. Besides pH and turbidity measurements, rheological behaviour of the solutions including viscosity was determined by CVO 100 Digital Rheometer (Bohlin Instrument).

Commercial Pyrex glass and Si (100) were used as substrates. The substrate surface was cleaned with acetone just before coating process. The solutions were used for coatings on the Pyrex glass and Si (100) substrates by spin coating system at ambient conditions. The spin coating system has been used for several decades in thin film preparation process. A typical process entails depositing a small puddle of a sol or gel onto the centre of a substrate and then spinning the substrate at high speed.

Centripetal acceleration causes the gel to spread to, and eventually off, the edge of the substrate leaving a thin film of gel on the surface. Hence the film thickness and their properties depend on the nature of the gel (viscosity, drying rate, percent solids, surface tension, etc.) and the parameters chosen for the spin process. Factors such as final rotational speed, acceleration, and fume exhaust contribute to how the properties of coated films are defined. For the most resin materials the final film thickness will be inversely proportional to the spin speed and spin time. Final thickness will be also be proportional to the exhaust volume although uniformity will suffer if the exhaust flow is too high since turbulence will cause non uniform drying of the film in the spin process [11]. Bi-YIG solutions were spin coated on the glass and Si (100) substrates with 3000 rpm for 20 s at room temperature. After spin coating process, the obtained gel coating was dried up to 500 °C with a heating rate 15°C/ min, at this temperature of the calcined films for 30 minutes in air. Same processes were applied for two times to obtained three layered Bi-YIG film. After coating and drying of three layers, the films were annealed at 800 °C for 2 hours in air to obtain Bi-YIG film structure in an electric furnace according to obtained DTA/TG result.

Thermal behavior of Bi-, Y- and Fe-based xerogels, which were dried at 200 °C for 30 minutes in air, was evaluated at a heating rate of 10 °C/min under oxygen atmosphere by using DTA/TG machine (DTG-60H Shimadzu) in order to gain decomposition and phase formation, and to obtain an optimum heat treatment regime. The experiments were conducted in oxygen flow (60 ml/min) in the temperature range from ambient to 700 °C at a heating rate of 10 °C/min. XRD patterns of the coatings were determined to identify phase structure by means of a Rigaku (D/MAX-2200/PC) diffractometer with a CuK_α irradiation (wavelength, λ=0.15418 nm) by both θ-2θ mode and 2θ scan mode. Thin-film XRD geometry where incident angle was fixed at 1° was used to collect data from only thin films. Thicknesses of the films were measured by spectrophotometer. The surface topographies of Bi-YIG coatings were examined by using SEM (JEOL JSM 6060) attached with energy dispersive spectroscopy (EDS). The magnetic properties of YIG films were measured at room temperature in a vibrating sample magnetometer (VSM, Lakeshore 736, 7400 Series) in a maximum applied field of 1000 Gauss. From the obtained hysteresis loops, the saturation magnetization (M_s) and coercivity (H_c) were determined.

3. Results and discussion

Turbidimetric measurements were performed to determine dissolution properties of solutions. It is interpreted that powder-based precursors are completely dissolved as turbidity value approaches to 0 ntu and they are not dissolved when some powder particles are suspended in a solution at near 1000 ntu. The fabrication of homogeneous and continuous thin film is directly related to turbidity value which is 0 ntu. According to these measurements results for solutions, BM1, BM2, BM3 and BM4 were obtained like; 280 ntu, 240 ntu, 200 ntu and 186 ntu respectively. Although we have homogenous and transparent solutions, the turbidity values are high because of iron colour. Due to the fact that iron has red colour, the turbidity values increase even if they are dissolved very well.

The pH values of BM1, BM2, BM3 and BM4 solutions were 5,20, 4,20, 3,38 and 3,19 respectively. Since the acidic content of BM4 is higher than that of the others, the pH value is lower. The pH value of the solution is an important factor influencing the formation of the polymeric three-dimensional structure of the gel during the gelation process; it should be taken into consideration during solution preparation.

A characteristic feature of many sol-gel solutions is the shear-rate or test time dependence of the viscosity. It was determined that the viscosity values of the BM1, BM2, BM3 and BM4 solutions were approximately equal to 2.8 mPa.s, 2.5 mPa.s, 2.3 mPa.s, and 2 mPa.s respectively. We have to point here that the viscosity values of the solutions are almost similar that shows a key factor in controlling film thickness. The obtained results are reasonable for sol-gel processing because the thicknesses of the films decrease from BM1 to BM4

Due to optimisation of drying, heat treatment and annealing regimes, DTA/TG curves were obtained from the xerogel which was dried at 200 °C for 30 minutes. As shown in Figure 1, the exothermic and endothermic reactions occur in a temperature range between 30 °C and 1000 °C. It was determined that there are four different phenomena including removal of solvent-based materials, combustion of carbon-based content, formation of oxides and garnet formation. The first phenomenon is removal of solvent. In this case, physical water adsorbed in gel was evaporated occurs at 100 °C. Secondly the combustion of carbon-based content occurs at 300 °C. Thirdly oxidation occurs at 400 °C and at last step formation of garnet phase starts at 700 °C and reaches a maximum point at 800 °C. According to TG results in Figure 1, weight loss of the Bi-YIG xerogel up to 500 °C was determined as ~64%. Based on these results, a heat regime for coating process was determined at temperatures such as 500°C and 700-1000°C which indicate respectively combustion or drying, oxidation or heat treatment and annealing processes prior to not using a sol-gel coating processes.

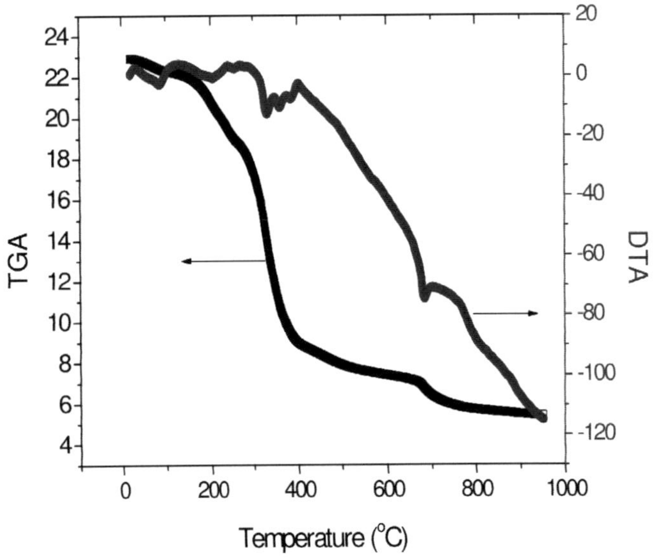

Figure 1. DTA and TG curves of Bi-YIG xerogel.

Figure 2 shows XRD patterns of selected samples from Bi-YIG coatings. It is clear from Figure 2 that all produced samples contains YIG phase. However, there are $FeYO_3$ (YIP), Fe_2O_3 and orthorhombic YIG or Bi-YIG phases in our films (shown with circle symbols). According to XRD results, the patterns show that there is partially YIG phase in all films. Even though YIG phase partially presents in the films Fe_2O_3 (shown with diamond symbols) and $FeYO_3$ (shown with square symbols) phases present at all specimens from BM1 to BM4. As shown in Figure 2, lattice constant of YIG increases with Bi substitution causing the peaks of YIG slight shift since radius values of Bi^{3+} ion is different from that of Y^{3+}. The radius of Bi^{3+} is 0.103 nm, whereas the radius of Y^{3+} is 0.09 nm. This results in a slight shift of YIG peak due to their characteristic properties. Studies of Haitao ve Shouhua [11] show that annealing process of Bi-YIG films at 800 °C causes to obtain Fe_2O_3, $FeYO_3$ ve Bi-YIG together. They reported that they increased annealing temperature above 800 °C and also Bi-YIG fraction was increased in the specimen. Similar results can be seen in our studies. X.W. Zhang [12] reported that YIG phase is not obtained at temperatures below 800 °C. Increasing temperature accelerates the reaction of Fe_2O_3 with $FeYO_3$ to form YIG as shown in Equation 1.

$$3FeYO_3 + Fe_2O_3 \rightarrow Y_3Fe_5O_{12} \ (T>800\ ^{\circ}C) \quad (1)$$

According to these results, it is obvious that time for transformation of YIG is not enough. Also many papers reports that Fe_2O_3 and $FeYO_3$ are inter-phases for YIG formation.

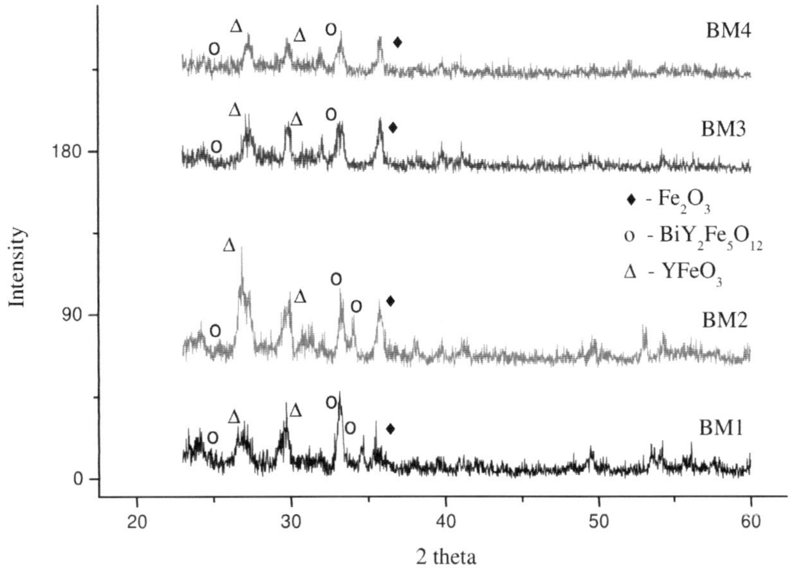

Figure 2. XRD patterns of the prepared specimens.

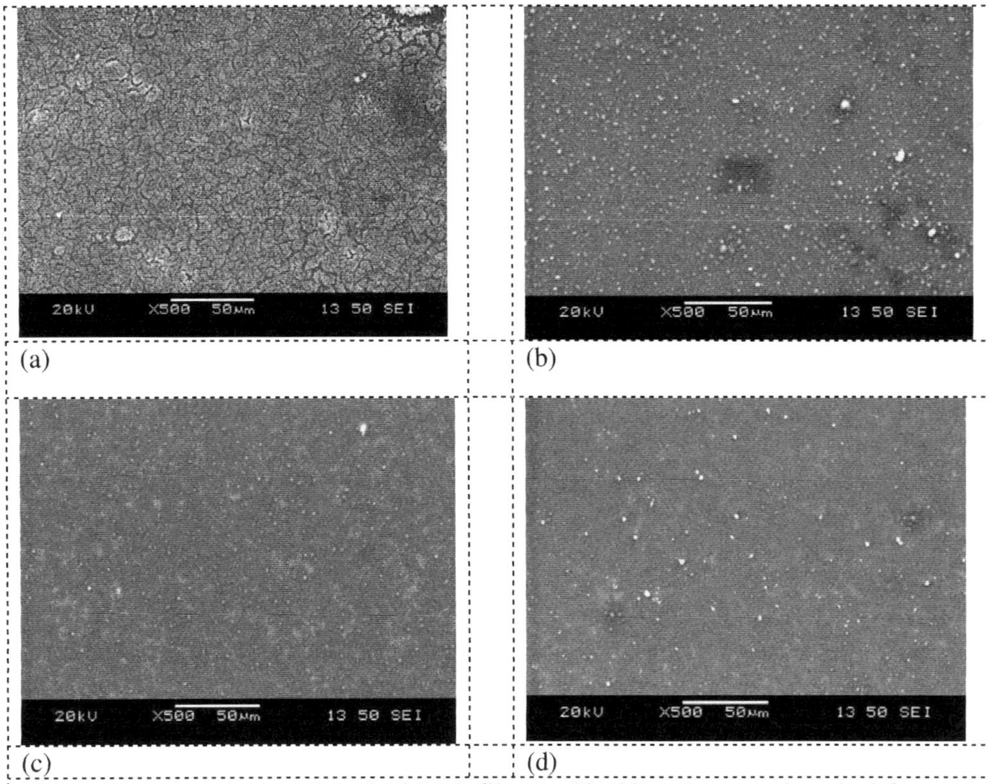

Figure. 3 SEM micrographs of Bi-YIG film on Si(100) substrates deposited from (a) BM1, (b) BM2, (c) BM3 and (b) BM4 solutions. The scale bars are 10 μm.

Microstructural properties of Bi-YIG films are shown in Figure 3 from BM1 to BM4. Since the molar concentration of the solutions decreases from BM1 to BM4 according to C.R. values, the fraction of the cracks, pinholes and deposition islands also decreases too. The obtained structure is directly related with deposition process. Spin coating technique involves a well spreading of the film. In addition, their microstructures are directly affected from pH value, heat treatment regime, and viscosity. Decreasing viscosity causes crack free deposition which influences good magneto-optic properties.

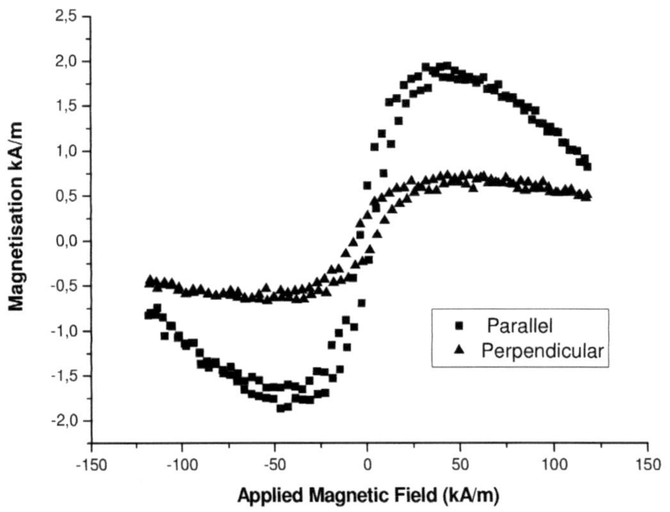

Figure 4. M-H curves of the Bi-YIG film at room temperature on a Si(100) substrate produced from BM2 solution.

The magnetization M versus applied magnetic field H curves was obtained from a VSM device where diamagnetic contribution is not substracted. We have measured for two configurations: the applied field parallel and perpendicular to the film plane. Since the results are similar for all our samples, we have shown here results for a selected sample. Prepared Bi-YIG specimens show magnetic properties in-plane and perpendicular configurations. Along with VSM results we should also regard XRD results: the XRD results indicate a small value of magnetization due to lower fraction of Bi-YIG is in comparison to other phases present. According to XRD result in Figure 2, the existing phases are Fe_2O_3 $YFeO_3$ and $Y_3Fe_5O_{12}$. Fe_2O_3 has paramagnetic behaviour and also $YFeO_3$ and $Y_3Fe_5O_{12}$ are ferrimagnetic materials but there is evidently a diamagnetic affect in our measurements dominating other phases. Si (100) was used as a substrate. Due to the fact that bulk Si is a diamagnetic material, the curve probably is affected from this condition as explained in Ref. [15]. However for perpendicular configurations diamagnetic effect is suppressed. This can be attributed to film perpendicular case has nonzero de-magnetization coefficient whereas for in plane case de-magnetization coefficient is almost zero.

4. Summary and Conclusions

Bi-YIG thin films were synthesized on Si(100) substrates from the solutions prepared from Bi, Y, and Fe based precursors, methanol, glacial acetic acid using sol–gel process for magneto-optical applications. The experiments were carried out at non-defined conditions just to obtain phase. At further studies we will perform experiments at refined conditions. The DTA-TG curve revealed that endothermic and exothermic reactions occur at temperatures between 30 °C and 1000 °C. The formation of Bi:YIG phase was occurred in the range of 700 °C and 800 °C. We decided the heat treatment regime as temperature versus time but the reaction kinetics carried out slower than it was thought. To optimize the heat treatment and to obtain exact phase and good magnetic properties, different experiments will be performed at different regimes.

Acknowledgments: The authors would like to acknowledge to Dr. S. Tari at Dept. of Physics of Izmir Institute of Technology, for supporting of the magnetic characterization studies and Ms. I. Birlik, Mr. M.F. Ebeoglugil, Mr. O. Culha and Mr. O. Sancakoglu at Dept. of Metallurgical and Materials Engineering of Dokuz Eylul University for helping structural characterization studies like XRD and SEM. We also thank to Dr. G. Kahraman for valuable discussions. This work has been supported by The Scientific and Technological Research Council of Turkey (TUBITAK).

References:

[1] Qinghui Y, Zhang H, Liu Y, and Wen Q 2007 *IEEE Trans. on Mag.* **43** 9
[2] Fujikawa R, Baryshev A V, Kim J, Uchida H and Inoue M 2008 *J. Appl. Phys* **10** 7
[3] Körner T, Heinrich A, Weckerle M, Roocks P and Stritzker B 2008 *J. App. Phy.* **103** 07B337
[4] Shintaku T and Uno T 1996 *Jpn. J. Appl. Phys.* **35** 4689
[5] Tate A, Uno T, Mino S, Shibukawa A, Shintaku T 1996 *Jpn. J. Appl. Phys.* **35** 3419
[6] Wickersheim K A and Buchanan R A 1967 *J. Appl. Phys.* **38** 1048
[7] DiBiccari A 2002 *Sol-gel processing of* $R_xY_{3-x}Al_yFe_{5-y}O_{12}$ *magneto-optical films* M.S. Thesis Virginia Polytechnic Institute and State University Materials Science and Engineering Department (Virginia)
[8] Inoue N and Yamasawa K 1996 *Elect. Eng. In Jpn.* **117** 1
[9] Sekijima T, Kishimoto H, Fujii T, Wakino K and Okada M 1999 *Jpn. J. Appl. Phys.* **38** 5874
[10] Hench L L and West J K 1990 *Principles of Electronic Ceramics* (New York: John Wiley & Sons)
[11] Xu H, Yang H, Xu W and Feng S 2008 *J. Matter. Proc. Technol.* **197** 296
[12] Zhang X-W 2008 *Journal of Crystal Growth* **310** 3235
[13] Zhou X, Cheng W, Lin F, Ma X and Shi W 2006 *Applied Surface Science* **253** 2108
[14] Kumar N, Misra D S, Venkataramani N, Prasad S, Krishman R 2004 *J. Mag. Mag. Matter.* **272** 899
[15] Kopnov G, Vager Z and Naaman R 2007 *Adv. Mater.* **19** 928

The Effect of NaCl prepared by ultra-sonic vibration on Sintering of annealed FePt nanoparticles

M. Farahmandjou[1, 4], S. A. Sebt[2], S. S. Parhizgar[2], P. Aberomand[2] and M. Akhavan[3]

[1]Department of Physics, Varamin (Pishva) Branch, Islamic Azad University, Varamin, Iran
[2]Physics Research Center, Science and Research Branch, Islamic Azad University, Tehran, Iran
[3]Magnet Research Laboratories (MRL), Department of Physics, Sharif University of Technology, Tehran, Iran

E-mail: farahmand_ph@iauvaramin.ac.ir

Abstract. FePt nanoparticles were synthesis via high-temperature solution phase. Monodispersed 3.5 nm FePt nanoparticles were first obtained by reduction of $FeCl_2.4H_2O$ and $Pt(acac)_2$. These preformed FePt nanoparticles were then combined with NaCl to prevent from sintering after annealing. The 1 micron salt particles were fabricated by ultra-sonic vibration method and used as the separating media. Annealing studies show that these salt-matrix FePt nanoparticles are stable after annealing at 700 °C for 4 h. The annealed FePt particles diameter increases from 4 to 25 nm as observed by TEM. The magnetic properties are largely affected by annealing temperature. The role of these salt particles on sintering of annealed FePt nanoparticles has been investigated.

1. Introduction

$L1_0$ FePt nanoparticles have a large magnetocrystalline anisotropy and, thus, exhibit large coercivity at room temperature, even when their size is as small as several nanometers. FePt nanoparticles have been intensively studied for many applications such as high-density recording media [1-3] and ferromagnetic nanocomposites [4]. Undesirable sintering between nanoparticles, which takes place when nanoparticles are annealed to transform the crystalline structure from the chemically disordered face-centered cubic (fcc) to the chemically ordered L10 phase (face-centered tetragonal), should be avoided [5]. Among currently available techniques for preparing FePt nanoparticles, wet chemical methods produce the nanoparticles with well controlled particle size and size distribution. Unfortunately, as-synthesized FePt nanoparticles always have fcc structure and have no magnetic ordering at room temperature. Heat treatments to convert the fcc structure to the face-centered tetragonal (fct) structure with high magnetic anisotropy must be performed at elevated temperatures (>600 °C) which undesirably lead to the particle agglomeration and sintering.

To solve this dilemma, extensive efforts have been made since 2000 to obtain monodisperse fct FePt nanoparticles. One of the approaches in such efforts is to lower the onset temperature for the fcc to fct phase transformation. Doping by elements such as Ag, Au, and Cu in the FePt phase was tested resulting in the onset of the fct phase at 400 °C [6, 7]. Direct synthesis of the fct structured particles by the polyol process using high boiling point solvents is another major approach [8–11]. In this approach, higher synthesis temperatures can be applied which allows partial formation of the fct phase. Recently, methods have been tested to coat the fcc FePt nanoparticles with high-melting-point materials and then to anneal the mixture to obtain the fct nanoparticles without sintering. Immiscible silica matrix [5], polyethyleneimine linker molecules [12], SiO shells [13,14] and TiO matrix [15] were used as the high-melt-point materials. The advantage of this coating approach is that high temperature annealing can be performed without causing particle sintering. The coating materials cannot be easily removed after the annealing so that these methods are not suitable in large quantity monodisperse magnetic nanoparticles free of nonmagnetic impurities. Other approaches to obtaining monodisperse fct nanoparticles include rapid thermal annealing [16] and pulse laser annealing [17]. To date, however, it is still a very challenging task to obtain both high coercivity and monodisperse morphology simultaneously of fct FePt nanoparticles. Most reported results show a trade-off between the two properties. It is, therefore, highly desirable to find processing techniques to produce completely transformed fct FePt monodisperse nanoparticles. A new technique has been adopted, based on salt-matrix annealing at temperatures up to 700 °C, with which monodisperse fct FePt nanoparticles have been successfully produced [18].

In this invited paper we present our recent research results in producing ferromagnetic FePt nanoparticles by salt-matrix method. We will demonstrate that the size of NaCl prepared by ultra-sonic vibration affects on diameter of FePt nanoparticles during the salt-matrix annealing.

2. Experimental details

The fcc FePt nanoparticles with size of 3.5 nm were prepared using the synthesis described by sun et al. [1]. Synthesis of the nanoparticles, involves the reduction of $Pt(acac)_2$ and $FeCl_2.4H_2O$, in phenyl ether solvent in presence of 1,2-hexadecanediol. The oleic acid and oleylamin surfactants were added to solvent, at 100 °C, as a protective agent in order to prevent agglomeration and oxidation. By adding superhydride ($LiBEt_3H$) under a blanket of N_2 at 200 °C, followed by refluxing, the FePt nanoparticles were formed. The refluxing temperature was fixed at 245 °C for 30 min. The black reaction mixture was cooled to room temperature and then combined with ethanol to remove the impurity. The product was precipitated and separated by centrifugation (8000 rpm, 10min). Any undissolved material was removed by centrifugation. Then nanoparticles were dispersed in hexane in presence of surfactants.

Sodium chloride (NaCl) was selected as a separating media in this investigation due to its chemical stability and high solubility in water. The salt particles were produced by two different methods: (i) The salt powder was crumbled by mechanical ball-milled for more than 24 h to reach particle size smaller than 20 microns. The ball-milled NaCl powder was then dispersed in hexane. (ii) Pure NaCl powder was solved in de-ionized water to reach saturation and then hexane solution was added to the saturated solution of salt water and ultra-sonic vibration was applied simultaneously. Ultra-sonic vibration led to formation and dispersion of the 1 micron salt crystallized particles in hexane phase.

The salt particles were mixed with hexane dispersion of 3.5 nm as-synthesized fcc FePt nanoparticles. The mixture was stirred until all the solvent evaporates at room temperature and then annealed in forming gas (93%Ar +7%H2) to complete the fcc to fct transition. The NaCl-to-FePt weight ratio 100:1 was chosen and annealing temperature fixed at 700 °C for 4 h. The annealed mixture which salts prepared using method (i) washed in de-ionized water and centrifuged for several times to remove all the NaCl particles and for nanoparticles which salts prepared by method (ii) washing process did not necessary because the fine salt particles removed from nanoparticles and dispersed in hexane [18]. The effect of NaCl on FePt particle size was investigated by each of the two salt preparation methods. Morphology of FePt nanoparticles before and after annealing was observed by a transmission electron microscope (TEM) analysis using a Philips EM 208 TEM (100KV) with resolution 200KX. To determine the nanoparticles structure, the X-ray diffraction (XRD) measurement was prepared after evaporation of hexane on a Silicon wafer using a Seifert with Cu-K_α (wavelength = 1.54 A°) radiation. The magnetization of the FePt samples in a variable magnetic field was measured using a vibrating sample magnetometer (VSM) model PAR-155.

3. Results and discussion

Figure 1 shows XRD patterns of the as-synthesized and annealed FePt nanopartcles; the annealing was done in forming gas at 700 °C for 4 h. The as-synthesized 3.5 nm particles exhibit a disordered fcc structure with lattice constant 3.871 A° which comply with the report made on FePt structures analysis [18]. The X-ray patterns confirm that the annealing in the salt-matrix transforms the disordered fcc structure into the tetragonal $L1_0$ structure. Because of the small particle size, the peaks are very broad and hence overlap each other. The average particle diameter estimated from scherrer's formula is consistent with that determined by statistical analysis of the TEM images (figure 3), indicating that each individual particle is a single crystal.

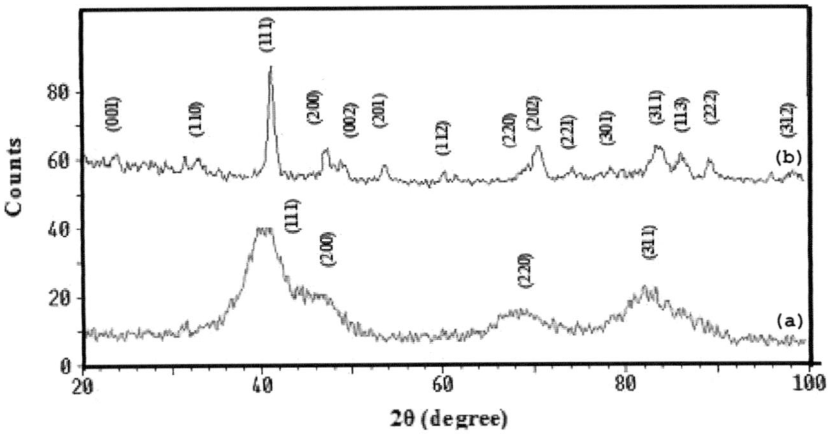

Figure 1. XRD patterns of the (a) as-synthesized and (b) annealed FePt nanoparticles at 700 °C for 4 h with NaCl : FePt ratio 100:1.

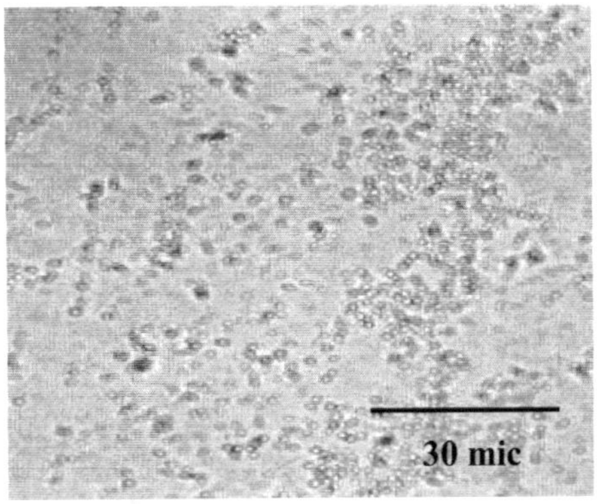

Figure 2. The image of 1 micron salt particles prepared by ultra-sonic vibration method.

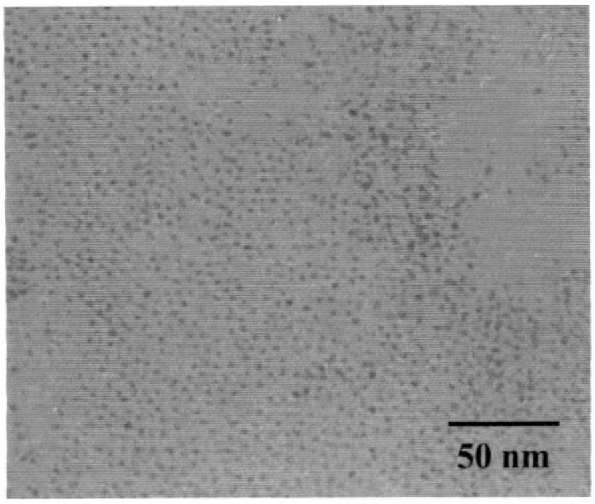

Figure 3. TEM image of the 3.5 nm as- synthesis FePt nanoparticles.

Figure 2 show the image of the salt particles fabricated by ultra-sonic vibration method with salt water concentration of 190 mg/mL and an average salt particle size about 1 micron and ultrasonic duration of 45 min. Figure 3 indicates TEM image of the as-synthesized fcc FePt nanoparticles. The particles prepared at 245 °C, have an average diameter of about 3.5 nm and narrow size distribution with standard deviation of about 8%.

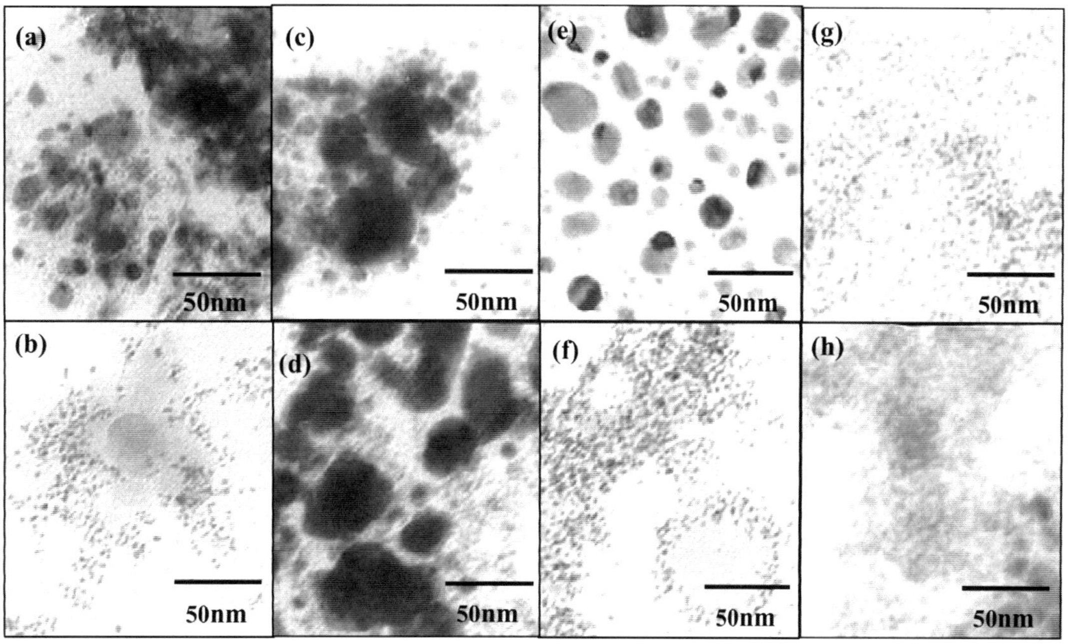

Figure 4. TEM images of fct FePt nanoparticles annealed at 700 °C for 4 h. The annealed (a) 15 and (b) 4.5 nm fct FePt particles before and after ultra-sonic vibration without NaCl; (c) 10 and (d) 4.5 nm with ultra-sonic duration of 15 and 30 min with NaCl; (e) 25 and (f) 4.5 nm with and without NaCl after ultra-sonic vibration. The TEM images of (g) and (h) show the effect of stabilizers on $L1_0$-FePt particles with and without surfactants respectively.

Figure 4 shows the TEM images of annealed FePt particles at 700 °C for 4h. Figures 4(a) and (b) show annealed FePt particles without salt media. In figure 4(a) the fct FePt nanoparticles were dispersed in hexane solvent and in figure 4(b) the ultra-sonic vibration was also done on dispersed particles for 30 min. As we can see the size of nanoparticles were decreased by ultra-sonic vibration.

Figures 4(c), (d), (e) and (f) show annealed salt-matrix FePt particles. The salt powder was crumbled by ball milled machine to reach particle size smaller than 20 microns and mixed with 3.5 nm as-synthesis FePt nanoparticles and then mixture annealed. In figures 4(c) and (d) the ultra-sonic time was increased from 15 to 30 min for mixture respectively. It is found that the diameter of fine FePt particles was decreased by increasing the ultra-sonic duration. In figures 4(e) and (f) the annealed mixture were washed without and with de-ionized water respectively. It is obvious that when the salts removed from nanoparticles and then ultra-sonic process applied the size of FePt particles decreases.

Figures 4(g) and (h) show annealed salt-matrix FePt nanopaticles which NaCl were fabricated by ultra-sonic vibration method. The 1 micron salt particles dispersed in hexane were mixed with 3.5 nm as-prepared FePt nanoparticles and then mixture annealed. For these annealed particles we also investigated the stability of nanoparticles during ultra-sonic vibration using oleic acid and oleylamine surfactants with 2:1 ratio. It can be seen that the stabilizers control the spaces between nanoparticles and the dispersity and stability of fct FePt particles increases.

Figure 5 shows the results of magnetic measurements before and after heat treatments. It indicates that as-synthesized FePt nanoparticles was first superparemagnetism at room temperature and after annealing under forming gas at 700 °C for 4 h, the saturation magnetization increases to 45 emu/g and coersivity raises to 7.5 kOe. These results confirm the suitable magnetic property of nanoparticles.

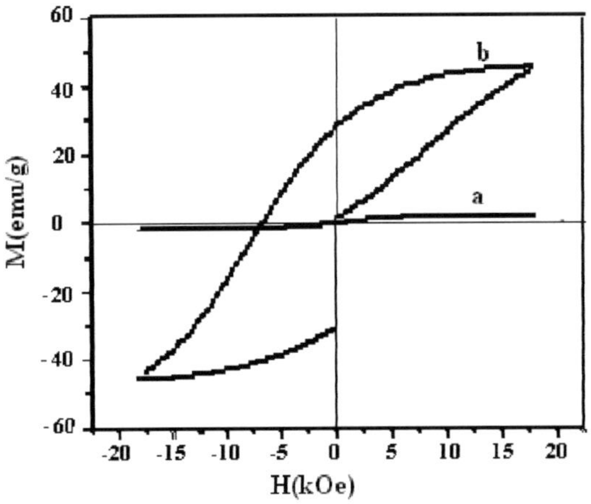

Figure 5. Magnetic hysteresis loops of (a) as-synthesized and (b) annealed FePt nanoparticles at 700 °C for 4 h.

4. Conclusions

Salt (NaCl)-matrix annealing, an easy-handling method, has been developed to produce fct FePt nanoparticles. It has been realized that the salt particles prepared by ultra-sonic vibration method make well-separated annealed FePt nanoparticles related to NaCl fabricated by ball milled method. The effects of ultra-sonic vibration, ultra-sonic duration, salt removing and surfactant stabilizers on fct FePt nanoparticles were investigated and realized that each of parameters can be important in size decreasing and stabilizing of the FePt nanoparticles. We found the following: (i) The size of $L1_0$-FePt particles was decreased from 25 nm to 5 nm by ultra-sonic vibration action. (ii) The diameter of annealed nanoparticles was decreased from 10 to 4.5 nm as a result of increasing the ultra-sonic duration from 15 to 30 min. (iii) Due to salt removing from fct FePt particles by de-ionized water the diameter of nanoparticles was decreased from 25 nm to 4.5 nm. (iv) The average distance between 4 nm FePt nanoparticles was controlled by using surfactant stabilizers. Finally (v) in purification stage the NaCl prepared by ultra-sonic vibration method was removed from fct FePt particles without washing. In conclusion, the NaCl fabricated by ultra-sonic vibration method make more advantageous annealed FePt nanoparticles in size and particles dispersion as compared to the ball milled method.

Acknowledgments

This work was supported in part by the Physics Research Center, Science and Research Campus at Islamic Azad University and the Centre of Excellent in Complex Systems and Condensed Matter (CSCM) at Sharif University of Technology.

References

[1] Sun S, Anders S, Thomas T, Baglin J E E, Toney M F, Hamann H F, Murray C B and Terris B D 2003 J. Phys. Chem. B, **107** 5419
[2] Kodama H, Momose S, Ihara N, Uzumaki T, Tanaka A 2003 *Appl. Phys. Lett.* ***83*** 5253
[3] Christodoulides J A, Zhang Y, Hadjipanayis G C, Fountzoulas C 2000 *IEEE Trans. Magn.* **36** 2333.
[4] Rong C B, Zhang H W, Du X. B, Zhang J, Zhang S Y, Shen B G 2004 *J. Appl. Phys.* ***96*** 3921
[5] Ding Y, Majetich S A, Kim J, Barmak K, Rollins H, Sides P 2004 *J. Magn. Magn. Mater.* **284** 336
[6] Kitakami O, Shimada Y, Oikawa K, Daimon H, Fukamichi K 2001 *Appl. Phys. Lett.* **78** 1104
[7] Platt C L, Wierman K W, Svedberg E B, Veerdonk R V, Howard J K, Roy A G, Laughlin D E 2002 *J. Appl. Phys.* **92** 6104
[8] Kang S, Jia Z, Shi S, Nikles D E, Harrell W 2005 J. *Appl. Phys. Lett.* **86** 62503
[9] Jeyadevan B, Urakawa K, Hobo A, Chinnasamy N, Shinoda K, Tohji K, Djayaprawira D D J, Tsunoda M, Takahashi M 2003 *Jpn J. Appl. Phys.* **42** 350
[10] Sato K, Jeyadevan B, Tojhi K 2003 *J. Magn. Magn. Mater* **266** 227
[11] Jeyadevan B, Hobo A, Urakawa K, Chinnasamy C N, Shinoda K, Tohji K 2003 *J. Appl. Phys.* **93** 7574
[12] Mizuno M, Sasak Yi, Yu A C C, Inoue M 2004 *Langmuir* **20** 11305–11307
[13] Yamamoto S, Morimoto Y, Ono T, Takano M 2005 *Appl. Phys. Lett.* **87** 032503
[14] Aslam M, Fu F, Li S, Dravid V P 2005 *J. Colloid Interface Sci.* **290** 444
[15] Sort J, Suriñach S, Baró M D, Muraviev D, Dzhardimalieva G I, Golubeva N D, Pomogailo S I, Pomogailo A D, Macedo W A A, Weller D, Skumryev V, Nogués J 2006 *Adv. Mater.* **18** 466
[16] Zeng H, Sun S, Sandstrom R L, Murray C B 2003 *J. Magn. Magn. Mater.* **266** 227
[17] Saita S, Maenosono S 2004 *J. Phys.: Condens. Matter.* **16** 6385
[18] Elkins K, Li D, Poudyal N, Nandwana V, Jin Z, Chen K, Liu J P 2005 *J. Phys. D: Appl. Phys* **38** 2306

The effect of milling speed on the magnetic properties of mechanically alloyed Fe-45%Ni powders

Kh Gheisari[1, 2,3], S Javadpour[1] and J T Oh[2]

[1] Dept. of Materials Science and Engineering, School of Engineering, Shiraz University, Shiraz, 7134851154, Iran
[2] School of Materials Science and Engineering, Nanyang Technological University, Nanyang Avenue, Singapore 639798, Singapore

Abstract. FeNi-based alloys are commonly used as soft magnetic materials, exhibiting high permeability, low coercivity and high saturation magnetization. The present study is focused on the effect of the milling speed (vial and disc rotation speed) on the magnetic properties of nanocrystalline Fe–45 wt. % Ni powders. The phase analysis and the structural properties were characterized by the use of X-ray diffraction. Structural parameters including lattice parameter, crystalline size and lattice strain were obtained from Rietveld's powder structure refinement analysis of X-ray powder diffraction data. The magnetic measurements were performed, using a vibration sample magnetometer (VSM) at room temperature. The results showed that the higher milling speed leads to larger lattice parameter, lower crystalline size, higher saturation magnetization and lower coercivity of the investigated system.

1. Introduction

Magnetic iron-nickel alloys, generally called permalloys, are industrially used due to high saturation magnetization and low coercivity [1]. Fe-45%Ni alloys are one of the major iron-nickel compositions, well-known for high flux density (Bs=1.6 T), low coercivity (H_c=0.3 Oe) and their responsiveness to the magnetic annealing [2].

Nanostructured materials have a significant influence on physical and especially magnetic properties. The coercivity of nanocrystalline soft magnetic materials strongly depends on crystal size. In accordance with the exchange anisotropy model, decrease in the grain size (D<40 nm) leads to a rapid decrease in coercivity, in proportion to the sixth power of the grain size [3]. Furthermore, such materials are of interest for magnetization saturation since the reduction of grain size to single magnetic domain size offers the possibility to eliminate the influence of magnetic walls [4, 5].

In order to improve such magnetic properties, nanostructured FeNi alloys have been produced by distinct methods such as thermal evaporation [6], electrodeposition [7], metal-plasma reaction [8] and mechanical alloying [9-13]. Among these methods, mechanical alloying (MA) is a powerful technique which allows the preparation of nanostructured FeNi alloys [9-13]. It is known that this process

[3] Corresponding author. Dept. of Materials Science and Engineering, School of Engineering, Shiraz University, Shiraz,7134851154, Iran. Tel: +98-913-3715597, Fax: +98-711-6287294
Email address: khgheisari@shirazu.ac.ir

involves a repeated welding, fracturing and rewelding of powder particles in a high-energy ball mill [14]. In this method, final structure and morphology of the powders are influenced by the process variables such as milling time, milling speed, ball-to-powder weight ratio and process control agent [14]. In the past, several attempts have been made to simulate the dynamics of the milling process in terms of ball velocity, frequency of impact and kinetic energy transferred to the powder charge during milling [15-18]. Abdellaoui and Gaffet have derived a set of kinetic equations to compute the collision velocity and kinetic energy released from the ball to the powders [17, 18].Based on their model and hypotheses, the collision velocity and kinetic energy were obtained in the following forms:

$$\left\|\vec{V}_c\right\|^2 = (R\Omega)^2 + (r - r_b)^2\omega^2(1 + \frac{2\omega}{\Omega}) \quad (1)$$

$$E_k = \frac{1}{2}m\left\|\vec{V}_c\right\|^2 \quad (2)$$

Where R is the distance between the disc center and the vial center, r is the vial radius, r_b is the ball radius, Ω is the disc rotation speed, ω is the vial rotation speed and m is the ball mass [18].

In our previous work, attention has been paid to the effects of the ball milling speed on the structural properties of nanocrystalline Fe-45 wt. % Ni powders [19]. The current study is focused on the effect of ball milling speed (vial and disc rotation speed) on the magnetic properties of nanocrystalline Fe-45 wt. % Ni powders.

2. Experimental method

Iron powder with an average particle size of 100 μm and the purity of 99% and nickel powder with an average particle size of 5 μm and the purity of 99.8% were used. The powders were mixed to give the composition of Fe-45 wt. % and introduced into a cylindrical tempered steel vial of capacity of 100 ml. Each vial was loaded with two type of tempered steel balls (D (12mm): D (10mm) =1:1). The ball-to-powder weight ratio was 15:1. In order to minimize oxidation, the entire operation was performed in a dry argon atmosphere and vials were sealed with a Teflon O-ring. The MA process was carried out at ambient temperature by using a planetary ball mill with different value of Ω and ω (R = 120 mm, r = 45 mm, $\Omega \leq 303$ rpm, $|\omega| \leq 500$ rpm, $\Omega/|\omega|=1.65$), where Ω is the rotation speed of the disc on which the vial holders were fixed. The vial holders turn at a rotation speed of ω. R is the distance between the disc center and the vial center and r is the internal radius of the vials. The duration of milling process was 24h. Ball milling conditions, collision velocity and kinetic energy are mentioned in table 1. These values are based on Abdellaoui and Gaffet's findings [5].

The phase analysis and the structural properties were characterized by the use of X-ray diffraction (XRD, Shimadzu XRD-6000 diffractometer with Cu-Kal radiation). Structural parameters including grain size, lattice parameter, lattice strain, and the quantitative amount of FCC γ-(Fe, Ni) phase were obtained from Rietveld's powder structure refinement analysis of X-ray powder diffraction data. The Rietveld calculations were performed by the TOPAS 3 software (from Bruker AXS). The crystallographic models of FCC FeNi (Fm-3m, a=0.35975 nm), BCC Fe (Im-3m, a=0.28664 nm) and FCC Ni (Fm-3m, a=0.35238 nm) were used as starting models for the refinements. For each refinement, the following variations were applied: the background parameter, scale factor, cell parameter, zero point correction, Lorentzian crystal size and Gaussian lattice strain. In order to judge the quality of the fitting on the structure model, the Bragg reliability factor (R_{Bragg}) was used. The refinements resulted in proper fits to the experimental data and average R_{Bragg} value was less than 6.5%. Magnetic measurements were obtained at room temperature using a vibrating sample magnetometer (VSM, Lakeshore 7404) with a saturating field of +/-10 kOe, field increment of 404 Oe and a field ramp rate of 40.4 Oe/s.

3. Results and discuss

3.1. Structure

Figure1 shows the X-ray patterns of ball milled powders which are given for different milling speeds. It is evident from this figure that in the course of ball milling the Fe and Ni mixture, the characteristic Ni and Fe lines decrease gradually and decay after 24h at vial rotation speeds higher than 350 rpm.

The X-ray diffraction data have been analyzed for the different milling speed. The average lattice strain, crystalline size, lattice parameter and quantitative amount of γ-(Fe, Ni) phase at different milling conditions are given in table 1.

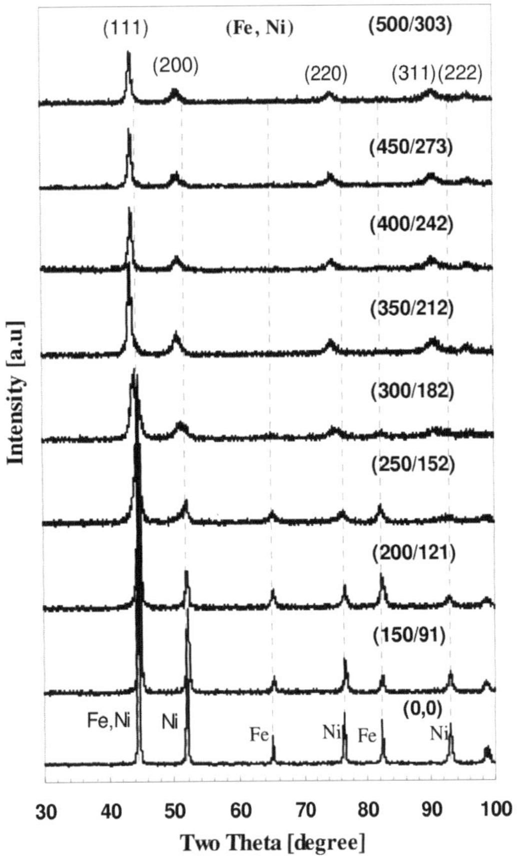

Figure 1. X-ray diffraction patterns of Fe–45 wt. % Ni powders milled for various milling conditions. (ω (rpm)/ Ω (rpm), milling time 24h).

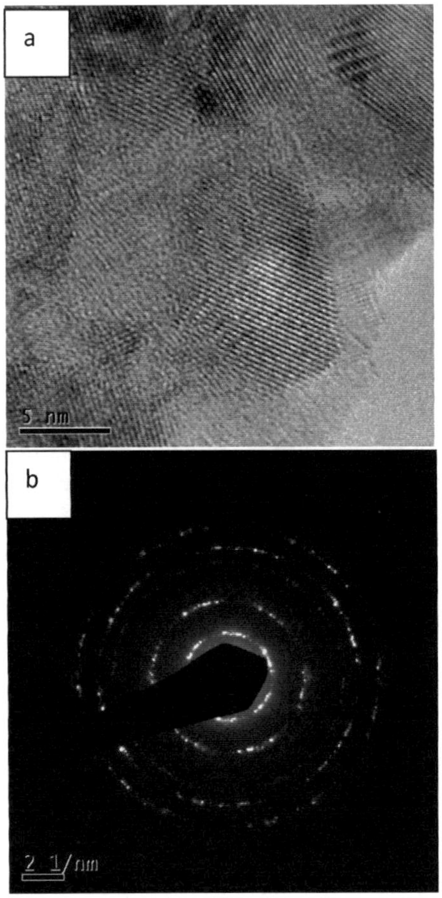

Figure 2. TEM micrograph of Fe–45 wt. % Ni powders milled at vial rotation speed of 350 rpm (a) bright field image, (b) corresponding diffraction pattern.

It is found that the faster milling speed transfers higher kinetic energy into the powders. Higher kinetic energy leads to the increase of lattice parameter and quantitative amount of γ-(Fe, Ni) phase. The crystalline size and lattice strain follow a different trend. The lattice strain of γ-(Fe, Ni) phase initially increases significantly with increasing milling speed up to 250 rpm (ω), after that, decreases slightly up to 350rpm (ω) and then remains constant at about 0.26 % for higher milling speeds. Unlikely, the

crystalline size of γ-(Fe, Ni) phase is found to fall moderately up to 250 rpm (ω), decreases slightly up to 350 rpm (ω) and then remains constant at about 15 nm for higher milling speeds.
The nanocrystalline structure of the powders is confirmed by the TEM image. Figure 2 shows the TEM bright field image and the corresponding diffraction pattern of the powder, milled at vial rotation speed of 350rpm. The average grain size is about 12 nm. This value is relatively consistent with the value obtained from XRD results. The diffraction pattern is indexed as FCC crystal structures. The corresponding diffraction pattern and the lattice fringes indicate an almost completely random orientation of the resulting nanocrystalline structure. TEM micrograph also shows some amorphous phase between nanocrystalline grains. Additional details of the structural properties have been discussed in our previous paper [19].

Table 1. Structural properties of γ-(Fe, Ni) phase obtained from Rietveld's powder structure refinement analysis of X-ray powder diffraction data at different milling conditions

Ball milling conditions ω (rpm)/ Ω (rpm)	Collision velocity Vc (m/s)	Kinetic energy per hit (J)	Lattice parameter (nm)	Crystalline size (nm)	Lattice strain (%)	Wt. (%)
150/91	0.2738	0.2099	0.353162	170	0.125	7.6
200/121	0.3651	0.3732	0.35224	230	0.154	22.8
250/152	0.4564	0.5832	0.3564	26.9	0.688	32.3
300/182	0.5476	0.8398	0.3597	20.3	0.492	71.8
350/212	0.6389	1.1431	0.3599	15	0.267	100
400/242	0.7302	1.4931	0.3606	16.19	0.267	100
450/273	0.8215	1.8897	0.3605	17.7	0.26	100
500/303	0.9128	2.3330	0.3613	15.19	0.25	100

3.2. Magnetic properties

Figure 3 shows examples of hysteresis curves (magnetization vs. applied field) for samples milled at the vial rotation speeds of 200 and 400 rpm (ω). Saturation magnetization, coercivity and retentivity can be obtained from these curves. It should be mentioned that the magnetic properties especially coercivity and retentivity are strongly dependent on magnetic measurement condition (increment field and ramp rate).
Figure 4 presents the saturation magnetization of the ball milled powders as a function of vial rotation speed. It is clear from the plot that the saturation magnetization increases nonlinearly with increasing the milling speed. The increase of the magnetization is found to be linked to the structural properties of ball milled powders at different milling speeds. With increasing milling speed from 150 rpm to 350 rpm, saturation magnetization increase from 140 to146 emu/g. In this range, there is a progressive disappearance of pure iron and nickel in such a way that γ-(Fe, Ni) phase completely is formed at vial rotation speed of 350 rpm. On the other hand, the larger the lattice parameter, the higher the saturation magnetization. It has been shown by Amils *et al.* [20] and R.Hamzaoui *et al.* [5] that there is a clear correlation between the saturation magnetization and the lattice parameter (increase of lattice parameter and the saturation magnetization). They have suggested that the observed magnetic transition may be partially related to the change in the density of states at Fermi level. Furthermore, crystalline size is decreased in this range from 150nm to 15nm. Each grain may be considered as being a single magnetic domain eliminating the influence from the magnetic walls [4, 5].

With increasing milling speed from 350 rpm to 500 rpm, no remarkable change of crystalline size and lattice strain is found and the powder is completely transformed into the γ -(Fe, Ni) phase (table 1). In addition to increase of lattice parameter, the probable phase transformation from nanocrystalline to amorphous might be another reason for increase in saturation magnetization in this region.
Figures 5 and 6 illustrate the variation of coercivity and retentivity as a function of milling speed, respectively. It is shown that coercivity decreases from 13 to 1.5 Oe as the milling speed increases from 150 to 400 rpm. Coercivity gradually increases at the higher milling speeds. Retentivity also changes similarly with increasing milling speed, in such a way the minimum value of coercivity and retentivity can be seen at the milling speed of 400 rpm. The low value of coercivity and retentivity at this milling speed can be due to the low value of crystalline size and lattice strain.

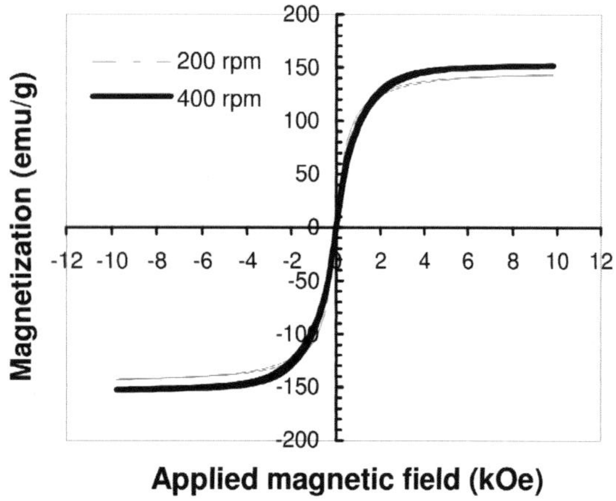

Figure 3. Hysteresis curves for samples milled at the milling speeds of 200 and 400 rpm.

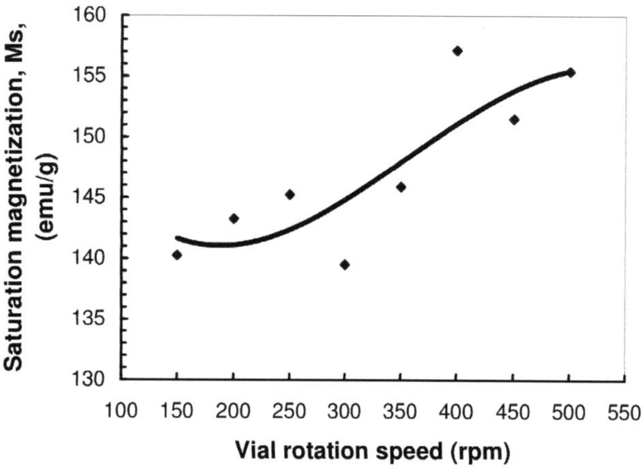

Figure 4. Saturation magnetization vs. vial rotation speed.

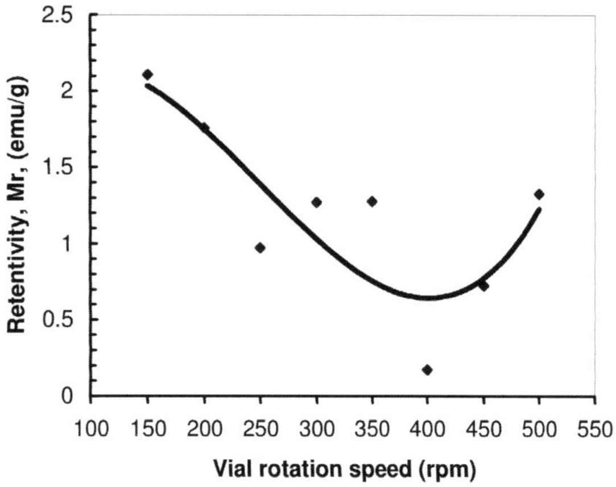

Figure 5. Coercivity vs. vial rotation speed.

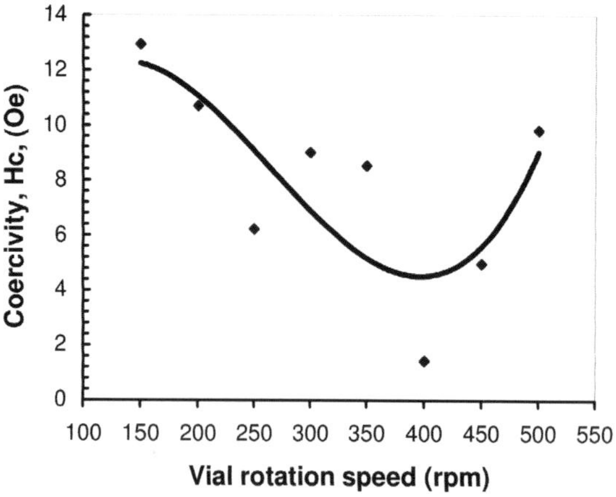

Figure 6. Retentivity vs. vial rotation speed.

In Figure 7, the coercivity values as a function of crystalline size are shown. According to Herzer' diagram [3], it can be concluded that the domain wall width of Fe-45%Ni alloy is about 15nm. The value of coercivity and retentivity increases at the higher milling speeds, while crystalline size and lattice strain remain constant. It seems that probably occurring amorphous phase at the high milling speed is responsible for this behavior.

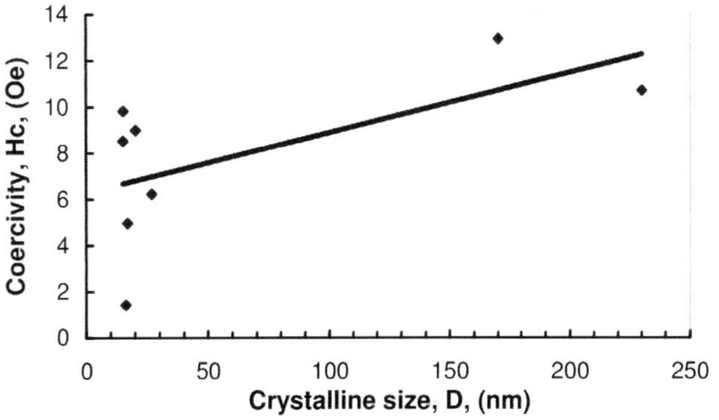

Figure 7. Coercivity vs. crystalline size for various ball milling conditions.

4. Conclusions

Nanostructured $Fe_{55}Ni_{45}$ powders have been prepared by high-energy ball milling of elemental Fe and Ni powders. The effect of ball milling speed on the magnetic properties has been studied. The important observations could be summarized as follow:

(i) Increasing milling speed led to an increase of the lattice parameter and a decrease of the grain size.

(ii) The saturation magnetization is increased and coercivity decreased with increasing milling speed up to 400 rpm. It is believed that the crystalline size reduction to single domain size and high value of lattice parameter could be the predominate factors in the magnetic properties in this region.

(iii) Unlike high saturation magnetization at the higher milling speeds (>400 rpm), coercivity and retentivity are decreased in this region. It seems, this was related to the formation of probably occurring amorphous phase at the high milling speed with the high kinetic energy per hit during milling process.

Acknowledgments

The authors would like to thank the Shiraz University Research Council and the Nanyang Technological University for providing financial support of this study.

References

[1] O'Handley R C 2002 *Modern Magnetic Materials* (New York: Wiley-Interscience)
[2] Bozorth R M 1964 *Ferromagnetism* (New York: D. Van Nostrand Co)
[3] Buschow K H J 2004 *Physics of Magnetism and Magnetic Materials* (New York: Kluwer Academic Publishers)
[4] Hamzaoui R, Elkedim O, Greneche J M and Gaffet E 2005 *J. Magn. Magn. Mater.* **294** 145–149
[5] Hamzaoui R, Elkedim O and Gaffet 2004 *Mater. Scie. Eng. A* **381** 363–371
[6] Baldokhin Y V, Kolotyrkin P Y, Petrov Y I and Shafranovsky E A 1994 *Phys. Lett. A* **189** 137
[7] Schirmer B and Wuttig M 1999 *Phys. Rev. B* **60** 945
[8] Li X G, Chiba A and Takahashi S 1997 *J. Magn. Magn. Mater.* **170** 339–345
[9] Valderruten J F, Perez Alcazar G A and Greneche J M 2006 *Phys. B* **384** 316–318
[10] Zhou P H, Deng L J, Xie J L, Liang D F, Chen L and Zhao X.Q 2005 *J. Magn. Magn. Mater.* **292** 325
[11] Kaloshkin S D, Tcherdyntsev V V and Tomilin I A 2001 *Phys. B* **299** 236–241
[12] Tcherdyntsev V V, Kaloshkin S D, Tomilin L A, Shelekhov E V and Baldokhin Yu V 1999 *Nanostruct Mater.***12** 139-142.
[13] Guittoum A, Layadi A, Bourzami A, Tafat H, Souami N, Boutarfaia S and Lacour D 2008 *J. Magn. Magn. Mater.* **320** 1385–92
[14] Suryanarayana C 2001 *Prog. Mater. Sci.* **46** 1-184
[15] Maurice D R and Courtney T H 1990 *Metall. Trans. A* **21A** 289–303
[16] Maurice D R and Courtney T H 1995 *Metall. Mater. Trans. A* **26A** 2431-35
[17] Abdellaoui M and Gaffet E 1995 *Acta. Metall. Mater.* **43** 1087–98
[18] Abdellaoui M and Gaffet E 1994 *J. Alloys Compd.* **209** 351-361
[19] Gheisari Kh, Javadpour S, Oh J T and Ghaffari M 2008 *J. Alloys Compd.* to be published
[20] Amils X, Nogués J, Suriñach S and Baró M.D 1999 *J. Magn. Magn. Mater.* **203** 129–131

EPR study of Mn-implanted single crystal TiO_2

S. Güler[1]*, B. Rameev[1,2], R.I. Khaibullin[2], O.N. Lopatin[3] and B. Aktaş[2]
1 Department of Physics, Gebze Institute of Technology, 41400 Gebze- Kocaeli, Turkey
2 Lab. of Radiation Physics, Kazan Physical-Technical Institute, 420029 Kazan, Russia
3 Faculty of Geology, Kazan State University, 420008 Kazan, Russia

E-mail: sumeyra@gyte.edu.tr

Abstract. Single crystals of manganese-implanted TiO_2 rutile have been investigated by electron paramagnetic resonance (EPR) technique at room temperature. ESR spectra have been interpreted to correspond to the transitions among the spin multiplet (S=3/2) of the paramagnetic Mn^{4+} ion. Characteristic six-line hyper-fine splitting of the ESR spectra resulting from the spin I=5/2 of the Mn^{55} nucleus has been observed. Analysis of EPR spectra shows that manganese in TiO_2 rutile host substitutes for Ti^{4+} ions. Two equivalent Mn^{4+} centers have been observed in the EPR spectra in correspondence with two equivalent octahedral positions of Ti ions in the rutile structure. Parameters of the crystal field of orthorhombic symmetry on the Mn^{4+} centers have been obtained as result of computer modelling.

1. Introduction

The discovery of high-Curie temperature diluted magnetic oxides (various oxides doped by magnetic atoms, such as TiO_2, ZnO, SnO_2 and others) has attracted much attention due to their potential value for the development of the spintronic devices. A number of experimental works claim the room temperature ferromagnetism observed in various oxides. Despite of essential progress achieved in this field there is still an issue about either intrinsic or extrinsic magnetism is realized in these materials. It has been shown that the local environment of doped magnetic ions and oxygen defects have a drastic effect on the magnetic properties of these materials. There are a few theoretical calculations showing that the magnetic state of the doped ions is expected to be very sensitive to the local crystalline structure [1,2]. Therefore, there is a strong need in the studies that probe the local structure near the doped magnetic ion. On the other hand, Electron Paramagnetic Resonance (EPR) is well known technique to probe the local magnetic and structural properties of crystals. In particular, it is a very efficient method for the investigations of the local environment around a substitutional magnetic ion in various crystals [3,4].

In this study, the local structure of Mn ions in the single crystalline TiO_2 rutile doped by ion-beam implantation has been studied by EPR technique. The samples under study were (100) and (001) substrates implanted by Mn and annealed at very high temperature (T=950°C) to prevent manganese clusterization and to obtain highly diluted concentration of Mn ions in the TiO_2 host. We have shown that implantation with 40 keV Mn^+ ions with subsequent annealing at high temperature results in formation of the paramagnetic Mn^{4+} centers substituting Ti^{4+} ions in the rutile structure [1,2,5,10,11]. The parameters of spin Hamiltonian of the Mn^{4+} ions in the implanted TiO_2 rutile crystals have been obtained and discussed.

2. Experimental procedures

Single crystalline (100) and (001) TiO_2 rutile substrates have been implanted on the ILU-3 ion accelerator (Kazan Physical-Technical Institute) with 40 keV Mn^+ ions to a fluence of 1.50×10^{17} ions/cm^2 at ion current density of 8 μA/cm^2. The sample holder was cooled by flowing water during the implantation to prevent the samples from overheating. The implanted samples have been annealed at T=950°C in air during 1 hour. Structural studies have revealed that annealing procedure results in the diffusion of the most of manganese impurity out of the implanted layer with thickness of about 30 nm [3]. As another result of the post-implanting thermal treatment the samples has became transparent and the colour of samples has changed from dark-grey to light-orange.

The annealed samples have been studied by electron paramagnetic resonance (EPR) technique using *Bruker EMX* X-band spectrometer (9.8 GHz) at room temperature. Angular dependences of EPR spectra have been recorded with the static magnetic field rotated either in the plane of the plate-like samples ("in-plane" geometry) or in the two perpendicular planes ("out-of-plane" geometry) nearly coinciding with crystallographic planes of a substrate. That is in total there are three rotational planes for each sample to obtain the angular dependences of the EPR spectra.

3. Experimental results and discussion

For the both (001) and (100) manganese- doped TiO_2 rutile single crystals multi-line EPR spectra have been observed. The EPR spectra are strongly anisotropic with the resonance fields of the components up to 16 kG.

The typical EPR spectra observed for the ***(001) sample*** for the DC field applied along the [100], [001] and [110] crystal directions are shown in Fig. 1. The angular dependences of the resonance fields for ***the (001) sample*** on the rotation of the magnetic field in the (001) plane (in-plane geometry) and in the (100) plane (out-of-plane geometry) are presented in Fig.2a and Fig.2b, respectively. The angular dependences of the resonance fields for the ***(100) sample*** on the rotation of the DC field in the (001) plane (out-of-plane geometry) and in the (100) plane (in-plane geometry) are presented in Fig.3a and Fig.3b, respectively.

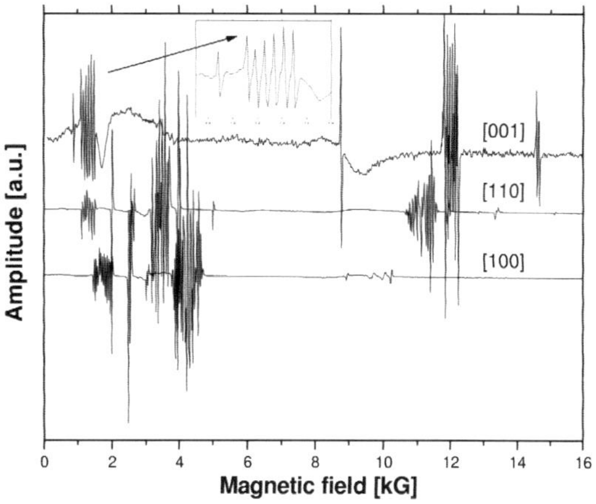

Figure 1. Experimental EPR spectra for three symmetry direction of the Mn-implanted TiO_2 ***(001) sample***. Narrow groups of lines belong to the Mn^{4+} centers, single lines are due to Fe^{3+} centers, and the broad lines is the background signal of the cavity and sample holder.

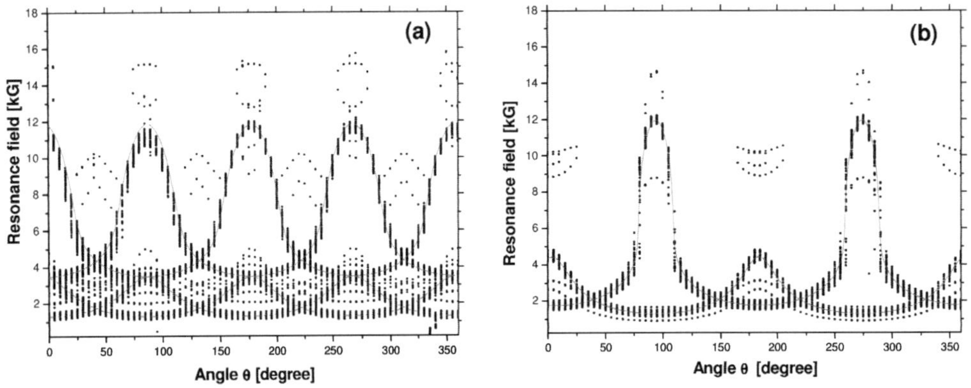

Figure 2. Experimental and modelled angular dependences of the EPR resonance fields of the Mn-doped TiO_2 ***(001) sample***. **(a)** – the static magnetic field (H) is rotated in the (001) plane (in-plane), **(b)** – H is in the (100) plane (out-of-plane). Symbols are the experimental data and lines are the theoretical fit results.

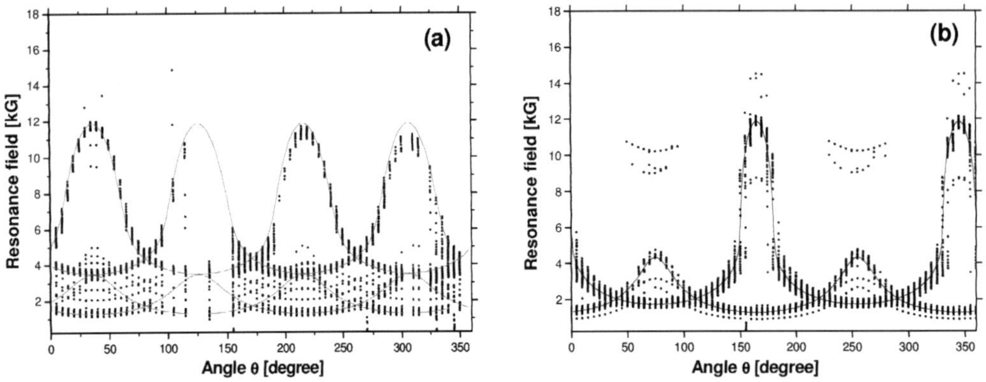

Figure 3. Experimental and modelled angular dependences of the EPR resonance fields of the Mn-doped TiO_2 ***(100) sample***. **(a)** – the static magnetic field (H) is rotated in the (001) plane (out-of-plane), **(b)** – H is in the (100) plane (in-plane). Symbols are the experimental data and lines are the theoretical fit results.

A strong anisotropy of the EPR lines as well as typical six-line pattern of the Mn_{55} hyperfine structure [1,2] clearly indicates that the observed EPR spectra are due to paramagnetic Mn ions. The (001) plane rotation patterns of both samples (Figs. 2&3) reveal a presence of at least two equivalent centers with symmetry axes perpendicular to each other. Besides, a set of additional EPR lines has been observed which EPR parameters and the resonance field dependence on angle unambiguously point to EPR of Fe^{3+} ions. It should be noted that room temperature EPR is very sensitive to any small concentrations of the paramagnetic Fe^{3+} centers [9]. Apparently the contamination by iron results from a non-perfect magnetic separation of ions in the implanter due to very close masses of the Fe and Mn atoms which are neighbours in the periodic table. EPR signal on Fe^{3+} centers in the Fe-implanted TiO_2 crystals have been studied in details in our previous work [3]. In this work we pay attention to the manganese resonance signal only. Manganese in rutile has a large fine structure splitting, so that for the most of the crystal orientations the hyperfine structure patterns of various electronic transitions are well

Journal of Physics: Conference Series **153** (2009) 012052 doi:10.1088/1742-6596/153/1/012052

separated from each other. Besides the manganese resonances could easily be distinguished from the Fe^{3+} signal by their characteristic hyperfine structure.

We found that the observed EPR manganese signal belongs to the Mn^{4+} centers substituting Ti atoms in the rutile lattice. Our theoretical analysis provides a set of the spin-Hamiltonian parameters which is very close to that of reported in the literature for Mn^{4+} ions in a bulk rutile crystal [1,2,5,10,11]. It should mentioned that both our and the literature data [1,2] indicate that in the recent works of D. Menzel *et al* and F. Iacomi et al [6,7] a wrong assignment of the single (without hyperfine structure!) EPR lines as the resonance signal of Mn^{4+} ions have been made. These lines should be attributed to other paramagnetic species (e.g. Fe^{3+}). Besides an assignment of the hyperfine lines at the central part of their spectra as due to Mn^{2+} ions is also suspicious.

It is known that the ground state of free Mn^{4+} ion ($3d^3$ electronic configuration) is ^4F. In the octahedral coordination the ground state of d^3 ions is an orbital singlet. However as result of the spin-orbital coupling a fine structure splitting of spin quartet of the ground state with nearly isotropic g-factor is observed in the EPR of d^3 ions [8]. Besides the hyperfine interaction of electron spin of Mn^{4+} with nuclear spin (I=5/2) results in the additional six-fold splitting of the spin levels. The unit cell of rutile contains two equivalent lattice sites of local orthorhombic (D_{2h}) symmetry. Thus, the spin Hamiltonian of the Mn^{4+} centers should include three terms which are: **1)** the nearly isotropic Zeeman term, **2)** the crystal field interaction of the orthorhombic symmetry and **3)** hyperfine interaction of the manganese nuclear and electron spins [8,12]. For S=3/2, only terms involving the second orders of the spin operators are necessary:

$$\mathrm{H} = \beta \mathbf{H}\tilde{\mathbf{g}}\mathbf{S} + D\left(S_z^2 - \frac{35}{12}\right) + E(S_x^2 - S_y^2) + \mathbf{I}\tilde{\mathbf{A}}\mathbf{S}\,, \tag{1}$$

where g, D, E and A are the experimentally determined spin-Hamiltonian parameters. Experimental data reveals a very small anisotropy of the A tensor of the hyperfine interaction. It is also obvious from experimental spectra that the hyperfine interaction is much smaller than the Zeeman one, therefore we can extract nearly isotropic hyperfine constant directly from the splitting of the resonance lines in hyperfine patterns. Our estimations provide a nearly isotropic A = 78 G.

To obtain parameters of the fine structure splitting parameters of Mn^{4+} ions a computer modelling has been performed to simulate digitally the angular dependences of the resonance fields at an arbitrary rotation plane. The central positions of the hyperfine groups in the rotational patterns presented in Figs. 2&3 have been fitted. In order to extract the crystal field parameters more accurately the "in-plane" experimental patterns for both samples have been preferred in modelling due to the smallest misalignment between the rotational and crystalline planes during the in-plane measurements (<1 degree).

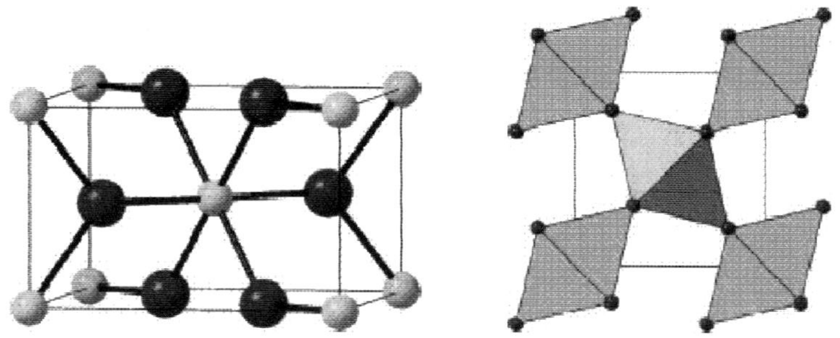

Fig. 4. On the left: the unit cell of the rutile lattice (O and Ti ions are shown by large dark and small bright spheres, respectively). On the right: mutual orientation of elongated TiO_6 octahedrons centered in the middle and corners of the TiO_2 unit cell.

Before we proceed with the simulation results it is useful to look at the rutile crystalline structure. The crystal structure of rutile is shown in Fig. 4. There are two equivalent sites per unit cell. The local symmetry of each site is D_{2h} and the principal axes of their spin-Hamiltonian parameters are perpendicular to each other and also perpendicular to the tetragonal axis of the crystal structure. Therefore, in the case of the substitutional Mn^{4+} ions located on the Ti^{4+} sites in the oxygen octahedrons centers one should expect two equivalent sets of paramagnetic centers in the crystal with principal axes given in Table 1. We observe 90° period of paramagnetic spectrum while turning the crystal around the c axis that proves that manganese atoms in fact substitute the Ti^{4+} ions in the rutile lattice.

Table 1 Principal axis directions of the Mn^{4+} centers in the rutile crystal.

Principal axes of the CF term	Crystalline directions	Sites
x	$[\bar{1}10]$	Site I
y	[110]	
z	[001]	
x	$[\bar{1}\bar{1}0]$	Site II
y	$[\bar{1}10]$	
z	[001]	

We have found that to simulate the experimental patterns of the most intense lines one have to use the same set CF parameters for each sort of the Mn^{4+} paramagnetic centers with their principal axes as given in the Table 1. Results of the modelling are presented in Figs. 2&3 by solid lines. The best agreement has been obtained with values of CF parameters as given in Table 2.

Table 2. Crystal field parameters of the Mn^{4+} centers

D [G]	E [G]	A [G]	g_x	g_y	g_z
4290±20	-1420±5	78±2	1.988±0.02	1.985±0.02	1.985±0.02

Thus, our analysis of EPR results with aid of computer modelling shows that high-temperature annealing treatment of the manganese-implanted TiO_2 rutile results in appearance of the paramagnetic Mn^{4+} centers which substitute for the Ti^{4+} ions in the rutile structure. Minor difference with the literature data on the bulk rutile crystal [1,2] may be attributed to the crystal structure imperfections of our Mn-implanted rutile samples (for instance, due to formation of oxygen vacancies and interstitial defects). We have to note also that in some recent publications a wrong assignment of the observed EPR lines as the resonance signal of Mn^{4+} have been done [6,7].

4. Conclusion

We have established that the paramagnetic Mn^{4+} ions formed as result of the Mn^{+} ion-beam implantation of the single crystalline TiO_2 rutile substrates with following high-temperature annealing procedure. The local environment of orthorhombic symmetry on the doped Mn^{4+} magnetic ions has been determined. Two equivalent Mn^{4+} centers have been observed in the EPR spectra in correspondence with two equivalent octahedral positions of the Ti ions in the rutile structure. The spin

Hamiltonian parameters of the Mn^{4+} centers in manganese-implanted rutile crystals have been obtained.

5. References

[1] Andresen H G 1961 *The Journal of Chemical Physics* **35** 1090
[2] Andresen H G 1960 *Phys. Rev.* **120** 1606
[3] Güler S, Bayrakdar H, Rameev B, Khaibullin R I, and Aktas B 2006, *phys. stat. sol.(a)* **203** 1533
[4] Rameev B Z, Yıldız F, Tagirov L R, Aktaş B, Park W K and J. S. Moodera 2003 *J. Magn. Magn. Mater.* **258-259** 361
[5] Sangaletti L, Mozzati M C, Galinetto P, Azzoni C B, Speghini A, Bettinelli M and Calestani G 2006 *Journal of Physics: Condensed Matter* **18** 7643
[6] Menzel D, Jursic I, Schoenes J, Iacomi F and Cacaina D 2006 *phys. stat. sol. (c)* **3** 4119
[7] Iacomi F, Apetroaei N, Calin G et al 2007 *Thin Solid Films* **515** 6402
[8] Abragam A and Bleaney B 1970 *Electron Paramagnetic Resonance of Transition Ions* (Oxford: Clarendon Press)
[9] Carter D L and Okaya A 1960 *Phys.Rev.* ***118*** 1485
[10] Cordischi D Valigi M Gazzolli D and Indovina V 1975 *Journal of Solid State Chemistry* **15** 82
[11] Yamaka E and Barnes R G 1964 *Phys Rev.* **135** A144
[12] H. J. Gerritsen, S. E. Harrison, H. R. Lewis, and J. P. Wittke J P 1959 *Phys. Rev. Letters* **2** 153

Effect of annealing temperature and substitution of Zr-Cu on magnetic properties of strontium hexaferrite nanoparticles

Muhammad Javed Iqbal[*] [1], Muhammad Naeem Ashiq[2] and Pablo Hernandez-Gomez[3]

[1]Surface and Solid State Chemistry Laboratory, Department of Chemistry, Quaid-i-Azam Islamabad-45320, Pakistan.
[2] Department of Chemistry, Bahauddin Zakariya University Multan, Pakistan.
[3] Dpto. Electricidad y Electrónica, Universidad de Valladolid, 47071 Valladolid, Spain.
E-mail: mjiqauchem@yahoo.com

Abstract: ZnTe and ZnTe:Cr films were prepared onto glass substrates using the thermal evaporation method. Structural properties of the prepared samples were analyzed using X-ray diffractometry, and the presence of a ZnCrTe phase was identified along with poor crystallinity. Composition analysis was done using XPS and the Cr content in the film was found to be 0.05 atomic percent. Transmittance spectra were recorded using UV-Vis spectrophotometry. The valence state of Cr in ZnTe:Cr film is determined to be +2 using electron spin resonance (ESR) spectroscopy. Magnetic moment data as a function of magnetic field were recorded using a Superconducting Quantum Interference Device (SQUID) magnetometer at temperatures of 5, 77 and 300 K. The results showed minority ferromagnetic behavior even at room temperature. Magnetic domains were observed using Magnetic Force Microscopy and the average domain size is 3.7 nm.

Keywords: Nanomaterials; Sr-hexaferrite; Saturation magnetization; Coercivity; Zr-Cu substitution

1. Introduction

Nano-materials have become a very active research area in the field of material chemistry. The surface effect is mainly responsible for deviation of the properties of nano-materials from that of the bulk [1-2]. Recently, attention has been drawn toward ferrite nanomaterials because of their great scientific and technological importance. Fine particles of strontium hexaferrites are suitable for use in high density magnetic recording media. One factor that limits the performance of high density recording media is media noise which results from coupling between magnetic grains. It is generally believed that the grain interaction is less significant when the grain size becomes small. In high density recording media, magnetic grains of the size of <50 nm are desirable for obtaining a suitable signal-to-noise ratio [3]. The fine particles of ferrites sintered at low temperature are also required for multilayer chip inductors [4]. It also has applications in the field of telecommunication, microwave devices and as permanent magnets due to its perfect chemical stability and high uniaxial magnetocrystalline anisotropy [5-7].

The solid state reaction method is commonly used for the fabrication of strontium hexaferrites but it requires high annealing temperature of ~ 1200 ^{0}C and it is also not easy to obtain very fine and mono-dispersed particles [8]. Recently, several techniques such as glass crystallization [9], sol-gel [10], microemulsion [11], aerosol pyrolysis [12], hydrothermal [13] and chemical co-precipitation [14] have been developed to synthesize nanosized and mono-dispersed particles of strontium hexagonal ferrites for utilization in recording and microwave devices. Some of these methods are complex and expensive. Sol-gel and microemulsion methods use low annealing temperatures, yet a much longer annealing time (~5-10) hours is generally required [11,15]. The chemical co-precipitation method [14] ensures a proper distribution of various metal ions to produce stoichiometric and smaller size particles as compared to the others. It also requires an annealing time of only one hour in order to obtain a single phase.

Requisites for application of materials in the high density recording media are high values of saturation magnetization and remanence but low coercivity (600 Oe) [16]. The physical, electrical and magnetic properties of the M-type hexaferrites depend upon the occupation of the substituted cations at five different sites [17]. In M-type hexaferrites, 12 ferric ions are distributed over five distinct sites i.e. 2a, 2b, $4f_1$, $4f_2$ and 12 k: Out of these, 2a, $4f_2$ and 12k are octahedral, $4f_1$ is tetrahedral and 2b is a trigonal bi-pyramid [18]. It has been reported that the substitution of Ti^{4+} along with Ni^{2+} and Co^{2+} at octahedral sites occupied by Fe^{3+} ions results in an improvement of its magnetic properties for application in the magnetic recording media [19-21]. Unfortunately, the annealing temperature required for single phase formation by substitution of these elements is higher than that of the unsubstituted hexaferrite[22]. Zr^{4+} has a similar electronic configuration and is in the same group of the periodic table in the present study it has been chosen to replace Ti^{4+}. Zirconium salts are relatively cheaper than those of titanium. In addition the Cu^{2+} has been selected to maintain the overall electroneutrality.

The aim of the present study is to synthesize nano-materials with large saturation magnetization and remanence and low coercivity. This paper presents the effect of the substitution of Zr-Cu and annealing temperature on the saturation magnetization (M_s), remanence (M_r) and coercivity (H_c) of Sr-hexaferrites.

2. Experimental Details

The chemicals used in the synthesis of samples were $Fe(NO_3)_3{\cdot}9H_2O$(Panreac Quimica SA, 98%), $Sr(NO_3)_2$ (Fluka, ≥99%), $Cu(CH_3COO)_2{\cdot}H_2O$ (Merck, 99%), $ZrOCl_2{\cdot}4H_2O$ (BDH, 96%) and NaOH (Fluka, ≥97%). The strontium hexaferrites substituted with Zr–Cu and having nominal composition of $SrZr_xCu_xFe_{12-2x}O_{19}$ (where x = 0.0–0.8) were prepared by a chemical coprecipitation method [14]. The metallic salt solutions of required molarities were prepared in distilled water and mixed together. The solution mixture was heated up to 70 ^{0}C with continuous stirring. When the temperature of the solution reached 70 ^{0}C, the 2M NaOH was added drop wise to form a precipitate. The pH of the solution was kept in the range 12.5–13 during precipitation. All the samples were stirred for 3 hours in order to control the crystallite size and homogeneity of the samples. The precipitates were washed with distilled water, dried at 100 ^{0}C and annealed at an optimized temperature of 920 ^{0}C for 1h in a temperature-programmed tube furnace at a heating rate of 5 ^{0}C min^{-1}.

Transmission electron microscopy (TEM) analysis was carried out with a Jeol JEM-1200EX microscope. The hysteresis loops of the samples were recorded by the standard AC induction method [23] where the magnetic field H and the magnetic induction B were both measured through two coils placed near the sample. The values of the saturation magnetization (M_s), remanence (M_r) and coercivity (H_c) were calculated from the loops. The magnetic moments of all the samples were also calculated from the saturation magnetization data.

3. Results and discussion

XRD data show that the synthesized materials are in the magnetoplumbite hexagonal phase as already reported [24]. Figure 1a shows the particles in the TEM micrograph of the sample annealed at 920 ^{0}C for one hour are in the range of 40-70 nm. The particle sizes increase to 150-250 nm as the annealing temperature is increased to 1120 ^{0}C as shown in figure 1b.

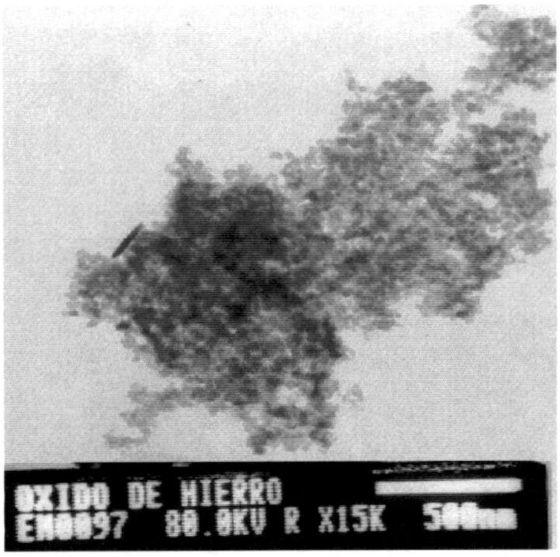

Fig. 1a

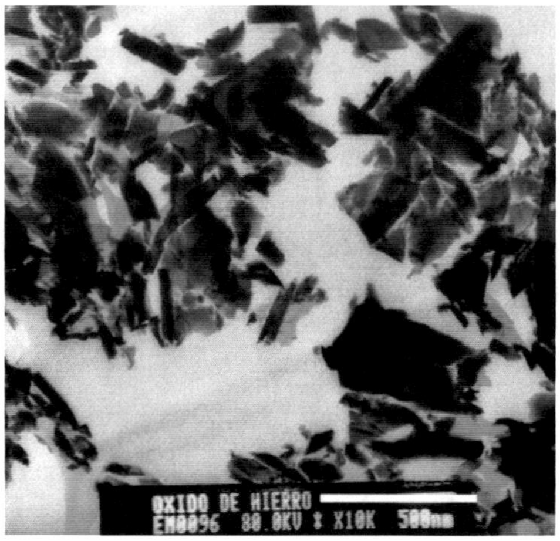

Fig. 1b

Figure 1. Transmission electron micrographs (TEM) of $SrZr_{0.4}Cu_{0.4}Fe_{11.2}O_{19}$ annealed at (a) 920 ^{0}C and (b) 1120 ^{0}C.

Figure 2 shows the effect of annealing temperature on the saturation magnetization (M_s), remanence (M_r) and coercivity (H_c) of $SrZr_{0.4}Cu_{0.4}Fe_{11.2}O_{19}$. Higher values of saturation magnetization and remanence are obtained at the annealing temperature of 1393K while the value of coercivity decreases continuously with temperature. It is well known that particle size has a significant effect on the magnetic properties of the ferrite materials [7]. As the annealing temperature is increased, the particle size is increased to reach the critical single domain size and as a result the remanence (M_r) and saturation magnetization (M_s) increase reaching eventually a maximum value at the single domain size because of the coherent rotation of spins. As the particles become larger than the single domain size at higher temperature, the values of M_r and M_s decrease. The coercivity of the samples decreases with increasing annealing temperature due to an increase in their particle size. This is also complemented by the TEM micrograph (fig. 1b). The sample of larger particle size has low coercivity and vice versa.

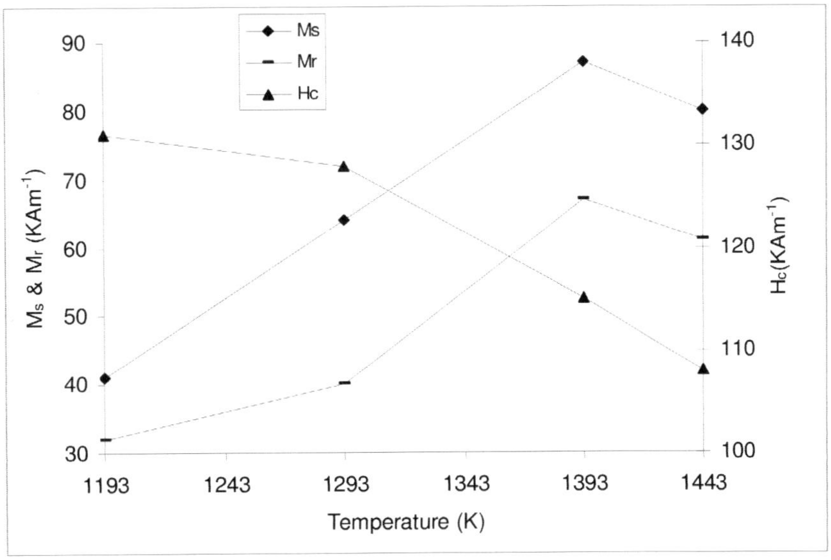

Figure 2. Effect of annealing temperature on the saturation magnetization (M_s), remanence (M_r) and coercivity (H_c) of $SrZr_{0.4}Cu_{0.4}Fe_{11.2}O_{19}$.

Figure 3 shows the variation of the saturation magnetization, coercivity and remanence with the Zr-Cu contents. The saturation magnetization increases from a value of 71 kAm^{-1} to 84 kAm^{-1} when Zr-Cu contents, x = 0.0-0.4 (table 1). However, the coercivity decreases regularly from 137 kAm^{-1} to 49 kAm^{-1} (table 1) in the entire concentration range. The value of remanence remains almost constant up to the Zr-Cu concentration of x = 0.4 but decreases above this concentration. The variation in the magnetic properties can be explained on the basis of occupancy by the substituted ions at different sites in the hexagonal structure. It has been reported that the Cu^{2+} occupies the $4f_2$ (octahedral site) which has spin in the downward direction [25] and Zr^{4+} occupies the site having spin in downward direction ($4f_1$) [26]. Both of these sites may contribute negatively to the total saturation magnetization of M-type hexaferrites. When both of the substituted ions replace the Fe^{3+} ions in a site having downward spin, the total number of spin in the upward direction increases and as a result the total magnitude of the magnetic moment increases. Consequently, the saturation magnetization increases and the abrupt fall in the value saturation magnetization and remanence (above x = 0.4) may be due to the following reasons: 1) the presence of a number of nonmagnetic Zr^{4+} ions and less magnetic Cu^{2+} ions probably decreases the superexchange interaction between $4f_1$-12k, $4f_2$-12k and $4f_1$-2a leading to a decrease in the saturation magnetization. 2) This abrupt fall in both the properties is due to the loss of magnetic collinearity leading to the spin canting. The observed decrease in coercivity is owing to a reduction in magnetocrystalline anisotropy.

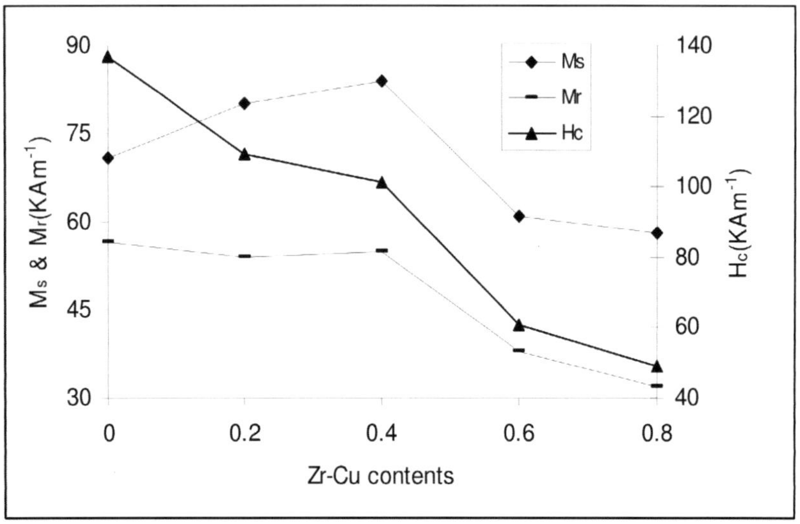

Figure 3. Variation of saturation magnetization (M_s), remanence (M_r) and coercivity (H_c) of Zr-Cu substituted strontium hexaferrites by the substitution of Zr-Cu.

In M-type hexaferrites the 12k, $4f_2$ (octahedral) and 2b (trigonal bipyramidal) sites are known as major contributors to the magnetocrystalline anisotropy [27]. As mentioned above, Cu^{2+} occupies the $4f_2$ and has a negative impact on the magnetocrystalline anisotropy, so consequently the coercivity decreases. This is one of the few examples in which the saturation magnetization increases and at the same time the coercivity decreases. The decrease in coercivity and increase in saturation magnetization suggest that the synthesized materials are promising materials for application to high density recording media.

Table 1 Saturation magnetization (M_s), remanence (M_r), coercivity (H_c) and magnetic moment (n_B) for the $SrZr_xCu_xFe_{12-2x}O_{19}$ (x = 0.0-0.8).

Parameters	**X = 0.0**	**X = 0.2**	**X = 0.4**	**X = 0.6**	**X = 0.8**
Saturation magnetization (M_s) kAm^{-1}	71	80	84	61	58
Remanence (M_r) kAm^{-1}	56	54	55	38	32
Coercivity (H_c) kAm^{-1}	137	109	101	61	49
Magnetic moment (n_B) μ_B	11.2	12.9	13.2	11.6	11.3

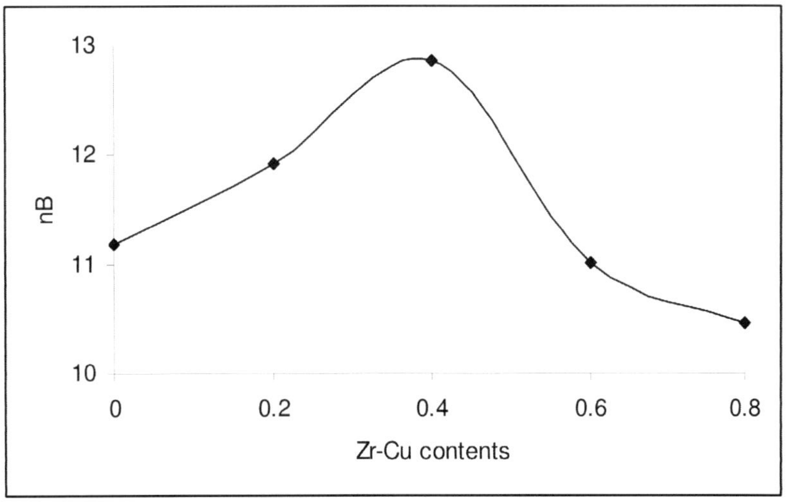

Figure 4. Dependence of magnetic moment on the Zr-Cu contents of substituted strontium hexaferrites.

The magnetic moment (n_B) for all the samples has been calculated using the following equation.

$$n_B = M.\ M_s/5.585 \quad (1)$$

where 'M' is the molar mass of the sample, 'M_s' the saturation magnetization and d_b is the measured density of the sample. The magnetic moment increases with the substitution of Zr-Cu content up to a value of x = 0.4 as shown in figure 4. The variations of magnetic moment can be explained on the basis of saturation magnetization which also increases up to this substitution level. From equation 1, it is clear that the saturation magnetization is directly proportional to the magnetic moment.

4. Conclusion

We have used an economic and simple method for the synthesis of nano-sized strontium hexaferrites doped with Zr-Cu. The particle sizes are in the range 40-70 nm and 150-250 nm for the samples annealed for one hour at 920 ^{0}C and 1120 ^{0}C, respectively. The observed particle size is small enough to be used in the high density recording media to minimize the signal to noise ratio. The maximum saturation magnetization and remanence are observed at the annealing temperature of 1120 ^{0}C. Up to a substitution level of x = 0.4 the saturation magnetization increases while the remanence remains almost constant, but the coercivity decreases from 137 to 49 KAm^{-1}. This improvement in both properties suggests that the synthesized material is quite significant, and makes the synthesized materials suitable for application in the high density recording media.

Acknowledgment

Financial support to one of the authors (M. M. A) for this work from the Higher Education Commission (HEC) of Pakistan under indigenous and IRSIP schemes is gratefully acknowledged.

Journal of Physics: Conference Series **153** (2009) 012053 doi:10.1088/1742-6596/153/1/012053

References

[1] Koch C C 2002 *Nanostructured materials: processing, properties and potential applications* (Norwich- New York: William Andrew Publishing)

[2] Ozin G A and Arsenault A C 2005 *Nanochemistry: A chemical approach to nanomaterials* (UK: The Royal Society of Chemistry)

[3] Che S, Wang J and Chen Q 2003 *J. Phys.: Cond. Mat.* **15** L335

[4] Nakamura T J 1997 *J. Magn. Magn. Mater.* **168** 285

[5] Sui X and Kryder M H 1994 *IEEE Trans. Magn.* **30** 4044

[6] Jin Z, Tang W, Zhang J, Lin H and Du Y 1998 *J. Magn. Magn. Mater.* **182** 231

[7] Iqbal M J, Ashiq M N, Gomez P H and Munoz J M, 2008 *J. Magn. Magn. Mater.* **320** 881

[8] Hussain S and Maqsood A 2007 *J. Magn. Magn. Mater.* **316** 73

[9] Rezlescu L, Rezlescu E, Popa P D and Rezlescu N 1999 *J. Magn. Magn. Mater.* **193** 288

[10] Zhong W, Ding W, Zhang N, Hong J, Yan Q and Du Y 1997 *J. Magn. Magn. Mater.* **168** 196

[11] Fang J, Wang J, Gan L, Ng S, Ding J and Liu X 2000 *J. Amer. Ceram. Soc.* **83** 1049

[12] Carreno T G, Morales M P and Serna C 2000 *J, Mater. Lett.* **43** 97

[13] Wang J F, Ponton C B and Harris I R 2004 *J. Alloy Compd.* **369** 170

[14] Leccabue F, Muzio O A, Kany M S E, Calestani G and Albanese G 1987 *J. Magn. Magn. Mater.* **68** 201

[15] Fang Q Q, Bao H W, Fang D M, Wang J Z and Li X G 2004 *J. Magn. Magn. Mater.* **278** 122

[16] Li Y, Liu R, Zhang Z and Xiong C 2000 *Mater. Chem. Phys.* **64** 256

[17] Suarez G M, Johal K K and Molinar H M 2001 *J. I. E. Garcia, M. M. C. Guerrero, Mater Res. Bull.* ***36*** *2597*

[18] Lechevallier L and Breton J M L 2005 *J. Magn. Magn. Mater.* **290-291** 1237

[19] Castellanos P A M and Jarque J C S 2005 *J. A. Rivera, Physica B: Cond. Mat.* **362** 95

[20] Gruskova A, Slama J, Dosudil R, Kevicka D, Jancarik V and Toth I 2002 *J.Magn. Magn Mater.* **242–245** 423

[21] Yong S, Shim I B and Kim C S 2002 *J Appl. Phys.* **91** 8465

[22] Sharma P, Rocha R A, Medeiros S N d and Paesano A 2007 *J. Alloys Compd.* **443** 37

[23] Kramer R B G, Egorov V S, Jansen A G M and Joss W 2007 *J. Magn. Magn. Mater.* **310** 1675

[24] Iqbal M J and Ashiq M N 2008 *Chem. Eng. J.* **136** 383

[25] Qiao L, You L, Zheng J, Jiang L and Sheng J 2007 *J. Magn. Magn. Mater.* **318** 74

[26] Rane M V, Bahadur D, Nigam A K and C. M. Srivastava C M 1999 *J. Magn. Magn. Mater.* **192** 288

[27] Yang Z, Wang C S, Li X H and Zang H X 2002 *Mater. Sci. Eng. B* **90** 142

Analysis of the series resistance and interface state densities in metal semiconductor structures

G. Güler[1], Ö. Güllü[2], Ş. Karataş[3*], Ö. F. Bakkaloğlu[4]
[1]Department of Physics, Faculty of Education, University of Adıyaman, 02100 Adıyaman, Turkey
[2]Department of Physics, Faculty of Sciences and Arts, Atatürk University, 25240 Erzurum, Turkey
[3]Department of Physics, Faculty of Sciences and Arts, University of Kahramanmaraş Sütçü İmam, 46100 Kahramanmaraş, Turkey
[4]Department of Engineering Physics, Faculty of Engineering Physics, University of Gaziantep, 27310 Gaziantep, Turkey

E-mail: skaratas@ksu.edu.tr

Abstract. The electrical properties of Co/*n*-Si metal-semiconductor (MS) Schottky structure investigated at room temperature using current-voltage (*I-V*) characteristics. The characteristic parameters of the structure such as barrier height, ideality factor and series resistance have been determined from the *I-V* measurements. The values of barrier height obtained from Norde's function were compared with those from Cheung functions, and it was seen that there was a good agreement between barrier heights from both methods. The series resistance values calculated with Cheung's two methods were compared and seen that there was an agreement with each other. However, the values of series resistance obtained from Cheung functions and Norde's functions are not agreeing with each other. Because, Cheung functions are only applied to the non-linear region (high voltage region) of the forward bias *I–V* characteristics. Furthermore, the energy distribution of interface state density was determined from the forward bias *I-V* characteristics by taking into account the bias dependence of the effective barrier height. The results show that the presence of thin interfacial layer between the metal and semiconductor.

1. Introduction

The metal-semiconductor (MS) structures have an important role in modern electronics, and MS structures are one of the most widely used rectifying contacts in the electronics industry. Due to the performance and stability of metal semiconductor structures which are of great importance to the electronic devices [1-3]. Although these devices have been studied extensively, satisfactory understanding in all details has still not been achieved. Because there is a continuing need for faster and more complex systems for the information age, existing semiconductor devices are being studied for improvement, and new ones are being invented [1-5]. Due to technological importance of metal-semiconductor structures, a full understanding of the nature of the electrical characteristics of Schottky barrier diodes in this system is of great interest. Electronic properties of a Schottky diode are characterized by its barrier height and ideality factor parameters. The interface states play an important role on determination of Schottky barrier height and other characteristic parameters and these can affect device performance, stability and reliability [3,6-8]. Thus, barrier height, ideality factor, series resistance (RS) and interface states density values (NSS) at metal-semiconductor structures play an important role in the determination of the main parameters of the structures. These parameters give useful information concerned with the nature of the device. When voltage is applied across the MS device, the combination of the interfacial insulator layer, depletion layer and the series resistance of the device will share applied voltage.

Usually, the forward bias I–V characteristics are linear in the semi-logarithmic scale at low voltages, but deviate considerably from linearity due to the effects of parameters such as the series resistance RS, the interfacial layer and interface states at sufficiently large applied voltage. The parameter RS is only effective in the downward-curvature region (non-linear region) of the forward I–V characteristics at sufficiently high applied voltage, but parameters such as the ideality factor n and barrier height □b, are effective in both the linear and non-linear regions of these characteristics, accompanying the change of the SBH [9,10]. The first studies on the interface layer in Schottky diodes were made by Cowley and Sze [11], who obtained their estimates from an analysis of barrier heights with different metallization as a function of the metal work function. Card and Rhoderick [12] examined the effects of the interface layer on the ideality factor of the forward bias I-V characteristics. Tseng and Wu [13] analysed the effect of the presence of an interfacial layer on the behaviour of Schottky barrier diodes.

It is well known that, unless specially fabricated, a Schottky barrier diode (SBD) possesses a thin interfacial native oxide layer between the metal and the semiconductor. The existence of such an insulating layer can have a strong influence on the diode characteristics as well as the interface states [1,3,14-16]. Therefore, the interfacial layer and the interface states play an important role in the determination of the barrier height. Consequently, it has been concluded that the barrier height determined from the I-V characteristics controlled by the interface states energy distribution in equilibrium with the semiconductor and the applied voltage under forward bias condition.

In the present study, we have investigated the electrical characteristics and interface states of the Co/n-Si Schottky structure. Au-Sb alloy and Co metal has been chosen as the ohmic contact and rectifier contact, respectively. The electronic parameters controlling the device performance, such as barrier height, ideality factor, series resistance and interface state energy distribution were evaluated by current–voltage measurements at room temperature and in the dark.

2. Experiment

The samples have been prepared using cleaned and polished n-Si (as received from the manufactured) with (100) orientation. The wafers were chemically cleaned using the RCA cleaning procedure [i.e., a 10 min boil in NH4OH+H2O2+6H2O followed by a 10 min boil in HCl+H2O2+6H2O]. The native oxide on the front surface of the substrates was removed in HF:H2O(1:10) solution and finally the wafer was rinsed in de-ionized water for 30 s. Then, low resistivity ohmic back barrier diode to n-type Si wafer was made by using Au-Sb alloy, followed by a temperature treatment at 420 oC for 3 min in N2 atmosphere. The Schottky barrier diode were formed on the front face of the n-Si as dots with diameter of about 1 mm by the galvanostatic electrodeposition of Co. Acid resistant adhesive tape was used to mask off all the substrate except for the deposition area. The electrodeposition of Co films on n-type Si substrate has been carried out at room temperature from an aqueous electrolyte containing 1M Co sulphate and 0.5M boric acid. The current–voltage characteristics were measured using a Keithley 487 Picoammeter/Voltage Source at room temperature and in the dark.

3. Results and discussion

3.1 Current-voltage (*I-V*) characteristics of Co/*n*-Si Schottky structure

The forward bias current-voltage characteristics of the Co/p-Si (MS) Schottky structure at 300 K (room temperature) are given in Fig. 1.

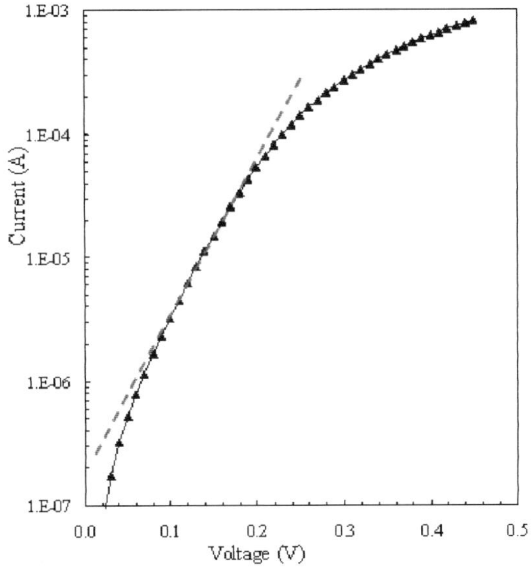

Figure 1. The experimental forward bias current-voltage characteristics of the Co/n-type Si Schottky structure at room temperatures

The forward current-voltage characteristics of the Schottky structure can be described by the following relation [1, 3,14-17],

$$I = I_o \exp\left(\frac{qV}{nkT}\right)\left[1 - \exp\left(-\frac{qV}{kT}\right)\right] \tag{1}$$

where,

$$I_o = AA^*T^2 \exp\left(-\frac{q\Phi_b}{kT}\right) \tag{2}$$

is the saturation density, V is the definite forward-bias voltage, A is the effective diode area, k is the Boltzmann constant, T is the absolute temperature, A* is the Richardson constant for n-type Si ($A^* = 112\ Acm^{-2}K^{-2}$) [1,18,19]. The barrier height was calculated using I_0 value and was found to be 0.70 eV. The ideality factor of Co/*n*-Si/AuSb Schottky contact from Eq. (1) is contained in the slope of straight line region of the forward-bias logarithmic characteristics of *I-V* through the relation;

$$n = \frac{q}{kT}\frac{dV}{d(\ln I)} \tag{3}$$

where n is a measure of conformity of the diode to pure thermionic emission, and was found to be 1.19. For an ideal Schottky barrier diode n=1. However, n has usually a value greater than unity. This value indicates that the effect of the series resistance in the linear region is important. Fig. 1 show the dark forward bias *I-V* characteristics of one of the electrodeposited Co/*n*-Si Schottky contacts respectively. From Eq. (2) the barrier height, Φ_b is given by;

$$\Phi_b = \frac{kT}{q} \ln\left(\frac{A^* A T^2}{I_0} \right) \tag{4}$$

Using Eqs (3) and (4), as explained above, the values of the barrier height (Φ_b) and the ideality factor (n) of the Co/n-type Si Schottky contact were found as 0.70 eV and 1.19, respectively. Moreover, at high currents there is forever a deviation which has been clearly shown to depend on parameters such as the interfacial layer thickness, the interface states density and series resistance, as one would expected [20,21].

The series resistance is a very important parameter of Schottky diode. The resistance of the Schottky contact is the sum total resistance value of the resistors in series and resistance in semiconductor device in the direction of current flow. The Schottky diode parameters as the barrier height Φ_b the ideality factor n and the series resistance R_S were also achieved using a method developed by Cheung and Cheung [22]. Cheung's functions can be written as follows:

$$\frac{\mathrm{dV}}{\mathrm{d(lnI)}} = IR_S + \mathrm{n}\frac{\mathrm{kT}}{\mathrm{q}} \tag{5}$$

$$H(I) = V - n\left(\frac{\mathrm{kT}}{\mathrm{q}}\right)\ln\left(\frac{I}{AA^*T^2}\right) \tag{6}$$

and

$$H(I) = IR_S + n\Phi_b \tag{7}$$

where Φ_{bo} is the real barrier height extracted from the lower-voltage part of forward I–V characteristics. The dV/d(ln(I)) plot is a straight line region where dominates the series resistance. Fig. 2 and Fig. 3, represent H(I) (a) and dV/dln(I) (b) vice versa current, respectively.

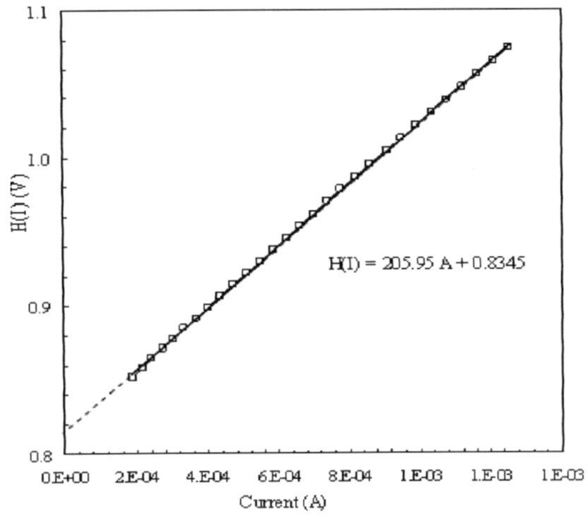

Figure 2. The experimental H(I) vs I plot obtained from forward bias current-voltage characteristics of the Co/n-Si Schottky structure.

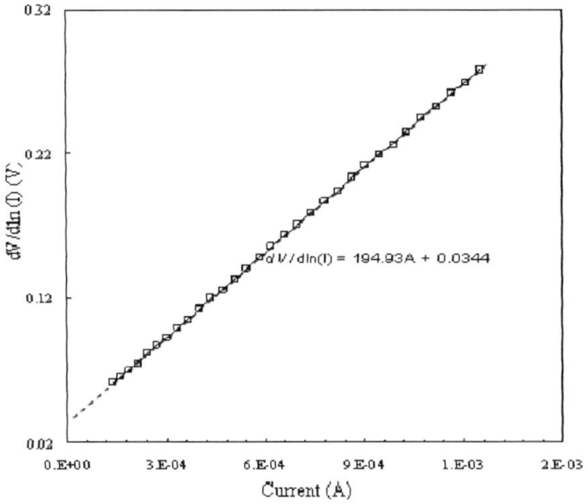

Figure 3. The experimental dV/dln(I) vs I plot obtained from forward bias current-voltage characteristics of the Co/n-Si schottky structure.

The values of barrier height and series resistance from Eq.7 were found to be 0.71 eV and 205.95 Ω. In the same way, the values of ideality factor and series resistance from Eq. 5 were found as 1.34 and 193.62 Ω, respectively. Thus, it can clearly be seen that there is relatively difference between the values of n obtained from the downward curvature regions of forward bias I–V plots and from the linear regions of the same characteristics. The reason of this difference can be attributed to the existence of effects such as the series resistance and the bias dependence of the Schottky contact according to the voltage drop across the interfacial layer and change of the interface states with bias in this concave region of the I–V plot. Furthermore, the R_S values obtained from the Cheung functions are in agreement with each other due to consistency of Cheung functions.

Norde proposed an alternative method to determine values of the series resistance and barrier height [23]. Norde method was also employed [23] because high series resistance (R_S) can hinder an accurate evaluation of barrier height from the standard ln$I$–V plot.

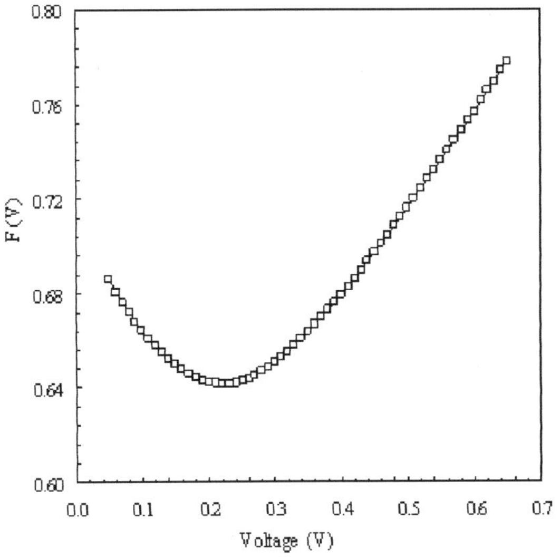

Figure 4. F(V) vs.V plot obtained from forward bias current-voltage characteristics of the Co/n-type Si Schottky structure at room temperatures.

The following function has been defined in the modified Norde's method:

$$F(V) = \frac{V}{2} - \frac{kT}{q}\ln\left(\frac{I}{AA^{*}T^{2}}\right) \tag{8}$$

The effective Schottky barrier height is given by

$$\Phi_b = F(V_{\min}) + \frac{V_{\min}}{2} - \frac{kT}{q} \tag{9}$$

and

$$R_S = \frac{kT}{qI_o} \tag{10}$$

where $F\ (V_{\min})$ is the minimum value of $F\ (V)$, $V_{\min}$ is the corresponding voltage, and I_0 is the corresponding current at V=$V_{\min}$, respectively. A plot of $F(V)$ versus V for the Co/n-Si Schottky structure at room temperature is as shown in Fig. 4. From the $F(V)$–V plot the values of Φ_b and R_S of the structure have been determined as 0.72 eV and 257.16 Ω, respectively. It is noted that these values are higher than those obtained by the *I–V* method. This means that the value of series resistance indicates that the series resistance is a current-limiting factor for the Co/*n*-Si Schottky diode.

3.2 Interface state density properties of Co/n-Si Schottky structure

As can be seen from Fig. 5, the interface state density N_{SS} has an exponential drop with bias from the midgap towards the top of the conduction band for Co/*n*-Si Schottky contact. The density distribution curves of the interface states can be thus obtained from experimental data of the forward bias *I-V* in Fig. 1. For a real Schottky diode having interface states in equilibrium with the semiconductor, the ideality factor, *n* becomes greater than unity as proposed by Card and Rhodercik [3,12] and is given by;

$$n = 1 + \frac{\delta}{\varepsilon_i}\left[\frac{\varepsilon_s}{W} + q.N_{SS}(V)\right] \tag{11}$$

where N_{SS} is the density of interface states in equilibrium with the semiconductor, W is the space charge width, ε_s and ε_i are the permittivity of the semiconductor and the interfacial layer, respectively. In n-type semiconductors, the energy of the interface states Ess with respect to the bottom of the conduction band at the surface of the semiconductor is given by:

$$E_C - E_{SS} = q\Phi_e - qV \tag{12}$$

where V is the applied voltage drop across the depletion layer and Φ_e is the effective barrier height. Eq. (11) can be used to obtain interface states densities (N_{SS}) values as a function of voltage. The interface states densities values were obtained from the Eq. (11) by obtaining the real parameters of the contact. As can be seen in Fig. 5, N_{SS} value decreases with increasing E_C-E_{SS} value. The increase in N_{SS} values forms from mid gap towards the bottom of the conduction band. This increase is probably due to the increase in the series resistance (R_S). The potential drop across the interfacial layer varies with bias because of the change in the charge in the interface states and thus the interface state energy distribution, and it alerts the diffusion potential and therefore the depletion capacitance [1,3, 15,16 and 24].

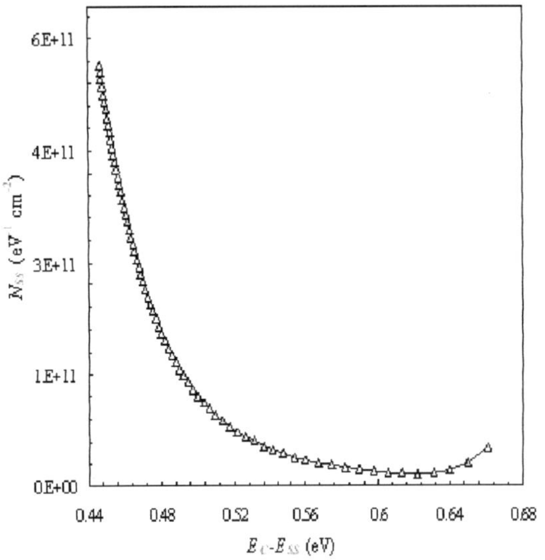

Figure 5. Density of interface states N_{SS} as a functions of E_{SS}-E_V obtained from the *I-V* measurements at room temperature

4. Conclusion

In this study, we have investigated electrical properties of Co/*n*-type Si Schottky structure (MS) using current-voltage (*I-V*) measurement at room temperature. The values of diode parameters such as ideality factor (n), series resistance (R_S) and barrier height (Φ_{IV}) were obtained from *I-V* measurements using Cheung and Norde's method. The ideality factor and barrier height values calculated using Cheung functions with different methods were compared and seen that there was an agreement with each other. However, the parameters obtained from Cheung functions and Norde's functions are not agreeing with each other. Because, Cheung functions are only applied to the non-linear region (high voltage region) of the forward bias *I–V* characteristics. Whereas, Norde's functions are applied to the full forward bias *I–V* characteristics of the junctions. The interface state density has an exponential rise with bias from the midgap towards the top of the conduction band, and gives a peak about 0.446 eV. It is clear that ignoring the role of interfacial insulator layer, interface state density and series resistance can lead to significant errors in the electrical characteristics of devices.

References

[1] S. M. Sze, Physics of Semiconductor Structures, second ed., New York **1981** 1-138
[2] P. W. Loscutoff, S. F. Bent, Annu. Rev. Phys. Chem. **2006** 57 467.
[3] E.H. Rhoderick., R.H.Williams, Metal-Semiconductor Contacts, Second ed. Oxford, **1988**.
[4] K.K. Kwok. , Complete Guide to Semiconductor Devices, New York: McGraw-Hill. **1995**
[5] M.S. Tyagi, Introduction to Semiconductor materials and Devices, New York: John Wiley, **1991**
[6] P.L. Hanselaer,W.H. Laflere, R.L. Van Meirhaeghe, F. Cardon , J. Appl. Phys. **1984** 56 2309.
[7] P.L. Hanselaer,W.H. Lafle´re, R.L. Van Meirhaeghe, F. Cardon , Appl. Phys. **1986** A 39 129.
[8] E.H. Rhoderick, Metal–Semiconductor Contacts, Oxford University Press, **1978**
[9] K. Akkılıç,, M.E. Aydin, I. Uzun, T. Kılıçoğlu, Synth. Met. **2006** 156 958.
[10] K. Akkılıç, I. Uzun, T. Kılıçoğlu, Synth. Met. **2007** 157 297.
[11] A.M. Cowley, S.M. Sze, J. Appl. Phys. **1965** 36 3212.
[12] J.C. Card, E.H. Rhoderick, J. Phys. **1971** D 4 1589
[13] H.H. Tseng, C.Y. Wu , Solid State Electron. **1987** 30 383.
[14] Ş. Karataş Solid State Comm. **2005** 135 500
[15] Ş. Karatas, Ş. Altındal, M. Çakar, Physica B **2005** 357 386
[16] Ş. Karatas, A. Türüt Physica B, **2006** 381 199.
[17] S. Kumar Y. S. Katharria, S.Kumar, and D. Kanjilal, J.of Appl. Phys. **2006** 100 113723.
[18] A. Keffous, M. Siad, S. Mamma, Y. Belkacem, C. Lakhdar Chaouch, H. Menari, A. Dahmani and W. Chergui , Applied Surf. Sci. **2003** 218 337
[19] G. Güler, Ö. Güllü, Ö. F. Bakkaloglu, A. Türüt, Physica B, **2008** 403 2211
[20] B. Şahin, H. Çetin, E. Ayyildiz Solid State Comm. **2005** 135 490
[21] M.A.Ebeoğlu, F. T. Temurtaş and Z. Z. Öztürk, Solid State. Electron. **1998** 42 23.
[22] S.K. Cheung, N.W. Cheung, Appl. Phys. Lett**. 1986** 49 85.
[23] H. Norde , J. Appl. Phys. **1979** 50 5052.
[24] A. Turut, M. Sağlam, Physica B. **1992** 179 285.

GALVANOMAGNETIC EFFECTS IN $A_2^VB_3^{VI}$

T.G.Kerimova, N.A.Abdullayev, K.M.Mustafayeva
Institute of Physics of Azerbaijan National Academy of Sciences
Baku AZ-1143, ave. G.Javid, 33

E-mail: ktaira@physics.ab.az

Abstract: Magnetoresistivity of $A_2^VB_3^{VI}$ –based compounds has been investigated (H~0÷89 kE, T=0,5÷4,2 K). Oscillation of magnetoresistivity, Shubnikov de Gaas oscillation has been observed at high magnetic fields (H>30 kE). The charge carrier concentration and area of extreme cross section of Fermi surface by plane, perpendicular to magnetic field have been evaluated from oscillation period.

$A^5_2B^6_3$-based compounds are used as high effective thermoelectric transformers [1]. Recently, the interest to these compounds increases due to the perspectivities of widening of working temperature range and increase of thermoelectric efficiency of $A^5_2B^6_3$ compounds, doped by different impurities [2-6]. Therefore the investigation of fundamental characteristics of these compounds is not only extended our representation on band structure, also has a practical meaning.

In the present work the results of investigations of galvanomagnetic effects in layered $A^5_2B^6_3$ compounds which have rhombohedral structure and crystallize in sp.gr. D_{3d} ($R\bar{3}m$) [1], are given. The layers in $A^5_2B^6_3$ compounds consist of five monoatomic hexagonal grids alternating in sequence in B(1)-A-B(2)-A-B(1). Atom B(2) has 6 nearest atoms B(1). The bond between B(1)-A-B(2)-A-B(1) layers-quintets is of weak Van-der-Vaals type, but in layers of B(1)-A and A-B(2) bonds have covalent character with small fraction of ionic bond.

Single crystals under investigation have been were grown by Bridgman method from components with stoichiometric relationship. It is known, that by synthesis of $A^5_2B^6_3$ compounds from the melt of stoichiometric structure are characterized by presence of a significant amount of own points defects conditioned by transition of A atoms into the position of B atoms. These antistructural defects are the acceptors. Concentration of holes is $p\sim10^{19}cm^{-3}$.

Investigations have been carried out in temperature range 0,5÷300K and magnetic field up to 8 Tl. At galvanomagnetic investigations the sample is placed in the centre of superconductive solenoid. Measurements are carried out by selective method under the alternating current by 20 Hz frequency. The current magnitude does not exceed 1mA. The current is directed along the layer plane, magnetic field is perpendicular to layer plane.

1. Theory

As it is known, at strong magnetic field (μB>1, μ-mobility of carriers is quanted on the plane perpendicular to magnetic field and takes discrete magnitudes (Landau levels). At definite conditions it is possible to observe these oscillations experimentally. These conditions are so:

1) $\hbar\omega_c$ >>kT- energy distance between Landau levels must be exceed many times thermal expansion of distribution of charge carriers on energy near Fermi level,

here $\omega_c = \dfrac{eH}{m^*_c c}$ cyclotron frequency,

2) $\omega_c\tau$ >>1 – broadening of Landau levels must be little many time than the distance between the levels,

3) $\varsigma > \hbar\omega_c$ - to observe the oscillation it is necessary Fermi level to be above for some Landau levels. At the same time in aggregate with the first condition, this condition provides fulfillment of a condition of strong degeneration.

As for as Landau levels are placed equidistantly, then observed oscillation of kinetic characteristics (in peculiar, magnetoresistivity) is periodical in reverse magnetic field (with the period $\dfrac{2\pi}{\omega_c} = \dfrac{2\pi m^*_c}{eH}$). The investigation of magnetoscillation effects are comfortable, therefore some characteristic magnitudes depend only on form of izoenergy surfaces in space of reverse lattice and are unsensitive to dissipation mechanism.

For example, from the p period of magnetoresistivity oscillation one can determine n the concentration of charge carriers. Generally the oscillation period for secluded Fermi surface of arbitrary form is determined by the equation [7].

$$P\left(\frac{1}{H}\right) = \frac{2\pi e}{\hbar c S_F} \tag{1}$$

here, S_F – extreme square of cross of Fermi surface $\varepsilon(\vec{k}) = \mu_F$ by plane, perpendicular to direction of magnet field.

In frameworks of sixellipsoidal model of Drebl-Volf for ellipsoid, centered on reflection plane, referred to center of a zone, dependence of energy with wave vector $\varepsilon(\vec{k})$ is described by:

$$\varepsilon(\vec{k}) = \frac{\hbar^2}{2m_0}\sum_{i,j}\alpha_{ij}k_i k_j \tag{2}$$

Here, α_{ij} - tenzor components of reverse effective mass $\alpha_{ij} = \dfrac{m_0}{m_{ij}}$, k_1 axis is perpendicular to reflection plane, k_2 axic is parallel to crystallography C_1 axis, k_3 axis is directed along tetrahedral C_3 axis. Then oscillation period [8]:

$$P\left(\frac{1}{H}\right) = \frac{e\hbar}{m_0 c\mu_F}\left[\left(\alpha_{22}\alpha_{33} - \alpha_{23}^2\right)Cos^2\alpha + \alpha_{11}\alpha_{33}Cos^2\beta + \alpha_{11}\alpha_{22}Cos^2\gamma + 2\alpha_{11}\alpha_{23}Cos\beta Cos\gamma\right]^{1/2} \tag{3}$$

In the case of magnet field is directed along C_3 tetrahedral axis (in the case Bi_2Te_3 is perpendicular to layers), then $\alpha = \beta =$ 90^0, γ =0, from (3) it is follow, that oscillation period is

$$P\left(\frac{1}{H}\right)=\frac{e\hbar}{m_0 c\mu_F}\sqrt{\alpha_{11}\alpha_{22}} \tag{4}$$

For sixellipsoidal model with square low of dispersion, Fermi level [9,10]:

$$\mu_F=\frac{\hbar^2}{2m^*}\left(3\pi^2\frac{n}{K_Ý}\right)^{2/3} \tag{5}$$

here, n – full hole concentration, k_e - number of ellipsoids,

$$m^*=\frac{m_0}{\sqrt[3]{\alpha_{11}\left(\alpha_{22}\alpha_{33}-\alpha_{23}^2\right)}} \tag{6}$$

Thus, oscillation period is:

$$P\left(\frac{1}{H}\right)=\frac{2e\sqrt{\alpha_{11}\alpha_{22}}}{c\hbar\left(3\pi^2 n/K_Ý\right)^{2/3}\sqrt[3]{\alpha_{11}\left(\alpha_{22}\alpha_{33}-\alpha_{23}^2\right)}} \tag{7}$$

Respectively, carrier concentration is determined as

$$n^{2/3}=\frac{2e\sqrt{a_{11}a_{22}}}{P(1/H)c\hbar(3\pi^2/k_E)^{2/3}\sqrt[3]{a_{11}(a_{22}a_{33}-a_{23}{}^2)}} \tag{8}$$

2. Results of investigations and their discussion

We reveal, that in $A^5{}_2B^6{}_3$ –based compounds the magnetresistivity oscillation is observed on magnetic field dependence of magnetresistivity in the range of high magnet field (H>30kE) at low temperature. By the special program prepared by us, oscillation part of magnetresistivity was secured, it is peripodical in reverse magnet field.

As for as, at orientation of magnet field along C_3 axis all six ellipsoids of Fermi surface are sited symmetrically and have same extreme cross section by plane, perpendicular to magnet field, on fig.1 and 2 oscillation of one period is observed.

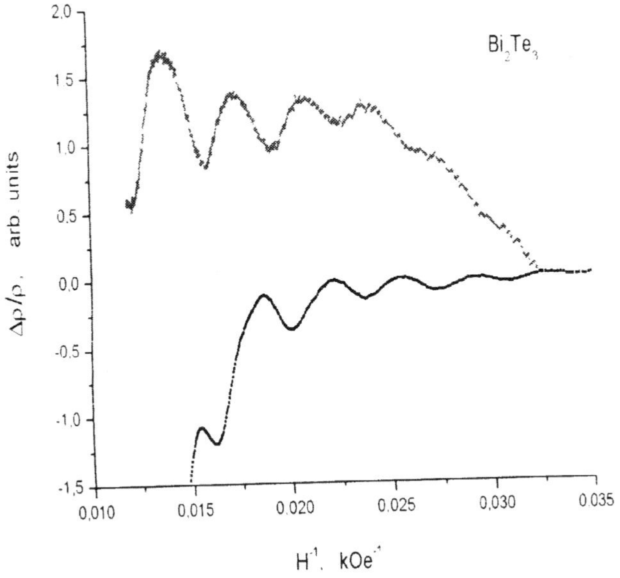

Figure1. Oscillation part of magnetoresistivity, shown in reverse magnet field for two different samples of Bi_2Te_3 single crystal.

In fig.1 oscillation part of magnet resistivity for two samples of Bi_2Te_3 single crystals is shown. Oscillation period of magnetresistivity in Bi_2Te_3 single crystal was within $(3\div3{,}5)\cdot10^{-6}$ E^{-1}. To take account in (8) tenzor components of reverse effective mass are a_{11}=2,8; a_{22}=20,8; a_{23}=4,65 and a_{13}=-1,05 [11], then determined value of charge carrier consentration appears to be equal $(6{,}5\div8)\cdot10^{18}$ cm^{-3}. Area of extreme cross section of Fermi surface by plane perpendicular to magnetic field is S=(28÷32))$\cdot10^{12}$ cm^{-2}.

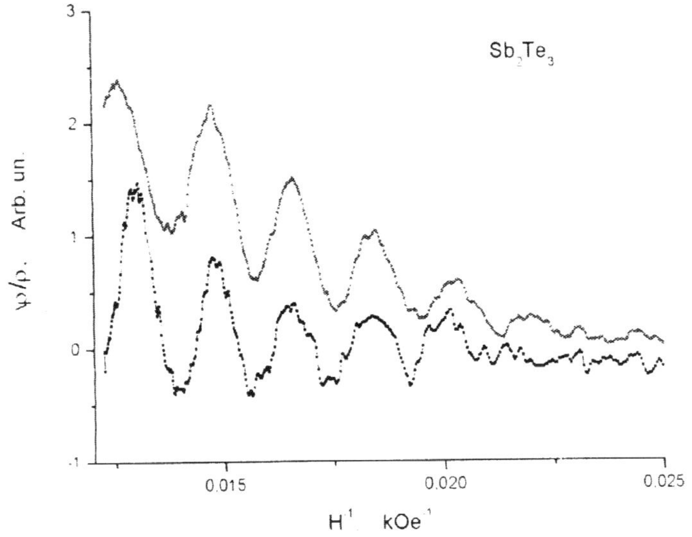

Figure 2. Oscillation part of magnet resistivity, shown in reverse magnet field for two different samples of Sb_2Te_3 single crystal.

In fig.2 oscillation part of magnetoresistivity for two samples of Sb_2Te_3 single crystals is shown. It is shown from figure the oscillation period of magnetoresistivity in these crystals appear to be less than in Bi_2Te_3 and is $(1,7\div1,9)\ )\cdot10^{-6}\ E^{-1}$. From (8) calculated value of charge carrier concentration is $(1,6\div1,9)\ )\cdot10^{19}\ cm^{-3}$. Area of extreme cross section of Fermi surface by plane perpendicular to direction of magnetic field and is $S=(50\div57)\cdot10^{12}\ cm^{-2}$. It is necessary to note the charge carrier concentration in Bi_2Te_3 and Sb_2Te_3 determined by Hall measurement, appears to be more, than the charge carrier concentration calculated by Shubnikov-de Gaas oscillation period in high magnet fields, in account of, six-ellipsoidal model of zone structure. Probably, it testifies for the existence of subband in valence band with the large effective mass, sited below the valence band top. Energy of holes in this band isn′t quanted and this band doesn′t give contribution to period of quant oscillation, whereas each bands give contribution to Hall effect. Authors of [12] investigated p-Bi_2Te_3 samples, supposed same disagreement by the existence of subband with the holes of large effective mass $m^*_p=2,4m_0$, sited below than upper subband on ~20 meV. .

References

[1] Golsman B M, Kudinov V A and Smirnov I A 1972 *Semiconductive thermoelectrical materials on Bi_2Te_3* base Nauka M
[2] Alekseyeva G T, Konstantinov P P, Kutasov V A, Lukyanova L N and Ravich Y I 1996 *Fizika tvordogo tela (Rassian)* **38** 2998
[3] Jitinskaya M K, Nemov S A and Ivanova L D, 2002 *Fizika tvordogo tela (Russian),* **44**(1) 41
[4] Kutasov V A, Lukyanova L H and Konstantinov P P 2000 *Fizika i texnika poluprovodnikov* **34**(4) 389
[5] Layxo R,Nemov S A, Lashkul A V, Laxderanta E, Svechnikova T E, and Dvornik D S 2007 *Fizika i texnika poluprovodnikov* (Russian) **41(5)** 565
[6] Stepanov N P, Nemov S A, Jitiyskaya M K and Svechnikova T E 2007 *Fizika i texnika poluprovodnikov (Russian)* **41(7)** 808
[7] Lifshits I.M and Kosevich A M 1955 *JETF* (Russian) **29(6)** 730
[8] Kulbachinski V A, Kovalyuk Z D and Pyrlya M N 1992 *Phys.Stat.Sol.(b)* **169** 157
[9] Parfenyev R V, Sologub V V and Goltsman B M 1968 *Fizika tvordogo tela* (Russian) **10(10)** 3087
[10] Schroder, VonMiddendorf A, Kohler H and Landwehr G 1973 *Phys.Stat.Sol. (b)* **59** 561
[11] Tastardi L R, Stiles P I and Burshtain E 1963 *Solid State communication* **1** 28
[12] Sologub V V, Goletskaya A D and Parfenyev RV 1972 *Fizika tvordogo tela* (Russian) **14(3)** 915

Structural, static and dynamic magnetic studies of evaporated Co_xCr_{1-x}/Si (100) and Co_xCr_{1-x}/glass thin films.

A Kharmouche

Faculté des Sciences de l'Ingénieur, Université Ferhat Abbas, Sétif 19000, Algeria.

E-mail: kharmouche_ahmed@yahoo.fr

Abstract.

We have evaporated a series of Co_xCr_{1-x} thin films under vacuum onto Si (100) and glass substrates, with a perpendicular incidence. The thickness of the magnetic layer ranged from 17 to 220 nm, and the content chromium, from 0.12 to 0.20, values determined by means of Rutherford Backscattering Spectrometry (R.B.S.) spectra using SIMNRA programme. Microscopic characterizations of the films were done with X-ray diffraction (XRD) measurements and infer that all the samples were polycrystalline, with an hcp structure and show a <0001> preferred orientation, and with the grain size increasing with the chromium content decrease. Atomic force microscopy (A.F.M.) observations reveal very smooth film surfaces. The static and dynamic magnetic properties have been investigated by means of Alternating Gradient Field Magnetometer (A.G.F.M.), and Brillouin Light Scattering (B.L.S.) measurements. The saturation magnetization M_s was found to decrease from 1200 emu/cm^3 to 220 emu/cm^3 as the chromium content increases from 12%at. to 20%at., whatever the thickness is.

From the fit of the B.L.S. spectra, we have computed effective magnetic anisotropy factors, as well. All the results are discussed and correlated.

Keywords: CoCr system; thin films; Hysteresis; Magnetic anisotropy; Brillouin Light Scattering (B.L.S.).

P.A.C.S.: 81.30.Dz; 75.70.-i; 75.60.-d; 75.30.Gw; 78.35.+c.

1. Introduction

Co and CoCr thin films have been intensively studied. They are the dominant materials media designed for magnetic longitudinal recording. As promising candidates for the perpendicular recording media, the CoCr thin films have got much attention [1-10].To achieve the perpendicular magnetic

anisotropy, the anisotropy constant **1. l** has to be greater than the demagnetizing energy. This depends, of course, on the comprehension of the microstructural and the magnetic properties, as well as on the performance of the fabrication techniques of the thin films media.

In this contribution, we study the effect of thickness of the magnetic layer on the magnetic properties of CoCr films, deposited onto Si and glass substrates.

The thickness measurements and atomic composition of the samples were performed by Rutherford Backscattering Spectrometry (RBS) technique (Section 3.1). The structural properties and film morphologies were investigated by x-ray diffraction (XRD) and atomic force microscopy (AFM). The static magnetic properties have been performed by alternating gradient field magnetometer (AGFM) (Section 3.2.1) experimental technique. The dynamic magnetic properties have been performed by Brillouin Light Scattering (B.L.S.) technique (Section 3.2.2.).

2. Experimental procedures

Two series of Co_xCr_{1-x} thin films were grown onto Si (100) and Corning glass substrates, under the same conditions. The CoCr samples have been prepared, at room temperature, by thermal evaporation, under vacuum, from a CoCr (70%at.-30%at) powder to which we added a quantity of Co powder of purity 99.99%. The pressure was 10^{-7} mbar before deposition; during the evaporation the vacuum achieved was better than 2 x 10^{-6} mbar.

The thickness measurements and the atomic composition of the samples were achieved by means of Rutherford Backscattering Spectrometry (R.B.S.) spectra using SIMNRA programme.

Each series consists of several samples with thickness ranging from 17 to 220 nm, and chromium content ranging from 12 at. % to 20 at. %.

The structural properties were monitored by x-ray diffraction, using a Siemens D-500 diffractometer with $\lambda = 1.789$ Å.

To study the macroscopic magnetic properties of the Co_xCr_{1-x} thin films, hysteresis loops were recorded by an Alternating Gradient Field Magnetometer (A.G.F.M.), with the external magnetic field H applied both perpendicular (polar configuration) and parallel to the film plane (longitudinal configuration). For the longitudinal configuration, the measures have been done applying the magnetic field in two different directions in order to study the plane magnetic anisotropy.

Brillouin Light Scattering (B.L.S.) experiments were used to investigate the magnetic anisotropy in Co_xCr_{1-x}/Si samples. The BLS measurements were done using a (2x3)–pass tandem Fabry-Perot interferometer. The samples were illuminated by a single-mode Ar^+ ion laser at the wavelength of $\lambda = 5145$ Å, with an incident power of 100 mW. The value of the transferred in-plane wave vector $q_{||}$ is connected to θ by the relation $q_{||} = (4\pi/\lambda) \sin\theta$. The experiments were done with different applied magnetic values in the 0 to 5 kOe range.

Finally the state of the surface of these Co_xCr_{1-x} films has been investigated by tapping mode atomic force microscopy (A.F.M.), using a Veeco 3100 apparatus.

3. Results and discussion

3.1 *Structural and morphological characterizations*

Many results concerning these Co_xCr_{1-x} films have been published elsewhere [10, 11] and are just summarized here. Table 1 gives the thickness and the composition of the samples, values determined by Rutherford backscattering (R.B.S.) spectrometry, using a 2MeV monoenergetic He^{2+} incident beam, and a backscattering angle equal to 160°. Figure 1 shows a specimen of the experimental and the simulated spectra; for the latter we used SIMNRA program [12, 13].

Table 1. Thickness and content chromium of the Co_xCr_{1-x} samples.

Thickness t (nm)	17	19	44	56	70	215	220
Cr (at. %)	18	18	12	20	12	15	20
Co (at. %)	82	82	88	80	88	85	80

Figure 1a shows the experimental (dotted curve) and simulated (solid curve) RBS spectra for a sample of thickness 17 nm.

The big peak corresponds to Co atoms and the small peak to Cr ones. The difference in the height between the two peaks is due, of course, essentially to the small amount of the Cr atoms in the composition of the CoCr sample.

In Figure 1b we observe a small overlapping of the chromium's peak with the cobalt one, for this film of 44 nm thickness.

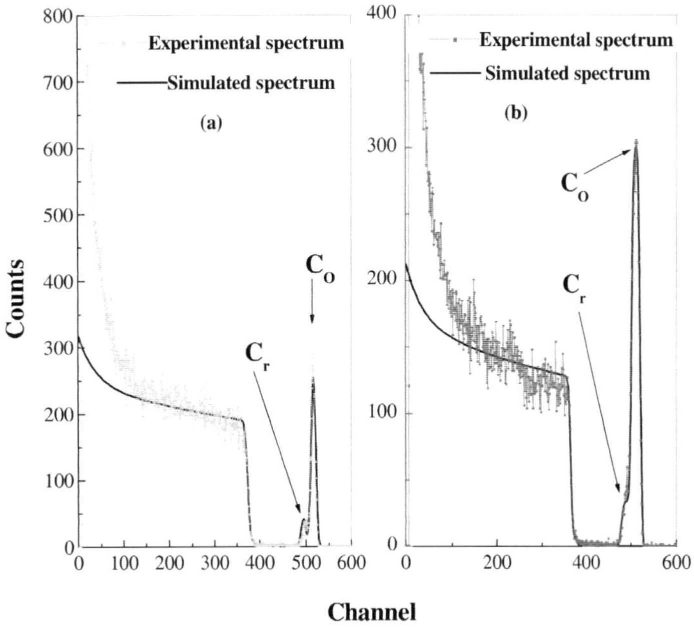

Figure 1 : Example of RBS spectra for samples of thickness 17 nm and 44 nm.

The x-ray diffraction patterns of CoCr/glass (not shown here) are very noisy, the reflections being weak and very broad. Nevertheless, the Bragg hexagonal close packed (hcp) peak (00.2) is clearly identified but its intensity is drastically reduced, the grain sizes being very small and their orientation random.

Examples of x-ray diffraction spectra are presented in figure 2 for CoCr/Si for t = 17 nm (the thinnest film), t = 44 nm (intermediate thickness) and t = 215 nm (among the thickest ones). All the spectra are characterized by three Bragg peaks, corresponding to the diffraction angles θ equal to 24.4°, 26.3° and 27.9° (i.e.2θ = 48.8, 52.5, and 55.8°) and the diffraction planes with "Bravais-Miller" indices (10.0), (00.2) and (10.1), respectively.

It is worth noting the great intensity of the hcp Co (00.2) Bragg peak, for all samples, which infer that the films crystallize in the hcp structure, are polycrystalline and with the (0001) texture. This undoubtedly indicates the preferential orientation of the crystal plane in parallel with the substrate surface.

The values of the lattice constants *c* and *a,* derived from the x-ray spectra and computed for different thicknesses and crystallographic orientations, range from 2.491 Å (for t = 70 nm) to 2.498 Å (for t = 17 nm) for *a*, and from 4.059Å (for t = 17nm) to 4.069Å (for t= 70 nm) for *c*. Compared to the bulk value parameters (*a* = 2.51 Å and *c* = 4.07 Å) these lattice parameters infer that the thinnest films are under a compressive stress.

Furthermore, table 2 displays the grain sizes computed using the x-ray diffraction patterns and the Scherer formula [14]:

$$L = \lambda / \Delta\theta \cos\theta$$

where L is the grain size for a grain with a particular orientation, λ the x-ray wavelength, θ the diffraction angle and Δθ the width at half height of the Bragg peak corresponding to this particular orientation. Table 2 gives the evolution of the grain size vs. the thickness t of the magnetic layer, for the three Bragg peaks present in the x-ray spectra. For the three Bragg peaks present, namely (10.0), (00.2) and (10.1), respectively, the most intense Bragg peak, which corresponds to hcp (00.2) peak of cobalt, shows the greater grain size (49, 70 and 70 nm) while the other two peaks display grain sizes with an average value of 20 nm.

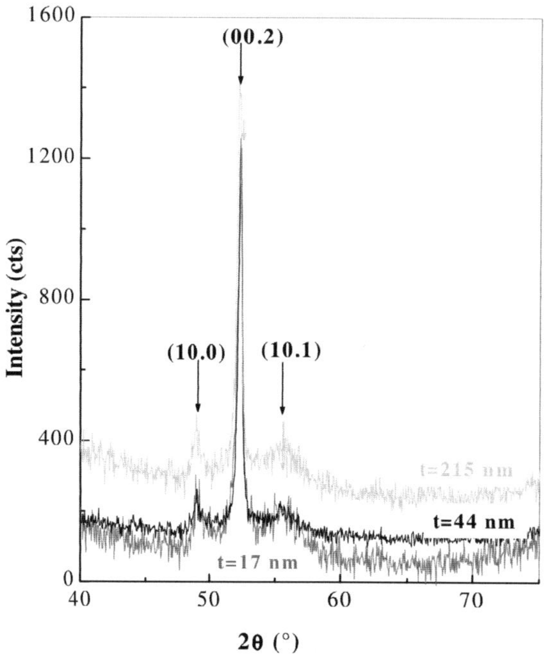

Figure 2 .Examples of x-ray diffraction spectra for CoCr/Si for t = 17 nm (the thinnest film), t = 44 nm (intermediate thickness) and t = 215 nm (among the thickest ones).

Table 2. Grain size vs. thickness.

t (nm)	17			44			215		
(hk.l)	(10.0)	(00.2)	(10.1)	(10.0)	(00.2)	(10.1)	(10.0)	(00.2)	(10.1)
L (nm)	26	49	20	19	70	13	16	70	10

For the most intense peak, a slightly increase of the grain size is noticed between the thinnest film, with t=17 nm, and the other films where the grain size is almost constant. Other researchers have reported an increase of the grain size with thickness [9, 15].

Moreover, topographic images given by the atomic force microscope (AFM) in the tapping mode reveal very smooth surfaces for most of the films, the root mean square (rms) roughness amplitude ranging from 0 (within the measures uncertainty) to 18 Å, the roughest film being the 70 nm thick film, and the 17 nm thick (the thinnest one) film displaying a very smooth surface (see table 3). So, no clear dependence of the roughness vs. the thickness of the magnetic layer is noticed. The AFM scans of some samples (among the thinnest (t = 17 nm) and the thickest (t = 215 nm) films) are presented in figure 3. Scan area is 5 x 5 μm^2 in all cases.

Table 3. Root-mean-square (rms) values of surface roughness as a function of thickness for Co_xCr_{1-x} thin films.

t (nm)	17	18	44	56	70	215	220
rms (Å)	----	15	----	14	18	-----	11

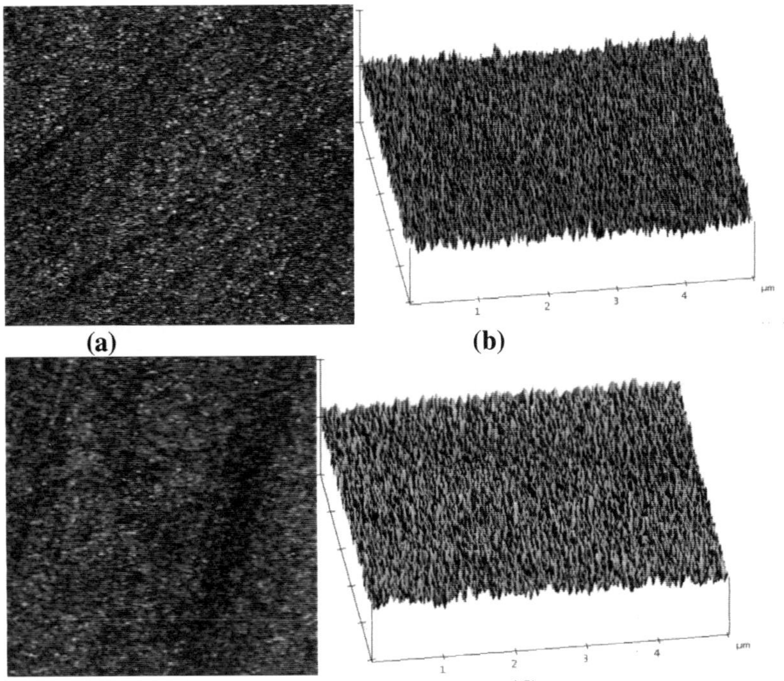

(a) **(b)** **(c)** **(d)**

Figure 3: 2D ((a) and (c)) and 3D ((b) and (d)) surface topography of CoCr thin films as observed by tapping mode AFM. Thicknesses are 18 nm (a,b) and 215 nm(c,d). Scan area is 5 x 5 μm^2 in all cases.

3.2. Magnetic characterizations

3.2.1 Static magnetic characterizations. In order to study the macromagnetic properties of the films, magnetic field dependence of magnetization curves are recorded, at room-temperature, using an alternating gradient field magnetometer (AGFM). The hysteresis loops depicted in figure 4 for two representative CoCr films, 18 nm thick, either on Si (100) and Corning glass substrates, are plotted both in the longitudinal as well as in the polar configurations.

From this loops it can be noticed clearly that the easy axis lies in the film plane: the longitudinal configuration is characterized by a quite square (rectangular) loops implying an easy axis lying in this direction, whereas the polar configuration, on the other hand, shows a small hysteresis and an extremely reduced remanent magnetization. Moreover, the loops in the polar configuration, representing the hard direction, is characterized at intermediate fields by a wide linear portion, followed at greater fields by a curvature going on up to the saturation. We may deduce at once that the demagnetizing filed exceeds the anisotropy constant K_u.

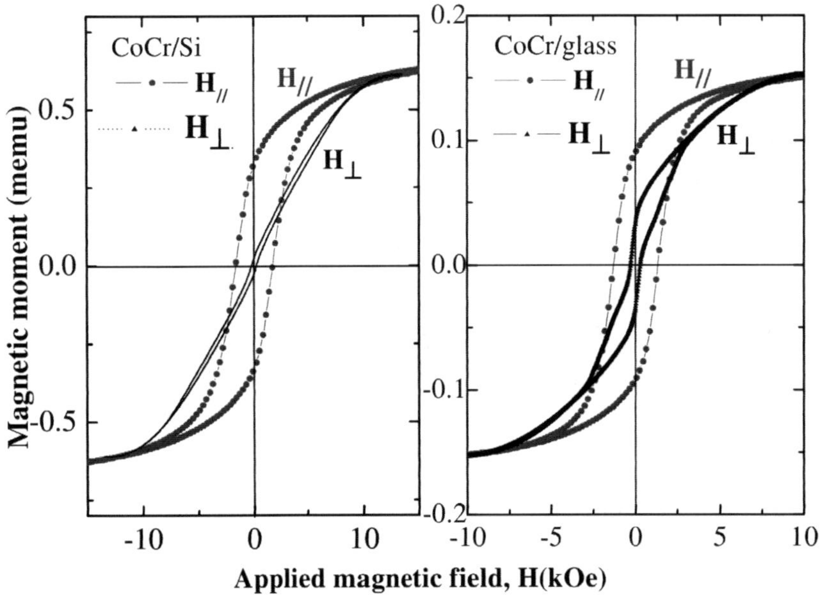

Figure 4 . Hysteresis loops for 17 nm CoCr/Si and CoCr/glass thick films. The external magnetic field is applied both in the longitudinal configuration ($H_{//}$) and the polar configuration $H_{\perp}$.

Table 4. Values of saturation magnetization M_s.

t(nm)		17	18	44	56	70	215	220
Cr(%at.)		18	18	12	20	12	15	20
M_s	Si	700	700	1200	220	1200	870	220
(emu/cm^3)	glass	500	500	900	170	900	840	170

No substantial change in the hysteresis loops was observed when rotating the substrate plane comparative to the applied field direction, indicating the absence of in-plane magnetic anisotropies.

On another hand, using the magnetization curves and the volume of the samples, we compute the saturation magnetization values. Table 4 displays value of M_s as a function of thickness and chromium content, the most prevailing factor being the Cr proportion. M_s values range from 220 emu/cm^3 for the $Co_{80}Cr_{20}$/Si, with t = 220 nm to 1200 emu/cm^3 for $Co_{88}Cr_{12}$/Si with t = 44 nm. For the glass substrate, the M_s values are somewhat smaller, ranging from 170 emu/cm^3 for $Co_{80}Cr_{20}$/glass, with t = 220 nm to 900 emu/cm^3 for $Co_{88}Cr_{12}$/glass with t = 44 nm. So, the effect of substrate on the saturation magnetization is significant.

Several authors found a similar dependence of M_s. Shun-ichi Iwasaki [2] reported values of M_s ranging from 300 emu/cm^3 to 700 emu/cm^3, when the percentage of chromium varies between 17 at.%Cr and 13 at.%Cr. J. Suzuki et al.[16] reported values of M_s ranging from 300 emu/cm^3 to 500 emu/cm^3, when the substrate temperature (T_s) rises from room temperature to 300°C, in their samples, prepared by RF magnetron sputtering. T. Takanashi et al. [17] have prepared $Co_{81}Cr_{19}$/Al multilayered films on glass-slide substrate at room temperature by facing targets sputtering (FTS) apparatus. They noticed a significant dependence of M_s on the thickness of $Co_{81}Cr_{19}$ layer and reported a maximum value of 660 emu/cm^3 when this thickness was 84 Å.

Figures 5 and 6 display the in-plane squareness, ratio of the remanent magnetization to the saturation magnetization $S = M_r/M_s$, and the coercive field H_c, vs. thickness of the magnetic layer, for both the substrates.

The same dependence of the squareness S on thickness is observed for both Si and glass substrates, the difference existing just on the amount of the value of this squareness. The thinnest film having the highest value, S=0.83 and 0.67, for Si and glass, respectively.

As for H_c, it is observed from figure 6 that the in-plane coercive field for CoCr/glass films decays with increasing thickness, following a law of the type $1/t^n$, reflecting a strong dependence on the surface effects. Moreover, for the CoCr/Si films, this behaviour is noticeable for films thicker than 70 nm. For films thinner than 70 nm, the relatively smaller values of H_c and their dependence on thickness may be related to the surface roughness. Indeed, the presence of roughness being directly associated to the tridimensional growth of a polycrystalline structure, possessing a great quantity of defects which work as pinning centres for domain walls, so, as the surface roughness values for the 17 nm and the 44 nm thick films are much lower than the 70 nm thick film one, a decrease of the coercive field for these films is observed comparatively to H_c for thicker films.

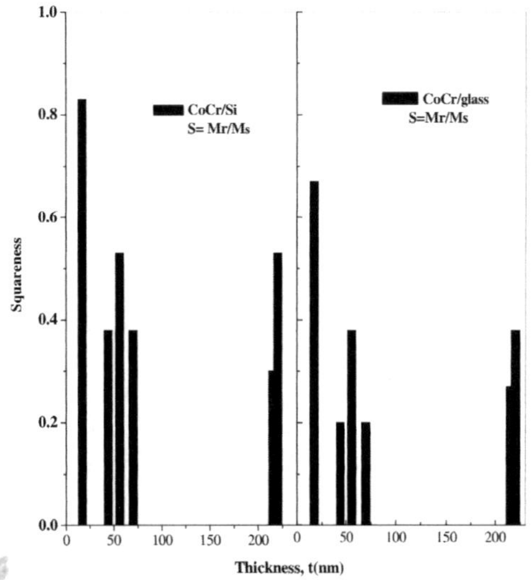

Figure 5. Squareness vs. thickness.

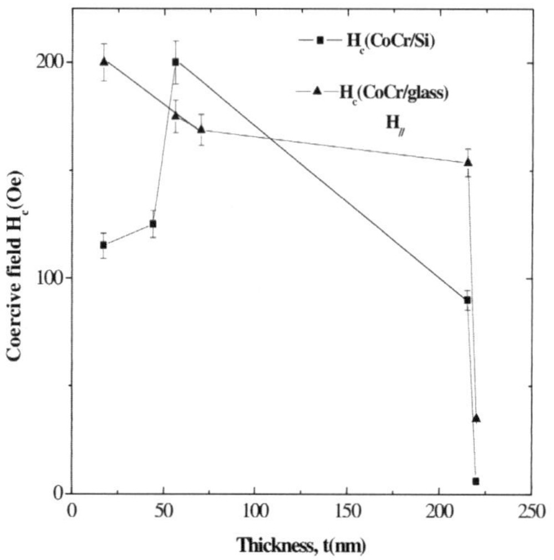

Figure 6. In-plane coercive field vs. thickness.

3.2.2 Dynamic magnetic characterizations. In the last decade, Brillouin light scattering has been used accurately to investigate magnetic materials [18-21].

BLS experiments were performed on the CoCr/Si series under the conditions described in Section 2. In figure 7, we show examples of experimental and calculated Brillouin spectra obtained with H = 1 kOe for CoCr(56nm)/Si sample, and with H = 5 kOe for CoCr(215nm)/Si sample. The peak with the highest intensity in the Stokes (low frequency) side of the spectra corresponds to the Damon-Eshbach (DE) surface spin wave. The DE [22] mode only appears in one side of the spectra because of its non-reciprocal nature and because the film is sufficiently thick that the light cannot interact with the DE mode localized on the lower surface of the film.

The adjustment of the theoretical and experimental spectra have been performed (taking $4\pi M_{eff} = 1.02\ T$, for CoCr (56nm) and $4\pi M_{eff} = 1.05\ T$ for CoCr (215nm), and the published values of the splitting factor and the spin wave exchange constant (g=2.16, and D = 2.6×10^{-9} erg.cm^2) and a negligible magnetocrystalline anisotropy.

Figure 8 displays the evolution of the frequency shift versus the applied magnetic field (for the Damon-Eshbach mode). This figure represents the experimental data (•, Δ) and the theoretical fit (—) computed for the dipolar mode using the dispersion relation:

$$\frac{\omega}{2\pi} = f = \frac{\gamma}{2\pi}\left\{H(H+4\pi M)+(2\pi M)^2(1-e^{-2q_{//}t})\right\}^{1/2}$$

where f is the frequency in GHz, γ is the magnetogyric ratio, related to the g factor (a value of 2.16 was taken for the splitting factor), $q_{//}$ is the transferred in-plane wave vector $q_{//}=1.73x10^5\,cm^{-1}$ *(for* $\theta = 45°$, *in this experiment).*

Finally, the study of the variation of the Damon-Eshbach mode frequency versus applied magnetic field intensity (figure 8) infers that the magnetic parameters determined for the Co_xCr_{1-x}/Si(100) investigated samples are satisfactory for a large range of the applied magnetic field variation.

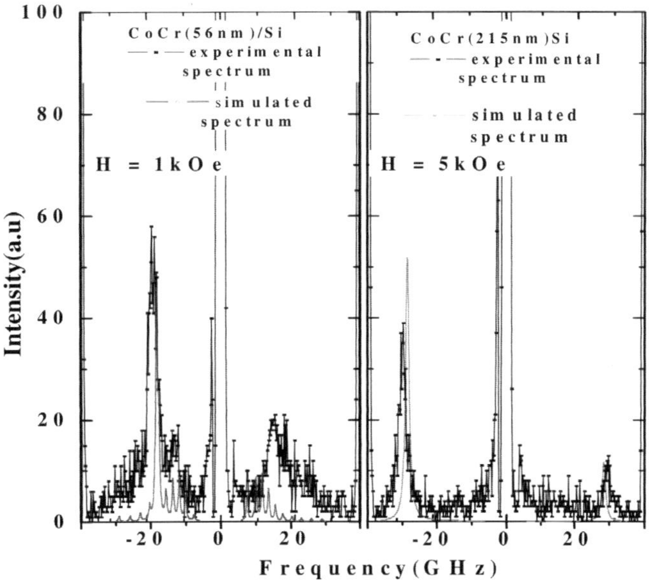

Figure 7 : Examples of BLS spectra for CoCr(56nm)/Si sample, for applied field H = 1kOe, and CoCr(215nm)/Si sample for applied field H = 5 kOe.

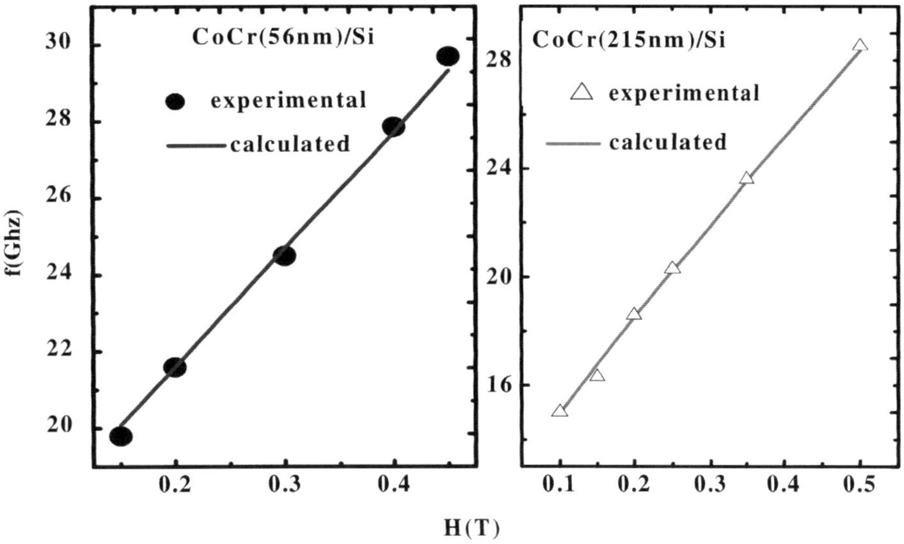

¶

Figure 8 : Damon-Eshbach mode frequency versus applied magnetic field intensity for the Co_xCr_{1-x}/Si samples. Indications are inside the figure.

From the fit of the BLS spectra, we extracted values of the anisotropy field H_a from which we computed the effective anisotropy magnetic factors K_{ueff}, these constants being connected to the anisotropy fields by the relation

$$K_{ueff} = \frac{1}{2} M_s H_a$$

where M_s is the saturation magnetization of the film.

The K_{ueff} values range from -0.7 x 10^6 to -3.2 x 10^6 erg/cm^3. These negative values, in the whole range of the chromium content, infer that the easy magnetization axis lies in the surface of the film for all samples and corroborate the conclusion found by means of the magnetization curves study (see section 3.2.1).

Unfortunately, due to the low thermal conductivity of glass, it was not possible to perform BLS measurements on CoCr/glass samples in order to point out the substrate effect. The experiments has to be performed using very low illuminating powers (less than 40 mW instead of the 100 mW used for CoCr/Si) in order to avoid undesirable heating. So, due to this experimental difficulty, no conclusive spectra were obtained about CoCr/glass samples.

4. Conclusion

The structural and magnetic properties of the Co_xCr_{1-x} series prepared by thermal evaporation under vacuum onto Si(100) and glass substrates have been studied as a function of the magnetic layer thickness, ranging from 17 nm to 220 nm. In this range, it has been observed that M_s decreases from 1200 emu/cm^3, to 220 emu/cm^3, for the CoCr/Si sample. The value of M_s was smaller for the CoCr/glass correspondent samples. It was also observed that the coercive field decreases with the thickness increase for CoCr/glass films, following a $1/t^n$ law, and hence, reflecting a clear dependence on the surface effects, whereas this behaviour is available only for CoCr/Si films thicker than 70 nm. For thinner films, the in-plane coercivity may be correlated to the surface roughness of the films. The anisotropy field values deduced from the fit of the BLS spectra give negative values for the effective anisotropy factors which infer that the easy magnetization axis lies in the film plane for all the samples, result corroborated by the magnetization curves study.

References

[1] Iwasaki S.-I and Nakamura Y 1977 *IEEE. Trans. Magn.* **13** 1272
[2] Iwasaki S.-I 1980 *IEEE. Trans. Magn.* **16** 71
[3] Schoenmaker J , Oppenheim I C, Chinaglia E F 2001 J. *Magn. Magn.Mater.* **226-230** 1653
[4] Dieny B, Speriosu V S, Gurney B A, Parkin S S, Wilhoit D R, Roche K P, Metin S, Peterson DT and Nadimi S 1991 *J. Magn. Magn. Mater* **93** 101
[5] Childress J R , Durand O, Schull A and George J M 1993 *Appl. Phys. Lett.* **63** 1996
[6] Chaiken A , Gutierrez C J, Krebs J and Prinz G A 1993 *J. Magn. Magn. Mater.* **125** 228
[7] Bottoni G , Candolfo D, Cecchetti A 1999 *J. Magn. Magn. Mater.* **193** 237
[8] Manago T , Kuramochi H, Akinaga H 2005 *J.Appl.Phys.* **97** 023907
[9] Kharmouche A, Chérif S-M, Bourzami A, Layadi A and Schmerber G 2004 *J.of Phys.D: Appl.Phys.* **37** 2583
[10] Kharmouche A, Chérif S-M, Schmerber G and Bourzami A 2007 *J. Magn. Magn. Mater.* **310** 152
[11] Kharmouche A and Djouada I 2008 *Appl. Surf. Sci.* **254** 5732
[12] Cliché L , Gujrathi S C and Hamel L A 1990 *Nucl. Instrum.Methods* B **45** 270

[13] Hösler W and Darji R 1994 *Nucl. Instrum.Methods* B **85** 602

[14] Eberhart J P 1989 '*Analyse structurale et chimiques des matériaux*', Bordas, Paris.

[15] Gil W, Görlitz D, Horisberger M and Kötzler J 2005 *Phys. Rev. B* **72** 134401
[16] Suzuki J, Takei K, Maeda Y and Morii Y 1998 *J. Magn. Magn. Mater.* **184** 116
[17] Takanashi T, Ikeda N and Naoe M 1991 *J.Appl.Phys.* **70** 6056
[18] Chérif S-M, Roussigné Y and Moch P 1999 *Phys. Rev.* B **59** 9482
[19] Chérif S-M, Roussigné Y, Dugautier C and Moch P 2000 *J. Magn. Magn. Mater.* **222** 337
[20] Hillebrands B, Mathieu C, Bauer M, Demokritov S O, Bartenlian B, Chappert C, Decanini D and Rousseaux F 1997 *J. Appl. Phys.* **81** 4993
[21] Gubbiotti G, Carlotti G, Ono T and Roussigné Y 2006 *J.Appl.Phys.* **100** 023906
[22] Damon R W and Eshbach J R 1961 *J. Phys. Chem. Solids* **19** 308

Superconductivity, antiferromagnetism and ferromagnetism in periodic table of D.I. Mendeleev

[1]**L Koroleva**[1,2]**, T.M. Khapaeva**[1]
M.V. Lomonosov Moscow State University, Moscow 119992, Russia

E-mail: koroleva@phys.msu.ru

Abstract. Definite regularity in the distribution of ferromagnetic, antiferromagnetic and superconducting elements is observed in the periodic table starting with the 4th period. Elements with superconductivity, by which *d*-shells start to fill up, are at the beginning of each period; then follow antiferromagnetics and ferromagnetics (in 4th period and lanthanides), or elements without any of the three listed order types (5th period and 6th period), in which the *d (f)*-shells continue to fill up almost exceedingly; then again appear superconductors by filling the *p*-shell up to the number is equal to 4. We calculated the radii of the external *d (f)*- and *p*-orbitals and the nearest to them orbitals with the Slater method. These trends were explained by distinction of degree of division of the external *d (f)*- or *p*-orbitals of the neighboring atoms in the crystal. Largest division occurs in ferromagnetics. In antiferromagnetics it is smaller than in ferromagnetics. It is demonstrated that in the superconducting crystals the external *d*- or *p*-shells approach the nucleus of neighboring atoms are much closely those for ferromagnetic or antiferromagnetic crystals. Furthermore the external *d*- or *p*-shells of some elements in the 5th and 6th periods approach the deeper shells of neighboring atoms. Hence the electron in this shell is situated in neighboring atoms in a different electric field from its own. This fact is open to speculation that the separation of spin and charge in electron, disposed on the external *d*- or *p*-orbitals, is quite possible. The charges without spin become bosons. Spins that have the magnetic moments are ordered antiparallel in pairs. Magnetic field transfers this pair in a parallel state and a magnetic flux component along of magnetic field from the pair is equal to one fluxon (the quant of the magnetic flux).

1. Introduction

The nature of ferromagnetism (FM), antiferromagnetism (AFM) and superconductivity (SC) is only partly clear at present, in spite of numerous researches in this area. At the same time, the clear regularity of arrangement of superconductors, antiferromagnetics and ferromagnets in the periodical table is amazing. (See figure 1 where allocation of FM, AFM and SC elements in periodic table is shown.) Recently the distribution of the elements with the indicated types of ordering was analyzed in

[2] To whom any correspondence should be addressed.

our short article [1]. This analysis was based on the supposition that the indicated type of ordering are connected with *d*-, *f*- and *p*-shells of atoms.

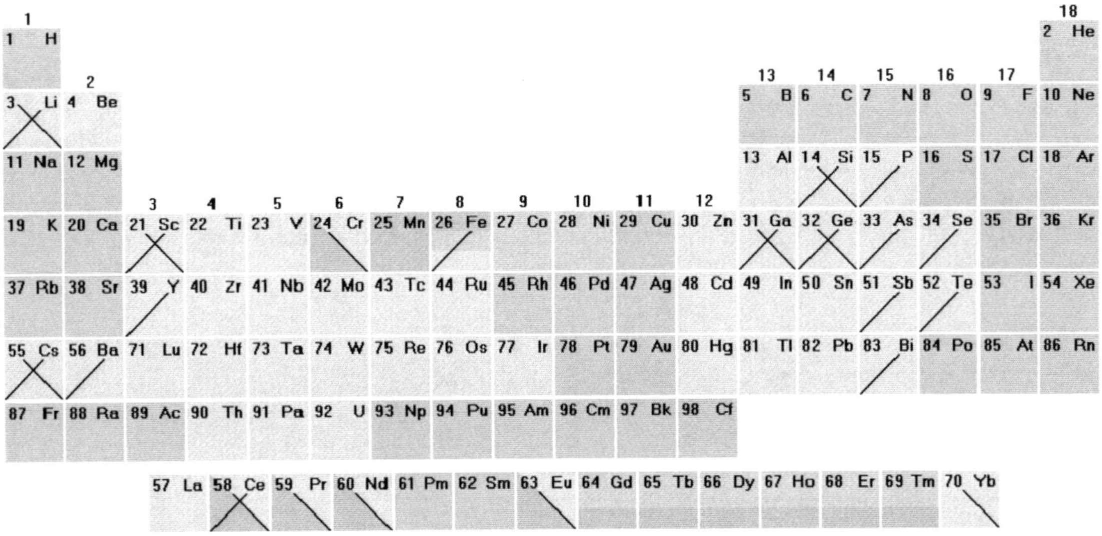

Figure 1. Periodic Table of D.I. Mendeleev. Ferromagnetics, antiferromagnetics and superconductors coloured Red, Blue and Green respectively. The signs / and \ indicate superconductivity under pressure and in thin films respectively.

Calculations of radii of external *d*-, *f*- and *p*-shells and the nearest to them shells of the elements of periodic table are submitted in the present article. These calculations were carried using a method of Slater [2] who proposed the next empirical rules. So the maximum of the radial charge density for one atom shell comes at

$$r = (n^*)^2/(Z - s). \qquad (1)$$

Here Z is the atom number and s is a shielding constant. The value n^* and $(Z - s)$ are given by the following rules:

(1) n^* is assigned by the following table, in terms of real principal quantum number n:

for n = 1, 2, 3, 4, 5, 6

n^* = 1, 2, 3, 3.7, 4.0, 4.2

(2) For determining $(Z - s)$, the electrons are divided into the following groups, each having a different shielding constant: 1*s*; 2*s*, *p*; 3*s*, *p*; 3*d*; 4*s*, *p*; 4*d*; 4*f*; 5*s*, *p*; 5*d*; etc. That is, the *s* and *p* of a given *n* are grouped together, but the *d* and *f* are separated. The shells are considered to be arranged from inside out in the order named.

(3) The shielding constant *s* is formed, for any group of electrons, from the following contributions:

(a) Nothing from any shell outside the one considered.

(b) An amount 0.35 from each other electron in the group considered (except in the 1*s* group, where 0.30 is used instead).

(c) If the shell considered is an *s, p* shell, an amount 0.85 from each electron with total quantum number less by one, and an amount 1.00 from each electron still further in; but if the shell is a *d* or *f*, an amount 1.00 from every electron inside it.

Though estimations of the radii of the atomic orbitals on Slater method are reasonably rough, they are in good agreement with the spectroscopy experiments.

Comparison of radii of external *d-*, *f-* and *p*-shells with the interatomic distances in crystals is permitted to reveal the connection of FM and AFM ordering and SC with the division degree of these shells in crystals. The largest division degree is observed in ferromagnetics and the least in superconductors up to penetration of the external *d-* and *p*-shells inside neighbor atoms. In last case the electron on this shell is situated in neighbor atom in a stronger electrical field than in its own and the separation of spin and charge in an electron is quite possible; in this case the charges without spin become bosons.

2. Connection of ferromagnetism, antiferromagnetism and superconductivity in elements of periodic table with filling degree of external *d-*, *f-* and *p*-shells

Figures 2 - 5 and table 1 show the radii *r* of the external *d-*, *f-* and *s*-shells of atoms as well as the intrinsic shells adjacent to them which were calculated of us using Slater method [2], the nearest interatomic distances in the real crystalline lattices of elements *b* with their types, according to the data from the monograph [3], and the differences *b- r*.

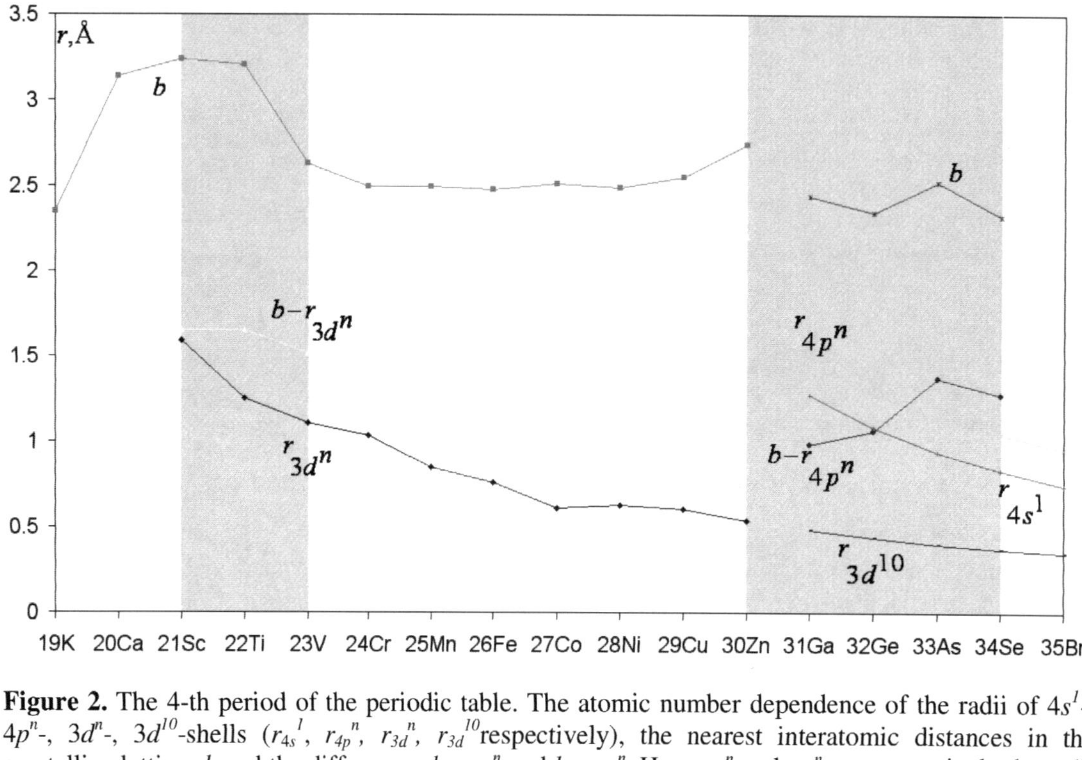

Figure 2. The 4-th period of the periodic table. The atomic number dependence of the radii of $4s^1$-, $4p^n$-, $3d^n$-, $3d^{10}$-shells ($r_{4s}{}^1$, $r_{4p}{}^n$, $r_{3d}{}^n$, $r_{3d}{}^{10}$respectively), the nearest interatomic distances in the crystalline lattices *b* and the differences b - $r_{3d}{}^n$ and b - $r_{4p}{}^n$. Here $r_{4p}{}^n$ and $r_{3d}{}^n$ are respectively the radii of the external shells of 4*p*- and 3*d* – layers of the atoms. The regions of superconducting elements are darkened.

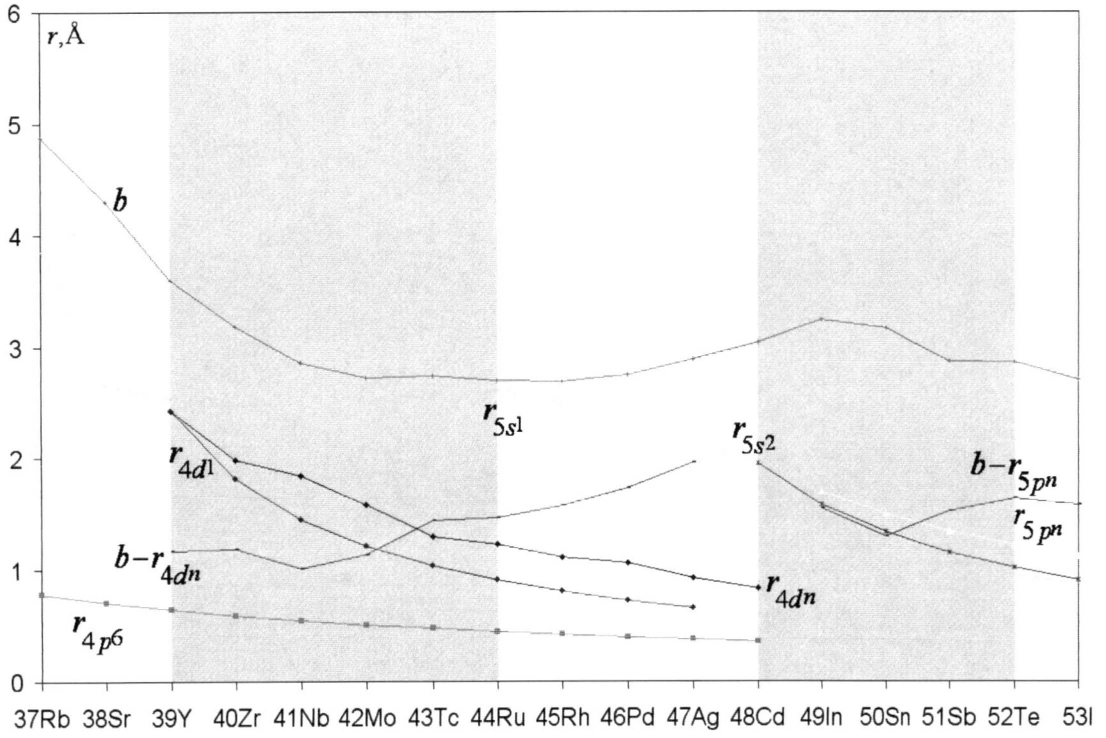

Figure 3. The 5-th period of the periodic table. The atomic number dependence of the radii of $5s^1$-, $5s^2$-, $5p^n$-, $4p^6$-, $4d^n$-, $4d^1$-shells ($r_{5s}{}^1$, $r_{5s}{}^2$, $r_{5p}{}^n$, $r_{4p}{}^6$, $r_{4d}{}^n$, $r_{4d}{}^1$ respectively), the nearest interatomic distances in the crystalline lattices b and the differences $b - r_{4d}{}^n$ and $b - r_{5p}{}^n$. Here $r_{5p}{}^n$ and $r_{4d}{}^n$ are respectively the radii of the external shells of the $5p$- and $4d$ – layers of atoms. Regions of superconducting elements are darkened.

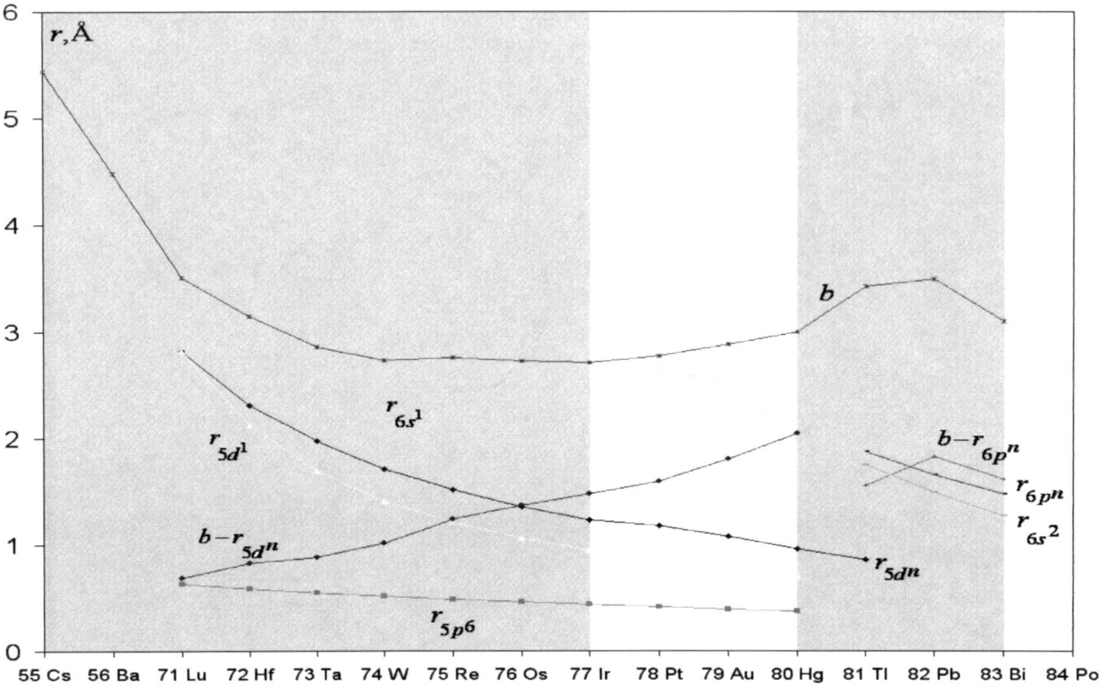

Figure 4. The 6-th period of the periodic table. The atom number dependence of the radii of $6s^1$-, $6s^2$-, $6p^n$-, $5\,p^6$,- $5d^n$-, $5d^1$-shells ($r_{6s}{}^1$, $r_{6s}{}^2$, $r_{6p}{}^n$, $r_{5p}{}^6$, $r_{5d}{}^n$, $r_{5d}{}^1$respectively), the nearest interatomic distances in the crystalline lattices b and the differences $b - r_{5d}{}^n$ and $b - r_{5p}{}^n$. Here $r_{6p}{}^n$ and $r_{5d}{}^n$ are respectively the radii of the external shells of $6p$- and $5d$ – layers of the atom. Regions of superconducting elements are darkened.

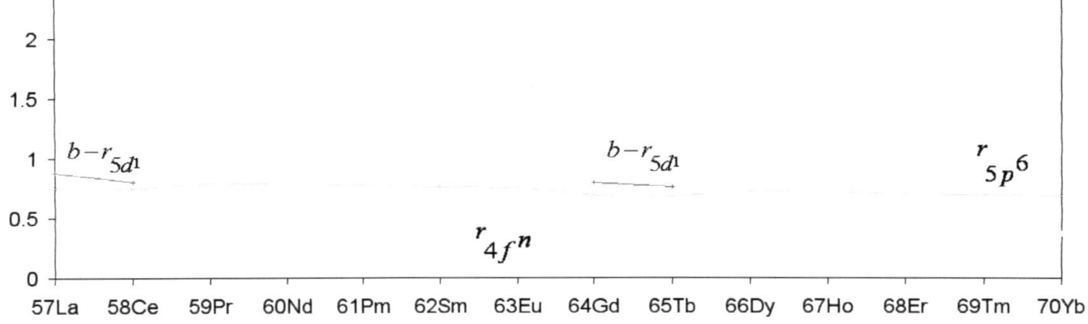

Figure 5. Lanthanides. The atomic number dependence of radii of $6s^2$-, $5p^6$-, $4f^n$-, $5d^1$-shells ($r_{6s}{}^2$, $r_{5p}{}^6$, $r_{4f}{}^n$, $r_{5d}{}^1$ respectively), the nearest interatomic distances in the crystalline lattices b and the differences $b - r_{5d}{}^1$ and $b - r_{4f}{}^n$. Here $r_{4f}{}^n$ is radii of the external shells of the $4f$ – layers of atoms.

Table 1. The radii of the external electronic p^n-, d^n-, f^n-shells (r_p^n, r_d^n, r_f^n respectively), the nearest interatomic distances in the crystalline lattices of elements *b*, their types, the electronic configuration of the external shells of the elements with the atomic number Z in the periodic table of Mendeleev and the temperatures of the superconducting transition T_C and Curie temperatures T_K of the elements. The sign / indicate the superconductivity under pressure.

The 4-th period (3d). Internal electronic shell of Ar: $1s^2 2s^2 2p^6 3s^2 3p^6$.

Z	Electronic configuration of external shells	r_{3d}^n (Å)	r_{4p}^n (Å)	b (Å)	Type of lattice	$b - r_{3d}^n$ (Å)	T_C / T_K (K)
19 K	$4s^1$			2.3500	b.c.c.		
20 Ca	$4s^2$			3.4000	f.c.c.		
21 Sc	$3d^1 4s^2$	1.590		3.2400	h.c.p. (α)	1.650	
				3.2040	f.c.c. (β)	1.610	
22 Ti	$3d^2 4s^2$	1.250		2.9000	h.c.p. (α)	1.650	0.39 (T_C)
				2.8750	b.c.c. (β)	1.625	
23 V	$3d^3 4s^2$	1.110		2.6274	b.c.c.	1.517	5.430 (T_C)
24 Cr	$3d^5 4s^1$	1.037		2.4930	b.c.c. (α)	1.456	
				2.7000	h.c.p. (β)	1.663	
25 Mn	$3d^5 4s^2$	0.850		2.4940	compl. (α)	1.644	
26 α-Fe	$3d^6 4s^2$	0.760		2.4778	b.c.c.	1.718	1043 (T_K)
26 γ-Fe	$3d^6 4s^2$			2.5800	f.c.c.	1.820	
27 Co	$3d^7 4s^2$	0.690		2.5070	h.c.p.	1.817	1404 (T_K)
28 Ni	$3d^8 4s^2$	0.630		2.4878	f.c.c.	1.858	636 (T_K)
29 Cu	$3d^{10} 4s^1$	0.608		2.5509	f.c.c.	1.943	
30 Zn	$3d^{10} 4s^2$	0.539		2.7400	h.c.p.	2.201	0.875 (T_C)
						$b - r_{4p}^n$	
31 Ga	$3d^{10} 4s^2 4p^1$	0.484	1.451	2.4370	rhombic	0.986	1.08 (T_C)
32 Ge	$3d^{10} 4s^2 4p^2$	0.440	1.284	2.3450	cub.diam.	1.061	
33 As	$3d^{10} 4s^2 4p^3$	0.403	1.147	2.5200	rhombohedral	1.373	0.31-0.5 (T_C)/
34 Se	$3d^{10} 4s^2 4p^4$	0.371	1.044	2.3200	h.chain	1.276	6.95 (T_C) /
35 Br	$3d^{10} 4s^2 4p^5$	0.355	0.955	2.2700	rhombic	1.315	
36 Kr	$3d^{10} 4s^2 4p^6$			3.9500	f.c.c.		

The 5-th period (4d). Internal electronic shell of Kr: $1s^2 2s^2 2p^6 3s^2 3p^6\ 3d^{10} 4s^2 4p^6$.

Z	Electronic configuration of external shells	r_{4d}^n (Å)	r_{5p}^n (Å)	b (Å)	Type of lattice	$b - r_{4d}^n$ (Å)	T_C (K)
37 Rb	$5s^1$			4.870	b.c.c.		
38 Sr	$5s^2$			4.290	f.c.c.		
39 Y	$4d^1 5s^2$	2.420		3.590	h.c.p.	1.170	1.2 – 2.7 /
40 Zr	$4d^2 5s^2$	1.990		3.180	h.c.p. (α)	1.190	0.53
				3.126	b.c.c. (β)	1.136	0.65
41 Nb	$4d^4 5s^1$	1.840		2.852	b.c.c. (β)	1.012	9.25
42 Mo	$4d^5 5s^1$	1.580		2.720	b.c.c. (β)	1.140	0.916
43 Tc	$4d^5 5s^2$	1.296		2.735	h.c.p.	1.439	7.78
44 Ru	$4d^7 5s^1$	1.230		2.693	h.c.p.	1.463	0.493
45 Rh	$4d^8 5s^1$	1.110		2.684	f.c.c. (β)	1.574	
46 Pd	$4d^{10}$	1.059		2.745	f.c.c. (β)	1.736	
47 Ag	$4d^{10} 5s^1$	0.924		2.883	f.c.c. (β)	1.959	
48 Cd	$4d^{10} 5s^2$	0.829		3.040	hexag.	2.211	0.52
						$b - r_{5p}^n$	
49 In	$4d^{10} 5s^2 5p^1$	0.744	1.696	3.240	tetragonal	1.544	3.40
50 Sn	$4d^{10} 5s^2 5p^2$	0.675	1.501	3.160	cub.diam.	1.300	3.721
51 Sb	$4d^{10} 5s^2 5p^3$		1.35	2.870	rhombohedral	1.520	2.6 – 2.7 /
52 Te	$4d^{10} 5s^2 5p^4$		1.22	2.860	h.chain	1.640	2.05 /
53 I	$4d^{10} 5s^2 5p^5$		1.12	2.700	rhombic	1.580	
54 Xe	$4d^{10} 5s^2 5p^6$				f.c.c.		

Rare earths group (4f). Internal electronic shell of Xe: $1s^2 2s^2 2p^6 3s^2 3p^6 3d^{10} 4s^2 4p^6 4d^{10} 5s^2 5p^6$.

Z	Electronic configuration of external shells	r_{4f}^n (Å)	r_{5d}^n (Å)	b (Å)	Type of lattice	$b - r_{5d}^n$ or $b - r_{4f}^n$ (Å)	T_C / T_K (K)
57 La	$5d^1 6s^2$		2.83	3.710	hexag. (α)	0.880	4.88 (T_C)
58 Ce	$4f^5 d6s^2$	0.605	2.83	3.634	f.c.c.	0.804	1.7 (T_C) /
				3.650	h.c.p. (α)	0.830	
59 Pr	$4f^3 6s^2$	0.590		3.662	hexag. (α)	3.072	
				3.642	b.c.c.	3.052	
60 Nd	$4f^4 6s^2$	0.560		3.657	h.c.p.	3.100	
61 Pm	$4f^5 6s^2$	0.534					
62 Sm	$4f^6 6s^2$	0.509					
63 Eu	$4f^7 6s^2$	0.487		4.084	b.c.c.	3.597	
64 Gd	$4f^7 5d^1 6s^2$	0.456		3.622	h.c.p.	0.792	310 (T_K)
65 Tb	$4f^8 5d^1 6s^2$	0.438	2.83	3.585	h.c.p.	0.755	236 (T_K)
66 Dy	$4f^{10} 6s^2$	0.431		3.578	h.c.p.	3.147	151 (T_K)
67 Ho	$4f^{11} 6s^2$	0.415		3.557	h.c.p.	3.142	88 (T_K)
68 Er	$4f^{12} 6s^2$	0.400		3.532	h.c.p.	3.132	50 (T_K)
69 Tu	$4f^{13} 6s^2$	0.386		3.523	h.c.p.	3.137	17.4 (T_K)
70 Yb	$4f^{14} 6s^2$	0.373		3.866	f.c.c.	3.493	

6-th period (5d). Internal electronic shell of Yb: $1s^22s^22p^63s^23p^63d^{10}4s^24p^64d^{10}4f^{14}5s^25p^66s^2$.

Z	Electronic configuration of external shells	r_{5d}^n (Å)	r_{6p}^n (Å)	b (Å)	Type of lattice	$b - r_{5d}^n$ (Å)	T_C (K)
55 Cs*	$6s^1$			3.302	h.c.p.		1.5 /
56 Ba*	$6s^2$			4.340	b.c.c.		1.3 /
71 Lu	$5d6s^2$	2.820		3.509	h.c.p.	0.689	0.1
72 Hf	$5d^26s^2$	2.310		3.140	h.c.p.	0.830	1.36
73 Ta	$5d^36s^2$	1.970		2.854	b.c.c.	0.884	4.47
74 W	$5d^46s^2$	1.720		2.735	b.c.c. (α)	1.015	1.59
75 Re	$5d^56s^2$	1.514		2.755	h.c.p.	1.241	1.697
76 Os	$5d^66s^2$	1.357		2.725	h.c.p.	1.368	0.655
77 Ir	$5d^76s^2$	1.229		2.708	b.c.c.	1.479	0.11-0.14
78 Pt	$5d^96s^1$	1.178		2.769	b.c.c.	1.591	
79 Au	$5d^{10}6s^1$	1.080		2.883	b.c.c.	1.803	
80 Hg	$5d^{10}6s^2$	0.958		2.999	rhombohedr.	2.041	4.15
						$b - r_{6d}^n$	
81 Tl	$6s^26p^1$	0.861	1.87	3.423	b.c.c.	1.553	1.39
82 Pb	$6s^26p^2$		1.66	3.490	b.c.c.	1.830	7.23
83 Bi	$6s^26p^3$		1.48	3.100	rhombohedr.	1.620	3.91 /
84 Po	$6s^26p^4$				monoclinic		

*Internal electronic shell of Xe: $1s^22s^22p^63s^23p^63d^{10}4s^24p^64d^{10}5s^25p^6$.

The 7-th period (6d). Internal electronic shell of Rn: $1s^22s^22p^63s^23p^63d^{10}4s^24p^64d^{10}4f^{14}5s^25p^66s^26p^6$.

Z	Electronic configuration of external shells	r_{5f}^n (Å)	r_{4p}^n (Å)	b (Å)	Type of lattice	$b - r_{3d}^n$ (Å)	T_C (K)
87 Fr	$7s^1$						
88 Ra	$7s^2$						
89 Ac	$6d^17s^2$						
90 Th	$6d^27s^2$		3.070	3.59	b.c.c.	0.520	1.39
91 Pa	$5f^26d^17s^2$		2.544	3.25	tetragonal	0.706	1.4
92 U	$5f^36d^17s^2$		2.170	2.77	compl.	0.600	2.3 /
				2.97	b.c.c. (β)	0.800	
93 Np	$5f^46d^17s^2$	0.53	3.460				
94 Pu	$5f^67s^2$						
95 Am	$5f^77s^2$	0.53					
96 Cm	$5f^76d7s^2$	0.56					
97 Bk	$5f^86d7s^2$						
98 Cf	$5f^{10}7s^2$						

1. Fourth period of D.I. Mendeleev periodic table

Figure 1 shows the periodic table. Let us examine how one type of ordering is substituted for another when moving from left to right along the 4-th period of the table. When the number of 3*d*-electrons *n* equals 1 (Sc), SC appears both in thin films and in bulk samples under pressure. SC takes place at $n = 2$ (Ti) and 3 (V) in bulk samples with the temperature of the phase transition T_C equal to

0.39 K and 5.43 K respectively [4]. The Cr bulk samples with $n = 5$ show AFM, SC also occurs in thin films. Then follow antiferromagnetic Mn ($n = 5$) and the ferromagnets Fe ($n = 6$), Co ($n = 7$) and Ni ($n = 8$), in which the indicated types of ordering take place in bulk samples. It is necessary to note that α-Fe, having a b.c.c. structure, is a ferromagnet while γ-Fe, having a f.c.c. structure is an antiferromagnet [5]. Under a pressure of ~15 GPa, α-Fe undergoes a structural transition from b.c.c. to f.c.c. that is accompanied by a FM-SC transition; at the same time SC is observed between 15 and 30 GPa [6]. Cu ($n = 10$), which follows Ni, shows none of the 3 types of ordering. However, in the Cu compounds, where Cu is bi- or trivalent (n = 9, 8 respectively), high-temperature SC is observed. Moving along this period, superconductors again occur: Zn (bulk form; the configuration of electrons is $3d^{10}4s^2$); Ga, Ge (under pressure in thin films); As, Se (only under pressure), in which the 4*p*-orbital is sequentially filled with the maximum number of electrons equal to 4. Thus, the following tendency exists in the 4-th period: SC occurs only when the number of 3*d*-electrons is small ($1 \le n \le 3$) or, vice versa, large ($8 \le n \le 10$). It should be note that elements at a boundary between different types of ordering possess both types at once: bulk Cr with a b.c.c. lattice is antiferromagnetic but, in thin films, it is a superconductor; Fe with a b.c.c. lattice is a ferromagnet and γ-Fe is an antiferromagnet. Furthermore, in the 4-th period SC begins to manifest only under pressure and in thin films (Sc) and ends at Se, only under pressure. So, the transition from one of 3 types of ordering to another one (or from absence of ordering to its appearance) occurs through the elements with both order types in the first situation and on evident "weakening" of SC in the second situation as appears only in thin films and under pressure. It is interesting to note that the elements with $n = 4$ and $n = 9$, looking favorable for the appearance of SC, are absent. This «absence» takes place after the most high-temperature superconductor in this period V and before Cu, in whose compounds high-temperature SC is found out.

2. *Fifth period of D.I. Mendeleev periodic table*

In the 5-th period the 4*d*-orbital is filled. The 5-th period in which the 4*d*-orbitals are filled, contains much more superconductors than the 4-th period but contains no FM or AFM elements those manifesting no ordering elements occupy their places. SC here appears at Y ($n = 1$) under pressure and lasts in the bulk samples from Zr ($n = 2$) to Ru ($n = 7$). Let us note that the «absence» in filling of 4*d*-orbital at $n = 3$ here occurs before the most high-temperature superconductor Nb ($T_C = 9.3$ K). Then 3 elements without any ordering (Rh, Pd, Ag) follow and then again SC: Cd ($n = 10$) and In, Sn, Sb, Te, at which 5*p*-orbital is filled up to the $5p^4$ state, at two last elements SC occurs only under pressure at that. «Absence» is observed before Pd element too because the element with $n = 9$ is absent.

3. *Sixth period of D.I. Mendeleev periodic table*

Let us consider the 6-th period without lanthanides at first. SC appears here under pressure and in thin films earlier than in the previous period. Namely, SC appears in thin films and under pressure at Cs (1*s*-shell is not completely filled) and under pressure at Ba (all the shells are filled). Then 7 superconductors from Lu ($n = 1$) to Ir ($n = 7$), at which the 5*d*-orbital is filled, follow. Thereafter the «absence» in filling of 5*d*-orbital (the element with $n = 8$ is absent) occurs and the noble metals Pt ($n = 9$) and Au ($n = 10$) follow. Then SC appears again at Hg ($5d^{10}6s^2$) and Tl, Pb, Bi (in the last one — under pressure), in which the 6*p*-states are sequentially filled up to p^3.

4. *Lanthanides*

Let us consider the group of lanthanides separately. This is the second area in the periodic table, where AFM and FM are observed. SC takes place at La with the number of 5*d*-electrons $n = 1$. When 4*f*-orbital is filled from Ce (the number of the 4*f*-electrons $k = 1$) to Nd ($k = 4$), the combination from SC (in Pr and Nd in thin films and in Ce in thin films and under pressure) and AFM occurs. Then two elements with $k = 5$ (Pm) and $k = 6$ (Sm), which don't show any of three listed above orderings, follow and Eu ($k = 7$), in which SC in thin films and AFM take place. Then 6 ferromagnets from Gd to

Tm (although with the admixture of the AFM interactions), in which 4*f*-orbital is filled ($7 \le k \le 13$), follow at the same time $n = 1$ at Gd and Tb and $n = 0$ at the rest elements. Yb with $k = 14$, $n = 0$, in whose films SC is found out, ends the row. There is the «absence» in filling of 4*f*-shell between Tb ($k = 8$) and Dy ($k = 10$), however at Tb the magnetic moment is defined by the sum $k + n$.

5. *Seventh period of D.I. Mendeleev periodic table*

In 7-th period, where only 6*d*- and 5*f*-shells are filled, merely SC takes place in Th ($n = 2$, $k = 0$; $T_C = 1.39$ K), Pa ($n = 1$, $k = 2$; $T_C = 1.4$ K) and U ($n = 1$, $k = 3$). At the last element SC is observed only under pressure. Hence SC occurs only at $2 \le n + k \le 4$. It should be remarked that for the 7-th period the rules of Slater do not apply since the effective quantum number n^* in the relation (1) does not adduces in [2]. Therefore data of radii of 6*d*- and 5*f*-shells in Table I is adduced from [7].

3. Base tendencies of existence of SC, FM and AFM in periodic table

Thus, some basic trends for the existence of SC, FM, AFM areas are found. Mention should be made of the phenomenon of lanthanide contraction and *d*-compression, implying that an increase in number of electrons in the *d*- and *f*-shells results in a decrease in their radii. This is apparent from figures 2-5 too. FM occurs for the elements, in which the 3*d*-shell is more than half-filled, or for those having a 4*f*-shell, in which the sum $k + n \ge 8$, that is, for the elements with the smallest radii of the 3*d*- and 4*f*-shells. AFM is observed in the elements, in which the 3*d*- or 4*f*-shell is precisely half-filled. Furthermore, for 4*f*-elements a combination of SC and AFM occurs at $2 \le k \le 4$ and a mixture of AFM and FM occurs at $8 \le k + n \le 13$. SC is observed for 3*d*-elements with a less than half-filled 3*d*-shell ($1 \le n \le 3$) and with a completely filled 3*d*-shell, at the same time the 4*p*-orbital is filled up to 4 and the 4*s*-orbital is completely filled. For lanthanides SC takes place only under pressure and in thin films with a less than half-filled 4*f*-shell ($1 \le k \le 4$) and for one element, Yb, with the completely filled 4*f*- and 6*s*-orbitals. That is, the 3*d*- and 4*f*-shells of SC elements are compressed less than in ferromagnets. As *d*-shells move away from the nucleus, the region of existence of SC expands. So, for the 4*d*- and 5*d*-elements SC occurs at $1 \le n \le 7$. This is due to the fact that the 4*d* and 5*d* wave functions are damped out at a distance from the nucleus longer than the 3*d* wave functions and consequently having a wide overlap with neighboring atoms (figures 3,4). For heavier elements, where the 6*d*- and 5*f*-orbitals are filled, SC occurs only for $2 \le n + k \le 4$. Further, SC is observed for the elements, in which the 3*p*-, 4*p*-, 5*p*- and 6*p*-orbitals are incompletely filled and contain no more than 4 electrons. Thus, SC occurs for elements with no more than 0.7-filled *d*- or *p*-orbitals and 0.5-filled *f*-orbitals. SC occurs for the elements with completely filled *d*-, *f*- and *s*-orbitals (Zn, Cd, Hg, Yb) too.

4. Explanation of superconductors, ferromagnetics and antiferromagnetics allocation in periodic table of D.I. Mendeleev

Let us remind that wave functions of *p*- and *d*-electrons are petal-shaped. As figures 2 - 5 and table 1 indicate, in the superconducting crystals the external *d*-shells approach the nucleus of neighboring atoms appreciably closer than in the ferromagnetic and antiferromagnetic crystals. It is obvious that the degree of the division of *d (f)*- or *p*-orbitals of the neighboring atoms in the crystals is written by the difference $b - r$. So, in the 4-th period $b - r_{3d}^{\,n}$ for SC's is within limits 1.52 – 1.62 Å whereas this difference for AFM's is equals to 1.63 Å at Cr and 1.65 Å at Mn. For FM's it is substantially greater: from 1.72 Å at Fe to 1.86 Å at Ni. At SC's in the 5-th period $1.19 \le b - r_{4d}^{\,n} \le 1.46$ Å, in the 6-th period $0.689 \le b - r_{5d}^{\,n} \le 1.48$ Å and in 7-th period $0.52 \le b - r_{6d}^{\,n} \le 0.71$ Å. What's more figure 5 shows that $(b - r_{5d}^{\,1}) \ll (b - r_{4f}^{\,n})$ and $(b - r_{5d}^{\,1}) \ge r_{4f}^{\,n}$. This clearly demonstrates that in lanthanides the $5d^1$-orbital is answerable for SC other than the $4f^n$-orbitals. Furthermore, in the number of the superconducting crystals of 5-th and 6-th periods the external *d*-shells of each atom penetrate inside neighboring atoms disposing between more deep the p^6-shells and the d^1-shells (for Y, Zr, Nb, Mo, Lu, Hf, Ta,W) (figures 3,4). The penetration of the external *p*-orbitals inside of neighboring atoms is also observed in the crystals of some superconducting elements disposed in the end periods. So,

$5p^n$- shells are disposed between more deep $5s^1$- and $5s^2$-shells of the neighboring atoms (for In and Sn) and $6p^1$-shell of Tl is disposed in neighboring atoms nearer to nuclear than $6s^1$ –shells.

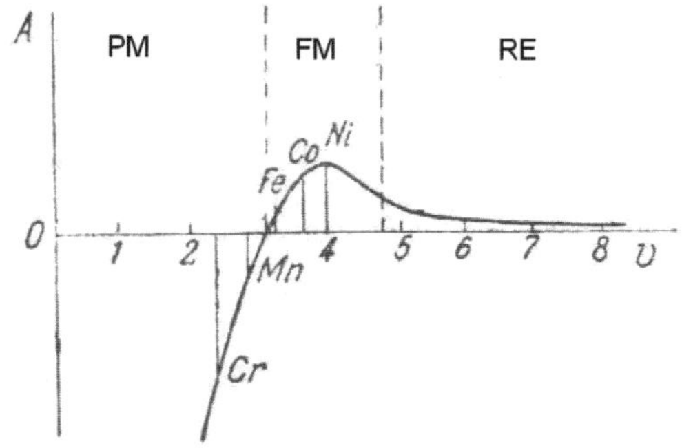

Figure 6. The dependence of the exchange integral (A) from $v = b/r$ is shown. Here r is the radius of the external $3d$ or $4f$-shell and PM is paramagnetism, FM is ferromagnetism and AFM is antiferromagnetism. (This figure took from [10]).

It is well known that atoms have to be enough separated for the FM interactions. Only in this case the exchange integral becomes positive. At the smaller distances between atoms the negative exchange interactions prevail and in crystal AFM order takes place [2, 5, 8]. This is apparent from figure 6, in which the dependence of the exchange integral (A) from $v = b/r$ is shown. Here r is the radius of the external $3d$ or $4f$-shell. (This figure took from [9]). As it can be seen from the table 1 and the figure 1, these rules are well carried out both for the elements of 4-th period, where the incompletely filled $3d$-shells are responsible for FM and AFM, and for the lanthanides, where the incompletely filled $4f$-shells are responsible for FM and AFM. At the $4f$-ferromagnets, where $4f$-electrons are responsible for FM, their orbitals are shielded by $5d$- and $6s$-orbitals and the $4f$-orbitals of the neighboring atoms are separated still stronger than in $3d$-ferromagnets as indicated by figures 2 and 5.

As it follows from the consideration above, in superconductors vice versa, the external d- and p-orbitals must approach enough closer to the neighboring atomic nucleus than in ferromagnetics and antiferromagnetics. As indicated above, in the number of superconductors the external d- and p-orbitals penetrate inside neighboring atoms disposing between the orbitals with the same principal quantum number but different l or s. It has been known that the states with the same principal quantum number but different l and s are obtained by the different energy. This energy difference is 0.1 – 1 eV. Hence in the superconducting crystals the electron on the external $d(p)$ shell is situated in the neighboring atom in the stronger electric field than in its own. Although in some superconductors the external d-/p-orbitals do not penetrate inside the external d-/p-orbitals of neighboring atoms they approach those enough closely. In this case external d-/p-orbitals are apparently subjected to the influence of the electric field of the neighboring atoms. That is the electrons on these orbitals are situated near neighboring atom in a different electric field from its own. This fact is open to speculation that the separation of spin and charge in electron, disposed on the external d-/p-orbitals, is quite possible.

It counts in favour this hypothesis that SC for a number of elements occurs only in thin films (Cr, Ga, Ge, Si, Sc, Nd, Eu, Yb) or under pressure (Y, Sb, Te and so on). It can be seen from the figure 1. Even such typical ferromagnetic as Fe founds out SC under pressure [6]. It is obvious, that under pressure, neighboring atoms are approached each other more closely (the distance b diminishes). It has been proposed another hypothesis too, at high pressures, $s \rightarrow d$ electron transfer occurs, for example, in trivalent rare earths Sc, Y and Lu and in nonmetals such as Te, I etc. [10,11]. In the thin films the compression of the crystalline lattice occurs by the connection with substrate.

The proposed hypothesis of the SC nature leans upon fact that the magnetic field penetrates in superconductor by the quantum of magnetic flux (fluxons). The fluxon value is equal to $hc/2e$, where h is the Plank constant, c is the light speed and e is the electron charge. The spin magnetic moment can submit in the form two magnetic charges with the different signs connected by the shoulder. In this case the value of the magnetic charge g is equal to the spin magnetic moment divided on the classic radius of the electron, i.e.

$$g = \hbar c/2e. \quad (2)$$

(Bohr magneton is equal to $e\hbar/2mc$ and classic radius of electron is equal to e^2/mc^2 where m is the electron mass.) Now put this magnetic charge in narrow tube and calculate the magnetic flux component along axis of tube on its end. This value is equal to half of fluxon. In other words the fluxon is equal to the magnetic flux component along axis of narrow tube on its end from two appointed magnetic charges of same sign, belonging to two spin magnetic moments parallel ordered. Hence it can be supposed the next explanation of Meissner effect. In superconductor the spins, deprived of the electrical charges, are antiparallel by pairs. It is obvious that in the absence of the magnetic field the chains of such oriented pairs of dipoles exist in the places, where the minima of free energy are observed. In order for the magnetic field penetrates to superconductor it can tear up this AFM connection and transfer the spins in FM state, that occurs at achievement of critical field H_{C1} in the superconductors of 2th type. At the same time, they combine with the electrical charges, deprived of spins, and the transition to the normal state occurs in this tube. The magnetic field is supported in it by the paramagnetic vortex currents of the electrons remained by in superconducting state. Diameter of the normal core is ~ 10^{-4} cm, i.e. it is approximately equal to the coherency length. In superconductors of the first type the situation is almost the same, but on account of more perfect lattice one tube, as it were, exists.

4. Conclusion

Basic tendencies in disposition of superconductors, antiferromagnetics and ferromagnetics are observed in the periodic table D.I. Mendeleev starting with 4th period. Elements with the superconductivity, in which the d-shells start to fill up, are at the beginning of each period; then follow antiferromagnetics and ferromagnetics (in the 4th period and lanthanides), or elements without any of the three listed order types (the 5th period and the 6th period), in which d (f)-shells continue to fill up almost exceedingly; then again appear superconductors with filling the p-shell up to 4 electrons.

In this connection, more superconductors are in those periods, where ferromagnetics and antiferromagnetics are absent. Furthermore, a transition from each of the three order types to another (or from the absence of the ordering to its appearance) occurs through elements with the both order types in the first situation and on evident "weakening" of superconductivity in the second situation that appears only in thin films or under pressure.

The calculation of the radii of the external d (f)- or p-orbitals and the nearest to them orbitals using the Slater method and the comparison of them with interatomic distances in the crystals allow to explain the specified above rule by the distinction of the degree of the division of the neighboring external d (f)- or p-orbitals in the crystal. As can be seen from figures 2-5, in the early each period the interatomic spacing between the nearest neighbors b drops with the growth of the element number Z, although slower than the external orbital's radius r_{d^n}; after that b stops to drop with growth of Z and then even starts to increase. As a result, distance $b-r_{d^n}$ increases with Z. So, in 4th period this difference for superconductors is within the limits 1.52–1.62 Å whereas for antiferromagnetics it is equal to 1.63 Å for Cr and 1.65 Å for Mn. For ferromagnetics it is substantially greater: from 1.72 Å at Fe to 1.86 Å at Ni. Thus the distance $b-r_{d^n}$ grows at filling of d^n-shells for $n \geq 5$, and in 4th period AFM (n = 5) and FM ($6 \leq n \leq 8$) occur. That is, the d-shells are strong separated in the crystal for $6 \leq n \leq 8$, what is necessary for FM [2,3,5,8]. In 7th period superconductivity takes place for $0.52 \leq b-r_{6d^n}$ ≤ 0.71 Å. What's more, in superconductors the external d-shells of some elements of 5th and 6th periods penetrate inside neighboring atoms disposing between the deeper p^6-shell and d^1-shell of neighboring atom (for Y, Zr, Nb, Mo, Lu, Hf, Ta, W) (figures 3, 4). Penetration of the external *p*-orbitals inside neighbor atoms is also observed in the crystals of the superconducting elements in the end of periods. So, the $5p^n$-shells are disposed between deeper $5s^1$ and $5s^2$-shells of the neighboring atoms (for In and Sn), and $6p^1$-shell of Tl is disposed in the neighboring atoms nearer to nuclear than their $6s^1$–shells. It is obvious that superconductivity, observed only in the thin films or under pressure in the number of the elements (see figure 1), is connected with the decrease of *b*-distance (in a thin film a compression of crystal lattice may be caused by connection with the substrate).

Hence in a superconductor an electron on the external *d(p)*-shell is situated in a neighboring atom in a different electric field from its own. This fact is open to speculation that the separation of spin and charge in electron, disposed on the external *d-/p*-orbitals, is quite possible. In this case, the charges without spin become bosons. The proposed hypothesis of the SC nature leans upon fact that the magnetic field penetrates in superconductor by the quantum of magnetic flux (fluxons). As is shown in Sec.3, fluxon is equal to the magnetic flux component along axis of a narrow tube on its end from two appointed magnetic charges of same sign, belonging to two spin magnetic moments parallel ordered.

Thus, it is shown that for superconductivity the atoms in the crystal must be greatly nearer to each other than in ferromagnetics and antiferromagnetics, and even their *d(p)*-shells must penetrate into the electronic cover of the neighboring atoms. Quit probably this statement is true not only for the elements from the periodical system, but for more complete compounds. Thereby, new ways for search of superconductors are opened.

References

[1] Koroleva L I and Khapaeva T M 2007 *Phys. Lett.* **A 371** 165
[2] Slater J C 1930 *Phys. Rev.* **36** 57
[3] Ormont B F 1950 *Structures of Inorganic Substances* (Moscow: Gostekhizdat)
[4] Vonsovskii S V, Izumov Yu A and Kurmaev E Z 1977 *Superconductivity of Transient Metals, Their Alloys and Compounds* Moscow: Nauka
[5] Weiss R J 1964 *Phil. Mag.* **9** 361
[6] Shimisu K, Kimura T, Furumoto Sh, Takeda K, Kontani K, Onuki Y and Amaya K 2001 *Nature* **412** 316
[7] Vonsovskii S V and Izumov Yu A 1962 *UFN* **77** 377
[8] Slater J C 1930 *Phys. Rev.* **35** 509
[9] Bethe G, Sommerfeld A 1938 *Electronentheorie der Metalls,* 1938 *Handb. D. Phys.* **24** 2 (Berlin: Shpringer)
[10] Brijman P W 1947 *Phys. Rev.* **72** 533
[11] Syassan K, Takemura K, Tups H and Otto A. 1981 *Physics of Solids under High Pressure*, eds. J.S. Shilling and L.N. Shelton (New York: Noth-Holland) p 125

Room-Temperature Ferromagnetism in Novel Mn-doped $ZnSiAs_2$ Chalcopyrite

S F Marenkin[1], V M Novotortsev[1], I V Fedorchenko[1], S A Varnavskiy[1], L I Koroleva[2,4] , D M Zashchirinskii[2], T M Khapaeva[2], R Szymczak[3], B Krzymanska[3], V Dobrowolski[3], L Kilanski[3].

[1]Institute of General and Inorganic Chemistry RAS, Leninskii str. 31, 119991 Moscow, Russia
[2]M.V. Lomonosov Moscow State University, Vorobyevy Gory, 119992 Moscow, Russia
[3]Institute of Physics PAS, Lotnicow al., Warsaw 02668, Poland

E-mail: koroleva@phys.msu.ru

Abstract. Based on Mn-doped chalcopyrite $ZnSiAs_2$ the new dilute magnetic semiconductor with *p*-type conductivity was produced. The Curie temperature behaviour of the produced semiconductor is distinctly dependent on the Mn concentration: 325 K for 1 wt.% and 337 K for 2 wt.% of Mn, consequently. Magnetization, electrical resistance, magnetic resistance and Hall effect of mentioned compositions were studied. Temperature dependence of magnetization $M(T)$ have complicate behaviour. For $T \le 15$ K the $M(T)$ dependence is characteristic for superparamagnetic and at $T > 15$ K magnetization is sum of magnetizations of ensemble of superparamagnetic clusters and ferromagnetic phase contained frustration regions.

1. Introduction

At the present time the new way of solid state physic is growing rapidly. This way relate with possibility of transfer oriented electronic spin from ferromagnetic to paramagnetic semiconductor [1, 2]. These investigations will allow produce one electron logic structure and spin-information systems with using electron spin as a memory cell: one spin – one information bit [3]. In order to achieve high polarized spin and to find solution of electrical contact between ferromagnetic and semiconductor, it is interesting to get ferromagnetic semiconductor with Curie temperature more that 300 K by dope of conventional semiconductor by element with the not fully filled 3d-shell. The best results were achieved on $Ga_{1-x}Mn_xAs$ films with Curie temperature $T_C \sim 170$ K [4, 5]. However for spintronic application ferromagnetic semiconductors with $T_C > 300$ K are desirable. Recently high temperature ferromagnetism was observed in Mn-doped $A^{II}B^{IV}C^{V}_2$ chalcopyrite. Its were $CdGeP_2$:Mn [6], $ZnGeP_2$:Mn [7] и $ZnSnAs_2$:Mn [8] compounds with T_C up to 350 K. In our papers [9, 10] new chalcopyrite $CdGeAs_2$:Mn with $T_C = 355$ K and $ZnGeAs_2$:Mn with $T_C = 367$ K were discussed. $A^{II}B^{IV}C^{V}_2$ ternary semiconductors are crystal-chemical and electronic twins of $A^{III}B^{V}$ semiconductors [11]. These compounds mostly have covalent type of chemical bond, small carrier effective mass, high

[4] To whom any correspondence should be addressed.

carrier mobility and absolute minima and maxima of conduction and valence bands in the centre of Brillouin zone. New compounds of $ZnSiAs_2$:Mn with $T_C > 300$ K were produced and observed in this paper. The value of these compositions is in the fact that they are combined with «siliceous technology», i.e. it is possibility to make on them of epitaxy and other technological processes allowed to create the devices of the solid state electronics [12]. Mentioned compound $ZnSiAs_2$ have following characteristics: band-gap energy is 2.1 eV and maximum carrier mobility is ~150 cm^2/V.s (for holes with $4\cdot10^{15}$ cm^{-3} concentration).

2. Experimental methods

At the present paper magnetization intensity, electrical resistance ρ, magnetic resistance $\Delta\rho/\rho = (\rho_H - \rho_{H=0})/\rho_{H=0}$ and Hall effect of Mn 1 and 2 wt % doped $ZnSiAs_2$:Mn polycrystalline were studied. Magnetization was measured by SQUID-magnetometer in a $5 \le T \le 300$ K temperature range and by weight method with electromagnetic compensation in a $T \ge 290$ K. The last method normally used to measure paramagnetic susceptibility, but due to low Mn consumption and low magnetization amount it was possible to apply it for measure magnetization. To measure ρ and Δρ/ρ a four-probe method was applied. Electrical contacts to samples were prepared on etched and polished surface using indium solder and gold wires.

3. Samples producing and analysis

Polycrystalline $ZnSiAs_2$:Mn samples have been prepared by direct fusion method from high purity powders of Si, $ZnAs_2$ and Mn with temperature 10 – 15^{0}C higher that $ZnSiAs_2$ melting point. Based on analysis of quasi-binary cut Si – $ZnAs_2$ this method looks more efficient [13, 14]. In order to increase Mn solubility in the alloy, crystals were cooled down with high speed 10°/c. Samples identification were done by x-ray phase analysis, thermal differential analysis (DTA), micrographic analysis, x-ray fluorescence analysis as well as by electron microscope.

X-ray phase analysis stated that samples consist of $ZnSiAs_2$ phase with following parameters: a =5.6084 Å; c = 10.8816 Å. These parameters good enough comply with dates from International center of diffraction data, 2007. X-ray fluorescence analysis also endorses x-ray phase analysis results.

4. Magnetic and electrical properties of $ZnSiAs_2$

Figure 1 and 2 show temperature dependence versus magnetization for composite with 2 wt.% of Mn measured in applied magnetic field 10 and 50 kOe respectively (curve 1).

As it shown on figures, $M(T)$ curve under $T > 50$ K have shape characteristic for ferromagnetic. However, in case of $T < 15$ K magnetization rapid grow with T-decreasing and $M(T)$-dependence obeys by Langevin function

$$M/M_0 = cth(\mu H/kT) - kT/\mu H, \tag{1}$$

where M_0 is equal to M at $T\rightarrow 0$ K and μ is magnetic moment of superparamagnetic cluster. We determined the M_0-value by extrapolation of $M(T)$ at $T \rightarrow 0$ K and μ-value by selection to experimental curve. These values are next: $M_0 = 0.321$ emu/g and $\mu = 55$ μ_B in $H = 11$ kOe and 11 μ_B in $H = 50$ kOe for compound with 2 wt %. Contrary to the conventional superparamagnets in this compound magnetic moments of clusters depend from the magnetic field value H: they diminish with grow H. Similar picture was observed for compound with 1 wt % of Mn. Here $\mu = 52$ μ_B in $H = 11$ kOe and 8.6 μ_B in $H = 50$ kOe. Difference between FC and ZFC magnetization and shift of the hysteresis loop of FC sample observed at $T = 5$ K confirms the superparamagnetic clusters presence (figure 3). We proposed that at $T > 15$ K magnetization M is sum of magnetization of ensemble of superparamagnetic clusters and magnetization from phase with long-range magnetic order. We separate $M(T)$-curve (curve 1 on figure 1 and 2) on curve of the superparamagnetic clusters magnetization (curve 2) and curve of T-dependence of magnetization from phase with long-range magnetic order $M_S(T)$ (curve 3) using a computer subtraction of curve 2 from curve 1. At $T > 15$ K appears a spontaneous magnetization as figure 1 and 2 shows. M_S-value does not almost depend from T at $100 \le T \le 250$ K. From highest value of M_S in this T-region was determined a magnetic moment on f.u. It is found this magnetic moment on f.u. very understand comparison with the one at total ferromagnetic ordering of spins of Mn^{2+} ions or antiferromagnetic ordering of spins Mn^{2+} and Mn^{3+}, that is these compounds are frustrated magnetic at $T > 15$ K.

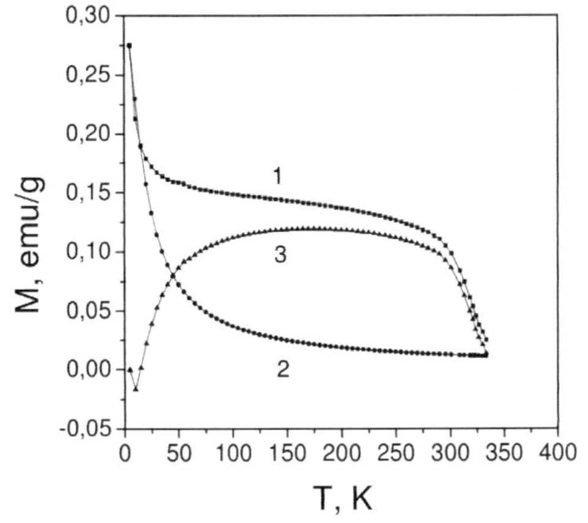

Figure 1. Temperature dependence of magnetization for composite with 2 wt.% of Mn measured in applied magnetic field 10 kOe {curve 1}. Curve 1 is separated on T-dependence of magnetization of superparamagnetic clusters ensemble {curve 2} and the one of ferromagnetic phase contained frustration regions {3}.

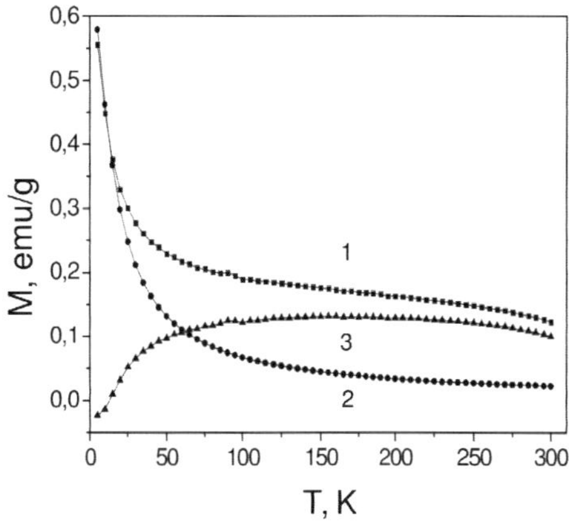

Figure 2. Temperature dependence of magnetization for composite with 2 wt.% of Mn measured in applied magnetic field 50 kOe {curve 1}. Curve 1 is separated on T-dependence of magnetization of superparamagnetic clusters ensemble {curve 2} and the one of ferromagnetic phase contained frustration regions {3}.

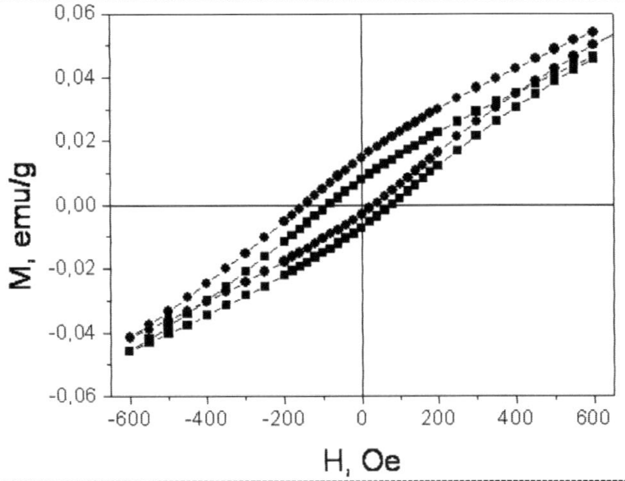

Figure 3. 2 wt.% Mn-doped $ZnSiAs_2$. Hysteresis loop measured with $T = 5$ K and in applied magnetic field from -0.1 kOe to 0.1 kOe after sample cooling from $T = 370$ K to $T = 5$ K under this field (circles, FC ssample) and without field (squares, ZFC sample). Shift of the loop can be observed on the axis H for FC sample, but for ZFC sample the loop is symmetrical.

Therefore their Curie points were determined by extrapolating the steepest part of $M_S(T)$ curve to intersection with T-axis. They are 325 and 337 K for samples with 1 and 2 wt % correspondingly.
The $\rho(T)$–dependence is semiconductive character with an activation energy 0.12 – 0.38 eV at $124 \le T \le 263$ K (both compounds). Mobility and concentration of the charge carriers (holes) are 1.33, 2.13

cm^2/Vs and 2.2×10^{16}, 8×10^{16} cm^{-3} at $T = 293$ K compounds with 1 and 2 wt % correspondingly. Magnetoresistance of both compounds does not exceed of 0.4 % in $H = 10$ kOe.

5. Conclusion

Up to highest magnetic field (50 kOe), magnetization have complicated behaviour. At $T \leq 15$ K sample magnetization is a result of existence of superparamagnetic clusters. Its magnetic moment is decreasing while magnetic field is growing. This dependence might be due to split big magnetic clusters to small one under magnetic field. Existence of superparamagnetic clusters is confirmed by difference of magnetizations of FC- and ZFC-samples and by shift in hysteresis loop for FC-sample. At $T > 50$ K ferromagnetic phase become play the key role. At $100 \leq T \leq 250$ K there is not big change of this magnetization. At $T > 300$ K $M(T)$ curve drop down what is an evidence of Curie temperature. Obviously, Curie temperature idea for such system is tentatively: this is Curie temperature of frustrated ferromagnetic phase. Due to magnetic heterogeneity of the system it's not possible to measure Curie temperature by using the thermodynamic coefficients method of Belov-Arrott. Also not true to measure T_C from $M(T)$ in low magnetic fields, because this magnetization is a result of demagnetizing factor. Demagnetizing factor of frustrated ferromagnetic phase depend of phase configuration, which can be changed with temperature. Therefore Curie temperatures were determined by extrapolating the steepest part of $M_S(T)$ curve measured under 50 kOe to intersection with T-axis. They are 325 and 337 K for samples with 1 and 2 wt % correspondingly.

In work [15] was shown that modified RKKI interaction of Mn atoms positioned in the neighbour sites of Ga-sublattice in $Ga_{1-x}Mn_xAs$ leads to their attraction and promotes their clustering. One would expect a conclusion of [15] can be justly for chalcopyrite structure since it is crystallochemical analog of $A^{III}B^{V}$. In this case the cluster formation occurs in the same manner as in Ga_{1-x} Mn_x As, as well as zinc to Mn replacement lead to cluster formation in $ZnSiAs_2$:Mn.

6. References

[1] Zutic I, Fabian J, Das Sarma S 2004 *Rev. Mod. Phys.* **323**, 76
[2] Ohno H 1998 *Science* **951** 281
[3] Matsukura F, Ohno H, Shen A, Sugawara Y 1998 *Phys. Rev.* B **57**, R2037
[4] Edmonds K M, Wang K Y, Campion R P, Neumann A C, Farley N R S, Gallagher B L, Foxon C T 2002 *Appl. Phys. Lett.* **4991** 81
[5] Edmonds K M, Boguslawski P, Wang K Y, Campion R P, Novikov S N, Farley N R S, Gallagher B L, Foxon C T, Sawicki M, Dietl T, Buongiorno Nardelli M, Bernholc J 2004 *Phys. Rev. Lett.* **92** 03720
[6] Medvedkin G A, Ishibashi T, Nishi T, Hayata K, Hasegawa Y, Sato K 2000 *Jpn J. Appl. Phys.* **10A** L949
[7] Medvedkin G A, Hirose K, Ishibashi T, Nishi T, Voevodin V G, Sato K, 2002 J Cryst *Growth* **609** 236
[8] Choi S, Cha G-B, Hong S C, Cho S, Kim Y, Ketterson J B, Jeong S-Y, Yi G-C, 2002 *Solid State Commun.* **165** 122
[9] Demin R V, Koroleva L I, Marenkin S F, Mihailov S G, Novotortsev V M, Kalinnikov V T, Aminov T G, Szymzak R, Szymzak G, Baran M 2004 *Letter in JSP* **81** 30
[10] Koroleva L I, Pavlov V Y, Zashchirinskii D M, Marenkin S F, Varnavskii S A, Szymzak R, Dobrowolski V, Kilanskii L 2007 *SSP* **49** 2022
[11] Gorynova N A *Ternary Diamond-like Semiconductors* (Soviet Radio, Moscow, 1968)
[12] Ivanov V A, Aminov T G, Novotortsev V M, Kalinnikov V T 2004 *Proceedings RAS. Series Chemical* **49** 116
[13] Novotortsev V M, Fedorchenko I V, Kuprianova T A, Koroleva L I, Szymzak R, Marenkin S F 2007 *Chemical technology* **9** 385
[14] Fedorchenko I V, Kuprianova T A, Marenkin S F, Kochura A V 2008 *JIC* **6** 15

[15] Meilikov E Z and Farzetdinova R M 2007 *Phys. Rev. B* **75** 052402

Journal of Physics: Conference Series **153** (2009) 012059 doi:10.1088/1742-6596/153/1/012059

First-principle Calculation of GaAs Electronic Structure by Doping with Mn and P

Z B Li[1], X Wang[2], Y Zhu[1], M W Ran[1], and K L Yao[3]

[1] Department of Physics, Tongren University, Tongren 554300, China;

[2] Department of Chemistry Tongren University, Tongren 554300, China;

[3] Department of Physics, Huazhong University of Science and Technology, Wuhan 430074, China

E-mail: zongbaoli1982@gmail.com

Abstract. The geometry structure of Mn, P doped 64-atom supercell of GaAs ($Ga_{1-x}Mn_xAs_{1-y}P_y$) was optimized and studied by the self-consistent full-potential linearized augmented plane wave method (FP_LAPW) based on the density functional theory (DFT). Cell parameters of both doped and undoped were calculated theoretically. Band structure, binding energies, partial density of states, mulliken charges, and electron density different of doped GaAs crystals were calculated and analyzed in detail. The results revealed that the both doped compounds are true half-metallic ferromagnets and the near distance doped one is a stable ground state.

1. Introduction

The interplay between the charge carriers in a semiconductor and the electron spins of a ferromagnetic metal doped into the semiconductor can be utilized for many magneto/spin-electronic devices. The desirable features for the successful application of a semiconductor spintronics material are that it should show ferromagnetism or half-metal ferromagnets with a reasonably high T_C, and a sizable concentration of magnetic sites. Interest in the family [1] of carrier induced dilute magnetic semiconductors (DMS) started with the discovery of ferromagnetism in ~5% Mn-doped InAs and GaAs with a Curie point of 110 K[1] and the study of ferromagnetism in (Ga, Mn)As and a theoretical prediction that Mn-doped GaN could be ferromagnetic(FM)at or above room temperature[2], have led to intensive research on transition metal elements doped (III, V) DMSs [3-18]. Among the various DMS materials, Mn-doped GaAs is particularly interesting because GaAs is a direct wide energy band gap semiconductor with high thermal, chemical, and mechanical stability. In addition, Mn-doped GaAs is one of the few materials where magnetism above room temperature has been reported and is extensively studied. Half-metallic ferromagnetism, especially dilute magnetic semiconductors [19,20], have attracted more and more attentions since its first discovery by De Groot et al. in 1983 [21]. Recently, there is considerable controversy in the reported results. The nature and origin of the magnetic coupling in this material continue to be hotly debated issues. The mechanism for the observed magnetic behaviour is complex and appears to depend on a number of factors.

The first attempt to explain carrier-controlled ferromagnetism in Mn-doped III-V semiconductors was made by Dietl *et al.*[2] who used the double-exchange Zener model[8] of ferromagnetism and

[1] E-mail: zongbaoli1982@gmail.com

predicted how the incorporation of Mn ions induces ferromagnetic behaviour via hole-mediated exchange interaction (Ruderman-Kittel-Kasuya-Yosida type). Litvinov and Dugaev[9] subsequently proposed the indirect exchange interaction caused by virtual electron excitations from magnetic impurity acceptor levels to the valence band. *Ab initio* calculations [10–13] have also been carried out to understand the underlying mechanism of carrier-induced ferromagnetism in transition-metal (TM) - doped III-V and II-VI semiconductors. Schilfgaarde and Mryasov[10] have performed a thorough investigation of the anomalous magnetic interactions in doped DMS systems and they had performed linearized muffin-tin orbital calculations on a zinc-blend structured supercell, in order to investigate the exchange interactions between 3*d* magnetic dopants (Cr, Mn, and Fe) in different III-V semiconductors (GaAs, GaN, and AlN). They had observed that doping caused deviations in the magnetic exchange interaction from what is predicted by simple models due to the formation of deep levels and concluded that short-range interaction between the dopant atoms could cause a clustering of these atoms. More recently, Ohno and co-workers has enabled its practical growth with stoichiometric amounts of $Mn_xGa_{1-x}As$ alloys, are inherently compatible with existing GaAs technology, resulting in the practical realization of device structures combining ferromagnetic and nonmagnetic layers.

In this paper, we present a detailed study of the electronic and magnetic properties of *Mn, P*-doped GaAs($Ga_{1-x}Mn_xAs_{1-y}P_y$) supercell and the wurtzite structure to demonstrate the effect of the compounds. We calculated using the WIEN2K package [22], a full potential linearized augment plane wave method within the density functional theory. The generalized gradient approximation (GGA) [23] proposed in 1996 by Perdew, Burke and Emzerhof is used for the exchange-correlation potential and the relativistic effect is taken into account in the scalar approximation, but the spin–orbit coupling is neglected because it does not destroy the half-metallicity. The radii R_{mt} of the muffin tins are chosen to be as large as possible according to the approximate ionic radii but not overlapped and $R_{mt}K_{max}$ are all taken to be 9.0. The self-consistency is better than 10^{-4} e/a.u. for charge density, and the stability is better than 10^{-5} Ry for the total energy per formula unit. Our *ab initio* investigations yield the electronic structure, energetics, and magnetism of *Mn* and *P*-doped GaAs in a cluster as well as in a bulk crystalline environment. In Sec. II we outline our computational results and discussed.

2. Results and discussion

The study presented in this work is based on $Ga_{1-x}Mn_xAs_{1-y}P_y$ materials. The crystalline structure used by the host semiconductor is a 64-atom cubic cell $Ga_{32}As_{32}$. In this cell one Ga atom has been substituted by one Mn atom and one As atom has been substituted by one P atom. Atoms of Mn and P were located as far as possible in Far Crystalline (FC) and in the Near Crystalline (NC) two atoms were located as immediate neighbors, which were shown in Fig.1. In this structure the atomic environment of the transition-metal atoms has a tetrahedral symmetry considering their immediate neighbors and the calculated bong lengths for Mn substituting the Ga site and P substituting As site are listed in Table 1. where they are compared with the ideal bond length in the pure GaAs. Substitution of Mn on the Ga site creates a longer bond in FC than in NC. Ignore the influence of atom P, the As atoms can be broken down into two groups: those directly bonded to the Mn atoms (As1) and those which are not (As2). Using the *ab initio* method described above, we have carried out an extensive analysis of the electronic structure, energetics and magnetism prosperities of FC and NC. We find both the doped compounds structures to be energetically stable, with the NC spin order structure lower in energy 0.002eV per supercell than FC which indicates that the NC structure is more stable. The reason due to bond strength of Mn-P in NC is smaller than that of Mn-As bond strength which is shown in Table 1.

Table 1. Calculated bong lengths for Mn substituting the Ga site and P substituting As site in the supercells of FC and NC

Bond length for FC(Å)				Bond length for NC(Å)					
Mn-As	Ideal	P-Ga	Ideal	Mn-As	Ideal	P-Ga	Ideal	Mn-P	Iedal
2.303	2.448	2.395	2.448	2.292	2.448	2.401	2.448	2.225	2.448

2.308	2.395	2.292	2.401
2.299	2.395	2.292	2.401
2.299	2.395		

From the FC structure, we find a global magnetic moment of $3.99409\mu_B$ per-supercell. As expected, the Mn dopant atom is found to be the main contributor to the magnetic moment of the structure, having a localized magnetic moment of $3.76882\mu_B$ per atom and in the NC structure the magnetic moment of Mn is bigger by $0.01214\mu_B$. The As neighbors of Ga together make a much smaller contribution of about $0.002\mu_B$, while the Mn second neighbors of As together contribute about $0.0008\mu_B$ to the magnetic moment, which due to the conjugation effect of the spin delocalization between manganese ions and arsenic atoms. Note that this nonintegral value of magnetic moment is due to the use of atomic sphere approximation in our self-consistent calculation.

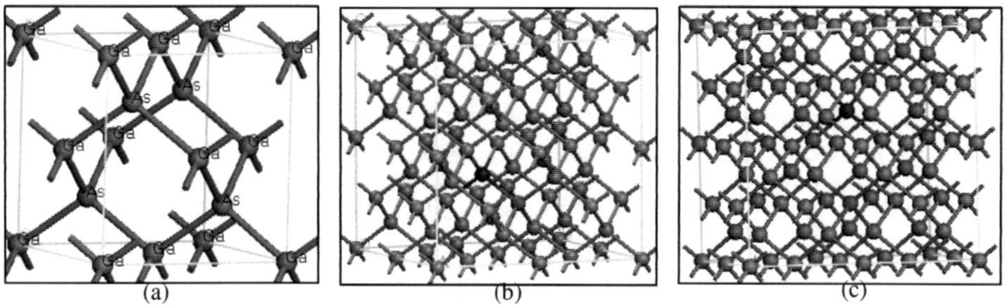

Figure 1. Cell of GaAs(a), Far Crystalline (b) and the Near Crystalline (c). the red and blue balls stand for Mn and P atoms, respectively

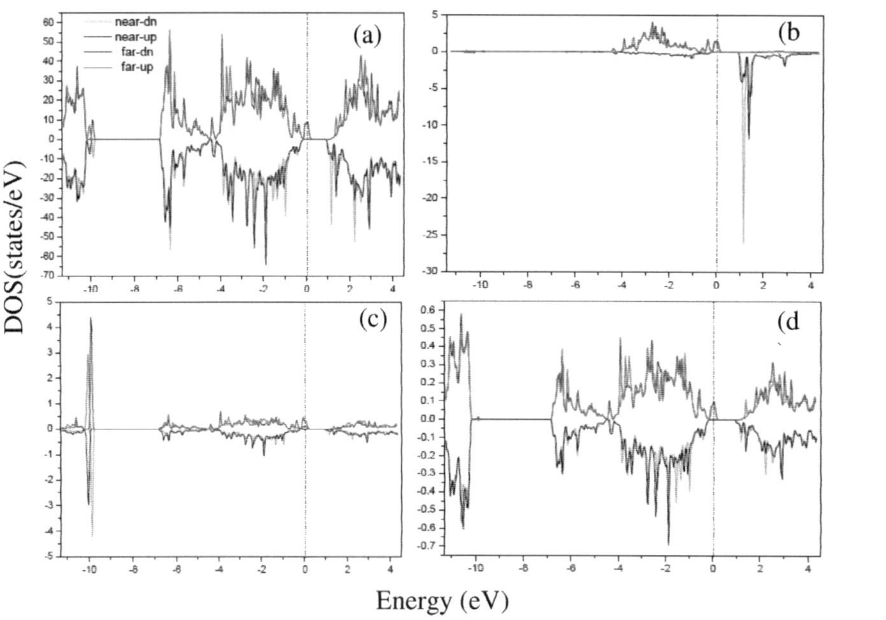

Figure 2. Cell of GaAs(a), Far Crystalline (b) and the Near Crystalline (c). the red and blue balls stand for Mn and P atoms, respectivelly

The main characteristic near the Fermi energy is the determination of the magnetic properties. The density of states (DOS) of the two $Ga_{1-x}Mn_xAs_{1-y}P_y$ (x=y=3.125%) compounds is illustrated in Fig.2. and Fig.3. The horizontal axis is energy relative to the Fermi energy (E_F). The host valence band consists of two parts. One is the low-lying Ga-*3d* band which appears at approximately 1.2 Ry below E_F as a peak whose width is 0.35 Ry. The other is the broad As-*2p* band which distributes from –4.0 Ry to –0.2 Ry. Between the valence band and the conduction band, which consists of Ga-4s states, there are impurity-*3d* bands. These impurity bands show large exchange splitting. Because the exchange splitting is larger than the crystal field splitting, as shown in figure 3, the TM is in the high-spin states. The half-metallic DOS means that the DOS is metallic for up-spin states and insulating for down-spin states, or vice versa. This kind of asymmetry in the DOS at E_F is shown in the other III–VI-based DMSs, and it is proposed that these III–VI compound-based DMSs are useful to realize spintronic devices such as a spin-injection device and a spin-polarized FET and so on. Aslo as Fig.3 shows, a half-metallic behaviour in the sense that the Fermi level state density is finite for the spin-up and zero for the spin-down, which is different with the ferromagnetic behaviour of Mn-doped GaAs[24]. This half-metallicity is due to the Mn-*3d* states hybridize well with the As-*2p* states, and the P-*2p* states. Due to this hybridization with the Mn-*3d* states, holes are itinerant with keeping their *d*-character, therefore the kinetic energy is lowered so efficiently that the ferromagnetic state is stabilized by the double-exchange mechanism. As shown in figure 2, the impurity bands appeared between the valence band and the conduction band of the host GaAs. The introduction of Mn, therefore, changes the semiconducting nature of GaAs. The result are similar to the obtained previously for Cr-doped GaN [24], where it was suggested that the half-metallicity is due to Cr-*3d* hybridizing with the N-*2p* state. In Fig.2(b), it is also significance that the Mn-*d* band of DOS is strongly spin split into three-peaked structure. The peaks for the Mn-*d* band spin-up just above the valence band edge, which is in tune with what Mayasov[24] have mentioned in their paper. This implies that Mn acts as a provider in the doped-structure.

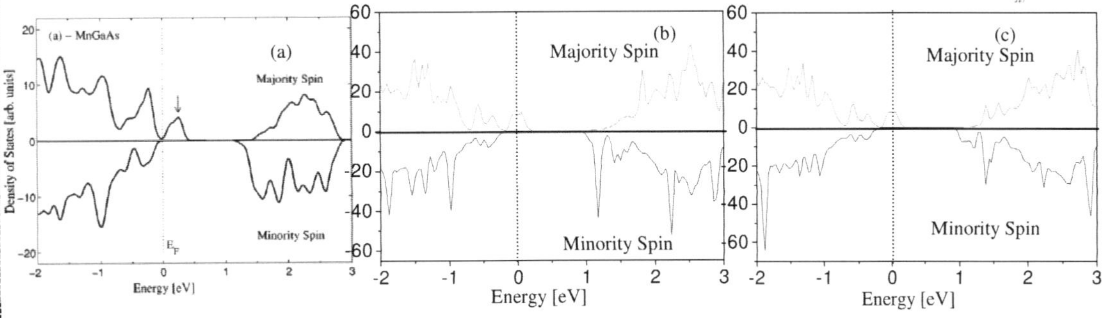

Figure 2. Density-of-states cover for: (a) $Mn_{0.063}Ga_{0.937}As$, (b) $Mn_{0.0375}Ga_{0.9625}P_{0.0375}As_{0.9625}$ for near distance, and (c) $Mn_{0.0375}Ga_{0.9625}P_{0.0375}As_{0.9625}$ for far distance

3. Conclusions

We have studied the structural, electronic, and magnetic properties of Mn and P doped GaAs with far and near distance through first-principles calculations. The result suggest that the two compounds are likely to order half-metallicity, leading to a dilute magnetically order semiconductor. The two compounds of $Ga_{1-x}Mn_xAs_{1-y}P_y$ with a realistic Mn and P content of $x=y=0.3125\%$ are all found to possess a magnetic moment of about $4.0\mu_B$ per-supercell and the Mn dopant atom is found to be the main contributor to the magnetic moment of the structure. But in the NC structure the magnetic moment of Mn is a smaller bigger than FC which due to the spin delocalization. Also, we found both the doped compounds structures to be energetically stable, with the NC spin order structure lower in energy 0.002eV per supercell than FC which indicates that the NC structure is more stable.

Acknowledgments
This work was supported by the Natural Science Foundation of Guizhou province (Nos. 2007039).

References

[1] Ohno H 1998 *Science* **281** 951
[2] Dietl T, Ohno H, Matsukura F, Cibert J and Ferrand D 2000 *Science* **287** 1019
[3] Dietl T 2002 *Semicond. Sci. Technol.* **17** 377
[4] Morkoc H 1999 *Nitride Semiconductors and Devices* 2nd ed. 2002 (Berlin: Springer-Verlag)
[5] Park S E, Lee H-J, Cho Y C, Jeong S-Y, Cho C R, and Cho S 2002 *Appl. Phys. Lett* **80** 4187
[6] Reed M L, El-Masry N A, Stadelmaier H H, Ritums M K, Reed M J, Parker C A, and Bedair S M 2001 *Appl. Phys. Lett.* **79** 3473
[7] Hashimoto M, Zhou Y-K, Kanamura M, and Asahi H 2002 *Solid State Commun.* **122** 37
[8] Zener C 1951 *Phys. Rev.* **82** 403
[9] Litvinov V I and Dugaev V K 2001 *Phys. Rev. Lett* **86** 5593
[10] VanSchilfgaarde M and Mryasov O N 2001 *Phys. Rev.* B **63** 233205
[11] Kronik L, Jain M, and Chelikowsky J R 2002 *Phys. Rev.* B **66** 041203(R)
[12] Sato K and Katayama H -Yoshida 2002 *Semicond. Sci. Technol.* **17** 367 ; Sato K *et al.* 2001 *Jpn. J. Appl. Phys.* Part 2 **40** L485
[13] Rao B K and Jena P 2002 *Phys. Rev. Lett.* **89** 185504
[14] Perdew J P and Wang Y 1986 *Phys. Rev.* B **33**, 8800 ; Perdew J P 1986 *ibid.* **33**, 8822 ; Perdew J P and Wang Y 1992 *ibid.* **45**, 13244
[15] Andersen O K and Jepsen O 1984 *Phys. Rev. Lett.* **53**, 2571 ; Andersen O K 1975 *Phys. Rev.* B **12**, 3060
[16] Christensen N E and Gorczyca I 1994 *Phys. Rev.* B **50** 4397
[17] Blochl P E, Jepsen O and Andersen O K 2001 *Phys. Rev. B* 245205
[18] Jepsen O and Andersen O K 1995 *Z. Phys. B: Condens. Matter* **97** 35
[19] Blochl P E, Jepsen O and Andersen O K 1994 *Phy. Rev.* B **49** 16223
[20] *GAUSSIAN 98* Revision A.7 Frisch M J *et al.* 1998 (Pittsburgh PA: Gaussian, Inc.,)
[21] Becke A. D, *Phys. Rev. A* **38**, 3098 (1988); Perdew J. P., Burke K., and Wang Y., *Phys. Rev. B* **54**, 16533 (1996); K. Burke, J. P. Perdew, and Y. Wang, *in Electronic Density Functional Theory: Recent Progress and New Directions*, edited by J. F. Dobson, G. Vignale, and M. P. Das (Plenum, New York, 1998).
[22] Blaha P, Schwarz K, Madsen G K H, Kvasnicka D and Luitz J 2002 WIEN2k,Vienna University of Technology improved and updated Unix version of the original copyrighted Wiencode, which was published by P. Blaha, K. Schwarz, P. Sorantin, S.B. Trickey, Comput. Phys. Commun. 59 (1990) 399
[23] Perdew J P, Burke K and Ernzerhof M 1996 *Phys. Rev. Lett.* 77 3865
[24] VanSchilfgaarde M and Mryasov O N 2001 *Phys. Rev* B 63, 233205

Correlation between crystallographic texture and the degree of $L1_0$-ordering in post-annealed Ag/CoPt bilayers and comparison with Ag/CoPt nanocomposites

E Manios[1], D Stamopoulos[1], I Panagiotopoulos[2] and D Niarchos[1]

[1] Institute of Materials Science, NCSR "Demokritos", 153-10, Aghia Paraskevi, Athens, Greece
[2] Department of Materials Science and Engineering, University of Ioannina, 451-10, Ioannina, Greece

E-mail: manios@ims.demokritos.gr (E Manios)

Abstract. CoPt and FePt compound films with $L1_0$ ordered structure have been intensively studied, due to their extremely high uniaxial magnetocrystalline anisotropy which makes them suitable for application in ultra high density magnetic recording media. A basic requirement in these type of media is the development of strong perpendicular magnetic anisotropy. The role of Ag underlayers in promoting strong (001) crystallographic texture and perpendicular magnetic anisotropy in post-annealed Ag/CoPt and Ag/FePt bilayers (BLs) has already been reported, along with a possible correlation between $L1_0$ formation and development of (001) crystallographic texture. In this work we present new data, which provide further evidence that there is indeed such a correlation during the annealing process of Ag/CoPt BLs. The most obvious manifestation of this correlation is the fact that the X-ray intensity ratios I_{001}/I_{002} (used as a measure of the degree of $L1_0$-ordering) and I_{002}/I_{111} (used as a measure of the crystallographic texture) and the coercivity H_c and relative remnant magnetization m_r (for field H normal to the surface of the films) exhibit the same kind of dependence from the thickness of the Ag underlayer. Comparison with respective crystallographic data from post-annealed Ag/CoPt nanocomposites (NCs) shows that in the case of NCs the (001) texture starts to degrade for lower total film thickness, compared to the case of BLs. This difference can be attributed to the structural incoherence in the growth of the CoPt grains imposed by the presence of Ag inside the Ag/CoPt NCs, while in BLs Ag is only used as an underlayer. Based on the above data and on detailed Heavy Ion Elastic Rutherford Back Scattering (HIRBS) measurements, performed on post-annealed Ag/CoPt NCs, we propose a possible mechanism for the interpretation of the observed correlation. The proposed mechanism is based on the reduction of total strain (residual strain of as-deposited film and transformation strain due to deformation of the unit cell as $L1_0$-CoPt is formed) throughout the annealing process.

1. Introduction

Co-Pt and Fe-Pt alloys with almost equiatomic stoichiometry (CoPt and FePt) have been studied in the past, for use as permanent magnets. CoPt and FePt are disordered single-phase solid solutions with FCC crystal structure above 820°C for CoPt and above 1300°C for FePt. *Below these critical temperatures they undergo a long-range ordering phase transformation* [1, 2, 3, 4] *to a structure known as the CuAu-I type* [3, 4, 5] *or face centered tetragonal (FCT) or* $L1_0$. The unit cell of this structure is tetragonal, with the (002) planes (normal to the [001]-axis or c-axis with distance $c/2$ between successive planes) occupied alternately by Co and Pt atoms for CoPt or by Fe and Pt atoms for FePt [1, 2, 5] and the c-axis smaller than a and b axes. The $L1_0$ phase of both CoPt and FePt is characterized by large uniaxial (along c-axis) magnetocrystalline anisotropy, with reported bulk values $K_u \sim 5\times10^7$ erg/cm^3 for CoPt and $K_u \sim 7\times10^7$ erg/cm^3 for FePt [6, 7, 8, 9, 10, 11, 12, 13]. This high anisotropy constant K_u is the reason why $L1_0$ structured CoPt and FePt are possible candidates for use in high density magnetic recording media [14, 15, 16, 17, 18, 19, 20], in the form of thin films [14, 21, 22, 23, 24, 25, 26, 27, 28] and nanocomposite thin films [25, 29, 30, 31, 32, 33, 34, 35, 36]. The reported K_u values for CoPt and FePt are at least 20 times higher than these of the currently used Co-alloy based media, like CoPtCr [17]. The minimum size D_p, for which the energy of exchange coupling between the spins is sufficiently larger than thermal energy and is able to maintain the ferromagnetic state of the particle/grain (superparamagnetic limit), is estimated to be approximately $D_p=(60kT/K_u)^{1/3}$ [16, 17]. This gives a critical particle size well below 10 nm: it is $D_p \sim 3$–4 nm for CoPt and FePt at room temperature [16, 17]. Due to this values of D_p for CoPt and FePt, it is expected that we can use at least 3 times smaller particle diameters D [17]. The result is a potential 10-fold areal density increase, since the areal density is analogous to $1/D^2$ [17]. Therefore the size of the CoPt or FePt grains can be reduced significantly without affecting the thermal stability of the recording medium.

Additional requirements (along with high K_u) that must be satisfied by a thin film with CoPt or FePt, in order to be suitable for use in magnetic recording, are:

(a) Grain isolation for reduction of the noise caused by inter-grain interactions and further increment the thermal stability.

(b) High coercivity H_c, which assures that small fields can not reverse the magnetization of grains. Coercivity is connected to uniaxial anisotropy, and therefore this requirement is not independent from the basic requirement for high anisotropy.

(c) Narrow width of grain size distribution, because in this case the width of the distribution of magnetic fields – that are needed for magnetization reversal in the grains – is also narrow (around H_c). The narrow width in the distribution of magnetic fields results in square-shaped hysteresis loops. This assures that the grain magnetization will not be affected from a field opposite to it, if the strength of this field is smaller than coercivity H_c.

(d) Crystallographic texture, since in this case the easy axis of magnetization of each grain is highly aligned towards a specific direction and produce a square-shaped loop in this direction. The most interesting case is the development of (001) texture, where the direction of alignment of the easy axis is normal to the plane of the film. This arrangement is magnetostatically favorable, since it reduces the magnetostatic interaction between successive bit cells, and allows the significant increment of the areal density of stored bits [37].

In this work we focus our attention on the development of (001) crystallographic texture. It has been reported that this kind of texture can be the result of suitable deposition and/or annealing conditions, which may be combined with the use of specific additives such as Ag, C, B_2O_3, SiO_2, BN and B_4C [23, 25, 27, 28, 30, 33, 34, 38, 39, 40, 41, 42, 43, 44, 45, 46, 47, 48, 49, 50]. In particular, it has been reported that the use of Ag underlayers [27, 28] is a very effective method for the development of strong (001) texture on CoPt or FePt films. In Ref. [25] we have reported on the influence of thin (of the order of 1.2–1.8 nm) Ag underlayers on the structural and magnetic properties of post-annealed Ag/CoPt BLs. The results of the X-ray measurements presented there, implied that there is a correlation between the degree of FCT-ordering and the degree of (001) crystallographic texture. Here we present additional X-ray data, which are combined with magnetization data. These combined crystallographic and magnetization measurements provide further evidence for the existence of this correlation. Based on these detailed data and on additional Heavy Ion Elastic Rutherford Back Scattering (HIRBS) measurements, performed on post-annealed Ag/CoPt NCs, we propose a possible interpretation for this correlation. The basic idea of the proposed mechanism is that the development of (001) texture results from the minimization of the elastic energy related to strain. Finally we present comparative data concerning the degree of (001) texture exhibited by BLs and NCs. These data clearly show that, while for small total film thickness (in the range of 9–12 nm) the NCs exhibit a higher degree of (001) texture compared to the BLs, for thicker films the degree of (001) texture of NCs reduces more sharply with the total film thickness. This difference is probably a result of the residual Ag in the interior of the NC films after the annealing process.

2. Preparation of samples and Experimental details

Below we briefly present information on the preparation of the samples used in this work. All the samples were initially deposited by magnetron sputtering on Si(001) substrates at ambient temperature, under an ultra pure Ar environment (99.999%), at a pressure of 3 mTorr. The as-deposited samples were magnetically soft, containing only the disordered FCC phase of CoPt. The base pressure (before inserting Ar) in the sputtering chamber was $BP=5\times10^{-7}$ Torr. In order to achieve the transformation from the disordered FCC phase to the highly anisotropic and magnetically hard $L1_0$ phase of CoPt, the films were annealed in high vacuum (10^{-6} Torr). High purity (99.99%) Ag and Co targets (with 2 in diameter) were used for the deposition of the various layers. In the case of the CoPt layers, the Co target was partially covered with Pt. X-ray Fluorescence (XRF) measurements were performed (not shown here), to verify that the atomic stoichiometry of the CoPt layers was close to 50% Co – 50% Pt. For the preparation of the BLs, the Ag and CoPt layers were sequentially deposited on the Si(001) substrates. The Ag layers were deposited by dc-sputtering at 5 W and at a rate of 1.5 Å/sec [25]. The CoPt layers were deposited by rf-sputtering at 30 W and at a rate of 1.2 Å/sec [25]. The as-deposited BLs were annealed for 20 min at 600°C [25], in order to produce post-annealed BLs with adequate magnetic properties. For annealing at lower temperatures or for shorter time periods, the maximum values of coercivity or relative remnant magnetization – exhibited by the produced BLs – were not satisfying. For the preparation of the NCs, the films were initially deposited in multilayer form and subsequently post-annealed at 600°C.

X-ray diffraction (XRD) spectra were collected with a SIEMENS D500 powder diffractometer using Cu-Ka radiation. The XRD intensity ratio I_{001}/I_{002} was used as a measure of the degree of $L1_0$-ordering of CoPt. The XRD intensity ratio I_{002}/I_{111} was used as a measure of the crystallographic texture of CoPt. We note here that it is I_{001}/I_{002}=1.286 and I_{002}/I_{111}=0.28 for single-phased, polycrystalline, bulk $L1_0$-CoPt with randomly oriented grains [51]. Comparison of the I_{001}/I_{002} and I_{002}/I_{111} ratios of the samples with these values serves as a measure of $L1_0$-ordering and texture respectively. Magnetic hysteresis loops were obtained at room temperatute with a Digital Measuring System 660 Magneto-Optical Kerr Effect (MOKE) Magnetometer for field H normal ($H\perp$film) and papallel ($H\|$film) to the plane of the films. For $H\|$film, the noise in MOKE magnetic measurements was comparable to the signal. Therefore the results of MOKE measurements for $H\|$film will not be presented. The maximum field that can be produced by the electromagnet of the magnetometer is: H_{max}^{MOKE}=1.2 Tesla. The MOKE hysteresis loops that are presented are normalized (magnetization $M(H)$ divided by the maximum magnetization M_{max}). The coercivity H_c, saturation magnetization M_s, remnant magnetization M_r and relative remnant magnetization $m_r=M_r/M_s$ were determined by performing calculations on the curves that were obtained from fitting equations of the form $M(H,z) = M_0 + M_{S1}\tanh[(H-H_0+zH_{C1})/W_1] + M_{S2}\tanh[(H-H_0+zH_{C2})/W_2]$ with $z=\pm1$ in the experimental hysteresis loops M/M_{max}=f(H) with the least squares method. The atomic stoichiometry of Ag/CoPt films, as function of depth, was determined by Heavy Ion Elastic Rutherford Back Scattering (HIRBS) measurements on selected samples. A 8 MeV $^{12}C^{+++}$ beam was used, with surface barrier Si-detector at 170° with respect to the beam direction,

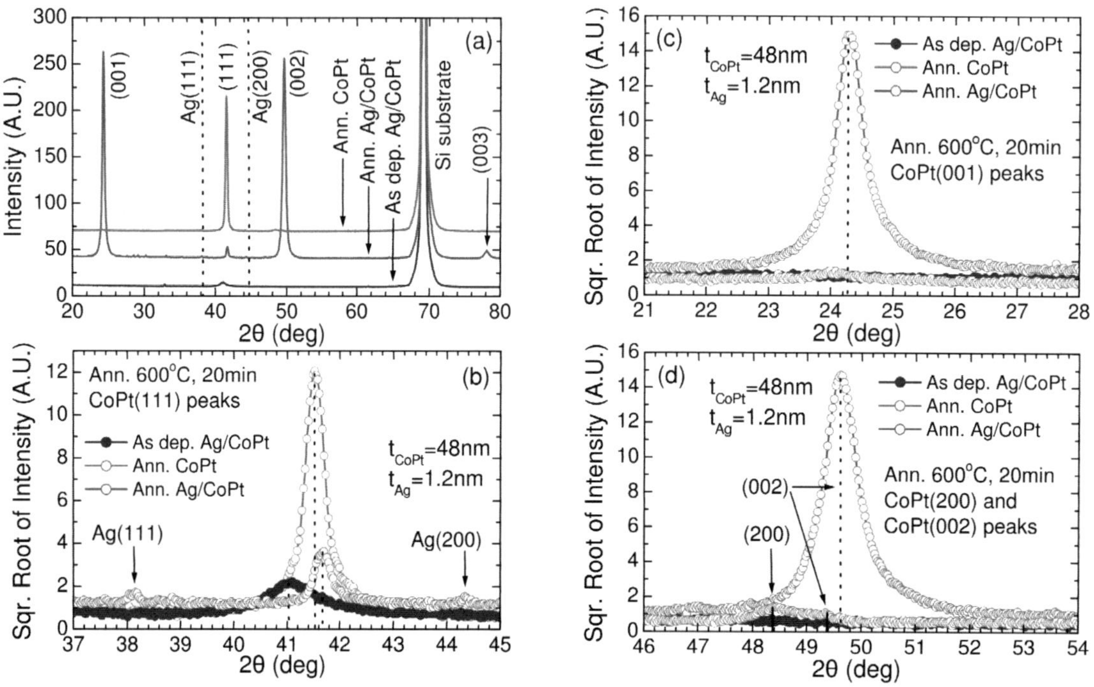

Figure 1. XRD spectra (panel (a)) of a Si(001)/Ag(1.2nm)/CoPt(48nm) BL, as-deposited (bottom) and annealed at 600°C for 20 min (middle). These spectra of the BL are compared to the XRD spectrum of a Si(001)/CoPt(48nm) MNL, annealed at 600°C for 20 min (top). The vertical dotted lines indicate the positions where the two strongest peaks of Ag are expected: $2\theta \approx 38.15^o$ for Ag(111) peak and $2\theta \approx 44.32^o$ for Ag(200) peak. Panels (b), (c) and (d) present the regions around the (111), (001) and (002) peaks of CoPt. In these panels we used the square root of the intensity, in order to make the weak peaks more clearly visible.

having a Full Width at Half Maximum (FWHM) equal to 67 keV, solid angle 2.54 msr (steradian is the SI unit of solid angle with symbol sr) and total beam charge 32 μCb. The pressure in the goniometer chamber was 2×10^{-7} Torr.

Note: Despite the fact that the annealing process may (partially or fully) destroy the layer-by-layer structure of an as-deposited film, we will use the layer structure of the as-deposited film as a label for the post-annealed film. For instance, a multilayer with structure Si(001)/[Ag(0.5nm)/CoPt(4.5nm)]×4 before annealing will be named "post-annealed Si(001)/[Ag(0.5nm)/CoPt(4.5nm)]×4 NC" after the annealing process.

3. Experimental results and discussion

3.1. Crystallographic data of post-annealed plain CoPt films and Ag/CoPt BLs

For completeness reasons, we present here in short a comparison between the XRD data of post-annealed plain CoPt monolayers (MNLs) and the XRD data of post-annealed Ag/CoPt BLs. Representative X-ray data are shown in Fig. 1, where we present XRD spectra of a Si/Ag(1.2nm)/CoPt(48nm) BL before (as-deposited) and after annealing at 600°C for 20 min (post-annealed). These values of annealing temperature and annealing time were chosen, because for annealing at lower temperatures or for shorter time periods the degree of $L1_0$-ordering and (001) texture were not satisfying. The XRD spectrum of a post-annealed Si/CoPt(48nm) MNL was added for comparison reasons. The as-deposited Ag/CoPt BL shows only a weak and broad

CoPt(111) peak, indicating that it is (111) textured and the CoPt grains are quite small. The absence of any other CoPt peaks shows that it contains only the disordered FCC phase of CoPt: the characteristic superstructure peaks of $L1_0$-CoPt [5, 51] are absent. The (111) texture is expected, because the (111) plane is the close-packed plane of FCC structure [28] with minimized surface energy. The subsequent annealing has as an effect not only the formation of the $L1_0$ phase of CoPt, but also typically leads to grain growth and sintering or coalescence [52, 53] which results in larger grains. Therefore, the post-annealed films should exhibit more intense and narrow peaks compared to the corresponding peaks of the as-deposited ones. The post-annealed Si/CoPt(48nm) MNL, which is a mixture of FCC-CoPt and $L1_0$-CoPt, does indeed show a strong and sharp CoPt(111) peak, but the other CoPt peaks are very weak. This MNL clearly shows a (111) texture, in agreement with Ref. [54] which reports that (in the case of FePt) the (111) plane has the lowest surface energy in both FCC and $L1_0$ structures. The fact that the (111), (001) and (002) peaks of CoPt are very weak in the post-annealed MNL indicates a relatively low degree of $L1_0$-ordering. The presence of $L1_0$-CoPt is mainly indicated by the fact that the CoPt(111) peak of the post-annealed MNL is placed at an angle 2θ relatively higher than the CoPt(111) peak of the as-deposited BL ($\Delta(2\theta)\approx 0.5^o$), as a result of the deformation of the unit cell during the FCC to $L1_0$ transformation.

In contrast to the behavior of the post-annealed MNL, the post-annealed BL shows a sufficient degree of $L1_0$-ordering, and a strong (001) texture. Indeed we observe strong and sharp CoPt(001) and CoPt(002) peaks and a clear CoPt(003) peak, while the CoPt(111) peak is weak. The presence of the CoPt(001) and CoPt(003) superstructure peaks is a clear evidence for the presence of the $L1_0$-phase of CoPt [5]. For the intensity ratio I_{001}/I_{002}, which is proportional to the square of the $L1_0$-ordering parameter [5, 13], we have $I_{001}/I_{002}\approx 1$. This shows a sufficient degree of $L1_0$-ordering, compared to the value 1.286 (see second paragraph of section 2). The development of the CoPt(00l) peaks, combined with the weak CoPt(111) peak, shows strong (001) texture. Using the measure I_{002}/I_{111} of the developed texture, we indeed see that $I_{002}/I_{111}\gg 1$. The absence of the CoPt(200) peak in the spectrum of the post-annealed BL is probably the result of the combination of the good $L1_0$-ordering with the strong (001) texture. We conclude that the presence of the Ag underlayer has as an effect the enhancement of $L1_0$-formation (and of the degree of $L1_0$-ordering) and simultaneously the development of (001) texture, compared to the (111) texture of the MNL.

3.2. HIRBS data on the post-annealed Ag/CoPt NCs

The annealing process at 600^oC, in order to achieve the formation of the $L1_0$ phase of CoPt, is expected to stimulate inter-diffusion between the two kinds of layer (CoPt magnetic layer, Ag layer) of the as-deposited film. The degree of this inter-diffusion was determined by HIRBS measurements on selected samples (mostly post-annealed Ag/CoPt NCs). Representative HIRBS data are shown in Fig. 2(a), where we present the HIRBS spectrum of a post-annealed Si/[Ag(0.5nm)/CoPt(4.5nm)]×4 NC (annealed at 600^oC for 40 min). Figures 2(b), 2(c) and 2(d) present the layer-by-layer structure after the annealing process, which was calculated by simulating the spectrum of Fig. 2(a) with the RUMP program: the simulation calculates the atomic stoichiometries of Ag, Pt and Co inside the NC film as functions of the distance from Si surface. The analysis of the spectrum showed that: (i) the film is continuous, (ii) there is indeed inter-diffusion between the Ag and CoPt layers during the annealing process. Although there is this inter-diffusion, the mixing between the two kinds of layers is not complete even for the elevated annealing temperature of 600^oC and the increased annealing time of 40 min. For this reason it is convenient to describe the structure of the post-annealed NC as a succession of "magnetic" (pure CoPt before annealing) and "spacer" (pure Ag before annealing) layers with thickness of 4.5 and 0.5 nm respectively, according to the as-deposited multilayer structure. The atomic stoichiometries X_{Ag}, X_{Pt} and X_{Co} of Ag, Pt and Co respectively (as functions of the

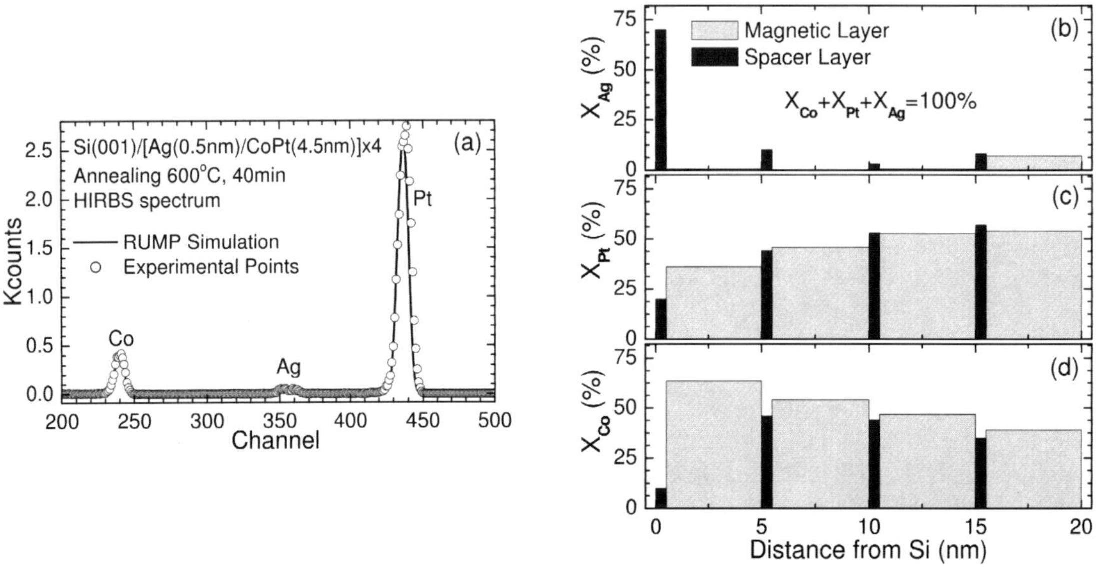

Figure 2. HIRBS spectrum (panel (a)) of a 20 nm thick Si(001)/[Ag(0.5nm)/CoPt(4.5nm)]×4 post-annealed NC. The continuous line is the result of the RUMP simulation that was used for the calculation of the chemical profile of the NC from the experimental points (open circles). Panels (b), (c) and (d) show the results of the calculation: the atomic stoichiometries X_{Ag}, X_{Pt} and X_{Co} of Ag, Pt and Co respectively as functions of the distance from Si substrate. Dark (bright) bars correspond to film areas that were 100% occupied by Ag (CoPt) in the initial as-deposited multilayer.

distance from Si substrate) are shown in Figs. 2(b), 2(c) and 2(d) in histogram form: the dark (bright) bars represent the "spacer" ("magnetic") layers of the post-annealed film. In these figures we observe that:

(a) Inside the magnetic layers it is $X_{\mathrm{Ag}}\approx 0$, except from the magnetic layer at the top of the film. In the inner spacer layers, X_{Ag} also takes small – but not vanishing – values. We conclude that Ag shows a strong tendency to remain away from the magnetic layers and to move near the film-substrate interface and the top region of the NC film. In the case of BLs, where Ag exists in the as-deposited film only as an underlayer, Ag is expected to remain near the substrate during annealing. In the case of NCs, the small amount of Ag at the inner regions of the post-annealed film (the not vanishing X_{Ag} at the inner spacer layers) may result in the structural incoherence of the growth of the CoPt grains during annealing.

(b) Deviations from the equiatomic CoPt composition of the as-deposited film are observed, with the relative Co concentration X_{Co} increasing in layers next to the Si substrate. This gradient can be explained by Co-Si alloying at the substrate-film interface.

(c) Pt diffuses in Ag layers more easily than Co. Thus the Pt stoichiometry X_{Pt} varies smoothly with depth, while Co stoichiometry X_{Co} is reduced in spacer regions. This difference can be a result of the fact that Co is almost immiscible in Ag [4].

3.3. $L1_0$ formation and development of (001) texture in post-annealed Ag/CoPt BLs and NCs

In Ref. [25] we presented crystallographic data which showed the dependence of I_{001}/I_{002} ($L1_0$-ordering) and I_{002}/I_{111} (texture) from the thickness t_{CoPt} of the CoPt layer of post-annealed Ag/CoPt BLs. These data showed that the samples which exhibited the higher (lower) degree of (001) texture were also characterized by better (worse) $L1_0$-ordering, which suggested that there

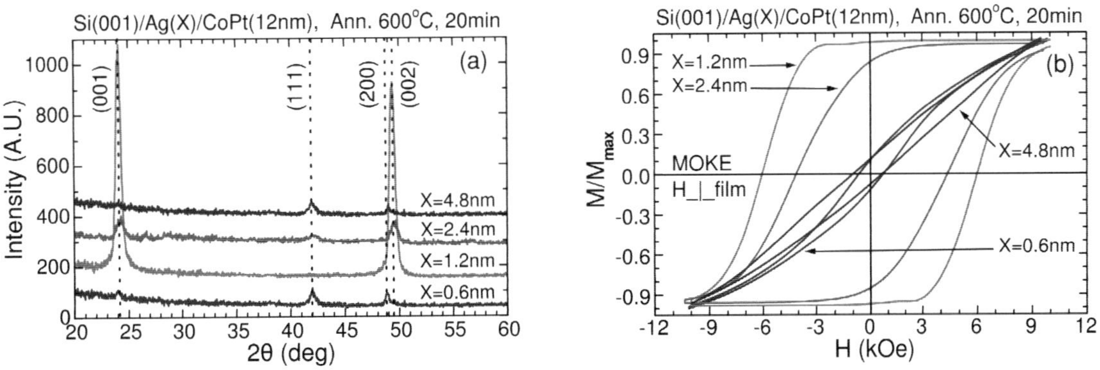

Figure 3. XRD spectra (panel (a)) and normalized MOKE hysteresis loops for $H\perp$film (panel (b)), of post-annealed Si(001)/Ag(X)/CoPt(12nm) BLs with X=0.6, 1.2, 2.4, 4.8 nm (annealing at 600°C for 20 min).

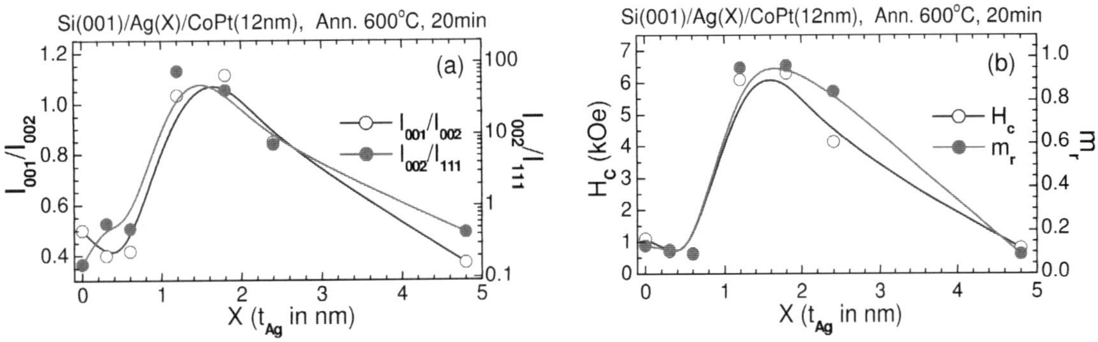

Figure 4. (a) $L1_0$-ordering (I_{001}/I_{002}, open circles) and crystallographic texture (I_{002}/I_{111}, solid circles), calculated from the XRD spectra of post-annealed Si(001)/Ag(X)/CoPt(12nm) BLs (600°C for 20 min), presented as functions of X. (b) Coercivity H_c (open circles) and relative remnant magnetization m_r=M_r/M_s (solid circles), calculated from MOKE measurements on the same post-annealed BLs (for $H\perp$film), presented as functions of X. X (Ag layer thickness) takes the values X=0, 0.3, 0.6, 1.2, 1.8, 2.4, 4.8 nm.

is a correlation between the mechanisms of texture and ordering. We also presented the results of VSM (Vibrating Sample Magnetometer) measurements on post-anneled BLs with strong (001) texture, which showed that (001)-textured samples exhibit perpendicular magnetic anisotropy. This is an expected behavior, since the CoPt grains are highly aligned in a direction with the c-axis (easy axis) normal to the plane of the film and the $L1_0$ phase of CoPt is characterized by large uniaxial anisotropy. Here we present new XRD data concerning the dependence of ordering and texture from the thickness t_{Ag} of the Ag layer. These data, combined with the results of magnetization (MOKE) measurements, provide further evidence for the existence of the the above mentioned correlation. For the purposes of this study we prepared two series of BL samples with varying t_{Ag}: a series with t_{CoPt}=12 nm and a series with t_{CoPt}=48 nm. Figure 3 presents characteristic XRD spectra and MOKE hysteresis loops, obtained from the measurements that were performed on the series of BL samples with t_{CoPt}=12 nm. It is clear that strong (001) texture results in high coercivity and relative remnant magnetization for $H\perp$film.

The results extracted from all the XRD and MOKE measurements, performed on the series of BL samples with t_{CoPt}=12 nm, are shown in Fig. 4: panel (a) presents I_{001}/I_{002} ($L1_0$-

ordering) and I_{002}/I_{111} (texture) as functions of t_{Ag}, panel (b) presents coercivity H_c and relative remnant magnetization $m_r=M_r/M_s$ as functions of t_{Ag}. Both panels of Fig. 4 clearly show a correlation between ordering and texture and between H_c and m_r. The correlation of the magnetic quantities is probably a manifestation of the correlation between ordering and texture. Indeed, it is expected [10, 12] that higher degree of formation of the L1_0 phase of CoPt corresponds to higher coercivity H_c, as long as the sample is not fully ordered (CoPt has not been completely transformed into its L1_0 phase). Also it is expected that, for stronger (001) texture, the relative remnant magnetization in a direction normal to the film's plane will tend to increase. For stronger (001) texture, the alignment of the easy axis of magnetization (c-axis) of grains along a direction normal to the film's plane is enhanced. Therefore the remnant magnetization normal to film's plane is expected to increase, since for zero field is along the easy axis. Here the correlation between the mechanisms of texture and L1_0-ordering becomes more evident and it is clear that the Ag underlayer is strongly involved in this correlation, since its thickness variation causes a variation to the values of both I_{001}/I_{002}-H_c and I_{002}/I_{111}-m_r.

From the HIRBS data, which were presented in subsection 3.2, we know that the Ag underlayers of the as-deposited BLs tend to remain at the film-Si(001) interface after the annealing process. This is a strong indication that the mechanism, through which Ag influences both ordering and texture, is of interfacial origin. However this does not indicate which is the nature of this mechanism, and does not explain the dependence of ordering and texture on t_{Ag}. In order to be able to make assumptions on the nature of this mechanism, we need a description of the Ag-Si(001) system. According to Refs. [55, 56, 57, 58, 59, 60, 61], this system exhibits the following properties:

(a) Although the Ag-Si(001) system exhibits high lattice mismatch [a_{Si}=5.431 Å (diamond), a_{Ag}=4.086 Å (FCC)], the ratio of Si to Ag lattice constants is almost exactly equal to 4:3. Therefore the two lattices can be adjusted with a very small remaining mismatch (2.2%).

(b) The Stranski-Krastanov (SK) mode of growth [62, 63] for Ag is well established. At the early stages of growth, Ag grows in 2D islands. In these 2D islands, the orientation relationship between Ag and Si is $(001)_{Ag}\|(001)_{Si}$ and Ag shows (001) texture. It continues to form a 2D Ag layer until Si coverage takes a critical value at about 0.5 ML, or equivalently at 2 nm [63, 64]. Above this value, the growth mode becomes gradually 3D. For 1 ML Ag forms 3D islands. The $\langle 111 \rangle$ becomes gradually the preferred growth orientation and Ag shows (111) texture, due to the lowest surface free energy for the (111) surface of the FCC crystals. This behavior is valid even at elevated temperatures (possibly with different critical value of thickness).

(c) With the sputtering process we can produce almost "epitaxial" Ag grains with (001) texture, in the initial stage of Ag growth, at a low sputtering power.

The proposed mechanism, for the explanation of the correlation between ordering and (001) texture development, is based on the reduction of total microstrain through the annealing process and is in good agreement (at least qualitatively) with the experimental data. Microstrain minimization in similar Ag/CoPt NCs has already been observed in Refs. [33, 34] (using XRD-peak profile analysis) and has been correlated to the optimization of (001) texture. In Ref. [28] Yu-Nu Hsu et al reported that the presence of (001) textured Ag underlayer is possible to induce (001) texturing in epitaxial FePt films and reduce the annealing temperature that is necessary for L1_0-formation. Since the Ag lattice parameter (a_{Ag}=4.086 Å) is slightly larger than that of FePt [65] (as also of CoPt [51]), the epitaxial growth is achieved via distortion of FePt (or Copt) unit cell and expansion of the unit cell constants along the film-Ag interface. On the other hand, during the cubic to tetragonal transformation (throughout the annealing process), the a and b axes expand while the c-axis contracts. This creates a mechanism by which the (001) texturing and ordering processes are correlated and the microstrain of FePt or CoPt is reduced, as a result of the reduction of the lattice mismatch with Ag. Also the surface energy must be minimized. In Ref. [54] Jae-Song Kim *et al* also studied the origin of (001) texture of FePt films.

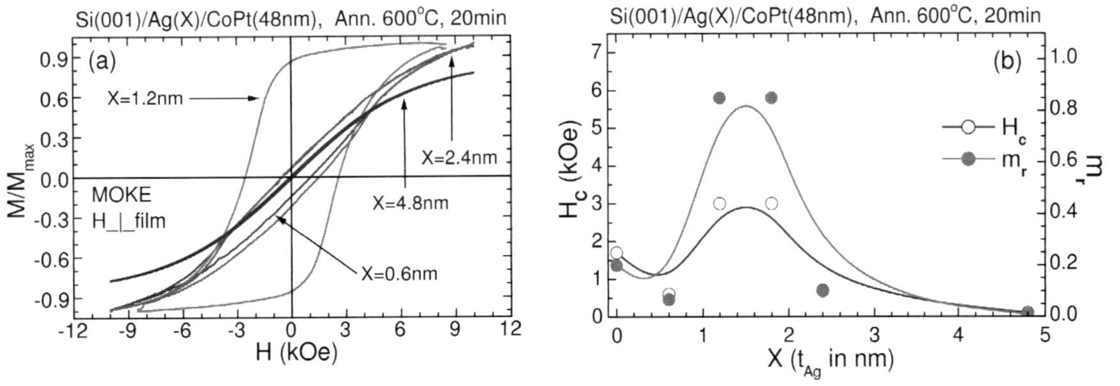

Figure 5. (a) Normalized MOKE hysteresis loops for $H\perp$film, of post-annealed Si(001)/Ag(X)/CoPt(48nm) BLs with X=0.6, 1.2, 2.4, 4.8 nm (annealing at 600°C for 20 min). (b) Coercivity H_c (open circles) and relative remnant magnetization $m_r=M_r/M_s$ (solid circles), presented as functions of X. H_c and m_r were calculated from the MOKE measurements presented in panel (a) (only the most representative loops are shown there). X (Ag layer thickness) takes the values X=0, 0.6, 1.2, 1.8, 2.4, 4.8 nm.

They distinguished the terms that contribute in the total energy of the polycrystalline material in three categories: (a) surface energy anisotropy [66], since surface energy depends on Miller indices (hkl), (b) biaxial in-plane strain caused by various sources such as residual strain of as-deposited film [66], difference in the thermal expansion coefficient between film and substate [67] (or in our case also between underlayer and top layer) and volume shrinkage in films by defect elimination and grain growth [68] during annealing, (c) anisotropic transormation strain due to the deformation of the unit cell during FCC to $L1_0$ transformation throughout annealing. According to Ref. [54], the formation of the (001) texture during annealing results from the minimization of the total energy, in which the 3rd term is the dominant term.

The above described mechanism, combined with the behavior of the Ag-Si(001) system can explain our experimental data. In Fig. 4 the region for Ag thickness lower than 1.2 nm should correspond to the first stage of Ag growth on Si(001), i.e. 2D almost "epitaxial" with (001) texture. The coverage of Si with Ag increases with t_{Ag}. As a result, larger regions of the CoPt layer are influenced by the described mechanism and the ordering and texture are increased with t_{Ag}. The region for $t_{Ag}>2$ nm should correspond to the stage where Ag growth becomes gradually 3D with (111) texture, and the mechanism becomes weaker with the increment of t_{Ag}. The coherence of growth of (001) textured CoPt is disrupted by the Ag island growth with (111) texture. The optimum Ag underlayer thickness is in the region 1.2–2 nm. Also the strain relaxation along film thickness can be more effectively realized in very thin films, through the formation of islands. This may be the reason for the relatively lower coercivity H_c and relative remnant magnetization m_r values of Fig. 5 when compared to the values of Fig. 4. Figure 5 presents the MOKE measurements performed on the series of BL samples with $t_{CoPt}=48$ nm. For this series of samples, the optimum Ag underlayer thickness is also in the region 1.2–2 nm.

In subsection 3.2, where we studied the HIRBS data of post-annealed NCs, we concluded that the small amount of Ag at the inner regions of a post-annealed Ag/CoPt NC (the not vanishing atomic percentage of Ag at the inner spacer layers) may result in the structural incoherence of the growth of the CoPt grains during annealing. Figure 6 shows crystallographic data that support this conclusion. Panel (a) shows texture and ordering as functions of the thickness of the CoPt layer for BLs with 1.2 nm of Ag underlayer. Correlation between texture and ordering is evident. Panel (b) shows texture as function of the total thickness for NCs with 10% Ag,

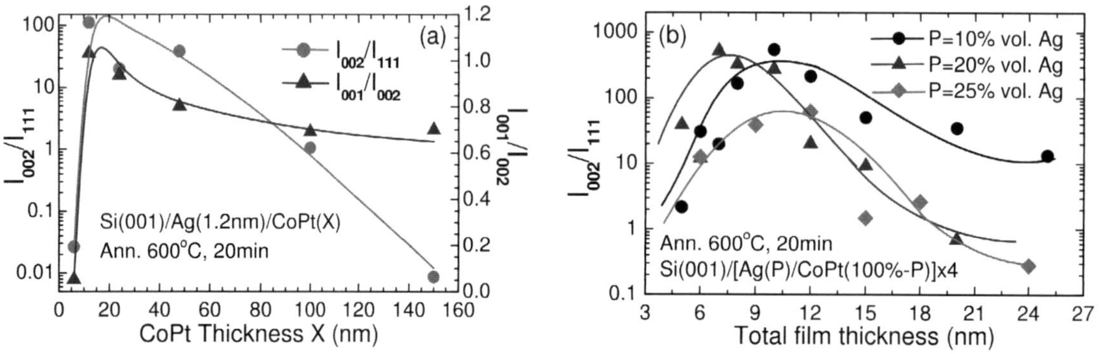

Figure 6. Comparison between the textures developed on post-annealed BLs and NCs. Panel (a) shows texture (circles) and ordering (triangles) as functions of the thickness of the CoPt layer for BLs with 1.2 nm of Ag underlayer. Panel (b) shows texture as function of the total thickness for NCs with 10% Ag (circles), 20% Ag (triangles) and 25% Ag (rhombs). Percentage refers to volume.

20% Ag and 25% Ag. Percentage refers to volume. First we observe that increment of the percentage of Ag results gradually in an overall reduction of the texture developed in the NC, which is compatible with the hypothesis of structural incoherence due to the presence of Ag in the inner regions of the film. Direct comparison of the degree of (001) texture between BLs and NCs, shows that in the case of NCs the (001) texture starts to degrade for lower total film thickness compared to the case of BLs. For NCs with total film thickness near 9–12 nm it is $I_{002}/I_{111}\approx500$, compared to the value $I_{002}/I_{111}\approx100$ exhibited by the BLs. In contrast to that, for NCs with total thickness above 15 nm, the degree of (001) texture is already lower than the value $I_{002}/I_{111}\approx40$, while for BLs it is $I_{002}/I_{111}\approx50$ even for 50 nm total thickness. This is also compatible with the hypothesis of structural incoherence.

3.4. Conclusions

The basic conclusions that can be deduced from the X-ray, magnetization and HIRBS data presented in this work, for post-annealed Ag/CoPt BLs and NCs, are the following:

- Ag underlayers can promote a high degree of $L1_0$-ordering and a strong (001) crystallographic texture in the post-annealed Ag/CoPt BLs. CoPt films without Ag underlayers have a (111) texture.
- The influence of Ag underlayers on post-annealed Ag/CoPt BLs is Ag and CoPt thickness dependent, as it is shown by both magnetic and XRD measurements. It was found that the optimum thickness of the Ag underlayer, for the development of strong (001) crystallographic texture in the post-annealed BLs, is $t_{Ag}\approx1.2$–2.0 nm. For Ag thickness within this range of values, the post-annealed BLs develop a strong (001) texture. For values of t_{Ag} outside this range, the post-annealed BLs tend to develop a (111) texture (this includes the CoPt MNLs, which correspond to $t_{Ag}=0$ nm).
- It was also found that the degree of (001) crystallographic texture (represented by I_{002}/I_{111}) and the degree of $L1_0$-ordering (represented by I_{001}/I_{002}), of the post-annealed BLs, are correlated. Indeed, XRD measurements for various values of t_{Ag} showed that: the maximum (medium or minimum) values of I_{002}/I_{111} correspond to maximum (medium or minimum) values of I_{001}/I_{002}. This correlation was also observed in the magnetic measurements.
- The HIRBS data of post annealed Ag/CoPt NCs show a very strong tendency of Ag to remain near film-substrate interface, as well as at the top region of the film, after the

annealing process. Especially in the case of BLs, Ag underlayers tend to remain near the Ag-Si(001) interface. This suggests that the mechanism responsible for the correlation between $L1_0$-formation and development of (001) texture is of interfacial origin.

- The mechanism, that is proposed for the explanation of this correlation, is based on the reduction of total strain (residual strain of as-deposited film and transformation strain due to deformation of the unit cell as $L1_0$-CoPt is formed) throughout the annealing process and the accompanied reduction of the lattice mismatch with the Ag underlayer.
- Comparison of the degree of (001) crystallographic texture between BLs and NCs, shows that in the case of NCs the (001) texture starts to degrade for lower total film thickness compared to the case of BLs. For NCs with total film thickness near 9–12 nm it is $I_{002}/I_{111}{\approx}500$, compared to the value $I_{002}/I_{111}{\approx}100$ exhibited by the BLs. In contrast to that, for NCs with total thickness above 15 nm, the degree of (001) texture is already lower than the value $I_{002}/I_{111}{\approx}40$, while for BLs it is $I_{002}/I_{111}{\approx}50$ even for 50 nm total thickness. This difference can be explained by the structural incoherence in the growth of the CoPt grains, imposed by the presence of intermediate spacer layers (which contain Ag) during the annealing process.

References

[1] Cullity B D 1972 *Introduction to Magnetic Materials* (Addison-Wesley Publishing Company)
[2] O'Handley R C 2000 *Modern Magnetic Materials: Principles and Applications* (John Wiley & Sons Inc.)
[3] Barrett C S and Massalski T B 1966 *Structure of Metals* (New York: McGraw-Hill)
[4] Massalski T B 1996 *Binary Alloy Phase Diagrams* (ASM International)
[5] Warren B E 1990 *X-ray Diffraction* (New York: Dover)
[6] Darling A S 1963 *Platinum Met. Rev.* **7** 96
[7] McCurrie R A and Gaunt P 1966 *Philos. Mag.* **13** 567
[8] Gaunt P 1966 *Philos. Mag.* **13** 579
[9] Brissonneau P, Blanchard A, Schlenker M and Laugier J 1968 *J. Appl. Phys.* **39** 1266
[10] Yermakov A Y and Maykov V V 1985 *Fiz. Met. Metalloved.* **60** 113
[11] Klemmer T, Hoydick D, Okumura H, Zhang B and Soffa W A 1995 *Scr. Metall. Mater.* **33** 1793
[12] Ristau R A, Barmak K, Lewis L H, Coffey K R and Howard J K 1999 *J. Appl. Phys.* **86** 4527
[13] Barmak K, Kim J, Lewis L H, Coffey K R, Toney M F, Kellock A J and Thiele J-U 2005 *J. Appl. Phys.* **98** 033904
[14] Coffey K R, Parker M A and Howard J K 1995 *IEEE Trans. Magn.* **31** 2737
[15] Yanagisawa M, Shiota N, Yamaguchi H and Suganuma Y 1983 *IEEE Trans. Magn.* **19** 1638
[16] Weller D and Moser A 1999 *IEEE Trans. Magn.* **35** 4423
[17] Weller D, Moser A, Folks L, Best M E, Lee W, Toney M F, Schwickert M, Thiele J-U and Doerner M F 2000 *IEEE Trans. Magn.* **36** 10
[18] Wang J-P, Qiu J-M, Taton T A and Kim B-S 2006 *IEEE Trans. Magn.* **42** 3042
[19] Jones B A, Dutson J D, OGrady K, Hickey B J, Li D, Poudyal N and Liu J P 2006 *IEEE Trans. Magn.* **42** 3066
[20] Sun A-C, Hsu J-H, Kuo P C and Huang H L 2007 *IEEE Trans. Magn.* **43** 2130
[21] Luo C P and Sellmyer D J 1995 *IEEE Trans. Magn.* **31** 2764
[22] Luo P, Shan Z S and Sellmyer D J 1996 *J. Appl. Phys.* **79** 4899
[23] Farrow R F C, Weller D, Marks R F, Toney M F, Cebollada A and Harp G R 1996 *J. Appl. Phys.* **79** 5967
[24] Liao W M, Lin Y P, Yuan F T and Chen S K, 2004 *J. Magn. Magn. Mater.* **272–276** 2175
[25] Manios E, Karanasos V, Niarchos D and Panagiotopoulos I 2004 *J. Magn. Magn. Mater.* **272–276** 2169
[26] Manios E, Alexandrakis V and Niarchos D 2007 *J. Magn. Magn. Mater.* **316** E166
[27] Hsu Y-N, Jeong S, Laughlin D E and Lambeth D N 2001 *J. Appl. Phys.* **89** 7068
[28] Hsu Y-N, Jeong S, Laughlin D E and Lambeth D N 2003 *J. Magn. Magn. Mater.* **260** 282
[29] Yu M, Liu Y and Sellmyer D J 1999 *Appl. Phys. Lett.* **75** 3992
[30] Luo C P, Liou S H, Gao L, Liu Y and Sellmyer D J 2000 *Appl. Phys. Lett.* **77** 2225
[31] Karanasos V, Panagiotopoulos I, Niarchos D, Okumura H and Hadjipanayis G C 2000 *J. Appl. Phys.* **88** 2740
[32] Karanasos V, Panagiotopoulos I, Niarchos D, Okumura H and Hadjipanayis G C 2001 *J. Appl. Phys.* **90** 3112

[33] Karanasos V, Panagiotopoulos I, Niarchos D, Okumura H and Hadjipanayis G C 2001 *Appl. Phys. Lett.* **79** 1255
[34] Karanasos V, Panagiotopoulos I and Niarchos D 2002 *J. Magn. Magn. Mater.* **249** 471
[35] Christodoulides J A, Bonder M J, Huang Y, Zhang Y, Stoyanov S, Hadjipanayis G C, Simopoulos A and Weller D 2003 *Phys. Rev. B* **68** 054428
[36] Tomou A, Panagiotopoulos I, Gournis D and Kooi B 2007 *J. Appl. Phys.* **102** 023910
[37] Plumer M L, van Ek J and Weller D (Editors) 2001 *The Physics of Ultra-High-Density Magnetic Recording* (Springer - Springer Series in Surface Sciences)
[38] Zeng H, Yan M L, Powers N and Sellmyer D J 2002 *Appl. Phys. Lett.* **80** 2350
[39] Luo C P and Sellmyer D J 1999 *Appl. Phys. Lett.* **75** 3162
[40] Sellmyer D J, Luo C P, Yan M L and Liu Y 2001 *IEEE Trans. Magn.* **37** 1286
[41] Yan M L, Zeng H, Powers N and Sellmyer D J 2002 *J. Appl. Phys.* **91** 8471
[42] Yan M L, Powers N and Sellmyer D J 2003 *J. Appl. Phys.* **93** 8292
[43] Yan M L, Li X Z, Gao L, Liou S H and Sellmyer D J 2003 *Appl. Phys. Lett.* **83** 3332
[44] Shao Y, Yan M L and Sellmyer D J 2003 *J. Appl. Phys.* **93** 8152
[45] Yan M L, Sabirianov R F, Xu Y F, Li X Z and Sellmyer D J 2004 *IEEE Trans. Magn.* **40** 2470
[46] Sellmyer D J, Yan M, Xu Y and R. Skomski R 2005 *IEEE Trans. Magn.* **41** 560
[47] Chen J S, Lim B C, Hu J F, Liu B, Chow G M and Ju G 2007 *Appl. Phys. Lett.* **91** 132506
[48] Wu Y-C, Wang L-W and Lai C-H 2007 *Appl. Phys. Lett.* **91** 072502
[49] Li B-H, Feng C, Gao X, Teng J, Yu G-H, Xing X and Liu Z-Y 2007 *Appl. Phys. Lett.* **91** 152502
[50] Yang F J, Wang H, Wang H B, Cao X, Yang C P, Li Q, Zhou M J, Chong Y M and Zhang W J 2007 *J. Appl. Phys.* **102** 106101
[51] $L1_0$-CoPt: JCPDS-ICDD database, card No 43-1358; Axes lengths of $L1_0$-CoPt unit cell: $a = 3.803$ Å and $c = 3.701$ Å.
[52] Dai Z R, Sun S and Wang Z L 2001 *Nano Lett.* **1** (8) 443
[53] Klemmer T J, Liu C, Shukla N, Wu X W, Weller D, Tanase M, Laughlin D E and Soffa W A 2003 *J. Magn. Magn. Mater.* **266** 79
[54] Kim J-S, Koo Y-M, Lee B-J and Lee S-R 2006 *J. Appl. Phys.* **99** 053906
[55] Lin X F, Wan K J and Nogami J 1993 *Phys. Rev. B* **47** 13491
[56] Lin X F, Wan K J and Nogami J 1994 *Phys. Rev. B* **49** 7385
[57] Doraiswamy N, Jayaram G and Marks L D 1995 *Phys. Rev. B* **51** 10167
[58] Kimura K, Ohshima K and Mannami M 1995 *Phys. Rev. B* **52** 5737
[59] von Hoegen M H, Schmidt T, Meyer G, Winau D and Rieder K H 1995 *Phys. Rev. B* **52** 10764
[60] Glueckstein J C, Evans M M R and Nogami J 1996 *Phys. Rev. B* **54** R11066
[61] Je J H, Kang T S and Noh D Y 1997 *J. Appl. Phys.* **81** 6716
[62] Zangwill A 1988 *Physics at Surfaces* (Cambridge University Press).
[63] Venables J A 2001 *Introduction to Surface and Thin Film Processes* (Cambridge University Press).
[64] The monolayer (ML) unit is a measure of areal density of atoms and it is expressed as atoms/m^2. 1ML corresponds to the areal density of one monolayer of atoms.
[65] FCC-FePt: JCPDS-ICDD database, card No 29-0717 and 29-0718; $L1_0$-FePt: JCPDS-ICDD database, card No 43-1359; Axes lengths of $L1_0$-FePt unit cell: $a = 3.852$ Å and $c = 3.713$ Å.
[66] Thompson C V and Carel R 1995 *Mater. Sci. Eng.* **B-32** 211
[67] Witt F and Vook R W 1968 *J. Appl. Phys.* **39** 2773
[68] Zeigera W, Brucknerb W, Schumannb J, Pitschkeb W and Worcha H 2000 *Thin Solid Films* **370** 315

Dielectric and magnetic properties of YIG/PMMA nanocomposites

Anjum Qureshi[1], Ayhan Mergen[1] and Bekir Aktaş[2]
[1]Department of Metallurgical and Materials Engineering, Marmara University, Istanbul-81040, Turkey
[2]Gebze Institute of Technology, 41400 Gebze, Kocaeli, Turkey

anjumqur@gmail.com

Abstract.

Yittrium iron garnet, $Y_3Fe_5O_{12}$ (YIG), is a material used widely in electronic devices for the microwave region as well as the magnetic bubble domain-type memories. Yittrium iron garnet ($Y_3Fe_5O_{12}$) was produced by mechanochemical synthesis from Y_2O_3 and Fe_2O_3 with particle size of around 150 nm. PMMA/YIG composite films were prepared by solution casting method at diffferent concentration (i.e. 10%, 20% and 40%) of YIG filler. Dielectric permitivity of composite materials were studied over a wide a range of frequency and temperature as a function of filler concentration. The electrical properties of composites were explained by in terms of molecular mobility and interfacial polarization.

1. Intoduction

Composite systems consisting of an insulating matrix and randomly dispersed fine inorganic particles have generated significant research interest, mostly, due to their electrical and electromagnetic performance [1-3]. The primary applications of these polymer composites refer to electromagnetic interference (EMI) shielding, radio frequency interference (RFI) shielding and electrostatic dissipation of charges (ESD). Composites consisting of inorganic particles embedded in a polymer matrix have attracted more and more attentions due to integrative advantages of the two constituents. Introducing high dielectric constant (ε) ceramic particles into polymer matrix will result in high ε composite with the processibility and flexibility similar to polymers [4–8]. Such kind of composites could contribute to the promising candidates for dielectric materials in flexible capacitors with high performance and as a microwave absorber.

The dielectric properties of polymer/ceramic composites are influenced not only by the dielectric constant of the polymer and the ceramic, but also by the dispersion and loading of the ceramic in the polymer matrix. Using higher dielectric constant particles and polymer matrix, and increasing the loading of particle fillers can increase the effective dielectric constant of a composite. As well, the morphological and structural optimization of particles will enhance their contribution to dielectric constant improvement [9]. Though lifting the filler content is a choice, however, too high loading of fillers would lead to poor quality of the composite. Polymers and polymer matrix composites are basically electrical insulators, due to their low concentration of free charge carriers. Thus their electrical response is, mainly, associated with relaxation phenomena occurring under the influence of ac field. The observed relaxation processes are related to dipolar orientation effects or space charge migration [10, 11]. Molecular mobility and interfacial polarization are regarded as the origin of dielectric effects. At sufficient high temperatures, in the vicinity of the glass transition temperature, large segments of the polymer chains are able to move trying to follow the alternation of the field, while at lower temperatures polar side groups are contributing to the electrical performance of the system. Interfacial polarization is the result of the heterogeneity of the system, such as mobile charges accumulated at the fillers–polymer matrix interface, form large dipoles. Recently, adding conductive fillers (with infinitively large ε) into polymer was reported on preparation of high ε composites [12–17], but the insulating properties of the composites will be reduced in this case. So, the inorganic fillers with super high ε should be the key consideration in obtaining higher ε composites with integrative properties needed in practical applications.

The ferromagnetic garnet, yttrium iron garnet (YIG), is a well-known material, and it has been widely used in electronic devices, such as circulators and phase shifters for microwave and magneto-optical devices [18, 19]. Additionally, YIG has controllable saturation magnetization, low dielectric loss tangent (tan δ) in microwave regions, and small line-width (ΔH) in ferromagnetic resonance [20]. Oscillations and waves of magnetization or spin waves in yttrium–iron garnet (YIG) ferrite films and mono-crystals have very interesting and useful properties such as a high Q-factor, a wide variety of dispersion laws dependent on the orientation of bias magnetic fields, and a possibility to tune their frequencies in the range of several GHz by changing the magnitude of the bias magnetic field. Another important property of ferrite films is low (of the order of several microwatts) power threshold of nonlinear spin wave processes. The nonlinear properties of ferrite films can be used for the development of novel microwave signal processing devices based on the parametric interaction of spin wave packets propagating in the films with localized electromagnetic fields of the microwave pumping. Excellent features of the ferrite/polymer composites, such as sharply reduced dielectric loss compared to in the bulk ferrites, while uninfluenced microwave absorption properties owing to the domination of natural ferromagnetic resonance absorption in the loss mechanism of the ferrite absorber materials [21, 22], make them quite attractive for applications not only as inductive and capacitive materials but also as microwave absorber materials.

In this work, Yittrium iron garnet, $Y_3Fe_5O_{12}$ (YIG) is synthesized using mechanochemical synthesis method and the polymethyl methacryalate polymer-based composites consisting of YIG nanoparticle fillers with varying concentrations were reported, aiming to investigate how nano fillers affects the electrical properties of such composites which could be explained by in terms of molecular mobility and interfacial polarization.

2. Experimental procedure

YIG ceramic was produced from Y_2O_3 and Fe_2O_3 by mechanochemical method. The synthesis was carried out in a Fritsch Pulverisette 5 Planetary high energy ball milling system. The following milling conditions were used: stainless steel vessel with a volume of 500 cm^3, stainless steel balls with diameter of 10 mm, ball-to-powder weight ratio was 40:1, air atmosphere, rotation speed of discs with vials was 320 min^{-1} and milling time was 8 h. Polymethyl methacrylate (PMMA) was used as the polymer matrix. Polymer-based nanocomposites were prepared by directly mixing YIG nanoparticles and PMMA together, by solution casting method. Tetrahydrofuran (THF) was used as a solvent to dissolve the polymer. The mixture of polymers and nanofillers were stirred with different volume fraction of YIG filler (i.e. 10%, 20% and 40%) at room temperature using magnetic stirrer at high speed of around 700 rpm until all polymer dissolved. After obtaining a homogeneous mixture, the mixture was poured into a clean glass trough. The solvent was evaporated at room temperature ($25^0C\pm 1^0C$) to get thin films of polymer nanocomposites which were then dried in vacuum oven at 30^0C. Powder X-ray diffraction analysis (Rigaku, Cu Ka, 2°/min) was used to determine the crystalline phases of milled powder. The particleshape and morphologies of YIG were investigated by SEM (JEOL-5910LV) after coating with gold. The particle sizes of milled powder were measured using Malvern particle size analyzer (Nano ZS Zeta Sizer). High frequency magnetic properties of the YIG particles were studied using Bruker MMX X-band ESR spectrometer. A small sample was located at the center of x-band cavity and field derivative FMR spectrum was recorded as a function of external field. The direct microwave absorption spectrum of the YIG particles was investigated using a Network analyzer operating as a function of microwave frequency between 1-50 GHz. A small amount of sample was placed in a cavity in the microwave transmission line. The cavity was located in between and producing an electromagnetic field in the range of 0-2.2 Tesla. Microwave signal with different frequencies was applied and the external field was scanned from 0 up to 2.2 Tesla. After coating silver layers on the two surfaces of the composite films as electrode, the dielectric and electrical properties were measured using HP4980A LCR meter.

3. Results and discussion

3.1 The structure and morphology of YIG nanoparticles

Fig.1 shows the particle size distribution of powders after 8 hrs of high energy milling indicating a nearly homogeneous distribution. The mean particle size of the powders was around 150 nm. Fig. 2 shows XRD patterns of the YIG powders milled for different times indicated that after 8h of high energy milling nearly all of the phases converted to $Y_3Fe_5O_{12}$ (JCPDS: 43-0507) through formation of $FeYO_3$ (JCPDS: 39-1489). After 2 h of high energy milling, iron and yittrium oxides started to react to form $FeYO_3$ and the first peaks of $Y_3Fe_5O_{12}$ were observed in samples milled for 4 hrs but there were still unreacted iron and yittrium oxides in addition to $FeYO_3$.

Fig. 3 indicates the SEM micrograph of mechanochemically activated powders after 8 hrs of milling which shows nanoscale particles. The particles had nearly uniform size distribution but contained also some agglomeration. After producing a pellet from mechanochemically treated powder, it was then sintered at 1425°C for 10 hrs which gave a YIG ceramic with a theoretical density of around 95%. This indicated that mechanochemical method gave a high density YIG ceramic after sintering without any calcination step.

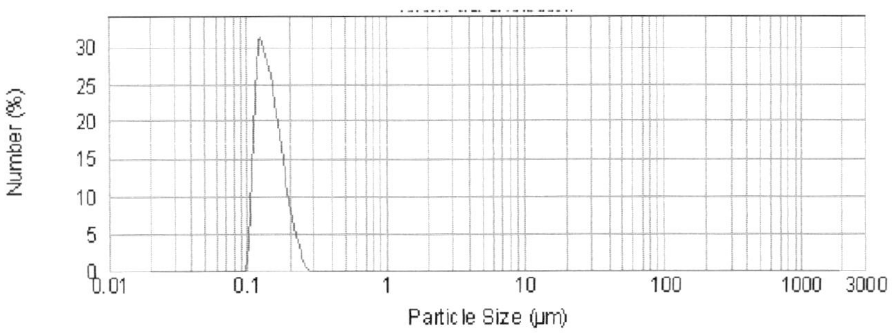

Figure 1. Particle size distribution of powders produced by mechanochemical synthesis after 8 hrs of high energy milling.

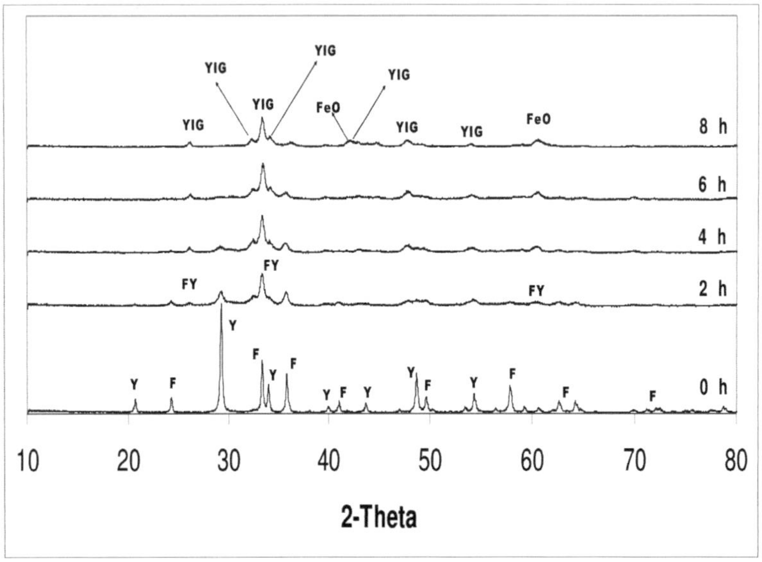

Figure 2. XRD patterns of the YIG powders milled for different times (Y: Y_2O_3, F: Fe_2O_3, FY: $FeYO_3$ (JCPDS Card No:39-1489), YIG: $Y_3Fe_5O_{12}$ (JCPDS Card No:21-1450).

Figure 3. SEM of mechanochemically activated powder with 8 hrs of milling.

3.2 Magnetic properties of YIG particles

Ferro-magnetic resonance (FMR) spectrum of YIG particles is given in Fig. 4. As seen from the spectrum, FMR signal is very strong even at possible minimum microwave energy. This means that the individual particles are strongly magnetic as expected from YIG particles. The FMR absorption curve is considerably wide as well. This behavior might be attributed partially to crystalline anisotropy of the individual particles and partly to the magnetostatic long range interactions among the magnetic particles. Since magnetic axes are randomly oriented, resonance field of individual particles might be distributed in the field up to some extent determined by the magnetic anisotropy field. The asymmetric character of the resonance peaks could be originated from the conducting properties of the sample at higher frequencies.

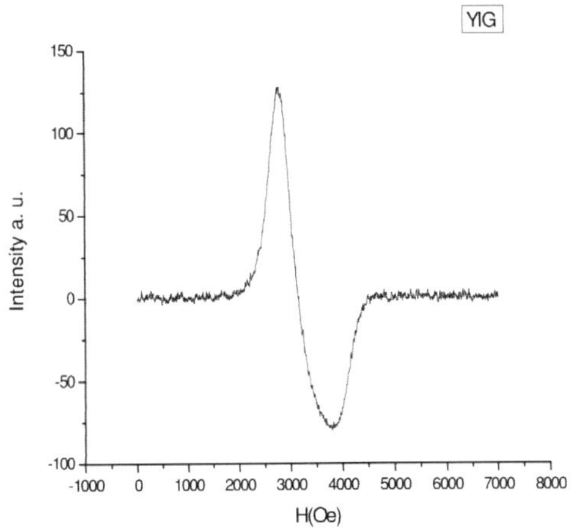

Figure 4. FMR spectrum of YIG nanoparticles

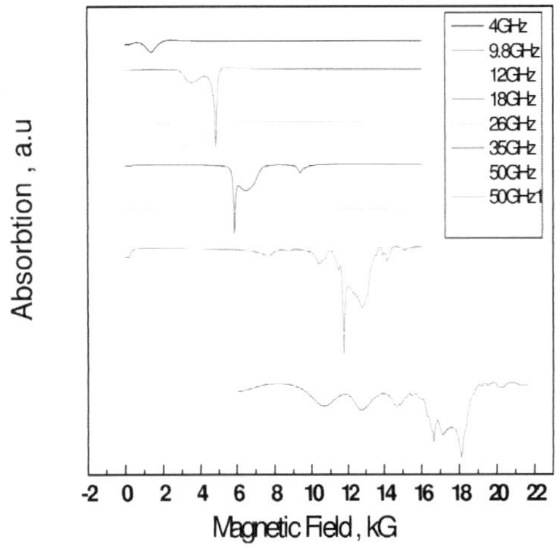

Figure 5. Microwave absorption spectrum of the sample at different frequencies and magnetic fields.

The FMR absorption spectrum with different applied magnetic fields is shown in Fig. 5. It was observed that the signal exhibits a resonance character at each frequency; however the location of the peak at the field axes monotonically shifts to higher field with increasing frequency. This behavior is consistent with FMR theory that the sample is ferromagnetic in nature even at very low amount of sample. Further additional peaks beside main FMR mode were observed and the number of these peaks increases with frequency. The shape and the number of these peaks depend on the external geometry of the sample in the cavity which is a basic property for spin wave resonance. It indicates that there is an existence of a long wave length spin wave i.e. magnetostatic wave modes throughout the sample.

3.2 The dielectric and conductivity properties of the PMMA/YIG nanocomposites

Fig.6 shows the dielectric constants and the electrical conductivities of the PMMA/YIG composites while varying the concentration of the YIG, frequency of applied fields and temperature. Both the dielectric constant and the electrical conductivity of the composites increased with the increase of YIG content. These phenomena could be interpreted from interfacial polarization inside the composites in applied alternating field [23]. The dielectric constant of the composites decreases with increasing the frequency of applied alternating filed. The reason is higher the frequency field, the shorter time the YIG particles have to polarize, and then the smaller the dielectric constant of the composite performs. Meanwhile, the conductivity of the composites increases with lifting the frequency. According to the percolation theory [24], this property is the result of the direct proportion of YIG particles with each other. The variations in the dielectric constant, dielectric loss and the electrical conductivity of the PMMA/YIG composites with temperature are shown in Fig. 7. Similar to the most case for other polymer/ceramic composites, which indicated a behavior that dielectric constant increased with temperature [25-27], the PMMA/YIG composites display an increasing tendency in the dielectric constant with increasing temperature. This phenomenon is explained from the occurrence of two kinds of behaviors, which would yield converse effects on dielectric constant of the composite by changing the temperature [28]. First, increasing the temperature would improve the segmental mobility of the polymer, facilitate the polarization of polar fillers and consequently, increase the dielectric constant. Secondly, the obvious differential in thermal expansion coefficient between polymer and filler would change the aggregations of polar components, and thus increase the dielectric constant. PMMA is a kind of polymer with a low glass transition temperature, and both behaviors are the main effects in the present case. Fig. 7c shows the plot of dielectric loss versus temperature at a frequency of 10 kHz, 1 MHz and 2 MHz. It is observed that dielectric loss increases with the increase in temperature. The dielectric loss is due to the perturbation of phonon system by an electric field, the energy transferred to the phonon is dissipated in the form of heat. The growth in tan δ and thus increase in conductivity is brought about by an increase in conduction of residual current and the conduction of absorption current.

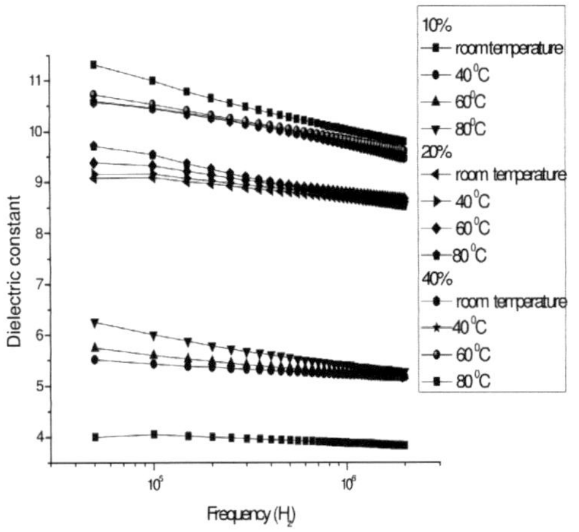

Figure 6(a). Frequency dependence of dielectric constant of the PMMA/YIG composite films on the different concentration of YIG at different temperature.

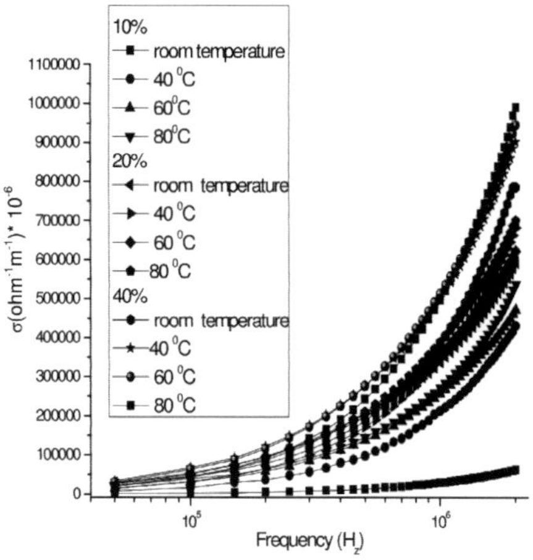

Figure 6(b). Frequency dependence of electrical conductivity of the PMMA/YIG composite films on the different concentration of YIG at different temperature.

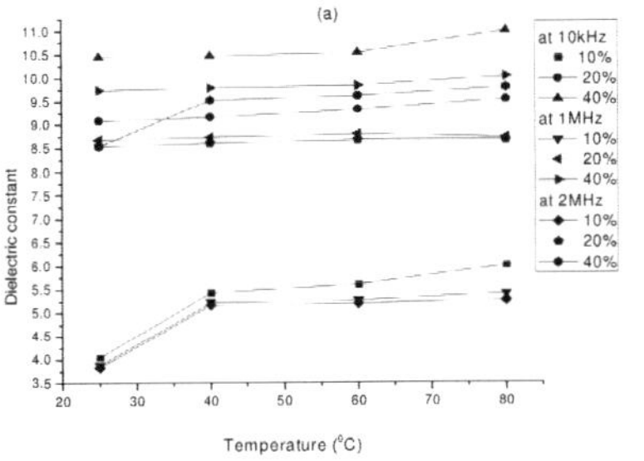

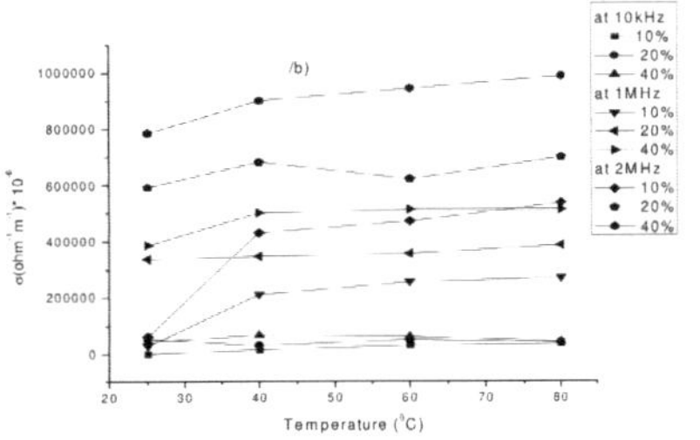

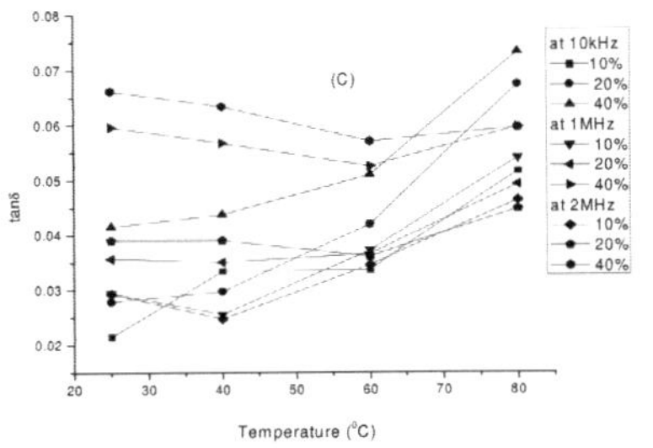

Figure 7. Influence of temperature on the (a) dielectric constant and (b) electrical conductivity (c) dielectric loss of PMMA/YIG composite films containing different contents of YIG (at 10 kHz, 1 MHz and 2 MHz).

4. Conclusion

In this paper, we prepared YIG material by mechanochemical synthesis from oxides of iron and yittrium. Electrical and dielectric properties of PMMA/YIG composites were studied over a wide range of frequencies and temperatures as a function of filler concentration. Both the dielectric constant and the electrical conductivity of the composites increased with the increase of YIG content. These phenomena could be interpreted from interfacial polarization inside the composites in applied alternating field. It was also observed that PMMA/YIG composites display an increasing tendency in dielectric constant and dielectric loss while increasing temperature. It is attributed to improving the segmental mobility of the polymer, facilitating the polarization of the polar fillers and increasing the dielectric constant.

5. Acknowledgements

We would like to give our great thanks to The Scientific and Technological Research Council of Turkey (TUBITAK) for financial support of this investigation under the Project grant number 107M372.

References

[1] Abbas S M, Chandra M, Verma A, Chatterjee R, Goel T C 2006 *Composites part A* **37** 2148.
[2] Peng C H, Hwang C C, Wan J, Tsai J S, Chen S Y 2005 *Mat. Sci. Eng.* B **11** 27.
[3] Abbas S M, Aiyar R, Prakash O, 1998 *Bull. Mater. Sci.* **21** 279.
[4] Bai Y, Cheng Z-Y, Bharti V, Xu H S, Zhang Q M 2000 *Appl. Phys. Lett.* **76** 3804.
[5] Vrejoiu I, Pedarnig J D, Dinescu M, Bauer-Gogonea S, Ba¨uerle D 2002 *Appl. Phys.* A **74** .
[6] Lam K H, Chan H L W, Luo H S, Yin Q R, Yin Z W, Choy C L 2003 *Microelectron Eng.* **66** 792.
[7] Kuo D-H, Chang C-C, Su T-Y, Wang W-K, Lin, B-Y 2001 *J. Eur. Ceram. Soc.* **21** 1171.
[8] Liou J W, Liou B S 1998 *J. Phys., Condens. Matter* **10** 2773.
[9] Bhattacharya S K, Tummala R R 2000 *J. Mater. Sci., Mater. Electron.* **11** 253.
[10] Stru¨mpler R, Glatz-Reichenbach J 1999 *J Electroceram* **3(4)** 329.
[11] Scho¨nhals A 1997 *Dielectric properties of amorphous polymers. In: Runt JP, Fitzgerald S, editors. Dielectric spectroscopy of polymeric materials* (Washington DC: American Chemical Society) p 81.
[12] Zois H, Apekis L, Omastova M 2001 *Macromol. Symp.***170** 249.
[13] Brosseau C, Boulic F, Queffelec P, Bourbigot C, Mest Y Le, Loaec J 1997 *J. Appl. Phys.* **81** 882.
[14] Po¨ tschkea P, Dudkinb S M, Aligb I 2003 *Polymer* **44** 5023.
[15] Zois H,. Mamunya Y P, Apekis L 2003 *Macromol. Symp.* **198** 461.
[16] Dang Z-M,. Lin Y-H, Nan C-W 2003 *Adv. Mater.* **15** 1625.
[17] Xiao M, Sun L, Liu J, Li Y, Gong K 2002 *Polymer* **43** 2245.
[18] Jeon Y H, Lee J W, Oh J H , Lee J C, Choi S C 2004 *Phys. Status Solidi A* **201 (8)** 1893.
[19] Kim T Y, Yamazaki Y, Hirano T 2004 *Phys. Status Solidi* B **241 (7)** 1601.
[20] Tsay C Y, Lin C Y, Liu K S, Lin I N, Hu L J, Yeh T S 2002 *J. Magn. Magn. Mater.* **239** 490.
[21] Ubizskii S B, Matkowskii A O, Mironova-Ulmane N, et al 2000 *Phys. Stat. Sol.* **A 177** 349.
[22] Ghasemi A, Hossienpour A, Morisako A, Saatchi A, Salehi M 2006 *J. Magn. Magn. Mater.* **302** 429.
[23] Van Beek L K H 1967 *Progress in Dielectric* (Heywood).
[24] Huang C, Zhang Q 2004 *Adv. Funct. Mater.* **14** 501.
[25] Dang Z-M, Wu J-B, Fan L-Z, Nan C-W 2003 *Chem. Phys. Lett.* **376** 389.
[26] Ardi M S, Dick W 1995 *Plast. Rubber Compos. Process. Appl.* **24** 157.
[27] Chiang C K, Popielarz R 2002 *Ferroelectrics* **275** 1.
[28] Berger M A, McCullough R L 1985 *Compos. Sci. Technol.* **22** 81.

Optically induced Hall effect in semiconductors

M. Idrish Miah[1,2] and E. MacA. Gray[1]

[1]Nanoscale Science and Technology Centre, Griffith University, Nathan, Brisbane, QLD 4111, Australia. [2]Department of Physics, University of Chittagong, Chittagong 4331, Bangladesh.

e-mail: m.miah@griffith.edu.au

Abstract: We describe an experiment which investigates the effect of a longitudinal electric field on the spin-polarized carriers generated by a circularly polarized light in semiconductors. Our experiment observes the effect as a Hall voltage resulting from nonequilibrium magnetization induced by the spin-carrier electrons accumulating at the transverse boundaries of the sample as a result of asymmetries in scattering for spin-up and spin-down electrons in the presence of spin-orbit interaction. It is found that the effect depends on the longitudinal electric field and doping density as well as on temperature. The results are presented by discussing the dominant spin relaxation mechanisms in semiconductors.

1. Introduction

The anomalous Hall effect (AHE) [1] and the most recently optically observed (predicted a long time ago [2]) spin Hall effect (SHE) [3] are the two types of spin-dependent Hall effects. The proper understanding of the spin-dependent Hall effects, that is of that part of the transverse resistance which depends on the magnetization, is still a challenging problem. However, it has been generally recognized that the spin-orbit (SO) interaction is responsible for the effects, and due to this interaction, electrons with their spin polarization parallel to the magnetization axis will be deflected at right angles to the directions of the longitudinal electric current and of the magnetization while electrons with antiparallel spin polarization will be deflected in the opposite direction via the skew-scattering (SS) [4] and side-jump (SJ) [5]. Thus, if the two spin species or spin populations (spin-up and spin-down) are unequal, there appears a net current in the transverse direction.

In the ordinary Hall effect (HE), the Lorentz force (F) causes a charge imbalance by deflecting like-charge carriers towards one boundary of the sample, creating a voltage transverse to the applied electric and magnetic fields (Fig. 1(a)). The AHE is the result of spin-dependent deflection of carrier (spin-polarized) motion, which produces a Hall voltage proportional to the spin polarization, or the magnetization, and spin accumulation at the transverse boundaries. In the AHE, the SO interaction plays the role of the force, so-called SO coupling force (F_{SO}), that deflects like-spin carriers to one boundary and opposite spins to the other boundary, which causes a spin imbalance as well as a charge imbalance and results in a Hall voltage and a spin accumulation (Fig. 1 (b)). In the SHE, the F_{SO} deflects like-spin carriers in the presence of the magnetization, which causes only the spin imbalance [because the spin-up ($\vec{j}_\uparrow$) and spin-down ($\vec{j}_\downarrow$) currents contain an equal number of charge carriers, i.e. $\vec{j}_\uparrow = \vec{j}_\downarrow$] and results in no Hall voltage but spin accumulation on the transverse boundaries (Fig.

1(c)). The difficulties associated with the absence of the Hall voltage in SHE make it a challenging problem for the electrical detection which is very desirable for spintronic device applications.

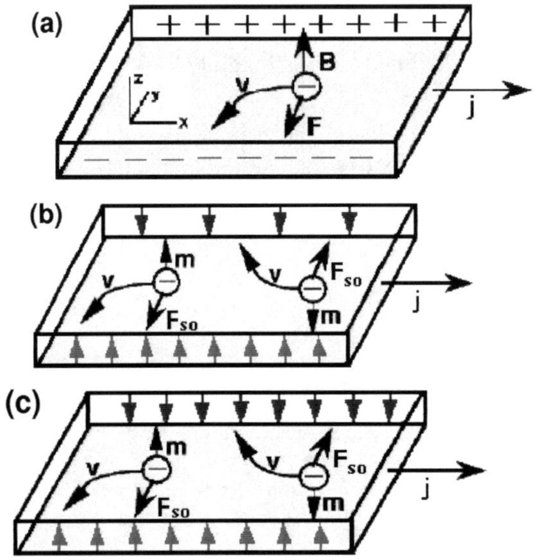

Fig. 1. (a) Hall effect in the presence of a magnetic field, **B** (Hall voltage but no spin accumulation), (b) anomalous Hall effect (Hall voltage as well as spin accumulation) and (c) spin Hall effect (no Hall voltage but spin accumulation). In the anomalous Hall effect and spin Hall effect, carriers with same charge but opposite spin are deflected (in the presence of the magnetization, **m**) by the spin-orbit coupling to opposite sides.

Although AHE in magnetic materials with the application of magnetic fields has been observed for a long time since its discovery in 1880 by Hall [1], AHE has attracted recently much attention because of its close theoretical connection to SHE and because of the rapidly growing interest in spintronics, particularly, spin-dependent transport phenomena, or spin transport in general [6]. In addition, AHE has been used in demonstrating the existence of a ferromagnetic state in several semimagnetic or diluted magnetic semiconductors while searching for materials for possible applications in spintronics.

While the HE results from the Lorentz force, AHE is due to the SO coupling in the presence of spin polarization. In a ferromagnetic material, the presence of the spontaneous magnetization leads to AHE. For a nonmagnetic semiconductor, there is no spontaneous magnetization, so one needs to apply a strong magnetic field to polarize the carriers. However, if one can generate (or inject into) a spin-polarized current ($\vec{j}_{\uparrow} \neq \vec{j}_{\downarrow}$) in a semiconductor such that it can be sustained on the required length or time scale, one could observe the AHE. In the presence of the generated or injected spin-polarised currents, the nonmagnetic semiconductor could behave more like a ferromagnet as far as, or as long as, the transport is concerned. Furthermore, since AHE often persists above the ferromagnetic Curie temperature, an analogous AHE should be observable in a paramagnetic system in which the charge carriers are spin polarized by an external magnetic field or by other methods, for instance, optically since circularly polarized light breaks time-reversal symmetry and could lead to AHE [7].

In this work, we experimentally observe a photo-induced Hall-type charge and spin transport in *n*-doped nonmagnetic semiconductor GaAs. Our paper describes an experiment which electrically investigates the effect, as a Hall voltage, of a longitudinal electric field on the spin-polarized carriers induced by a circularly polarized light [8]. Since the photo-generated hole spins relaxation is extremely fast [9], the experiment observes only the effect resulting from spin-polarized electrons (not holes) accumulating at the transverse boundaries of the sample as a result of left-right asymmetries in scattering for spin-up and spin-down electrons in the presence of SO interaction. It is found that the effect depends on the longitudinal electric field and doping density as well as on temperature. The results obtained from the experiment are discussed in details.

2. Device fabrication and measurements

Investigated samples were fabricated on *n*-doped (with various doping densities) GaAs. Prior to contact deposition, the substrates were cleaned using conventional organic solvents. The native surface oxide was then removed using $HCl:H_2O$ (1:1 vol.) followed by a de-ionized water rinse and blown dry with nitrogen before loading the substrates in the evaporation chamber. Au/Ge/Pd contacts were deposited on the substrates, with Pd layers adjacent to the GaAs substrates, using an e-beam evaporator with a base pressure of $\sim 5\text{x }10^{-8}$ Torr . The contact metallization was annealed in a tube furnace in flowing nitrogen at temperature 180 ^{0}C for 1 h to achieve reliable or transparent contacts with low contact resistance. The specific contact resistivity was assessed using transmission line model and was found to be $\sim 10^{-6}\ \Omega\ cm^2$, which confirms that the device contains ohmic conatcts. Gold wires were bonded from the sample-holder to the contact pads. Experiments were performed at various sample temperatures from low (3 K) to room temperature (300 K) by placing the samples in a cryostat. For optical excitation, a mode-locked Ti:sapphire laser which generates ps pulses at 76 MHz repetition rate was used. The excitation photon energy was tuned at $\lambda \approx 800$ nm, i.e. slightly above the energy bandgap of GaAs. A neutral density wheel was used to vary the optical power level. The average power on the sample was ~5 mW. The polarization of the optical beam was modulated using a photo-elastic modulator at a lock-in reference frequency of 42 kHz. The laser beam was focused to a ~90 μm (FWHM) spot of the sample with a lens. Care was taken not to illuminate any of the electrical contacts to avoid the generation of any artefacts. For this purpose (i.e. to check that the beam hits the sample on the desired location), a microscope was used. The lens was designed for minimum spherical aberrations. The spot size was measured by knife-edge scans and the spatial (or the sequence focus) profile of the pulse was found to be Gaussian. The laser beam was found to be stable at the power level we used. A regulated electric power supply was used as a bias source. The signal was measured by a lock-in amplifier coupled to a computer. A schema of the experimental set-up is shown in Fig. 2.

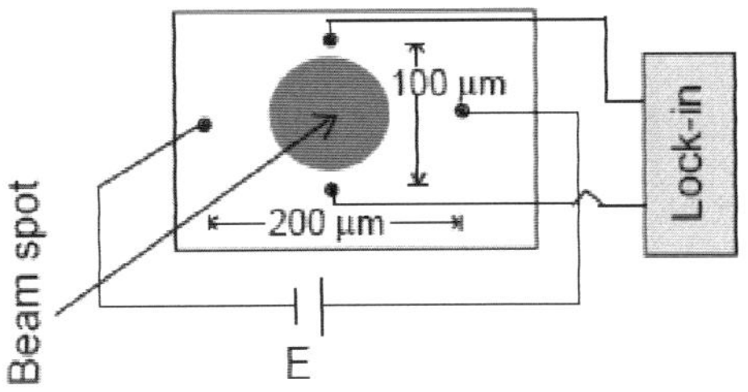

Fig. 2. A schema of the experimental set-up along with an illustration of the geometry of the sample.

3. Results and discussion

We observed a Hall-type effect in a nonmagnetic semiconductor without the application of an external magnetic field, but in the presence of a circularly polarized light. The circularly polarized light created a spin-polarized longitudinal current. If the longitudinal current in a nonmagnetic semiconductor is spin-polarized, e.g. current carriers contain more spin-up electrons than spin-down electrons, there would be more electrons scattered to the right than to the left via SS and SJ in the presence of the SO interaction. This leads to both spin and charge accumulations in the transverse direction of the sample. When the photo-induced spin-polarized electrons are dragged by an external bias in a sample, a light-induced Hall voltage (V_{LH}) proportional to the net spins or spin-polarized current is observed. The V_{LH} is a measure of the net charge accumulation on the transverse boundaries of the sample for the generated spin-polarized electrons. The origin of the V_{LH} is similar to the AHE, or the extraordinary Hall effect, in ferromagnetic materials in the presence of magnetic fields.

The results obtained from the experiments are presented in Figs. 3-5. In Fig. 3, we plot V_{LH} as a function E for a sample at sample temperature T=300 K, where an average of the data taken from several experiments for a field strength was calculated. As can be seen, V_{LH} increases quickly to a maximum value and remains almost constant, and then suppresses almost exponentially. Above 150 mV μm^{-1}, V_{LH} was found to be decayed, and it disappeared very quickly after 180 mV μm^{-1}. The decay of V_{LH} with E might be due to the enhanced electron spin relaxation at higher current densities.

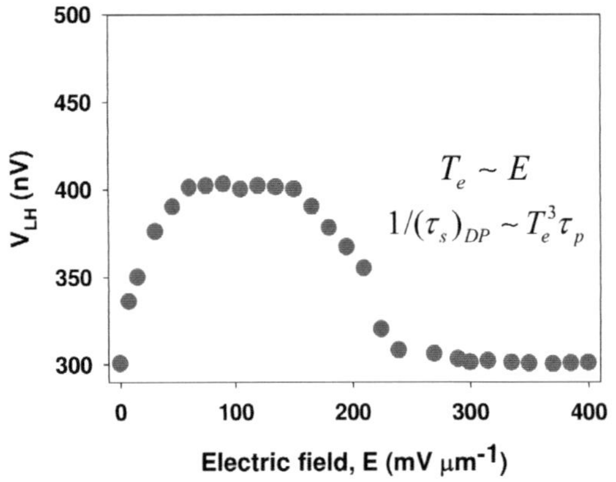

Fig. 3. Field dependence of V_{LH} for a sample with n=1 x10^{22} m^{-3} at sample temperature T=300 K. An average of the data taken from several experiments for a field-strength was calculated. The resulting high temperature (T_e) of the field-drifting electrons together with the momentum relaxation (τ_p) leads to enhanced DP spin relaxation ($1/(\tau_s)_{DP} \sim T_e^3 \tau_p$).

As seen from Fig. 3, V_{LH} and thus polarization, or spin lifetimes (τ_s), is preserved during transport in *n*-doped GaAs in moderate electric fields, which is very important for implementing spin-sensitive or spintronic new device ideas. Although the spin for moderate electric field is preserved, the spin relaxation rate becomes considerably larger for higher fields. When the electrons were injected with a high E, a significant reduction in the spin relaxation is observed. Increasing E leads to larger charge and spin accumulations near sample boundaries, but polarization saturates because of shorter τ_s for larger E. The suppression of τ_s with increasing E from saturation implies that spin decay increases with

E is consistent with other observations [10,11]. The decay of polarization was found to be exponential (Fig. 3) and the fields (for all the samples) at which spin relaxation or depolarization occurs was found to be within a moderate range. A similar feature can be seen in a theoretical investigation based on a spin drift-diffusion model [12], where it was shown that a significant increase in the spin current could not be achieved by increasing the field owing to the competing effect of field on diffusion because the spin drift and spin diffusion currents contribute additively to the spin current and because the spin diffusion current decreases with the field while the spin drift current increases.

As found, the photo-generated spin can travel without losing their initial orientation as long as their field is below a threshold value. Above the threshold, spin depolarization is considerably enhanced due to the enhanced spin relaxation as a result of increase of the electron temperature (T_e). Our results agree with those of the field-dependent spin transport experiment by D. Hägele et al. [13]. These authors studied spin relaxation of photo-generated electrons during drift transport in GaAs at low temperatures by photoluminescence measurements. They found almost complete conservation of the orientation of the electron spin during transport in GaAs over a distance as long as 4 μm, but fields up to a couple of hundreds mV μm^{-1}.

The results obtained by us are also in good agreement with those of the spin transport experiments [14], similar to one employed in Ref. 13. They showed that the photo-generated spins could travel without losing their initial spin orientation as long as E was below 100 mV μm^{-1} and the spin relaxation rate increased rapidly with E, and polarization disappeared at ~350 mV μm^{-1}. It is noted here that the photo-generated spins in their study were generated with a higher optical power.

The origin of the field-dependent efficient electron spin relaxation is discussed based on the D'yakonov–Perel (DP) spin relaxation mechanism [2]. The DP mechanism is due to SO coupling in semiconductors lacking inversion symmetry. In III-V semiconductors, such as GaAs, the degeneracy in the conduction band is lifted for $\vec{k} \neq 0$ due to the absence of inversion symmetry. Without inversion symmetry the momentum states of spin-up (↑) and spin-down (↓) electrons are not degenerate, i.e. $E_{\vec{k}\uparrow} \neq E_{\vec{k}\downarrow}$, where $E_{\vec{k}\uparrow}(E_{\vec{k}\downarrow})$ is the momentum-dependent electron energy with spin ↑(↓). The resulting energy difference, for electrons with the same $\vec{k}$ but different spin states, plays the role of an intrinsic $\vec{k}$-dependent magnetic field [15], $\vec{h}(\vec{k}) = \alpha\hbar^2(2emE_g)^{-1/2}[k_x(k_y^2 - k_z^2)\hat{x} + c.p.]$, where α is a dimensionless, material-specific parameter which gives the magnitude of the SO splitting and is approximately given by $\alpha \simeq 4\eta(m/m_{cv})/\sqrt{3-\eta}$ (where $\eta = \Delta/(E_g + \Delta)$, E_g is the bandgap, Δ is the SO splitting of the valence band, m is the electron's effective mass and m_{cv} is a constant close in magnitude to free electron mass m_0) induced by the presence of the Dresselhaus [16] SO interaction in a zinc-blende structure, acting on the spin with its magnitude and orientation depending on $\vec{k}$ and results in spin precession (spin relaxation) with intrinsic Larmor frequency $\Omega_s(\vec{k}) = (e/m)\vec{h}(\vec{k})$ during the time between collisions.

The corresponding Hamiltonian term (DP Hamiltonian) describing the precession of electrons in the conduction band is $H_{SO}(\vec{k}) = (\hbar/2)\vec{\sigma}\bullet\Omega_s(\vec{k})$, where $\vec{\sigma}$ is the vector of Pauli spin matrices. In a quantum well, for example, both the bulk and structural inversion asymmetries are present. So the DP term is composed of the Dresselhaus [16] and Rashba [17] terms.

The increased electron momentum at higher electric fields brings about a stronger $\vec{h}(\vec{k})$ and consequently, the electron precession frequency $\Omega_s(\vec{k})$ becomes higher. The intrinsic magnetic field depends on the underlying material, on the geometry of the device, and on $\vec{k}$. Momentum-dependent spin procession described by the DP Hamiltonian, together with momentum scattering characterized by momentum relaxation time ($\tau_p(E_{\vec{k}})$) leads to the DP spin relaxation. Since the magnitude and the direction of $\vec{k}$ changes in an uncontrolled way due to electron scattering with the environment, this process contributes to (DP) spin relaxation given by $1/(\tau_s)_{DP} = \gamma(\alpha/\hbar)^2 E_{\vec{k}}^3/(E_g)\tau_p(E_{\vec{k}})$ [15], where $E_{\vec{k}} = k_B T_e$ and γ is a dimensionless factor that ranges from 0.8 to 2.7 depending on the dominant momentum relaxation process.

During transport in the electric field, electrons are accelerated to higher velocities at higher fields, where T_e increases sharply due to the energy-independent nature of the dominant energy relaxation process via the longitudinal polar optical phonon scattering [15]. The resulting high T_e leads to enhanced DP spin relaxation ($1/(\tau_s)_{DP} \sim T_e^3 \tau_p$) because they have large kinetic energy between successive collisions.

The V_{LH} decreases rapidly with E, which is consistent with the results of the Monte Carlo simulation performed by Barry et al. [18]. They have shown that for relatively low fields up to 100 mV μm^{-1} a substantial amount of spin polarization is preserved for several microns at room temperature and the DP spin relaxation rate increased rapidly for fields higher than 150 mV μm^{-1} and became infinite for E higher than 250 mV μm^{-1}. For GaAs, the Elliot-Yafet spin relaxation is less effective due to the large E_g and low scattering rate [6]. For *n*-doped materials, as holes are rapidly recombined with electrons due to the presence of a large number of electrons, spin relaxation due to the Bir-Aronov-Pikus mechanism is usually blocked [6]. The spin relaxation in *n*-doped GaAs is therefore dominated by the DP mechanism.

Fig. 4 shows the doping-density dependence of the signal, where V_{LH} is plotted as a function of n. As seen, an enhancement of V_{LH} with increasing doping is observed. The introduction of *n*-type dopants in semiconductors increases τ_s, because the electronic spin polarization in these systems survives for longer times. Studies of spin precession in *n*-GaAs reveal that moderately *n*-doping yields significantly extended τ_s [11,19].

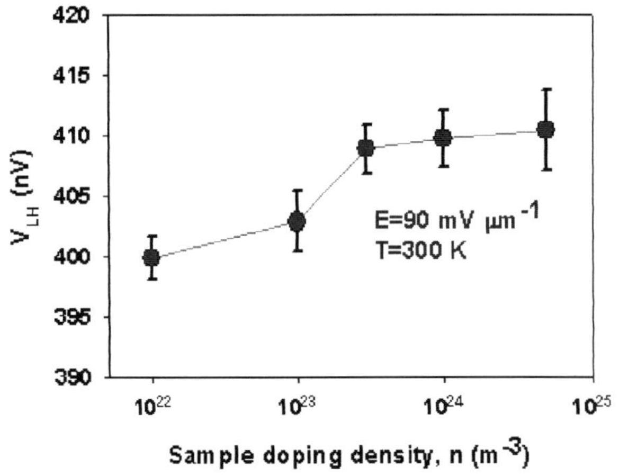

Fig. 4. Doping-density dependence of V_{LH}.

As known from the literature [6], the spin-dependent Hall effects originate from a variety of mechanisms and consists of the contribution due to an asymmetric SS of charge carriers, the contribution due to a SJ that the charge carriers undergo at each scattering event and the contribution not related to scattering (but arising as a result of the band structure), i.e. the intrinsic contribution [7]. As they originate from the same mechanisms, the same physics underlies both AHE and SHE. For the less-doped sample, the spin-dependent Hall conductivity is low because of the higher compensation ratio [20], and the SJ contribution dominates the effect. Although the SJ contribution is independent of τ_s, SS contribution will grow with τ_s, resulting in a higher contribution to the effect for higher n [21], but would be cut off when τ_s becomes comparable to $\triangle^{-1}$ at the Fermi level due to the k^3-Dresselhaus term in the DP expression. For the higher-doped samples, SS increases with n, agreeing with the theoretical predictions performed in Refs. [22,23].

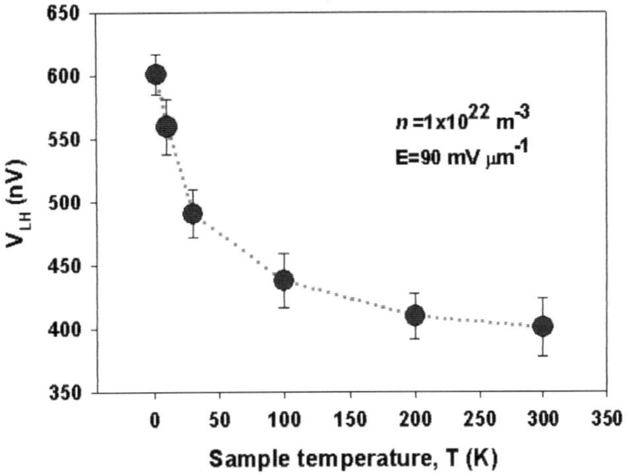

Fig. 5. Temperature dependence of V_{LH} for a sample.

The dependence of the Hall signal on temperature is shown in Fig. 5. As can be seen, V_{LH} increases with decreasing temperature, in consistence with the studies by others [14]. These authors studied spin relaxation of photo-generated electrons during transport in GaAs at low temperatures by the time-resolved and photoluminescence polarization measurements. They showed that the polarization under drift increased with decreasing temperature. The result is also consistent with the time-resolved Faraday rotation experimental findings for the τ_s measurements [10,11].

4. Conclusions

We experimentally studied the circularly polarized light-induced Hall effect in n-doped GaAs at temperatures ranging from 3 K to 300 K. Our experiment investigated the effect of a longitudinal electric field on the optically spin-polarized electrons accumulating at the transverse boundaries of the sample via the measurements of V_{LH}. It was found that the effect depends on the longitudinal electric field and doping density as well as on temperature. The results obtained from the investigation were discussed based on the dominant spin relaxation mechanisms for the semiconductor.

References

[1] Hall E. H, Philos. Mag. **12**, 157 (1881).

[2] D'yakonov M. I, Perel V. I, Zh. Eksp. Teor. Fiz. **60**, 1954 (1971) [Sov. Phys. JETP **33**, 1053 (1971)].
[3] Kato Y, Myers R. C, Gossard A. C, Awschalom D. D, Science **306**, 1910 (2004).
[4] Berger I, Phys. Rev. B **2,** 4559 (1970).
[5] Mott N. F, Massey H. S. W, The Theory of Atomic Collisions (Clarendon Press, 3rd ed., Oxford, 1965).
[6] Žutić I, Fabian J, Sarma S. D, Rev. Mod. Phys. **76**, 323 (2004).
[7] Karplus R, Luttinger J. M, Phys. Rev. **95**, 1154 (1954).
[8] Miah, M. I, Mater. Lett. **60**, 2863 (2006).
[9] Pierce D. T, Meier F, Phys. Rev. B **13**, 5484 (1976).
[10] Kikkawa J. M, Awschalom D. D, Phys. Rev. Lett. **80**, 4313 (1998).
[11] Semiconductor Spintronics and Quantum Computation, Awschalom D. D, Loss, Samarth N, eds. (Springer, Berlin, 2002).
[12] Miah M. I, J. Appl. Phys. **103**, 063718 (2008).
[13] Hägele D, Oestreich M, Rühle W. W, Nestle N, Eberl K, Appl. Phys. Lett. **73**, 1580 (1998).
[14] Sanada H, Arata I, Ohno Y, Chen Z, Kayanuma K, Oka Y, Matsukura F, Ohno H, Appl. Phys. Lett. **81**, 2788 (2002).
[15] Pikus G. E, Titkov A. N, in Optical Orientation, Modern Problems in Condensed Matter Science (Meier F, Zakharchenya B. P, eds.), Vol. **8** (North-Holland, Amsterdam, 1984).
[16] Dresselhaus G, Phys. Rev. **100**, 580 (1955).
[17] Bychkov Y. A, Rashba E. I, J. Phys. C: Solid State Phys. **17**, 6039 (1984).
[18] Barry E. A, Kiselev A. A, Kim K. W, Appl. Phys. Lett. **82**, 3686 (2003).
[19] Miah M. I, Appl. Phys. Lett. **103**, 092104 (2008).
[20] Blakemore J. S, J. Appl. Phys. **53**, R123 (1982).
[21] Chazalviel J.-N, Phys. Rev. B **11**, 3918 (1975).
[22] Engel H.- A, Halperin B. I, Rashba E. I, Phys. Rev. Lett. **95**, 166605 (2005).
[23] Tse W.-K, Sarma D. S, Phys. Rev. Lett. **96**, 056601 (2006).

Observation of magnetocaloric effect in the $LaMn_{1.9}Fe_{0.1}Si_2$ compound at low fields in the vicinity of phase transition around room temperature

Onur Nane[(a)], Hasan Ağıl[(b)], Berdan Özkurt[(c)], Bekir Özçelik[(a)]*, Ali Gencer[(b)]

[(a)] *Department of Physics, Faculty of Sciences and Letters, Cukurova University 01330 Adana, Turkey*
[(b)] *Department of Physics, Faculty of Sciences, Ankara University, Turkey*
[(c)] *Department Electronic and Computer Education,* Tarsus *Technical Education Faculty, Mersin University, İçel, Turkey*

e-mail: ozcelik@cu.edu.tr

Abstract:Low field magnetic and magnetocaloric properties of the compound $LaMn_{1.9}Fe_{0.1}Si_2$ have been investigated. Using the M-H data measured at various temperatures, we have derived the magnetic entropy change $|\Delta S|$ as a function of temperature and magnetic field, by using the Maxwell relation. The temperature dependence of the Landau coefficient has been derived by fitting the magnetization, using the Landau expansion of magnetic free energy and Arrott plots. The results show that $LaMn_{1.9}Fe_{0.1}Si_2$ is ferromagnetically ordered around room temperature (285 K) and the nature of the phase transition at this temperature is found to be second order.

1.Introduction

The magnetic properties and structures of the ternary rare-earth compounds RT_2X_2, where R is the rare earth metal, T is the transition metal and X is Si or Ge, with tetragonal $ThCr_2Si_2$-type structure (space group I4/mmm) have recently been studied extensively [1-6]. In early studies, the $LaMn_2Si_2$ compound was considered as a simple ferromagnet with a Curie temperature Tc=305 K [1,2]. Based on the neutron diffraction and Mössbauer studies, it was reported that $LaMn_2Si_2$ has a canted ferromagnetic structure (F_{mc}) at temperatures $T<Tc$ and in-plane antiferromagnetic structure at $T>Tc$ up to T_N=470 K [3,4]. In addition to the canted ferromagnetic structure, the conical magnetic arrangement (F_{mi}) was observed around $T_h \approx 45$ K. In the last decade, the magnetic properties of the $R(1)_{1-x}R(2)_xMn_2X_2$ systems were extensively studied [7-11]. The substitution of Fe or Co for Mn on $LaMn_{2-x}T_xSi_2$ systems results in a decrease of the value of d^a_{Mn-Mn} due to the size effects. This substitution should induce the transition from ferromagnetic to antiferromagnetic alignment in adjacent Mn layers [6]. The X-ray diffraction patterns showed that $LaMn_{2-x}Fe_xSi_2$ compounds are single phase and the lattice parameters a, c and unit cell volume V of $LaMn_{1.9}Fe_{0.1}Si_2$ were found to be 4.1088 Å, 10.6034 Å and 179 $Å^3$, respectively [5]. The nature of the magnetic coupling in these compounds within and between the Mn layers is closely related to the in-plane Mn-Mn spacing d^a_{Mn-Mn}. Roughly, if d^a_{Mn-Mn} >2.87 Å, the intra-layer in-plane coupling is antiferromagnetic and interlayer coupling is ferromagnetic (F_{mi} or F_{mc}). If 2.84Å< d^a_{Mn-Mn} <2.87Å, both the intra-layer in-plane coupling and the interlayer coupling are antiferromagnetic (AF_{mi} or AF_{mc}). In the case when d^a_{Mn-Mn} <2.84 Å, there is no intra-layer in-plane spin component, and the interlayer coupling remains antiferromagnetic (AF_{il}) [5].

Since the discovery of the giant magnetocaloric effect (MCE) in $Gd_5Si_2Ge_2$ ($\Delta S = 18.5 J/kg-K$ under 5 T field at Tc=276 K) [12], the number of studies in magnetocaloric materials that can be used around room temperature has increased considerably. The MCE is defined as the heating or cooling (i.e. the temperature change) of a magnetic material due to the application of a magnetic field. This effect has been called adiabatic demagnetization for years, though this phenomenon is just one practical application of the MCE in magnetic materials [13,14]. The magnetocaloric effect was discovered in 1881, during an experiment performed on iron [15]. The origin of the MCE was explained independently by Debye [16] and Giauque [17]. Nowadays, there is a great deal of interest in using the MCE as an alternative technology for refrigeration, from room temperature to the temperatures of hydrogen and helium liquefaction (~4.2-20 K). An ideal magnetic refrigerant should work over a wide range of temperatures (e.g 220-330 K). The magnetic refrigeration offers the prospect of an energy-efficient and environmentally friendly alternative to the common vapor-cycle refrigeration technology in use today [18,19].

In this paper, we have investigated the low field magnetic and magnetocaloric properties of $LaMn_{1.9}Fe_{0.1}Si_2$ with the expectation that this will exhibit MCE. The Curie temperature is measured as 285 K which is very close to the room temperature range consistent with the values reported by other workers [5,6]. If high Curie temperature and magnetic behavior in the low field are considered the $LaMn_{1.9}Fe_{0.1}Si_2$ compound appears to show the MCE with second order phase transition, thus being an appropriate candidate for magnetic refrigerators.

2.Experimental procedure

The preparation method of the compound was given in detail elsewhere [5]. The magnetic properties were measured by a Closed Cycle Vibrating Sample Magnetometer (Model-7304, Lake Shore).

3.Results and Discussion

The zero field cooled (ZFC) magnetization of $LaMn_{1.9}Fe_{0.1}Si_2$ with respect to temperature is given in Fig.1, for different applied magnetic fields. As can be seen, the magnetization increases rapidly from room temperature up to 270 K, then goes to slightly so-called saturation values for the individual field values. On the other hand, the decrease around 90 K for the field value of 200 Oe is probably due to a reorientation of a sub-lattice magnetization, and the conical ferromagnetic structure (F_{mi}) together with canted ferromagnetic structure (F_{mc}) known to establish below 50 K which is in agreement with the ac susceptibility results [5].

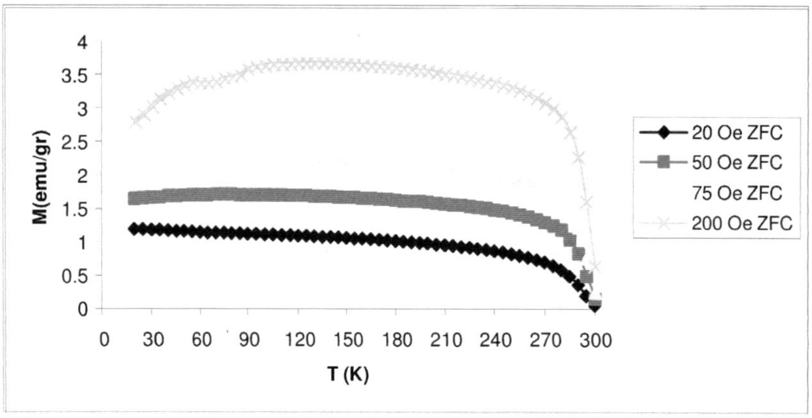

Fig.1. Temperature dependence of magnetization of the $LaMn_{1.9}Fe_{0.1}Si_2$ compound, measured for different field values.

We have shown the zero field cooled (ZFC) and field cooled (FC) magnetization results, measured in a field of 200 Oe, in Fig.2. It is seen that there is a large difference between the FC and ZFC results. In the FC measurement, the sample undergoes a phase change (from an antiferromagnetic to a ferromagnetic regime) during the cooling process with an external magnetic field. This gives a preferred orientation of the ferromagnetic components. On the other hand, in the ZFC process a preferred orientation is not formed and so the magnetization for ZFC becomes smaller than the FC case. The decreasing behavior around 90 K at the ZFC magnetization is not observed in the FC magnetization measurements. This is probably due to pinning of the canted ferromagnetic phase by conical ferromagnetic phase, and means the canted ferromagnetic phase is suppressed by the external magnetic field during FC process.

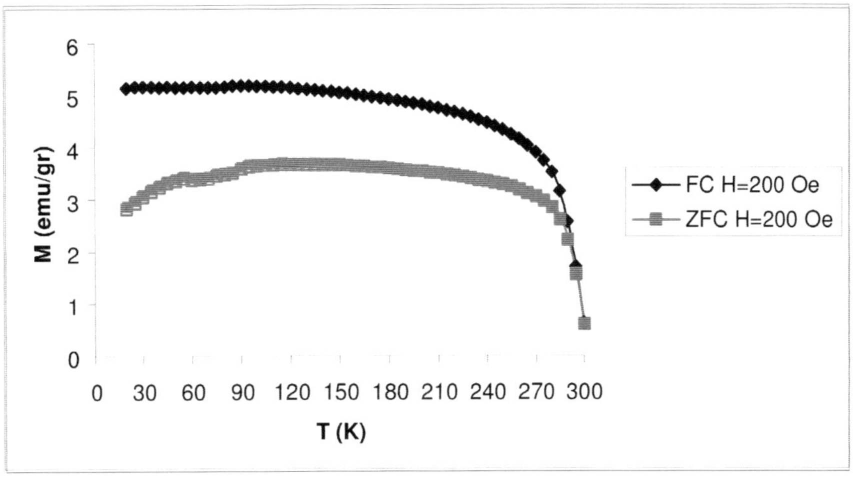

Fig.2. Temperature dependence of magnetization of the $LaMn_{1.9}Fe_{0.1}Si_2$ compound, measured for *H*=200 Oe

For the temperature value of 40 K, the hysteresis loop of $LaMn_{1.9}Fe_{0.1}Si_2$ is given in Fig.3. The variations in the magnetization curves exhibit typical ferromagnetic behavior. The coercivity field and saturation magnetization are measured to be about 540 Oe and 29 emu/g, respectively.

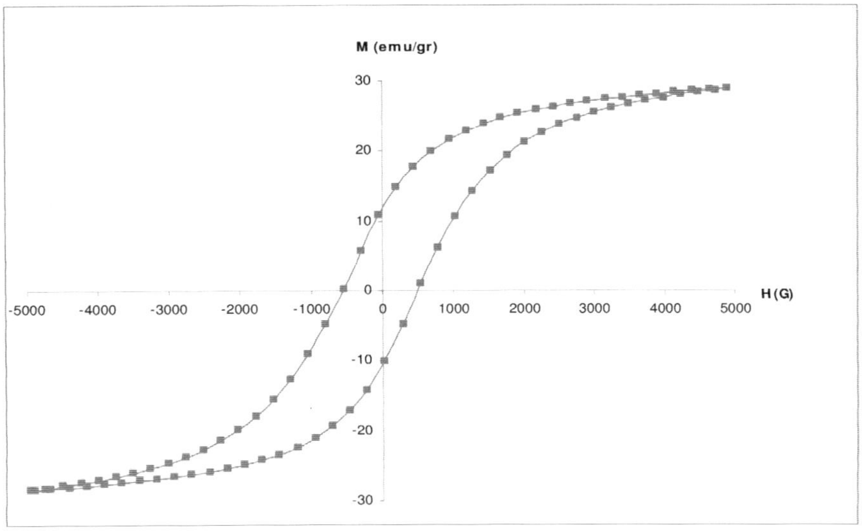

Fig.3. Hysteresis loops of the $LaMn_{1.9}Fe_{0.1}Si_2$ compound, for *T*=40 K

From the *M-H* data, the magnetic entropy change associated with the magnetocaloric effect can be calculated from the thermodynamic Maxwell relation:

$$\left(\frac{\partial S}{\partial H}\right)_T = \left(\frac{\partial M}{\partial T}\right)_H \tag{1}$$

or integrating over the magnetic field

$$\Delta S = \int_0^H \left(\frac{\partial M}{\partial T}\right)_H dH \tag{2}$$

which gives the magnetic entropy change, ΔS. The integral in eq.2 may be computed numerically, and $|\Delta S|$ can be approximated as

$$|\Delta S| = \sum_i \frac{M_i - M_{i+1}}{T_{i+1} - T_i} \Delta H \tag{3}$$

where, M_i and M_{i+1} are the experimental values of the magnetization at the temperatures of T_i and T_{i+1}, respectively, under an applied magnetic field H_i. Fig. 4 shows the magnetization isotherms for $LaMn_{1.9}Fe_{0.1}Si_2$, which were measured around its transition temperature from canted antiferromagnetic state to canted ferromagnetic state. The magnetization behavior is similar to that of typical ferromagnetic materials. It is observed that the magnetization increases rapidly with magnetic field and then almost saturates in a small magnetic field indicating a canted ferromagnetic state. The magnetizations around room temperature (above *Tc*) are much lower than those at lower temperatures (below *Tc*) and increase almost linearly with magnetic field.

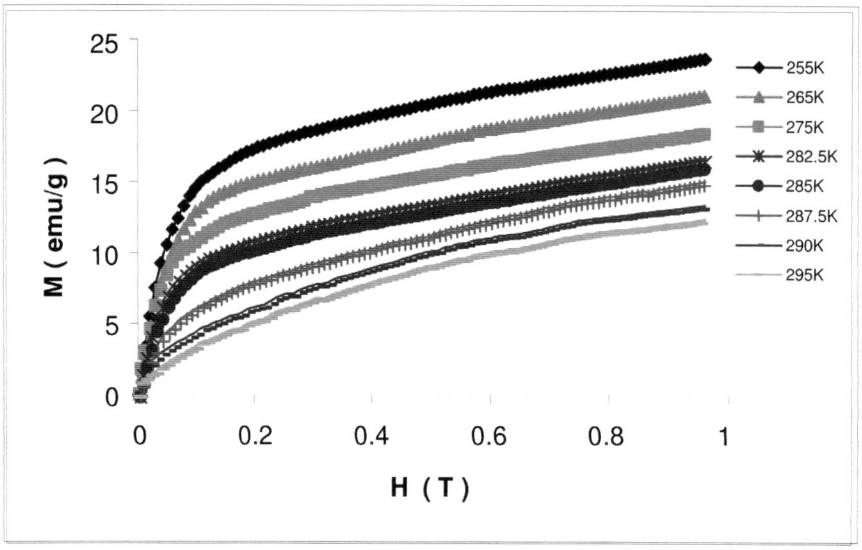

Fig.4. Magnetization isotherm of the $LaMn_{1.9}Fe_{0.1}Si_2$ compound as a function of magnetic field

The magnetic entropy change calculated using eq. 3 as a function of temperature is illustrated in Fig.5. It can be seen that the magnetic entropy change (ΔS) exhibits a single peak around 286 K which is somewhat above *Tc*. It is predominantly negative (positive MCE) and the magnitude increases with applied field. Furthermore, the temperatures of the entropy peaks stay almost constant with increasing field within experimental accuracy. The peak values of $|\Delta S|$, for the fields of 0.2 and 0.9 T are about 0.2 J/kg-K and 0.53 J/kg-K, respectively. The small values of magnetic entropy in $LaMn_{1.9}Fe_{0.1}Si_2$ are probably due to the small value of saturation magnetization and the nature of the complex phase transition. This is in agreement with the results given in literature that the materials having second order magnetic phase transitions exhibit very small magnetocaloric effects [20-22].

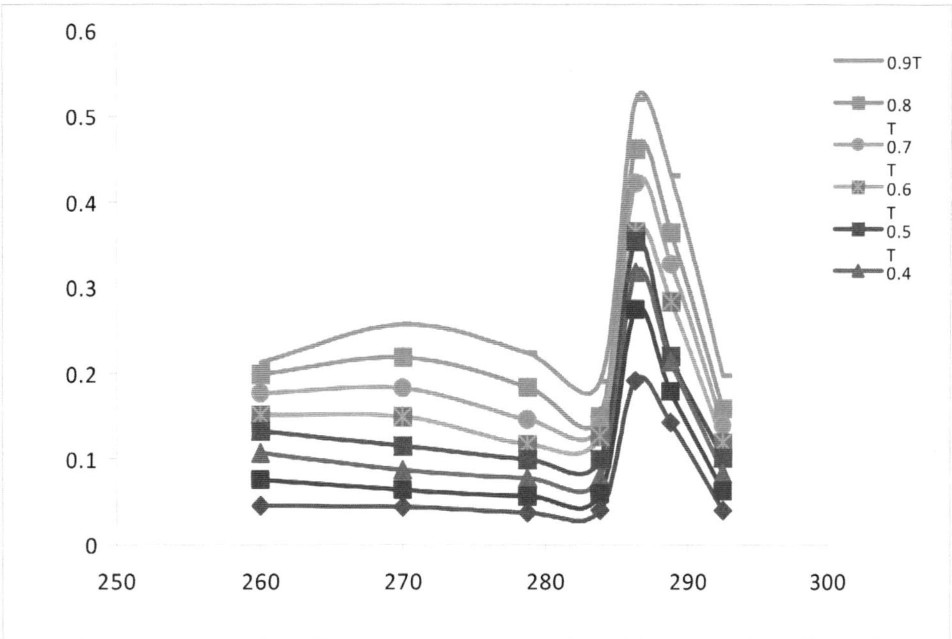

Fig.5. Magnetic entropy change versus temperature at various magnetic fields, the lines are drawn by simply connecting the data points.

The magnetic free energy generally can be expressed in powers of magnetization, neglecting higher order terms as:

$$F(M,T) = \frac{1}{2}c_1(T)M^2 + \frac{1}{4}c_3(T)M^4 - MH \quad (4)$$

where c_1 and c_3 are known as Landau coefficients. The sign of the $c_3(T)$ for the M^4 term allows us to identify the characteristic of the magnetic phase transition. The first order phase transition is expected in the case $c_3(T) < 0$ and second order phase transition in the case $c_3(T) > 0$ [20,23,24].
From the condition of equilibrium $\partial F(M,T)/\partial M = 0$, one can obtain around *Tc*:

$$H = c_1(T)M + c_3(T)M^3 \tag{5}$$

Eq. 5 can be arranged into the form

$$\frac{H}{M} = c_1(T) + c_3(T)M^2 \tag{6}$$

Using eq. 6, one can plot M^2 versus H/M in the critical region which is known as the Arrott plot. The slope of the curves points out whether the nature of the magnetic transition is of first- or second order. It is possible to deduce that if all curves have positive slopes, the magnetic transition is second order. On the other hand, if the curves have negative slopes at some point or inflection points, then the magnetic transition is first order. We have plotted the Arrott plots of $LaMn_{1.9}Fe_{0.1}Si_2$ compound in fig. 6. The Arrott curves have positive slope and no inflection points, in agreement with a positive $c_3(T)$ around the transition temperature, which is indicative of a second order magnetic phase transition. We have calculated the value of $c_3(T)$ approximately as $3\text{x}10^{-4}$ $T.gr^3/emu^3$ at 285 K, which is somewhat smaller than the value mentioned in the literature [20,23,24].We attribute this to our limited experimental condition in which the maximum applied magnetic field is 1 Tesla, and the small value of saturation magnetization is inconsistent with the nature of the magnetic transition.

In conclusion, we have used a known technique [13] to determine MCE. To best of our knowledge, this kind of experiment is the first to be applied to the $LaMn_{1.9}Fe_{0.1}Si_2$ compound which deserves a further investigation in terms of potential applications in refrigerators.

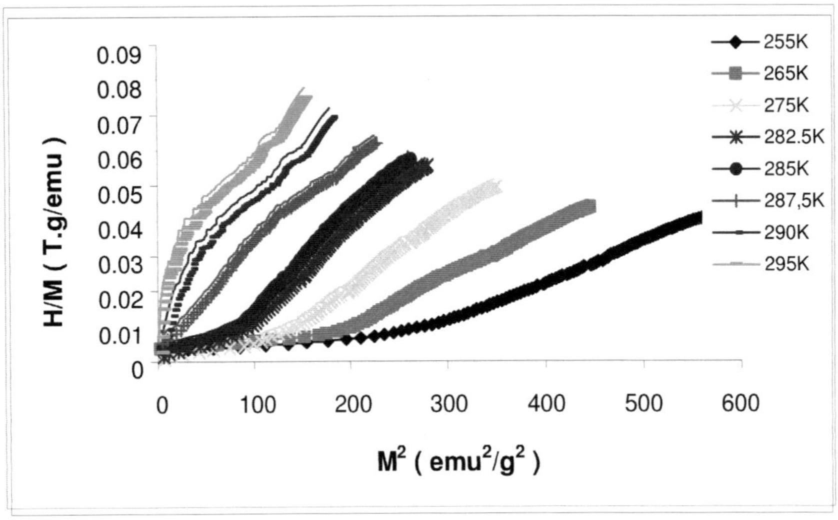

Fig.6. Arrott plots of the $LaMn_{1.9}Fe_{0.1}Si_2$ compound at different temperatures

Acknowledgements,
This work is partially supported under contract no: BOREN 2006-Ç0060 and FEF2008BAP4 (Research Fund of Cukurova University, ADANA)

References

[1] Fujii H, Okamota T, Shigeoka T and Iwata N 1985 *Solid State Comm.* **53 (8)** 715
[2] Szytula A 1991 *Handbook of Mag.Mat.* vol 6,ed K H J Buschow (Amsterdam:Elsevier Science) **85**

[3] Hofmann M, Campbell S J, Kennedy S J and Zhao X L 1997 *J.Magn.Magn.Mater.* **176** 279
[4] Nowik I, Levi Y, Felner I and Bauminger E.R 1995 *J.Magn.Magn.Mate*r. **147** 373
[5] Kervan S, Kılıç A, Özcan Ş and Gencer A 2004 *Solid State Communications* **130** 101
[6] Dincer I., Elmali A, Elerman Y, Ehrenberg H, Fuess H and Daoud-Aladine A 2005 *J. Alloys Compd.* **403** 53
[7] Welter R, Venturini G, Ressouche E and Malaman B 1995 *J.Alloys Compd.* **218** 204
[8] Venturini G, Welter R, Ressouche E and Malaman B 1995 *J.Magn.Magn.Mater* **150** 197
[9] Venturini G, Welter R, Ressouche E and Malaman B 1995 *J.Alloys Compd.* **223** 101
[10] Venturini G, Malaman B and Ressouche E 1996 *J.Alloys Compd.* **241** 135
[11] Ijjaali I, Venturini G, Malaman B and Ressouche E 1998 *J.Alloys Compd.* **266** 61
[12] Pecharsky V K and Gschneidner K A 1997 *Appl.Phys.Lett.* **70** 3299
[13] Tishin A M 1999 *in Handbook of Magnetic Materials* vol 12, ed K H J Buschow (Amsterdam), **395**
[14] Pecharsky V K and Gschneidner K A Jr. 1999 *J. Magn. Magn. Mater.* **200** 44
[15] Warburg E 1881 *Ann. Phys.* **13** 141
[16] Debye P 1926 *Ann. Phys.* **81** 1154
[17] Giauque W F 1927 *J. Amer. Chem. Soc.* **49** 1864
[18] Pecharsky V K and Gschneidner K A Jr. 1999 *J. Appl. Phys.* **85** 5365
[19] Gschneidner K A Jr. and Pecharsky V K 2000*Annu. Rev. Mater. Sci.* 30387
[20] Gencer H, Kervan N, Gencer A, Güneş M and S.Atalay 2007 *J.Alloys Compd.* **466** 1-4
[21] Pecharsky V K and Gschneidner K A Jr. 1999 *J.Magn.Magn.Mater.,* **200** 44
[22] Hou D L, Bai B, Xu J, Tang G D and Nie X F 2004 *J.Alloys Compd.* **384** 62
[23] Liu X B, Ryan D H and Altounian Z 2004 *J.Magn.Magn.Mater.***270** 305
[24] Dunhui W, Shaolong T, Songling H, Jianrong Z and Youwei D 2004 *J.Magn.Magn.Mater.***268** 70

International Conference On Superconductivity and Magnetism (ICSM2008) IOP Publishing
Journal of Physics: Conference Series **153** (2009) 012064 doi:10.1088/1742-6596/153/1/012064

Magnetostriction in orthorhombic manganites

Iliya Radulov, Vassil Lovchinov, Dimitar Dimitrov and Andrey Apostolov

Institute of Solid State Physics "Georgi Nadjakov", Bulgarian Academy of Sciences, 72 Tzarigradsko Shaussee Blvd., 1874 Sofia, Bulgaria

E-mail: `radulov@issp.bas.bg`

Abstract.
From the physical point of view multiferroics present an extremely interesting class of systems and problems. In the present article we report our results from magnetic properties measurements on orthorhombic $ReMn_2O_5$ single crystals (Re = Ho, Tb). The inherent magnetic frustration in these materials is lifted by a small lattice distortion, primarily involving shifts of the Mn^{3+} cations and giving rise to a canted antiferroelectric phase. In both compounds a colossal magnetostriction effect was observed and a novel phase transition diagrams, which indicate the correlation between magnetostriction and polarization for these two compounds were build.

1. Introduction

Multiferroics are single phase compounds in which magnetism and ferroelectricity coexist. Surprisingly, orthorhombic and hexagonal manganites with general formula $ReMnO_3$ and $ReMn_2O_5$ (Re=rare earth) belong also to the group of multiferroics , which are currently of great interest because of the possibility to use them for the design of different kind magneto-electric devices. Among the multiferroic materials displaying the coexistence of magnetic and ferroelectric (FE) orders the rare earth manganites, $ReMn_2O_5$, have attracted attention because of a cascade of phase transitions originating from the wealth of fundamental physical phenomena observed in these compounds. The physical origin of this phase complexity lies in the partially competing interactions between the Mn^{4+}/Mn^{3+} spins, the rare earth magnetic moments, and the lattice. Geometric magnetic frustration among the Mn^{3+} - Mn^{4+} spins leads to a ground state degeneracy of the magnetic states that can be lifted by a distortion of the lattice (magnetic Jahn-Teller effect) [1]. This is believed to be the origin of ferroelectricity in $ReMn_2O_5$ compounds. Strong spin-lattice coupling is needed to explain the ionic displacements in the FE state. From a fundamental point of view, however, these multiferroic manganites contain a puzzle.

2. Samples and experimental

The orthorhombic single crystals were grown by the high temperature solution growth method from pure polycrystalline $ReMn_2O_5$, as described elsewhere [2]. Before investigation all single crystals were characterized and oriented by using EDAX and XDR techniques. The investigated samples are of two types: cubical type with dimensions 1.2 x 1.4 x 1.5 mm^3 and weights 9.8 -

13.5 mg and thin rectangular specimens with typical surface area 3 - 4 mm^2, thickness 0.3 mm and weight 7.4 - 9.2 mg.

3. Results and discussion

At room temperature orthorhombic $ReMn_2O_5$ manganites have space group *Pbam*, but it is expected that in their multiferroic state the actual symmetry group is $Pb2_1m$, which allows for a macroscopic electric polarization along the **b** axis [2, 3, 4, 5, 6]. The structure consists of edge-sharing $Mn^{4+}O_6$ octahedra, forming chains along the **c** axis, cross linked via $Mn^{3+}O_5$ pyramidal units. Recently *ab initio* calculations have shown [7], that the experimental magnetic structure of $HoMn_2O_5$ with $Pb2_1m$ symmetry is indeed the magnetic groundstate. Mn and Re ions in $ReMn_2O_5$ systems form three magnetic subsystems: Re^{3+} , Mn^{3+} and Mn^{4+} with crystallographic position 4g, 4h and 4f respectively. Magnetic interactions in the Mn subsystem are geometrically frustrated and $ReMn_2O_5$ shows complicated series of magnetic transitions involving the Mn and Re ions on cooling below T_N. The Mn^{3+} ions have a decisive importance for the ferroelectric ordering of the $ReMn_2O_5$ system.

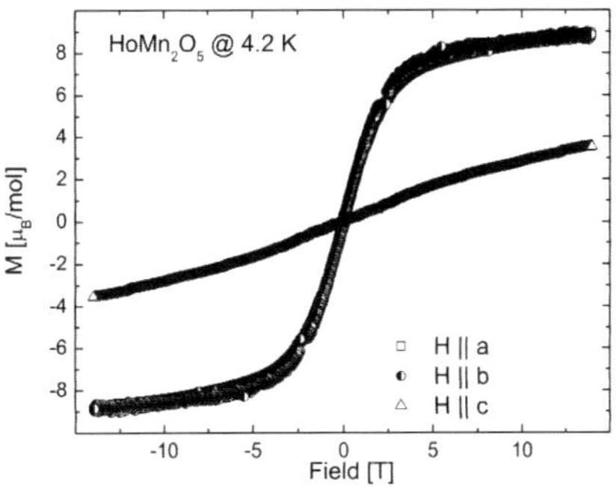

Figure 1. Field dependencies of magnetization in $HoMn_2O_5$ single crystals along the three principal crystallographic directions: **a** - □ **b** - ◐ **c** - △.

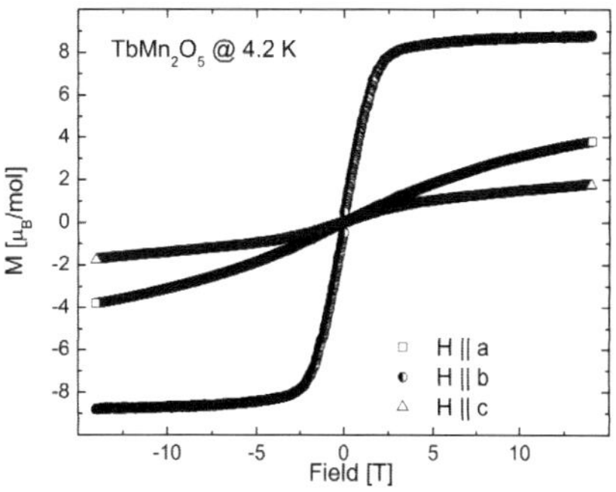

Figure 2. Field dependencies of magnetization in $TbMn_2O_5$ single crystals along the three principal crystallographic directions: **a** - □ **b** - ◐ **c** - △.

Our magnetization measurements were realized by a vibrating sample magnetometer. Magnetization dependencies along the three principal crystallographic directions for $HoMn_2O_5$ and $TbMn_2O_5$ single crystals were measured in the temperature range 4.2 - 120 K and in magnetic fields up to ± 14 T. Typical field dependence curves for $HoMn_2O_5$ and $TbMn_2O_5$ are shown on Fig.1 and Fig.2 respectively. In both compounds a significant magnetic anisotropy is present. In $HoMn_2O_5$ monocrystal two directions of easy magnetization, along the **a** and **b** axes, were found, while for the $TbMn_2O_5$ monocrystal only one easy magnetization direction along the **b** axis was found. Saturation of the sample magnetization was not observed in fields up to 14 T. According to the results from our magnetostriction measurements, presented below, some peculiarities in the magnetization curves of both compounds were expected, but in fact not observed.

Aiming to eliminate any possibility for equipment fault or malfunction, a series of test magnetization and magnetostriction measurements on orthorhombic $NdMnO_3$ single crystal were performed. The obtained typical magnetization and magnetostriction curves are presented in Fig.3. From the presented results can be clearly seen, that in $NdMnO_3$ single crystal a strong correlation between magnetization and magnetostriction was observed, which eliminates the possibility for equipment fault or malfunction.

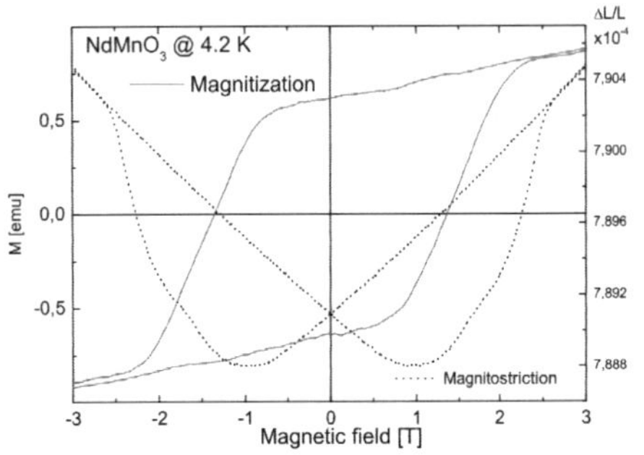

Figure 3. Field dependencies of magnetization (solid line) and magnetostriction (dashed line) in $NdMnO_3$ single crystals along the **b** - axis.

It is characteristic for all $ReMn_2O_5$ compounds that the various magnetic phase changes are reflected in sharp and distinct anomalies of the dielectric constant, as shown for $HoMn_2O_5$ and $TbMn_2O_5$ in Fig.4, which exhibit distinct anomalies at all magnetic transitions temperatures, with the strongest peaks observed at the FE transitions. Our data provide striking evidence for the existence of extraordinarily large spin-lattice interactions in $ReMn_2O_5$.

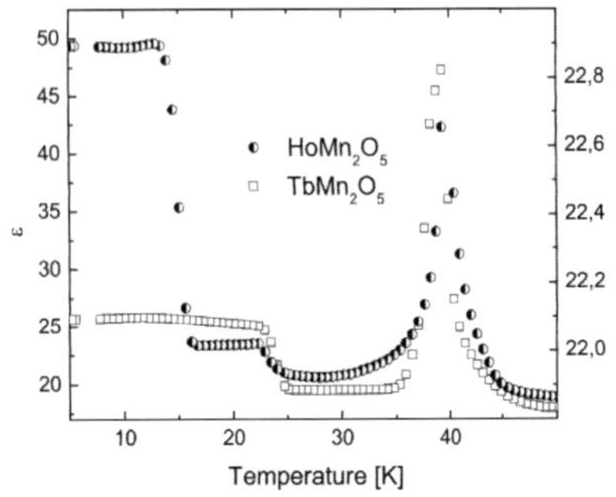

Figure 4. Temperature dependency of dielectric constant for $TbMn_2O_5$ - □ and $HoMn_2O_5$ - ◐ single crystals at 1 kHz along the **b** axis measured by sample cooling the in presence of 1 T magnetic field.

This is a clear indication of strong magneto-electric coupling due to large spin-lattice interactions. Long-range antiferromagnetic (AFM) ordering of the Mn^{3+}/Mn^{4+} spins occur at T_N = 43 K. This transition into a high Néel temperature phase is a common features for all $ReMn_2O_5$. Subsequently the FE transition takes place at T_C slightly below T_N. The pure ferroelectric lock-in transition is not influenced by magnetic fields. With further temperature decreasing, another magnetic transition takes place at T'_N (22.4 K for $HoMn_2O_5$ and 24 K for $TbMn_2O_5$), where commensurate AFM ordering becomes low temperature incommensurate. This transition is accompanied by a significant decrease of the FE polarization and is often referred to a second FE phase transition. T'_N turns out to be associated with a commensurate - incommensurate transition. This is very surprising in that the typical incommensurate-commensurate transition results in the commensurate state being the low-T ground state. This is a consequence of the "locking in" of magnetic and lattice modulations that is energetically favored at low T. Under T'_N the spin wave vector remains unchanged and the transition involves an increase in the ordered moments of the Mn^{4+}/Mn^{3+} sublattice. It has been assumed that the long-range magnetic ordering of Mn^{3+}/Mn^{4+} induces the FE transition via an additional Jahn-Teller distortion of Mn^{3+} ions [8]. The FE state exhibits canted antiferromagnetic displacements

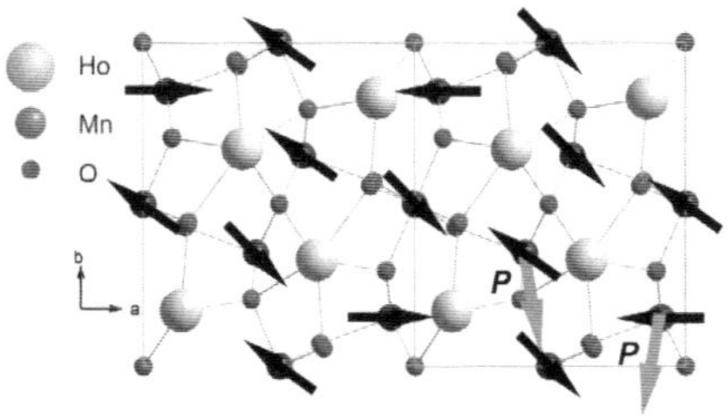

Figure 5. Spin and dipolar moment (black and white arrows) orientation in doubled along the **a** axis $ReMn_2O_5$ unit cell.

of the Mn^{3+} ions. These displacements lift the magnetic degeneracy by lowering the crystal symmetry to $Pb2_1m$, thus stabilizing the FE state via the magnetic Jahn-Teller effect.

The spontaneous polarization along the **b** axes appears by cooling the sample below T_N, where a long-range anti-ferromagnetic (AFM) ordering of the Mn^{3+} /Mn^{4+} spins occurs. As shown in Fig.5 the spins of two Mn^{3+} per unit cell are each frustrated with two neighboring Mn^{4+} with the same spin direction. Reducing this frustration by moving the Mn^{3+} away from the Mn^{4+} generates a dipolar moment **P** [1] between the Mn^{3+} and the surrounding oxygen ions. The $\mathbf{P_a}$ components of this dipolar moments cancel out while the $\mathbf{P_b}$-components add up to the macroscopic polarization and ferroelectricity along the **b** axis. The proposed displacement lowers the symmetry to the space group $Pb2_1m$. The AFM modulation along the **a** axis with $\mathbf{q}_x$=0.5 leads to the frustration and displacement of both Mn^{3+} and the net polarization along the **b** axis. In the temperature range 4.2 − 43 K a number of magneto-elasto-electric phase transitions was observed. It was pointed out that all the ferroelectric phases are strongly tied to the antiferromagnetic Mn^{3+}/Mn^{4+} spin structure, with the latter being dominated by the f-d exchange interaction [9]. It is important to emphasize that the described above mechanism leads to the appearance of a spontaneous polarization in both investigated compounds $HoMn_2O_5$ and $TbMn_2O_5$.

In general the values of the macroscopic polarization P, for nonmagnetic ferroelectrics, computed by *ab initio* band-structure methods [7] are in good agreement with the experimentally observed ones [10]. According to the big difference between this both values, the class of multiferroic manganites can be considered as an exception to this rule. They are two possible explanations for this large discrepancy: On the one hand, it can be due to experimental artifacts, for instance the formation of ferroelectric domains, but on the other hand it can be due to an incompleteness in our understanding of the physical properties of these magnetic ferroelectric materials.

As it is well known, the total polarization **P** for a given material is the sum of the ionic and electronic polarization which originates from the lattice displacements and the valence electrons respectively. The ionic contribution $\mathbf{P}_i$ to the total polarization is easily obtained by summing the product of the ionic displacements from the centrosymmetric to the ferroelectric structure with the nominal charge of the ions rigid core. The calculation of the electronic contribution $\mathbf{P}_e$, which is is induced by the magnetic superstructure is more complicated. The electronic polarization $\mathbf{P}_e$ arises from a reorganization of valence charges caused by both ferroelectric lattice distortions and magnetic ordering. Both these cause changes in covalency, which in turn cause a flow of valence electron charge across the material. Therefore the effective charge that is displaced by a distortion can be much larger than just the bare ionic value. In the magnetically ordered high symmetry phase the ionic contribution is $\mathbf{P}_i$=0, but the electronic part is not required to vanish. In fact, the material is bound to have a magnetically induced ferroelectric polarization as it is a dislocated spin density wave system in which the center of symmetry of the magnetic and lattice structure does not coincide, providing a basic mechanism for multiferroicity [11, 12]. An equivalent point of view is that in this oxide symmetry allows for a purely electronic part to the magnetostriction [13, 14]. As a consequence the spin ordering induces a redistribution of charge on crystallographically inequivalent manganese sites.

Recent calculations have shown [7] that the Hubbard U causes a near cancellation of the electronic and ionic contributions to the polarization and effectively reduces the polarization in the multiferroic manganites by over an order of magnitude. Therefore we can conclude that the electron-electron interactions decimate the polarization in the multiferroic $ReMn_2O_5$ manganites. Electronic correlation effects are thus of prime importance and quantitatively dominate the physical properties of these multiferroic transition metal compounds.

As recently reported [2] a colossal magnetostriction effect in $HoMn_2O_5$ single crystals was observed. According to our assumption about the existing correlation between polarization and magnetostriction, we carried out a number of experiments on $HoMn_2O_5$ and $TbMn_2O_5$ single crystals in the temperature range 4.2 - 10 K. In Fig.6 and Fig.7 typical polarization and magnetostriction dependence curves from our measurements on $HoMn_2O_5$ and $TbMn_2O_5$ single crystals are shown. As clearly can be seen in Fig.6, for $HoMn_2O_5$ the place of the peculiarities in polarization field dependencies and the magnetostriction phase transition at 4.2 K is nearly the same. For $TbMn_2O_5$ (Fig.7) where the peculiarities in polarization are not so clear indicated, the exact transition fields can be determinate by dielectric constant derivative. More detailed measurements corroborated our assumption for possible correlation between the peculiarities in polarization and the magnetostriction phase transition. In our opinion, the observation of polarization peculiarities below 10 K only is a direct consequence of the RE (Ho,Tb) spin reorientation.

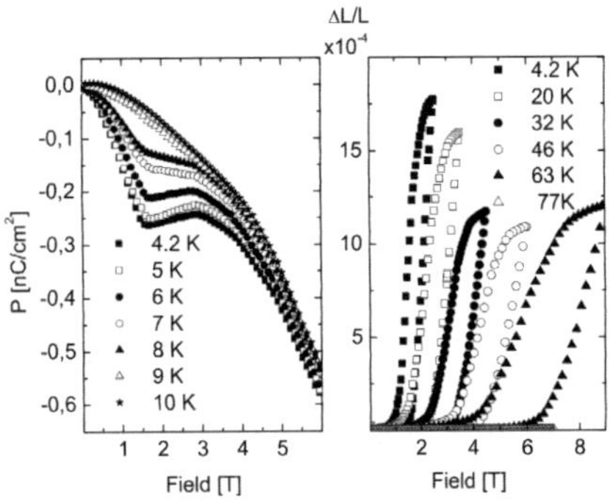

Figure 6. Polarization (left block) and magnetostriction (right block) field dependencies measured along the **b** axis of $HoMn_2O_5$ single crystals.

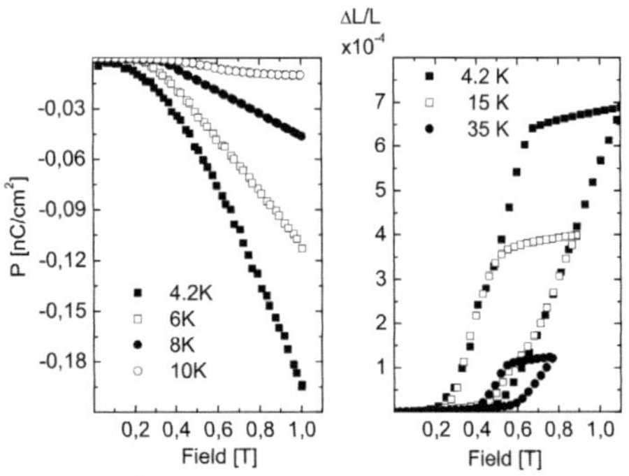

Figure 7. Polarization (left block) and magnetostriction (right block) field dependencies measured along the **b** axis of $TbMn_2O_5$ single crystals.

As discussed in [2] the reason for observing such a colossal magnetostriction effect is the total effect of the exchange magnetostriction of the manganese ions and the holmium single ion magnetostriction. Holmium and Terbium in metal state showed gigantic single-ion magnetostriction , which is due to both the strong spin-orbit coupling between orbital magnetic moment M_L and the non-spherical charge cloud of 4f - electron shell (which is highly anisotropic), and strong spin-lattice interactions. When a Ho ion is placed in the crystal lattice the anisotropy of the 4f -electron shell remains practically unchanged. In an external magnetic field the spin moment MS changes its orientation and this leads to reorientation of M_L. This causes a strong perturbative effect on the crystal field (the spin - lattice interactions in the $HoMn_2O_5$ and $TbMn_2O_5$ compound are strong) and a colossal magnetostriction effect appears.

The phase diagram build from magnetostriction (line) and polarization (circle) measurements on $HoMn_2O_5$ and $TbMn_2O_5$ single crystals is shown in Fig.8.

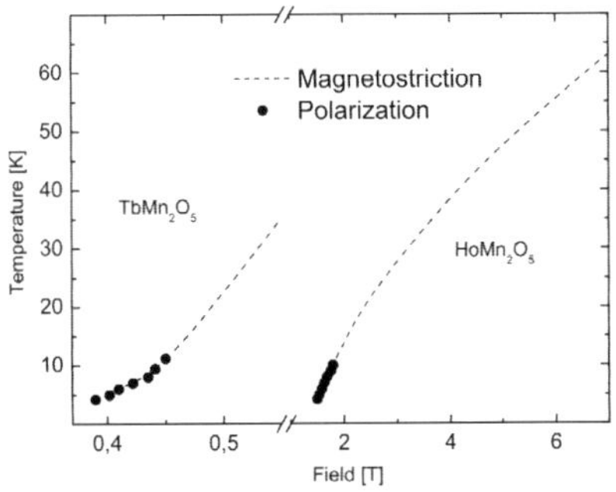

Figure 8. Phase diagram built from data received by magnetostriction (solid line) and polarization (circles) measurements along the **b** axis in $HoMn_2O_5$ and $TbMn_2O_5$ single crystals .

4. Conclusions

In the present article the temperature dependencies of the magnetization and the dielectric constant, as well the field dependencies of the polarization and magnetostriction for two series of $HoMn_2O_5$ and $TbMn_2O_5$ monocrystals are discussed. As is evident, by temperature lowering these orthorhombic manganites undergo cascade phase transitions, whose complexity originates from in the partially competing interactions between Mn^{3+}/Mn^{4+} spins, rare earth magnetic moments and the lattice [1]. We also observed a colossal magnetostriction effect in both $HoMn_2O_5$and $TbMn_2O_5$ monocrystals. This effect results from the reorientation of the spin moments in an external magnetic field and its strong perturbative effect on the crystal field. Compared to $HoMn_2O_5$ monocrystals the observed in $TbMn_2O_5$ monocrystals magnetostriction effect appears at lower magnetic fields (as shown in Fig. 6 and fig. 7). The same magnetostrictive behavior was observed in pure Ho/Tb monocrystals in metal state [15]. The more detailed measurements of magnetization, polarization and magnetostriction of $HoMn_2O_5$ and $TbMn_2O_5$ monocrystals in the temperature range 4.2 - 10 K gave us the possibility to build the phase diagrams for both monocrystals. These diagrams clearly demonstrate the correlation between the peculiarities in polarization and the magnetostriction phase transition. In our opinion, the field couples to the magnetic order resulting in field-induced spin reorientations and magnetic phase transitions, which in turn should affect the lattice via the spin-lattice interaction. In this way they realize the correlation observed between the polarization and magnetostriction. Knowing the signature and the nature of the magnetoelastic effect we acquit essential information about the intrinsic magnetoelastic and magnetoelectric interactions. To elucidate the role of the rare earth ions in this processes further investigations are necessary.

5. Acknowledgements

The work of I. Radulov is supported by NATO EAP.RIG.981824 and BUX Grant - 308/2007

References

[1] I. Radulov, V. Lovchinov and M. Daszkiewicz , *JMMM*, **320**, (2008), 43-46
[2] I. Radulov, V.I. Nizhankovskii ,V. Lovchinov, D. Dimitrov and A. Apostolov , *Eur. Phys. J. B* , **52**, (2006), 361-364
[3] J. A. Alonso, M. T. Casais, M. J. Martinez-Lope, J. L. Martinez and M. T. Fernandez-Diaz., *J. Phys. Condens. Mat.* **9**, (1997), 8515
[4] N. Hur, S. Park, P. A. Sharma, J. S. Ahn, S. Guha and S-W. Cheong , *Nature* **429**, (2004), 392.
[5] N. Hur et al., *Phys. Rev. Lett.* **93**, (2004), 107207
[6] L.C. Chapon et al., *Phys. Rev. Lett.* **93**, (2004), 177402
[7] G. Giovannetti, and J. van den Brink, *Phys. Rev. Lett.*, **100**, (2008), 227603
[8] M. Kenzelmann et al., *Phys. Rev. Lett.* **95**, (2005), 087206
[9] M. Fiebig, *J. Phys. D: Appl. Phys.*, **38**, (2005), R123-R152

[10] K.M. Rabe, C.H. Ahn and J.-M. Triscone, eds., *Physics of ferroelectrics: a modern perspective*,(Berlin: Springer)(2007)
[11] L.C. Chapon et. al., *Phys. Rev. Lett.*, **96**, (2006), 097602
[12] J.J. Betouras, G. Giovannetti and J. van den Brink, *Phys. Rev. Lett.*, **98**, (2007), 257601
[13] A.B. Harris, *Phys. Rev. B*, **76**, (2007), 054447
[14] S.W. Cheong and M. Mostovoy , *Nature Mater*, **6**, (2007), 13
[15] S. A. Nikitin, *JETF*, **43**, (1962), 31-36

Study of Ferromagnetism in Mn Doped ZnO Dilute Semiconductor System

R K Singhal[1*], M S Dhawan[1], S K Gaur[1], S N Dolia[1], Sudhish Kumar[2], T Shripathi[3], U P Deshpande[3], Y T Xing[4], Elisa Saitovitch[4] and K B Garg[1]

[1]Department of Physics, University of Rajasthan, Jaipur 302004, India
[2]Department of Physics, M L Sukhadia University, Udaipur 313 002, India
[3] UGC-DAE CSR, University Campus, Indore 452 017, India
[4]CBPF, Rua Xavier Siguad 150, Urca, Rio de Janeiro, RJ, Brazil

Abstract
The wide band gap semiconductor ZnO when doped with a very low percent of some transition metal ions can exhibit above room temperature ferromagnetism, transforming it into a unique compound for spin-electronic applications. In the present work we have compared the electronic structure of two polycrystalline $Zn_{1-x}Mn_xO$ pellets (for x=0.02 and 0.04), prepared by low temperature processing, and carefully characterized. The Rietveld refinement of the XRD patters established that the samples have the ZnO lattice with ZnS type Wurtzite hexagonal symmetry and no detectable impurities. The samples exhibit distinctly different magnetic properties. The pure ZnO pellet shows a diamagnetic behaviour, the 2% sample displayed a clear FM ordering at 300 K while the 4% sample did not show any ordering even upon cooling. Their electronic structure has been investigated using x-ray absorption and x-ray photoemission spectroscopy with an aim to find out how the changes in the electronic structure can correlate to the magnetic properties in such diluted magnetic semiconductor materials. The results show that most of the Mn ions of the ferromagnetic sample are in the divalent state. For the higher Mn percent nonmagnetic sample, a larger contribution of higher oxidation Mn states are dominant and the oxygen content also increases. The two factors can be correlated to the suppressed ferromagnetism, though it is hard to exactly predict that which of these two factors weighs more.

1. Introduction
Dietl's prediction, that the ferromagnetism at high temperature can be obtained in many wide band gap semiconductors such as ZnO, GaAs, GaN when doped with diluted transition metals can show

*Corresponding author: singhal46@yahoo.co.in, Ph: 91 141 2545931 ferromagnetism above room temperature, has attracted considerable research interest [2-14]. Following it, there has been intense searching for ferromagnetic ordering in doped dilute magnetic semiconductors (DMS's) focusing on possible spin-transport properties, which has many potentially interesting device applications. Spintronic devices such as spin-valve transistors spin light-emitting diodes, non-volatile memory, logic devices, optical isolators and ultra-fast optical switches are some of the areas of interest for introducing the ferromagnetic properties at room temperature in a semiconductor [15-18].While the DMS's based on III–V semiconductors show ferromagnetism only at very low temperatures, the oxide-based DMSs may be ferromagnetic at room temperature or at even higher temperatures.

The occurrence of ferromagnetism in DMS's has been long debated due to confronting reports. Considerable doubts and fears have been expressed that even though the Mn clusters are expected to give an antiferromagnetic state, rather than a ferromagnetic (FM) order, very small clusters can be ferromagnetic [2]. The first reports of room temperature FM ordering in Mn-doped ZnO powder, bulk pellets as well as thin films with Curie temperatures far above 300K came from Sharma et al [19,20]. This was confirmed by Blythe et al [21] who made a detailed study of dependence of the FM ordering on the processing temperature. On contrary, Kane et al [22] claimed no evidence of an FM ordering in $Zn_xMn_{1-x}O$, $Zn_xCo_{1-x}O$ bulk single crystals. Instead, they claimed a paramagnetic behaviour in case of the former and hysteresis in the latter owing to super-paramagnetic Co clusters embedded in a diamagnetic ZnO matrix. Kittilstved et al [23] have, on the other hand, not only observed high- T_c ferromagnetism in ZnMnO and ZnCoO systems but also show how their T_c can be manipulated through *in situ* as well as *ex situ* chemical manipulation. Long et al [24] studied the effects of annealing atmosphere on FM behaviour for the ZnCoO bulk samples. They show that the air-annealed samples has weak FM ordering, but the samples annealed in vacuum or Ar/H_2 show a strong FM behaviour, indicating that the strong ferromagnetism is associated with high oxygen vacancies density. The enhanced room-temperature FM behaviour is also found in the samples with high carrier concentration controlled by doping interstitials Zn.

Various theoretical attempts have been put forward to explain the ferromagnetism in these DMS systems. In ZnO system it has been proposed that the ferromagnetism can be obtained by the mediation of the shallow donors or acceptors [23] or the holes in the valence band [1]. Garcia etal [6] suggested that the ferromagnetism in this system should be associated with the co-existence of Mn^{3+} and Mn^{4+} via a double-exchange mechanism. Based on x-ray-magnetic-circular dichroism studies of diluted $Zn_{1-x}Co_xO$ films, it is shown that Co ions substituted to Zn in the ZnO matrix are not responsible for the origin of FM ordering [6-8] but the anion sublattice may be instead responsible of the ferromagnetism [8]. Occurrence of ferromagnetism in undoped HfO_2 thin films have urged the researchers to re-judge the real role that doping may play in tailoring the magnetic properties of these oxides [25]. Some reports [25-28] on ferromagnetism in various undoped oxides claimed that magnetism is certainly possible in undoped semiconducting and insulating oxide thin films, that originates due to oxygen vacancies. But despite a lot of theoretical and experimental attempts, the origin of ferromagnetism in semiconductors remains an issue of debate. When assigning the origin of ferromagnetism in Mn-doped ZnO DMS, it is important to distinguish the contributions of the transition metal with different valence bonds, providing evidence to explore whether the magnetization originates from the magnetic precipitates or not.

In this paper we have studied the electronic structure of two single phase polycrystalline ZnMnO samples with 2% and 4% Mn concentration. The two samples exhibit entirely different magnetic behaviour as shown by SQUID. The 2% sample displayed clear FM ordering at 300 K, the 4% sample did not show any ordering even upon cooling. The electronic structure of these two samples has been investigated using x-ray absorption and photoemission spectroscopy to find out how the changes in the electronic structure can correlate to the observed magnetic properties.

2. Experimental techniques

High purity (99.999%) ZnO and MnO_2 powders from Johnson Matthey Company (JMC) were mixed and ground for long time to get a homogeneous mixture. The ground mixture was calcined at (450±5)°C for nearly 14 hrs raising the temperature very slowly above 100°C in a microprocessor controlled furnace and similarly cooling it very slowly after completion of the calcinations. The resulting lumps were ground for ~5 hrs and pellets of 10 mm diameter using a hydraulic pressure of nearly 7 tons. The pellets were then sintered at (500±5)°C for 18 hrs. *The important feature of our sample preparation was the low-temperature processing. When high-temperature (T > 700°C) methods were used, samples were found to exhibit no ferromagnetic ordering at room temperature.*

The hysteresis loops were measured at different temperatures by a dc-SQUID magnetometer (MPMS-XL, Quantum Design) at CBPF, Brazil. The x-ray diffraction (XRD) patterns of the samples were collected on a Rigaku x-ray diffractometer equipped with CuK_α radiation. The 2θ scan was from 10^0 to 90^0 with a step size of 0.05^0. Rietveld profile refinements of the XRD patterns were carried out using the FULLPROF Program [29].

The Mn L_{23} edge x-ray absorption spectra in total electron yield mode were recorded at the BACH beamline of Elettra in Trieste, Italy. The spectra were recorded at 300 K at pressure better than $5x10^{-10}$ mbar. The measurements were performed on $Zn_{0.98}Mn_{0.02}O$ and $Zn_{0.96}Mn_{0.04}O$ pellets (referred to as the 2% sample and 4% sample, respectively, in the following) and on some reference samples, i.e. MnO, $La_{1.2}Sr_{1.65}Ca_{0.15}Mn_2O_7$ bilayer manganite, all scraped in UHV before each measurement.

The core level x-ray photoemission spectra were measured using an x-ray photoelectron spectrometer at IUC, Indore, India using the Al $K\alpha$ radiation. The samples were scraped uniformly with diamond files before carrying out the measurements. Final spectra were taken after minimizing the feature coming from carbon contamination of the surface (C 1s peak). The vacuum in the chamber was ~4.4×10^{-10} Torr. We kept scraping the samples *in situ* to get uncontaminated surface throughout the measurements. No shift was observed due to charging of the samples.

3. Results and discussion

3.1. Characterization results

The XRD patterns of pure ZnO with Wurtzite hexagonal structure and no impurity peaks are displayed in figure 1a. The cell parameters were refined with the help of a least square refinement program that are in good agreement with the reported values. Figures 1 b,c show the fitted XRD patterns of the 2% and 4% Mn doped ZnO sample, respectively, recorded at 300 K. The patterns show the single phase ZnS type Wurtzite hexagonal symmetry with no signature of any peak from any other phase. We used the FULLPROF Program [29] for Rietveld profile refinements of the XRD patterns. The Rietveld refinement were carried out using the $P6_3mc$ space group (No.186, Z=3), with each atom in the Wurtzite hexagonal structure residing on the 2b Wyckoff position with the Zn atoms at (1/3,2/3,0) and the O atoms at (1/3,2/3,z) co-ordinates. In the refinement process the Mn occupancy was varied for the two Zn and O sites to locate the exact site of Mn. The best fit was obtained when Mn atoms occupy the Zn site with total preference while the Mn atoms occupying the O site gave very poor fits. In fact, the atomic scattering lengths of the three atoms i.e. Zn (Z= 30), Mn (Z= 25) and O (Z= 8) are quite different as these are far away in the periodic table, hence all the three atoms can be easily distinguished in the profile refinement. Our Rietveld analysis shows that Mn atoms occupy the Zn site with total preference. The cell parameters are shown in Table 1. The refined parameters shown in the table further confirm the stoichiometry and the single phase nature of the samples.

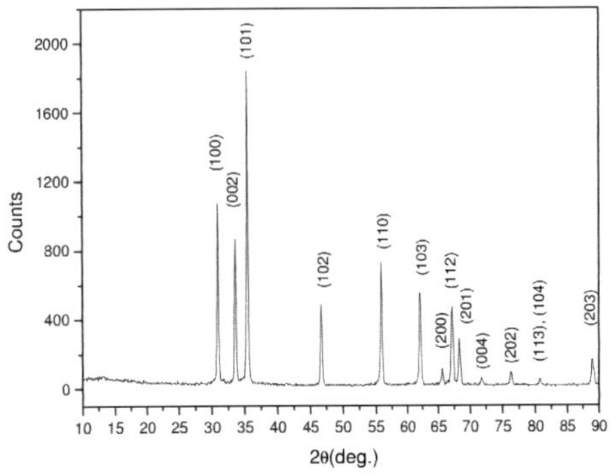

Fig. 1a

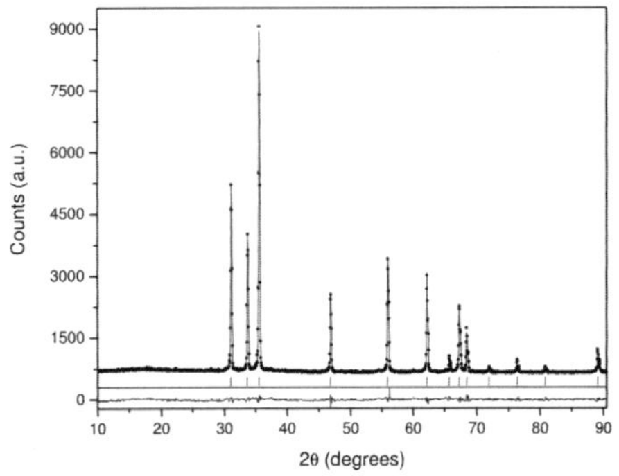

Fig. 1b

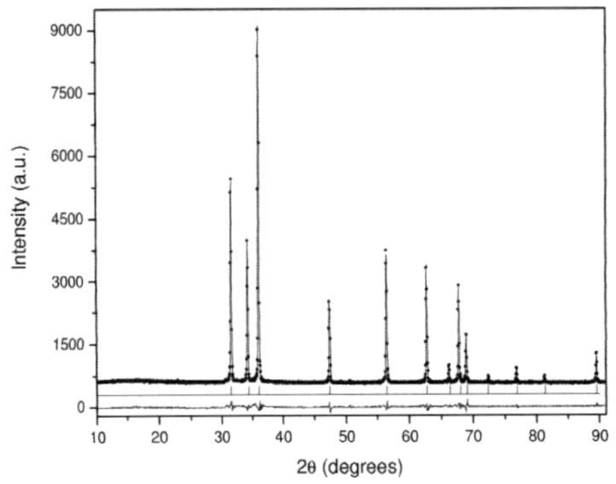

Fig. 1c

Fig. 1a. The XRD pattern of ZnO powder at 300K **b**. the fitted XRD pattern of $Zn_{0.98}Mn_{0.02}O$ sample. **c.** the fitted XRD pattern of $Zn_{0.96}Mn_{0.04}O$ at 300K. Observed (calculated) profiles are shown by dotted (solid) curves in case of fig. **b** and **c**. The short vertical marks represent Bragg reflections. The lower curves are the difference plot.

Table-1. Crystallographic data for $Zn_{1-x}Mn_xO$ samples at 300K obtained using the Rietveld refinement of the powder XRD patterns. Space group is $P6_3mc$ (No. 186, Z = 2). Each atom in the Wurtzite hexagonal structure resides on the 2b Wyckoff position with Zn atoms at (1/3,2/3,0) and O atoms at (1/3,2/3,z).

Sample	Cell parameters				Discrepancy factors		
	a [Å]	c [Å]	V [Å^3]	O [positional parameter z]	χ^2	R_p [%]	R_{wp} [%]
ZnO	3.249	5.207	47.69	0.612			
$Zn_{0.98}Mn_{0.02}O$	3.252(3)	5.212(5)	47.73	0.628	1.76	6.15	7.92
$Zn_{0.96}Mn_{0.04}O$	3.253(2)	5.213(4)	47.76	0.637	1.76	6.34	8.63

3.2. Magnetization measurements

The magnetization versus field (hysteresis) curves at 300 K for $Zn_{0.98}Mn_{0.02}O$ and $Zn_{0.96}Mn_{0.04}O$ samples are shown in figure 2a. All the measurements were repeated several times with the time interval of several weeks to minimize the experimental error. The curves clearly reveal a room temperature FM ordering of the 2% sample but no such signature for the 4% sample. In fact, the 4% sample shows a purely a paramagnetic state (see inset). The shape of the loop does not rule out the presence of some para-magnetic (PM) contribution even for the 2% sample. Figure 2b displays the separated PM and the FM contributions along with the experimental curve and the fit made to it. The saturation magnetization and coercivity estimated come out to be ~ 0.0019 emu/g and ~ 19.8 Oe respectively. Figure 3 displays the comparison of the curves at 300K and 50K for the FM sample. The sample shows an appreciable enhancement in saturation magnetization as well as the coercivity upon cooling (~0.0029 emu/g and ~ 27.9 Oe respectively at 50K). The sample with 4% Mn did not show any FM ordering even upon cooling down to 5K (not shown here). Pure ZnO pellet prepared in same conditions was also measured at room temperature and 50 K but it showed a diamagnetic behaviour at both these temperatures (not shown).

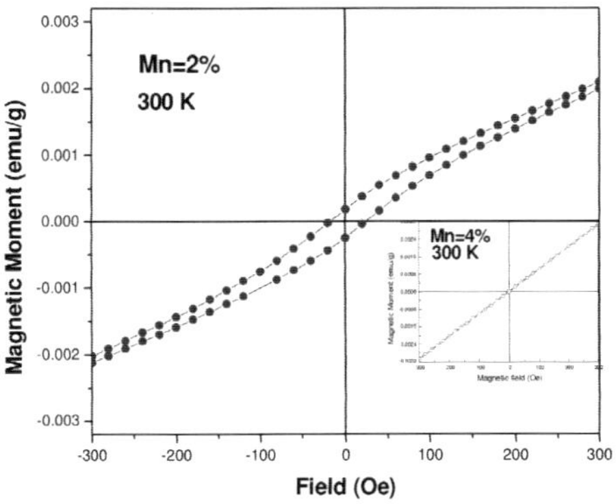

Fig. 2a

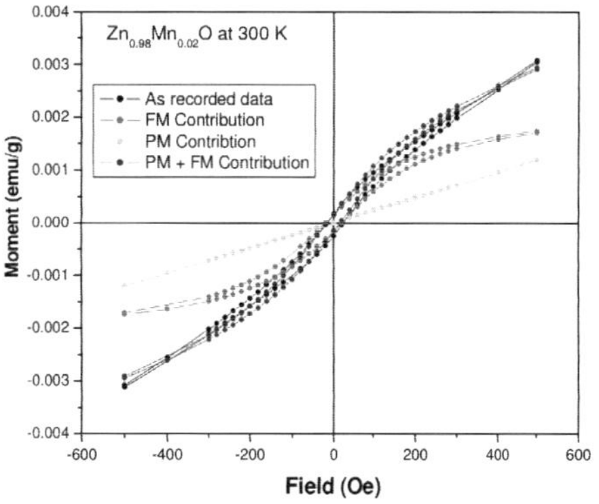

Fig. 2b

Fig. 2a. Magnetization vs. Field curve for $Zn_{0.98}Mn_{0.02}O$ at 300K. The inset shows the curve for $Zn_{0.96}Mn_{0.04}O$ at 300K **2b.** Magnetization vs. Field curve for $Zn_{0.98}Mn_{0.02}O$ at 300K alogwith the paramagnetic and FM contributions.

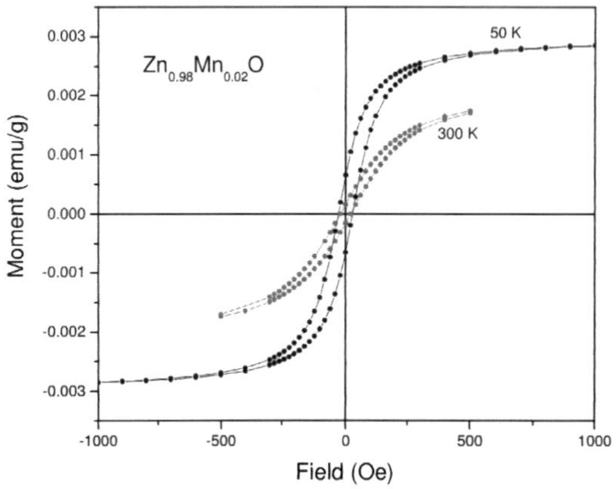

Fig. 3. Comparison of hysterisis curves for $Zn_{0.98}Mn_{0.02}O$ sample at 50K and 300K.

3.3. XAS m,easurements

The Mn L_{23} edge x-ray absorption spectrum measured for the 2% sample is shown in figure 4a. Spectra for MnO (Mn^{2+} reference system) and the $La_{1.2}Sr_{1.65}Ca_{0.15}Mn_2O_7$ bilayer manganite (reference system of a mixed valent Mn^{3+}/Mn^{4+} compound) are also shown for a comparison. The spectrum of the 2% FM sample is very different from that of the bilayer manganite, which has the main peak at nearly 2.5 eV higher energy than the main peak of the 2% sample (at 640 eV). This indicates a significant change in the charge state and crystal field of Mn in the two systems. The spectrum of the 2% sample is also different from the spectrum of Mn metal [our ref. 10], which does not display multiplet structures. This rules out the possibility of Mn impurities getting segregated in metallic clusters form. In fact, the spectrum from of the 2% sample shows multiplet structures qualitatively similar to those of MnO, suggesting that Mn ions in the sample $Zn_{0.98}Mn_{0.02}O$ are nearly divalent. However, the Mn L_3 edge reveals several differences between the two spectra, if wee look at figure 4 b carefully. Firstly, the lower energy multiplet lines at 639 eV do not appear for $Zn_{0.98}Mn_{.02}O$. Secondly, there is an increased spectral weight around 642 eV in $Zn_{0.98}Mn_{.02}O$. These differences indicate that the local environments of Mn in $Zn_{0.98}Mn_{.02}O$ and MnO are different. The absence of the multiplet lines at 639 eV can be assigned to a lower absolute value of the crystal field parameter 10Dq for $Zn_{0.98}Mn_{.02}O$ with respect to the corresponding value for MnO. In fact the difference spectrum between the $Zn_{0.98}Mn_{0.02}O$ and MnO spectra (figure 4b) does show similarities with the Mn L_3 spectra of the bilayer manganite.

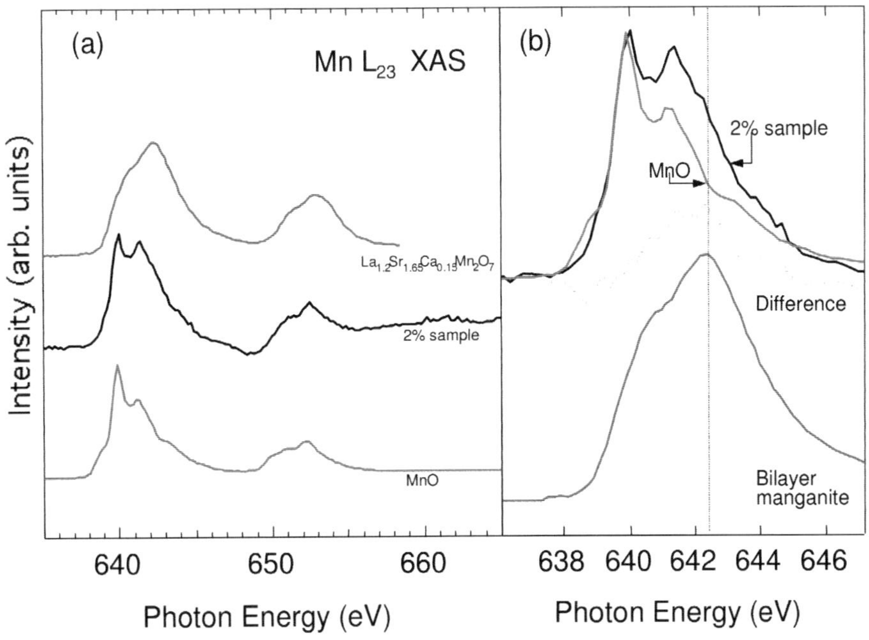

Fig. 4 (a) Mn L_{23} XAS spectra of $La_{1.2}Sr_{1.65}Ca_{0.15}Mn_2O_7$ bilayer manganite, $Zn_{0.98}Mn_{0.02}O$, MnO systems **(b)** Mn L_3 XAS of the $Zn_{0.98}Mn_{0.02}O$ compared to Mn L_3 XAS of MnO, difference spectrum between the Mn L_3 XAS of MnO and the Mn L_3 XAS of $Zn_{0.98}Mn_{0.02}O$, Mn L_3 XAS of the bilayer manganite shown for comparison.

3.4. XPS measurements

Zn 2p, 3p, Mn 2p and O1score level XPS spectra were measured for pure ZnO and the two doped samples to study the electronic structure in each case. The results are presented for each in the following subsections.

3.4.1. Zn 2p and 3p XPS data.

The Zn $2p_{3/2}$ XPS spectra for the pure ZnO and the Mn doped samples are displayed in figure 5a. All the three spectra show a symmetric featured single peak at ~1023 eV. A single Gaussian was nicely fit to all the spectra and no change in the peak positions was noticed upon Mn doping. The single Gaussian rules out the possibility of multiple components of Zn in any of these samples. Figure 5b shows the Zn 3p core level XPS spectra. The spectra show the $3p_{3/2}$ and $3p_{1/2}$ bands located at ~88.6 and 91.6 eV, respectively. The two Gaussians were fit to the Zn 3p spectra but again no shift observed in the positions of two bands ($3p_{3/2}$, $3p_{1/2}$), that reconfirms the absence of any other multiple Zn components in these samples. The area under the peaks of $2p_{3/2}$, $3p_{3/2}$ and $3p_{1/2}$ progressively decreases as the concentration of Mn at Zn site increases that was natural. Thus no electronic structural change is observed at the Zn site by Mn doping.

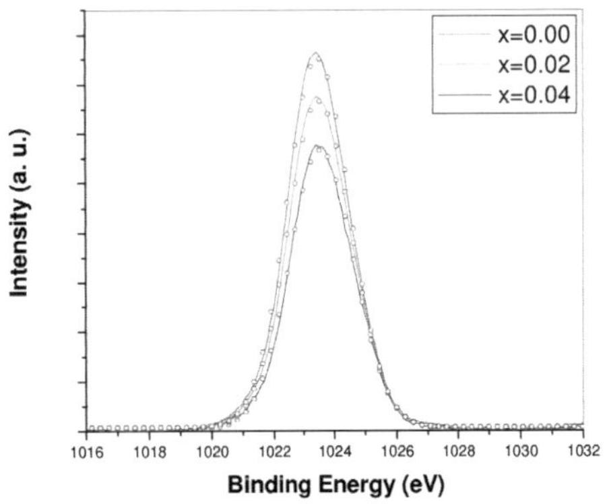

Fig. 5a.

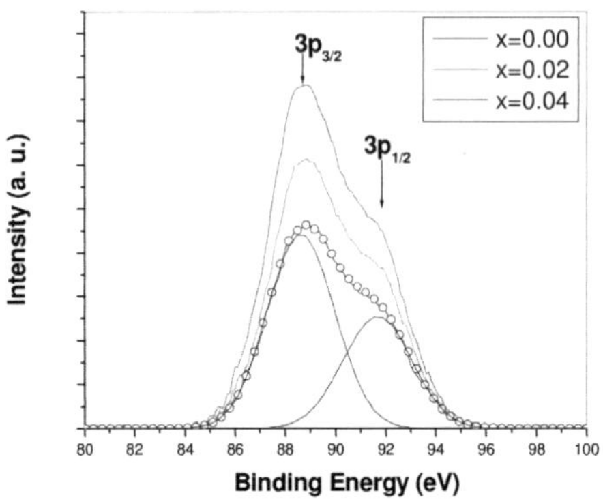

Fig. 5b.

Fig. 5a. The Zn $2p_{3/2}$ XPS spectra for the $Zn_{1-x}Mn_xO$ [x=0, 0.02, 0.04] samples along with the Gaussian fits **b.** The Zn 3p XPS spectra along with the Gaussian fits.

3.4.2. Mn $2p_{3/2}$ XPS data.
The Mn $2p_{3/2}$ XPS spectra are shown in figure 6. The three Gaussian peaks located at 640.0, 642.6 and 645.0 eV, can be attributed to Mn^{2+}, Mn^{3+} and Mn^{4+} valence states, respectively. To make a quantitative estimation of atomic percent of Mn^{2+} and the other valence states, the area of the different valence bonds of Mn were calculated and the results are summarized in the table 2. It has been found that the largest Mn^{2+} atomic concentration appeared in the FM 2% sample. In the non-FM 4% sample the Mn^{2+} percent declines and the higher valence states Mn^{3+} and Mn^{4+} dominate.

3.4.3. Oxygen 1s data.
Asymmetric Oxygen 1s spectra, shown in figure 7 indicate that multi-component oxygen species are present in the near-surface region of these samples. The high energy peaks located at 532.7 and 533.5 eV are due to the chemisorbed oxygen of the surface hydroxyl, $-CO_3$, absorbed H_2O, absorbed O_2 or the surface contamination. But the most intense first peak at ~ 531.2 eV is due to the actual O1s contribution from the samples. Area under the peak O 1s progressively increases by the doping of Mn. The first Gaussian peak that is corresponding to Oxygen 1s feature has been highlighted in the inset of figure 7. This enhancement of oxygen and the increase of Mn valence state with increase in Mn concentration can be attributed to the charge neutrality of the samples similar to the reports of magnetic ordering in ZnO thin films [30] and may be causing the suppression of ferromagnetism. This is in agreement to reports [24] that have shown that the air-annealed sample has weak FM ordering, but the sample annealed in vacuum or Ar/H_2 show a strong FM behaviour, indicating that the strong magnetic ordering might be associated with high oxygen vacancies density.

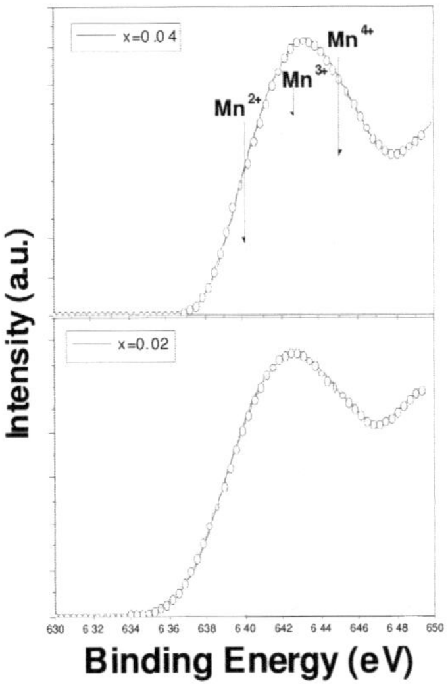

Fig. 6

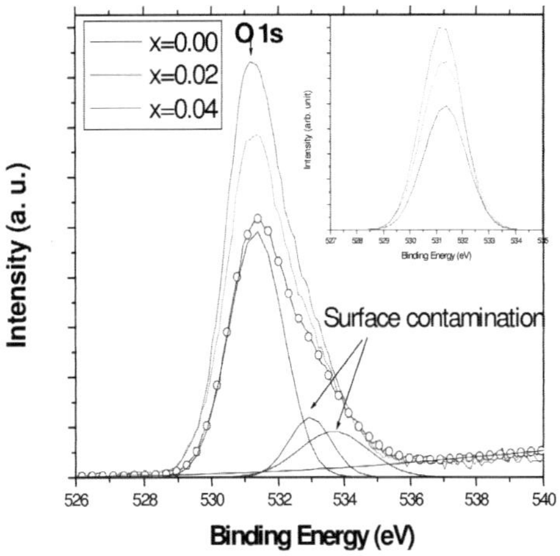

Fig. 7

Fig. 6. Gaussian fits in the Mn $2p_{3/2}$ XPS for the $Zn_{1-x}Mn_xO$ [x=0.02, 0.04] samples. **Fig. 7.** The O 1s XPS spectra for the two samples along with the Gaussian fits. The Gaussian corresponding to Oxygen 1s feature is shown in the inset for clarity.

Table-2. Area of the peaks under the Mn^{2+}, Mn^{3+} and Mn^{4+} Gaussians and the absolute quantity of Mn^{2+} in the 2 and 4% doped samples.

Sample	Area of peaks [in arb.units] 640.0eV Mn^{2+}	642.6eV Mn^{3+}	645.0eV Mn^{4+}	Quantity [%] of Mn^{2+} in all Mn valence bonds	Quantity [%)] of Mn^{2+} in $Zn_{1-x}Mn_xO$ sample
$Zn_{0.98}Mn_{0.02}O$	3672.1	2521.3	1681.3	46.63	0.93
$Zn_{0.96}Mn_{0.04}O$	1011.8	4073.9	3494.2	11.79	0.47

The observed FM ordering in these materials has been speculated to be due to presence of some secondary phases or clustering effect. In the present samples the possible secondary phases could be manganese oxide phases, such as MnO, MnO_2 and Mn_3O_4. Here, first of all, we should mention that we had followed the low-temperature processing to avoid any secondary phases which are reported to be evolved only when high-temperature (T > 700 °C) methods were used. However, of the manganese oxides stated above, the MnO and the MnO_2 may be safely ruled out as they are anti-ferromagnetic. Ferromagnetic Mn_3O_4 has a Curie temperature at 43K that is far below all these samples [31]. In addition, another possible phase $ZnMnO_3$ is established to show a spin-glass behaviour [32,33]. If the formation of a secondary Mn-related phase were responsible for the FM behaviour, an increase in Mn concentration would presumably increase the secondary phase volume fraction and related magnetization signature, instead, the opposite behaviour is observed that by increasing the Mn content from 2 to 4 at% caused a complete destruction of FM order. This rules out the magnetization due to any precipitating secondary phase in our samples.

Many theoretical attempts have been put forward to explain the observed magnetic ordering in these materials. Some are based upon mediation of the charge carriers [23] or the holes in the valence band [1], others are based upon the co-existence of Mn^{3+} and Mn^{4+} via a double-exchange mechanism. Also it has been argued, on the basis of experimental data, that the FM ordering in the ZnO system should not originate from oxygen vacancies as in the case of TiO_2 and HfO_2 films, but from defects on Zn sites [30,34]. A theoretical work stating that for ZnO, oxygen vacancies or Zn interstitials should result in anti-ferromagnetism, while ZnO with Zn vacancies has more chance to be ferromagnetic [34], supports this argument. It was also reported that magnetization of films of 10-50 nm-thick is much larger than that of the thicker films. This reveals that defects must be located mostly at the surface and/or the interface between the film and the substrate, or possibly nanometer-size layers at the interface/surface. According to Ruderman-Kittel-Kasuya-Yoshida theory [35,36] the ferromagnetism is a carrier-induced mechanism. This arises due to the exchange interaction between local spin-polarized electrons like the electrons of Mn^{2+} ions, and conductive electrons. This interaction leads to the spin polarization of conductive electrons. Subsequently, the spin-polarized conductive electrons perform an exchange interaction with local spin-polarized electrons of other Mn^{2+} ions. Thus, after the long-range exchange interaction, almost all Mn^{2+} ions exhibit the same spin direction. The conductive electrons are regarded as a media to contact all Mn^{2+} ions. As a result, the material exhibits FM ordering. More local spin-polarized electrons are produced with increasing Mn^{2+} concentration. Hence the best FM properties should be occurred for the low percent Mn sample as seen in our case, a plausible and explanation of the ferromagnetic ordering in the Mn-ZnO system.

4. Summary

We have studied the electronic structure of two polycrystalline ZnMnO pellets doped with diluted Mn concentration (2% and 4%). The important feature of our sample preparation was the low-temperature processing to avoid the secondary phases. When high-temperature (T > 700 °C) methods were used, samples were found to exhibit no ferromagnetic ordering at room temperature. The samples were characterized by SQUID and XRD. The XRD patterns and the Rietveld refinement confirmed that the samples are single phase with no detectable impurities having the ZnO lattice with Wurtzite hexagonal symmetry. The SQUID measurements show that the samples exhibit different magnetic properties. The 2% sample shows a clear room temperature FM ordering, the 4% sample did not show any ordering down to 5K. Pure ZnO pellet prepared in same conditions showed rather a diamagnetic behaviour from 300 to 50K.The electronic structure of these samples was investigated using x-ray absorption and x-ray photoemission spectroscopy. The XPS, XAS results show that most of the Mn ions of the ferromagnetic sample are in the divalent state. For the nonmagnetic sample, a larger contribution of higher oxidation Mn states (3+, 4+) is present and the oxygen content also shows an increment, which can be correlated to the suppressed ferromagnetism. Our studies reveal that the suppression of ferromagnetism can be associated with higher oxidation Mn state and the higher oxygen vacancies density. However, these studies are not enough to conclude which of these two factors weighs more in the suppression mechanism or some more factors are important.

5. Acknowledgments

RKS is grateful to the TWAS, Italy and CNPq Brazil for the associate membership and CBPF Brazil for providing the hospitality and all the experimental facilities during this work. We are grateful to DST, New Delhi and ELETTRA Sincrotrone, Trieste, Italy for funding this work under the Indo-Italian POC in S&T. Part of this work was performed at UGC-DAE Consortium for Scientific Research, Indore, India.

References

[1] Dietl T, Ohno H, Matsukura F, Cibert J and Ferrand D 2000 *Science* **287** 12019
[2] Bobadova-Parvanova P, Jackson K A, Srinivas S and Horoi M 2005 *J. Chem. Phys.* **122** 014310
[3] Rode K, Mattana R, Anane A, Cros V, Jacquet E, Contour J-P, Petroff F and Fert A 2008 *Appl. Phys. Lett.* **92** 012509
[4] Ueda K, Tabata H, Kawai T 2001 *Appl. Phys. Lett.* **79** 988
[5] Chambers S A, Droubay T, Wang C M, Lea A S, Farrow R F C, Folks L, Deline V and Anders S 2003 *Appl. Phys. Lett.* **82** 1257
[6] García M A, Ruiz-González M L, Quesada A, Costa-Krämer J L,Fernández J F, Khatib S J, Wennberg A, Caballero A C, Martín-González M S, Villegas, Briones F, González-Calbet J M, and Hernando A 2005 *Phys. Rev. Lett.* **94** 217206
[7] Gacic M, Jakob G, Herbort C, Adrian H, Tietze T, Brueck S and Goering E 2007 *Phys. Rev. B* **75** 205206
[8] Barla A, Schmerber G, Beaurepaire E, Dinia A, Bieber H, Colis S, Scheurer S, Kappler J P, Imperia P, Nolting F, Wilhelm F, Rogalev A, Müller D and Grob J J 2007 *Phys. Rev. B* **76** 125201
[9] Park Y D, Hanbicki A T, Erwin S C, Hellberg C S, Sullivan J M, Mattson J E, Ambrose T F, Wilson A, Spanos G and Jonker B T 2002 *Science* **295** 651
[10] Bondino F, Garg K B, Magnano E, Carleschi E, Heinonen M, Singhal R K, Gaur S and Parmigiani F 2008 *J. Phys.: Condens. Matter* **20** 275205
[11] Fukumora T, Yamada Y, Toyosaki H, Hasegawa T, Koinuma H, Kawasaki M 2004 *Appl. Surface Sci.* **223** 62
[12] Jung S W, An S J, Gyu-Chul Yi, Jung C U, Sung-ik Lee and Choe S 2002 *Appl. Phys. Lett.* **80** 4561
[13] Singhal R K, Dolia S N, Dhawan M, Kumar S, Garg K B, Xing Y, Elisa Saitovitch 2008 *Int. J. Mod. Phys. B (in press)*
[14] Hong N H 2006 *J. Magn. Magn. Mater.* **303** 338
[15] Chambers S April 2002 *Mater: Today* 34
[16] Ohno H Matsukura F, Ohno Y 2002 *JSAP Int.* **5** 4
[17] Wolf S A, Awschalom D D, Buhrman R A, Daughton J M, Von Molnar S, Roukes M L, Chtchelkanova A Y, Treger D M 2001 *Science* **294** 1488
[18] Chiba D, Yamanouchi M, Matsukara F, Ohno H 2003 *Science* **301** 943
[19] Sharma P, Gupta A, Rao K V, Owens F J, Sharma R, Ahuja R, Osro J M Guillen, Johansson B, Gehring G A 2003 *Nature Materials* **2** 673
[20] Sharma P, Gupta A, Owens F J, Inoue A, Rao K V 2004 *J. Magn. Magn. Mater* **282** 115
[21] Blythe H J, Ibrahim R M, Gehring G A, Neal J R, Fox A M 2004 *J. Magn. Magn. Mater.* **283** 117
[22] Kane M H, Shalini K, Summers C J, Vartharajan R, Nause J, Vestal C R, Zhang Z L, Ferguson I T 2005 *J. Appl. Phys.* **97** 23906
[23] Kittilstved K R, Norberg N S, Gamelin D R 2005 *Phys. Rev. Lett.* **94** 147209
[24] Peng Long, Z H-Wu, Wen Qi-Ye, Song Yuan-Qiang, Su Hua and John Q. Xiao 2008 *Chinese Phys. Lett.* **25** 1438
[25] Venkatesan M, Fitzgerald C B and Coey J M D 2004 *Nature* **430** 630

[26] Pemmaraju D P Sanvito S 2005 *Phys. Rev. Lett.* **94** 217205
[27] Hong N H, Sakai J, Poirot N and Brizé V 2006 *Phys. Rev. B* **73** 132404
[28] Sundaresan A, Bhagravi R, Rangarajan N, Siddesh U and Rao C N R 2006 *Phys. Rev. B* **74** 161306
[29] Rodriguez -Carvajal J., *FULLPROF version 3.0.0. Laboratorie Leon Brillouin,CEA-CNRS* (2003)
[30] Hong N H, Sakai J, and Brizé V 2007 *J. Phys.: Condens. Matter* **19** 036219
[31] S. J. Pearton, D. P. Norton, K. Ip 2005 Prog. Mater. Sci. **50** 293
[32] Kolesnik S, Dabrowski B and Mais J 2002 *J. Super.* **15** 251
[33] Cong C J and Zhang K L 2006 *Phys. Stat. Sol. (b)* **243, No. 12** 2764
[34] Iusan D, Sanyal B and Eriksson O 2006 *Phys. Rev. B.* **74** 235209
[35] Ruderman M A and Kittel C 1954 *Phys. Rev.* **96** 99
[36] Yosida K 1957 *Phys. Rev.* **106** 893, also in 1958 *Phys. Rev.* **112** 1649 and also in 1962 *Phys. Rev.* **126** 1470

Another alternative to the solid state reaction method to synthesize nanocrystalline YIG: Ammonium nitrate melt technique

H Sozeri[1] and N Ghazanfari[2]
[1] TUBITAK-UME, National Metrology Institute, PO Box 54, TR-41470, Gebze Kocaeli, TURKEY
[2] Middle East Technical University, Physics Department, 06531 Ankara, TURKEY

E-mail: huseyin.sozeri@ume.tubitak.gov.tr

Abstract. The structural and magnetic properties of yttrium iron garnet particles, synthesized in an ammonium nitrate melt (ANM), were investigated by magnetization measurements, XRD analysis and SEM microscopy and compared with that of the samples prepared using a solid state reaction (SSR) route. The phase formation of YIG starts at lower temperatures with the ANM technique and then develops with increasing temperature and sintering time. An almost single-phase sample was obtained by annealing for 2 hours at 1300°C, after which the YIG fraction in the SSR sample was only 0.34. Similarly, saturation magnetization of the samples sintered in the same conditions is always higher in the samples sintered via the ANM technique. SEM micrographs were used to determine particle sizes of the ANM samples which vary from the sub-micron to the micron range depending on the sintering temperature. These samples have uniform structures, small grains, better magnetic properties and do not contain massive agglomerates. As a result, the synthesis of YIG via the ANM technique represents another alternative to the SSR route.

1. Introduction

Ytrrium iron garnet (YIG) is a ferrimagnetic material with unique features which make this material suitable for various device applications including circulators, oscillators, phase shifters for the microwave region, sensors and lasers. In addition, epitaxially grown YIG crystals on GGG (Gadolinium Gallium Garnet) substrates have been used in studies of the vortex dynamics in high-temperature superconductors for observation of magnetic field penetration [1-4].

Synthesis of this material with strict control of the composition, the homogeneity, the size and the particle shape is crucially important for both magnetic and structural properties of these garnets. The samples prepared by using a SSR route generally include $YFeO_3$ and Fe_2O_3 as impurity phases and these remain in the sample unless heated to high temperatures [5]. Generally, the final product has a limited degree of homogeneity. The sol-gel, co-precipitation, and glycothermal synthesis have been developed for obtaining finer particles [6-10].

[1] To whom any correspondence should be addressed.

In this report, we introduce a combustion method to synthesize YIG as an alternative method to the SSR route, namely the ANM technique. The method has been successfully applied for sintering ceramic high-T_c superconductors, like $YBa_2Cu_3O_{7-x}$ [11] and BSCCO [12]. The magnetic and structural properties of the samples were compared with those from materials synthesized via the SSR route.

2. Experimental

Appropriate amounts of high-purity Y_2O_3 and Fe_2O_3 were mixed at a 3:5 molar ratio and ground for 15 minutes in ethyl alcohol. The mixture was added to an ammonium nitrate melt, kept at 185°C, and a reddish liquid was formed. Next, the temperature of the mixture was gradually raised to 300°C over a period of 1 hour, and kept at this temperature for 24 hours to obtain the precursor. Milling was performed for 5 hours at 100 rpm with a ball-mixture weight ratio of 10:1. Finally, the powder obtained was sintered at different temperatures, starting from 800°C up to 1300°C, for varying durations, ranging from 1 to 12 hours.

The structural properties of the samples were investigated using an X-ray powder diffractometer (Rigaku Miniflex, Cu-K_α radiation). Quantitative analysis software was used to calculate the phase fractions. The surface morphology and microstructure of the samples were examined with a scanning electron microscope (JEOL 6335F Field Emission Gun). The magnetic characterization of the samples was performed at room temperature using a vibrating-sample magnetometer (LDJ Electronics, Inc. Model 9600) and an applied field of 2 kOe.

3. Results and Discussion

3.1. Structural Characterization

The X-Ray diffraction patterns of the YIG samples synthesized by both the ANM technique and the SSR route are shown in Fig.1. At 900°C the impurities, $YFeO_3$ (also known as YIP) and Fe_2O_3, are in the majority and the YIG does not appear, even as a minor phase. However, annealing at 1000°C for 1 hour revealed the conversion of the YIP into the YIG phase with a fraction of 0.35. When the sample was annealed for 2 hours or more at this temperature, the YIG became the major phase. As the sintering temperature and time were increased, the fraction of YIG was found to increase. For instance, at 1200 °C with 2 hours annealing, it was 0.95 with small traces of YIP retained. An almost pure YIG sample was sintered at 1300 °C with 2 hours annealing; it had a cubic structure with lattice parameter of 12.3585 ± 0.0015 Å. With the SSR method, on the other hand, YIG formation starts at 1200°C with a very small fraction of 0.08 and reaches 0.70 with 12 hours of annealing at 1300°C. Sintering at 1350°C for 12 hours did not result in a pure YIG phase.

The scanning electron microscopy (SEM) micrographs of the YIG samples synthesized with the ANM method show that sub-micron particles were formed at 1000°C with 6 hrs of annealing (Fig.2a). It seems that there is a homogeneous grain structure, i.e., there are no agglomerates or relatively big grains. The average particle size can be estimated as 0.3-0.4 μm. In a sample sintered at 1200°C, the particle size increased to approximately 1 μm and the particles were stuck to each other in a regular and homogeneous manner to form a network-like structure, see Fig.2b. Heat treatment at 1300°C for 2 hours resulted in an average particle size of 1.5 μm (Fig.2c). On the other hand, the samples sintered via the SSR route have grains with sizes that vary over a wide range (from 1 μm up to 10 μm), see Fig.2d-e.

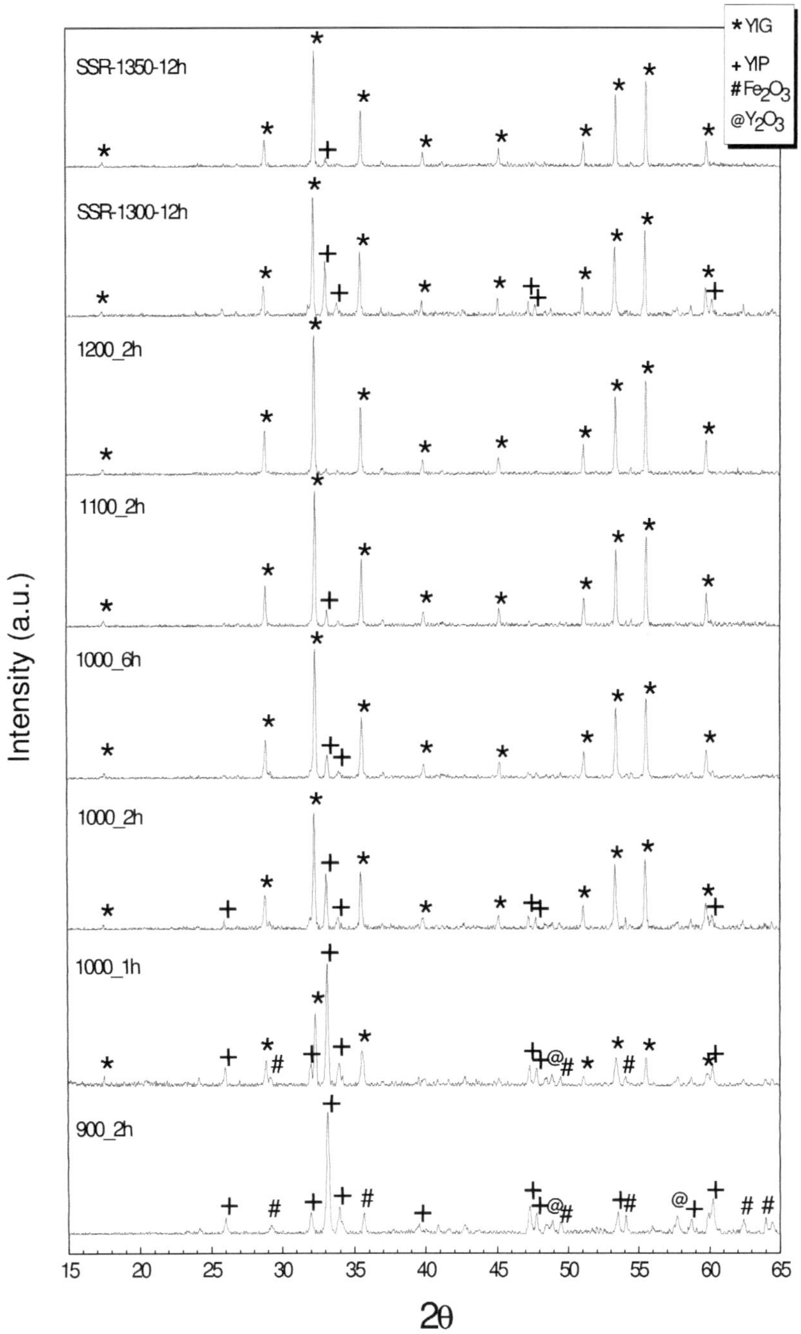

Fig.1: XRD patterns of the samples sintered by ANM and SSR.

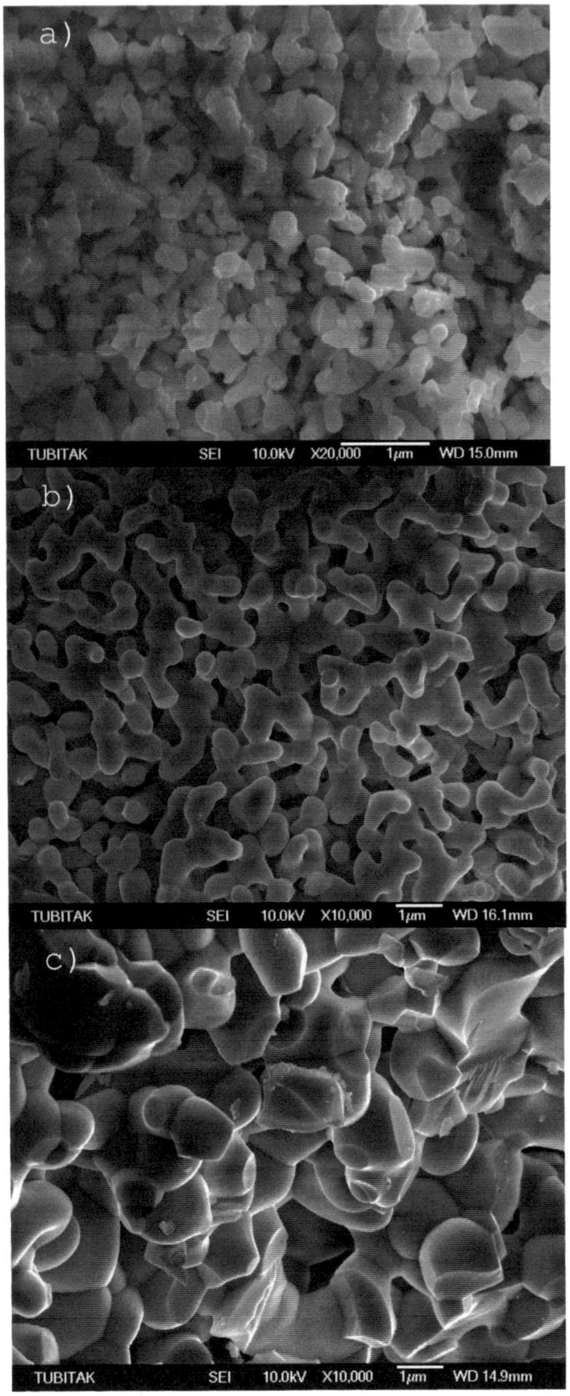

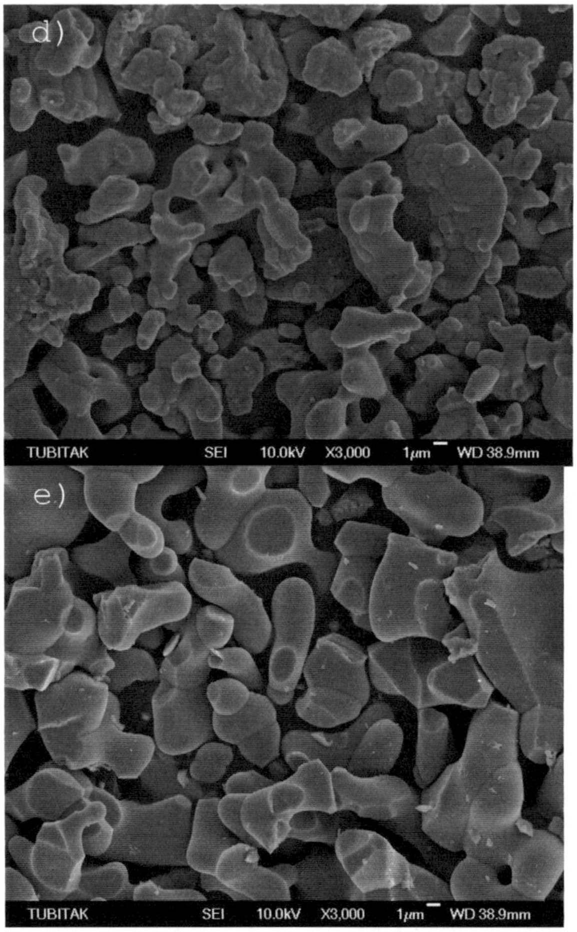

Fig.2: SEM micrographs of the ANM samples sintered at a) 1000°C for 6 hours, b) 1200°C, c) 1300°C for 2 hours and SSR samples annealed at d) 1300°C, e) 1350°C for 2 hours.

3.2. Magnetization Measurements

Room-temperature magnetization curves of some selected YIG samples synthesized by the ANM technique and the SSR route are shown in Fig. 3a for a comparison. It is clear that the ANM sample sintered at 1200ºC for 2 hours has a higher M_s value compared to the SSR samples sintered at higher temperatures for longer times. The saturation magnetization of the ANM samples starts with 9.39 emu/g for the sample annealed at 1000°C for 1 hour and increases at higher sintering temperatures due to the increasing YIG fraction and the crystallinity of the particles, see Fig3b. Then, it saturates at approximately 23 emu/g for the temperatures above 1200ºC.

4. Conclusion

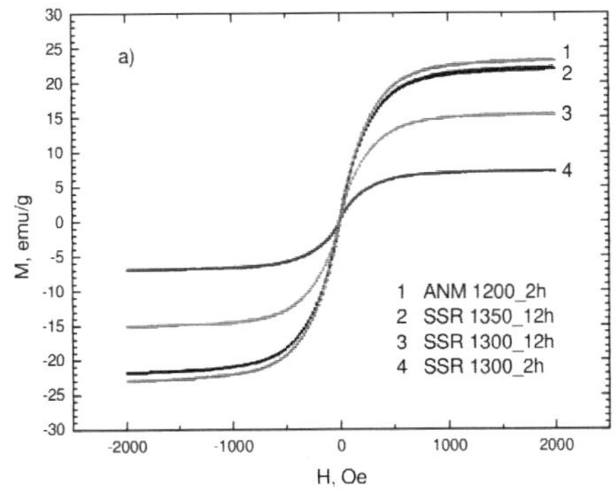

Figure 3a. a) M-H hysterisis curves of some of the selected ANM and SSR samples.

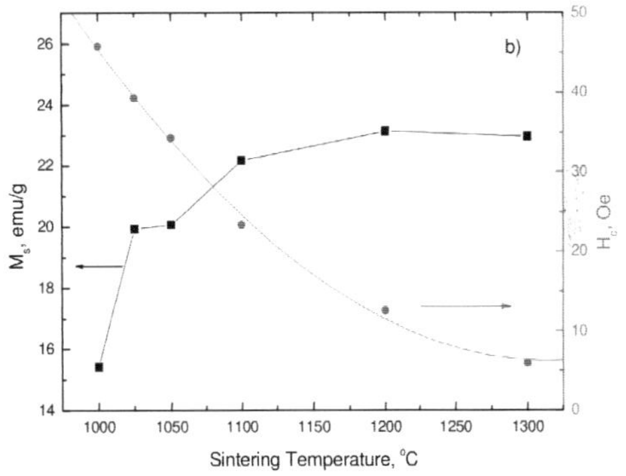

Figure 3b. b) Sintering-temperature dependence of M_s and the coercivity.

YIG was synthesized using the ANM technique. The structural and magnetic properties appeared to be better than those of the SSR samples. It was observed that the phase formation starts at 1000°C and increases with increasing annealing temperature and duration. M_s first increases sharply and then saturates at approximately 23 emu/g for temperatures higher than 1100°C. Meanwhile, H_c decreases from 45 to 5 Oe due to the increasing particle size. This behavior of H_c indicated that the average particle sizes of our garnets were greater than 200 nm, above which the particles have a multi-domain structure. However, the crystallite sizes of the samples determined using the Scherrer formula are between 50-60 nm. In contrast to the SSR samples, the ANM garnets have homogeneous structures, and no big agglomerates are observed in the SEM micrographs. This combustion technique appears to be another alternative to the SSR route, although further work is needed to decrease the average particle size, which will be subject of our next investigations.

5. References

[1] Dorosinskii L, Bocuk H and Topal U, 2001 *Supercond. Science & Tech.* **14** 839
[2] Bocuk H, Dorosinskii L, and Topal U, 2002 *Physica C* **377** 561
[3] Sozeri H, Dorosinskii L, Topal U, 2006 *Physica C* **434/1** 95
[4] Sozeri H and Dorosinskii L, 2003 *Physica C* **388** 745
[5] Grosseau P, Bachiorrini A, Guilhot B, 1997 *Powder Tech.* **93** 247.
[6] Vaqueiro P, Lopez-Quintela M A, Rivas J, Greneche J M, 1997 *J. Mag. Mag. Mater.* **169** 56.

[7] Sanchez R D, Rivas J, Vaqueiro P, Lopez-Quintela M A, Caeiro D, 2002 *J. Mag. Mag. Mater.* **247** 92
[8] Grosseau P, Bachiorrini A, Guilhot B, 1996 *J. Thermal Anal.* **46** 1633
[9] Taketomi S, Ozaki Y, Kawasaki K, Yuasa S, Miyajima H, 1993 *J. Mag. Mag. Mater.* **122** 6
[10] Inoue M, Nishikawa T, Inui T, 1998 *J. Mater. Chem.* **13** 856
[11] Sozeri H, Ozkan H, Ghazanfari N, 2007 *J. Alloys and Comp.* **428/1-2** 1
[12] Celebi S, Karaca I, Ozturk A and Nezir S, 1998 *J. Alloys and Comp.* **268** 256

Magnetic and fluorescence properties of cobalt implanted hydrogels

H Sozeri[1], A Gelir[2], E Alveroğlu[2], M Tulun[2], Y Yılmaz[2]

[1]TUBITAK-UME, National Metrology Institute, PO Box 54 TR-41470, Gebze Kocaeli, Turkey
[2]Department of Physics, Istanbul Technical University, Maslak, 34469 Istanbul, Turkey

E-mail: huseyin.sozeri@ume.tubitak.gov.tr

Abstract. Co-doped plain gels (PAAm) were prepared with and without pyranine fluoroprobe (8-hydroxypyrene-1, 3,6-trisulfonic acid, trisodium salt; POH), which have a OH^- reactive group and three SO_3^- ionic groups. Magnetization and fluorescence measurements were used for characterization of the gels. It was observed that both the plain gel itself and the gel including only POH (PAAm-POH) have a diamagnetic response to the external field. When the gels were synthesized together with cobalt (PAAm-Co) and cobalt/POH (PAAm-Co-POH), both gels magnetized in the direction of the magnetic field, as expected. However, the magnitude of the magnetization was different in the gels which were synthesized in the presence of POH and the one into which POH molecules were diffused after the synthesis, even though they have the same cobalt concentration. Fluorescence measurements and SEM micrographs were used to understand this surprising difference in the magnetization of PAAm-Co-POH gels. It was proposed that the electrostatic interaction between SO_3^- groups of POH and free electrons of cobalt (Co^{+2}) atoms led to the formation of Co-POH clusters having fewer unpaired electrons, and this results in a decrease in overall magnetization. It was also observed from the SEM images that the size of clusters depends on the Co-POH composition.

1. Introduction

The synthesis of hydrogels which are sensitive to external stimuli like pH, temperature, light intensity and magnetic field have become an interesting subject of study. These gels may have applications in medicine, drug-delivery, bio-sensors, etc.[1,2] More recently, dual response polymers like pH/thermo, thermal/magnetic and pH/magnetic have been developed to increase the number of possible applications [3-7].

Hydrogels which are sensitive to magnetic fields can be synthesized by introducing ferro- [8,9] or paramagnetic [10] nanoparticles either during the gelation or by diffusing them into previously formed gels. Magnetic gels, especially ferrogels which include ferromagnetic particles, are typical examples of smart materials for mechanical actuators and the subject of many studies in recent years. For

[1] To whom any correspondence should be addressed.

instance, it was also reported that hydrogels cross-linked by paramagnetic ions can be used as colorimetric pH sensors or used to remove organic materials from aqueous solution [10].

In our work, cobalt particles were embedded in hydrogels during the gelation and pyranine (POH) was used as a fluoroprobe which can be used for monitoring gelation, phase transitions and the fractal nature of the gels [11]. Magnetic and fluorescence measurements together with SEM images were used to characterize the magnetic hydrogels.

2. Experimental

The PAAm gels were synthesized as described in Ref [16]. The gels prepared with Co, PAAm-Co gels (sample **2**), were prepared for two different concentrations of Co(II)acetate, 10^{-3}M, and 0.3M. The gels prepared with POH, PAAm-POH gels (sample **3**), were prepared with 10^{-4}M concentration of POH. The gels prepared with both Co and POH, PAAm-Co-POH gels (sample **4**), were prepared with two different concentration of Co(II) acetate, 10^{-3}M (sample **4L**), 0.3M (sample **4H**), and 10^{-4}M POH for each sample.

Once prepared, the four types of gels were divided into three different groups and the first one used without washing, the second and the third ones were washed with pure and high pH water (pH=11), respectively. Then, samples, **1**, **2**, and **4**, were swelled in sufficient amounts of pure water and the sample **2** was swelled in similar way in 10^{-4}M POH solution (10^{-4}M POH dissolved in pure water, sample **2D**) to bring them to certain swelling ratios.

The fluorescence spectra of the gels were recorded by using a charged-coupled-device (CCD) array camera (Ocean Optics USB2000) [12]. The fluorescence measurements were carried out at a 90° angle. POH molecules were excited at a wavelength of 400 nm. Magnetic measurements were performed by using a SQUID magnetometer (Quantum Design, MPMS XL).

3. Results and Discussion

3.1. Magnetic Characterization of the gels

Magnetization measurements, performed at several different applied fields but only two of them were included here, of the samples 1, 2, 3 and 4 were shown in Fig.1a as a function of increasing mass ratio, m/m_0. Here, m represents the mass after swelling and m_0 is the initial mass (i.e. dry mass) of the gels. The swelling process was performed in such a way that the diffusion of both POH molecules and Co-atoms were prevented. It was observed from the figure that the plain gel, PAAm, and the gel including POH, PAAm-POH have a diamagmetic response to the applied field and it increases with increasing water content (i.e. swelling ratio) which is also diamagnetic [13]. The magnetic behavior of Co-doped samples appeared to be more interesting, in that they have different positive magnetic moments despite both gels (2 and 4) having the same cobalt concentration. In order to understand this surprising result, we performed fluorescence measurements and used SEM images which were taken from the surface of the gels.

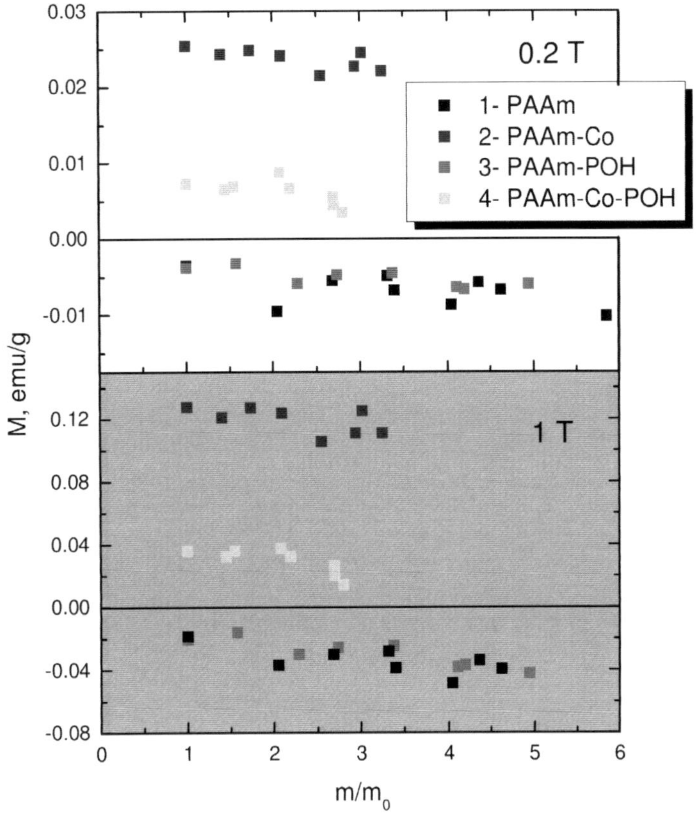

Figure 1. Magnetic behavior of the gels as a function of swelling ratio and at two different applied fields, 0.2 T and 1T.

3.2. Fluorescence Measurements

The fluorescence spectra of PAAm-Co-POH gels (sample **4**) in the collapsed and swollen states and the one which was washed with high pH water, are presented in Fig2. Almost no fluorescence intensity was observed for the gel including high Co content (**4H**), but the intensity became considerable for the gel including low Co content (**4L**) in the collapsed state (Fig. 2a). When these gels were swelled with sufficient water (pH=6), the fluorescence intensity became apparent for **4H** and it increased for the gel **4L** (Fig. 2b). The peak around 435 nm corresponds to the POH molecules which are chemically bonded to the PAAm strands, and the other peak around 500 nm (broad peak) is the superposition of the fluorescence peaks of free (515 nm) and excimer forms (480 nm) of POH molecules, details of which can be found in Ref. [14].

Fig.2c corresponds to the fluorescence spectra of samples **4H** and **4L** after washing with high pH (pH=11) water. The intensity disappears almost completely for the sample **4H**; in contrast, there is no significant change for the sample **4L** except for some decrease in the 515 nm side of the broad peak, which is due to the diffusion of free POH molecules out of the gel. The same results were found when the gels were washed with pure water. The peak around 480 nm which becomes apparent after washing indicates that POH molecules did not leave the gel. The fluorescence spectra presented in Fig.2c were also supported by the magnetization measurements performed in these gels, which were washed with high pH water and then brought to the collapsed state, in the following way. The change in the magnetization of these samples after washing was completely different from each other, ~0% for sample **4L** and ~86% for **4H**. This observation clearly shows that the Co atoms in sample **4H** were washed out of the gel together with the POH molecules during the washing process.

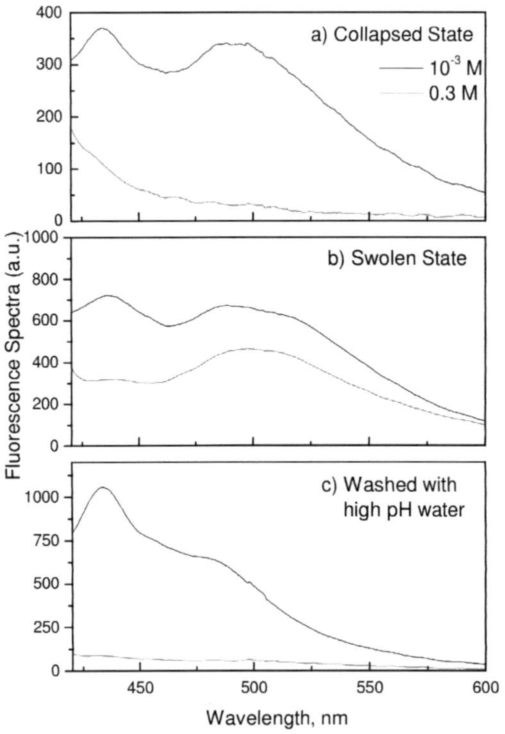

Figure 2. Swelling and washing effect on the fluorescence spectra of POH in the PAAm gel with Co

3.3. Microstructure of the gels: SEM Micrographs

SEM images of the samples 2, 4 and 2D (POH diffused) in the collapsed state are shown in Fig3; these images were taken from the non-magnetized samples. In other words, no magnetic field was applied before the SEM analysis. Cobalt clusters are formed in all PAAm gels, however the mechanism of the formation is quite different in samples 2 and 4. In sample 2, Co-atoms are connected by ponderamotive forces, while in sample 4 it is the electrostatic interaction between the empty electronic states of Co atoms and the free electrons of SO_3 groups in POH molecules which keeps the Co-atoms and POH molecules together. SEM-EDX analysis showed that these big and small clusters include Co-atoms which can not be due to the magnetic interactions, since magnetic fields were not applied either during the synthesis or the SEM analysis.

The average size of these cobalt clusters depends on the composition of Co and POH. Synthesizing the composites in the presence of Co and POH together (sample **4**) gave the largest average cluster size, ~ 5600 ± 500 nm. In addition, the density of these clusters is higher than the others formed in the composites prepared without POH. If the composites are synthesized with only Co, the average size of the clusters decreases to ~ 1550 ± 260 nm. When the POH molecules are diffused into previously formed PAAm-Co gel (**2D**), the average size surprisingly gets smaller than that of the others.

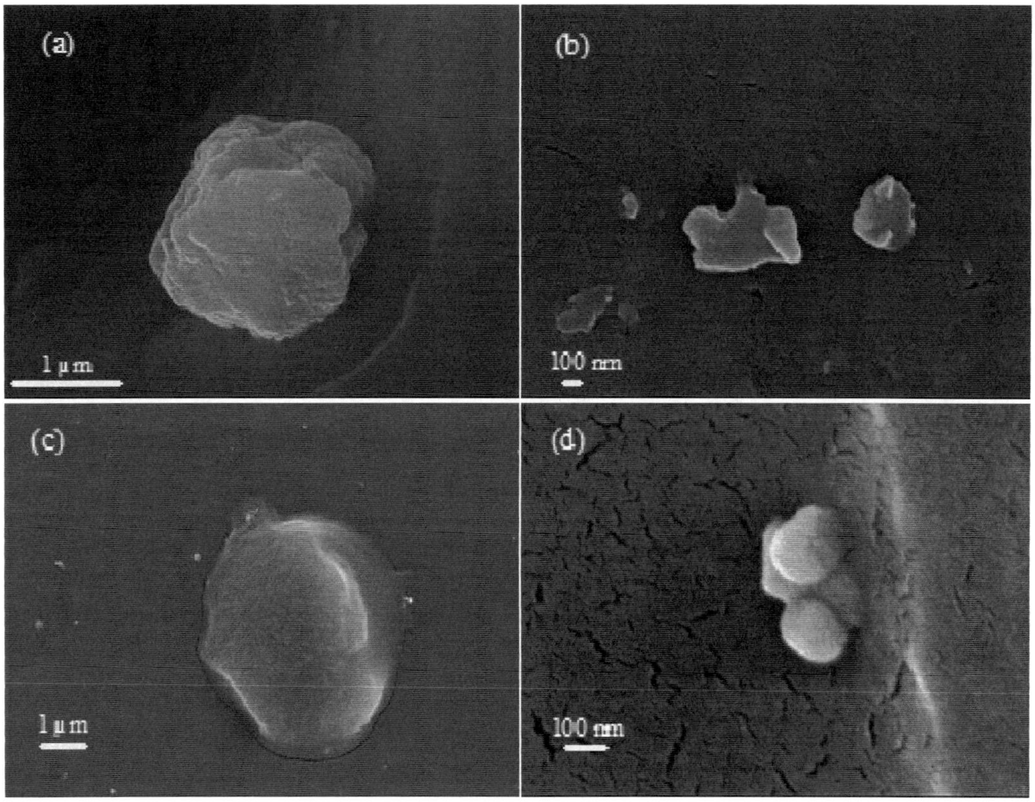

Figure 3. SEM micrographs of samples **2** (a) and (b), **4** (c) and **2D** (d) in collapsed states.

In summary, our results showed that synthesis of the gel together with POH molecules results in stable and large clusters between Co and POH molecules via only ionic interactions, i.e. electrostatic interactions. The cluster formation with Co atoms and POH molecules can occur only during the polymerization and these clusters can not be created by the magnetic interactions between the Co-atoms, since there was no applied field during the synthesis. In other words, Co-atoms were not magnetized during the synthesis which may cause an attractive interaction. When the POH molecules were diffused into the PAAm-Co gel after the preparation, no change in the magnetization was observed. They behave like flocculants and decrease the average size of the Co clusters dispersing them via attractive electrostatic interactions. Therefore, the stability and size of Co clusters formed in the gel composites are controlled by the anionic agents, POH.

Journal of Physics: Conference Series **153** (2009) 012067 doi:10.1088/1742-6596/153/1/012067

4. References

[1] Qiu Y, Park K 2001 *Adv. Drug Delivery Rev.* **53**, 321
[2] Miyata T, Uragami T, Nakamea K 2002 *Adv. Drug Delivery Rev.* **54**, 79
[3] Kim J Y, Lee S B, Kim S J, Lee Y M 2002 *Polymer* **43**, 7549
[4] Furukawa H, Shimojyo R, Ohnishi N, Fukuda H 2003 *App. Microbiol. Biotech.* **62**, 478
[5] Deng Y, Tang W, Wang C, Fu S 2003 *Adv. Mater.* **15**, 1729
[6] Pich A, Bhattacharya S, Lu Y, Boyko V, Adler H. 2004 *Langmuir* **20**, 10706
[7] Fernandes R, Wu L Q, Chen T, Yi H, Rubloff G W, Ghodssi R, Bentley W E, Payne G F 2003 *Langmuir* **19**, 4058
[8] Herna´ndez R, Sarafian A, Lo´pez D, Mijangos C 2004 *Polymer* **46**, 5543
[9] Chatterlee J, Haik Y, Chen C J 2003 *Colloid Polym. Sci.* **281,** 892
[10] Winkleman A, Bracher P J, Gitlin I, Whitesides G M 2007 *Chem. Mater.* **19** , 1362
[11] Pekcan Ö, Yılmaz Y, Okay O 1997 *Polymer* **38**, 1693
[12] Yilmaz Y, Gelir A, Alveroglu E and Uysal N 2008 *Physical Review E* **77**, 051121
[13] Kochmarsky V, 1996 *Magnetic and Electrical Separation* **7**, 77
[14] Valeur B, 2002 *Molecular Fluorescence Principles and Applications* Weinheim Germany, Wiley-Vch
[15] Gelir A, Yilmaz I and Yilmaz Y, 2007 *Journal of Physical Chemistry B* **111**, 478
[16] Yılmaz Y, Gelir A, Salehli F, Nigmatullin R R and Arbuzov A A, 2006 *J. Chemical Physics* **125**, 234705

FMR studies of half metallic ferromagnetic thin films Co_2MnSn and Co_2MnGe

R Yilgin[1,*], S Kazan[1], B Rameev[1,2], B Aktas[1], K Westerholt[3]

[1] Department of Physics, Gebze Institute of Technology, Gebze, 41400, Turkey.
[2] Kazan Physical Technical Institute, 420029 Kazan, Tatastan, Russian Federation.
[3] Institute for Experimental Physics/Condensed Matter Physics, Rurh-University Bochum.

E-mail: resul_yilgin@yahoo.com

Abstract. Thin films of Heusler alloys Co_2MnSn and Co_2MnGe prepared on sapphire *a*-plane substrates by RF-sputtering have been investigated by ferromagnetic resonance (FMR) technique. The magnetic anisotropies of the Co_2MnSn and Co_2MnGe thin films have been probed and the effective magnetization values have been extracted. Weak in-plane anisotropies observed for the Co_2MnGe sample indicate a textured growth of the films. Non-uniform (spin-wave and interface/surface) modes have been observed and the exchange stiffness constants for the Co_2MnSn and Co_2MnGe films have been estimated.

1. Introduction

The spin polarization of the conduction electrons is the most important parameter determining the performance of various spintronic devices [1]. For that reason, half-metallic Ferro magnets, which have completely spin-polarized electrons at the Fermi level, are very attractive for the spintronic applications. However in order to construct the multilayered MTJ device one need to know in details the magnetic anisotropies of a thin film material to tailor them in a desirable way. Besides any effects of surfaces/interfaces and bulk in-homogeneities should be studied thoroughly to understand physics involved and realize in practice a half-metallic MTJ device.

It is well known that ferromagnetic resonance (FMR) is a very efficient technique to study the nanoscale magnetic properties of various ferromagnetic materials. In this work thin films of various half metallic Heusler alloys Co_2MnSn and Co_2MnGe have been investigated using FMR technique.

2. Experimental procedures

Thin films of Co_2AlGe and Co_2AlSn (both with thickness of 100 nm) were prepared by RF-sputtering on sapphire a-plane substrates covered by a 5 nm thick V buffer layer and were finally capped by protecting overlayer of amorphous Al_2O_3 with thickness of about 5 nm, see details elsewhere [2,3].

* R Yilgin, resul_yilgin@yahoo.com, Gebze Institute of Technology, Science Faculty, Department of Physics, Cayirova, 41400 Gebze, Kocaeli, Turkey. Tel:+90(262)605 1321, Fax:+90(262)653 8490.

FMR spectra were recorded on Bruker EMX spectrometer (9.8 GHz) at room temperature. Angular dependences of EPR spectra were recorded with the magnetic field rotated either in the film plane ("in-plane" geometry) or in the plane which is perpendicular to the film plane ("out-of-plane" geometry). It should be noted that due to restrictions of the experimental setup in the in-plane geometry the microwave magnetic field is perpendicular to the film plane, while for the out-of-plane geometry this field is in the plane of film.

3. Experimental results and discussion

Splitting of the FMR signal due to excitation of non-uniform modes are observed (figure 1). Noticeable in-plane anisotropy for the Co_2MnGe film is evident from the angular dependence of FMR resonance fields H_{res} (Figure 2), while in-plane angular dependence for the Co_2MnSn film having nearly the same magnitude is masked by very broad line-width in the parallel orientation (about 1 kOe).

A very high anisotropy is observed in the out-of-plane measurements, while the magnitude of the in-plane anisotropy (figure 2) is very small. Therefore, it is possible to extract the effective magnetization M_{eff} from the out-of-plane dependences of H_{res} (not presented) using in theoretical modelling only the shape anisotropy contribution. Here, the term "effective" means that M_{eff} contains in general a small contribution due to various anisotropies, e.g. the surface or strain anisotropies. Computer simulations provide rather high values: M_{eff} = 892 emu/cm^3 for the Co_2MnGe film, and M_{eff} = 790 emu/cm^3 for the Co_2MnSn sample. Both results are very close to the bulk values, indicating an absence of sizable B2-type disorder [3, 4, 5], and in the excellent agreement with the saturation magnetization values, reported previously [6].

Discussing FMR results for Co_2MnGe Heusler film it should be noted that the sample is rather uniform magnetically as evident from narrow signal line-widths (about 30 Oe near the perpendicular position), while for the Co_2MnSn sample the FMR signal in the parallel orientation is very broad (up to 1 kOe). This indicates larger magnetic inhomogeneity of the Co_2MnSn film comparable with the Co_2MnGe sample.

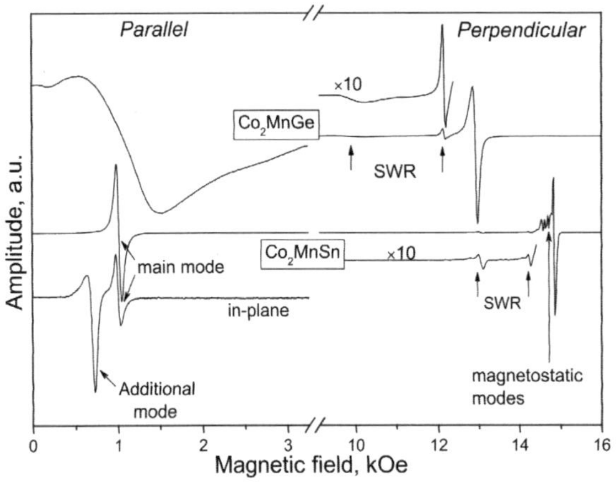

Figure 1. FMR spectra of Co_2MnGe and Co_2MnSn films for the DC magnetic field applied in parallel (left) and perpendicular (right) to the film plane.

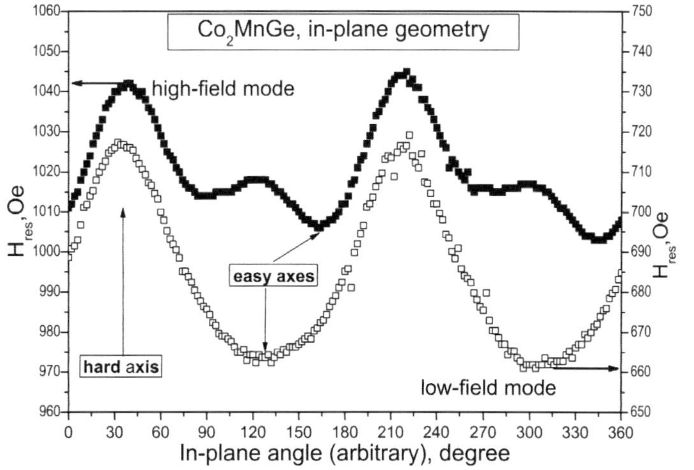

Figure 2. In-plane angular dependence of resonance fields for Co_2MnGe film.

Spin-wave (exchange) modes are observed for both Co_2MnGe and Co_2MnSn samples, that allows us to estimate the exchange stiffness constant D. Our estimations similar to the procedure applied in ref. [5] result in $D = 1.5\times10^{-9}$ Oe cm^2 for the Co_2MnSn sample, while for Co_2MnGe sample the estimations are very rough ($D = 0.8\div1.8\times10^{-9}$ Oe cm^2) due to overlapping between the SWR and magneto-static modes. It is remarkable that the above values are close to our estimation of $D= 1.9\times10^{-9}$ Oe cm^2 for Co_2MnSi system [5]. Besides, for the Co_2MnGe thin film the multiple magneto-static (or "magneto-exchange") modes (figure 1) are revealed in FMR spectra. These modes may appear, for instance, as result of exchange decoupling at grain boundaries, as proposed in refs. [7, 8].

The only main mode at higher field is observed in the parallel orientation of the out-of-plane geometry, while an additional lower-field mode is revealed in the in-plane geometry (figure 1). The first one is the bulk mode with nearly uniform spin precession, while an appearance of the additional mode is attributed to the "unconventional" in-plane geometry with non-uniform radiofrequency field distribution through the sample. Therefore, this mode is apparently some kind of non-uniform precession of magnetization (e.g. surface/ interface mode). Remarkably, there is a difference between the in-plane behaviour of the low- and high- field modes (figure 2). The low-field mode possesses dominant two-fold behaviour, which is attributed to a growth induced uniaxial anisotropy due to the stepped surface of the a-plane sapphire substrate [2, 3]. However, the main mode exhibits the dominant cubic anisotropy with symmetry expected for the DC field rotated in the (110) crystalline plane (some small uniaxial term is probably due to interface-induced anisotropy). We estimated the cubic anisotropy term (K_1/M_S) to be about ~10 Oe. A weak cubic anisotropy for this mode points to a partial in-plane (110) texture for the Co_2MnGe film. Formation of out-of-plane (110) texture for Co_2MnGe and Co_2MnSn films grown on V buffer has been reported previously in ref. [6].

Figure 3 shows the both experimental and theoretical angular dependence of resonance field results in the in-plane (a) and out-of-plane (b) geometry of Co_2MnGe film obtained at 35K. We observed contribution of both four-fold and two-fold anisotropies effect for Co_2MnGe film at low temperature, uniaxial anisotropy is much higher. We also attributed M_{eff} value for the Co_2MnGe film as 1000 emu/cm^3, at 35 K.

In conclusion, we have shown that Co_2MnGe and Co_2MnSn films grown on a-plane Al_2O_3 substrate reveal weak in-plane anisotropies. Rather large magnetic in-homogeneity of the Co_2MnSn film comparing with the Co_2MnGe sample has been observed. Cubic symmetry of the in-plane anisotropy for the Co_2MnGe sample indicates a partially textured growth of this film. The effective magnetization values and the exchange stiffness constants for the Co_2MnSn and Co_2MnGe films have been estimated.

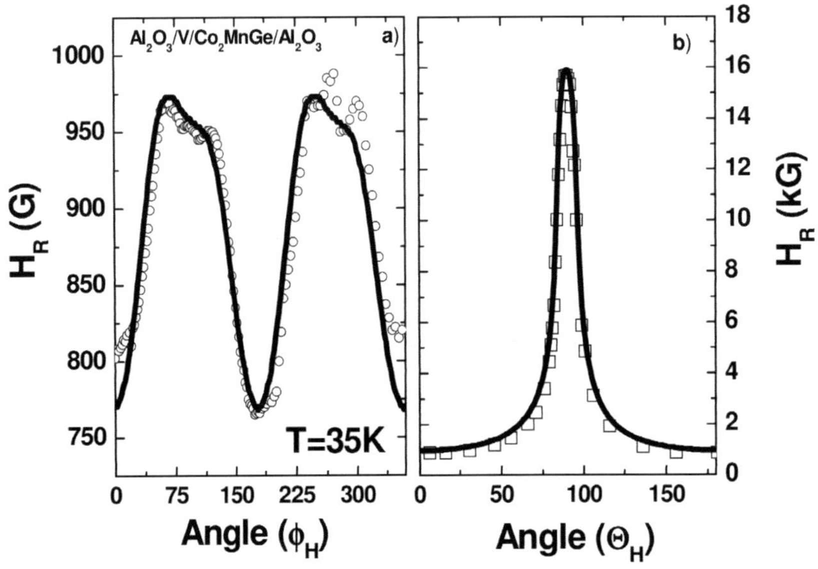

Figure 3. In-plane (a) and out-of-plane (b) angular dependences of resonance fields of FMR spectra recorded at 35K for Co_2MnGe film. The open circles and squares denote the experimental results in the in-plane and out-of-plane geometries, respectively. Continuous lines represent the theoretical fitting results.

4. References

[1] Wolf S A, Awschalom D D, Buhrman R A, Daughton J M, von Molnar S, Roukes M L, Chtchelkanova A Y and Treger D M 2001 *Science* **294** 1488

[2] Verdujin E A and Westerholt K 2006 *J. Appl. Phys.* **99** 084502

[3] Geiersbach U, Bergmann A and Westerholt K 2002 *J. Magn. Magn. Mater.* **240** 546

[4] Ziebeck P J and Webster K R A 1986 *Landolt-Börnstein New Series* III/19c

[5] Rameev B, Yildiz F, Kazan S, Aktas B, Gupta A, Tagirov L R, Rata D, Buergler A, Gruenberg P, Schneider C M, Kammerer S, Reiss G and Hutten A 2006 *Phys. stat. sol. (a)* **203** 1503

[6] Geiersbach U, Bergmann A and Westerholt K 2003 *Thin Solids Films* **425** 225

[7] Rameev B Z, Gupta A, Anguelouch A, Xiao G, Yildiz F, Tagirov L R and Aktas B 2004 *J. Magn. Magn. Mater.* **272-276** 1167

[8] Bergmann A, Grabis J, Toperverg B P, Leiner V, Wolff M, Zabel H and Westerholt K 2005 *Phys. Rev. B* **72** 214403

Ferromagnetism in Mn-implanted Ge/Si nanostructure material

I. T. Yoon[1], C. J. Park, and T. W. Kang

Quantum Functional Semiconductor Research Center, Dongguk University, Seoul, 100-715 Korea

ityoon@dongguk.edu

Abstract. Multistacked Ge quantum dots (QDs) with Si spacers of different thicknesses have been grown on (100) Si substrates by rapid thermal chemical vapor deposition followed by Mn ion implantation and post-annealing. The presence of a ferromagnetic structure was confirmed in the insulating $(Si_{0.45}Ge_{0.55})Mn_{0.03}$ diluted magnetic quantum dots (DMQD) and semiconducting $(Si_{0.45}Ge_{0.55})Mn_{0.05}$ DMQD. The DMQD materials were found to be homogeneous, and to exhibit p-type conductivity and ferromagnetic ordering with Curie temperatures $T_C = 350$ and 160 K respectively. The x-ray diffraction (XRD) data show that there is a phase separation of Mn_5Ge_3 from the MnGe nanostructure. Temperature-dependent electrical resistivity in the semiconducting DMQD material indicates that manganese introduces two acceptor levels in germanium at 0.14 eV from the valence band and 0.41 eV from the conduction band implying that Mn substitutes for Ge. Therefore, it is likely that the ferromagnetic exchange coupling of DMQD material with T_C = 160 K is hole-mediated due to formation of bound magnetic polarons and the ferromagnetism in the sample with $T_C > 300$ K is due to Mn_5Ge_3 phase.

1. Introduction

Diluted magnetic semiconductors (DMSs), often referred to as semi-magnetic semiconductors, based on III-V semiconductors have attracted a great deal of attention recently because of their potential application in spintronic devices that exploit the charge and spin of electrons [1]. The lattices of these materials consist of magnetic ions partially substituting for some of the cations, thereby inducing a local magnetic moment in the lattice and donating carriers into the system. The ferromagnetic nature of these materials is caused by the exchange interaction between localized magnetic moments introduced by the magnetic ions and the carrier spins. Therefore, for device applications, it is desirable to find materials that exhibit ferromagnetism at as high a temperature as possible. Since Dietl et al. [2] predicted that wide band-gap DMS materials, such as GaMnN and ZnMnO, can have Curie temperatures (T_c) above room temperature, based on the mean-field Zener model of ferromagnetism, several materials with Curie temperatures greatly above room temperature have been found in many studies [3-8]. In particular, group-IV-based DMS materials [9-13] with high Curie temperatures have attracted much attention because of their compatibility with conventional semiconductor integrated circuit manufacturing techniques since Park et al. reported an epitaxially grown single crystal of

[1]To whom any correspondence should be addressed.

Mn_xGe_{1-x} ($x = 3.5\%$) with a magnetically ordered phase and a Curie temperature up to 116 K because of a long-range ferromagnetic interaction [9]. The structural, electronic, and magnetic properties for Si_xGe_{1-x} alloys doped with Mn have been investigated theoretically and experimentally [14-17]. In this paper, we report on the magnetic and transport properties of $(Si_{0.45}Ge_{0.55})Mn_{0.05}$ diluted magnetic quantum dots (DMQD), formed through ion implantation and post-annealing. Energy dispersive x-ray fluorescence (EDX), x-ray diffraction (XRD), atomic force microscopy (AFM), high-resolution transmission electron microscopy (HRTEM), superconducting quantum interference device (SQUID), Hall effect and Raman measurements were performed to characterize the samples.

2. Experimental

Ten and twenty-stacked Ge QDs with Si spacers of different thicknesses were grown on *p*-type (100) Si substrates by rapid thermal chemical vapor deposition (RTCVD) with a base pressure of 2×10^{-7} Torr using SiH_4 and GeH_4 as Si and Ge source gases. The QD multilayers consisted of ten and twenty bilayers, each of which was composed of 6.5 monolayers of Ge and a Si-spacer layer of 14–100 nm in thickness. The samples underwent the following growth procedures: the Ge quantum dots (QDs) were deposited in the Stranski-Krastanow (SK) growth mode. After baking at 1000 °C in hydrogen, ten and twenty layers of self assembled Ge QDs with 44 and 59 nm-thick Si barrier (hereafter referred as samples A and B) were grown on a 200 nm-thick Si buffer layer at 700 °C with a pressure of 100 mTorr. The Ge QD samples A and B grown by RTCVD were uniformly implanted with Mn^+ ions at an energy of 200 keV and a dose of 2.5×10^{16} and 5×10^{16} cm^{-2}, respectively. After implantation, thermal annealing was performed at 650 °C under Ar gas at ambient pressure for 10 minutes in a rapid thermal annealing (RTA) furnace to recover the damage. XRD and AFM were employed to analyze the structural and morphological properties of the annealed samples. Hysteresis loops were measured at temperatures in the range 4 - 300 K using a SQUID, and a high-sensitivity (10^{-8} emu) alternating gradient magnetometer (AGM). In all the measurements, the magnetic field was applied parallel to the sample surface. The concentration of Mn in the sample A and B using an EDX was determined at around 3.0 and 5.0%, respectively. The grown sample doped with manganese exhibited p-type conductivity in Hall effect measurements. The hole concentration and resistivity of the $(Si_{0.45}Ge_{0.55})Mn_{0.03}$ sample A was determined to be 1×10^{14} cm^{-3} and 1.7×10^4 Ω cm, respectively. The hole concentration and resistivity of the $(Si_{0.45}Ge_{0.55})Mn_{0.05}$ sample B was determined to be 1×10^{15} cm^{-3} and 1.2×10^3 Ω cm, respectively.

3. Results and discussion

Fig. 1 shows the XRD patterns of the as-grown and annealed samples A and B. Si(002) and Si(004) substrate peaks can be observed from the as-grown sample. The XRD spectra for the annealed samples A and B are similar, containing a dominant Mn_5Ge_3 phase. This phase is known to have a theoretical Curie temperature above room temperature [18].

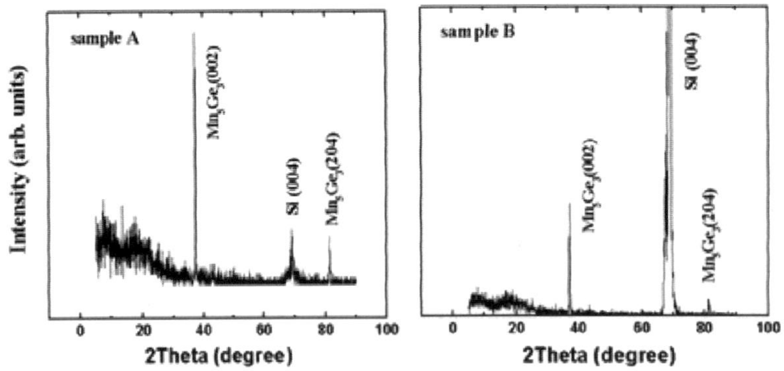

Fig. 1. XRD spectrum for the samples A and B after thermal annealing for 10 min.

Fig. 2 shows cross-sectional HRTEM images of ten and twenty stacked Ge QDs with 44 and 59 nm-thick Si barriers. The density of the dots was approximately 2×10^{8} cm^{-2}. They demonstrate that the Ge QDs are vertically ordered and fully contrasted with the Si spacer. No misfit dislocations were observed from the TEM images. The height and the base of the dots range from 8 to 15 nm and from 140 to 200 nm, respectively.

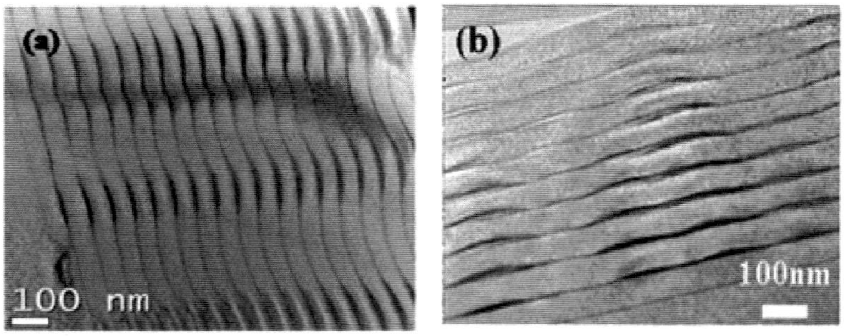

Fig. 2. Cross-sectional HRTEM images of twenty and ten-stacked Ge QDs with Si spacers of (a) 59 nm and (b) 44 nm in thickness.

Fig. 3(a) and (b) shows the AFM 2 μm image of samples A and B, respectively. It is noted that some DMQDs seem to merge into a single large QD. The DMQDs can still be seen after the thermal annealing process.

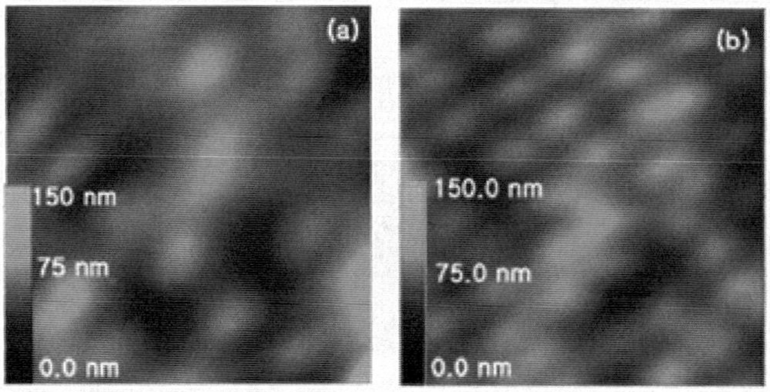

Fig. 3. Fig. 3(a) and (b) shows the AFM 2 μm image of samples A and B, respectively.

Fig. 4 shows the magnetic field-dependent magnetization, M, at 5 K for DMQD samples A and B after RTA for 10 min, with the magnetic field applied parallel to the plane of the film using an AGM. Clearly elongated hysteresis loops are observed in Fig. 4, indicating that ferromagnetic ordering is present; the diamagnetic background of the substrate was subtracted from the data. The samples showed coercivities in the range 40 ~ 60 Oe. The magnitude of hysteresis of sample B is greater than that of the sample A because of the increased Mn concentration.

The temperature dependence of the magnetization curve, M(T), for DMQD samples A and B are presented in Fig. 5. A negative magnetization signal is shown up to 350 K for sample A. Squares in Fig. 5 denote the magnetization of the insulating sample A. It can be determined that the Curie temperature for the sample A is around 350 K whereas the Curie temperature for the semiconducting sample B is around 160 K. The observed temperature-dependent magnetization curve M(T) for samples A and B appears to be very different from the usual convex M(T) curve expected from the

Weiss mean-field theory [19]. The temperature-dependent magnetization shows an unusual outwardly concave shape that is consistent with the behaviour of a GeMn system [9], suggesting a signature of charge carrier localization. It has been shown that such a non-mean-field like behaviour in M(T) curves is much more typical of semi-insulating DMS systems than metallic ones [20-21].

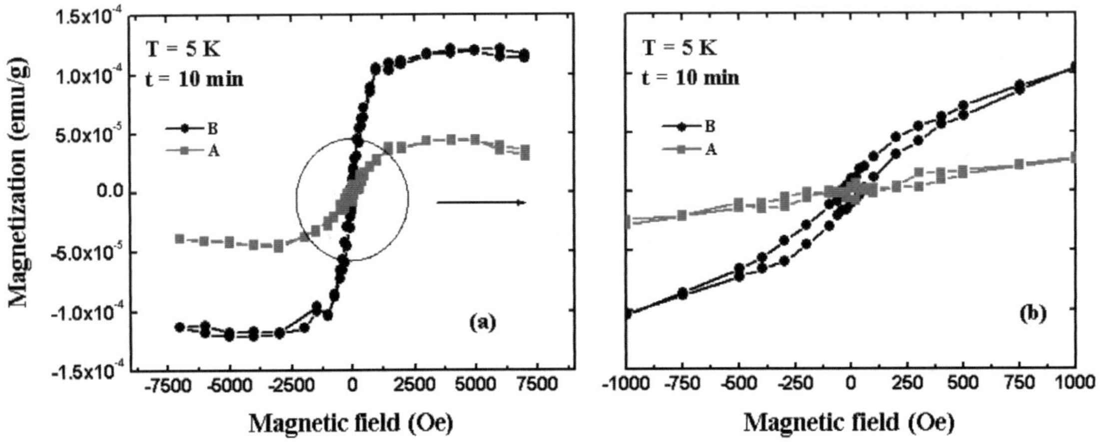

Fig. 4. (a) A ferromagnetic hysteresis loop at 5 K for DMQD samples A and B after thermal annealing for 10 min. The solid line serves as a visual guide (b) Magnified view of the hysteresis loop.

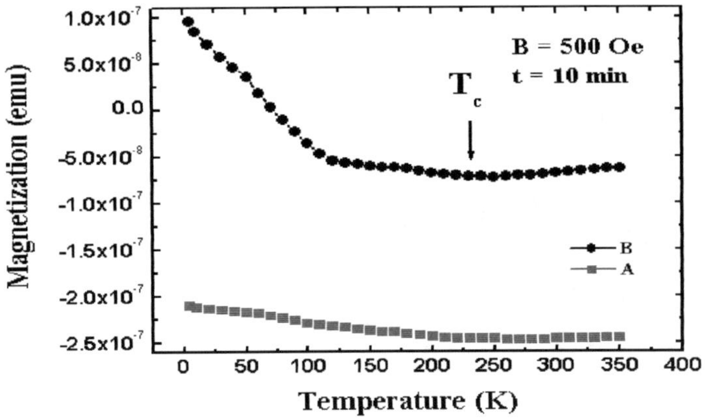

Fig. 5. Temperature dependence of magnetization for DMQD sample A and B after thermal annealing for 10 min. obtained by SQUID magnetometry with a magnetic field of 500 Oe.

Fig. 6 shows the electrical resistivity as a function of temperature for the sample A. As can be seen from Fig. 6, the resistance increases with decreasing temperature, which is also typical semi-insulating behaviour. The rate of resistance change can be divided into two regions. The temperature-dependent resistance was fitted with an Arrhenius function $\rho = \rho_0 \exp(\varepsilon_a/kT)$ where ε_a is the activation energy. Low lying levels at 0.05 and 0.01 eV from the valence band could be observed in Fig. 6. The source of these levels could not be ascertained; it does not seem to be directly related to the manganese [22] except clusters. We assumed that these differences in magnetic behaviour between samples A and B were due to the differences in electrical properties. They show clearly that both the semiconducting and insulating samples were ferromagnetic. Given the highly insulating nature of sample A, no itinerant hole based picture is possible, and so the RKKY-type scenario discussed extensively in the framework of GaMnAs magnetization [1,2] simply does not apply to sample A. However, the XRD spectra for the annealed sample A containing a dominant Mn_5Ge_3 phase indicates that the ferromagnetism of the sample is due to this phase.

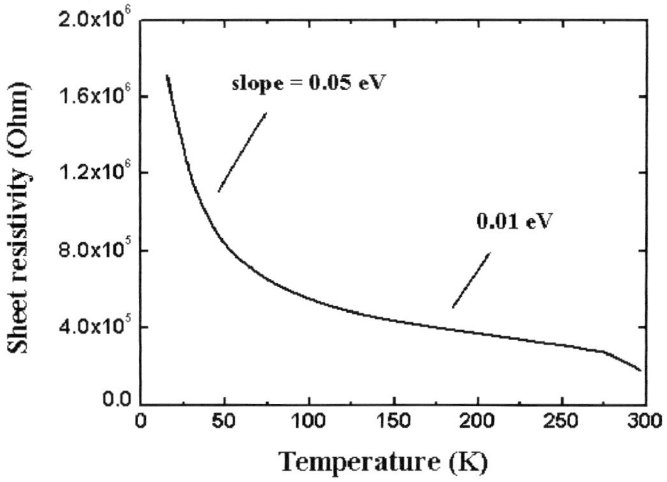

Fig. 6. Electrical sheet resistivity of DMQD sample A as a function of temperature.

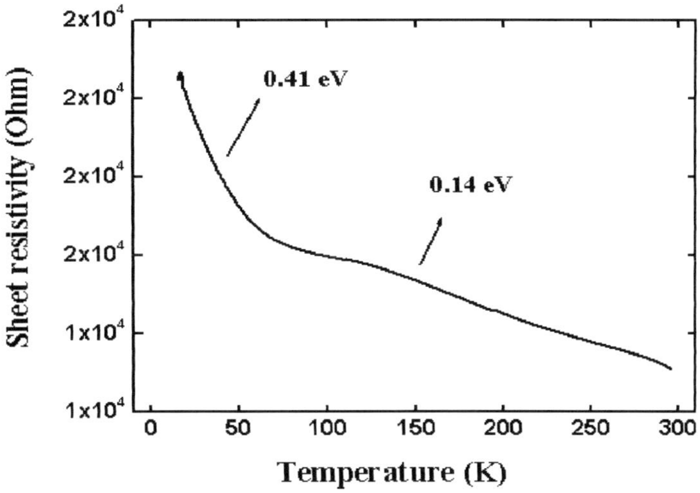

Fig. 7. Electrical sheet resistivity of DMQD sample B as a function of temperature.

Fig. 7 represents the electrical resistivity as a function of temperature for sample B. As can be seen from Fig. 7, the resistance increases with decreasing temperature, which is a typical semiconducting characteristic. The strong divergence at low temperature indicates strong carrier localization behaviour. The temperature-dependent resistance was also fitted with an Arrhenius function. Low lying levels at 0.41 eV and 0.14 eV from the valence band could be observed in Fig. 7. The activation barrier associated with the separation between the Mn acceptor level and Ge valence band is known to be ~ 160 meV in p-type Mn-doped Ge crystal and the activation barrier between the Mn acceptor and conduction band is known to be ~ 390 meV in n-type Mn-doped Ge crystals from the temperature-dependent resistivity measurements [22]. It is possible that Mn is also incorporated into the Si matrix. The reason why the activation energies are not associated with Si is that the Mn-related double donor level, which comes from Mn atoms occupying Si sites in p-type conditions, is around

0.29 eV above the valence band [23].According to Woodbury et al.[22], at low temperatures the exponential regimes with activation energies of 140 meV and 410 meV are most likely associated with a shallow acceptor level of Mn substituting Ge suggesting that the transport is dominated by a thermal activation mechanism. Therefore, the ferromagnetism observed in sample B with T_C = 160 K is associated with the carriers produced by Mn atoms occupying Ge sites (Mn_{Ge}), as confirmed by transport measurements. This behaviour further suggests that the ferromagnetic exchange coupling is hole-mediated, but the hole concentration of sample B was only ~10^{15} cm^{-3}, which is too low to show hole-mediated ferromagnetism. However, the concentration of localized carriers residing in the impurity band is not too low. Therefore, the bound magnetic polaron (BMP) percolation picture [20,21] as discussed in the framework of strongly semi-insulating DMS materials may be relevant. In this picture, the spin of localized carriers polarizes the surrounding magnetic impurities, leading to the so-called bound magnetic polarons [24]. Alignment of the polarization is mediated by the localized holes in the system via interaction according to the RKKY theory [2]. We consider that the ferromagnetism of sample B is due to the magnetic polarons, indicating that room temperature ferromagnetism in IV group semiconductors doped with Mn can be realized in the future.

4. Conclusion

In summary, DMQD has been formed through Mn ion implantation and thermal annealing. It was found that manganese introduces two acceptor levels in germanium at 0.14 eV from the valence band and 0.41 eV from the conduction band, implying Mn substituting Ge. Therefore, it can be concluded that the ferromagnetic exchange coupling of the semiconducting sample with a Curie temperature around 160 K is hole-mediated due to formation of bound magnetic polarons. On the other hand, the ferromagnetism in the insulating sample (with a Curie temperature over 300 K) is associated with the dominant Mn_5Ge_3 phase.

Acknowledgements

This work was supported in part by funds from the Quantum-functional Semiconductor Research Center and Dongguk University and by the Korea Science and Engineering Foundation (KOSEF) grant funded by the Korea government (MOST) in 2008.

References

[1] Ohno H 1998 *Science* **281** 951
[2] Dietl T, Ohno H, Matsukura H, Cibert J and Ferrand D 2000 *Science* **287** 1019
[3] Matsumoto Y, Murakami M, Shono T, Hasegawa T, Fukumura T, Kawasaki M, Ahmet P, Chihyow T, Koshihara S and Koinuma H 2001 *Science* **291** 854
[4] Hartman T, Lampalzer M, Klar P J, Stolz W, Heimbrodt W, Keug von Nidda H–A, Loids A and Svistov L 2002 *Physica E* **13** 572
[5] Theodoropoulou N, Hebard A F, Overberg M E, Abernathy C R, Pearton S J, Chu S N G and Wilson R G 2002 *Phys. Rev. Lett.* **89** 107203-1
[6] Medvedkin G A, Ishibashi T, Nishi T, Hayata K, Hasegawa Y and Sato K 2000 *Jpn. J. Appl. Phys*. Part 2 **39** L949
[7] Reed M L, El-Masry N A, Stadelmaier H H, Ritums M K, Reed M J, Parker C A, Roberts J C and Bedair S M 2001 *Appl. Phys. Lett.* **79** 3473
[8] Sonoda S, Shimizu S, Sasaki T, Yamamoto Y and Hori H 2002 *J. Cryst. Growth* **237-239** 1358
[9] Park Y D, Hanbicki A T, Erwin S C, Hellberg C S, Sullivan J M, Mattson J E, Ambrose T F, Wilson A, Spanos G and Jonker B T 2002 *Science* **295** 651
[10] Cho S, Choi S, Hong S C, Kim Y, Ketterson J B, Kim Y C and Jung J H 2002 *Phys. Rev. B* **66** 033303
[11] Nakayama H, Ohta H and Kulatov E 2001 *Physica B* **302–303** 419
[12] Zhang F M, Liu X C, Gao J, Wu X S, Du Y W, Zhu H., Xiao J Q and Chen P 2005 *Appl.*

Phys. Lett. **85** 786
[13] Bolduc M, Awo-Affouda C, Stollenwerk A, Huang M B, Ramos F G, Agnello G and LaBella V P 2005 *Phys. Rev. B* **71** 033302
[14] Peressi M, Debernardi A, Picozzi S, Antoniella F and Continenza A 2005 *Computational MaterialsScience* **33** 125
[15] Picozzi S, Antoniella F, Continenza A, MoscaConte A, Debernardi A and Peressi M 2004 *Phys. Rev B* **70** 165205
[16] Yu S S, Cho Y M, Ihm Y E, Kim D J, Kim H J, Hong S K, Oh S J, Woo B C and Kim C S 2006 *Current Appl. Phys.* **6** 478
[17] Stropa A, Picozzi S, Continenza A and Freeman A J 2003 *Phys. Rev. B* **68** 155203
[18] Yamada N, Maeda K, Usami Y and Ohoyama T 1986 *J. Phys. Soc. Jpn.* **55** 3721
[19] Kittel C *Introduction to Solid State Physics* Wiley, New York (1996)
[20] Sarma S D, Hwang E H and Kaminski A 2003 *Phys. Rev B* **67** 155201
[21] Kaminski A J and Sarma S D 2002 *Phys. Rev. Lett.* **88** 247202
[22] Woodbury H H and Tyler W W 1955 *Phys. Rev.* **100** 659
[23] Bernardini F, Picozzi S and Continenza A 2004 *Appl Phys Lett.* **84** 2289
[24] Chen J, Wang K L and Galatsis K 2007 *Appl. Phys. Lett.* **90** 012501

Meissner effect in films of ropes of boron-doped single-walled carbon nanotubes; Correlation with applied pressure and boron-doped multi-walled nanotubes

M. Matsudaira[1], J. Nakamura[1], T. Shimizu[1], J. Haruyama[1, 2, 4], J. Reppert[3], A. M. Rao[3], T. Eguchi[2], T. Nishio[2], Y. Hasegawa[2], H. Sano[2], Y. Iye[2]

[1]*School of Science and Engineerings, Material Science course, Aoyama Gakuin University, 5-10-1 Fuchinobe, Sagamihara, Kanagawa 229-8558, Japan*

[2]*Institute for Solid State Physics, University of Tokyo, Kashiwanoha 5-1-5, Kashiwa, Chiba 277-8581, Japan*

[3]*Department of Physics and Astronomy, Center for Optical Materials Science and Engineering Technologies, Clemson University, Clemson, SC 29634, USA*

[4]*JST-CREST, 4-1-8 Hon-machi, Kawaguchi, Saitama 332-0012, Japan*

J-haru@ee.aoyama.ac.jp

Abstract

Superconductivity in carbon nanotubes (CNTs) is attracting considerable attention. Recently, we have reported successful boron doping into single-walled CNTs (SWNTs) and also revealed its correlation with superconductivity. In the present study, we report results of pressure-applied magnetization measurements in films consisting of ropes of boron-doped SWNTs. It reveals that T_c and magnitude for Meissner effect is mostly independent of applied pressure, while magnitude of graphite diamagnetism drastically increases as pressure increases. We also report result of resistance measurements in the samples and also correlation of boron doped SWNTs with multi-walled CNTs, in which we reported superconductivity previously.

1. Introduction

It is known that the small mass of carbon can promote high transition temperature (T_c) in Bardeen -Cooper-Schrieffer (BCS) -type superconductivity (SC) and novel behaviors of SC can also be expected. Recently, new carbon-based superconductors with order of T_c of ~10K (calcium- intercalated graphite (C_6Ca) and highly boron *(B)*- doped diamond [*1-3*]) have been discovered and higher T_c has been expected. In particular, the SC in a carbon nanotube (CNT), which is a rolled-up graphene sheet and a typical one-dimensional (1D) carbon conductor, is attracting considerable attention [*4-8*] for the following reasons: (1) The curvature resulting from a small diameter (~1 nm) can lead to formation of sp^3 hybrid orbitals and σ-π mixed bands. Coupling of this σ-π electrons with radial breathing mode of phonon might bring high T_c [*9*]. (2) The alignment of the Fermi level (E_F) to a van Hove singularity (VHS) can also lead to high T_c due to the presence of an extremely large density of states (DOS) [*9*]. (3) It can provide an insight into the 1D electron correlation [*7, 10-14*]. Indeed, we have reported T_c as high as 12K in entirely end-bonded multi-walled CNTs (MWNTs) from viewpoint of an abrupt resistance drop.

Here, effective carrier doping is crucial for realizing high-T_c SC in any material. In fact, SCs in C_6Ca and *B*-diamond have been associated with it. Carrier doping into CNTs has also been studied in many previous works [*15-17*], including our present methods [15]. Nevertheless, correlation with carrier-doping has been never reported. Indeed, even in our previous finding of SC, presence of carriers in the MWNTs has not been clarified to date [7].

Only possibility of non-intentional boron doping through catalyst during CVD process has been discussed [7,8]. On the other hand, we have reported successful boron doping into single-walled CNTs (SWNTs) by mixing elemental boron into catalyst and realized Meissner effect with T_c o f 12K by assembling the boron-doped SWNTs (*B*-SWNTs) to thin film structures [18]. We found that 1. Boron concentration of the catalyst (N_B) > 4 at.% destroyed tube structure, 2. Lower N_B values are favorable to occurrence of evident and higher T_c, and 3. High uniformity of the thin film is necessary for the SC.

Here, in this study, we report results of some new experiments of the *B*-SWNT thin films, i.e., high pressure magnetic measurement of the thin films consisting of ropes of *B*-SWNTs, electrical measurement of the films by fabricating FET structure on them, and also nuclear magnetic resonance (NMR) measurements of the MWNTs, in which we reported SC [7]. In particular, it is well known that applying high pressure drastically changes behaviors of SC. In the present thin films consisting of *B*-SWNTs, applying pressure enhances interaction among neighboring B-SWNTs. If loop current for Meissner effect flows across assembled *B*-SWNTs, this enhancement makes magnitude of the current flow stronger. Thus, it is expected that it changes behaviors of Meissner effect.

2. Sample preparation and characterization

We have used mostly the same high-uniformity *B*-SWNT thin films with N_B = 1.5 at. %, which were reported in ref.[18], for the present experiments. *B*-SWNT bundles(ropes) were synthesized using the pulsed laser vaporization technique [15]. Elemental boron and Co:Ni catalysts (0.5:0.5 at. %) were homogenously mixed with carbon paste and hot-pressed into 1" diameter targets. The YAG laser-generated products were characterized by high resolution transmission electron microscopy, NMR, and Raman spectroscopy as well as those in ref.[18].

Thin films consisting of these *B*-SWNTs were prepared by solubilizing the *B*-SWNTs in 5 mg/mL dichloroethane. Then, the solution was centrifuged for 48 *h* by Tomy, low-speed centrifuge in order to purify the SWNTs, and ultrasonication (As One, US cleaner) was carried out also for 20 *h* to separate the bundles (ropes) into individual *B*-SWNTs. Next, the solution was spin-coated at 500 rpm on a Si substrate to produce a highly homogeneous film consisting of assembled *B*-SWNTs. Finally, the thin films were annealed at 1600 °C for 1 *h* in order to entirely remove the residual catalyst.

Figure 1 shows AFM image of the film. It exhibits that the film consists of closely assembled ropes of the *B*-SWNTs with a rope diameter of about 10nm. This result is different from previous case in ref.[18], in which ropes were sufficiently dispersed to individual SWNTs. This is due to four-times shorter ultrasonication time (i.e., it was 80 *h* in [18]). However, because the sample also showed Meissner effect, we have used this condition in the present measurements.

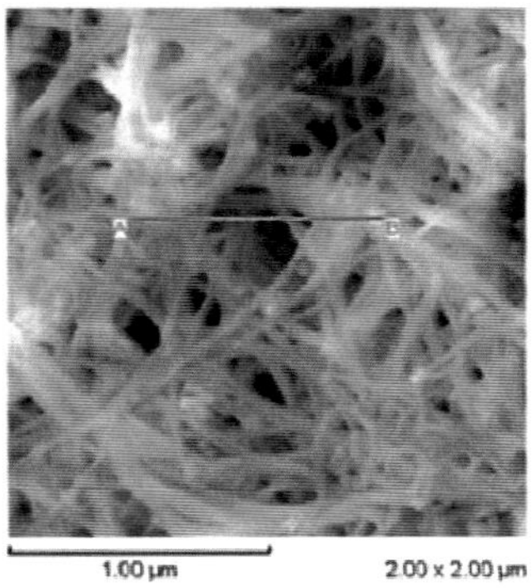

Figure 1. Room temperature AFM image of a thin film consisting of assembled *B*-SWNT ropes prepared on Si substrates by spin coating (500 rpm). Short ultrasonication time for 20 *h* resulted in insufficient dissolving of ropes and presence of the ropes of *B*-SWNTs in the film.

3. Measurement results and discussions

3.1. High-pressure magnetization measurement

The magnetization measurements of the thin films consisting of *B*-SWNTs were performed using a superconducting quantum interference device (Quantum Design, MPMS), applying pressure between 0.3 – 1.1 GPa. At first, Fig. 2 shows normalized magnetization ($M_N = M(T) - M(T{=}40K)$) as a function of the temperature (T) (M_N–T relationship) under zero-pressure in field-cooled (FC) and zero-field-cooled (ZFC) regimes in the highly uniform film with N_B ~ 1.5 at. % reported in ref. [18]. Evident drop in M_N with a large slope value is observed below T_c = 12 K in the ZFC regime, and below T = ~8 K in the FC regime. Interestingly, this value of T_c = 12 K exactly agrees with the T_c value for an abrupt resistance drop observed in the array of entirely end-bonded MWNTs [*7, 8*].

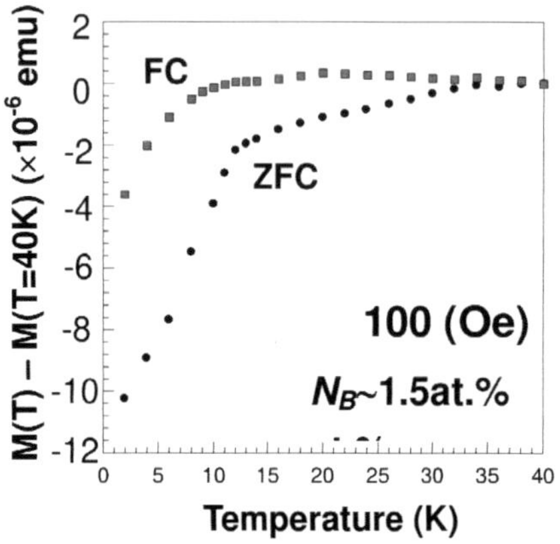

Figure 2. Normalized magnetization as a function of temperature at magnetic fields (*H*) of 100 Oe in field-cooled (FC) and zero-field-cooled (ZFC) regimes in thin films consisting of well dissolved *B*-SWNTs, which were synthesized from a target with N_B = ~1.5 at. % and prepared on a Si substrate by spin coating at 500 rpm [18].

In contrast, the small and gradual M_N drop observable at T_c > 12 K in the ZFC regime is due to diamagnetism of graphite structure of assembled SWNTs and not associate with Meissner effect. Graphite diamagnetism in ensemble of SWNTs very gradually appears in *M-T* curves from high temperatures (e.g., > 50K) even in undoped samples and the amplitude does not decrease in *M- H* features even at high magnetic fields (e.g., > 2000 Oe). They are very different from those in Meissner effect observed here. However, because the component is very small as the back ground (i.e., magnetization at T > 12K in Fig.2) in the samples shown here and different by respective samples, we could not entirely delete it in this study.

It is found that the magnitude of the drops observed in the M_N–*T* relationship becomes considerable at magnetic field (*H*) < ~1400 Oe as *H* values increase. In contrast, it decreases for *H* > ~1400 Oe, and at *H* = ~3500 Oe, the magnitude becomes almost zero. This is typical behavior of type II SC as reported in refs.[8, 18]

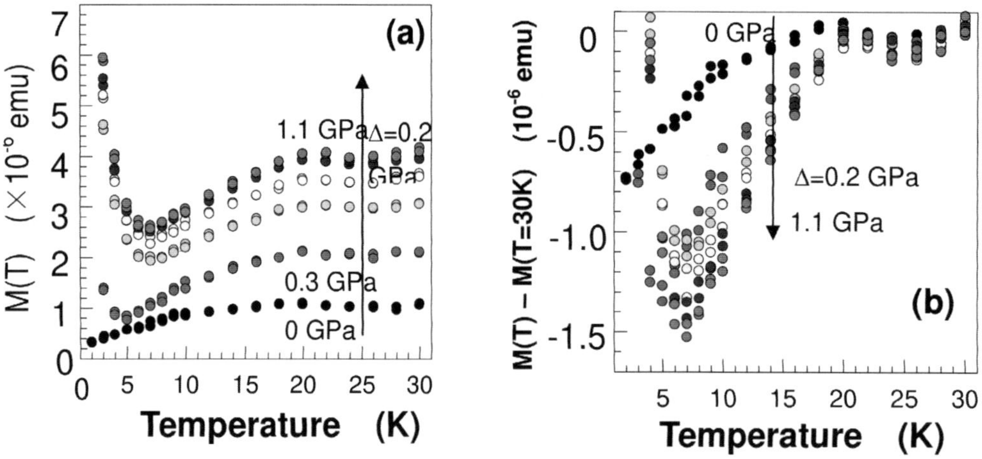

Figure 3. **(a)** No-normalized and **(b)** normalized magnetizations as functions of temperature for various pressures in the sample shown in Fig. 1, which consist of ropes of *B*-SWNTs. In order to apply pressure, the thin films were immersed into oil-filled cells.

Here, Fig. 3(a) shows pressure dependence of *M-T* curves in ZFC regime in a thin film consisting of ropes of the *B*-SWNTs as shown in Fig.1. The *M-T* feature similar to Fig.2 is observable at zero-pressure and T_c is mostly 10 K in this sample. As magnitude of pressure increases, positive *M* value itself increases in the entire temperature range. It should be noticed that magnetization drop in low temperature regime becomes significant with increasing pressure. Figure 3(b) shows pressure dependence of the M_N–*T* relationship of Fig.3 (a). It is found that amplitude of magnetization drops drastically increases at entire temperatures < 20K and T_c of ~10K for Meissner effect becomes ambiguous, as pressure increases. In particular, change in the M_N–T relationship is significant under 0.3 GPa pressure Although the T_c becomes unclear, it still remains T_c = ~10K even under this high pressure. The T_c entirely disappears under 0.7 GPa. On the other hand, magnetization drop at T > 10K becomes significant, as magnetic field increases. It eliminates Meissner effect at T < ~10K under pressure > 0.7 GPa. As mentioned above, this magnetization drop at T > 10K is attributed to graphite diamagnetism. These results imply the following pressure dependence of magnetization and that interpretation.

Amplitude for Meissner effect at T < 10 K is mostly independent of applied pressure, while amplitude of graphite diamagnetism at T < 20 K drastically increases as magnitude of the pressure increases. Here, interaction among neighboring SWNTs in the film increases with increasing pressure magnitude, when individual SWNTs are well dispersed. It should lead to enhancement of Meissner effect around ensemble of SWNTs including boron, while it should result in enhancement of graphite diamagnetism in ensemble of SWNTs without boron. However, as shown in Fig.2, amplitude of Meissner diamagnetism (volume fraction of SC) is very small in the present *B*-SWNT film (i.e., 10^{-6} emu). This is four order smaller than those for conventional Meissner effect (i.e., 10^{-2} emu). Indeed, we estimate superconducting coherence length of ~ 10 nm [18].

These two things suggest a possibility that the loop current for Meissner effect is confirmed in one rope of *B*-SWNT, because the present film consist of ropes with the diameter of ~10nm as shown in Fig.1. Because interaction among neighboring *B*-SWNTs in a rope is conventionally very strong as it was synthesized, it does not mostly change even under high pressure. This well explains the present result (i.e., mostly constant amplitude of Meissner effect at T < 10K as mentioned above).

In contrast, loop current for graphite diamagnetism can occur even among neighboring ropes of no-boron-doped SWNTs within a large volume faction. Interaction among these ropes in a thin film should drastically increase as magnitude of the pressure increases. This is consistent with drastic increase in graphite diamagnetism at T < 20 K. On the other hand, the critical temperature of 20 K for the graphite diamagnetism is mostly constant even with the increase in magnitude. This implies that magnitude of loop current both for graphite diamagnetism has correlation with only magnitude of those diamagnetisms.

In order to clarify correlation of Meissner diamagnetism with applying pressure more evidently, thin films consisting of further dispersed SWNTs (e.g., in rope with diameter < 5nm) and also increase in volume fraction of *B*-SWNTs are required.

3.2. Resistance measurements

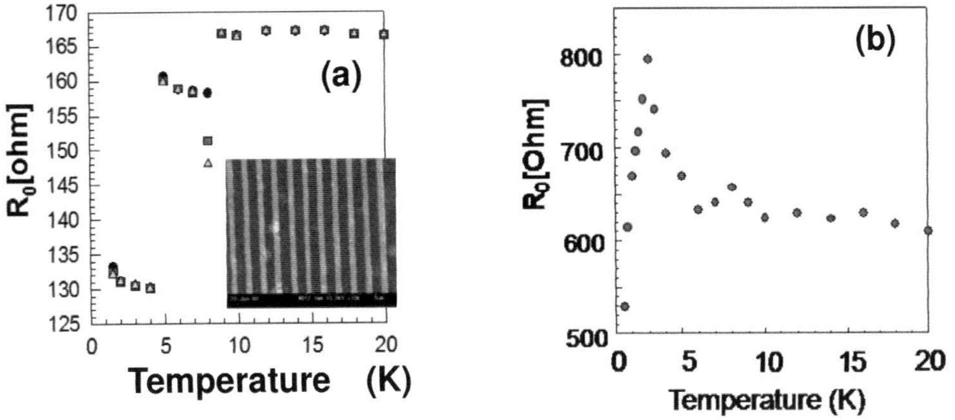

Figure 4. Result of zero-voltage resistance measurements of FETs fabricated on *B*-SWNT thin films with N_B=1.5 at.% used for Fig.3 **(a)** and the other sample **(b)**. Resistance was measured three times at each temperature and noted by different symbols in (a). **Inset of (a):** SEM top view of Au/Ti electrode pattern of a FET fabricated on the film. In order to detect very small superconducting area, many source and drain electrode fingers were patterned within 300nm spacing.

Here, it is indispensable to detect resistance drop down to zero ohm as well as detecting Meissner effect for identification of SC. However, the volume fraction of superconducting region is extremely small as mentioned above. This suggests that the area of SC might be only a few μm scales. Thus, in order to detect this SC region, we patterned source and drain finger electrodes (Ti/Au) within 300 nm spacing on the *B*-SWNT film as many as possible and investigated electrical features. Figure 4 shows two examples of the resistance measurement results in the sample used for Fig. 3 (Fig.4(a)) and the other sample(Fig.4(b)).

In Fig.4(a), the resistance slightly drops at T = 10 K and becomes unstable (i.e., it shows different values for three-times measurements). Then, it drastically and abruptly drops to zero ohm at T_c = 5 K. This implies possibility of superconducting transition, because T = 10 K is in good agreement with T_c of Meissner effect in Fig.3 and also the magnitude of the drop is as large as ~1/5. The instability might originate from poor contact between electrodes and the B-SWNTs. Also in Fig.4(b), we can confirm an abrupt resistance drop starting from T_c = 3K with the magnitude of the drop as large as ~1/3. Only SC can be an origin for such a large and abrupt resistance drop in our system.

However, when we repeated measurements with applying further voltages or magnetic fields, the resistance drops disappeared and only high resistance at T > T_c remained in both samples. This might mean that small superconducting region was destroyed by applied high energy or concentration of superconducting current. At present stage, most of samples are very weak for repeated measurements with applying electrical and magnetic fields. Thus, increase of superconducting volume fraction and improvement of FET structures are indispensable for detecting resistance drops.

3.3. Correlation with superconductivity in MWNTs

As mentioned above, importantly the observed highest T_c of 12K for Fig.2 is exactly the same as that in the entirely end-bonded MWNTs, which we reported SC previously [7]. But we could not show evidence of carrier doping in ref.[7] and suggested only a possibility of boron doping via catalyst during CVD process [8]. "The T_c = 12K" is also in good agreement with that in C_6Ca [1, 2]. Moreover, Ishii et al. have recently reported a resistance drop at T_c = 12K in a FET using an isolated *B*-MWNT as the current channel. Hence, it can be a "Magic Number".

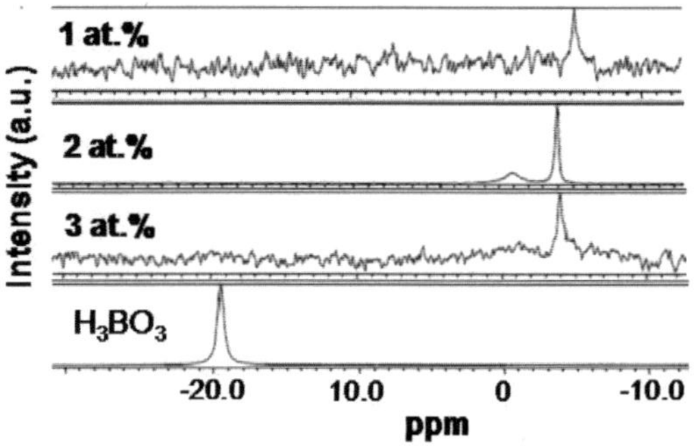

Figure 5. Results of NMR measurement of *B*-MWNTs synthesized with N_B = 1 ~ 3 at. % catalyst (upper three) reported in ref. [7, 8] and H_3BO_3 (lowest) as a reference. The Fe/Co catalyst was electrochemically deposited into the bottom ends of alumina membrane. H_3BO_3 was occasionally used for only activation of chemical reaction for Fe/Co catalyst.

Here, Fig.5 shows result of NMR measurement of the MWNTs, which were synthesized from catalyst with different N_B values, reported in Ref.[7, 8]. It evidently shows NMR peaks, which are similar to those in the B-SWNTs reported in ref.[18] and implies presence of boron-carbon chemical bonds in the MWNTs. In particular, the MWNTs with N_B = ~2 at.%, which showed the most evident SC with the highest possibility, exhibited the highest peak height. This is mostly consistent with the result reported in ref.[18], which implied that lower N_B value was favorable to appearance of evident SC. Thus, the common T_c = 12K for the present SC in *B*-SWNT films and the previous SC in *B*-MWNTs [7, 8] implies presence of common origins for ensemble of *B*-CNTs. One of the candidates for the mechanism can be the best alignment of E_F to a VHS under such lows N_B values [18, 20] and suppression of one-dimensional quantum effect (e.g., repulsive Coulomb interaction in Tomonaga-Luttinger liquid) [19], which tends to obstruct occurrence of SC, by assembled structures. In contrast, because C_6Ca has a different electronic structure (e.g.,difference in sp^2 and sp^3 orbitals), the agreement of T_c may be occasional.

5. Conclusion

We have shown the results of high-pressure magnetization measurements in the films consisting of the ropes of boron-doped SWNTs. It revealed that T_c and magnitude for Meissner effect was mostly independent of applied pressure, while magnitude of graphite diamagnetism drastically increased as pressure increases. We have also reported the result of resistance measurement and correlation of boron doping with SC in MWNTs. Consequently, we suggested that assembling of well dispersed *B*-CNTs was important.

The present results assure that applying high pressure to thin films consisting of sufficiently dispersed *B*-SWNTs with a larger volume fraction could lead to considerably high T_c. Homogeneously assembled *B*-CNTs are promising as a novel structure which is expected to open doors to the fields of carbon-based SC.

6. Acknowledgement

The authors acknowledge H. Fukuyama, J. Akimitsu, H. Shinohara, S. Saito, T.Koretsune, T. Ando, J. Gonzalez, H. Bouchiat, K. McGuire, and M.Dresselhaus for the encouragement provided and for fruitful discussions, and also O. Kamo for technical assistance for NMR measurement. This research was partially supported by Japanese ministry, Grant-in-Aid for Scientific Research (A), 18204033.

References

[1] T. E. Weller et al., 2005 *Nature Physics* **1**, 39
[2] N. Emery et al., 2005 *Phys. Rev. Lett.* **95**, 87003
[3] E. A. Ekimov et al., 2004 *Nature* **428**, 542
[4] M. Kociak et al., 2001 *Phys. Rev. Lett.* **86**, 2416
[5] Z. K. Tang et al., 2001 *Science* **292**, 2462
[6] M. Ferrier et al., 2006 *Phys. Rev. B* **73**, 094520
[7] I. Takesue, J.Haruyama H.Shinohara et al., 2006 *Phys. Rev. Lett.* **96**,057001
[8] N. Murata, J.Haruyama, S.Maruyama, H.Shinohara et al., 2007 *Phys. Rev. B* **76**, 245424
[9] R. Barnett, E.Demler et al., 2005 *Phys. Rev. B* **71**, 035429
[10] M. Matsudaira, J.Haruyama., S.Maruyama, H.Shinohara et al., 2008 *Physica E* **40/7**, 2299
[11] M. Matsudaira, J.Haruyama., et al., to be published *on Phys. Rev. B* (2009)
[12] I.Takesue, et al., 2004 *Physica E* **24**, 32; P. Recher, D. Loss, 2002 *Phys. Rev. B* **65**, 165327; C. Bena et al., 2002 *Phys. Rev. Lett.* **89**, 037901
[13] J. Gonzalez, 2002 *Phys. Rev. Lett.* **88**, 076403
[14] E. Perfetto, J. Gonzalez, 2006 *Phys. Rev. B* **74**, 201403(R)
[15] K. McGuire et al., 2005 *Carbon* **43**, 219
[16] K. Liu et al., 2001 *Phys. Rev. B* **63**, 161404R
[17] S. Bandow, S. Numao, S. Iijima, 2007 *J. Phys. Chem. C* **111**, 11763
[18] N.Murata, J.Haruyama et al., 2008 *Phys.Rev.Lett.* **101,** 027002
[19] J.Gonzalez, 2002 *Phys.Rev.Lett.* **88,** 076403
[20] T. Koretsune, S. Saito, 2008 *Phys. Rev. B* **77**, 165417

AUTHOR INDEX

Abdullayev, N A 434
Aberomand, P 402
Agamirzoeva, G M 323
Ağil, Hasan 501
Akdogan, M 169
Akhavan, M 402
Aksan, M A 1, 24
Aktaş, B 414, 485, 539
Albiss, B A 68
Aliyev, Y I 323
Ali-zade, R A 329
Altin, S 1, 24
Altug, D Turkoz 343
Alveroğlu, E 533
Anis-ur-Rehman, M 192
Antal, V 62
Apostolov, Andrey 508
Ashiq, Muhammad Naeem 420
Askerbeyli (Tagiyeva), R T 336
Aslan, Ö 6
Avgin, I 395
Awana, V P S 102, 142
Babayev, A M 343
Bagwe, V C 240
Baik, S K 15, 129
Bakkaloğlu, Ö F 427
Balci, Y 1, 24
Barbaran, J Hasani 323
Basoglu, M 31
Belenli, I 82, 212
Bern, F 354
Borges, R P 354, 360
Bozbey, Ali 121
Cai, C B 318
Cakmaktepe, S 343
Çam, B Ates 39
Cansiz, A 39
Cao, S X 318
Casaca, A 360
Celik, E 395
Çelik, Ş 31, 47, 160
Cetinkara, H A 306
Cheikhrouhou, A 370
Chen, M Q 318
Costa, A T 354
Cruz, M M 354
Da Silva, R C 354, 360
De Luca, R 55
Deshpande, U P 515
Dhawan, M S 515
Diez, J C 149
Diko, P 62, 106
Dimitrov, Dimitar 508
Dobrowolski, V 463
Doğan, N 375
Dolia, S N 515
Dorosinskii, L 383
Eguchi, T 550
ElMassalami, M 208
Erol, M 395
Farahmandjou, M 402
Fardmanesh, Mehdi 121
Fedorchenko, I V 463
Feng, Xiao 291
Ferreira, P 360
Garg, K B 515
Gaur, S K 515
Gelir, A 533
Gencer, A 169
Gencer, Ali 501
Gharaibeh, M 68
Ghazanfari, N 527
Gheisari, K H 408
Godinho, M 354, 360
Gokay, U S 73
Golod, T 204
Gömöry, F 234
Gorur, O 82
Gray, E Mac A 493
Grovenor, C R M 31
Güler, G 427
Güler, S 414
Güllü, Ö 427
Gundogdu, Ö 39
Guseinov, G G 323
Hallet, X 95
Haruyama, J 550
Hasan (Qaseer), M K 68
Hasar, U C 39
Hasegawa, Y 550
Hernandez-Gomez, Pablo 420
Hirata, K 208
Hoseini, Seyed Mehdi 121
Huhtinen, H 102
Iqbal, Muhammad Javed 420
Irjala, M 102
Iye, Y 550
Javadpour, S 408
Jin, G 318
Jirsa, M 106
John, J. 240
Kang, T W 543
Kaňuchová, M 62
Karataş, Ş 427
Karci, A B 113
Kasahara, S 208
Kazan, S 539
Kerimova, T G 434
Keskin, S S 6
Khaibullin, R I 414

AUTHOR INDEX

Khapaeva, T M ... 450, 463
Kharmouche, A ... 439
Khoshaman, Amirhosein ... 121
Kilanski, L ... 463
Kim, H J ... 129
Kim, H M ... 15
Kim, Y C ... 15
Kishan, H ... 142
Kokabi, Alireza ... 121
Koroleva, L ... 450, 463
Koubaa, M ... 370
Koubaa, W Cheikh-Rouhou ... 370
Kováč, J ... 62
Kováč, P ... 136
Krasnov, V M ... 204
Krzymanska, B ... 463
Kumar, Anuj ... 102, 142
Kumar, Shiva ... 142
Kumar, Sudhish ... 515
Kwon, W S ... 15
Kwon, Y K ... 15
Laiho, R ... 102
Lee, J D ... 15
Li, Yanrong ... 291
Li, Z B ... 468
Li, Zhuyong ... 129
Lomonosov, M V ... 450
Lopatin, O N ... 414
Lovchinov, Vassil ... 508
Ma, YongHu ... 129
Madre, M A ... 149
Magalhães, S ... 360
Mangalaraj, D ... 383
Manios, E ... 473
Manolache, S ... 186
Marconi, D ... 155
Marenkin, S F ... 463
Mátéfi-Tempfli, M ... 95
Mátéfi-Tempfli, S ... 95
Matei, I ... 186
Matsudaira, M ... 550
Mergen, Ayhan ... 485
Miah, M Idrish ... 493
Michotte, S ... 95
Moftakharzadeh, Ali ... 121
Moreira, M Dionízio ... 354
Moshchalkov, V V ... 95
Murakami, M ... 283
Muralidhar, M ... 106
Mustafayeva, K M ... 434
Nakamura, J ... 550
Nane, Onur ... 501
Narlikar, A V ... 102
Narsale, A M ... 240
Nataraj, D ... 383
Niarchos, D ... 473
Nishio, T ... 550
Novotortsev, V M ... 463
Nunes, W C ... 360
Nursoy, M ... 82, 306
Obaidat, I M ... 68
Oh, J T ... 408
Onbaşli, Ü ... 6
Özabaci, M ... 1, 550
Özçelik, Bekir ... 501
Özdemir, Z Güven ... 6
Özkurt, Berdan ... 501
Öztürk, K ... 31, 47, 160
Ozturk, O ... 169, 306
Ozturk, Y ... 395
Pai, S P ... 240
Panagiotopoulos, I ... 473
Parhizgar, S S ... 402
Park, C J ... 543
Park, G S ... 15
Park, H J ... 15
Paturi, P ... 102
Pinto, R ... 240
Piraux, L ... 95
Pop, A V ... 155, 186
Pop, V ... 186
Qin, Wenfeng ... 291
Qin, Z P ... 318
Qureshi, Anjum ... 485
Radulov, Iliya ... 508
Rameev, B ... 375, 414, 539
Rames, M ... 106
Ran, M W ... 468
Rao, A M ... 550
Rasekh, Sh ... 149
Razmkhah, Sasan ... 121
Reppert, J ... 550
Rosa, M A ... 360
Rosenstein, B ... 222
Rydh, A ... 204
Ryu, K ... 129
Saitovitch, Elisa ... 515
Sano, H ... 550
Santoyo-Salazar, J ... 383
Saraiva, A ... 360
Savaskan, Burcu ... 160
Sebt, S A ... 402
Šefčiková, M ... 62
Seki, H ... 283
Seong, K C ... 129
Shimizu, T ... 550
Shripathi, T ... 515
Singhal, R K ... 515
Song, Xiaoke ... 291
Sotelo, A ... 149

AUTHOR INDEX

Šouc, J 234
Soundararajan, D....383
Sozeri, H....113, 527, 533
Stamopoulos, D....473
Stiufiuc, G....155
Subaşi, H....375
Suzuki, A....283
Szymczak, R....463
Takeya, H....208
Tanatar, B....336
Tao, Bowan....291
Tepe, M....73, 113
Terzioglu, C....82, 169, 212, 306
Tinh, B D....222
Togano, K....208
Topal, Ugur....229
Topkaya, R....375
Tulun, M....533
Vajpayee, Arpita....142
Vanacken, J....95
Varilci, A....82, 212
Varnavskiy, S A....463
Venezuela, P....354
Vojenčiak, M....234
Wanchoo, S K....240
Wang, X....468
Westerholt, K....539
Winkler, W....247
Wongsatanawarid, A....283
Xing, Y T....515
Xiong, Jie....291
Yahya, A K....300
Yakinci, M E....1, 24
Yakinci, Z D....1
Yanmaz, E....31, 47, 160
Yao, K L....468
Yao, X....318
Yegen, D....212
Yerli, Y....375
Yilgin, R....539
Yilmaz, Y....533
Yilmazlar, M....306
Yoon, I T....543
Zashchirinskii, D M....463
Zhang, Fei....291
Zhang, J C....318
Zhu, Y....468